# Ostracoda and
# Global Events

## British Micropalaeontological Society Publication Series

Series Editor
**Ronald L. Austin**
*Reader in Palaeontology*
*University of Southampton*

This series, published on behalf of the British Micropalaeontological Society by Chapman and Hall, aims to provide a synthesis of the current state of knowledge of all microfossil groups. The stratigraphic indexes detail the distribution of these groups in British sequences. Supported by notes on the systematics and identification criteria of the various taxa, these volumes are a unique compilation of data crucial to the work of those geologists concerned with stratigraphic correlation. The Series also includes the proceedings of selected conferences and edited volumes on specialist themes.

Books in the series provide essential reading for all micropalaeontologists and palaeontologists in academia and in industry.

Forthcoming titles

**Stratigraphic Index of Dinoflagellate Cysts**
Edited by A.J. Powell
**Stratigraphic Index of Acritarchs and Other Palaeozoic Microflora**
Edited by K. Dorning and S. Molyneux

# Ostracoda and Global Events

Edited by
**Robin Whatley**
*Professor of Micropalaeontology,*
*University College of Wales*

and

**Caroline Maybury**
*Post-doctoral Researcher, and Honorary Lecturer,*
*University College of Wales*

CHAPMAN AND HALL
LONDON • NEW YORK • TOKYO • MELBOURNE • MADRAS

| UK | Chapman and Hall, 11 New Fetter Lane, London EC4P 4EE |
|---|---|
| USA | Van Nostrand Reinhold, 115 5th Avenue, New York NY10003 |
| JAPAN | Chapman and Hall Japan, Thomson Publishing Japan, Hirakawacho Nemoto Building, 7F, 1-7-11 Hirakawa-cho, Chiyoda-ku, Tokyo 102 |
| AUSTRALIA | Chapman and Hall Australia, Thomas Nelson Australia, 480 La Trobe Street, PO Box 4725, Melbourne 3000 |
| INDIA | Chapman and Hall India, R. Sheshadri, 32 Second Main Road, CIT East, Madras 600 035 |

First edition 1990

© 1990  Chapman and Hall

Printed in Great Britain at the
University Press, Cambridge

ISBN 0 412 36300 3     0 442 31167 2 (USA)

**British Library Cataloguing in Publication Data**

Ostracoda and global events.
  1. Ostracoda 2. Fossil Ostracoda
  I. Whatley, Robin C.  II. Maybury, Caroline
  595.33
  ISBN 0-412-36300-3
  ISBN 0-442-31167-2 pbk


**Library of Congress Cataloging-in-Publication Data**

Ostracoda and global events / edited by Robin C. Whatley
  and Caroline Maybury. - 1st ed.
    p.  cm.
  Includes bibliographical references.
  ISBN 0-412-36300-3
    1. Ostracoda, Fossil-Congresses. 2. Ostracoda-Congresses.
  3. Paleoecology-Congresses. 4. Ecology-Congresses.
  I. Whatley, Robin C. (Robin Charles), 1936- . II. Maybury,
  Caroline, 1955- .
  QE817.O8075  1990
  565'.33-dc20                                    90-30457
                                                        CIP

# Contents

# Contents

# Contributors

**Katsumi Abe**
Geological Institute,
University of Tokyo,
Japan

**Bernard Andreu**
Résidence Joffre,
59640 Jeumont,
France

**Martin V. Angel**
Institute of Oceanographic Sciences,
Deacon Laboratory,
Wormley,
Godalming,
Surrey GU8 5UB,
UK

**Michael Ayress**
Geochem Laboratories Ltd,
Chester Street,
Saltney,
Chester CH4 8RD,
UK

**Jean-François Babinot**
Centre for Sedimentology and Palaeontology,
University of Provence,
Marseilles,
France

**Sara C. Ballent**
Division of Invertebrate Palaeozoology,
Museum of Natural Sciences,
La Plata,
Argentina

**Olivier Beckaert**
Department of Geology and Oceanography,
University of Bordeaux I,
France

**Gerhard Becker**
Geological-Palaeontological Institute,
Frankfurt,
West Germany

**Richard H. Benson**
Smithsonian Institution,
Washington DC,
USA

**Martin J.M. Bless**
The Natural History Museum,
Maastricht,
The Netherlands

**Willem A. van den Bold**
Department of Geology and Geophysics,
Louisiana State University,
USA

**Gioacchino Bonaduce**
Palaeontological Institute,
University of Naples,
Italy

**Willi K. Braun**
Department of Geological Sciences,
University of Saskatchewan,
Saskatoon,
Canada,
S7N OWO

**P. Carbonel**
Department of Geology and Oceanography,
University of Bordeaux I,
France

**Anne C. Cohen**
Department of Biology,
University of California at Los Angeles,
Los Angeles,
California 90024,
USA

**Graham Coles**
Institute of Earth Studies,
University College of Wales,
Aberystwyth,
Dyfed,
UK
(Chap 22):
Geochem Laboratories Ltd,
Chester Street,
Saltney,
Chester CH4 8RD,
UK

**Jean-Paul Colin**
Esso Rep.,
33321 Bègles,
France,
and c/o Exxon Production Research
Company,
Houston,
Texas,
USA

**August Coomans**
Rijksuniversiteit Gent,
Laboratoria voor Morfologie en Systematiek
der Dieren,
K.L. Ledeganckstraat 35,
9000 Gent,
Belgium

**Christine Crumière-Airaud**
U.R.A.,
1208 CNRS,
France

**Dan L. Danielopol**
Limnological Institute,
Austrian Academy of Sciences,
A-5310 Mondsee,
Austria

**Hugh B. Devery**
Department of Geology and Geography,
Mississippi State University,
USA

**Chris P. Dewey**
Department of Geology and Geography,
Mississippi State University,
USA

**Odette Ducasse**
Department of Geology and Oceanography,
University of Bordeaux I,
France

**Yoran Eshet**
Geological Survey of Israel,
Jerusalem,
Israel

**Ephraim Gerry**
The Israel Institute of Petroleum and Energy,
Tel Aviv,
Israel

**Helga Groos-Uffenorde**
Institute and Museum for Geology and
Palaeontology,
University of Göttingen,
Goldschmidt-Str. 3,
D-3400 Göttingen,
FRG

**Joseph E. Hazel**
Department of Geology and Geophysics,
Louisiana State University,
USA

**Francis Hirsch**
Geological Survey of Israel,
Jerusalem,
Israel

**Avraham Honigstein**
Department of Geophysics and Planetary
Sciences,
Tel-Aviv University,
Israel

**David J. Horne**
Thames Polytechnic,
London,
UK

**Kunihiro Ishizaki**
Institute of Geology and Paleontology,
Tohoku University,
Japan

**Ian Jarvis**
Kingston Polytechnic,
Surrey,
UK

**Roger L. Kaesler**
Department of Geology,
Museum of Invertebrate Paleontology and
Paleontological Institute,
The University of Kansas,
USA

**Michael Keen**
Department of Geology and Applied
Geology,
University of Glasgow,
Glasgow G12 8QQ,
UK

**Dietmar Keyser**
Zoological Institute and Zoological Museum,
University of Hamburg,
West Germany

**Larry W. Knox**
Department of Earth Sciences,
Tennessee Technological University,
USA

**Mervin Kontrovitz**
Department of Geosciences,
Northeast Louisiana University Monroe,
USA

**Edith Kristan-Tollmann**
A-1180 Wien,
Scheibenbergstrasse 53,
Austria

**Miroslav Kruta**
Thalmannova 4,
Prague 6 160 00,
Czechoslovakia

**María Luisa Machain-Castillo**
Instituto de Ciencias del Mar y Limnología,
Universidad Nacional Autónoma de México,
Ap. Postal 70-305,
México

**Rosalie F. Maddocks**
Department of Geosciences,
University of Houston,
Texas,
USA

**Koen Martens**
Koninklijk Belgisch Instituut voor
Natuurwetenschappen,
Hydrobiology,
Vautierstraat 29,
1040 Brussels,
Belgium

**John Milne**
Department of Geology,
Saint Mary's University,
Halifax,
Nova Scotia,
Canada B3H 3C3

**Alicia Moguilevsky**
Institute of Earth Studies,
University College of Wales,
Aberystwyth,
Dyfed,
UK

**James C. Morin**
Department of Biology,
University of California at Los Angeles,
Los Angeles,
California 90024,
USA

**Nasser Mostafawi**
Geological-Palaeontological Institute and
Museum,
Kiel,
West Germany

**Ph. Mourguiart**
Department of Geology and Oceanography,
University of Bordeaux,
France

**Eduardo A. Musacchio**
National University of Patagonia San Juan
Bosco,
Comodoro Rivadavia,
Argentina

**Tomohide Nohara**
Department of Earth Sciences,
College of Education,
University of the Ryukyus,
Okinawa,
Japan

**Henri J. Oertli**
F-64320 Bizanos,
France

**Pang Qiqing**
Hebei College of Geology,
Hebei,
China

**Ana María Pérez-Guzmán**
Department of Geosciences,
University of Houston,
Texas,
USA

**J.-P. Peypouquet**
Department of Geology and Oceanography,
University of Bordeaux,
France

**James A. Pilch**
Department of Geology,
Museum of Invertebrate Paleontology and
Paleontological Institute,
The University of Kansas,
USA

**T. Mark Puckett**
Department of Geology,
University of Alabama,
USA

**N.J. Riley**
British Geological Survey,
Keyworth,
Nottingham NG12 5GG
UK,

**Amnon Rosenfeld**
Geological Survey of Israel,
Jerusalem,
Israel

**Lucienne Rousselle**
Department of Geology and Oceanography,
University of Bordeaux I,
France

**Eberhard Schindler**
Institute and Museum for Geology and
Palaeontology,
University of Göttingen,
Goldschmidt-Str. 3,
D-3400 Göttingen,
FRG

**Qadeer Siddiqui**
Department of Geology,
Saint Mary's University,
Halifax,
Nova Scotia,
Canada B3H 3C3

**I.G. Sohn**
US Geological Survey,
Room E-308,
National Museum of Natural History,
Washington DC,
20560,
USA

**Jonathan C. Sporleder**
Department of Geology,
Museum of Invertebrate Paleontology and
Paleontological Institute,
The University of Kansas,
USA

**P. Lewis Steineck**
Division of Natural Sciences,
SUNY College at Purchase,
New York,
USA

**Ryoichi Tabuki**
Department of Earth Sciences,
College of Education,
University of the Ryukyus,
Okinawa,
Japan

**Klaus Trier**
Glandwr,
Mathry,
Haverfordwest,
Dyfed SA62 5HG,
UK

**Ruth D. Turner**
Museum of Comparative Zoology,
Harvard University,
USA

**Anne Van Frausum**
Instituut voor Aardwetenschappen,
University of Leuven,
Leuven,
Belgium

# Contributors

**Dick Van Harten**
Institute of Earth Sciences,
Free University,
Amsterdam,
The Netherlands

**Robin Whatley**
Institute of Earth Sciences,
University College of Wales,
Aberystwyth,
Dyfed,
UK

**Ian P. Wilkinson**
British Geological Survey,
Keyworth,
Nottingham NG12 5GG,
UK

**Karel Wouters**
Koninklijk Belgisch Instituut voor
Natuurwetenschappen,
Brussels,
Belgium

**Ye Dequan**
Daqing Petroleum Administrative Bureau,
Heilongjiang,
China

**Jaromir Zelenka**
Czechoslovakian Geological Survey,
Prague

# Acknowledgements

Gratefully acknowledged is the support of the following in the production of this volume:

Britoil
Enterprise Oil
Esso Exploration and Production U.K. Limited
Gearhart Geodata Limited
Mid-Wales Development
Robertson Research International Limited
The British Micropalaeontological Society
The British Petroleum Company p.l.c.
The International Paleontological Association
The Stereo-Atlas of Ostracod Shells.

Without the assistance of the staff of the Computer Unit, University College of Wales, Aberystwyth this volume could not have been produced in its present form. Mohammed Jalloq, Roger Mathews, Jeremy Perkins and David Roberts have helped and advised us in a multiplicity of ways. We are especially indebted to the latter whose advice and assistance with the conversion of wordprocessing packages, the Software Bridge and the Aldus Pagemaker desktop publishing application is greatly appreciated. This volume owes him much. Paul Badcock and Les Dean repaired and maintained the hardware. Christine Hughes, Jean Mathews and Lilian Prosser-Evans retyped numerous manuscripts. Their help is gratefully acknowledged.

Staff of the Institute of Earth Studies, University College of Wales, Aberystwyth have also provided invaluable help. We particularly thank Alicia Moguilevsky who has proof read all the manuscripts and remade a number of plates. In addition, Arnold Thawley and Ian Gulley have re-drawn or improved much of the original artwork.

John Whittaker (British Museum (Natural History), London) is thanked for his help and support in many ways, especially with respect to the 'World List of Scientific Periodicals'.

Denis Bates, Elaine Maybury and Guy Whatley are thanked for their support.

We also thank the referees who performed their task with speed, accuracy and diligence.

Finally, we extend our thanks to all the contributory authors of this volume.

# Preface

This volume represents a selection of the papers presented to the Tenth International Symposium on Ostracoda held at Aberystwyth in the summer of 1988 on the theme of 'Ostracoda and Global Events'. It is the latest volume in a formidable line of British Micropalaeontological Society publications.

The first symposium in this series was held in Naples in 1963 and subsequent meetings, approximately triennially, have been held in Hull, Newark (Delaware), Pau, Hamburg, Saalfelden, Houston and Shizuolka. The return of the symposium to the United Kingdom, twenty-one years after the Hull meeting, is very much a reflection of the current strength of ostracod studies in this country. The worldwide expansion of research into this group, by earth, life and environmental scientists continues apace and during the course of organizing the symposium we had correspondence with more than 800 specialists. The Aberystwyth meeting attracted some 150 participants, almost 50% more than any previous symposium.

The major theme of both the symposium and the volume is highly topical and reflects the current interest in both the detection and the consequences of past global change. In the section of the book, various leading authors, whose interests range from the Devonian to the present day, consider ways in which Ostracoda can be used as indices and monitors of past environmental change. From our increasing knowledge of the biology and ecology of this group and their known usefulness in evaluating past environments, surely it is but a short step to their employment in environmental modelling for the future?

Although the major section of this volume is concerned with global events, papers of major significance under a variety of other section headings indicate both the diversity and vibrancy of research into fossil and Recent Ostracoda and the complexity and utility of the group. Other sections of the book are entitled Biology and Genetics; Biostratigraphy; Deep Sea; Ecology; Morphology;

Palaeoecology; Palaeogeography, Zoogeography and Palaeozoogeography and Educational. These have all attracted major contributions from leading authorities in those fields.

This volume clearly demonstrates the strength of ostracod studies and also presents a vision for the future.

Robin Whatley & Caroline Maybury

# GLOBAL EVENTS

1

# Ostracoda and global events

Robin Whatley

Institute of Earth Studies, University College
of Wales, Aberystwyth, Dyfed, U.K.

## ABSTRACT

Some of the often overlooked considerations in
the relationship between organic evolution and
global events are examined. Catastrophism is
considered as an alternative to gradualism and is
shown to be unacceptable unless it is uniformitarian. It is argued that terrestrial causation, for
which evidence can be obtained, should be preferred to speculative extraterrestrial events in the
search for answers to the reasons for faunal and
floral change. Similarly, intrinsic causes of biotic
change should not be marginalized in favour of
totally extrinsic hypotheses.

Transgressions and regressions are shown to be
of importance in the evolution of Mesozoic Ostracoda.
However, the end-Cretaceous extinctions were
probably brought about by a Maastrichtian global
regression and massive and sustained volcanic activity
in the Deccan. Eruptions of this magnitude,
introduced huge amounts of gas, $H_2SO_4$ aerosols
and particulate matter into the upper atmosphere
and brought about a prolonged 'nuclear winter'
with global reduction in solar radiation and
temperature.

While global geological changes can be correlated with changes in Mesozoic and Cainozoic
ostracod faunas, these correlations should be
made with great caution. Morphological and
physiological evolutionary innovations are also
slow to influence diversity and origination/extinc-
tion rates. Advances in the means of carapace
articulation, enhanced sensorial capability and the
occupation of new niches, such as the infaunal, are
all shown to be of importance in influencing the
patterns of evolution of Mesozoic Ostracoda.

In the case of the huge non-marine ostracod
diversity increase over the Jurassic - Cretaceous
transition, this is shown to be not simply associated
with the wholesale global provision of 'Wealden'

environments, but also with pre-adaptive evolution of reproductive and dispersal strategies in the Cypridacea during the late Jurassic.

Lastly, the vexed question of extinction periodicity is extended to the Ostracoda for the first time. Although 2 (and possibly 3) of the 7 peaks of Mesozoic ostracod extinction do not correspond to those of other authors, the remainder do. More interesting is that the mean interval between the 7 peaks is 25.7 MY! No such correlation between the major ostracod extinction peaks in the Cainozoic can however be demonstrated.

## INTRODUCTION

Event hunting has become a fashionable pastime for many earth scientists in recent years. I am not immune from the fashion myself in that three International Geological Correlation Programme (IGCP) projects with which I am associated all have 'event' in the title. (IGCP No. 216, Global Bio-Events; No. 199, Rare Events in Geology and No. 246, Pacific Neogene Events). A brief survey of recent and current literature indicates that geologists and palaeontologists of all persuasions are increasingly publishing on events of one sort or another.

Fashions have always ebbed and flowed in science and both the title of this chapter and the major theme of this book demonstrates that ostracod workers are as much subject to it as are any of our peers. Although at present in second place to the search for the 'Philosophers' Stone' which will elucidate the problem of polymorphism in Ostracoda, I confidently predict that 'event studies' through the medium of Ostracoda will become increasingly popular. This is already evidenced by the growing amount of literature which seeks to relate observable fluctuations in the composition and character of successive fossil ostracod populations to so-called 'global events', be they palaeoceanographical, climatic, tectonic or extraterrestrial (Babinot & Airaud, this volume; Benson, this volume; Bless, 1988; Horne *et al.*, this volume; Jarvis *et al.*, 1988; Lethiers, 1988; Van Harten, 1984; Whatley, 1986; 1988; Whatley & Coles, 1987, in press).

Another part of current fashion demands that both the changes in the fortunes of animal and plant communities through time and the events which allegedly brought these changes about, must be sudden and dramatic. Gradualism, to an extent uniformitarianism and, in some instances, one might argue common sense and rationality, seem to be in retreat. While debates range as to whether these extrinsic events are extraterrestrial or geological or, in some cases both, the fact that intrinsic causes may play a part is largely overlooked. This is the case, despite the increasing amount of molecular evidence to the contrary. Little or no attention is paid to those few in the Earth Sciences who preach caution or dissent. There are some voices left, but they are all too muted and of the 'crying in the wilderness' variety; so that it becomes a pleasure to read the sceptic's viewpoint (Hoffman, 1989) for a change.

While it is, of course, remotely possible that all major faunal and floral changes through time are catastrophic in character and invariably linked to cataclysmic geological or astronomical happenings, as Stenseth & Maynard-Smith's (1984) Stationary Model of Evolution would largely have it, the very remoteness of the possibility renders it too implausible to consider seriously. The antithesis (in its extreme form), the Red Queen Hypothesis (Van Valen, 1973), which invokes biotic causation for all evolution is almost equally improbable. However, one wonders whether it is possible to stem the gadarene tide for long enough to reflect on a number of increasingly overlooked or marginalized considerations? For example, why so suddenly has almost every geological and palaeobiological phenomenon become an event? Why are our explanations of these phenomena more likely to be catastrophic (as opposed to gradualistic or uniformitarian) today than they were even 10 years ago? Could this perhaps be a reflection of the societies in which we live?

Perhaps we should ask what is an event? How can events be distinguished from non-events? To what extent can one be certain that an event had global, rather than regional or local consequences? Is it not possible that at least some bio-events were the product of 'normal' biological and evolutionary processes? If, however, we conclude that environmental perturbations, whatever their cause, are responsible for these bio-events, exactly *how* are they responsible? Also, if we do concede that there has always been an inevitable relationship between biotic and abiotic events in the past, must we also concede, as some would wish, that these events occurred periodically or cyclically? Lastly, we must consider how ostracods fit into all this?

What is an event? In attempting to address this question, one encounters an interesting problem which seems to arise from the very inadequate definition of many phenomena. In 'event science' there seems to be no clear distinction made in much of the literature between cause and effect. Both are frequently referred to as 'events'. For example, an 'impact event' is presumably causal, while 'extinction events' are presumably consequential. One would not always know this from reading the literature. The term 'event' has become an almost essential suffix for many writers who wax nonsensical about such things as 'limestone events' (or substitute any other lithology) or *Yoldia* (or any other taxon) events, etc. Perhaps, however, so called 'anoxic events' should take the prize for attracting the woolly-minded and for totally confusing cause and effect? To these I shall return later.

## EVENTS: THE CAUSES

### Terrestrial

Plate tectonics allows us to better understand the consequences of many geological processes. We now appreciate why many transgressions and regressions are synchronous and global and why, for example, major volcanic or earthquake activity is regional or local. We have now mapped the breaking up and coming together of continental blocks and have documented the history of oceans and seas throughout the Phanerozoic.

Such phenomena as the break up of Gondwanaland, the opening and subsequent closure of the Iapetus, the opening of the Atlantic and the Drake Passage and the closure of the Straits of

Panama and the Tethys etc., are all now dated to more or less acceptable limits and are no longer geo-myths. This archive of data, based on our understanding of plate tectonics, can be used to demonstrate that many events are the product of normal geological processes and must by definition, therefore, be uniformitarian in their consequences.

Transgressions and regressions can be correlated with changes in the nature of both animal and plant communities throughout the Phanerozoic. The extent of the effect is dependent on the biology of the organisms concerned. In the case of Ostracoda, I have shown (Whatley, 1986; 1988; in press a; b; Whatley & Coles, in press; Whatley & Stephens, 1976) that changes in their diversity and their rate of evolution during the Mesozoic and Cainozoic can arguably be related to these phenomena. I will return to this later.

Changes in the chemistry of the world's oceans and seas, particularly with respect to dissolved gases (notably oxygen) seem to have taken place from time to time in the past. These phenomena have come to be termed 'anoxic events' (Schlanger & Jenkyns, 1976). This is usually a misnomer since only some elements of the fauna or flora are deleteriously affected and those elements which either survive or flourish are certainly not anaerobic. The Greek prefix *'an'* means 'without' and it is very evident that most of these so-called anoxic events are merely times of reduced oxygen levels. They should, therefore, be renamed 'oxygen minimum' or 'oxygen reduction' events. Perhaps, however, we should use the term 'kenoxic' as introduced by Cepek and Kemper (1981) which they define as times when there were 'moderate concentrations of oxygen' rather than anaerobia. Such kenoxic events seem to be much more common than those involving true anoxia. Certainly the type of phenomenon described in this volume by Horne *et al.* are the product of a kenoxic event but the authors persist in referring to it as an 'anoxic event', despite advice to the contrary. The same comment could validly be made of most of the so-called 'anoxic events' described by a whole variety of authors (Arthur & Schlanger, 1979; Arthur *et al.*, 1987; Jarvis *et al.*, 1988; Jenkyns, 1980; Schlanger *et al.*, 1987). One wonders whether their ignorance is of science or Greek?

Some 'anoxic events' seem certainly to have been non-events. The pyritization of benthos alone is not good evidence for anoxia above the sediment/water interface. It is rather, indicative of strongly reducing conditions in early diagenesis or of later diagenetic chemical changes due to high temperature consequent upon deep burial or contact metamorphism. Notwithstanding this, the meerest whiff of pyrite and most black clays or shales (including those with a rich benthos) are deemed to represent 'anoxic events'.

Oxygen reduction seems to be brought about by initially subtle changes in the balance of organisms in the oceanic ecosystem, probably mainly the product of changes in circulation. This is likely to have been initially triggered by oceanographical changes brought about by tectonic or climatic changes. These processes were likely to have been gradual rather than rapid and also they are rather unlikely to have taken place in more than one ocean basin contemporaneously.

### Extraterrestrial

My first concern is with climatic changes. These were traditionally considered to be gradual and progressive but more recent research has suggested that some climatic changes, particularly it is argued those taking place in high latitudes during the past 2 MY or so, may have have come about relatively abruptly (Shackleton & Kennet, 1975; Shackleton & Opdyke, 1973; Porter, 1980).

Climatic changes seem to be an amalgam of the effects of both terrestrial and extraterrestrial processes. The most striking climatic changes in the past, particularly those resulting in glaciation must have been the consequence primarily of changes in the relationship of the planet to the sun. Similarly, the large scale shifts which result in the long term presence or absence of polar ice, cannot be explained by invoking terrestrial causes alone.

The consequences of the presence or absence of polar ice on oceanic circulation, on the very nature of the oceans and on global climatic belts

is fundamental and far-reaching and this is mirrored in animal and plant communities (Pianka, 1966). Compare, for example, those of the Mesozoic and Tertiary.

Geological processes, however, can also bring about far-reaching effects on climates and organisms. The elevation of mountain chains can create rain shadows which create deserts and bring about major faunal changes. Witness the post-Miocene faunal and floral changes in western North America and Patagonia. The opening and closure of the Straits of Panama had fundamental consequences for the faunas of South America, and the Southern Ocean must have been very different prior to the opening of the Drake Passage. Similarly, the Miocene closure of both ends of the Tethys had well-known major effects on Atlantic, Mediterranean and Indo-Pacific ostracod faunas (Coles *et al.*, this volume; Titterton & Whatley, 1988; Whatley, 1986; Whatley & Ayress, 1988; Whatley & Coles, 1987; in press).

In some cases global climatic changes seem to have had only regional or local effects on the biota. Relatively few marine animals became extinct during the Pleistocene and virtually no ostracods. With the onset of glacial episodes, marine organisms merely migrate down latitude (and also, downslope because of the drop in sea level) and return up latitude upon amelioration of the climate. The consequences of Pleistocene temperature changes on terrestrial faunas was much more marked (Martin & Klein, 1984).

In some cases climatic changes are brought about by the drift of continents through a range of latitudes. For example, during its migration northwards during the Cainozoic Australia has become progressively more arid with concomitant changes in its terrestrial fauna and flora (Hope, 1984; Horton, 1984; Kershaw, 1984).

Even the effects of the inception of the psychrosphere on the deep ocean benthos is debatable. While some authors argue that fundamental faunal changes took place at about 38 MY (Benson, 1975; this volume), others are unable to find any evidence for faunal change at this time, dramatic or otherwise (Coles, this volume; Millson, 1988; Whatley & Coles, in press; Whatley, *et al.*,

1983).

Solely extraterrestrial (and really non-uniformitarian) are bollide impacts. That they have occurred during the Earth's history is indisputable. Given both terrestrial and recent planetary evidence, it would be absurd to suggest that the Earth had escaped such impacts during its history. However, what is not known is what effects (if any) the impact, of even a relatively large body, would have on animal and plant communities and whether or not these would be global rather than local or regional. All the evidence for such a relationship is at best circumstantial although, to read some authors one would not imagine this.

To some who witness the demise of the ammonoids, the belemnites and the dinosaurs, together with major changes in the character and diversity of other groups, including some foraminifera and some shallow water ostracods at the Cretaceous-Tertiary boundary, this is evidence of a catastrophic event. Many and various are the postulated causes of these extinctions and faunal readjustments. However, what could be more suitable to the philosophy of the neo-catastrophists than a comet, asteroid or large meteor, crashing into the earth with enormous velocity and releasing iridium (an element most of us have never heard of in any other context) and dust into the upper atmosphere to blot out the sun and cause a 'nuclear winter'. Some authors seem to imagine that it is the iridium (which they surely would prefer to have been named *'eradicum'* or *'erasium'*) itself that is the agent of extinction, falling like some deadly cloud and wiping out entire types and communities of animals and plants. So much balderdash has been written on this subject that it almost defies belief.

The literature on bollide and asteroid impacts and their relationship to extinction has mushroomed since the publication of Alvarez *et al.* (1980) to include entire volumes (Silver & Schultz, 1982). Two relatively recent reviews (Hallam, 1988 and Raup, 1988) give all the subsequent essential literature. Van Valen (1984), is particularly articulate and thoughtful on the subject.

The zealous protagonists of the end-Cretaceous bollide impact extinction catastrophe either overlook or ignore such non-catastrophic considerations as the near terminal decline of the ammonoids and the major diversity decline of the dinosaurs long before the alleged impact event took place. Similarly, they are not at all distracted from their singleminded views by the survival into the Tertiary of many groups of reptiles (lizards, snakes, tortoises, turtles etc.) and many amphibians. Of the Cretaceous mammals, the placentals survived virtually intact but marsupials suffered some considerable extinction losses (Clemens, 1986). These, however, took place at some considerable time *before* the 'impact event'! Unfortunately, neither this evidence nor the fact that many deep-sea ostracods and benthonic foraminifera, Chinese non-marine ostracods, bivalves and gastropods cross the boundary largely unaffected, seems to influence them.

Neither the occurrence of other iridium enrichment levels elsewhere in the geological column which are not associated with actual or alleged mass extinctions nor the growing amount of evidence that much (why not all?) iridium is of terrestrial volcanic origin (Hansen *et al.*, 1988) seems to militate against bollide impact as the catastrophic cause of the end-Cretaceous mass extinctions.

What we must be clear about is that, if an impact had been responsible for global rather than regional or local extinctions, then it had to have been of a very large body. Certainly nothing like it has occurred in historical times and very doubtfully during the Tertiary. My own opinion is that no such impact with attendant mass extinctions can be certainly demonstrated to have occurred during the Phanerozoic.

None of the known impacts in history, or Holocene pre-history can be argued to have had anything but local or small scale regional effect, and that very transitory. Only the hypothetical impacts are claimed to have had major consequences and, given their alleged intensity and frequency, it is perhaps more surprising to encounter survivors rather than casualties.

It should not be forgotten that what is popularly thought to be the most powerful historical explosion, natural or man made, was the eruption of Krakatoa in 1883. This was an event of great power and destruction, the sound was heard as far away as St. Helena in the South Atlantic, almost 6,000 miles away, and the shock wave passed 7 times round the Earth. Actually, what was probably the most powerful explosion in history was ten times the magnitude of Krakatoa. This was the eruption of Tambora, a near neighbour of Krakatoa, in 1815. This was not only spectacular in its explosive force but, more importantly, the huge amount of effluvia discharged into the upper atmosphere was responsible for a 'nuclear winter' or 'volcanic summer' in the northern hemisphere. This was characterized by its very low temperatures, low sunlight and darkness at midday. The summer was exceptionally cold; enough for frost and snow in Bengal! It is now known that it is not so much the particulate matter which reflects the sun's rays, but aerosol droplets of $H_2SO_4$ in the stratosphere and above the ash levels. I know of no suggestion, however, despite the strength and global effects of these two explosions and the consequences of the large amounts of dust and aerosols introduced into the upper atmosphere, of any global effect on plant or animal species. The most abiding human memory seems to be of the magnificent red sunsets which the explosions are said to have caused, rather than the lack of summer.

It would be interesting to ascertain whether in, for example, polar ice cores, there could be recorded a Tambora, Krakatoa or a Mt. St. Helens iridium 'event'. Given that we have a rule of thumb measure of both the force of these explosion and their consequences, if dateable iridium levels could be encountered, their degree of enrichment could be used as an approximate scale by which to judge the intensity of the event which produced the iridium fallout at the Cretaceous-Tertiary boundary.

In the author's opinion, it is not necessary to invoke anything other than 'normal' uniformitarian terrestrial causes for the end Maastrichtian extinctions. Almost certainly the prime cause was the global regression of the time which was probably all the more effective because of the large scale series of transgressions, begun in the Aptian and

climaxing in the Upper Cretaceous, which preceded it. This transgressive phase, had for more than 50 MY, inundated a greater proportion of the continental crust than at any time in the Phanerozoic. This, together with the climatic cooling, for which there is good evidence (Savin, 1982), would produce a sufficiently significant environmental perturbation to account for the faunal and floral extinctions, crashes and changes at the Mesozoic-Tertiary boundary. The presence of an iridium enrichment level at the boundary could be of volcanic origin as could the associated shocked quartz (Carter *et al.*, 1986). There may, of course, have been a bollide impact at this time but only the geochemical and petrological evidence can be used to argue this; certainly not the palaeontological.

I personally do not find it difficult to rationalize the end Cretaceous extinctions, using straightforward uniformitarian arguments, invoking sea level and climatic changes. The latter were probably brought about by the 'nuclear winter' effect of the enormous volcanic outpourings of the Deccan. Recent research by Hammer and his colleagues, studying the stable isotopes, particulate matter and acidity of ice cores from Greenland, has demonstrated a direct relationship between volcanic activity and climatic deterioration (Hammer, 1980; 1984; 1985; Hammer *et al.*, 1980; 1987). Their cores clearly show the cooler years which follow such major eruptions as that of Santorini (1645 B.C.), Hekla 3 (1120 B.C.), Krakatoa (1883 A.D.) and Tambora (1815 A.D.). The end-Cretaceous eruption of the Deccan lavas seems to have been of an order of magnitude greater and may, therefore, have had a direct effect on extinction rates, especially among such vulnerable groups as the plankton. It certainly had to be much more powerful than any of the eruptions cited above since no global extinctions were caused by then and it is doubtful whether local or regional consequences were anything other than transitory. Much more difficult, I believe not yet possible to explain, is the mysterious extinction in the late Pleistocene (approximately 11000 years B.P.) of large mammals. In North America 73.7% of large mammals became extinct and the percentages for South America, Australia and Africa are 78.9%, 86.4% and 19.3%

respectively (Martin, 1967; 1984a; b; 1985). Among the numerous and varied suggestions as to the cause of this event, the author has not encountered one invoking bollide impact! Surprisingly, much less attention is paid to this event despite the fact that its consequences are much more striking, such as the complete extinction of the Equiidae in both North and South America, and the possibility that our ancestors may have had a hand in it!

## THE EFFECTS

The most obvious of these, as they apply to fossil organisms, are extinctions, originations, changes in diversity patterns, migrations etc. Of these, predictably, the former has captured popular imagination more than any other. The significance of the appearance in the fossil record of organisms with new feeding or reproductive strategies, such as the first infaunal bivalves, the first non-marine ostracods or the first marine mammals, for example, is usually not of popular concern. Dramatic changes in diversity patterns consequent upon the invasion of new environments such as the deep sea (Whatley, 1983) or the adaptive radiation of large ungulates, especially the Bovidae, in open grassland environments in the Miocene (Webb, 1969) are equally ignored.

If we believe that sudden, more or less mass extinctions followed by more or less mass originations can be correlated with global events of some sort or another, then we are consciously or not protagonists of the Stationary Model of (Stenseth and Maynard-Smith, 1984), which in effect, relates all evolutionary change to extrinsic causes.

If it could be demonstrated that mass extinctions involved numerous heterogeneous taxa which died out suddenly on a global scale, then perhaps they would warrant catastrophic explanations. However, an examination of most end-epoch mass extinction 'events' reveals a number of disturbing factors. Firstly, mass extinctions only seem to have occurred if one is considering higher taxa. Consider the end-Palaeozoic and end-Mesozoic extinctions as examples. If analysed at the ordinal or familial level, then terminal epoch

mass extinctions seem to have occurred. Consideration at the generic level, however, reveals that they are stepped while species level extinctions will appear more or less stepped according to the number of hiatuses. The fewer there are, the more likely the extinctions plotted against time will occur as a straight line, which is probably equivalent to the background level of extinction (Fig. 1). This figure shows that both the nature of the fossil record, as well as the level at which we consider it, can to an extent determine whether end epoch extinction events are sudden, stepped or gradual. (see also Kauffman, 1988).

There is little doubt, however, that major global changes in the nature of continental and shallow marine communities have been brought about by global transgressions and regressions. Since transgressions seem to take place much more slowly and cumulatively than regressions, arguably the latter should be responsible for the more dramatic consequences to animal and plant communities. Neither, however, should exert a major influence on oceanic biota unless, in their course, they were responsible for such phenomena as the opening of the Drake Passage, the closing of the Straits of Panama or the Iberian Portal etc. and, thereby, bringing about major changes in the patterns of oceanic circulation and oxygen levels.

It should be of universal concern that the current trend in deep-sea studies to correlate 'events' has sometimes seemingly led to the 'bending' of otherwise hard biostratigraphical data. To some, the 'event' has become the datum and, if necessary, all other evidence must be sublimated to proving its oceanic if not global extent. The search for 'anoxic' events is a rich ground for these phenomena and the present vogue is to look to 'anoxia' as the cause of most perturbations in oceanic biota.

One wonders to what extent it is possible for other, quite different causes to be responsible for observed biotic changes, such as short term but radical changes in ocean currents, or changes in the position of the boundaries between major and distinctive oceanic water masses. Is it not possible that these could lead to what, in the fossil record,

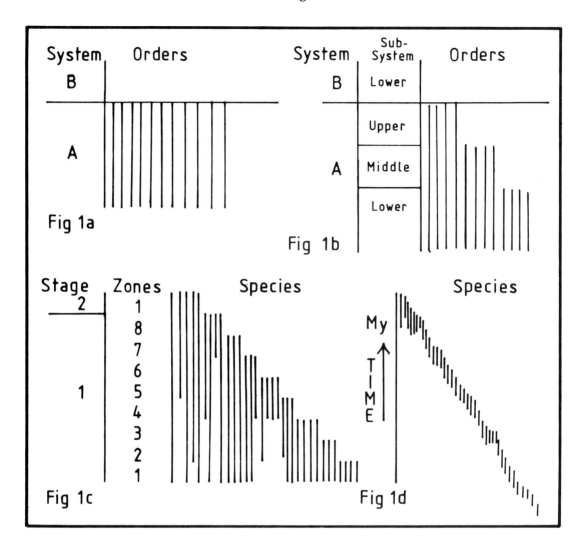

Fig. 1. Four ways of depicting the same feature: Fig. 1a. An extinction at a system boundary with taxonomic discrimination to ordinal level. Fig. 1b. An extinction within a system and at a system boundary at the ordinal level, but with refined stratigraphy. Fig. 1c. An extinction over a stage boundary with taxonomic and stratigraphical discrimination at the specific and zonal levels respectively. Note the stepped nature of the extinction model in this and the preceding figure. Fig. 1d. The probable true extinction of species given total stratigraphical coverage.

could be interpreted as the consequences of oxygen depletion.

The periodically devastating effects of El Niño to the total marine biota off the Pacific Coast of South America (Arntz, 1985) could, in a DSDP core, resemble an oxygen depletion event and could easily be quite incorrectly correlated with a similar local or regional event elsewhere in the world's oceans which was neither related nor contemporary.

## OSTRACODA AND EVENTS

Many papers in this volume are concerned with the global, regional or local changes in ostracod faunas, brought about by environmental changes. This has been a major preoccupation among students of the Ostracoda for several decades, although the local rather than the regional or global scale of research has predominated. On all three geographical scales, short-term changes in faunas are likely to be claimed to be the consequence of extrinsic (environmental) perturbations. This is simply because any intrinsic (evolutionary) changes taking place in the fauna may be too slow to detect. Generally these latter, while a possible factor to be considered in all studies, can only be isolated with certainty when longer-term studies are undertaken.

Notwithstanding this, it is always easier to attribute faunal or floral change to environmental rather than evolutionary causes. No doubt this is partly a function of the difficulty of obtaining total global evolutionary data and also to the simplistic philosophy which claims that evolution always actually takes place somewhere else. For example, it would be very difficult (and probably entirely erroneous) to demonstrate that the changeover within the Cenomanian *Plenus* Marls of Dover, from a fauna dominated by podocopids to one dominated by platycopids, (Horne *et al.*, this volume) was the product of global evolutionary changes within the Ostracoda, involving increased tolerance of dysaerobia by the genus *Cytherella*, rather than the response of the local/regional ostracod communities to local/regional environmental change, particularly kenoxia. So difficult to demonstrate are such possibilities that they are generally ignored. Although Horne *et al.* do not consider the biology of the two groups of ostracods concerned in their study, there is a simple biological explanation for the survival of the filter-feeding platycopids in reduced oxygen environments which cannot be tolerated by the Cytheracea or Cypridacea which feed in other ways. The Platycopida, in order to facilitate the circulation of water over the ventral surface of the body and from which they extract the particulate matter which constitutes their food, have more 'branchial' plates on their appendages than do the podocopids. The platycopids can, therefore, cope with reduced oxygen levels by virtue of the fact that they are able to extract a sufficiency from the larger volumes of water which they circulate.

Only those studies which are global (or deal with large regions) and which are long-term enough in scope can hope to detect major evolutionary changes in faunas but even these are often masked by (or mistaken for) those induced by global change.

Whatley & Stephens (1976), Whatley (1986, 1988), Whatley & Coles (in press) and Coles (this volume) have all studied changes in diversity and in origination and extinction rates over extended intervals of time. Whatley & Stephens (1976) studied fortunes of the Cytheracea throughout the Mesozoic, while Whatley (1986, 1988) studied all groups of Ostracoda through the same 183 MY interval. Coles (this volume) and Whatley & Coles (in press), studied the entire marine and brackish Cainozoic ostracod fauna (except the Myodocopina) from Europe, the North Atlantic and eastern North America and only the deep water Cainozoic faunas of the North Atlantic respectively.

All these studies reveal major changes in diversity levels and in the rates of both origination and extinction during the Triassic to Recent interval.

**The Mesozoic**

Whatley (1986, 1988) working with a database of 739 genera and 6797 species, demonstrated that it was possible to correlate diversity, extinction and origination maxima and minima for the Mesozoic with major global events, particularly transgression and regressions. It should be stressed that these were *possible correlations* and do not necessarily imply a cause and effect relationship. Indeed, he noted that some peaks and troughs on his histograms and graphs seemed not to correspond to known global changes. Figs 2-5, unlike earlier figures in Whatley (1986, 1988) and Whatley & Stephens (1976), are normalized for time. Fig. 2 shows the huge fluctuations in simple species diversity which took place during the Mesozoic and also, as ascertained by a comparison between inherited and new species, the considerable changes which took place in the rate of evolution. Fig. 3 shows that generic diversity and the rate of evolution at generic level, also fluctuated considerably. Apart from their fewer numbers, the generic

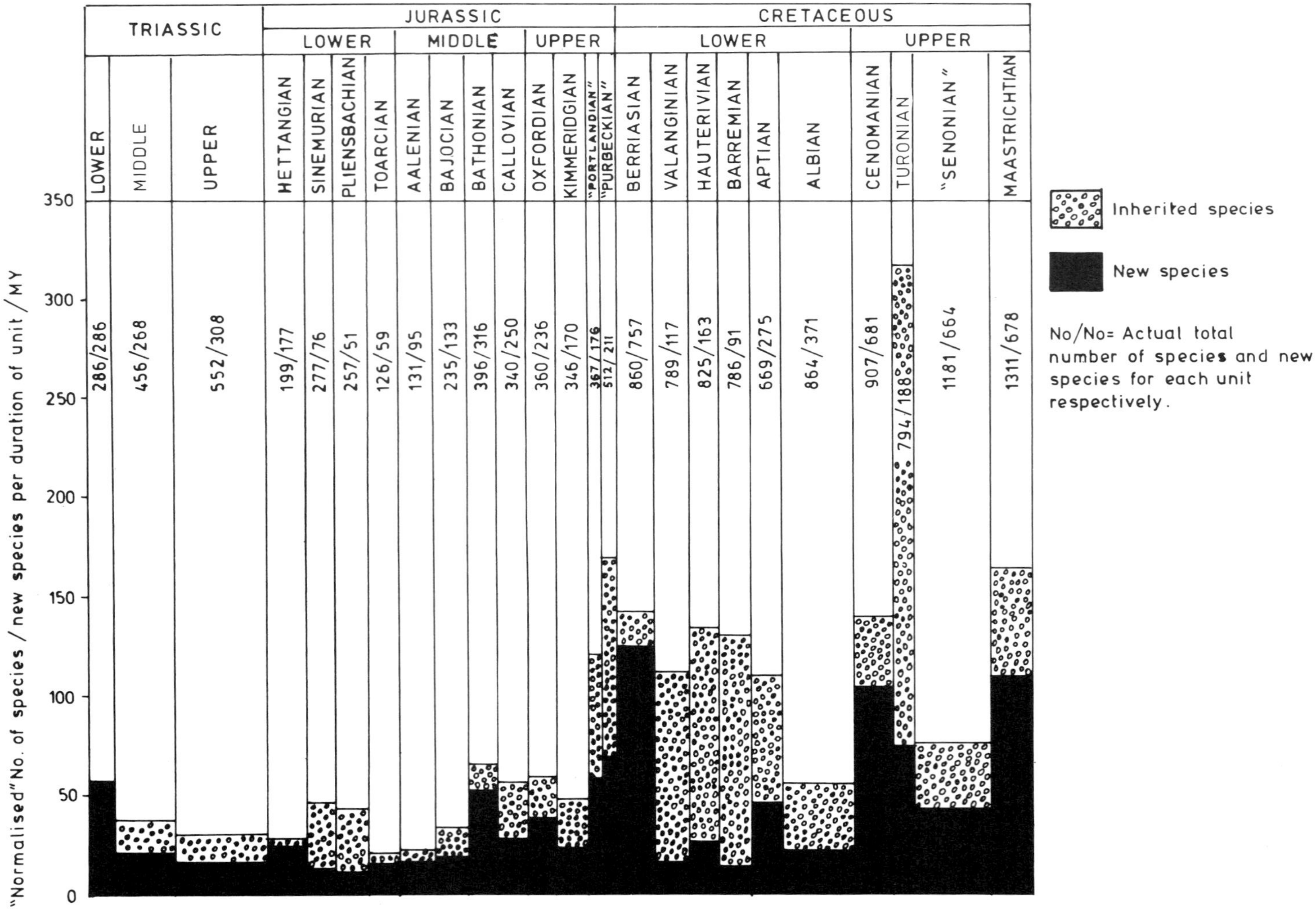

Fig. 2. Simple species diversity and the relationship between new and inherited species for the major stratigraphical units of the Mesozoic. Number of species for each unit normalized according to the duration of that unit.

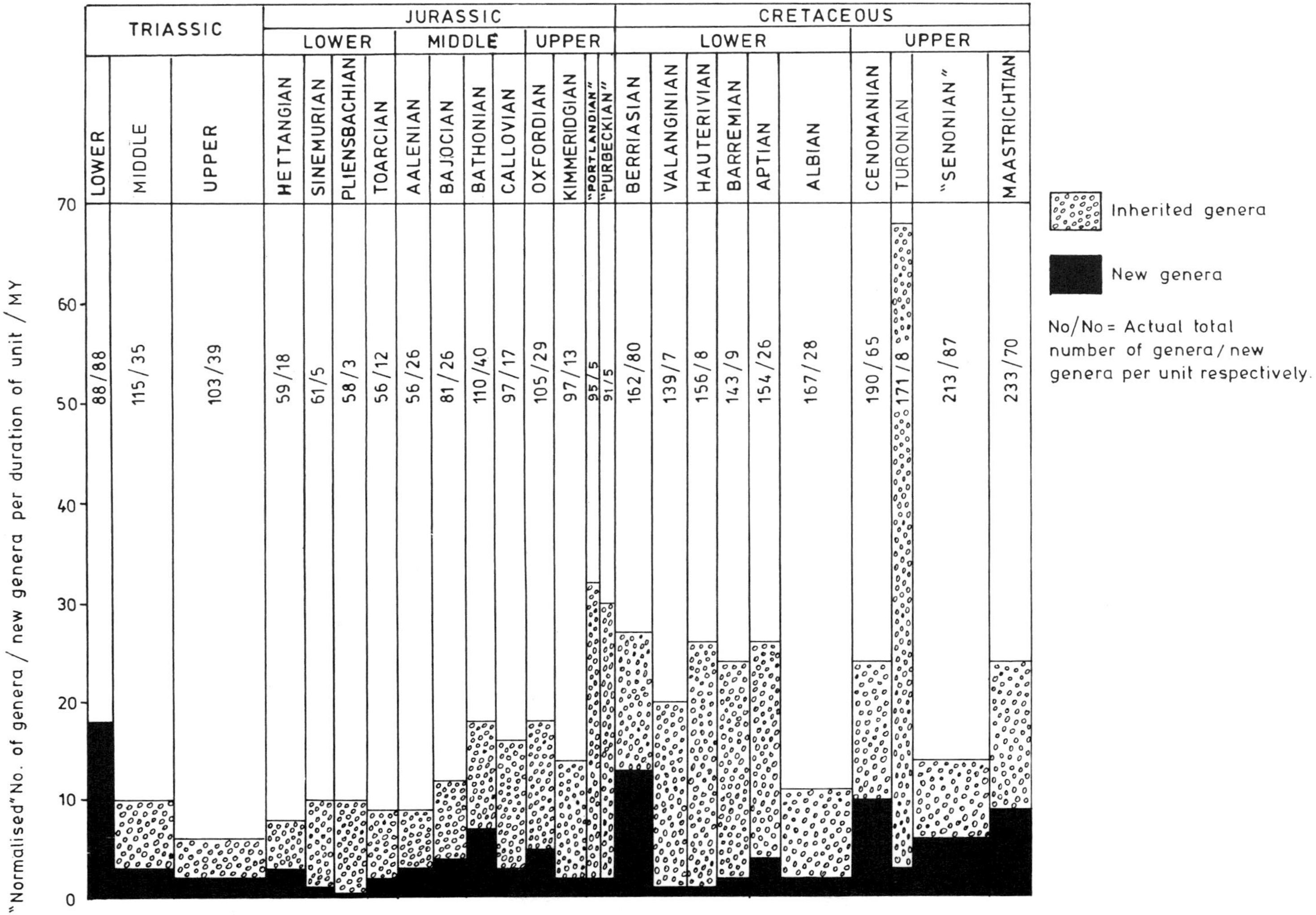

Fig. 3. Simple generic diversity and the relationship between new and inherited genera for the major stratigraphical units of the Mesozoic. Number of genera for each unit normalized according to the duration of that unit.

histograms closely mirror those for species and both diagrams show clearly the marked changes in ostracod faunas which took place at the Triassic-Jurassic, Jurassic-Cretaceous and Lower Cretaceous-Upper Cretaceous boundaries.

Figs 4 and 5 plot the normalized graphs for the origination and extinction of species and genera respectively. They illustrate a number of clear extinction peaks, especially with respect to species. These occur in the Upper Triassic, Pliensbachian, Bathonian, 'Purbeckian', Barremian and Turonian.

Whatley (1986, 1988) has previously argued that most of these peaks, both extinctions and originations, can be correlated with major Mesozoic global geological changes. For example, the Upper Triassic saw, with the Rhaetian transgression, the precursor of the Lower Liassic Transgression, while the Pliensbachian was something of a period of retrenchment prior to the Toarcian transgression. The Bathonian represents a Middle Jurassic regressive phase (at least in its upper part) in many parts of the world although the end-Jurassic-Lower Cretaceous regression was of several orders of magnitude more widespread and enduring. The 'Wealden' basins on most continents (not Asia) were terminated by the Aptian transgression, the precursor of the more major transgressions of the Upper Cretaceous. The Turonian extinction peak is arguably the product of a phase of oxygen depletion which began in the Cenomanian but whose consequences are most clearly seen in the succeeding stage. The short time interval of the Turonian tends to distort the figures but, the extinction level for species and (most importantly) genera is very high, indicating the severity of this event. Had the event been truly worldwide and had it involved anoxia, rather than kenoxia, then many more of the 71 genera known from the Turonian would have become extinct instead of just 19. The survival of the majority of the Ostracoda of the Turonian, the great majority of which were not filter feeders (Whatley, 1988, Fig. 8a) attests to the existence of large refugia where there was either no diminution in oxygen levels or very little.

There is, therefore, no lack of possible extrinsic causes which one can correlate with the various peaks in Figs 2-5. Whether such correlations between cause and effect are valid or not is another matter. Why do some transgressions or regressions correlate with ostracod extinction or origination peaks and not others? Is it reasonable to expect different causes (transgressions, stillstands, regressions, oxygen depletion) to produce the same effects? One can always justify these correlations if one believes implicitly that only major physical changes can produce such biological reactions.

Considerable evolutionary changes took place in the Ostracoda during the 183 MY of the Mesozoic. Triassic, Jurassic and Cretaceous faunas are all very distinctive and some of the morphological innovations evolved during the interval allowed their possessors to undertake quite distinct life styles, often in entirely new environments. Whatley (1987, 1988) has argued, therefore, that many of the diversity, origination and extinction peaks and troughs can be shown to be the product of biotic rather than abiotic phenomena.

The Bathonian is notable for peaks in all three criteria (Figs 2-5). Its ostracod fauna is interesting for a number of reasons:

a) The Cypridacea, previously virtually entirely marine, invade freshwater permanently for the first time.

b) Brackish water ostracod faunas are recognizable, using the same criteria by which we recognize modern ones, from this time onwards.

c) The entomodont hinge achieves its acme of development here, where it articulates numerous genera and species. This new hinge type allowed the carapace to become much thicker, heavier and more heavily ornamented than hitherto and thereby increased the range of habitats available to those ostracods which bore it to include those of high energy, for example. This was the time of the main adaptive radiation of the Progonocytheridae.

d) Certain groups of Ostracoda, notably the Schulerideidae and Cytherideidae, but also others, began at his time to increase their number of anterior radial pore canals. This allowed greatly increased sensorial ability anteriorly (in the direction of movement) and may have facilitated entry into infaunal and other environments, given more early warning of predators, improved location of food and/or better ability to compete with those groups which did not follow this trend.

My belief is that, at least with respect to the Bathonian, it is such factors as those outlined above which were responsible for the great diversity and evolutionary activity exhibited by Bathonian ostracods. Similarly, one of the reasons for the outstanding success of the entirely marine Trachyleberididae from the Neocomian (despite the regression) to the present day cannot be unrelated to their acquisition of the amphidont hinge, and their development of very numerous anterior radial pore canals. Similarly, the principal reason why the mean species diversity of Cretaceous marine faunas exceeds that of the Jurassic lies in the community structure of these faunas being very different. That of the Cretaceous is likely to be much more complete and to include a diverse infaunal population which had yet to evolve in the Jurassic. Perhaps the most obvious change in Mesozoic faunas took place around the time of the Jurassic-Cretaceous transition. The principal reason for this change is due to a reversal in the relative positions of dominance of the largely marine Cytheracea and the (from this time onwards) largely freshwater Cypridacea (Whatley, 1988, Fig. 8b). Since the origins of both superfamilies in the Lower Palaeozoic until the Neocomian; the cytherids had always dominated the cyprids and have always done so since the Aptian. For anyone seeking a relationship between bio- and geo-events, this correlation between the wholesale provision of non-marine aquatic environments consequent upon the end-Cretaceous regression and the rise of the freshwater Cypridacea seems to be ideal. Whatley (in press a, b) has considered this phenomenon in some detail and has shown that this relationship is not as simple as it seems at first sight.

Fig. 6 shows the simple species diversity of the Cypridacea and all Cytheracea and Darwinulacea throughout the Mesozoic and demonstrates the major change in dominance in the relationship between the cyprids and cytherids in the Neocomian-Aptian interval. Fig. 7 plots the simple species diversity of the Cypridacea, the Limnocytheridae (the only Mesozoic non-marine cytheraceans) and the Darwinulacea over the same interval.

Both figures show why, if the reason for the step increase in the number of freshwater ostracod taxa at the Jurassic-Cretacous boundary is a function of the wholesale global availability of 'Wealden' basins, of the three groups of freshwater ostracods, the Cypridacea were the only ones to take advantage of this provision. Also, how did they almost instantaneously expand their range from North West Europe in the late Turonian to world-wide in the Berriasian? Why did the darwinulids and limnocytherids signally fail to respond in the same way as the cyprids?

Whatley (in press a, b) has suggested that the answer lies in the different reproductive and dispersal strategies of the three groups.

The Darwinulacea have been in freshwater the longest; since the Devonian. They brood their young within the carapace, they cannot swim and, in the Mesozoic they seem to have reproduced both sexually and parthenogenetically. The limnocytherids appeared as freshwater ostracods in the Triassic although they may have existed with a different name in the Permian (Whatley, in press a) They have always reproduced sexually and have brooded their young within the carapace. They are incapable of swimming. Neither the darwinulids nor the limnocytherids, because of their brood care strategy have evolved desiccation resistant eggs.

The Cypridacea are very different. They seem not to have entered freshwater in a sustained and permanent manner until the late Bathonian and this seems to have happened in North West Europe (Whatley, in press a, b). Their subsequent history in Jurassic freshwater environments is one of important evolutionary innovation following, of course on the major evolutionary changes in their osmoregulatory physiology which allowed them to adapt to freshwater in the first place.

The marine ancestors of these pioneer Bathonian cyprids all reproduced sexually, as do all marine cyprids today. These ancestors were also rather poor swimmers or non-swimmers. Between the late Bathonian and the late Tithonian, some of

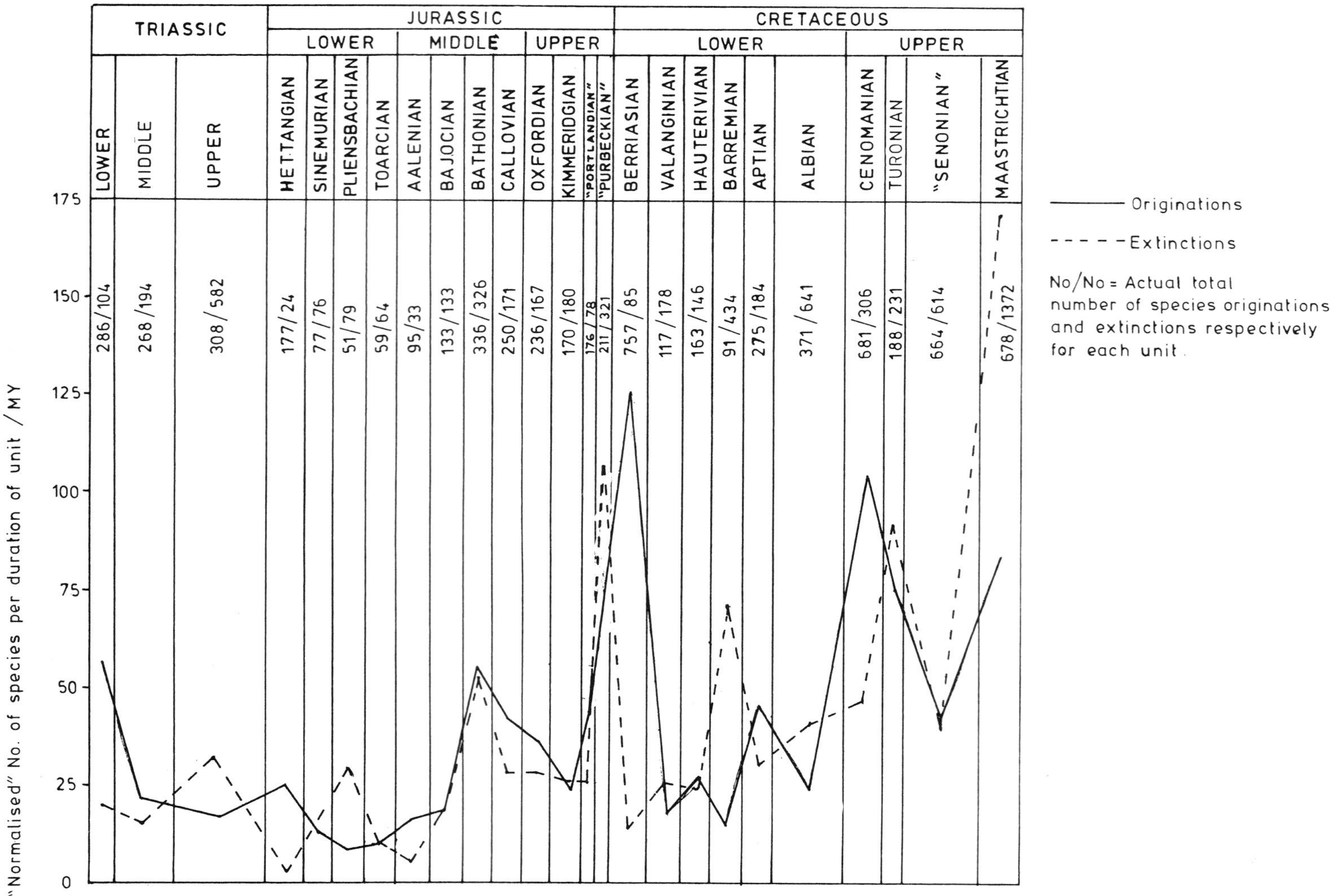

— Originations

– – – – Extinctions

No/No = Actual total
number of species originations
and extinctions respectively
for each unit

| | | | Actual total No/No |
|---|---|---|---|
| CRETACEOUS | UPPER | MAASTRICHTIAN | 678/1372 |
| | | "SENONIAN" | 664/614 |
| | | TURONIAN | 188/231 |
| | | CENOMANIAN | 681/306 |
| | LOWER | ALBIAN | 371/641 |
| | | APTIAN | 275/184 |
| | | BARREMIAN | 91/434 |
| | | HAUTERIVIAN | 163/146 |
| | | VALANGINIAN | 117/178 |
| | | BERRIASIAN | 757/85 |
| JURASSIC | UPPER | "PURBECKIAN" | 211/321 |
| | | "PORTLANDIAN" | 176/78 |
| | | KIMMERIDGIAN | 170/180 |
| | | OXFORDIAN | 236/167 |
| | MIDDLE | CALLOVIAN | 250/171 |
| | | BATHONIAN | 336/326 |
| | | BAJOCIAN | 133/133 |
| | | AALENIAN | 95/33 |
| | LOWER | TOARCIAN | 59/64 |
| | | PLIENSBACHIAN | 51/79 |
| | | SINEMURIAN | 77/76 |
| | | HETTANGIAN | 177/24 |
| TRIASSIC | UPPER | | 308/582 |
| | MIDDLE | | 268/194 |
| | LOWER | | 286/104 |

"Normalised" No. of species per duration of unit / MY

Fig. 4. Originations and extinctions of species through the Mesozoic. The data on which the graphs are plotted is normalized according to the duration of each unit; the actual numbers of species appearing and disappearing in each stratigraphical unit is given above the graph.

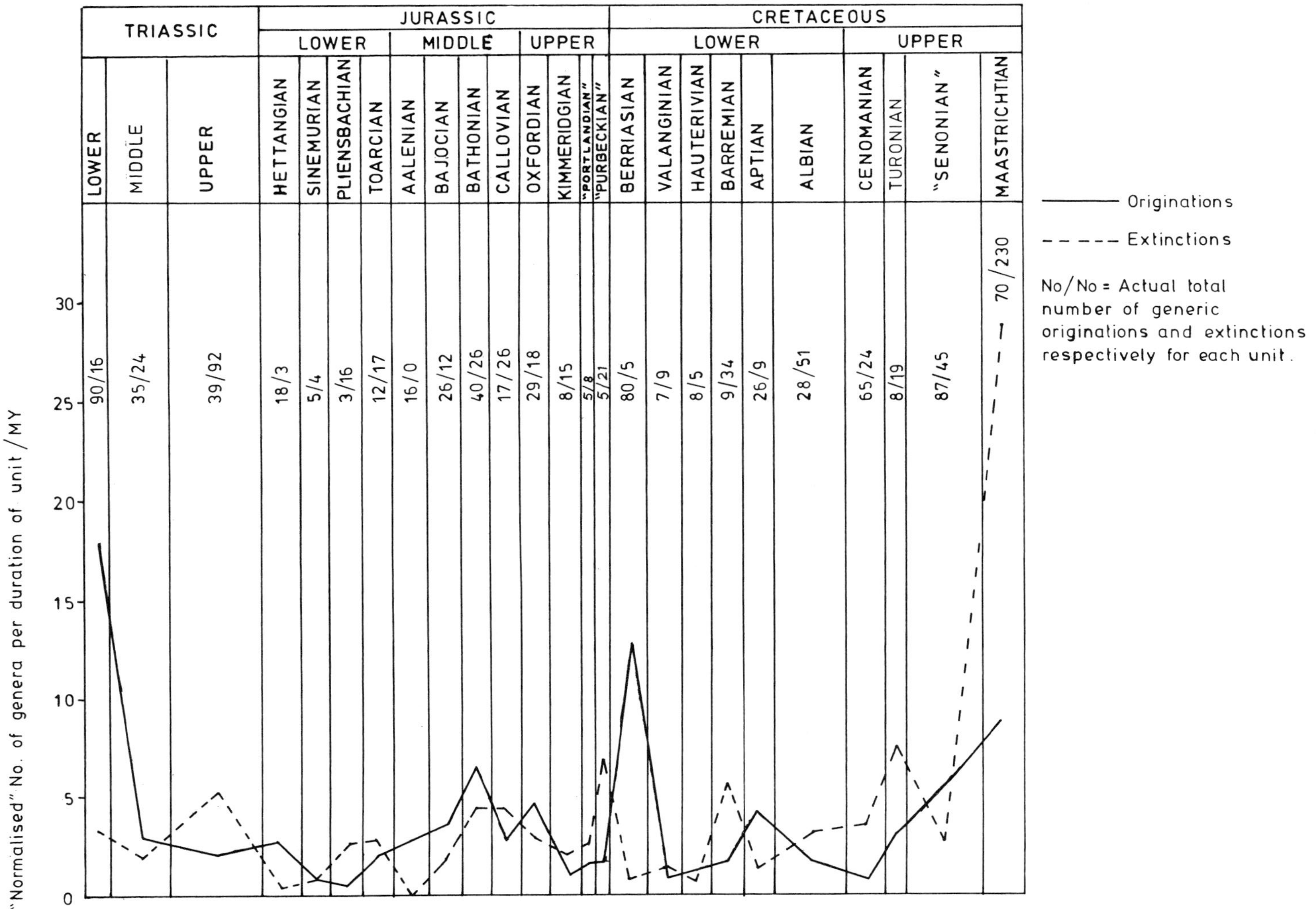

| Period | | | Unit | No/No |
|---|---|---|---|---|
| CRETACEOUS | UPPER | | MAASTRICHTIAN | 70/230 |
| | | | "SENONIAN" | 87/45 |
| | | | TURONIAN | 8/19 |
| | | | CENOMANIAN | 65/24 |
| | LOWER | | ALBIAN | 28/51 |
| | | | APTIAN | 26/9 |
| | | | BARREMIAN | 9/34 |
| | | | HAUTERIVIAN | 8/5 |
| | | | VALANGINIAN | 7/9 |
| | | | BERRIASIAN | 80/5 |
| JURASSIC | UPPER | | "PURBECKIAN" | 5/21 |
| | | | "PORTLANDIAN" | 5/8 |
| | | | KIMMERIDGIAN | 8/15 |
| | | | OXFORDIAN | 29/18 |
| | MIDDLE | | CALLOVIAN | 17/26 |
| | | | BATHONIAN | 40/26 |
| | | | BAJOCIAN | 26/12 |
| | | | AALENIAN | 16/0 |
| | LOWER | | TOARCIAN | 12/17 |
| | | | PLIENSBACHIAN | 3/16 |
| | | | SINEMURIAN | 5/4 |
| | | | HETTANGIAN | 18/3 |
| TRIASSIC | UPPER | | | 39/92 |
| | MIDDLE | | | 35/24 |
| | LOWER | | | 90/16 |

"Normalised" No. of genera per duration of unit /MY

30  25  20  15  10  5  0

Fig. 5. Originations and extinctions of genera through the Mesozoic. the data on which the graphs are plotted is normalized according to the duration of each unit; the actual numbers of genera appearing and disappearing in each stratigraphical unit is given above the graph. Note the difference in vertical scale for Figs 4 and 5.

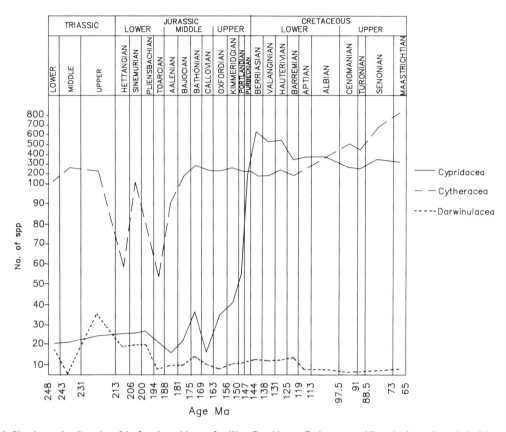

Fig. 6. Simple species diversity of the 3 podocopid superfamilies: Cypridacea, Cytheracea and Darwinulacea through the Mesozoic. Note the Barriesian-Aptian reversal of normal dominance relationships between the Cypridacea and Cytheracea. Data not normalized.

these freshwater cyprids had evolved parthenogenesis and this included the important *Cypridea* stock. At about the same time they also evolved an egg which was desiccation and freezing resistant and many of them, from their carapace morphology and phosphatized appendages, became accomplished swimmers.

Parthenogenetic reproduction and a desiccation and freezing resistant egg are formidable dispersal assets alone; in combination they brought about the remarkable dispersion of the cyprids. The egg could be wind-blown and even transported in the upper atmosphere and, with a parthenogenetic species, only one egg needed to be transported to a new locality to have the possibility of successful colonization. Their ability to swim also helped the cyprids to both dominate and maximize the environment (Whatley, in press a, b).

For all these reasons the cyprids were able to inherit the bounty of the world-wide 'Wealden' environments; for the same reasons the darwinulids and limnocytherids were unable to emulate them. Call the late Jurassic cyprids 'Hopeful Monsters' if you will. They certainly provide a classic example of the success of preadaptive evolution.

If this story has a moral, then that must be that in considering the way in which animals and plants respond to environmental change, one must never loose sight of the evolutionary dimension.

## PERIODICITY OF EXTINCTIONS

If the current debate on mass extinctions can be laid at the door of Alvarez *et al.* (1980), then discussion

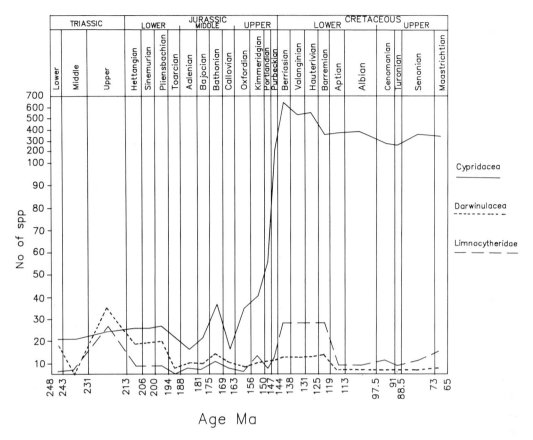

Fig. 7. The relationships between the Cypridacea (marine and freshwater), Darwinulacea and the Limnocytheridea during the Mesozoic in terms of simple species diversity. Note the reaction of the Cypridacea and, to a much more limited extent, the Limnocytheridae to the global availability of 'Wealden' environments and the virtual inability of the Darwinulacea to join this response.

of the periodicity of extinction events can be similarly shown to stem from a paper by Raup & Sepkoski (1984), in which they claim to demonstrate a 26 MY periodicity in extinctions through the Phanerozoic. The original hypothesis, however, seems to be attributable to Fischer & Arthur (1977), who argued a periodicity of extinction among Mesozoic and Tertiary pelagic ecosystems of 32 MY. A modern synopsis of the state of thinking on the subject, which includes all the essential and considerable literature, is given by Sepkoski (1989).

Much of this literature produced by the data manipulators relies on the various parts of the *Treatise on Invertebrate Paleontology* for generic and familial ranges. We students of the Ostracoda

are well aware of the extent to which such data in the ostracod volume (Part Q) is now out of date and misleading, and would be entitled to consider conclusions of extinction periodicity based upon these data with extreme scepticism. Some authors, notably Patterson & Smith (1987) have argued that, in general, *Treatise* data are so biased as to render the results void.

In the Mesozoic and Cainozoic, Sepkoski & Raup (1986) recognized the following extinction events. Rhaetian, Pliensbachian, Tithonian, Cenomanian, Maastrichtian, late Eocene and Middle Miocene. Sepkoski (1989) refines this somewhat by considering the main part of the late Triassic extinctions to have occurred in the Upper Norian and he also refers to the late Jurassic extinction as

Table 1. The interval between ostracod extinction peaks in the Mesozoic, as measured from the mid-point of each unit.

| Extinction Peak | Duration of Unit MY | Interval between Peaks MY |
|---|---|---|
| End-Permian | | |
| | | 26.0 |
| Upper Triassic | 18.0 | |
| | | 25.0 |
| Pliensbachian | 6.0 | |
| | | 26.0 |
| Bathonian | 6.0 | |
| | | 26.5 |
| Upper Tithonian | 3.0 | |
| | | 23.5 |
| Barremian | 6.0 | |
| | | 32.25 |
| Turonian | 2.5 | |
| | | 20.75 |
| Maastrichtian | 8.0 | |

Upper Tithonian. The Aptian is recognized as an additional stage with an extinction event.

The present author's data for the extinction of Mesozoic Ostracoda is second to none and it is considered at the specific and generic levels. These data clearly illustrate the following extinction peaks (Figs 4-5): Upper Triassic, Pliensbachian, Bathonian, 'Purbeckian', Barremian, Turonian and Maastrichtian. In attempting to determine the interval between these peaks, one is presented with a number of possibilities. For example, should the mid-point or the end of the unit in which the extinction peak occurs be used to measure the interval? In Tables 1-4, the mid-point has been used; a rather different result will be obtained if the end of units is used.

The intervals between the Jurassic extinction peaks for Ostracoda are very close to Raup & Sepkoski's 26 MY interval, although they do not recognize a Bathonian event, which is very clearly

Table 2. The interval between ostracod extinction peaks in the Cretaceous, measured from the mid-point of units and taking the Cenomanian rather than the Turonian data to represent an extinction peak.

| Extinction Peak | Duration of Unit MY | Interval between Peaks MY |
|---|---|---|
| Upper Tithonian | 3.0 | |
| | | 23.5 |
| Barremian | 6.0 | |
| | | 27.75 |
| Cenomanian | 6.5 | |
| | | 25.25 |
| Maastrichtian | 8.0 | |

Table 3. The interval between ostracod extinction peaks in the Cainozoic, measured from mid-point of each unit. (Data from Coles, this volume, entire fauna).

| Extinction Peak | Duration of Unit MY | Interval between peaks MY |
|---|---|---|
| Cretaceous-Tertiary | | |
| | | 27.0 |
| Upper Eocene | 6.0 | |
| | | 13.4 |
| Upper Oligocene | 3.2 | |
| | | 19.5 |
| Upper Miocene | 6.2 | |
| | | 5.5 |
| Upper Pliocene | 1.3 | |
| | | 2.0 |
| Upper Quaternary | 0.88 | |

depicted in the ostracod data. In the Cretaceous, the interval is less regular. The Ostracoda record distinct Barremian and Turonian extinctions while the Aptian is an extinction trough. The Maastrichtian extinction is artificially high because all taxa are assumed not to cross the Cretaceous-Tertiary boundary.

If, however, the Turonian extinction, which as Whatley & Stephens (1976) and Whatley (1986, 1988) have noted, is a strange one, and largely the product of events which took place in the Cenomanian, is largely due to oxygen diminution, perhaps the intervals should be measured to the mid-Cenomanian. This is done in Table 2. With these 'massaged' Cretaceous figures, the mean interval between the 7 ostracod extinction peaks in the Mesozoic is 25.71; quite remarkably close to Raup & Sepkoski's 26 MY periodicity! I have made no attempt to sophisticate these statistics; there seems little point since, in a sense, the

Table 4. The interval between ostracod extinction peaks in the Cainozoic deep water of the North Atlantic. Intervals measured from the mid-point of each unit.

| Extinction Peak | Duration of Unit MY | Interval between Peaks MY |
|---|---|---|
| Cretaceous-Tertiary Boundary | | |
| | | 17.75 |
| Middle Eocene | 6.5 | |
| | | 18.55 |
| Upper Oligocene | 8.2 | |
| | | 20.6 |
| Upper Miocene | 6.2 | |
| | | 5.55 |
| Upper Pliocene | 1.8 | |
| | | 2.0 |
| Upper Quaternary | 0.88 | |

quoted figure for the duration of unit is a measure of the experimental error.

While these figures for the Mesozoic are based on global data (Whatley, 1986; 1987) those for the Cainozoic are from western Europe, the North Atlantic and eastern North America (Coles, this volume). From Fig. 4 of Coles, it is possible to detect the extinction peaks for Ostracoda based on this large regional data which, one hopes will reflect global patterns.

These peaks are different from those of Sepkoski & Raup (1986), in that they recognize an extinction peak in the Middle rather than the Upper Miocene. The Upper Eocene peak is the same, however, and this occurs some 27 MY after the inception of the Tertiary.

When only the ostracods from the deep water Cainozoic of the North Atlantic are considered (Whatley & Coles, in press; Coles, this volume) a rather different extinction periodicity is produced.

While the mean interval between the 5 extinction peaks for the entire Cainozoic fauna of the region (Table 3) is 13 MY, for the deep water Atlantic only (Table 4) it is 12.29.

While on the whole the Mesozoic ostracod extinction data support an hypothesis of 26 MY extinction periodicity, no such clear support is forthcoming from the Cainozoic data. No doubt the data could be 'massaged' in order to improve their fit to existing dogma. The author, however, prefers to leave this task to a professional data manipulator.

With respect to the amplitude and relative magnitude of extinctions, Sepkoski (1989, Table 3) shows that the Upper Permian, Upper Triassic and Maastrichtian saw the disappearance of the highest number of specific and generic taxa. Of these, the former is much the largest; the latter two being 30% less and about equal. The remainder of the Mesozoic peaks are only about 1/2 the magnitude of the Upper Triassic and Maastrichtian and 1/3 that of the Upper Permian. The amplitude of the Tertiary extinctions is 1/2 to 1/3 smaller again.

Although the present data, in its present form cannot be directly compared to that of Sepkoski, it is interesting to note that the relative amplitudes

of the various peaks are quite different. In declining order, as Fig. 4. shows, they are as follows: Maastrichtian (but remember that this is hypothetical), Upper Tithonian, Turonian, Barremian, Bathonian, Upper Triassic and Pliensbachian.

## ENVOI

With the increasing emphasis on and funding for projects concerning global change, the present author expects to see something of a shift in ostracod studies in the next decade. These organisms with their sensitivity to environmental change are ideal indices and one hopes that they can be used increasingly and imaginatively in the future to detect and evaluate past global changes and thereby help us to prepare for those to come.

## ACKNOWLEDGEMENTS

The author wishes to thank Dr Caroline Maybury for reading, discussing and improving the manuscript. He also gratefully acknowledges access to data made available by Graham Coles. Arnold Thawley produced the diagrams and many authors have inspired the author to write this review.

## REFERENCES

Alvarez, L. W., Alvarez, W., Asaro, F. & Michel, H. V. 1980. Extraterrestrial cause for the Cretaceous-Tertiary extinction. *Nature*, London, **308**, 718-720.

Arntz, W. 1985. Zur Entstehung von Organisme-nansammlungen: "El Niño", 1982-1983 vor Peru. *Nutur. Mus., Frankf.*, Frankfurt, **114**, 134-151.

Arthur, M. A. & Schlanger, S. O. 1979. Cretaceous "oceanic anoxic events" as causal factors in development of reef-reservoired giant oil fields. *Bull. Am. Ass. Petrol. Geol.*, Chicago, **63**, 870-885.

Arthur, M. A., Schlanger, S. O. & Jenkyns, H. C. 1987. The Cenomanian-Turonian oceanic anoxic event. II. Palaeoceanographic controls on organic-matter production and preservation. *In* Brooks, J. & Fleet, A. J. (Eds), *Marine Petroleum Source Rocks. Geol. Soc. spec. Publ.*, **26**, 401-420. Blackwell Scientific, Oxford.

Benson, R. H. 1975 The origin of the psychrosphere as recorded in changes in deep-sea ostracode assemblages. *Lethaia*, Oslo, **8**, 69-83.

Benson, R. H. 1988. Ostracods and Paleoceanography. *In* De Deckker, P., Colin, J.-P. & Peypouquet, J.-P. (Eds),

*Ostracoda in the Earth Sciences*, 1-26. Elsevier, Amsterdam, Oxford, New York, Tokyo.

Bless, M. J. M. 1988. Possible causes for the change in ostracod assemblages at the Maastrichtian-Palaeocene boundary in southern Limburg, The Netherlands. *Meded. Wkgrp. Tert. Kwart. Geol.*, Rotterdam, **25**(2-3), 197-211.

Carter, N. L., Officer, C. B., Chesner, C. A., & Rose, W. I. 1986. Dynamic deformation of volcanic ejecta from the Toba Caldera: Possible relevance to Cretaceous-Tertiary boundary phenomena. *Geology*, Boulder, Co., **14**, 380-383.

Cepek, P. & Kemper, E. 1981. Der Blättertonstein des nordwestdeutschen Barrême und die Bedeutung des Nannoplanktons für die fein laminierten, anoxisch entstandenen Gesteine. *Geol. Jb.*, Hannover, A 58, 3-13.

Clemens, W. A. 1986. Evolution of the terrestrial vertebrate fauna during the Cretaceous-Tertiary transition. *In* Elliot, D. K. (Ed.), *Dynamics of Extinction*, 63-85. Wiley, New York, Chichester, Brisbane, Toronto, Singapore.

Fischer, A. G. & Arthur, M. A. 1977. Secular variations in the pelagic realm. *In* Cook, H. E. & Enos, P. (Eds), *Deep-Water Carbonate Environments*. *Soc. Econ. Paleont. Mineralogists Spec. Pub.* **25**, 19-50.

Hallam, A. 1988. A compound scenario for the end-Cretaceous mass extinctions. *In* Lamolda, M. A., Kauffman, E. G. & Walliser, O. H. (Eds), *Palaeontology and Extinction: Extinction Events III Jornadas de Paleontología. Revta esp. Paleont.*, Madrid, No. Extraordinario, 7-20.

Hammer, C. U. 1980. Acidity of polar ice cores in relation to absolute dating, past volcanism, and radio echoes. *J. Glaciol.*, London, **25**(93), 359-372.

Hammer, C. U. 1984. Traces of Icelandic eruptions in the Greenland Ice Sheet. *Jokull*, **34**, 51-65.

Hammer, C. U. 1985. The influence on atmospheric composition of volcanic eruptions as derived from ice-core analysis. *Ann. Glaciol.*, **7**, 125-129.

Hammer, C. U., Clausen, H. B. & Dansgaard, W. 1980. Greenland ice sheet evidence of post-glacial volcanism and its climatic impact. *Nature*, London, **288**(5788), 230-235.

Hammer, C. U. Clausen, H. B., Friedrich, W. L. & Tauber, H. 1987. The Minoan eruption of Santorini in Greece dated to 1645 BC? *Nature*, London, **328**(6130), 517-519.

Hansen, H. J., Gwozdz, R. & Rasmussen, K. L. 1988. High - resolution trace element chemistry across the Cretaceous-Tertiary boundary in Denmark. *In* Lamolda, M. A., Kauffman, E. G. & Walliser, O. H. (Eds), *Palaeontology and Extinction: Extinction Events III Jornadas de Paleontología. Revta esp. Paleont.*, Madrid, No. Extraordinario, 21-30.

Hoffman, A. 1989. Mass extinctions: the view of a sceptic. *J. Geol. Soc.* London, **146**, 21-35.

Hope, G. 1984. Australian environmental change. Timings, directions, magnitudes and rates. *In* Martin, P. S. & Klein, R. G. (Eds), *Quaternary Extinctions. A prehistoric revolution*, 681-690. University of Arizona Press, Tucson.

Horton, D. R. 1984. Red kangaroos; last of the Australian Megafauna. *In* Martin, P. S. & Klein, R. G. (Eds), *Quaternary Extinctions. A prehistoric revolution*, 639-680. University of Arizona Press, Tucson.

Jarvis, I., Carson, G. A., Cooper, M. K. E., Hart, M. B., Leary, P.

N., Tocher, B. A., Horne, D. & Rosenfeld, A. 1988. Microfossil assemblages and the Cenomanian-Turonian (late Cretaceous) Oceanic Anoxic Event. *Cret. Res.*, Academic Press, London, **9**, 3-103.

Jenkyns, H.C. 1980. Cretaceous anoxic events - from continents to oceans. *J. Geol. Soc. London*, **137**, 171-188.

Kauffmann, E. G. 1988. The dynamics of marine stepwise mass extinction. *In* Lamolda, M. A., Kauffman, E. G. & Walliser, O. H. (Eds), *Palaeontology and Extinction: Extinction Events III Jornadas de Paleontología. Revta esp. Paleont.*, Madrid, No. Extraordinario, 51-71.

Kershaw, P. 1984. Late Cenozoic plant extinctions in Australia. *In* Martin, P. S. & Klein, R. G. (Eds), *Quaternary extinctions. A prehistoric revolution*, 691-707. University of Arizona Press, Tucson.

Lethiers, F. 1988. La Moyenne des Durees des Especes (MDE): une approche nouvelle de l'Evolution. Application aux Ostracodes. *C. r. hebd. Seanc. Acad. Sci.*, Paris, **307**, serie II, 871-877.

Martin, P. S. 1967. Prehistoric overkill. *In* Martin, P. S. & Wright, H. E. (Eds), *Pleistocene Extinctions*, 75-120. Yale University Press, New Haven.

Martin, P. S. 1984a. Catastrophic extinctions and late Pleistocene blitzkrieg: two radiocarbon tests. *In* Nitecki, M. H. (Ed.), *Extinctions*, 17-35. University of Chicago Press, Chicago.

Martin, P. S. 1984b. Prehistoric overkill: the global model. *In* Martin, P. S. & Klein, R. G. (Eds), *Quaternary extinctions: a prehistoric revolution*, 354-403. University of Arizona Press, Tucson.

Martin, P. S. 1985. Refuting late Pleistocene extinction models. *In* Elliot, D. K. (Ed.), *Dynamics of Extinction*, 107-130. Wiley, New York, Chichester, Brisbane, Toronto, Singapore, 107-130.

Martin, P. S. & Klein, R. G. (Eds), 1984. *Quaternary extinctions. A prehistoric revolution*, i-x, + 892 pp. University of Arizona Press, Tucson.

Millson, K. J. 1988. *The palaeobiology of Palaeogene Ostracoda from Deep Sea Drilling Project Cores in the South West Pacific*. Unpub. Ph.D. thesis, University of Wales, Aberystwyth, 2 vols, 769 pp., 48 pls.

Patterson, C. & Smith, A. B. 1987. Is the periodicity a taxonomic artefact? *Nature*, London, **330**, 248-251.

Pianka, E. R. 1966. Latitudinal gradient in species diversity: a review of concepts. *Am. Nat.*, Salem, Mass., **100**, 33-46.

Porter, S. C. 1980. Rapid deglaciation of alpine regions at the end of the last glaciation. *Am. Quat. Assoc. 6th. Bien. Mtg. Abstract*, 157.

Raup, D. M. 1988. The role of extraterrestrial phenomena in extinction. *In* Lamolda, M. A., Kauffman, E. G. & Walliser, O. H. (Eds). *Palaeontology and Extinction: Extinction Events II Jornadas de Paleontología. Revta esp. Paleont.*, Madrid, No. Extraordinario, 99-106.

Raup, D. M. & Sepkoski, J. J. 1984. Periodicity of extinctions in the geologic past. *Proc. Natl Acad. Sci. U.S.A.*, **81**, 801-805.

Savin, S. M. 1982. Stable isotopes in climatic reconstructions. *In* *Climate in Earth History*, 164-171. National Academy Press, Washington DC.

Schlanger, S. O., Arthur, M. A., Jenkyns, H. C. & Scolle, P. A.

1987. The Cenomanian-Turonian oceanic anoxic event, I. Stratigraphy and distribution of organic carbon-rich beds and the marine 13C excursion. *In* Brooks, J. & Fleet, A. J. (Eds). *Marine Petroleum Source Rocks. Geol. Soc. spec. Publ.*, **26**, 371-399. Blackwell Scientific, Oxford.

Schlanger, S. O. & Jenkyns, H. C. 1976. Cretaceous oceanic anoxic events: causes and consequences. *Geologie Mijnb.*, Den Haag, **55**, 179-184.

Sepkoski, J. J. 1989. Periodicity of extinction and the problem of catastrophism in the history of life. *J. Geol. Soc. London*, **146**, 7-19.

Sepkoski, J. J. & Raup, D. M. 1986. Periodicity in marine extinction events. *In* Elliott, D. K. (Ed.), *Dynamics of Extinction*, 3-36. Wiley, New York.

Shackleton, N. J. & Opdyke, N. D. 1973. Oxygen isotope and palaeomagnetic stratigraphy of equatorial Pacific core V28-239; oxygen isotope temperatures and ice volumes on a $10^5$ - $10^6$ year scale. *Quaternary Res.*, New York and London, **3**, 39-55.

Shackleton, N. J. & Kennet, J. P. 1975. Palaeo-temperature history of the Cainozoic and the initiation of the Antarctic glaciation; oxygen and carbon isotope analyses in DSDP sites 277, 279, and 281. *In* Kennet, J. P. *et al.*, *Init. Reports DSDP*, **29**, 743-755. U.S. Govt. Printing Office, Washington, **29**, 743-755.

Silver, L. T. & Schultz, P. H. (Eds), 1982. *Geological implications of impact of large asteroids and comets on the Earth*, 528 pp. *Spec. Pap. geol. Soc. Am.*, Boulder.

Stenseth, N. C. & Maynard-Smith, J. 1984. Coevolution in ecosystems: red queen evolution or stasis? *Evolution*, **38**(4), 179-184.

Titterton, R. & Whatley, R. C. 1988. The provincial distribution of shallow water Indo-Pacific marine Ostracoda: origin, antiquity, dispersal routes and mechanisms. *In* Hanai, T., Ikeya, N. & Ishizaki, K. (Eds), *Evolutionary biology of Ostracoda, its fundamentals and applications*, proceedings of the Ninth International Symposium on Ostracoda, held in Shizuoka, Japan, 29 July - 2 August 1985, Developments in palaeontology and stratigraphy, **11**, 759-786, Kodansha Ltd., Tokyo and Elsevier, Amsterdam, Oxford, New York, Tokyo.

Van Harten, D. 1984. A model of estuarine circulation in the Pliocene Mediterranean based on new ostracode evidence. *Nature*, London, **312**, 359-361.

Van Valen, L. 1973. A new evolutionary law. *Evolut. Theory*, Chicago, **1**, 1-30.

Van Valen, L. 1984. Catastrophes, expectations, and the evidence. *Palaeobiol.*, **10**, 121-137.

Webb, S. D. 1969. Extinction-origination equilibrium in late Cenozoic land mammals of North America. *Evolution*, **23**, 688-702.

Whatley, R. C. 1983. Some aspects of the palaeobiology of Tertiary deep-sea Ostracoda from the S.W. Pacific. *J. micropalaeontol.*, London, **2**, 83-104.

Whatley, R. C. 1986. Biological events in the evolution of Mesozoic Ostracoda. *In* Walliser, O. H. (Ed.), *Lecture Notes in Earth Sciences 8, Global Bio-Events*, 257-265. Springer-Verlag, Berlin, Heidelberg.

Whatley, R. C. 1987. The southern end of Tethys: an important locus for the origin and evolution of both deep and shallow water Ostracoda. *In* McKenzie, K. G. (Ed.), *Shallow Tethys 2*, proceedings of the International Symposium on Shallow Tethys 2 Wagga Wagga/ 15-17 September 1986, 461-475. A. A. Balkema, Rotterdam, Boston.

Whatley, R. C. 1988. Patterns and rates of evolution among Mesozoic Ostracoda. *In* Hanai, T., Ikeya, N. & Ishizaki, K. (Eds), *Evolutionary biology of Ostracoda, its fundamentals and applications*, proceedings of the Ninth International Symposium on Ostracoda, held in Shizuoka, Japan, 29 July - 2 August 1985, Developments in palaeontology and stratigraphy, **11**, 1021-1040, Kodansha Ltd., Tokyo and Elsevier, Amsterdam, Oxford, New York, Tokyo.

Whatley, R. C. In press a. The reproductive and dispersal strategies of Cretaceous non marine Ostracoda: The key to pandemism. *Proc. 1st Int. Symp. on non marine Cretaceous correlations*, IGCP Project 245, Urumqi, China, August 1987.

Whatley, R. C. In press b. The relationship between extrinsic and intrinsic events in the evolution of Mesozoic non marine Ostracoda. *Proc. 2nd International Conference on Global Bio-Events*, IGCP Project 216, Bilbao, Spain, October, 1987.

Whatley, R. C. & Ayress, M. 1988. Pandemic and endemic distribution patterns in Quaternary bathyal and abyssal Ostracoda. *In* Hanai, T., Ikeya, N. & Ishizaki, K. (Eds), *Evolutionary biology of Ostracoda, its fundamentals and applications*, proceedings of the Ninth International Symposium on Ostracoda, held in Shizuoka, Japan, 29 July - 2 August 1985, Developments in palaeontology and stratigraphy, **11**, 739-758, Kodansha Ltd., Tokyo and Elsevier, Amsterdam, Oxford, New York, Tokyo.

Whatley, R. C. & Coles, G. 1987. The late Miocene to Quaternary Ostracoda of Leg 94, Deep Sea Drilling Project. *Revta esp. Micropaleont.*, Madrid, **19**, 33-97.

Whatley, R. C. & Coles, G. In press. Global change and the biostratigraphy of Cainozoic deep water Ostracoda from the North Atlantic. *J. micropalaeontol.*, London.

Whatley, R. C. Harlow, C. J., Downing, S. E. & Kesler, K. J. 1983. Some observations on the origin, evolution, dispersion, and ecology of the genera *Poseidonamicus*, Benson and *Bradleya* Hornibrook. *In* Maddocks, R. F. (Ed.), *Applications of Ostracoda*, proceedings of the Eighth International Symposium on Ostracoda, July 26-29, 1982, 581-590. Univ. Houston Geos., Houston, Texas.

Whatley, R. C. & Stephens, J. M. 1976. The Mesozoic explosion of the Cytheracea. *Abh. Verh. naturw. Ver. Hamburg.*, N.F. 18/19, (Suppl.), 63-76.

# 2

# The effect of global events on the evolution of Cenomanian and Turonian marginal Tethyan ostracod faunas in the Mediterranean region

**Jean-François Babinot & Christine Crumière-Airaud**

Centre for Sedimentology and Palaeontology, University of Provence, Marseilles and U.R.A. 1208 C.N.R.S., France

## ABSTRACT

During the Cenomanian-Turonian interval, on both northern and southern flanks of the Tethys in the Mediterranean region, major changes took place in the composition of the ostracod faunas. The early to mid-Cenomanian and the Cenomanian-Turonian boundary, seem to have been times of particular significance. The early Cenomanian saw a major renewal of Albian faunas, a diversification of associations and a progressive increase in species diversity, especially in shelf environments. A major crisis in the ostracod faunas occurred in the uppermost Cenomanian-early Turonian, when 80-90% of species disappeared.

After an early Turonian time of low diversity, the mid-Turonian saw the reappearance of diverse faunas. These faunas comprise the same genera as the pre-crisis associations, but have different species.

A detailed study across the Cenomanian-Turonian boundary is made, using Provence as a model and employing both palaeontological and sedimentological data from both platform and basin environments. The aim is to better understand the nature of the events which occurred at this boundary, including oceanic anoxic events and to attempt to achieve a correlation between eustatic sea level changes and tectonic control.

## INTRODUCTION

Tethyan marine Ostracoda have been analysed from Cenomanian and Turonian strata in the Mediterranean and adjacent areas. Cenomanian ostracod assemblages can be divided into three main bioprovinces (Fig. 1). Two of these, the Southwestern European and the North African-Middle East bioprovinces are well defined while the third, temporarily referred to as the 'dinaro-hellenic' bioprovince, which occurs on the Apulian Plate, is much less well-known. Recent studies (Babinot, 1988a; b) have demonstrated the strongly endemic character of late Cenomanian ostracod assemblages which seems to be the consequence of palaeobiogeographical barriers. In open marine areas, depth is probably the most important such barrier. The first two bioprovinces remain essentially unchanged into the Turonian; no data are available for the Apulian area.

Much valuable data are now available on the composition and distribution of these faunas through space and time and these can be used for comparison and correlation as previously recorded, for example, in a more global context by Whatley (1988). The data are patchy, however, with very little known about the Albian-Cenomanian boundary in many areas, or the faunas of Southeastern Europe, except for some recent work on Cenomanian assemblages from the Apulian region. Conversely, mid-Cenomanian to late Turonian faunas are much better known and detailed monographs exist for those of southern France, the Iberian Peninsula, The Mahgreb and both the Far and Middle East.

This study takes into consideration all the available data for the Cenomanian and Turonian throughout the Mediterranean region. Its main purpose is to consider the following:

1) The qualitative and quantitative evolution of global assemblages within particularly well-known areas and to propose a comparison.

2) The Cenomanian-Turonian boundary was a time of important renewal of faunas and many problems are identified with respect to the recognition of such an event. The work is based on data from both platform and basin environments, in particular on sections studied in detail in Provence.

The nature of the fauna is reviewed, the boundary is tentatively defined and the results are discussed within a global framework, with particular reference to oceanic anoxic events.

## THE EVOLUTION OF CENOMANIAN-TURONIAN OSTRACODA IN TETHYAN MARGINAL FACIES IN THE MEDITERRANEAN REGION

### Material and methods

Ostracod assemblages of early Cenomanian to late Turonian age have been studied in detail from a number of areas around the Mediterranean. A detailed database has been compiled from the literature, taking into account the qualitative composition of faunas at the species level. Only well-established species were considered. Also, an analysis of the number of species at certain stratigraphical levels is plotted against recent ammonite zonations (Wright & Kennedy, 1981; Birkelund et al., 1984; Cobban, 1984). At any interval, the number of inherited species is compared to the number of newly appearing species to give a measure of both evolutionary trends and rates (Fig. 2). Also, using the total number of species belonging to successive levels, it is possible to calculate a similarity index. The Sorensen Index (Legendre & Legendre, 1979) is produced by plotting: a) the number of common species between two successive levels; b) The number of species of the lower horizon and c) the number of species of the upper horizon, using the following formula:

$$Is = \frac{2a}{2a + b + c}$$

Values obtained range between 0.0 (no similarity) and 0.50 (complete similarity, i.e. the same fauna) (Fig. 3).

### Albian microfaunas as precursors of Cenomanian-Turonian assemblages

Major changes in the taxonomic composition of ostracod faunas took place during the Albian, with the disappearance of many early Cretaceous

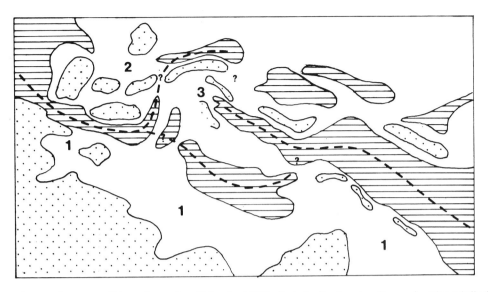

Fig. 1. Cenomanian ostracod bioprovinces. 1 = Mahgreb - Middle-East, 2 = Southwestern Europe, 3 = Dinaro-hellenic area, dotted = emergent areas, horizontal lines = deeper marine areas, blank areas = epicontinental seas, thick dashed lines = hypothetical palaeogeographical barriers. Palaeogeographical map schematized and slightly modified after Ricou, 1987; Dercourt *et al.*, 1985; Barron *et al.*, 1981.

genera belonging to the Protocytherinae and (less spectacularly) the Schulerideinae. This is accompanied by the progressive appearance of new trachyleberid genera, this family becoming very well represented in later Tethyan assemblages. In southern Europe (including the USSR) the diversification of the Trachyleberididae occurs precisely at the end of the Albian with the appearance of *Mauritsina, Limburgina, Dumontina, Oertliella, Planileberis* etc. (Babinot & Colin, 1988). In North Africa and the Middle East, typical associations also occur in the late Albian and flourish during the Cenomanian.

Unfortunately, within many Tethyan basins, data on Albian ostracods are very scarce. The exceptions are Aragon and Portugal (Andreu, 1981), the eastern Mahgreb (Bismuth *et al.*, 1981) and Israel (Honigstein *et al.*, 1985). This inhibits the extent to which the faunas can be compared geographically.

## Cenomanian-Turonian assemblages

The Cenomanian, particularly its mid-part is characterized by major faunal changes, already begun in the Albian. The Trachyleberididae and the Hemicytheridae become the dominant ostracod families. The relationship between local and global evolution of these faunas is now considered for each area.

## The northern margin of Tethys

**Southern France.** This region is considered as a model to which the others can be compared. The Upper Cretaceous Tethyan region of southern France has been defined by virtue of the strong regional differences (at the species level) in the ostracod faunas. The detailed stratigraphy of type sections and the overall palaeogeographical framework are also taken into account. Detailed studies of the area have been made by Babinot *et al.* (1978), Babinot & Colin (1983) and Babinot (1985). Data are sufficient to accurately analyse the evolution of assemblages from both the Cenomanian and Turonian throughout the area. In all other areas, results are less complete.

Ostracod faunas are recorded from the Aquitaine Basin (including Touraine) to the northern Pyrenees and Provence (Babinot *et al.*, 1978; Babinot, 1980; Babinot *et al.*, 1982; 1985; Tambareau *et al.*, 1985 and Floquet *et al.*, 1987).

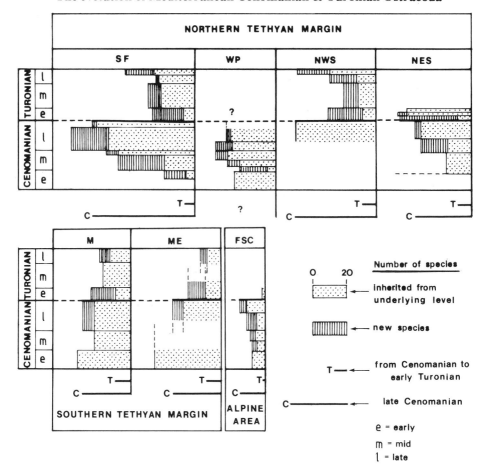

Fig. 2. Evolution of global ostracod assemblages within the Cenomanian-Turonian of the Mediterranean Tethyan margins. Northern Tethyan margin: SF = Southern France, WP = Western Portuguese Basin, NWS = Northwestern Spain, NES = Northeastern Spain. Alpine area: FSC = French Subalpine Chains. Southern Tethyan margin: M = Maghreb, ME = Middle-East.

The Cenomanian shows a progressive increase in the number of species (Fig. 2); this being strongest in the late (but not the uppermost) part of the stage. However, no major changes are apparent in the similarity curve (Fig. 3).

An important faunal change takes place at the top of the Cenomanian and in the lowermost levels of the Turonian. Here, both species diversity and the similarity curve strongly decrease. Very significant is the small percentage of inherited Cenomanian species in the early Turonian (10-15%), but it is worth noting that a number of new species appear at this time. In the mid-late Turonian, assemblage evolution closely parallels that of the Cenomanian in taxonomic composition

at the generic level and similarity indices.

**Spain**. Distinct areas have been taken into consideration. The Basco-Cantabrian region has yielded significant, well studied faunas in the late Cenomanian and Turonian (Ramírez del Pozo, 1972; Swain, 1978; Colin *et al.*, 1982; Méndez & Swain, 1983; Rodríguez-Lázaro, 1985) and northeastern Spain (Aragón and the eastern part of the Cantabrian region) where only Cenomanian and early Turonian records are available, predominantly from platform environments (Grekoff & Deroo, 1956; Andreu, 1983; Breman, 1976; Reyment, 1982a; 1984; Floquet, litt. comm.) The same general trends as those observed in Southern France can be

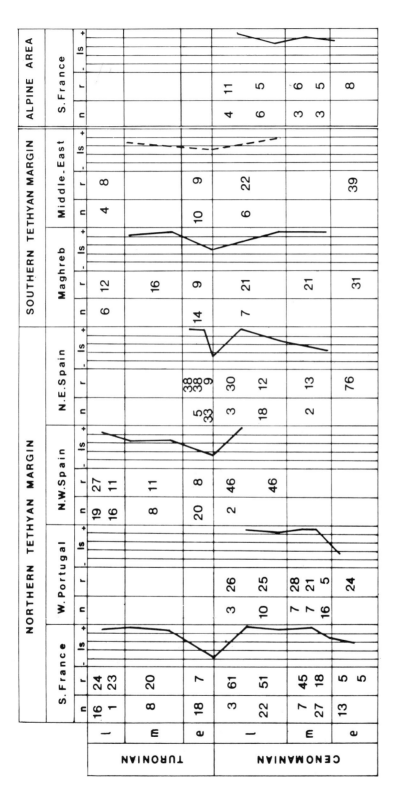

Fig. 3. Evolution of global ostracod assemblages within the Cenomanian-Turonian of the Mediterranean Tethyan margins, Sorensen Similarity Index. n = number of new species (appearance, origination); r = inherited species from underlying level; Is = Sorenson Index values (- = 0.0, + = 0.50).

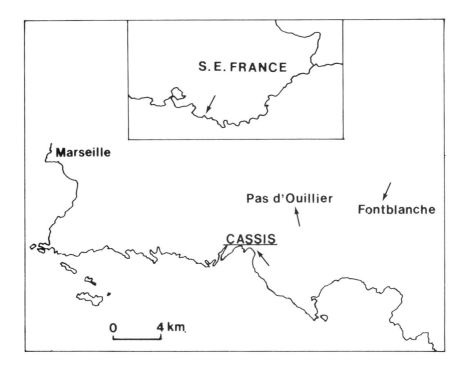

Fig. 4. The Southeastern France (Provence) model, location map of main sections. Cassis: proximal basin, Le Pas d'Ouiller; Fontblanche: carbonate platform.

seen in the faunas and, moreover, recent studies on the early Turonian of north-central Spain, particularly in sheltered (locally carbonate) platform areas, have yielded additional detailed information on the Cenomanian-Turonian boundary event. There is also some similarity in the evolution of the Ostracoda of the Turonian basin deposits of northwestern Spain and those of the platforms, especially of Provence.

**The Western Portuguese Basin.** From Estremadura to the Río Mongego region in the north, Cenomanian (especially mid-Cenomanian) deposits exhibit considerable lithological and biostratigraphical similarities with Aquitaine and Provence (Berthou, 1973; Babinot *et al.*, 1978; Andreu, 1981). Nevertheless, a degree of endemism is apparent in the early Cenomanian (Andreu, 1981). No data exist on the ostracods of the Cenomanian-Turonian boundary or of the Turonian, except for a few species described by Lauverjat (1972), which

are similar to French and Spanish species of the same age.

**The French Subalpine Chains (Alpine area).** The relationship between the faunas of Provence and the southern part of the French Subalpine Chains has been analysed for the mid and late Cenomanian and (less precisely) for the early Turonian. A few species are common between the two areas (Donze & Porthault, 1972; Donze & Thomel, 1972; Babinot, 1980) although the subalpine region is considered as a subprovince of the western European Realm (Babinot, 1985). The scarcity of taxa in the subalpine area is attributed to the prevailing hemipelagic conditions.

### The southern margin of Tethys

**The Mahgreb.** An analysis of the Ostracoda of the Cenomanian and Turonian of the southern margin of the Tethys can be obtained by using only the

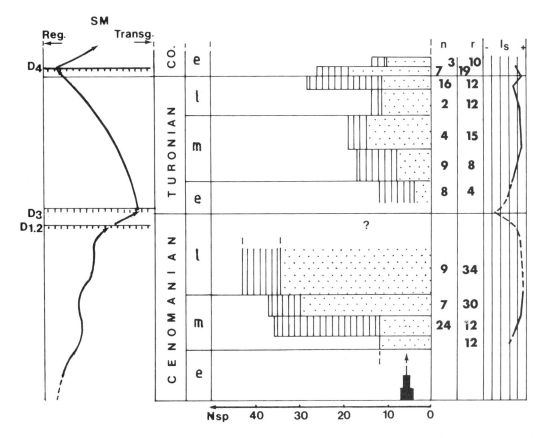

Fig. 5. Evolution of ostracod assemblages within the Cenomanian and Turonian platforms of Provence (Southeastern France). SM = Sequential megasequences, Reg. = regressive trend, Transg. = transgressive trend; D 1.2, D 3, D 4 = main discontinuities. Black histogram (representing the lower part of the Cenomanaian) = inherited microfaunas from earlier Cretaceous times (Albian inheritance). For further explanation see Figs 2-3.

considerable data from Algeria and Tunisia (Glintzboeckel & Magné, 1959; Grekoff, 1968; Bassoullet & Damotte, 1969; Ben Youssef, 1980; Bismuth et al., 1981; Bellion et al., 1973; Gargouri-Razgallah, 1983; Damotte, 1985; Vivière, 1985) and from the Tarfaya Basin, Morocco (Oertli, 1966; Reyment, 1978; 1979; 1982b). Some Moroccan species which have been placed in European taxa should be carefully revised. Despite this, the overall evolution of the fauna and the similarity curve exhibit considerable correspondence with those of the northern margins.

**The Middle East.** At present, data are still too scattered and imprecise to allow detailed comparisons with other regions. In Israel, both the Cenomanian and Turonian have been studied in various areas and environments (Rosenfeld & Raab, 1974; Honigstein et al., 1985) but in other countries in the region there exist only patchy records which are summarized below: Cenomanian of Lebanon (Damotte & Saint Marc, 1972) Kuwait (Al Abdul Razzaq & Grosdidier, 1981), Saudi Arabia (Wasia Formation) (Al Furiah, 1983), early Cenomanian of Jordan (Babinot & Basha, 1985), late Cenomanian of Egypt (Sinai), Colin & El Dakkak, 1975), Cenomanian-Turonian of Abu-Rawash, Egypt (Bold, 1964). However, zonation within stages is not exact enough for certainty. The similarity curve for this region, therefore, is drawn approximately. Areas further east (Iran, Oman etc.) are not considered here.

## Conclusions

The evolution of Cenomanian-Turonian ostracod assemblages over the areas discussed above and especially between Provence, the Iberian Peninsula and even the Maghreb, exhibit similar trends. This is detectable on either side of the Tethys and is also largely independent of latitude. The possible palaeoenvironmental causes of the similarities in the faunas is considered below using Provence as a model (Babinot, 1980; Tronchetti, 1981; Philip *et al.*, in press).

## THE OSTRACOD ASSEMBLAGES OF THE CENOMANIAN-TURONIAN BOUNDARY INTERVAL IN PROVENCE

### Carbonate Platforms

Previous studies on Cenomanian-Turonian ostracod faunas from Provence (Babinot, 1980) provided biostratigraphical range charts and palaeogeographical data, especially in the Cassis-Le Pas d'Ouillier region. Recent work (Floquet *et al.*, 1987) has demonstrated a general but variable transgressive trend throughout the Cenomanian, strongest during the uppermost Cenomanian and early Turonian and followed by a regressive episode in the late Turonian. This seems to correlate well with changes in the ostracod faunas over the same interval. There is a gradual increase in species diversity from mid to late Cenomanian and from the later part of the early to the late Turonian, this is also reflected in the higher values of the similarity index (Fig. 3). This seems to be a good example of the relationship between faunal change and transgressive (Cenomanian) and regressive (Turonian) episodes.

Unfortunately, the analysis of the Ostracoda at the Cenomanian-Turonian boundary is rendered difficult by the local induration of the late Cenomanian limestones (which makes the ostracods difficult to extract) and the rare and poorly preserved nature of the faunas in the calcareous marls of the earliest Turonian. However, the data clearly show a dramatic decrease in ostracod species diversity and, with the sole exception of a few ubiquitous species, the virtual extinction of the Cenomanian fauna. Detailed recent studies, particularly in the Pas d'Ouillier and Fontblanche sections (Fig. 4.) (Philip, 1978; Philip *et al.*, in press) have yielded important new palaeontological and sedimentological data which are summarized below:

a) Occurrence of discontinuities: the first discontinuity, in the late Cenomanian, occurs above a limestone with rudists. This feature is well-marked by bioturbation and by the sudden appearance of planktonic foraminifera, bryozoans, algae (Gymnocodiaceae), echinoid spines etc. The second discontinuity occurs 3m above and is similar but less strongly marked. The third discontinuity, 20m higher in the section, is a hardground with burrows, ferruginization and incrusting oysters.

b) A calcareous deposit, which occurs between the second and third discontinuities, is characterized by a high percentage of planktonic forms (foraminifera, calcispheres) probably belonging to the *Geslianianum* Zone (Tronchetti, pers. comm.). The uppermost part yields specimens of the hippuritid rudist *Vaccinites fontalbensis* (Philip, 1978).

c) Occurrence of nodular glauconitic limestones with sparse ferruginization and ammonites (*Nodosoides* Zone, early Turonian).

Given the above, the Cenomanian-Turonian boundary could be situated between the *Geslianianum* and the *Nodosoides* zones and, therefore, associated with the third discontinuity. However, the most recent evidence would locate the boundary below the horizons with *Vaccinites*, this form being indicative of (at least) the Turonian.

### Proximal basin

The Cassis section, located in the proximal part of the basin (Fig. 6), is also used as a model. Its ostracod faunas have been described and analysed by Babinot (1980).

Unfortunately, no ostracods were found in the few metres of sandy beds at the base of the Cenomanian nor in the mid-Cenomanian so-called 'Banc des Lombards', some 0.30m of ferruginous phosphatized sandstones with ammonites and molluscan shell beds. From the base of the

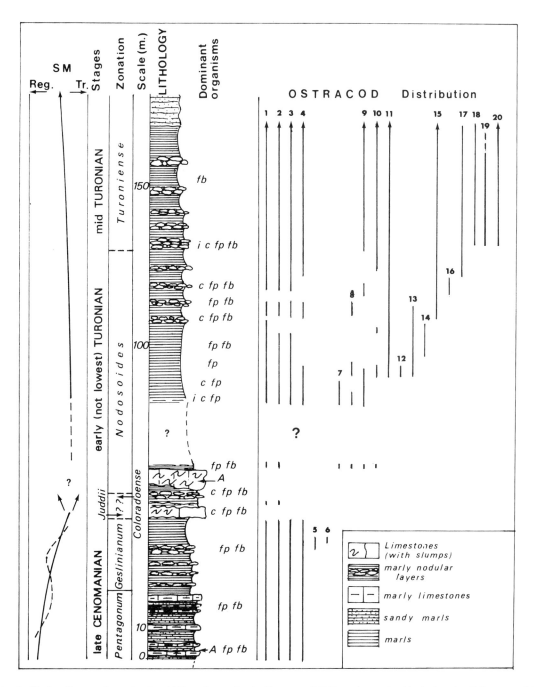

Fig. 6.  The late Cenomanian and early mid-Turonian at Cassis (Provence, S.E. France) - proximal basin domain.  SM = sequential megasequences.  Dominant organisms:  A = ammonites,  i = inoceramids, c = calcispheres, fp = planctonic foraminifera, fb = benthonic foraminifera. Ostracods: 1 *Cytherella ovata* (Roemer), 2 *Cytherella parallela* (Reuss), 3 *Bairdia* sp., 4  *Pontocyprella* sp.,  5 *Veenia* aff.  *ballonensis* Damotte & Grosdidier,  6 *Quasihermanites gruendeli* Colin, 7 *Phodeucythere*? sp., 8 *Polycope* sp., 9 *Pterygocythere* gr. *pulvinata* Damotte, 10 *Parvacythereis* cf. *plenata* Gründel, 11 *Paracypris* sp., 12 *Amphicytherura* sp., 13 Cytherurinae indet., 14 *Macrocypris* sp., 15 *Spinoleberis ectypus* Babinot, 16 *Bythoceratina* sp., 17 *Mauritsina provencialis* Babinot, 18 *Trachyleberidea* gr. *geinitzi* (Reuss), 19 *Paracyprideis* sp., 20  *Asciocythere polita* Damotte.

late Cenomanian (*Pentagonum* Zone) very rapidly deposited terrigenous deposits were laid down and the ostracods are of low diversity and do not record any major changes.

A re-evaluation of existing data and additional information from Airaud as a result of recent field excursions to Cassis, allows us to better understand the palaeoenvironmental conditions obtaining in the region at the time of the Cenomanian-Turonian boundary. This can be summarized by considering aspects of the Cassis section, in stratigraphical sequence, as follows:

a) Grey blue marls (10m) with mainly planktonic foraminifera, with *Rotalipora cushmani* (Morrow), belonging to the *Geslinianum* Zone (Tronchetti, 1981) and with muddy calcareous nodules at various levels in the lower part. This unit overlies alternations of marly sandstones and marls with *Eucalycoceras pentagonum* Jukes-Brown.

b) (i) Limestones with occasional slumped structures characterized by numerous calcispheres and both planktonic and benthonic foraminifera. Both glauconite and quartz are also recorded from this unit. Towards the top the number of planktonic foraminifera decreases and small benthonic forms predominate. The unit is barren of ostracods.

ii) This unit has marls and calcareous marls at the base with quartz and a poorly preserved microfauna of small foraminifera and a few ostracods. At the top are soft pebbles and another occurrence of abundant calcispheres and a very diverse fauna of planktonic foraminifera including *Dicarinella, Guembelitria, Hedbergella, Whiteinella* and *Heterohelix* etc., but no ostracods.

iii) Limestones with mainly benthonic bioclasts, lithoclasts and benthonic foraminifera. There are no ostracods and the microfacies suggests very rapid sedimentation.

c) Marly layers of 1-2m with mainly planktonic foraminifera, especially *Praeglobotruncana helvetica* (Bolli), *P. praehelvetica* (Trujillo), *Dicarinella* spp. and *Whiteinella* spp. No ostracods occur in this unit which belongs to the early (but not lowermost) Turonian *Nodosoides* Zone.

d) A unit of slumped carbonates of 5-7m with (according to Amedro, pers. comm.) the ammonoid *Pseudaspidoceras* cf. *pseudonosoides*. Slumps seem

to be related to the reworking of uppermost Cenomanian or earliest Turonian deposits.

e) A marly unit with planktonic foraminifera (early to mid-Turonian) up to 50m thick. The lower part belongs to the *Nodosoides* Zone. Ostracods appear at the interface between this unit and the one below and progressively increase in diversity and abundance.

While only advanced as an hypothesis, unit b) may be located between the *Gesilianum* and *Nodosoides* zones, i.e. *Juddi* (uppermost Cenomanian) and *Coloradense* (early Turonian) zones.

The ostracods from the section are few in species but often occur in great abundance. In the highest part of the *Pentagonum* Zone, the platycopids dominate with *Cytherella ovata* (Roemer) making up 70-80% of the total fauna; other taxa are *Bairdia* gr. *pseudoseptentrionalis* Mertens and some specimens of *Pontocyprella*.

At the top of the *Geslinianum* Zone, the ostracod fauna is essentially very similar but a few podocopids appear such as: '*Veenia*' aff. *ballonensis* Damotte & Grosdidier and *Quasihermanites gruendeli* Colin. A significant change is noted in the overlying unit b), where all samples are barren of ostracods. They reappear again, however, at the top of the slumped unit d) and in the first marly beds of unit e) of the early Turonian. Here the cytherellids dominate with *Cytherella ovata* and *C. parallela* (Reuss), some specimens of *Polycope* sp., *Bythoceratina* sp. and *Phodeucythere?* sp. and the alate species *Pterygocythere* gr. *pulvinata* Damotte.

In the overlying marls, which range up to the mid-Turonian, *Pontocyprella* reappears followed by *Parvacythereis?* cf. *plenata* Gründel, Cytherurinae indet. sp. and *Mauritsina provencialis* Babinot.

The above analysis, therefore, shows evidence of a decrease in ostracod diversity and abundance at the Cenomanian-Turonian boundary. From the earliest to the mid-late Turonian, a progressive enrichment in species can be observed, especially in adjacent areas of perireefal structures, such as the interbedded carbonates with rudists. These faunal changes correlate well with the observed palaeogeographical changes, such as the decrease in depth of the basin associated with the progressive outgrowth of deltas. In such conditions, reefal carbonates may reappear onlapping the platform deposits (Philip, 1970).

## DISCUSSION AND CONCLUSIONS

By using Provence as a regional model, it is possible to obtain information concerning the evolution of ostracod faunas during the Cenomanian-Turonian which has wider implications. There are two main observations which can be made:

1) The relative diversity and the progressive increase in ostracod diversity during the mid-late Cenomanian and early late Turonian, mainly in shallow carbonate platform environments, seems related to relatively stable palaeoceanographical conditions which were conducive to high organic productivity. While some faunal turnover does take place, a large proportion of inherited species is always apparent.

2) Cumulative diversity decline and eventual disappearance of ostracods during the uppermost Cenomanian and the early Turonian. This apparent crisis can, in fact, be explained by major palaeogeographical changes.

In all the platform areas, all evidence points to considerable instability and rapid rise in sea level. This is evidenced by the nature of the sediments which overlie the 1st and 2nd discontinuities, the large planktonic fauna of the *Geslinianum* Zone and periods of non-deposition. Under these transgressive conditions, assemblages are destroyed unless they are able to migrate landwards. The ostracods eventually disappear under these conditions.

Recent studies (Philip *et al.*, in press) suggest that eustatic sea level rise (deepening phase) was locally (Pyrenees and Provence) enhanced by local tectonics (Boillot *et al.*, 1984).

In the proximal part of the basin, the following 5 stages of palaeoenvironmental evolution are thought to have occurred:

**First stage.** Stable conditions with rich and abundant planktonic assemblages.

**Second stage.** Initially a phase of instability (unit b (i)), with heterogeneous bedding and local slumping, followed by a return to more stable conditions (unit b (ii)), with the reappearance of planktonic forms. The Cenomanian-Turonian boundary is located at the top of this stage.

**Third stage.** This is equivalent to unit c), in which there is clear evidence of increasing depth or oceanic influence.

**Fourth stage.** This is equivalent to unit d) and is indicative of very unstable conditions and of the reworking of earlier deposits.

**Fifth stage.** Return to stable, low energy conditions and progressive recolonization by benthos.

As indicated previously, the fortunes of the Ostracoda closely follow this pattern. They suffer cumulative impoverishment through the late Cenomanian to disappear from its uppermost levels and those of the early Turonian.

These studies demonstrate the relationship between sediment type (a function of the dynamics of substrate and sea bottom physiography) and changes in eustatic sea level. It is possible, therefore, to correlate the disappearance of ostracods with increasing water depth which is related to transgressive pulses and/or increasing oceanic influence. The so-called late Cenomanian-early Turonian 'ostracod crisis' seems to occur between two 'transgressive spikes'. It is not yet possible to know whether this is the product of a heightened transgressive pulse or of a time of lower relative sea level.

This late Cenomanian-early Cenomanian 'event' has been the subject of considerable study and speculation since the early 1980s. It is claimed that the 'event' is in part the consequence of an 'oceanic anoxic event', under which the biota was disturbed by important physicochemical changes in the water masses of the ocean basin. However, in spite of this event, many elements of the fauna, notably the ostracods, survived into the Turonian. Those faunal changes which did take place were at the specific rather than the generic level.

Considerable literature has accummulated on this O.A.E. (oceanic anoxic event) with respect to both Boreal and Tethyan regions: Anderson & Arthur, 1983; Arthur *et al.*, 1987; Crumière, in press; Graciansky de *et al.*, 1986; Jarvis *et al.*,

1988; Jenkyns, 1980; Schlanger *et al.*, 1987).

Geochemical analyses taken at the level of the O.A.E., indicate a major worldwide positive 13C spike (or excursion) (Jarvis, *et al.*, 1988). According to these authors, the O.A.E. was associated with a phase of increased upwelling which led to a widespread expansion and intensification of the oxygen minimum zone in the world's oceans. Thurow *et al.* (1982) came to a similar conclusion with respect to the western Tethys, with particular reference to changes in ocean currents and falling water temperature.

Geochemical analyses of samples from the Cassis section will be carried out shortly. However, it is possible on the basis of palaeontological data to achieve some degree of correlation with the Chalk Sea of N.W. Europe (Jarvis *et al.*, 1988). It seems possible to correlate changes in the fortunes of the foraminifera and calcispheres in the two areas and the evolution of the ostracods through the late Cenomanian *Plenus* Marls of Dover, England, agrees closely with those of Provence, particularly at Cassis.

Ostracod diversity increases strongly at the Cenomanian-Turonian boundary in Dover and the faunas at this level are dominated by platycopids, especially *Cytherella ovata*. On either side of the level dominated by platycopids, the same genera occur (*Cytherella*, *Phodeucythere*, *Bairdia*, *Pontocyprella*, *Parvacythereis*, *Pterygocythere*), but are represented by different species.

In their discussion, Horne and Rosenfeld (in Jarvis *et al.*, 1988) speculate that 'platycopids (which are filter-feeders) were more tolerant of low oxygen levels than podocopids'.

Also, at Cassis, there is a decline in the diversity of the benthonic foraminifera while the Buliminidae which, according to Tronchetti (pers. comm.) can survive low oxygen levels, flourish. The synchronous proliferation of calcispheres, also noted by Kuhnt *et al.* (1986) in the Tethyan realm, is considered to reflect 'an increasing nutrient supply and a decline in the zooplankton' (Jarvis *et al.*, 1988).

Especially in comparison with what Jarvis *et al.* (1988) describe from Dover, the Cassis section shows some evidence of 'anoxia'. More probably, however, the product of a degree of stratification of water masses, or what Philip *et al.* (in press) refer to as 'hypoxic conditions' or a 'Kenoxic event' as defined by Cepek & Kemper (1981). Changes in the ostracod, foraminiferal and calcisphere faunas tend to support this possibility. Furthermore, similarities in the evolution of the ostracod faunas of both the northern and southern margins of the Tethys during the Cenomanian-Turonian interval can be explained as being influenced by major global events in the following two ways:

1) **Eustatism**. a) Small to moderate sea level changes (positive or negative) are responsible for the relative stability of ostracod faunas and their ability to adapt to change.

b) Sudden increases in sea level enhanced by changes in the water mass and increase in its stratification leading to the instability and even the disappearance of ostracod faunas.

2) **Dynamics**. Tectonic control such as that which seems to have taken place in Provence. These tectonic events can be related to the geodynamic evolution of the Tethyan realm, i.e. relationships between major plates in the Northern Hemisphere. This global tectonism could be responsible for changes in the strength and direction of ocean currents and changes in other palaeoceanographical parameters. More research is necessary, however, to elucidate the relationship between these 'events' and the evolution of ostracods and other organisms.

## ACKNOWLEDGEMENTS

The authors wish to acknowledge Drs J. Philip, J. P. Masse and G. Tronchetti, all of the University of Provence, for their encouragement and valuable comments. They are also indebted to A. Mermighis, of the same University, who kindly provided documentation on the Apulian (Grecian) area and to M. Haufeurt for typing the manuscript. Furthermore, they thank especially Professor R. C. Whatley (University of Wales, Aberystwyth) for his revision of the English text and other comments.

# REFERENCES

Al Abdul Razzaq, S. & Grosdidier, E. 1981. Ostracode index species from the Cenomanian of the South shelf of the Tethys Sea. *Bull. Centres Rech. Explor.-Prod., Elf-Aquitaine*, Pau, **5**(2), 173-191.

Al Furiah, A. A. 1983. Middle Cretaceous (Cenomanian) Ostracoda from the Wasia Formation of Saudi Arabia. *Paleont. Contr. Univ. Kans.*, University of Kansas, Lawrence, **108**, 1-6.

Anderson, T. F. & Arthur, M. A. 1983. Stable isotopes of oxygen and carbon and their application to sedimentologic and paleoenvironmental problems. *In* Arthur, M. A., Anderson, T. F., Katlan, I. R., Veizer, J. & Land, L. S. (Eds), *Stable Isotopes in Sedimentary Geology*, Soc. *Economic Palaeontologists and Mineralogists Short Course*, **10**(1), 1-151.

Andreu, B. 1981. Nouvelles espèces d'Ostracodes de l'Albien et du Cénomanien d'Estremadura (Portugal). *Ciencias da Terra* (UNL), Lisboa, **6**, 117-152.

Andreu, B. 1983. Nouvelles espèces d'Ostracodes de l'Albien et du Cénomanien sud-pyreneen (Sierra d'Aulet, Espagne). *Bull. Centres Rech. Explor.-Prod., Elf-Aquitaine*, Pau, **7**, 1-43.

Babinot, J.-F. 1980. Les Ostracodes du Crétacé supérieur de Provence. Systématique - Biostratigraphie - Paléoécologie, Paléogéographie. *Trav. Lab. Géol. hist. Paléont. Univ. Provence*, Marseille, **10**, 634 pp.

Babinot, J.-F. 1985. Paléobiogéographie des ostracodes du Crétacé supérieur des marges ouest-européennes et nord-africaines de la Tethys. *Bull. Soc. géol. Fr.*, Paris, (8), **1**(5), 739-745.

Babinot, J.-F. 1988a. Premières données sur les Ostracodes du Cénomanien de Yougoslavie (Istrie du Sud). *Géobios*, Lyon, **21**(1), 5-15.

Babinot, J.-F. 1988b. Études préliminaires sur les asociations d'ostracodes du Cénomanien de Grèce (zone pelagienne, Argolide). Implications paléogéographiques et géodynamiques. *Géobios*, Lyon, **21**(4), 435-463.

Babinot, J.-F. & Basha, S. A. 1985. Ostracodes from the Early Cenomanian of Jordan. A preliminary report. *Géobios*, Lyon, **18**(2), 257-262.

Babinot, J.-F., Berthou, P. Y., Colin, J.-P. & Lauverjat, J. 1978. Les Ostracodes du Cénomanien du bassin occidental portugais: biostratigraphie et affinités paléogéographiques. *Cah. Micropaléontol.*, Paris, **3**, 11-23.

Babinot, J.-F. & Colin, J.-P. 1983. Marine late Cretaceous ostracode faunas from southwestern Europe: a paleoecological synthesis. *In* Maddocks, R. F. (Ed.), *Applications of Ostracoda*, proceedings of the Eighth International Symposium on Ostracoda, July 26-29, 1982, 182-205. Univ. Houston Geos., Houston, Texas.

Babinot, J.-F. & Colin, J.-P. 1988. Paleobiogeography of Tethyan Cretaceous marine ostracodes. *In* Hanai, T., Ikeya, N. & Ishizaki, K. (Eds), *Evolutionary biology of Ostracoda, its fundamentals and applications*, proceedings of the Ninth International Symposium on Ostracoda, held in Shizuoka, Japan, 29 July - 2 August 1985, Developments in palaeontology and stratigraphy, **11**, 823-839,

Kodansha Ltd., Tokyo and Elsevier, Amsterdam, Oxford, New York, Tokyo.

Barron, E. J., Harrison, C. G. A., Sloan, J. L. II & Hay, W. N. 1981. Paleograraphy, 180 million years ago to the present. *Eclog. geol. Helv.*, Lausanne, Basel, **74**/2, 443-470.

Bassoullet, J. P. & Damotte R. 1969. Quelques Ostracodes nouveaux du Cénomano-Turonien de l'Atlas saharien occidental (Algérie). *Revue Micropaléont.*, Paris, **12**(3), 130-144.

Bellion, Y., Donze, P. & Guiraud, R. 1973. Répartition stratigraphique des principaux Ostracodes (Cytheracea) dans le Crétacé supérieur du Sud-Ouest constantinois (confins Hodna-Aures, Algérie du Nord). *Publs Serv. géol. Algér.*, n. sér., **44**, 7-44.

Ben Youssef, M. 1980. *Étude stratigraphique et micropaléontologique du Crétacé des Djebels Koumine et Kharroub (Tunisie centrale)*. These 3° cycle, Univ. de Nice, 104 pp.

Berthou, P. Y. 1973. Le Cénomanien de l'Estremadure portugais. *Méms Servs géol. Port.*, Lisbon, N. S., **23**, 1-164.

Birkelund, T., Hancock, J. M., Hart, M. B, Rawson, P. F., Remane, J., Robaszynski, F., Schmid, F. & Surlyck, F. 1984. Cretaceous stage boundaries: proposals. *Bull. geol. Soc. Denm.*, Copenhagen, **33**, 3-20.

Bismuth, H., Boltenhangen, C., Donze, P., Le Fevre, J. & Saint-Marc P. 1981. Le Crétacé, moyen et supérieur du Djebel Semmama (Tunisie du Centre Nord): microstratigraphie et évolution sedimentologique. *Bull. Centres Rech. Explor.-Prod., Elf-Aquitaine*, Pau, **5**(2), 193-267.

Boillot, G., Monadert, L., Lemoine, M. & Biju-Duval, B. 1984. *Les marges continentales actuelles et fossiles autour de la France*. 342 pp. Masson Ed., Paris.

Bold, W. A. van den. 1964. Ostracoden aus der Oberkreide von Abu Rawash, Ägypten. *Palaeontographica*, Kassel, Abt. A, **123**, 111-136.

Breman, E. 1976. Paleoecology and Systematics of Cenomanian and Turonian Ostracodes from Guadalajara and Soria (Central Spain). *Revta esp. Micropaleont.*, Madrid, **8**(1), 71-122.

Cepek, P. & Kemper, E. 1981. Der Blättertonstein des nordwestdeutschen Barrême und die Bedeutung des Nannoplanktons für die fein laminierten anoxisch entstandenen Gesteine. *Geol. Jb.*, Hannover, **58**, 3-13, 1-13.

Cobban, W. A. 1984. Mid-Cretaceous ammonite zones, Western Interior United States. *Bull. geol. Soc. Denm.*, Copenhagen, **33**, 71-89.

Colin, J.-P. & El Dakkak, N. W. 1975. Quelques ostracodes du Cénomanien du Djebel Nezzazat, Sinaï, Egypte. *Revta esp. Micropaleont.*,, Madrid, n° special 1975, 49-60.

Colin, J.-P., Lamolda, M. A. & Rodríguez-Lázaro, J. 1982. Los Ostrácodos del Cenomaniense superior y Turoniense de la cuenca vasco-cantábrica. *Revta esp. Micropaleont.*, Madrid, **14**, 187-220.

Crumière, J. P. In press. Crise anoxique à la limite Cénomanien-Turonien dans le Sud-Est de la France. Relation avec l'eustatisme. *Géobios*, Lyon, sp. publ.

Damotte, R. 1985. Les Ostracodes du Crétacé moyen sud-mésogéen et leur répartition paléogéographique. *Bull. Soc. géol. Fr.*, Paris, (8), **1**(5), 733-737.

Damotte, R. & Saint-Marc, P. 1972. Contribution à la connaissance des Ostracodes crétacés du Liban. *Revta esp. Micropaleont.*, Madrid, **3**, 273-296.

Dercourt, J., Zoneshain, L. P., Ricou, L. E., Kazmin, V. G., Le Pichon, X., Knipper, A. L., Grandjacquet, Cl., Sborshchikov, I. M., Boulin, J., Sorokhtin, O., Geyssant, J., Lepvrier, Cl., Biju-Duval, B., Sibuet, J. C., Savostin, L. A., Westphal, M. & Lauer, J. P. 1985. Présentation de 9 cartes paléogéographiques au 1/20.000.000° s'étendant de l'Atlantique au Pamir pour la période du Lias à l'Actuel. *Bull. Soc. géol. Fr.*, Paris, (8), **1**(5), 637-652.

Donze, P. & Porthault, B. 1972. Les Ostracoda de la sous-famille des Trachyleberidinae dans quelques coupes de référence du Cénomanien du Sud-Est de la France. *Revta esp. Micropaleont.*, Madrid, **4**(3), 355-376.

Donze, P. & Thomel, G. 1972. Le Cénomanien de la Foux (Alpes de Haute-Provence). Biostratigraphie et faunes nouvelles d'ostracodes. *Eclog. geol. Helv.*, Lausanne, Basel, **65**(2), 369-389.

Floquet, M., Philip, J., Babinot, J.-F., Tronchetti, G. & Bilotte, M. 1987. Transgressions-régressions marines et évènements biosédimentaires sur les marges pyrénéo-provençales et nord-ibériques au Crétacé supérieur. *Mém. géol. Univ. Dijon*, **11**, 245-258.

Gargouri-Razgallah, S. 1983. Le Cénomanien de Tunisie centrale. Étude paléoécologique, stratigraphique, micropaléontologique et paléogéographique. *Docums Trav. I.G.A.L.*, Paris, **6**, 215 pp.

Glintzboeckel, C. & Magné, J. 1959. Répartition des microfaunes à plancton et a Ostracodes dans le Crétacé supérieur de la Tunisie et de l'Est algérien. *Revue Micropaléont.*, Paris, **2**(2), 57-67.

Graciansky, P. C. de, Deroo, G., Herbin, J. P., Jaquin, T., Magni, F., Montadert, L. & Müller, C. 1986. Ocean-wide stagnation episodes in the Late Cretaceous. *Geol. Rdsch.*, Leipzig, **75**, 17-41.

Grekoff, N. 1968. Sur la valeur stratigraphique et les relations paléogéographiques de quelques Ostracodes du Crétacé, du Paléocène et de l'Eocène inférieur d'Algérie orientale. *Proc. III° Afr. Micropal. Coll.*, Le Caire, 227-248.

Grekoff, N. & Deroo, G. 1956. Algunos Ostrácodos del Cretácico medio del Norte de España. *Estudios geol. Inst. Invest. geol. Lucas Mallada*, Madrid, **XII**, 31-32, 215-235.

Honigstein, A., Raab, M. & Rosenfeld, A. 1985. Manual of cretaceous Ostracodes from Israel. *Geol. Surv. Israel*, spec. publ. n°5, Jerusalem, 1-25.

Jarvis, I., Carson, G. A., Coope, M. K. E., Hart, M. B., Leary, P. N., Tocher, B. A., Horne, D. J. & Rosenfeld, A. 1988. Microfossil assemblages and the Cenomanian-Turonian (late Cretaceous) oceanic anoxic event. *Cret. Res.*, Academic Press, London, **9**(1), 3-103.

Jenkyns, H. C. 1980. Cretaceous anoxic event: from continents to oceans. *Journ. Geol. Soc.*, **137**, 171-188.

Kuhnt, W., Thurow, J., Wiedmann, J. & Herbin, J. P. 1986. Oceanic anoxic conditions around the Cenomanian/Turonian boundary and the response of the biota. *In* Degens, E. T., Meyers, P. A. & Brassen, S. C. (Eds), Biogeochemistry of Black Shales, *Mitt. geol.-paläont. Inst. Univ. Hamburg*, Hamburg, **60**, 205-246.

Lauverjat, J. 1972. *Le Crétacé supérieur dans le Nord du Bassin occidental portugais*. Unpub. thèse, Univ. P. & M. Curie (Paris VI), 715 pp.

Legendre, L. & Legendre, P. 1979. *Ecologie numérique. 2. La structure des données écologiques*, 247 pp. Masson Ed., Paris.

Méndez, C. A. & Swain, F. M. 1983. Ostrácodos cenomanenses de dos secciones en los alrededores de Oviedo, Asturias. *Revta esp. Micropaleont.*, Madrid, **15**(3), 467-496.

Oertli, H. J. 1966. Étude des Ostracodes du Crétacé supérieur du bassin côtier de Tarfaya. *Notes Mém. Serv. géol. Maroc*, **173**, 267-278.

Philip, J. 1970. *Les formations calcaires à Rudistes du Crétacé supérieur provençal et rhodanien*. Thèse Fac. Sc. Marseille, 438 pp.

Philip, J. 1978. Stratigraphic et paléoécologie des formations à Rudistes du Cénomanien: l'exemple de la Provence. *Géol. Méditerr.*, **5**(1), 155-167.

Philip, J., Airaud, Ch. & Tronchetti, G. In press. Évènements paléogéographiques en Provence (S.E. France) au passage Cénomanien-Turonien. Modifications biosédimentaires. Causes géodynamiques. *Géobios*, sp. publ., Lyon.

Ramírez Del Pozo, J. 1972. Algunas precisions sobre la bioestratigrafía, paleogeografía y micropaleontología del Cretácico Asturiano (Zona de Oviedo-Infiesta-Villaciosa-Gijón). *Boln Geol. Min.*, **33**(2), 122-162.

Reyment, R. A. 1978. Quantitative biostratigraphical analysis examplified by Moroccan Cretaceous Ostracods. *Micropaleontology*, New York, **24**(1), 24-43.

Reyment, R. A. 1979. Signification paléobiogéographique de la répartition de *Oertliella tarfayensis* au Maroc. *Revue Micropaléont.*, Paris, **22**(3), 185-190.

Reyment, R. A. 1982a. Preliminary account of Middle Cretaceous ostracods from Segovia and Guadalajara provinces (Spain). *Cuadernos Geol. Ibérica*, Madrid, **8**, 195-218.

Reyment, R. A. 1982b. Note on Upper Cretaceous Ostracodes from South-western Morocco. *Cret. Res.*, Academic Press, London, **3**, 405-414.

Reyment, R. A. 1984. Upper Cretaceous Ostracoda of North Central Spain. *Bull. geol. Instn. Univ. Uppsala*, n. s., **10**, 67-110.

Ricou, L. E. 1987. The Tethyan oceanic gates: a tectonic approach to major sedimentary changes within Tethys. *Geodinamica Acta*, Paris, **1**(415), 225-232.

Rodríguez-Lázaro, J. 1985. *Los ostrácodos del Coniaciense y Santoniense de la Cuenca Vasco-cantábrica occidental*. Unpub. Ph.D. thesis, Univ. Bilbao, Spain, 527 pp.

Rosenfeld, A. & Raab, M. 1974. Cenomanian-Turonian Ostracodes from the Judea Group in Israel. *Geol. Surv. Israel.*, Jerusalem, **62**, 1-64.

Schlanger, S. O., Arthur, M. A., Jenkyns, M.E. & Scholle, P. A. 1987. The Cenomanian-Turonian oceanic anoxic event, I: Stratigraphy and distribution of organic carbon-rich beds and the marine 13C excursion. *In* Brooks, E. J. & Fleet, A. (Eds), Marine petroleum Source Rocks, *Geol. Soc. spec. Publ.*, **36**, 371-399. Blackwell Scientific, Oxford.

Swain, F. M. 1978. Some middle Cretaceous Ostracoda from northern Spain and their interregional relationships. *Revta esp. Micropaleont.*, Madrid, **10**(2), 245-265.

Tambareau, Y., Bessière, G., Bilotte, M., Villatte, J., Babinot, J.-F. & Lethiers, F. 1985. Journée d'étude des Ostracodologistes de langue française dans les Hautes-Corbières. *Bull. Soc. Etud. Scient. Aude*, Carcassonne, **85**, 13-47.

Thurow, J., Kuhnt, W. & Wiedmann, J. 1982. Zeitlicher und palaontogeographischer Rahmen der Phthanit und Black Shale-sedimentation in Marokko. *Neues Jb. Geol. Paläont. Abh.*, Stuttgart, **165**(1), 147-176.

Tronchetti, G. 1981. Les Foraminifères crétacés de Provence (Aptien-Santonien). *Trav. Lab. Géol. hist. Paléont. Univ. Provence*, Marseille, **12**, 559 pp.

Vivière, J. L. 1985. Les Ostracodes du Crétacé supérieur (Vraconien a Campanien basal) de la région de Tébessa (Algérie du Nord-Est). Unpub. thèse 3° Cycle, Univ. P. & M. Curie (Paris VI), 261 pp.

Whatley, R. C. 1988. Patterns and rates of evolution among Mesozoic Ostracoda. *In* Hanai, T., Ikeya, N. & Ishizaki, K. (Eds), *Evolutionary biology of Ostracoda, its fundamentals and applications*, proceedings of the Ninth International Symposium on Ostracoda, held in Shizuoka, Japan, 29 July - 2 August 1985, Developments in palaeontology and stratigraphy, **11**, 1021-1040, Kodansha Ltd., Tokyo and Elsevier, Amsterdam, Oxford, New York, Tokyo.

Wright, C. W. & Kennedy, W. J. 1981. The Ammonoidea of the Plenus Marls and the Middle Chalk. *Palaeontogr. Soc. (Monogr.)*, London, **134**(560), 1-148.

## DISCUSSION

Pierre Donze: Do 'black shales' exist in the Cassis area in between the Cenomanian and Turonian as they do on the southern Tethyan margin in central Tunisia?

Jean-François Babinot: In the Cassis area, no 'black shales' occur. At a few horizons, however, facies analysis shows the existence of an enhanced organic matter content. This has been observed mainly in marly beds underlying and immediately above the slumped carbonate bank (unit d of the Cassis section), i.e. in the transitional zone between late Cenomanian and early (but not lowest) Turonian. More precise records (quantitative data, origin of the organic matter) will be given in the near future by Christine Crumière-Airaud (thesis in preparation).

David Horne: Why did *Cytherella* species survive across the Cenomanian/Turonian boundary?

Jean-François Babinot: No available data can be found to explain the survival of *Cytherella* species, especially across this boundary, but *Cytherella* is a very ubiquitous genus. Its occurrence, therefore, is clearly demonstrated in very particular environments such as coarse sandy, transitional marine to brackish and even ferruginous (reduced) deposits. In these cases, *Cytherella* species may be well represented in oligo- or monospecific assemblages.

# Ostracoda and the discovery of global Cainozoic palaeoceanographical events

**Richard H. Benson**

Smithsonian Institution, Washington, D.C., USA.

## ABSTRACT

Studies of changes in the generic diversity of fossil deep-sea ostracods, their evolution and faunal invasions into newly developed water-mass or ocean-basin systems, have helped to discover and demonstrate at least five global palaeoceanographical 'events' that have occurred over the last 80 MY. One of these reflects a system boundary, another approaches a series boundary, two occur within series and one defines a series boundary. These events took place:

1) at 65 Ma, the end of the Cretaceous; suggesting external influences, a 15 percent deep water ostracod generic extinction, without replacement until;

2) between 40 and 38 Ma, during the late Eocene; the formation of cold Antarctic bottom water, the breakdown of the Tethyan 'seaway' and latitudinal current system into independent basins, and the establishment of the modern thermally driven water-mass structure of the World Ocean and the developement of the 'psychrospheric' ostracod fauna;

3) about 16 to 14 Ma, during the Middle Miocene; changes in the Indo-Pacific and South Atlantic faunas, formation of the East Antarctic Ice Cap, closure of the eastern end of Tethys and completion of the formation of the gyres; and

4) at 6.3 Ma lasting until 4.9 Ma, the end of the Miocene, a glacial maximum in the Antarctic, thinning and acceleration of the gyres, and the Messinian Salinity Crisis. Discoveries in Morocco of the psychrospheric ostracod fauna in shallow depths with 'conoidal' planktonic foraminifera in the Rifian Corridor indicate an important reversal of currents and the cessation of the flow of Tethyan water into the North Atlantic in the early Messinian.

The Pliocene 'boundary event' (4.9 Ma; following the Messinian Salinity Crisis), is a global event only in the sense of the coincidence of the global Boundary Stratotype with the origin of the Mediterranean. It is in part identified with the invasion of *Agrenocythere pliocenica* and *Oblitacythereis mediterranea*, ostracods typical of the Lower Pliocene in the Mediterranean, but found in the bathyal Miocene of the Atlantic. These species have now been traced from the Atlantic through the tectonic divide in the Rifian Corridor in Morocco towards the Mediterranean.

5) and at 3.5 Ma, during the Pliocene, probably related to the closure of the Balboa Portal, increased ocean cooling and the end of low latitude Atlantic-Pacific warm-water transfer.

## INTRODUCTION

Global marine stratigraphy is directly related to oceanic water-mass stratification and history. Events of global importance, so far as microfossil distribution are concerned, are those in which the faunas reflect major water-mass formation or boundary shifts. Of course, the terms 'global' and 'event' have special stratigraphical meanings, indicating exceptional geographical extension and the sudden appearance (associated with chronohorizons) of new environmental conditions.

The underlying theme of the 10th International Symposium on Ostracoda (1988) and the 28th International Geological Congress in Washington (1989) was the 'global perspective'. This was an opportunity for those of us interested in palaeoceanography to take stock of how the ostracod fossil record in the deep sea has contributed to our world view of past oceanic events, and in particular their correspondance to our preconceptions of major divisions of Cainozoic history.

As the Chairman of the Stratigraphic Panel of the Deep Sea Drilling Project (1976-1981), the author participated in the development of 'global event stratigraphy', especially Cainozoic global event stratigraphy. The need to confirm classical stratigraphical boundaries and fill gaps turned our attention from onshore sections with their numerous unconformities to longer sections from the ocean floor sampled by continuous hydraulic piston cores. The need for correlation ultimately led

to the search for 'catastrophic' environmental changes; if they existed (Berggren & Van Couvering, 1984a; Benson, 1984a). Any global event could imply some sort of catastrophic change, using this term in the sense of a system's upset.

There was a need then, as now, to conduct a dialogue between global historical models and stratigraphical control. The ostracods are not yet likely to displace planktonic microfossils as biostratigraphic markers (FADs; First Appearance Datums) for inter-oceanic Cainozoic correlation in deep-sea cores. However, they have had an influence on the construction of historical models. As van den Bold's many works in the Caribbean, or that of the IGCP 124 Project (Gramann, 1988) in the Palaeogene of western Europe have shown, ostracods can have an influence on local, zonal correlation, if studied in sufficient detail. Perhaps even in the deep sea the ostracods could be more stratigraphically useful. Specimens in deep-sea sediments are not that much rarer than those of many critical planktonic foraminifera species.

This chapter, which is in large part a synthesis of previous work, is written in the context of the search by the author for sudden global change in deep-sea ostracods to aid in the construction of palaeoceanographical models. The five global events described here vary in duration, intensity and certainty. With any such descriptive effort, one must accept the limitations that come with initial approximations and the considerable ignorance of a fossil record we have just begun to study. A bias toward the ornate ostracods is admitted as these are the most easily recognized in a preliminary survey. The statistical data, which extend from late Cretaceous through the Pleistocene, have been partitioned in 1 MY time slices and time averaged throughout each major ocean (Figs 1 and 2). This sample interval may begin to compromise chronostratigraphical limits as the Recent is approached. The database (Benson *et al.*, 1985) consisted of 156 genera from 155 DSDP sites spanning a stratigraphical interval of 80 MY. Of the 1600 (50cc) samples examined, 1044 yielded 30,000 specimens and these were determined taxonomically by the author during a six month interval. The average number of genera per time-slice was between eight and nine, with

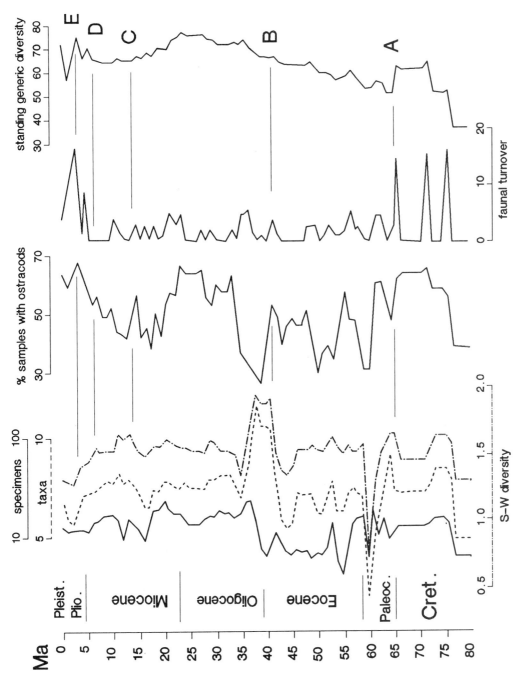

Fig. 1. Global census of deep-sea ostracod populations per million year time-slices in 50cc DSDP core samples; average number of specimens and genera; Shannon–Wiener diversity, standing generic diversity; and percent of samples with Ostracoda. The events indicated are A, the K/T Boundary Event; B, the origin of the psychrosphere; C, the Middle Miocene 'Event' (*circa* 16-14 Ma); D, the Messinian Salinity Crisis; and the 3.5 Ma Event.

variations somewhere between two and twelve. One is reminded that the events are the relatively large and rapid changes in the diversity, population, etc., and not just the peaks.

Considering the size of the database, which for any single time-slice after the Eocene was from 50 to 150 samples (rarely more than one per DSDP site), an average variation from the norm approaching two genera would be significant. An increase approaching four genera, such as took place after the Eocene, would signal a major event. It is noted that even for 20 MY after the crash of the Palaeocene, when ostracod samples ranged from 15 to 20, the average number of genera was only slightly less, but the variance increased. The number of genera extant at any one time throughout the world varied between 45 and 75 with a norm of 62 over a 50 MY period.

The study of Moroccan Neogene ostracods, now in progress, is a special case of a study of a global event. It began as an examination of a controlling threshold between an ocean and a sea related to the origin of that sea and correlation problems associated with it. Widespread water-mass changes that cause 'global events' are usually the direct extension of local threshold histories. In the Mediterranean, the global chronostratigraphic control datum for the Pliocene (the Global Stratotype Section and Point) is inextricably involved in the history of faunal invasions crossing that threshold.

## PALAEOCEANOGRAPHICAL STRUCTURE AND OSTRACOD DISTRIBUTION

The ability of sequential changes in the deep-sea fossil record to reflect global events resides in the integrity and stratification of water masses maintained over great distances. Often the same or similar ostracod fossil records can be found in deep facies that touch the edges of several continents, and even extend into the interior basins of some.

From the simplest palaeoecological point of view, the most important oceanic water-mass change begins at bathyal depths on the continental slope near the major pycnocline (where water density changes rapidly because of sudden, stratified changes

in temperature and salinity, and currents are indicated). The major pycnocline, which as a dynamic interface also defines the barostrophical surface beneath the gyres (as differentiated from minor pycnoclines separating intermediate water masses), is a significant modern ecological and hydrographical barrier where it intercepts the slope. It has been an historical frontier separating shelf and deeper faunas, especially in low latitudes since the Palaeogene. The colder underlying water-masses (the psychrosphere) originate in high latitudes, where the pycnocline (now a permanent thermocline) and psychrospheric faunas approach the surface. Warmer surface, or near surface water-masses (constituting the present thermosphere) originate in open ocean near the equator and in restricted basins or epeiric seas in lower latitudes. Gyres, or holoceanic surface current systems, are formed in the thermosphere as cybernetic response mechanisms in order to balance latitudinal temperature differences. It was the sudden shifts in gyral boundaries, their pycnoclines, that in the past resulted in biostratigraphical events.

Before a strong, thermally controlled circulatory structure existed in the World Ocean (before 40 Ma), the major pycnocline, if it existed in force, was controlled principally by stratification caused by salinity differences (Benson, et al., 1985). Gyral motion must have been sluggish and barostrophical surfaces weak. Whether ocean circulation was brought about by strong salinity differences, by temperature differences, or both; the tectonics of the thresholds of the separate ocean and sea basins, combined with the dominance of meridional or latitudinal seas, determined the water-mass structure of the World Ocean and strongly influenced ostracod distribution.

Now, temperature gradients or thermoclines strongly affect the distribution of those deeper benthonic marine ostracods that live beyond such coastal influences as large sediment and salinity differences. In the past, before there were strong thermally induced pycnoclines, the deep populations living in the oceanic basins may have suffered somewhat because of poor circulation in salinity controlled water-mass systems (such as the Albian-Aptian faunas of the South and equatorial Atlantic or the present Mediterranean). However, in general, ostracod taxa could easily move down slope across these open ocean salinity gradients. This was not so after the oceans became cold at depth.

Several Cainozoic deep-sea ostracod genera have been studied in detail and have revealed both gradual and punctuated evolutionary histories (Whatley, 1983a; Benson, 1984c). The difference in the speed of population replacement is probably due to rapid changes in water-masses and also new migratory routes opened by such changes. Very deep species may be morphologically determined by slight changes in the carapace reticulation patterns, after long periods of structural stability. Studies of variation in carapace shape have demonstrated 'speciation events', phyletic linkage over 10,000km ranges, and quite different morphotypes within a single genus (Benson, 1982; 1984a; also see Whatley, 1983a). Generally, the more radical were the morphological changes, the steeper the pycnocline associated with these changes was thought to have been.

The extreme temperature and pressure gradients extant in the deep sea have a strong effect on carbonate metabolism rates, which in turn affect ostracod carapace design, relative strength and adaptive reaction rates (Benson, 1988). By design is meant the ability to interpret the carapace as a functional static framework, whose strength is maintained for support and protection. The need for strength within the carapace and its selective value has not yet been sufficiently explored. This has nothing to do with internal-external, hydrostatic-pressure differences. It is more likely to be an adjunct to protection against predation, particularly important in the deep sea.

As the production of calcareous carapace mass becomes increasingly energy expensive, the carapace design must become more efficient and less redundant in order to retain the same functional effectiveness. Near the permanent thermocline (viewed separately as the thermal component of the pycnocline), there is a lag between metabolic demands and inherited design, and vestigial or relic morphological structures are not uncommon (in *Pterygocythereis*, for example; Benson, 1981). Populations of common inhabitants of former thermospheric seas and oceans (such as *Cytherella*, a dominant form of the Mesozoic) may be found living on the upper slope not far under the thermocline, but with larger, thinner shells.

More should be said about the partitioning of the ostracod faunas between the thermosphere and the psychrosphere, but we are just beginning to measure the effect of the frontier between these two worlds. It is sufficient to say that it was probably easier to migrate 10,000km along the continental shelf than it was to descend 200m down the continental slope. Most of the ostracods living in the greater depths of the deep sea after the formation of the psychrosphere were far beneath this interface. Their speciation events reflect much subtler pycnoclinal shifts.

Changes in oceanic thermal structure during the Palaeogene, from a World Ocean dominated by warm, saline-rich waters (coming from circumglobal 'seas' of low latitudes, collectively known as Tethys), to one dominated by cold, thermally driven currents (originating in polar regions and channeled by oceans separated by north-south continental barriers), fundamentally altered the ostracods already living in the deep sea (Benson, 1975; Benson *et al.*, 1985). The ostracods that could adapt to this change were, in a sense, condemned to this new but isolated world. They seem to have been left as either ornate or 'smooth' relics, compared to the morphologically varied, often transitional species of the faster changing world of the continental shelves and epeiric seas.

## PALAEOCEANOGRAPHICAL EVENTS

Exposed onshore outcrops or cored sections of upper bathyal sections are unfortunately rare. As stated above, here is where one is most likely to find the sharpest record of changes after the modern oceanic water-mass structure was formed. Many of the changes in the deep-basin benthonic data after this time are apt to be dulled, secondary reflections of the actual events. Older events, especially in the Atlantic, were controlled by deeper threshold changes as the basins were formed, the Tethys Ocean became more and more restricted, and the Southern Ocean 'heat sink' was formed. The deep-basin data of this time are probably more reliable, because the events concerned with thermal stratification, as are recorded and listed below, were felt more strongly.

Since its beginning during the Jurassic, the fossil record of the deep Atlantic Ocean ostracods (reflected more or less in the other ocean faunas as well; Fig. 2) has been marked by three historical episodes or phases in oceanic evolution, punctuated by five pycnoclinal shifts or global events:

1) The restricted basin phase, with faunal elements similar to those of the Permo-Triassic and probably coming from Tethys. This is the base-line stage, remaining until the Albian and Aptian in the South Atlantic. The pycnoclines were weak and deeply spread; the basins were poorly ventilated. The first invaders of restricted basins seem to include many of the survivors of the Phanerozoic Crisis (dominance of bairdiids, cytherellids, monoceratinids in 'off-shelf' deposits; Bate *et al* ., 1984; Benson, 1984a);

2) The thermospheric ocean phase with no strong taxonomic distinction between deep and neritic forms. This episode lasted until the Palaeogene, when colder circulation began to ventilate the restricted basins effectively. It was perpetuated in the isolated basins of Tethys (and exists today in the Mediterranean) after it had ended in open ocean. The pycnoclines were saline driven and probably not ecologically determinant for the ostracods. This phase was suddenly and temporarily interrupted by a loss of about 15% of the known deep sea ostracod genera near the Cretaceous/Tertiary boundary; a relatively high and sudden loss when compared with changes since that time. This is the oldest global 'event' now evident in ostracods found in deep-sea cores, but one suspects that there may have been another during the Turonian.

3) The psychrospheric phase, which began as an event in the late Eocene with the movement of Australia away from Antarctica far enough to make the Southern Ocean circuit thermally confining, the restriction of the warm saline driven outflow of Tethys as continental collision began to form shallower thresholds (Ricou, 1987), and the consequent filling of the deeper oceanic basins with cold water. The pycnoclines were becoming strong and ultimately dominated by thermal differences. They were compressed toward the surface by the basin infilling. This progressive alteration in water-mass structure was recorded by the evolution of the psychrospheric ostracods. It lasted more than several million years, and its effects are diachronous at different palaeodepths (Benson, 1975).

4) The psychrospheric phase, continuing even today, but augmented due to the restriction of the Drake Passage (22-20 Ma), the enlargement of the East Antarctic Ice Cap during the Middle Miocene (16-14 Ma) and the closure of the eastern entrance of Tethys. All of these events caused the punctuated acceleration of the filling of the Atlantic basins by cold water-masses. The rapid development of the Southern Ocean cooling system created an even sharper contrast between the cold supporting underlayers of water-mass structure and the remnants of the warm surface and intermediate layers derived from the Tethys. A deepening of the carbonate compensation depth occurred with a noticeable change in the deep faunas of the southern hemisphere. This third global event was felt most strongly in the South Atlantic, but is not clearly registered by the ostracods (Benson, 1984c).

5) The Messinian Salinity Crisis beginning with accentuated southern hemisphere glaciation and near closure of the western end of Tethys (the Palaeo-Mediterranean). A water deficit in the developing Mediterranean region, in part caused by damming of eastern continental drainage (Paratethys), caused a current reversal in the Rifian Corridor at the time of the Global Carbon Shift (6.3 Ma). Sea level dropped, the North Atlantic gyre thinned and accelerated, and upwelling increased along the coast of North Africa. The Palaeo-Mediterranean dried up, or at least was critically severed from the World Ocean. This is the fourth 'global' event to which Mediterranean and North Atlantic ostracods, deep, shallow and estuarine, theoretically bore witness. The lack of precise stratigraphical control has previously hindered the search for its effects. The Zanclean Deluge ended the 'crisis' and originated the Neo-Mediterranean (Benson, 1973b). Its global significance seems to be more stratigraphical than historical.

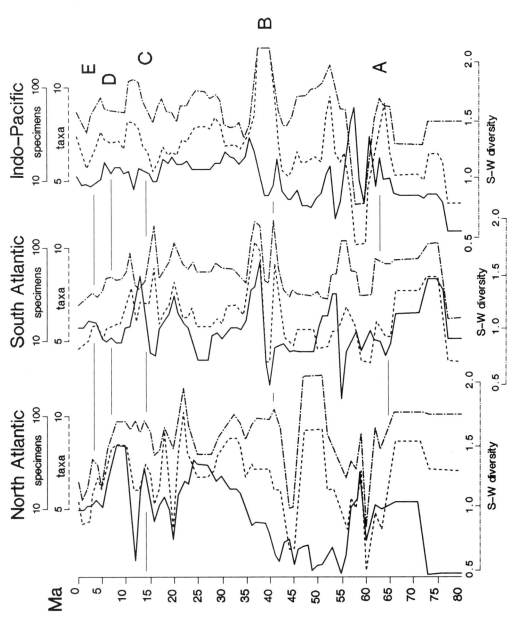

Fig. 2. Census of deep-sea ostracod populations in the North Atlantic, South Atlantic and Indo-Pacific Oceans per million year time-slices in 50cc DSDP core samples; average number of specimens and genera; Shannon–Wiener diversity, standing generic diversity; and percent of samples with Ostracoda. The events indicated are A, the K/T Boundary Event; B, the origin of the psychrosphere; C, the Middle Miocene 'Event' (*circa* 16-14 Ma); D, the Messinian Salinity Crisis; and the 3.5 Ma Event.

6) The Pliocene 'crash', which began with a surface warming trend in the first part of the Pliocene, followed by a general cooling and another series of glacial maxima. The deep and shallow water-mass flow from the Atlantic into the Pacific through the Balboa Portal stopped abruptly. There is a general deterioration of faunal diversity in the deep sea (lowest at 3.5 Ma). This 'event' is not finished. It is the strongest negative faunal change since the Palaeocene (Benson *et al.*, 1985; Whatley, 1983b; Whatley & Coles, 1987).

## THE K/T BOUNDARY EVENT - THE BOLIDE COLLISION

The end of the Cretaceous is followed by a sharp decline in deep-sea ostracod population density and a sudden change in generic diversity (about 15% extinctions, without replacement). Only in the Pliocene is the faunal turnover greater. This change is also found in deep-sea benthonic foraminifera (Emiliani *et al*., 1981; Beckmann, pers. comm.). The decline 'bottomed-out' in the late Palaeocene. Of the samples available for study, which for the 25 MY interval following the Cretaceous is at a minimum (10 to 30 sites per timeslice), the proportion of samples containing ostracods remained the same and only slightly lower than average.

At first, the level of extinctions does not seem to be very great but there was no replacement for a long time. No strong evolutionary changes are noticed in the survivors. Extinctions of such magnitude are common in shallow-water faunas. The generic diversity seems to recover, but it oscillates suggesting that the extant genera spread out and that there may have been unmeasured increases in species diversity (for Mesozoic diversity estimates, see Whatley, 1988). These changes occurred in a relatively buffered environment after a long time of faunal stability and high diversity (Campanian to Maastrichtian). It is followed by a relatively long interval of oscillating, generally depressed diversities and much lower populations. It is striking, to this author at least, how many of the survivors of this event, and that of the Phanerozoic Crisis, are numbered among the first invaders of

newly opening deep-sea basins (cavellinid-cytherellids, bairdiids and monoceratinids; Benson, 1984b).

The early Palaeocene 'crash' and following oscillations are difficult to explain by changes in terms of modern ocean water-mass structure. Perhaps the best analogue is the eastern Mediterranean. This was also a time of thermospheric deep ocean circulation, with bottom temperatures in excess of 12°C dominated by outflow from low latitude circum-global Tethys. No presently demonstrated pycnoclinal or tectonic threshold change could have suddenly and so strongly affected the deep faunas in such a manner. The possibility that it began because of something happening beyond the scope of oceanographic explanations seems reasonable by default. Yet those who suggest conventional gradual causes (geological catastrophes have uniformitarian explanations) must also explain why such a downturn in data is reflected globally, and not just in one or two basins.

The coincidence of its commencement with the iridium fall-out, suggesting a bolide collision (Alvarez *et al*., 1980; Hsu, 1980; Smit & Hertogen, 1980; Officer & Drake, 1983; for arguments for entirely terrestrial causation, see Hallam, 1987; Whatley, this volume), cannot be discounted. Although, ostracod evidence for such a spectacular event is indirect and delayed. The ostracod event could be a secondary food chain effect of the catastrophic change that took place in the plankton.

## THE ORIGIN OF THE PSYCHROSPHERE AND THE BEGINNING OF MODERN OCEANIC CIRCULATION

Modern oceans are characterized by thin, warm rotating surface gyres supported by thick, cold underlying water-masses originating in high latitudes with relatively weak stratification at depth. Today, relatively little outflow comes from low-latitude epeiric seas or oceans, as compared to that of the Mesozoic and early Palaeogene. The deep Atlantic now serves as a north-south, cold-water conduit. This conduit is the passage between northern, threshold controlled spillover and a southern refrigerating ring or heat sink. By comparison, the rest of the World Ocean is passive and less interesting, in terms of developing ocean systems.

When the present Antarctic continent was joined with Australia, the region of the South Pole was open to World Ocean surface circulation and the influence of warm water inflow. Heat exchange with the low latitudes was rapid. As Australia moved away from Antarctica and opened up the circumglobal circulation of the Southern Ocean in the polar region, the longer entrapment of water-masses became a 'heat sink', that is, heat-loss *versus* insolation-gain began to dominate. The Southern Ocean began to concentrate the cooling effect and feed cold, dense waters northward into the deep basins of the rest of the World Ocean. At the same time, the Tethys Ocean was becoming more restricted due to northward movement of the southern continents forming, shallower thresholds and diminishing the outward flow of thermospheric saline driven water masses. As a consequence, the World Ocean circulation was transformed from a dominantly saline driven to a dominantly thermal driven system. This series of changes formed the 'modern' psychrosphere (assuming that there could have been psychrospheric conditions before), which had global and lasting biological effects in the deep sea.

Ostracods show the cooling effects of this event worldwide by a major increase in diversity. When it subsides, after the first effects of the flushing of concentrated nutrients throughout the World Ocean, it is still maintained at a much higher level than before. This is most strongly felt in the Indo-Pacific and especially in the South Atlantic (Benson, 1975; 1984a; Benson *et al.*, 1985). Here the deep circulation to the north was controlled by aseismic ridges across a spreading, sinking ocean floor. The Vema Channel now forms a passage and threshold through the Rio Grande Rise to the west, while the Walvis Ridge forms a barrier to the east. The history of faunal change from thermospheric to psychrospheric is, in effect, the history of the origin of modern deep-sea circulation. In the North Atlantic the diversity peaked in the Middle Eocene (for reasons not yet understood) followed by the expected peak in the late Eocene.

The suggestion and discovery of this event came from three independent sources. The first, an indication of change, came from a study of the benthonic foraminifera of the northwest Pacific (Douglas, 1973); the second came from the discovery of the change in the ostracod faunas in the South Atlantic (Benson, 1975); the third, from changes in stable isotopes in the foraminiferal faunas south of the Tasman Sea (DSDP Site 277; Shackleton & Kennett, 1975).

A difference of opinion has existed about the date when this event happened (Corliss, 1981; Keller, 1982; Shackleton & Kennett, 1975; Benson, 1975; Boersma, 1986; Boersma & Premoli-Silva, 1986). Stable isotope data from Site 277 suggested that it took place at 38 Ma. Ostracod data suggested that it began at 40 Ma. However, if one examines the palaeodepth at which the data for the younger date were collected, the stable isotopes south of Tasman Sea, represent water-masses much shallower (present site depth 1214m) than those possible in the South Atlantic (present site depths 4000m+).

Both dates are probably correct. The two dates describe different stages of the filling of the deep ocean basins with cold water. A battle for dominance between the effects of polar cooling and the warming influence of Tethyan intermediate waters occurred. The ensuing change was not immediate. It is possible, as Whatley *et al.* (1983, 1984) suggest, that local tectonic subsidence during this time caused the submergence and introduction of individual genera (such as *Bradleya* ) into the psychrosphere. Polar submergence can explain the entrance of many new forms into the deep sea. The general faunal change evident in the world census of ostracod data is an event culminating in the late Middle Eocene, and lasting into the early Oligocene.

This interval marks the establishment of the modern deep-sea fauna characterized by ornate, reticulate ostracod genera, such as *Bradleya*, *Poseidonamicus*, and *Agrenocythere*. Phyletic lineages such as *Paleoabyssocythere* -*Abyssocythere* and *Rocaleberis* -*Henryhowella* developed less massive aspects in their carapace morphology in the South Atlantic. *Bradleya* came from New Zealand, *Poseidonamicus* from the Australian region of the eastern Indian Ocean and southwestern Pacific, *Agrenocythere* evolved from *Oertliella* in the North Atlantic and Tethys. The change in architecture of these and other deep-sea forms, most noticeable in late Middle and late Eocene, is generally from smaller more massive carapaces to larger thinner more spinose, more deterministic designs. The increase in metabolic stress seems evident in the elimination of excess or redundant structure (Benson, 1984b).

## THE MID-MIOCENE EVENT

The Middle Miocene event (16-14 MY; Srinivasan, & Kennett, 1981; Savin, *et al* ., 1981, 1985; Vail & Hardenbol, 1979) appears weakly in the world-wide census of deep-sea ostracod data as a recovery from a negative peak after a long period of stability following the origin of the modern psychrosphere. Strong but temporary increases in population abundance developed in the North and South Atlantic. This occurred during a time of increased diversity in the North Atlantic and between two very notable peaks of diversity in the South Atlantic, where *Poseidonamicus* undergoes a significant change in the pattern of the reticulum on the Rio Grande Rise and Walvis Ridge (Benson, 1984c). The event is clearest in the Indo-Pacific as a significant increase in taxa and diversity.

The formation of the East Antarctic Ice Cap occurred at about this time, as did the closure of the eastern end of Tethys (estimates of the date of these events vary in the order of 5 MY as neither were simple events). Together they increased the relative thickness of the psychrosphere, and the extant gyres significantly increased their rotational velocity to maintain their effectiveness as heat return mechanisms. The thinner the gyres became (the major pycnocline approaching the surface); the more defined they became (better stratified with stronger currents). This was the interval when *Globorotalia menardii*, a tropical epipelagic foraminifer, extended far into the high latitudes of the North Atlantic (Cifelli, 1976). The warm neritic faunas of the west migrated poleward, while the deeper faunas mounted the eastern continental slopes.

Widespread changes in hydrography,

sedimentation, surface productivity and carbonate dissolution level took place in the Pacific during this event. Only minor changes were noted in the ostracod faunas of the central equatorial region by Steineck, *et al.* (1988), but our data, based on a selection of 20 DSDP sites, show a definite temporary negative change in diversity, even in this relatively open ocean system.

## LATE MIOCENE OSTRACOD FAUNAS NEAR THE STRAIT OF GIBRALTAR

The Bou Regreg Section at Rabat in Morocco (Fig. 3) is the only onshore section known to be a continuous marine record of the interval between 7 and 4.5 Ma straddling the Miocene-Pliocene boundary (Feinberg & Lorenz, 1970; Bossio *et al.* 1976; Wernli, 1977; Cita & Ryan, 1978; Benson *et al* ., 1989b). The palaeodepth of this sedimentary sequence was sufficient (5-600m minimum) to escape the effects of the terminal Miocene regression (70-100m estimated; Adams *et al.*, 1977) and to register an important change in the pycnocline. The section begins with very shallow sands containing *Loculicytheretta*, deepens, and ends with typical sublittoral forms such as *Chrysocythere* and *Aurila. Pterygocythereis, Ruggieria, Buntonia* and *Oblitacythereis mediterranea* Benson are typical of the uppermost bathyal facies, immediately before and after the appearance of the psychrospheric fauna (Fig.4). The psychrospheric fauna contains *Agrenocythere pliocenica* (Seguenza), *Bythoceratina* sp., *Cytherella* spp. and *Oblitacythereis ruggierii* Russo. The fauna contains some 100 bathyal and outer neritic ostracod species.

Few of the Neogene ostracods of Morocco are yet formally described although, with Bonaduce, the author has this in hand. Those of the Bou Regreg section, however, have been used in a stratigraphical study by Bossio *et al.* (1976), in which Russo identified the ostracods. A similar study was made in the Gharb Basin by Peypouquet, *et al.*, 1980). As one might expect, there are species known from the Atlantic and the Mediterranean, and species of the fauna of the Guadalquivir Basin in Spain.

From borehole cores in the Bou Regreg Valley

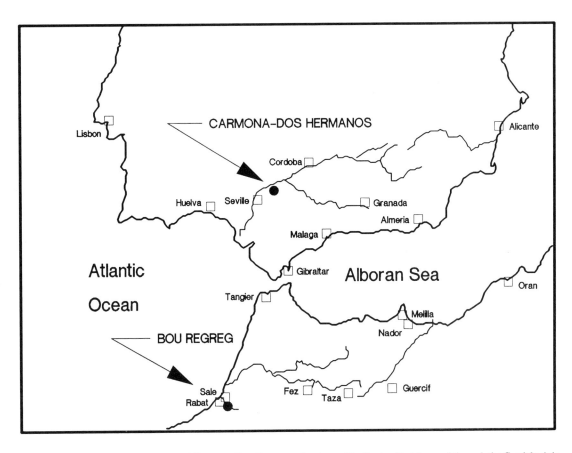

Fig. 3. Locations of the Bou Regreg and Carmona-Dos Hermanos Sections. The Iberian Portal passed through the Guadalquivir Valley (Seville, Cordoba) north of Alicante. The Rifian Corridor follows the Oued Sebou north of Rabat east to Fez, Taza, Guercif and then the Oued Moulouya east of Nador.

near Rabat, we have identified a zone of predominantly psychrospheric ostracods, referred to as the '*Agrenocythere pliocenica* Zone' (beginning at the Tortonian-Messinian boundary). This zone, with *Oblitacythereis ruggierii*, has been found deep in the Rifian Corridor eastward past regional tectonic and physiographical divisions separating the Atlantic from the Palaeo-Mediterranean. We have found *A. pliocenica* well within the Corridor east of Fez.

In the Guadalquivir valley in the northern Pre-Betic region of Andalusian Spain, a second, but probably less important passage existed during the Miocene. Psychrospheric ostracods were found there in Serravallian and Tortonian deposits

extending to the Balearic Islands (Benson, 1976; Berggren *et al.*, 1976). The Rifian Corridor, in the Pre-Rif region of Morocco, 200km south of Gibraltar was discovered by Gentil in 1918, which he thought was Burdigalian and 'Helvetian' in age. Nappes of olistostromes closed both of these passageways before the beginning of the Pliocene.

The changes in ostracod biofacies at Bou Regreg correspond to several important changes in the fossil microplankton. The *Agrenocythere pliocenica* Zone (6.3-5.9 Ma) begins with the first appearance of the very conoidal *Globorotalia conomiozea*, and a change in dominance from the epipelagic *G. menardii* to the mesopelagic *G. miotumida* (Sierro, 1985). A major nannofloral change occurs just

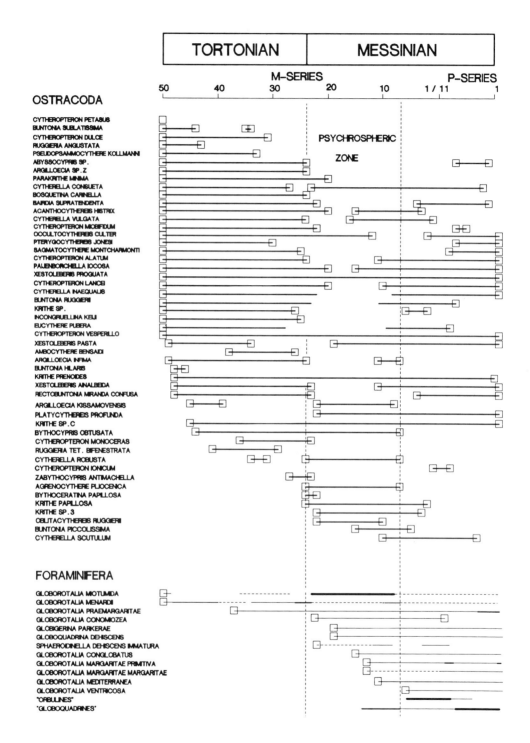

Fig. 4. The ranges of Ostracoda and planktonic foraminifera of the Tortonian-Messinian part of the cored section at Bou Regreg (Ain el Bieda; samples P1 through M50) showing the biofacies change caused by the intrusion of the psychrosphere. Single appearances and ubiquitous species have been omitted.

before the base of this zone (*fide* M.-P. Aubry). The global 6.3 Ma Carbon Shift occurs immediately after the beginning of this change. Benson *et al* ., (1989a) have interpreted this change to indicate a reversal in current direction that lifted the eastern barostrophical surface of the thinning Atlantic gyre and the psychrosphere with its fauna into the Rifian passage to form an inflow, 'siphon', or semipermanent upwelling (similar to the 'estuarine circulation' of Van Harten, 1984). This event lasted about 400,000 years.

As the pycnocline compressed in thickness, the contrast in the temperature between the thinning gyre and the cold upper supporting layers increased. The planktonic foraminifera required more mass in order to be able to descend to accustomed depths in their bathymetric migrations. To achieve this, they became encrusted or more conical or both. The association of 'conoidal' globoconellids with psychrospheric ostracods in upper bathyal palaeodepths is very significant, especially as it coincides with the 6.3 Ma Global Carbon Shift.

**THE MESSINIAN SALINITY CRISIS**

During Middle and early Upper Miocene, the western, salinity-controlled, thermospheric outflow of Tethys into the Atlantic was confined to an ever thinner, but far reaching, intermediate water-mass. This outflow became the principal source of the cooling deep waters of the Norwegian Basin, which overflowed to the west and eventually passed through the Balboa Portal into the deep basins of the Pacific (Ried, 1979). At 6.3 Ma (the Tortonian-Messinian Boundary), the deep current in the Rifian Corridor, which was then the largest of two passages for the outflow from the Palaeo-Mediterranean into the Atlantic, reversed its direction (Benson, *et al.*, 1989a). This was the first stage of the series of global events associated with the Messinian Salinity Crisis (Ruggieri, 1967; Hsu *et al.*, 1973; Hsu, 1986; Cita & McKenzie, 1986; Drooger, 1973; Benson, 1976).

Whether this event was caused by tectonic changes between the Alboran Microplate and the subducted edge of the African Plate, increased desiccation rates in the Palaeo-Mediterranean, or

acceleration in the formation of thickening psychrospheric water by increased glaciation in the Antarctic, is uncertain. What is certain is that the near isolation of the deep Mediterranean basins did ultimately occur, leading to the annihilation of the normal marine ostracod faunas of Tethys. With the desiccation phase of the 'salinity crisis' and the lowering of the Mediterranean sea level, came intermittent invasions of the caspi-brackish Paratethyan *Cyprideis* faunas.

If the contribution to the Atlantic of Tethyan Intermediate Waters did stop, the decrease in force of the deep Labrador Current of the western North Atlantic would have allowed North American bathyal faunas to move north. One can possibly see residual effects of this mechanism (which was repeated during the Pleistocene) in the modern deep water faunas off Newfoundland (Benson, *et al.*, 1983).

Many global events concurrent with that of the Salinity Crisis are now well established (Kennett *et al.*, 1985). Among these is the 6.3 Ma permanent negative Global Carbon Shift (Vincent, *et al.*, 1980; Hodell & Kennett, 1984), which represents a significant change in the carbon isotope balance and, therefore, a change in the speed of replenishment of carbon in the ocean reservoir. The effect of increased glaciation in the south polar region is also well established (Hodell *et al.*, 1986). A lowering of sea level of from 70 to 100m has been generally accepted (Adams *et al.*, 1977; Berggren & Haq, 1976; Haq *et al.*, 1988). It has been suggested that oceanic salinity possibly dropped by some four ppm (Ryan, 1973; Benson, 1984c) throughout the world, to compensate for evaporites formed in the Mediterranean.

**THE ORIGIN OF THE NEO-MEDITERRANEAN: A PLIOCENE EVENT**

The Messinian Salinity Crisis ended suddenly with restoration of normal marine conditions in the Neo-Mediterranean (Crescenti, 1971; Cita, 1975) by the sudden inundation of the basins by upper bathyal North Atlantic waters with species of the psychrospheric ostracod fauna (Benson & Sylvester-Bradley, 1971). Characteristic elements of this fauna, *Agrenocythere*, *Quasibuntonia*, *Oblitacythereis* and

*Bythoceratina*, remained in the Mediterranean until the early Pleistocene. They were replaced by the present submerged neritic and residual upper bathyal faunas with a few new invaders, such as deeper species of *Cytheropteron* and *Paijenborchella*. Not a few pandemic taxa (Whatley & Ayress, 1988) endured, and generally those with broad tolerance to near anoxic conditions (Van Harten & Droste, 1988). From the ostracod point of view, the Mediterranean did not become a 'sea' (in the hydrographical sense) until the Pleistocene.

The lowermost Pliocene (Zanclean) ostracod fauna is characterized by *Agrenocythere pliocenica* in the deep-basin 'Trubi' facies (Capo Rossello section in Sicily; Benson, 1972a), and *Oblitacythereis mediterranea* in upper bathyal facies (Santorno section of northern Italy and in the Myrtou Formation of Cyprus; Benson, 1977) immediately above Messinian *Cyprideis* bearing beds (remaining until the Pleistocene, Benson & Sylvester-Bradley, 1971; Benson, 1973a,b, 1978). Thus far they have not been found in the Atlantic in faunas of this age, but they have been found in bathyal deposits in the Rifian Corridor in Morocco in the lower Messinian (4.7 Ma in Sicily *versus* 6.3 Ma at Bou Regreg in Morocco). The evolution of *Agrenocythere pliocenica* can now be traced along the northwest African continental slope toward Morocco from the south during the Miocene.

The appearance in the Mediterranean of Zanclean psychrospheric ostracods required a deeper and broader passage to the Atlantic than now exists at the Strait of Gibraltar. The existence at that time of a passage at Gibraltar, and not elsewhere, is inferred from an hiatus from DSDP Site 121 (Nesteroff & Ryan, 1973; also see Durand-Delga, 1971; Andrieux & Mattauer, 1973; Bonini, *et al.*, 1973; Arana & Vegas, 1974; Hatzfeld, 1976; Reyment, 1981 for discussions of the continuity of the Rif-Betic Arc across the Strait). The presently known structural and physiographical configuration of the Pliocene Gibraltar threshold would not have allowed maintenance of psychrospheric conditions in the Mediterranean basins against the warming influence of local climatic conditions for 3 MY.

The abrupt entry of the psychrospheric ostracod fauna also has special global stratigraphical significance in that it, with other elements of the microfauna, indicates the restoration of '... open marine conditions in the Mediterranean...' after the Messinian Salinity Crisis. This is part of the definition by Cita (1975, p. 17) for the Boundary Stratotype of the Pliocene.

This is a case of a regional event becoming of global significance because of stratigraphical tradition and practice. Yet its history and demonstration as an event depends on the testimony of deep-sea ostracods, especially those that lived close to the major pycnocline and the thresholds of partially land-locked basins. These are forms that are most likely to have the greatest impact in event stratigraphy.

At present, there is no direct geological or fossil evidence in Spain, Morocco or the Gibraltar region for the passage of the Zanclean Deluge that formed the Mediterranean Sea after its desiccation by the Salinity Crisis.

## THE EVOLUTION OF *AGRENOCYTHERE* IN THE ATLANTIC

When *Agrenocythere* was first recognized as a genus (Benson, 1972b), *A. pliocenica* was known only from the Mediterranean particularly from Italy (Sicily and Calabria) from where Seguenza (1880) had originally described it. Later, it was frequently encountered in the Zanclean of DSDP cores (Legs 13 and 42A; Benson, 1973a,b, 1978), and confirmed (Van Harten, 1984) to have penetrated into the eastern Mediterranean with the 'Zanclean Deluge'.

Living species of *Agrenocythere* are principally known from the Indian Ocean, but its fossil record is global. Known from the Atlantic since the Eocene, it flourished in the Oligocene and Miocene to become extinct there in the Pliocene (youngest occurrences, *A. hazelae* van den Bold, DSDP sites 354, 526A; also see Whatley & Coles, 1987). Its exact depth limits are, of course unknowable, but it is safe to say that its presence is strongly indicative of waters deep enough to have originated at high latitudes, and not in the thermospheric Tethys or some other restricted basin with little or no access to the deeper open ocean.

*Agrenocythere pliocenica*, from its name, was thought to be Pliocene and indigenous to the Mediterranean. If this proved to be true, its interpretation as a psychrospheric species could be held in doubt.

The ancestral lineage of *Agrenocythere pliocenica* has been found in DSDP cores along the African Slope (Sites 366, 368 and 547) and at Lomo Pardo, Spain. This demonstrates that its origins are in the Atlantic in the Miocene and not the Mediterranean. It was found at Site 366 in the *Globorotalia acostaensis* Zone (core 14, sec. 5; Tortonian, 8 Ma) and later, still in the Miocene (core 3, sec. 3; early Messinian?), at Site 547A in the *Discoaster hamatus* Zone (core 8, sec. 4; NN9, N16; Tortonian) and in Spain in the Burdigalian.

The stock of *A. pliocenica* can now be clearly traced back into the Middle Miocene of the southeastern North Atlantic, where its evolution may have resulted from the changes that climaxed in the so-called '14 Ma event'. It is inferred to have arisen from an Oligocene form near the equator; but evidence for this is largely circumstantial due to the lack of any other possible connection.

There seem to be three lineages of *Agrenocythere* in the Atlantic of which the *A. pliocenica* branch is restricted to the eastern part before the desiccation phase of the Salinity Crisis and subsequently in the Mediterranean. *A. pliocenica* was abundant in the core at Ain al Beida (M25 to P8) and in the Oued Akrech section of Bou Regreg near Rabat (6.3 to 5.9 Ma; Lower Messinian).

*Agrenocythere pliocenica* is, like most of the other microfossil markers of important events in Mediterranean Pliocene history, an obvious invader from the Atlantic; and like the other invaders, it required oceanic conditions initially to survive in the Mediterranean. It could have adapted later, but we have no morphological or phylogenetic evidence of such a change. It left no Pliocene descendants in the Atlantic. The last examples found in the Atlantic are in the Messinian (5.9 Ma). The first examples found in the Mediterranean are a million years younger. There is no known overlap of the ranges. Like Lazarus, the Mediterranean *Agrenocythere pliocenica* seems to have risen from the dead (the dust of the Messinian Salinity Crisis).

## THE PLIOCENE 'CRASH'

Our data (Benson *et al.*, 1985) of the ostracod generic standing diversity in the deep sea, shows a general sharp increase after the Salinity Crisis, whereas the Shannon-Weiner diversity index shows a more general decline since the Middle Miocene (a Cainozoic maximum). This was true in all of the oceans. This is at a time when samples with ostracods are at a maximum. A sharp decline at about 3.5 Ma is followed by a noticeable recovery in the North Atlantic and the Indo-Pacific (in agreement with Whatley, 1983b). Only during the crisis at the end of the Cretaceous has the faunal turnover and the decrease in standing diversity been as great. Our data, based on million year time-slices, could have averaged Pleistocene diversity estimates that otherwise would show a wide range of local and short term variation.

The 3.5 Ma event seems strongest in the North Atlantic and the Pacific. It may be an effect of the closure of the Balboa Portal ('Panamanian Straits'), the return to glacial conditions, and the final refinements of the North Atlantic gyral system. The strong peak in ostracod faunal turnover with extinctions is especially noticeable in the North Atlantic (Benson *et al.*, 1985; Whatley & Coles, 1987). *Agrenocythere*, a common, if not dominant component of Miocene faunas and an important invader during the Zanclean Deluge, is extremely rare in Pliocene samples from the North Atlantic.

## ACKNOWLEDGEMENTS

Remembered are Peter Sylvester-Bradley and Giuliano Ruggieri, who started me on the trail of deep-sea ostracods. Thanks to Ralph E. Chapman for his analytical assistance, to G. Bonaduce for our work together on the ostracods of Morocco and to K. Rakic-El Bied for information on the distribution of foraminifera. This work has been a part of IPOM, sponsored by the Smithsonian Institution in co-operation with the National Geographic Society.

## REFERENCES

Adams, G. C., Benson, R. H., Kidd, R. B., Ryan, W. F. B. & Wright, R. C. 1977. The Messinian Salinity Crisis and evidence of late Miocene eustatic changes in the world ocean. *Nature*, London, **269**, 383-386.

Alvarez, L. W., Alvarez, W., Asaro, F. & Michel, H. V. 1980. Extraterrestrial Cause for the Cretaceous-Tertiary extinction. *Science*, New York, **208**, 1095-1108.

Andrieux, J. & Mattauer, M. 1973. Précisions sur un model explicatif de l'Arc de Gibraltar. *Bull. Soc. géol. Fr.*, Paris, Ser. 7, **15**, 115-118.

Arana, V. & Vegas, R. 1974. Plate tectonics and volcanism in the Gibraltar Arc. *Tectonophysics*, Amsterdam, **24**, 197-212.

Bate, R. H., Lord, A. R. & Riegraf, W. 1984. Jurassic Ostracoda from Leg 79, Site 574, *In* Hinz, K., Winterer, E. L. *et al.* (Eds), *Init. Repts DSDP*, **79**, 703-710. U.S. Govt Printing Office, Washington.

Benson, R. H. 1972a. Ostracodes as indicators of threshold depth in the Mediterranean during the Pliocene. *In* Stanley, D. J. (Ed.), *The Mediterranean Sea*, 63-73. Dowden Hutchinson & Ross, Stroudsburg, Pa.

Benson, R. H. 1972b. The *Bradleya* Problem with descriptions of two new psychrospheric ostracode genera, *Agrenocythere* and *Poseidonamicus* (Ostracoda, Crustacea). *Smithson. Contr. Paleobiol.*, Washington, **12**, 1-138.

Benson, R. H. 1973a. Psychrospheric and continental Ostracoda from ancient sediments in the floor of the Mediterranean. *In* Ryan, W. F. B., Hsu, K. J., *et al.* (Eds), *Init. Repts DSDP*, **13**, 1002-1008. U.S. Govt Printing Office, Washington.

Benson, R. H. 1973b. An Ostracodal View of the Messinian Salinity Crisis. *In* Drooger, C. W. (Ed.), *Messinian Events in the Mediterranean*, 235-242. North-Holland Publ. Co., Amsterdam.

Benson, R. H. 1975. The origin of the psychrosphere as recorded in changes in the deep-sea ostracode assemblages. *Lethaia*, Oslo, **8**, 69-83.

Benson, R. H. 1976. The Biodynamic Effects of the Messinian Salinity Crisis. *Palaeogeogr. Palaeoclimat. Palaeoecol.*, Amsterdam, **20**, 1-170.

Benson, R. H. 1977. Evolution of *Oblitacythereis* from *Paleocosta* (Ostracoda, Trachyleberididae) during the Cenozoic in the Mediterranean and Atlantic. *Smithsonian Contr., Paleobiol.*, Washington, **33**, 1-47.

Benson, R. H. 1978. The paleoecology of the ostracodes of DSDP Leg 42A. *In* Hsu, K. J., Montadert, L., *et al.* (Eds), *Init. Repts DSDP*, **42**, 777-787. U.S. Govt Printing Office, Washington.

Benson, R. H. 1981. Form, Function and Architecture in Ostracode Shells. *A. Rev. Earth planet. Sci.*, Palo Alto, Calif., **9**, 59-80.

Benson, R. H. 1982. Comparative transformation on shape in a rapidly evolving series of structural morphotypes of the ostracod *Bradleya*. *In* Bate, R. H., Robinson, E. & Sheppard, L. M., (Eds), *Fossil and Recent Ostracods*, 147-164. Ellis Horwood Ltd., Chichester for British Micropalaeontological Society.

Benson, R. H. 1984a. The Phanerozoic "Crisis" as viewed from the Miocene. *In* Berggren, W. A., & Van Couvering, J. A., (Eds), *Catastrophes and Earth History, The New Uniformitarianism*, 437-446. Princeton Univ. Press, Princeton, N.J.

Benson, R. H. 1984b. Estimating greater paleodepths with ostracodes, especially in past thermospheric oceans. *Palaeogeogr. Palaeoclimatol. Palaeoecol.*, Amsterdam, **48**, 107-141.

Benson, R. H. 1984c. Biomechanical stability and sudden change in the evolution of the deep-sea ostracode *Poseidonamicus*. *Paleobiol.*, **9**(4), 398-413.

Benson, R. H, 1985, Perfection, Continuity, and Common Sense in Historical Geology, *In* Berggren, W. A., & Van Couvering, J. A. (Eds), *Catastrophes and Earth History, the New Uniformitarianism*, 35-75. Princeton Univ. Press, Princeton, N.J.

Benson, R. H. 1988. Ostracods and Paleoceanography. *In* De Deckker, P., Colin, J.-P., Peypouquet, J.-P., (Eds), *Ostracoda in the Earth Sciences*, 1-26. Elsevier, Amsterdam, Oxford, New York, Tokyo.

Benson, R. H., Chapman, R. E. & Deck, L. T. 1985. Evidence from the Ostracoda of major events in the South Atlantic and world-wide over the past 80 million years. *In* Hsu, K. J., & Weissert, H. J. (Eds), *South Atlantic Paleoceanography*, 325-350. Cambridge University Press.

Benson, R. H., DelGrosso, R. M. & Steineck, P. L. 1983. Ostracode biofacies and distribution, Newfoundland, Continental Slope and Rise. *Micropaleontology*, New York, **29**(4), 430-453.

Benson, R. H., Rakic-El Bied, K. & Bonaduce, G. 1989a. In press. An important water-mass reversal in the Rifian Corridor (Morocco) at the Tortonian-Messinian boundary, *Continental Paleoenvironments*.

Benson, R. H., Rakic-El Bied, K., Bonaduce, G., Berggren, W. A., Aubry, M.-P., Hodell, D. A. & Napoleone, G. 1989b. In preparation. The value of the Bou Regreg Section, Morocco, as the Boundary Stratotype of the Pliocene.

Benson, R. H. & Sylvester-Bradley, P. C. 1971. Deep-sea ostracodes and the transformation of ocean to sea in the Tethys. *In* Oertli, H. J. (Ed.), *Paléoécologie des Ostracodes*, Pau 1970, *Bull. Centre Rech. Pau - SNPA*, **5** suppl., 63-91, 15 figs, 1 tab., 1 pl.

Berggren, W. A., Benson, R. H. & Haq., B. L. 1976. The El Cuervo Section (Andalusia, Spain): Micropaleontology of an Early Late Miocene Bathyal Deposit. *Mar. Micropaleontol.*, Amsterdam, **1**(3), 195-247.

Berggren, W. A. & Haq, B. U. 1976. The Andalusian Stage (late Miocene): biostratigraphy, biochronology, and paleoecology. *Palaeogeogr. Palaeoclimat. Palaeoecol.*, Amsterdam, **20**, 67-129.

Berggren, W. A. & Van Couvering, J. A. (Eds), 1984. *Catastrophes and Earth History, the New Uniformitarianism*, 464 pp. Princeton Univ. Press, Princeton, N.J.

Boersma, A. 1986. Eocene/Oligocene Atlantic Paleo-Oceanography using benthic Foraminifera. *In* Pomerol, C. & Premoli-Silva, I. (Eds), *Terminal Eocene Events*, in *Devs Paleont. Stratigr.*, Amsterdam, **9**, 225-236.

Boersma, A. & Premoli-Silva, I. 1986. Terminal Eocene events, Planktonic Foraminifera and isotopic evidence. *In* Pomerol, C. & Premoli-Silva, I. (Eds), *Terminal Eocene Events*, in *Devs Paleont. Stratigr.*, Amsterdam, **9**, 213-223.

Bonini, W. E., Loomis, T. P. & Robertson, J. D. 1973. Gravity anomalies, ultramafic intrusions and the Tectonics of the region around the Strait of Gibraltar. *J. geophys. Res.*, Richmond, Va., **73**(8), 1372-1382.

Bossio, A., Rakic-El Bied, K., Gianelli, L., Mazzei, R., Russo, A. & Salvatorini, G. 1976. Correlation de quelques sections stratigraphiques du bassin Méditerranéen sur la base des Foraminifères planktoniques, Nannoplankton calcaire et Ostracodes. *Memorie Soc. tosc. Sci. nat.*, Pisa, Ser. A, **83**, 121-137.

Cifelli, R. 1976. Evolution of ocean climate and the record of planktonic Foraminifera. *Nature*, London, **264**(5585), 431-432.

Cita, M. B. 1975. The Miocene/Pliocene boundary. History and definition. *In* Saito, T. & Burckle, L. (Eds), *Late Neogene Epoch Boundaries*, 1-30. Micropaleontology Press, Spec. Publ. 1, New York.

Cita, M. B. & McKenzie, J. A. 1986. The terminal Miocene event. *In* Hsu, K. J. (Ed.), *Mesozoic and Cenozoic Oceans*, Geodyn. Ser., AGU, Washington, D.C., **15**, 123-140.

Cita, M. B & Ryan, W. B. F. 1978. The Bou Regreg section of the Atlantic coast of Morocco. Evidence, timing and significance of a Late Miocene regressive phase. *Riv. ital. Paleont.*, Milano, **84**(4), 1051-1082.

Corliss, B. H. 1981. Deep-sea benthic foraminiferal turnover near the Eocene/Oligocene boundary. *Mar. Micropaleontol.*, Amsterdam, **6**, 367-384.

Crescenti, U. 1971. Sul limite Mio-Pliocene in Italia. *Geologica romana*, Roma, **10**, 1-22.

Douglas, R. G. 1973. Evolution and bathymetric distribution of Tertiary deepsea benthic Foraminifera. *Geol. Soc. Amer., Abstracts*, **5**(7), 603.

Drooger, C. W. (Ed.), 1973. *Messinian Events in the Mediterranean. Geodynamics Scientific Report No. 7*, 272. North-Holland Publ. Co., Amsterdam.

Durand-Delga, M. 1971. Hypotheses sur la genèse de la courbure de Gibraltar. *Bull. Soc. géol. Fr.*, Paris, Ser. 7, **15**, 119-120.

Emiliani, C., Kraus, E. B. & Shoemacker, E. M. 1981. Sudden death at the end of the Mesozoic. *Earth and planet. Sci. Lett.*, Amsterdam, **55**, 317-334.

Feinberg, H. & Lorenz, H. G. 1970. Nouvelles données stratigraphiques sur le Miocène supérieur et le Pliocène du Maroc nord-occidental. *Notes Serv. géol. Maroc.* **30**(225), 21-26.

Gramann, F. 1988, The Northwest European Tertiary Basin, Results of the International Geological Correlation Program, Project No 124, 4.3 Ostracods, *Geol. Jb.*, Hannover, **A100**, 225-252.

Hallam, A. 1987. End-Cretaceous Mass extinction event, Argument for Terrestrial Causation. *Science*, New York, **238**, 1237-1242.

Haq, B. U., Hardenbol, J. & Vail, P. R. 1988. Sea Level History, Technical comments, Response. *Science*, **241**, 596-597.

Hatzfeld, D. 1976. Étude de sismicité dans la région de l'arc de Gibraltar. *Ann. de Geophys.*, **32**, 71-85.

Hodell, D. A. & Kennett, J. P. 1984. Late Miocene Carbon Shift in DSDP Site 516A, western South Atlantic. *Geol. Soc. Amer., Abstracts*, **16**(6), 540.

Hodell, D. A., Elstrom, K. M. & Kennett, J. P. 1986. Late Miocene delta $^{18}O$ variability, global ice volume, sea level, and the "Messinian salinity crisis". *Nature*, London, **20**(320), 411-414.

Hsu, K. J. 1980. Terrestrial catastrophe caused by cometary impact at the end of Cretaceous. *Nature*, London, **285**, 201-203.

Hsu, K. J. 1986. Unresolved problems concerning the Messinian salinity crisis. *J. Geol.*, Chicago, Ser. 3a, **47**(1-2), 203-212.

Hsu, K. J, Ryan, W. B. F. & Cita, M. B. 1973. Late Miocene desiccation of the Mediterranean. *Nature*, London, **24**(5395), 240-244.

Keller, G. 1982. Biochronology and paleoclimatic implications of Middle Eocene to Oligocene planktic foraminiferal faunas. *Mar. Micropaleontol.*, Amsterdam, **7**, 463-486.

Kennett, J. P., Keller, G. & Srinivasan, M. S. 1985. Miocene planktonic foraminiferal biogeography and paleoceanographic development of the Indo-Pacific region. *Mem. geol. Soc. Am.*, Boulder, **163**, 197-236.

Nesteroff, W. D. & Ryan, W. B. F. 1973. Séries stratigraphic et implications tectoniques du forage Joides 121 en mer d'Alboran. *Bull. Soc. géol. Fr.*, Paris, Ser. 7, **15**, 113-114.

Officer, C. B. & Drake, C. L. 1983. The Cretaceous-Tertiary transition. *Science*, New York, **219**, 1383-1390.

Peypouquet, J.-P., Carbonel, P. & Cirac, P. 1980. Les ostracodes et les paléoénvironments profond du sillon sud-rifain au Mio-Pliocène. *8e Réunion Ann. Sciences de la Terre*, Marseille, 1980, 282.

Reyment, R. A. 1981. Asymmetry analysis of geologic homologues on both sides of the Strait of Gibraltar. *Math. Geol.*, **13**(6), 523-533.

Ricou, L.-E. 1987. The tethyan oceanic gates: A tectonic approach to major sedimentary changes within Tethys. *Geodynamica Acta*, Paris, **1**(4/5), 225-232.

Ried, J. L. 1979. On the contribution of the Mediterranean outflow to the Norwegian-Greenland Sea. *Deep-Sea Res.*, Oxford, **26A**, 17-91.

Ruggieri, G. 1967. The Miocene and later evolution of the Mediterranean Sea. *In* Adams, C. G., & Ager, D. V. (Eds), *Aspects of Tethyan Biogeography*, 283-290. Systematics Assoc. Publs., London, **7**.

Ryan, W. B. F. 1973. Geodynamic implications of the Messinian crisis of salinity. *In* Drooger, C. W. (Ed.), *Messinian Events in the Mediterranean*, 26-38. North-Holland Publ. Co., Amsterdam.

Savin, S. M., Douglas, R. G., Keller, G., Killingly, J. S., Shaughnessy, L., Sommer, M. A., Vincent, E. & Woodruff, F. 1981. Miocene benthic foraminiferal isotope records, a synthesis. *Mar. Micropaleontol.*, Amsterdam, **6**, 423-450.

Savin, S. M., Able, L., Barrera, E., Hodell, D. A., Kennett, J. P., Murphy, M., Keller, G., Killingly, J. S. & Vincent, E. 1985. The evolution of Miocene surface and near surface temperatures, Oxygen isotope evidence. *Mem. geol. Soc. Am.*, Boulder, Co., **163**, 49-82.

Seguenza, G. 1880. Le formatazioni tertiare nella provincia de Reggio (Calabria). *Atti Accad. naz. Lincei Memorie*, Roma, **4**(6), 1-406.

Shackelton, N. J. & Kennett, J. P. 1975. Paleotemperature history of the Cenozoic initiation of Antarctic glacial glaciation:

oxygen and carbon isotope analysis in DSDP Sites 277, 279, and 281. *In* Kennett, J. P. & Houtz, R. E. (Eds), *Init. Repts DSDP*, **29**, 743-755. U.S. Govt Printing Office, Washington.

Sierro, F. J. 1985. The replacement of the "*Globorotalia menardii* " Group by the *Globorotalia miotumida*, Group: An aid to recognizing the Tortonian-Messinian Boundary in the Mediterranean and adjacent Atlantic. *Mar. Micropaleontol.*, Amsterdam, **9**, 525-535.

Smit, J. & Hertogen, J. 1980. An extraterrestrial event at the Cretaceous/Tertiary boundary. *Nature*, London, **285**, 198-200.

Srinivasan, M. S. & Kennett, J. P. 1981. A Review of Neogene planktonic foraminiferal biostratigraphy and evolution: Applications in the Equatorial and South Pacific. *SEPM Spec. Publ.*, **32**, 395-492.

Steineck, P. L., Dehler, D., Hoose, E. M. & McCalla, D. 1988. Oligocene to Quaternary ostracods of the central equatorial Pacific (Leg 85, DSDP-IPOD). *In* Hanai, T., Ikeya, N. & Ishizaki, K. (Eds), *Evolutionary biology of Ostracoda, its fundamentals and applications*, proceedings of the Ninth International Symposium on Ostracoda, held in Shizuoka, Japan, 29 July - 2 August 1985, Developments in palaeontology and stratigraphy, **11**, 597-618, Kodansha Ltd., Tokyo and Elsevier, Amsterdam, Oxford, New York, Tokyo.

Vail, P. R. & Hardenbol, J. 1979. Sea level changes during the Tertiary. *Oceanus*, Woods Hole, Mass., **22**, 71-79.

Van Harten, D. 1984. A model of estuarine circulation in the Pliocene Mediterranean based on new ostracode evidence. *Nature*, London, **312**, 359-361.

Van Harten, D. & Droste, H. J. 1988. Mediterranean deep-sea ostracods, the species poorness of the Eastern Basin, as a legacy of an early Holocene anoxic event. *In* Hanai, T., Ikeya, N. & Ishizaki, K. (Eds), *Evolutionary biology of Ostracoda, its fundamentals and applications*, proceedings of the Ninth International Symposium on Ostracoda, held in Shizuoka, Japan, 29 July - 2 August 1985, Developments in palaeontology and stratigraphy, **11**, 721-738, Kodansha Ltd., Tokyo and Elsevier, Amsterdam, Oxford, New York, Tokyo.

Vincent, E., Killingly, J. S. & Berger, W. H. 1980. Magnetic Epoch-6 carbon shift: a change in the ocean's 13C/12C ratio 6.2 million years ago. *Mar. Micropaleontol.*, Amsterdam, **5**, 185-203.

Wernli, R. 1977. Les Foraminifères planktoniques de la limité mio-pliocène dans les environs de Rabat (Maroc). *Eclog. geol. Helv.*, Basel, **70**(1), 143-191.

Whatley, R. C. 1983a. Some Aspects of the evolution of the ostracode genera *Bradleya* and *Poseidonamicus* in the deep-sea Tertiary and Quaternary of the South Pacific. *In* Cope, J. C. W., & Skelton, P. W. (Eds), Evolutionary case histories from the Fossil Record, *Spec. Pap. Palaeont.*, London, **33,** 103-116.

Whatley, R. C. 1983b. Some aspects of the palaeobiology of Tertiary deep-sea Ostracoda from the S.W. Pacific. *J. micropalaeontol.*, London, **2**, 83-104.

Whatley, R. C. 1988. Patterns and rates of evolution among Mesozoic Ostracoda. *In* Hanai, T., Ikeya, N. & Ishizaki, K.

(Eds), *Evolutionary biology of Ostracoda, its fundamentals and applications*, proceedings of the Ninth International Symposium on Ostracoda, held in Shizuoka, Japan, 29 July - 2 August 1985, Developments in palaeontology and stratigraphy, **11**, 1021-1040, Kodansha Ltd., Tokyo and Elsevier, Amsterdam, Oxford, New York, Tokyo.

Whatley, R. C. & Ayress, M. 1988, Pandemic and Endemic Distribution of Deep Sea Ostracoda. *In* Hanai, T., Ikeya, N. & Ishizaki, K. (Eds), *Evolutionary biology of Ostracoda, its fundamentals and applications*, proceedings of the Ninth International Symposium on Ostracoda, held in Shizuoka, Japan, 29 July - 2 August 1985, Developments in palaeontology and stratigraphy, **11**, 739-755, Kodansha Ltd., Tokyo and Elsevier, Amsterdam, Oxford, New York, Tokyo.

Whatley, R. C. & G. Coles. 1987, The late Miocene to Quaternary Ostracoda of Leg 94, Deep-Sea Drilling Project. *Revta esp. Micropaleont.*, Madrid, **19**, 33-97

Whatley, R. C., Downing, S. E. Kesler, K. & Harlow, C. J. 1984. New species of the ostracod *Bradleya* from the Tertiary and Quaternary of DSDP sites in the Southwest Pacific. *Revta esp. Micropaleont.*, Madrid, **16**, 265-298.

Whatley, R. C., Harlow, C. J., Downing, S. E. & Kesler, K. 1983. Observations on the origin, evolution, dispersion and ecology of the genera *Poseidonamicus* Benson and *Bradleya* Hornibrook. *In* Maddocks, R. F. (Ed.), *Applications of Ostracoda*, proceedings of the Eighth International Symposium on Ostracoda, July 26-29, 1982, 492-509. Univ. Houston Geos., Houston, Texas.

# 4

# Devonian ostracod faunas of western Canada: their evolution, biostratigraphical potential and environmental settings

**Willi K. Braun**

Department of Geological Sciences, University of Saskatchewan Saskatoon, Saskatchewan, Canada, S7N 0W0

## ABSTRACT

The evolution of Devonian ostracods in western and northern Canada proceded in distinct bursts and lapses, reflecting major transgression and regression events punctuated by regional discontinuities. Lower Devonian ostracods are known only from scattered localities and stratigraphical horizons of the wilderness regions of Alaska and the Yukon, and seem to bridge Silurian and Emsian to earliest Eifelian faunas.

With the rapid marine transgression in the Middle Devonian, major parts of northern Canada and the Prairie Provinces were inundated, and ostracods spread widely. Two peaks in their diversity and abundance were reached in about the late Eifelian and mid Givetian. However, relative homogeneity at the generic level indicates a common origin from the same evolutionary stock, allowing the recognition of a well-defined Middle Devonian cycle in ostracod evolution.

Most ostracods of the Middle Devonian Fauna vanished from western Canada in the late Givetian, and only a few nondescript species cross the Frasnian boundary. The Frasnian transgressions introduced an entirely new suite of ostracods, referred to as the 'Frasnian Fauna', which reached its first peak in diversity and abundance in the early Frasnian, and its major development at about the mid-late Frasnian boundary. After a steady decline throughout late Frasnian time, another regional regression produced the third major reduction in taxa at the Frasnian-Famennian boundary. The distribution pattern of the Famennian Fauna is the result of the Famennian transgression and its oscillations, and the evolutionary cycle of the Famennian ostracods has been elucidated by Lethiers (1981). They vanished with the retreating seas from western Canada in late Famennian time, to be succeeded by an entirely different early Mississippian Fauna described by Green (1963), and as a result of the development of the North American Mississippian seaways.

A detailed biostratigraphical framework of 19 zones and 18 subzones has been established for Eifelian to Famennian strata, based on about 600 species of ostracods. This zonal scheme is equal to or, in many cases, more refined than those based on other fossil groups traditionally used in Devonian correlations. In terms of duration, some of the subzones represent time intervals of 600,000 years or less.

With respect to their usefulness as palaeoenvironmental indicators, Devonian ostracods are of limited value in western Canada mainly because they are known only from shallow-water marine or marginal marine deposits. They thrived on the vast carbonate platforms that are so characteristic of western Canada. Speciation was essentially allopatric, with the highest diversities coinciding with transgressive peaks and the early to mid-phases of the regressions.

## INTRODUCTION

Ostracods flourished in the warm and relatively shallow waters of a vast sea that covered major parts of western Canada in Middle and late Devonian times. Extensive carbonate platforms with many reef complexes formed south of, but in close proximity to the palaeoequator, which graded into sizable evaporite deposits in the southeastern part of the seaway. Both the reefs, with their numerous and often substantial oil and gas occurrences, and

the evaporites with the largest known potash deposits are of great economic value. It is this economic potential that has greatly stimulated geological exploration of the Devonian sequences in western Canada for the past four decades.

Despite the hydrocarbon exploration boom, palaeontological studies have not kept pace for a variety of reasons, although the ostracods were the least neglected of the fossils. Ostracoda were found to be more numerous in species and occur in higher individual abundances than any of the other Devonian groups used in western Canadian stratigraphical studies.

Over the past 25 years, more than 6,000 rock and microfossil samples have been collected from across western Canada and in particular from the regions indicated on the map (Fig. 1). The numerous surface and subsurface sections sampled (only cores were used from the latter) included most of the lithostratigraphical type and reference localities, with the largest number of samples collected personally in order to guarantee uniformity in sampling procedures. Also, much effort was expended in deeply trenching the surface sections and taking closely spaced and more or less continuous samples. The same applies to the core samples from the many boreholes analysed, especially those drilled during the early phases of the exploration boom, when large diameter cores were taken more frequently and more continuously than they are today. This ambitious programme of sampling and data collection from such a large country is considered essential for the construction of a detailed biostratigraphy and taxonomic evaluation of the ostracod faunas.

To date, more than 700 species of Eifelian to Frasnian ostracods have been discovered, evaluated and carefully recorded. Of these, some 500 species are utilized in the zonation. The others are either endemic or rare forms. The zones are essentially assemblage zones and their boundaries are drawn on the basis of first and last appearances of key taxa. The zones are denominated with letters and numbers rather than using the name of a single taxon. This system avoids confusion with schemes based on megafossils and conodonts,

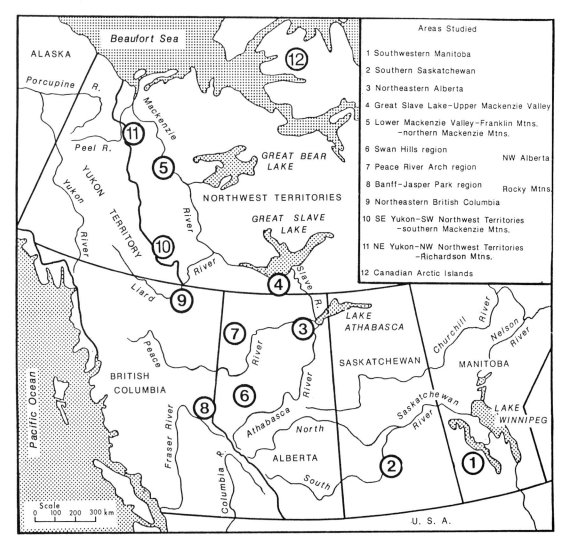

Fig. 1. Map of western Canada showing regions studied.

and provides flexibility for adjustments.

## GEOLOGICAL FRAMEWORK AND DISTRIBUTION OF THE OSTRACODS

The Devonian strata of western Canada are arranged in broad facies belts, changing from predominantly shales in the north and northwest to reef carbonates in the more central parts of the seaway, and giving way to dolomitic limestones, dolomites, and evaporites (anhydrite, halite, potash) in the southeast. This autochthonous pattern is the dominant one, and only negligible amounts of coarser terrigenous clastics and allochthonous material are associated with the carbonates and evaporites. Only in the northern third of the seaway were thick clastic sequences deposited in the late Devonian as erosional products from the uplifted Caledonides in the region of the present-day Canadian Arctic Islands.

The seaway itself was dissected by a number of intermittently active arches and basement highs (Fig. 2) which divided it into a series of minor basins and depocentres, but which influenced the

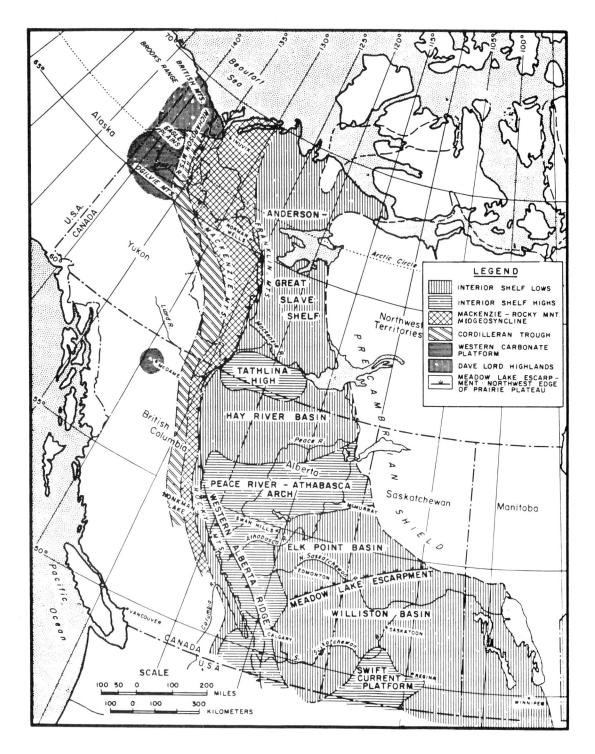

Fig. 2. Regional tectonic framework and extent of the western Canadian seaway, after Bassett & Stout, 1968.

sedimentation and environmental conditions only locally and temporarily. The Devonian transgressions and regressions, however, especially four which affected the entire western Canadian seaway, exerted the most profound influence on the facies patterns and their lateral migrations, as well as on the course and direction of ostracod evolution.

The direction of the transgressions and regressions can be readily deduced from a number of observations. There is the elongate shape of the seaway (Fig. 2) and the stacking pattern of rock facies within, with the evaporites essentially confined to the landlocked, southeastern portion of the seaway, and with the most open-marine sequences concentrated in the northwest. The same progression can be recognized with respect to the variety and abundance of fossils, including the ostracods. They are most varied and most abundant in the shales and carbonates of the northwest (Alaska, Yukon, western Northwest Territories) but become rare, or are absent, in the restricted environments and evaporites of the southeast (southern Saskatchewan mainly). It is only logical to assume, therefore, that the migration route for the ostracods and other animals was essentially along the same northwest-southeast axis. There may have been some exchanges of faunas and marine waters with the areas to the west and the eugeocline, but a string of islands or geanticlines, such as the West Alberta Ridge and lesser known structures to the north, seem to have provided effective barriers during most of Devonian time. Connections with the eastern and midcontinent seaways were confined to brief episodes and resulted in minor faunal exchanges only. For these reasons, the western Canadian Devonian seaway can be viewed as an essentially closed system largely independent of major external influences, developing ideal conditions for textbook examples of an autochthonous sedimentary pattern and its corresponding faunal distribution.

Despite the relative simplicity and predictability in the stacking pattern of the major facies belts, the local sedimentary systems and facies are highly varied and notoriously unpredictable, reflecting the effective interplay of a number of minor and local controlling factors. No major deviations would have been required, however, for even the smallest of change could have produced large consequences in the shallow basins with their seemingly flat-bottom topography.

A similar simplicity - complexity duality can be reconstructed for the various palaeoenvironments and distribution patterns of the ostracods. There were many minor influences that affected local populations, and which seem to have facilitated the evolution of many endemic species. Notwithstanding this, the relatively uniform regional setting and in particular the transgressions, provided ideal conditions for rapid and widespread distribution of the cosmopolitan species. It was this ideal mixture of conditions and opportunities that allowed the seaward parts of the carbonate platforms and their immediately adjacent, slightly deeper water areas to become important centres for the evolution of the ostracods from where they spread widely and efficiently. These were their favoured breeding grounds also, judging by their great individual numbers in these sediments and localities.

One other factor beneficial to the proliferation of Devonian ostracods in western Canada was the relative uniformity in water temperature. The western Canadian seaway lay either immediately south of the palaeoequator, or was crossed by it, especially in late Devonian times, as the reconstructions by Heckel & Witzke (1979) show convincingly. Considering that the seaway was some 3,000km long, there must have been a temperature gradient, but its effect on the biota would have been minimal so close to the palaeoequator. Precipitation would have been highest in the palaeoequatorial rain belts, but much lower in the adjacent dry belts from about 10 to 30 degrees north and south of the equator. The southeastern parts of the seaway were within this arid zone which, together with the shallowness of the sea, the lack of any appreciable runoff, and a stagnant current regime were all factors conducive to the formation of not only extensive but also highest grade evaporite (potash) deposits in southern Saskatchewan, especially during regressive phases.

## LIMITING FACTORS IN OSTRACOD DISTRIBUTION

The instabilities and frequent changes in the salinity regime may best explain the highly erratic distribution of ostracods in the Williston Basin, in the southeastern part of the seaway. Many of the carbonates there are products of a restricted environment and contain only scattered fossiliferous horizons with relatively few specimens of ostracods. The extensive evaporite deposits are barren altogether. Only during transgressions, which were especially strong in the Frasnian, did ostracods and other faunas spread widely and re-colonize the hitherto chemically hostile areas. The pattern of distribution is consequently one of thick, barren sequences interspersed with widely scattered and relatively thin fossiliferous units.

There is another type of salinity control, more difficult to assess however, which affected three assemblages in the northern and north-central parts of the seaway. It is impossible to correlate these monotypic or low diversity ostracod faunas precisely with their marine counterparts, for nowhere were they found interfingering. They are shown, therefore, as independent zones in the biostratigraphical scheme and in a relative position which had to be deduced on general stratigraphical considerations. Judging from the nature of the host and surrounding rock sequences, these lagoonal faunas developed during the end phases of regional regressions, and they are spread over sizable areas. Their low diversity and high dominance of a single species or, more rarely, two or three species indicates highly stressed conditions, but there is no firm clue in the rock record nor in the fauna that would point to either lower or elevated salinity levels. Assuming that temperature and water depth were comparable to those in the open-marine setting, slightly lower salinity levels are considered more likely than hypersaline conditions to have controlled the distribution of the ostracods, for there are no evaporites, or evaporitic minerals or their associated pseudomorphs.

While salinity probably exerted a major control on the distribution of Devonian ostracods in major parts of western Canada, the effect of depth is more difficult to assess. No ostracods have yet been recovered from the extensive, miogeoclinal Lower and Middle Devonian black to dark grey shales of the northern portion of the seaway. They are traditionally referred to as deep water deposits although there is no agreement whatever of what this term means in absolute figures. It does not follow, however, that there are no ostracods in the black shales, for no efforts were made to collect them. The corresponding eugeoclinal pile to the west is completely metamorphosed.

## BIOSTRATIGRAPHICAL POTENTIAL OF THE OSTRACODS

The uniformly warm and moderately shallow waters of the western Canadian Devonian seaway were populated by widely distributed and largely uniform faunas. The frequent regressions and transgressions brought about the elimination of older faunas and dispersed new ones in rapid succession. This applies to all organisms but in particular to the ostracods which are the most abundant and most varied of the invertebrate groups represented. All these factors in concert provide ideal prerequisites for a refined biostratigraphical zonation.

The biozones are identified and catalogued under a code which contains some letters and a number. The first letter 'D' stands for Devonian and those which follow are abbreviations for the periods and stages. The numbers are repeated within the Eifelian-Givetian, Frasnian and the Famennian faunas.

In the Eifelian-Givetian, including the transitional parts with the Lower Devonian (Emsian), 9 major ostracod zones (DLM 1 to DM 9) were discriminated, with 8 subzones (see Fig. 3). The Frasnian is characterized by 6 zones (DFr 1 to DFr 6) and 10 subzones, and Lethiers (1981) divided the Famennian into four ostracod zones. Most of these zones are based on ostracod assemblages which are open-marine, but three zones were created in the restricted facies. The latter biozones, however, are confined to the Middle Devonian, and none is known from Frasnian or Famennian

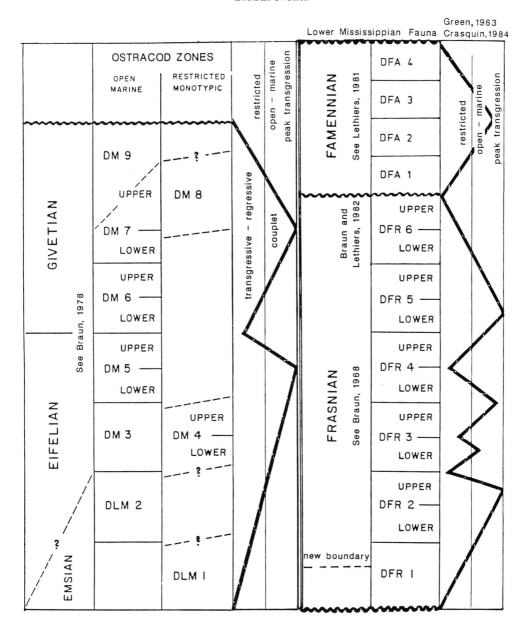

Fig. 3. Devonian ostracod biozones of western Canada, adapted from Braun 1968 and 1978.

deposits.

Fig. 4 shows the type of accuracy in correlation which can be achieved between two important Frasnian reference localities using ostracods, and the range chart compiled from the species involved is given in Fig. 5. These two areas represent in the Frasnian of western Canada the most fossiliferous, most open-marine, most shaly and stratigraphically most complete sections known.

The refinement of the zonation can be emphasized differently by using absolute ages. Assuming a duration for the Frasnian of 7 million years (374-367 MY, according to the Geological Society of America 1983 Time Scale), divided by the 11

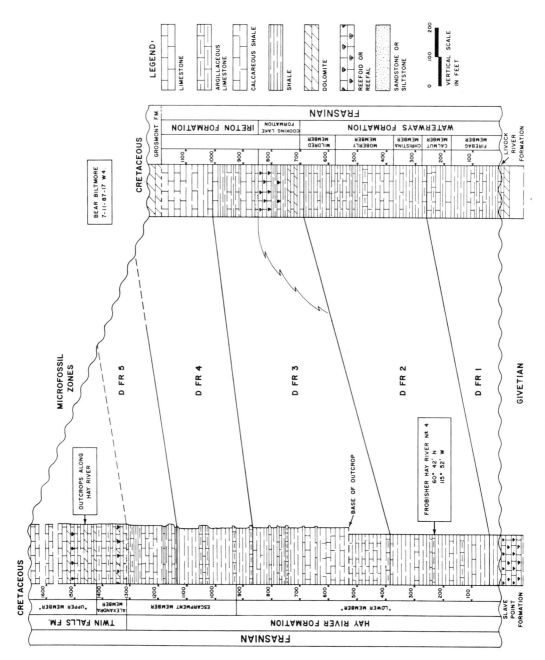

Fig. 4. Frasnian biostratigraphical correlations and ostracod biozones, Great Slave Lake to northeastern Alberta (areas 4 and 3 on map, respectively), after Braun, 1968.

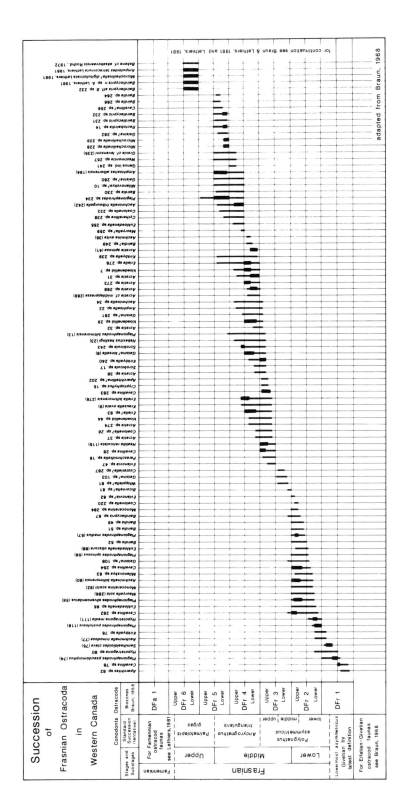

Fig. 5. Range chart of Frasnian ostracods between Great Slave Lake and northeastern Alberta (see also Fig. 4), adapted from Braun, 1968.

biostratigraphical units, a number of *circa* 600,000(+) years per zone and subzone emerges which is close to the value quoted for the duration of Jurassic ammonite zones, supposedly one of the best groups of index fossils. The same range, of between one half and one million years at most, can be calculated for the Middle Devonian ostracod zones, a remarkable refinement for the Palaeozoic.

## TRANSGRESSIONS, REGRESSIONS AND OSTRACOD EVOLUTION

One striking feature of the overall sequence of the ostracod species (an obvious example is the Frasnian succession shown in Fig. 5) are the many discontinuities indicated by major breaks or minor shifts in the faunal spectrum. There are four major and regional transgressive-regressive cycles, and the evolution of the ostracods followed approximately the same patterns of three major diversity bursts and declines. The Lower Devonian ostracods are poorly known and therefore not counted in the tabulation. Since the faunal and lithological data can be closely correlated or matched, this indicates a probable cause-consequence relationship. The oldest of the faunas studied in detail span the mainly Eifelian and Givetian, but most likely have their roots in Emsian or even older strata. This Middle Devonian Fauna came to an abrupt end with the full retreat of the late Givetian seas from all of western Canada.

Using conodonts as chronostratigraphical markers, this regional regression seems to have lasted for only a short time, however, and was followed by what used to be considered the first pulse of the Frasnian transgression. The pronounced faunal break (which affected all invertebrate groups and not just the ostracods), as well as karstic topography, weathered horizons, and a change in the regional stratigraphical grain, were used traditionally to demarcate the Givetian-Frasnian boundary and, by implication, the Middle-Upper Devonian boundary in western Canada. However, an international adjustment emerging over the past few years, now places this boundary at a higher stratigraphical level. Accordingly, the retreat of the last Middle Devonian sea from western Canada

would have taken place in late but not latest Givetian time; the beginning of the first of the Frasnian transgressions.

Semantics aside, the distinction between the Middle Devonian Fauna and the Frasnian Fauna is nearly total, with only a few species crossing the boundary. Impressive as this changeover may be on species level, however, a qualitative change at the generic level is even more striking, giving the two faunas their entirely different character.

The break between the Frasnian Fauna and the Famennian ostracods, in contrast, is mainly at the species level with very few changes in the dominant genera. Nevertheless, a sudden and pronounced break is indicated and obvious on the charts produced by Lethiers (1981) and by Braun & Lethiers (1982). Once again, there is lithological, local and regional stratigraphical and, above all, faunal evidence for a basin-wide regression at the end of the Frasnian, followed by the first of the Famennian transgressive pulses which brought with it new species of ostracods. Unlike the Givetian-Frasnian break, which is more clearly expressed in the changeover in the ostracods and in the demise of the novakias (cricoconarids), all microfossils, the Frasnian-Famennian event is most famous for a near total change of its megafaunas. The impact of a large meteorite is blamed for this apparently global mass extinction event by many palaeontologists.

The Famennian Fauna was similarly terminated by a regional regression, and the early Mississippian transgression brought along a new stock of ostracods with different species and a number of distinctive new genera. These faunas were treated monographically by Green (1963) and more recently by Crasquin (1984).

## THE SUCCESSION OF THE FRASNIAN OSTRACODS: SOME EVOLUTIONARY IMPLICATIONS

The punctuated nature of the faunal record is clearly evident in the range chart of the Frasnian ostracods from two important sections (Fig. 5). These are the outcrops along the Hay River where it empties into Great Slave Lake (Fig. 1, location 4) and one borehole in the vicinity of the delta. The second is a completely cored (large diameter core) well, Bear Biltmore, in northeastern Alberta (location 3 on the map) and about 300km south of the town of Hay River. As mentioned earlier, these sections are considered the most complete and most fossiliferous of all the Frasnian sequences of western Canada.

The range chart (Fig. 5) shows the shorter-ranging ostracods only (95 species); the longer-ranging ones (31 species) are excluded. However, even the latter species span no more than 3 to 4 biozones within the Frasnian, with only few species extending into the Famennian or downwards into the Middle Devonian.

Two features stand out: the short ranges of most of the ostracod species and the step-like, punctuated progression of first and last appearances of species throughout the Frasnian. With respect to the first point, a comparison was suggested earlier with the ammonites of the Jurassic. This does not, however, imply that ostracods are as useful in continent-wide and intercontinental correlations or in chronostratigraphical studies, although they have never been adequately tested in this respect. Ostracods are probably too substrate controlled and too endemic, like most other benthonic invertebrates, to serve as intercontinental index taxa. Their application, obviously, lies in more regional and especially local settings and studies, and regional in the western Canadian context refers to an area about the size of Western Europe.

In 1983, Lethiers & Braun summarized and illustrated the effects that two main variables, transgressions and regressions, have on the overall evolutionary path of Devonian ostracods. The result is a predictable pattern, the evolutionary sigmoid, which are strung together in a chain if a number of transgressive-regressive cycles are involved. The step-like nature of the range table conforms to such a multiple pattern. Each sigmoid is composed of three parts, and in each cycle, the new faunas are installed during the initial pulse of the transgression. A few short-ranging, pioneering forms appear and disappear in rapid succession, as is the case with the ostracods of the DFr1

zone and lower DFr2 subzone. During the second developmental phase, species diversity increases notably as the transgression reaches its climax. The species are longer ranging, because the environment has become much more stable relative to the initiating phase when changes were frequent and often profound. During the last phase, conditions deteriorated with the accelerating regression, successively rendering species extinct but also producing a few short-ranging, opportunistic species capable of adapting to the rapidly changing environment. The cycle and thus the sigmoid ends as a result of the basin-wide withdrawal of the sea, and new ones begin with the subsequent transgression. Each of these evolutionary cycles thus corresponds to a transgressive-regressive couplet, and the shape of the sigmoid itself may give a hint where to look for discontinuities which otherwise may remain unnoticed, and for which there are no clear indications in the rock record.

This first cycle has a noticeable asymmetry, with the upper part of the sigmoid distinctly truncated. This may indicate that the regression was very rapid and not lingering as in that ideal case when the duration of both transgression and regression are about equal and producing, therefore, a symmetrical sigmoid. The same truncation, however, could also be indicative of erosion, with part of the regressive sequences and its faunas removed. Notwithstanding this, the sharply delineated faunal break can be traced throughout the Prairie provinces and into the southwestern Northwest Territories pointing strongly to a regional discontinuity. The presence and implications of this unconformity will be published shortly.

The second cycle, spanning the DFr3 to Lower DFr4 assemblages, seems to be broken up into two sub-cycles, possibly due to local and certainly shorter-lived influences. This cycle is also asymmetrical with another pronounced discontinuity at the Lower-Upper DFr4 boundary. Across this boundary, there seems to be an additional irregularity because the basal part of the succeeding sigmoid is missing. This may be due to a delayed transgression and thus to non deposition at these localities, followed by a near instantaneous inundation or, alternatively, due to erosion having destroyed all or most of the basal transgressive pile. The sharp and conspicuous break at the Lower-Upper DFr5 boundary is more of an artefact until details with respect to the upper DFr5 and DFr6 ostracods are worked out.

Demarcating the position and speculating on the cause of a faunal break is one thing; recognizing these discontinuities in the rock record is quite another. There is ample evidence and general acceptance with respect to the major unconformities that parallel the system, series, stage, and first order faunal boundaries, and which mark the acme of the regressions. What will be more trying undoubtedly, is to convince the geologists of the presence of other discontinuities, such as the ones pointed out above, for which the evidence in the rock record is ambiguous at best, or not recognized as yet. The problem is compounded, for many geologists in western Canada still cling to deeply ingrained, gradualistic ideas and are most hesistant, therefore, to accept a highly punctuated rock record. Yet the message from the Frasnian ostracod record is unmistakable: it is punctuated, as Lethiers (1981) has also demonstrated for the Famennian Fauna, and as is also true for the Middle Devonian Fauna still to be published.

## REFERENCES

Bassett, H. G. & Stout, J. G. 1968. Devonian in western Canada. *In* Oswald, D. H. (Ed.), International Symposium on the Devonian System, Calgary. *Alberta Society of Petroleum Geologists*, Calgary, **I**, 717-752, 11 figs.

Braun, W. K. 1968. Upper Devonian ostracode faunas of Great Slave Lake and northeastern Alberta, Canada. *In* Oswald, D. H. (Ed.), International Symposium on the Devonian System, Calgary. *Alberta Society Petroleum of Geologists*, Calgary, **II**, 617-652, 9 pls, 8 figs, 6 charts.

Braun, W. K. 1978. Devonian ostracodes and biostratigraphy of western Canada. *In* Stelck, C. R. & Chatterton, B. D. E. (Eds), Western and Arctic Canadian Biostratigraphy. *Geological Association of Canada*, Special Paper **18**, Waterloo, 259-288, 6 pls, 2 figs.

Braun, W. K. & Lethiers, F. 1982. A new Late Devonian ostracode fauna and its bearing on the Frasnian-Famennian boundary in western Canada. *Can. J. Earth Sci.*, Ottawa, **19**, 1953-1962, 3 pls, 3 tables, 1 chart.

Crasquin, S. 1984. *Ostracodes du Dinantien: Systématique-Biostratigraphie-Paléoécologie (France, Belgique, Canada)*. Thèse de Docteur de Troisième Cycle, Lille, 236 pp,

23 pls, 109 figs.

Green, R. 1963. Lower Mississippian ostracodes from the Banff Formation, Alberta. *Research Council of Alberta*, Bulletin **11**, Edmonton, 237 pp, 17 pls, 13 tables, 23 figs.

Heckel, P. H. & Witzke, B. J. 1979. Devonian World Palaeogeography determined from distribution of carbonates and related lithic palaeoclimatic indicators. *In* House, M. R., Scrutton, C. T. & Bassett, M. G. (Eds), The Devonian System - A Palaeontological Association International Symposium. *Spec. Pap. Palaeont.*, Palaeontological Association, London, **23**, 99-123, 8 figs.

Lethiers, F. 1981. Ostracodes du Devonian Terminal de l'Ouest du Canada: Systématique, Biostratigraphie et Paléoécologie. *Géobios*, Lyon, Mémoire Special **5**, 1-234, 26 pls, 14 tables, 73 figs.

Lethiers, F. & Braun, W. K. 1983. Les extensions stratigraphique des espèces d'Ostracodes paléozoïque sur les platesformes. Abstract, thème 9, Premier Congres International de Paléoécologie, Lyon 1983.

## DISCUSSION

Dan Danielopol: Could some of the small-sized species live in an interstitial marine environment as one finds in present-day situations?

Willi Braun: It is possible that some did, but the majority of my ostracod species, especially the Middle Devonian and Frasnian faunas are quite small. The small size seems to be the rule rather than the exception and, therefore, not all could have lived in such limiting environments.

# 5

# A comparison of the evolution, diversity and composition of the Cainozoic Ostracoda in the deep water North Atlantic and shallow water environments of North America and Europe

**Graham Coles**

Institute of Earth Studies, University College
of Wales, Aberystwyth, Dyfed, U.K.

## ABSTRACT

The evolution and taxonomic diversity of Cainozoic marine and brackish water Ostracoda from the deep water North Atlantic and the shallow water environments of eastern North America and Europe are documented from a database abstracted from all available published and unpublished sources. The changes in the ostracod faunas over time at the species, generic and higher taxonomic levels are described and quantified and the patterns of evolution and composition of both shallow and deep water faunas are compared. Possible correlations between the observed patterns of diversity, evolution and composition, with global events such as climatic, palaeoceanographical or eustatic sea level changes are discussed.

## INTRODUCTION

The aims of this study are to document the diversity and evolution of Cainozoic marine and brackish water Ostracoda from the North Atlantic Ocean, eastern North America and Europe, to describe the taxonomic changes over time at the species, generic and higher taxonomic levels and attempt to account for the observed changes in diversity and in origination and extinction rates.

## THE STUDY AREA

This comprises the deep (over 1000m) bathyal and abyssal waters of the North Atlantic Ocean and the bordering continental areas of eastern North America, Europe and northwest Africa, between the Tropic of Cancer (latitude 23°N) and latitude 85° N. To the west it is bounded by the landward limit of Cainozoic marine sediments of the Atlantic and Gulf coastal plains, but includes small areas of Quaternary marine sediments in eastern Canada. To the east it is bounded by longitude 70°E, the western border of the U.S.S.R. and the easternmost limit of the Cainozoic deposits of the Mediterranean basin. The Ostracoda of the U.S.S.R., Bulgaria and the Black Sea were considered beyond the scope of this study. Only the Miocene to Recent Ostracoda of Africa north of latitude 33°N were studied, comprising the northern coastal regions of Morocco, Algeria and Tunisia. This area is considered sufficiently large to show the general trends of ostracod evolution over the past 65 million years, minimizing the effects of smaller scale, local events on the overall pattern. The marine and brackish water environments represented are sufficient to enable a comparison to be made between evolution in shallow and deep water regions. The Cainozoic geological history of eastern North America and Europe is relatively well known, as is that of the North Atlantic compared to other deep ocean basins. The Ostracoda of the shallow water regions are also by far the most intensively sampled and studied in the world; most of the early research on Ostracoda was carried out here. The relative lack of information on Cainozoic, particularly Palaeogene, deep water North Atlantic faunas has been improved by an extensive study by the author (Coles MS. in prep; Coles & Whatley, in press; Whatley & Coles, in press).

## THE OSTRACODA STUDIED

All benthonic marine and brackish water Ostracoda, comprising stenohaline and euryhaline marine and stenohaline brackish water Ostracoda are included. Planktonic Myodocopina and exclusively fresh water groups were excluded. Inclusion of these groups would have considerably increased the size and complexity of this study and have complicated the interpretation of the results. The following groups of Ostracoda have been considered:

Podocopina: Bairdiacea: all families.
Cypridacea: Marine cyprids only, i.e. Paracyprididae and Pontocyprididae.
Cytheracea: all families, except the non-marine Limnocytheridae.
Platycopina: Cytherellidae only.
Metacopina: *Cardobairdia* only.
Cladocopina: Polycopidae only.

The classification employed is essentially that of the *Treatise* (Moore, 1961), with certain necessary modifications.

## COMPILATION OF THE DATABASE

Information on the stratigraphical and geographical occurrence of all species was abstracted from all available published and unpublished sources. In total, over 300 references were consulted for this study; these are listed in Coles (in preparation) with the full database for the stratigraphical and geographical distribution of all species and genera from the Cainozoic of the study area. Open nomenclature species were also considered where there was evidence that they represented undescribed species. Most subspecies were excluded, unless they have been subsequently raised to specific rank by a consensus of authors. All known homonyms, synonyms and unrecognizable species were excluded, as were species based on the instars or sexual dimorphs of other described species.

## THE STRATIGRAPHICAL TIME SCALE

The interval studied comprises the entire Cainozoic

Table 1. Diversity and evolution statistics for the Cainozoic of the entire area studied (whole), shelf environments of North America and Europe (shallow) and North Atlantic (deep).

| Age | LPA | UPA | LE | ME | UE | LO | UO | LM | MM | UM | LP | UP | LQ | UQ | R |
|---|---|---|---|---|---|---|---|---|---|---|---|---|---|---|---|
| Duration of age (MY) | 4.8 | 5.3 | 4.4 | 6.5 | 6.0 | 5.2 | 8.2 | 10.2 | 3.1 | 6.2 | 1.8 | 1.3 | 1.1 | 0.8 | .01 |
| Species diversity: whole | 147 | 174 | 328 | 545 | 593 | 511 | 462 | 509 | 428 | 755 | 620 | 860 | 740 | 707 | 990 |
| Species diversity: shallow | 118 | 133 | 267 | 420 | 490 | 406 | 367 | 451 | 374 | 660 | 544 | 791 | 674 | 628 | 919 |
| Species diversity: deep | 16 | 23 | 58 | 99 | 85 | 101 | 107 | 59 | 53 | 70 | 76 | 74 | 79 | 124 | 124 |
| Generic diversity: whole | 51 | 66 | 94 | 111 | 114 | 122 | 123 | 131 | 128 | 149 | 144 | 161 | 165 | 170 | 180 |
| Generic diversity: shallow | 49 | 61 | 79 | 94 | 99 | 106 | 101 | 120 | 119 | 139 | 124 | 144 | 152 | 155 | 168 |
| Generic diversity: deep | 13 | 15 | 31 | 45 | 40 | 46 | 44 | 28 | 27 | 31 | 36 | 36 | 35 | 50 | 57 |
| Familial diversity: whole | 15 | 18 | 18 | 19 | 20 | 21 | 21 | 21 | 21 | 23 | 24 | 24 | 22 | 22 | 23 |
| Species originations: whole | 121 | 98 | 255 | 275 | 236 | 226 | 137 | 300 | 126 | 374 | 312 | 362 | 257 | 59 | 371 |
| Species originations: shallow | 118 | 85 | 231 | 235 | 223 | 203 | 122 | 295 | 118 | 361 | 294 | 359 | 252 | 47 | 361 |
| Species originations: deep | 16 | 15 | 36 | 56 | 16 | 24 | 19 | 8 | 8 | 18 | 23 | 6 | 18 | 38 | 53 |
| Species extinctions: whole | 62 | 78 | 98 | 183 | 291 | 189 | 237 | 160 | 78 | 428 | 66 | 396 | 84 | 178 | |
| Species extinctions: shallow | 59 | 75 | 85 | 159 | 287 | 181 | 204 | 152 | 71 | 413 | 63 | 390 | 84 | 144 | |
| Species extinctions: deep | 4 | 3 | 13 | 27 | 8 | 13 | 44 | 10 | 7 | 18 | 10 | 8 | 1 | 59 | |
| Generic originations: whole | 5 | 11 | 24 | 18 | 8 | 19 | 12 | 30 | 9 | 15 | 15 | 14 | 15 | 3 | 10 |
| Generic extinctions: whole | 2 | 3 | 2 | 5 | 7 | 6 | 13 | 3 | 2 | 15 | 2 | 14 | 3 | 9 | |
| Total species/MY | 31 | 33 | 75 | 84 | 99 | 98 | 56 | 50 | 138 | 122 | 344 | 662 | 661 | 803 | |
| Total genera/MY | 10 | 13 | 21 | 17 | 19 | 24 | 15 | 13 | 41 | 24 | 80 | 124 | 147 | 193 | |
| Species originations/MY | 25 | 19 | 58 | 42 | 39 | 44 | 17 | 29 | 41 | 60 | 173 | 279 | 230 | 67 | |
| Species extinctions/MY | 13 | 15 | 22 | 28 | 49 | 36 | 29 | 16 | 25 | 69 | 37 | 305 | 75 | 202 | |
| % of new species | 92 | 64 | 81 | 56 | 42 | 46 | 31 | 60 | 30 | 52 | 51 | 43 | 35 | 8 | 38 |
| % of new genera | 10 | 17 | 26 | 16 | 7 | 16 | 10 | 23 | 7 | 10 | 10 | 9 | 9 | 2 | 6 |

from the Palaeocene to the Recent, a time span of approximately 65 million years. To ensure consistency in the definition of the various stratigraphical divisions, the timescale of Harland *et al.* (1982) was used, although with some minor modifications, (the Bartonian is included within the Upper Eocene, following the usage of most previous authors) to divide the Cainozoic into 15 time periods. These comprise the Lower and Upper Palaeocene; Lower, Middle and Upper Eocene; Lower and Upper Oligocene; Lower, Middle and Upper Miocene; Lower and Upper Pliocene; Lower and Upper Quaternary and Recent. Further subdivision of these time periods could not be consistently or accurately made throughout the area studied. Ostracoda recorded from the 'Recent' are not confined to living species, but include those recorded from the Holocene, up to 10,000 years before the present.

## RESULTS

From the compiled species index, numerous statistics on diversity and evolution rates can be calculated, some of which are shown in Table 1. Throughout this study, the term 'diversity' refers to simple species or generic diversity, i.e. the total number of taxa countable within a particular time interval or geographical area. 'Evolution rates' are estimated from the total number of species or genera appearing (originations) or disappearing (extinctions) in a time interval within the area studied. Species and generic diversity is shown for the whole area studied (whole), the deep water North

Atlantic (deep) and shallow water areas (shallow). Familial diversity is given for the entire study area. 'Originations' and 'extinctions' indicate the first and last apparent records of species or genera. In the case of many genera, the apparent first or last appearance may not represent its true global origination or extinction. However, the recorded originations and extinctions of species broadly reflect the true pattern of appearances and disappearances since the vast majority (over 90%) of species are endemic to the area studied. However, many, if not most, deep-sea species occur outside the North Atlantic (Whatley & Ayress, 1988; Coles, Ayress & Whatley, this volume), as do many polar shelf species and warm water species which extend south of the study area along the continental shelves of North America or Africa. The percentage of new species and genera for each time interval is given as an indication of faunal turnover. (The percentage of inherited species in any interval is equal to 100% minus the percentage of new species.) The figures for total species and generic diversity, species originations and extinctions are also normalized for time by dividing the relevant figure by the time span of the relevant time period, to give diversity or originations/extinctions per MY. This practice eliminates any bias caused by the variable duration of the 15 time periods.

## DIVERSITY

A total of 3521 species and 273 genera have been recognized in the whole study area. Of these, only 12 species are definitely known to range back to the Cretaceous, although taxonomy has undoubtedly created artificial distinctions between Upper Cretaceous and Palaeocene species. Nevertheless, Cainozoic faunas are almost completely new at the species level. In contrast, 72 genera also occur in the Mesozoic, mostly in the Upper Cretaceous, representing 28% of all Cainozoic genera. The similarity between Mesozoic and Cainozoic faunas increases at successively higher taxonomic levels, so that almost all Cainozoic families and all superfamilies range back at least to the Cretaceous. The diversity of species, genera and

families over time for the whole area is shown in Fig. 1.

## SPECIES DIVERSITY

Overall species diversity rises steeply but irregularly through the Cainozoic. There are, however, several episodes of falling diversity. Species diversity is lowest in the Palaeocene, much lower than in the Maastrichtian, as calculated by Whatley (1988) for global Ostracoda. Diversity rises steeply in the Eocene, particularly in the Lower and Middle Eocene, to reach a peak in the Upper Eocene, so that Eocene diversity is 2 to 4 times higher than that of the Palaeocene. Diversity declines in the Oligocene to a level just below that of the Middle Eocene. A slight rise occurs in the Lower Miocene, followed by a fall in the Middle Miocene; Lower to Middle Miocene diversity is similar to that of the Oligocene. A steep rise in species diversity occurs in the Upper Miocene to a level 57.6% higher than in the Middle Miocene. Upper Miocene to Quaternary diversity remains very high and exceeds that recorded at any time in the Palaeogene; much of this reflects the improved taphonomy of successively younger Cainozoic faunas. Diversity falls slightly in the Lower Pliocene but recovers strongly in the Upper Pliocene to its maximum level before the Recent. Further slight falls occur during the Quaternary, although diversity remains very high, being comparable to Upper Miocene levels. Recent species diversity is the highest in the Cainozoic, with a total of 990 species. This compares to the global Maastrichtian total, the highest in the Mesozoic, of 1311 species recorded by Whatley (1988).

A somewhat distinct pattern is observed when diversity is normalized for time. Fig. 2 shows specific and generic diversity per MY, calculated by dividing the total species or generic diversity of each time period by the duration of the time period in millions of years. There are consistently fewer than 100 species per MY during the Lower Palaeocene to Lower Miocene interval, while there are always more than 100 species per MY from the Middle Miocene to the Recent. Diversity

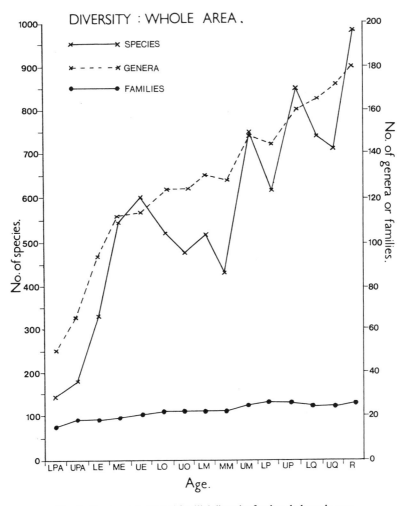

Fig. 1. Species, generic and familial diversity for the whole study area.

per MY rises steadily through the Palaeogene, but falls during the Upper Oligocene and Lower Miocene. A steep rise occurs in the Middle Miocene, reflecting the short duration (3.1 MY) of that interval, but a fall occurs in the Upper Miocene. The higher species diversity of the Upper Miocene is shown in part to be a function of its duration (6.2 MY). Very steep rises also occur in the Lower and Upper Pliocene, with a lesser rise in the Upper Quaternary, clearly reflecting the very high diversity of later Neogene and Quaternary faunas and the short duration of these periods. Recent diversity per MY greatly exceeds that recorded at any other time in the Cainozoic,

reflecting intensive research on Recent faunas, much more complete preservation and the very short (10,000 years) duration of the 'Recent'.

## GENERIC DIVERSITY

From Fig. 1, it is clear that generic diversity essentially mirrors the pattern of species diversity, although diversity fluctuations are less marked. There is a steady, almost linear increase in diversity from 51 genera in the Lower Palaeocene to 180 genera in the Recent, a rise of 353%. However, some differences between the patterns of species and generic diversity are apparent. Generic diversity

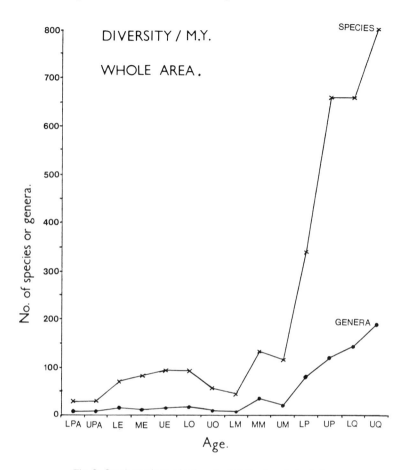

Fig. 2. Species and generic diversity/MY for the whole study area.

does not fall from the Upper Eocene to the Lower Miocene as does species diversity, but steadily increases throughout that interval. Slight falls in generic diversity occur in the Middle Miocene and Lower Pliocene, but these are less dramatic than the corresponding falls in species diversity. Generic diversity does not drop in the Quaternary, but rises steadily from the Upper Pliocene to the Recent. The differences between the patterns of species and generic diversity principally reflect the greater longevity of genera, as well as the more completely known stratigraphical ranges of genera compared to species.

The pattern of normalized generic diversity shown in Fig. 2 closely resembles the trend for species, although the two diverge dramatically from the Pliocene to Recent. The species:genus ratio is lowest when species diversity is lowest, most markedly so in the Palaeocene. The main differences between species and generic diversity per MY are that the rise in Palaeogene normalized species diversity is not matched by genera to the same extent, and that the dramatic rise for species from the Upper Miocene to Recent is much more subdued for genera.

## FAMILIAL DIVERSITY

Fig. 1 shows that familial diversity rises very slowly through the Cainozoic to a maximum of 24 families in the Pliocene, then falls very slightly to 22 in the Quaternary, with 23 families known from the Recent. The slight falls and rises are mainly due to rare families which include very few species.

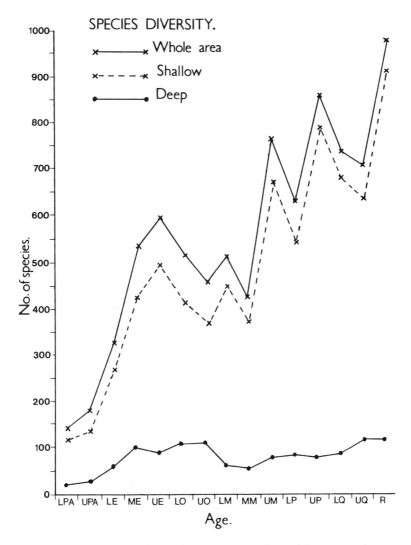

Fig. 3. Species diversity for the whole study area, shallow and deep water regions.

## COMPARISON OF SHALLOW AND DEEP WATER DIVERSITY

Species diversity over time for the whole study area, shallow water areas and the deep water North Atlantic are shown in Fig. 3. A total of 3303 species and 265 genera are recorded from the predominantly shallow water regions of North America and Europe, while only 351 species and 76 genera are recorded from the deep water North Atlantic. Total species diversity in the shallow environments is, therefore, more than 9 times higher than it is in the deep water North Atlantic, while total generic diversity is just over 3 times that of the North Atlantic. Only 133 species (mostly slope and bathyal taxa inhabiting depths between 200 and 1000m), but 68 genera are common to both shallow and deep water realms, while 218 species and only 8 genera are confined to the North Atlantic.

The pattern of shallow species diversity essentially mirrors that for the whole area, comprising the total number of species in the whole area minus the number of species endemic to the deep water

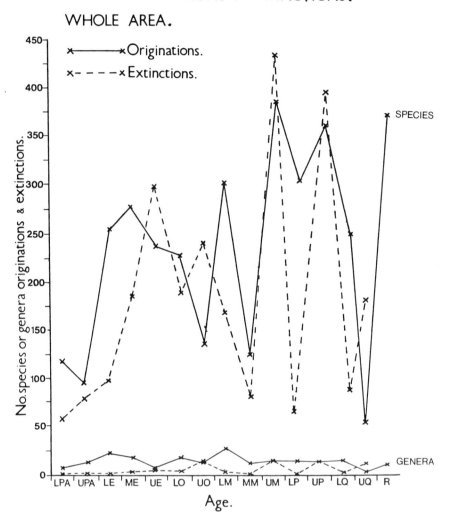

Fig. 4. Species and generic originations and extinctions in the whole study area.

North Atlantic in any one time period. All rises and falls in species diversity for the whole area are discernible in the shallow water pattern, although shallow species diversity is between 15 and 81 species less than that of the whole area. The difference between the shallow and the whole area species diversities is greatest in the Middle Eocene to Upper Oligocene and Quaternary.

Diversity in the deep water North Atlantic is always much lower than in contemporaneous shallow water environments, ranging from 9 to 25% of the total of shallow species. In this realm, species diversity is very low in the Palaeocene, rises during the Lower and Middle Eocene, falls slightly in the Upper Eocene, rises slowly in the Oligocene only to fall sharply in the Lower Miocene, but rises steadily from the Middle Miocene to the Recent, except for a small drop in the Upper Pliocene. Maximum diversity (124 species) occurs in the Upper Quaternary and Recent.

The patterns of species diversity in shallow and deep water environments share certain features,

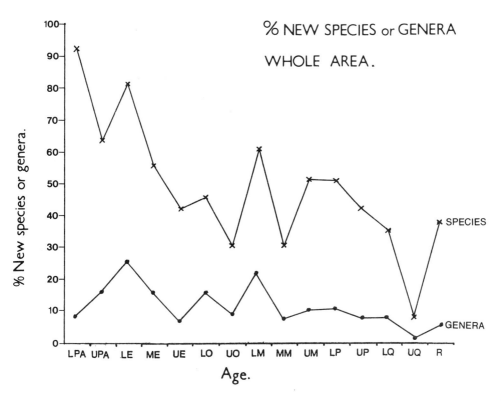

Fig. 5. Percentage of new species and genera for the whole study area.

but significant differences are also apparent. Both deep and shallow water realms show a net rise in species diversity through the Cainozoic, although the rise is much less marked in the deep water North Atlantic than it is in the shelf regions. Recent species diversity in both realms is 7 to 8 times greater than that of the Lower Palaeocene, but the actual numerical increase in species diversity during the Cainozoic is 837 species in the shallow environments, compared to only 108 species in the deep water North Atlantic. Both realms show a large increase in species diversity during the Eocene and peak diversity occurs in the Recent. The former reflects the common Eocene recovery after the Palaeocene diversity low; the latter is partially due to the intensity of research activity on Recent faunas and taphonomic bias in favour of Recent Ostracoda.

Shallow water species diversity rises steeply from the Middle Miocene to the Recent, so that Upper Miocene to Recent diversity is higher than

at any time in the Lower Palaeocene to Middle Miocene interval. In contrast, Lower Miocene to Lower Quaternary species diversity in the deep water North Atlantic is lower than that of the Middle Eocene to Upper Oligocene interval; here there is no Neogene taxonomic explosion. Instead, a significant drop in diversity occurs from the Upper Oligocene to the Lower Miocene due to high extinctions at the Palaeogene-Neogene boundary. This contrasts with the slight rise in species diversity from the Upper Oligocene to Lower Miocene in shallow water environments.

## EVOLUTION: ORIGINATIONS AND EXTINCTIONS

The total number of originations and extinctions (first and last appearances) of both species and genera over time for the whole study area are shown in Fig. 4, while Fig. 5 shows the percentage of new species or genera over time, again for the whole study area. When both graphs are examined, several features are apparent concerning the evolution of species.

## SPECIES EVOLUTION

The highest levels of species originations occur in the Lower and Middle Eocene, Lower and Upper Miocene, Pliocene and Recent. The highest extinction peaks occur in the Upper Eocene, Upper Oligocene, Upper Miocene and Upper Pliocene, at which times extinctions exceed originations. The principal faunal turnovers tend to occur at the end of epochs, particularly at the Eocene-Oligocene, Oligocene-Miocene and Miocene-Pliocene boundaries. The magnitude of originations and extinctions is greatest in the Miocene to Recent, when species diversity is highest. In contrast, very low origination rates occur in the Palaeocene (linked to low diversity), Upper Oligocene, Middle Miocene and Upper Quaternary, while low extinction rates occur in the Palaeocene to Lower Eocene, Middle Miocene, Lower Pliocene and Lower Quaternary.

Fig. 5 shows that there is a general, but irregular trend of a decreasing percentage of new species in the faunas of the successive Cainozoic time intervals. This contrasts with the trend of increasing species diversity of the whole study area. However, the general trend is interrupted by several peaks, marking a high percentage of new species. The most notable peak is in the Lower Miocene, when just over 60% of species are new, reflecting the great faunal renewal at the Palaeogene-Neogene boundary. Very few (8.4%) of Upper Quaternary species are new, but over 38% of Recent species are new, probably reflecting the large number of Recent species which do not preserve well and lack a fossil record.

## GENERIC EVOLUTION

The levels of generic originations and extinctions for the entire area shown in Fig. 4 are obviously much lower than the corresponding rates for species. Most of the evolutionary trends expressed by species are also evident for genera, but are less

dramatically expressed. Most notable are the generic origination highs in the Lower Eocene and Lower Miocene, and the extinction peaks in the Upper Oligocene (when extinctions exceed originations by 13 to 12), Upper Miocene and Upper Pliocene (when extinctions equal originations). The major difference between the pattern of generic originations and extinctions and that of species is that the Miocene to Recent generic evolutionary rates are broadly equivalent to those of the Palaeogene, rather than being much greater, as is the case for species.

## EVOLUTION IN SHALLOW WATER ENVIRONMENTS

The rates of species originations and extinctions for the shallow and deep water environments are compared in Fig. 6. The pattern of Cainozoic shallow water species evolution is very similar to the pattern shown for the whole area, except that both origination and extinction rates are slightly lower for the shallow water species. The greatest reductions in apparent origination rates occur in the Lower and Middle Eocene, while the greatest reductions in apparent extinction rates occur in the Upper Oligocene and Upper Quaternary, which correspond to origination and extinction highs respectively in the North Atlantic.

## EVOLUTION IN THE DEEP WATER NORTH ATLANTIC

The pattern of species evolution in the North Atlantic is quite distinct from that of shallow regions. Fluctuations in origination and extinction rates are naturally less marked than those of shallow environments because of the much lower species diversity. Nevertheless, some significant trends can be identified in the Cainozoic evolution of deep water North Atlantic Ostracoda.

Originations and extinctions are very low in the Palaeocene, ut rise in the Lower and Middle Eocene, fall in the Upper Eocene and rise slightly in the Lower Oligocene. No marked origination or extinction peaks occur at the Eocene-Oligocene boundary; indeed Upper Eocene extinction rates

## SPECIES ORIGINATIONS & EXTINCTIONS.

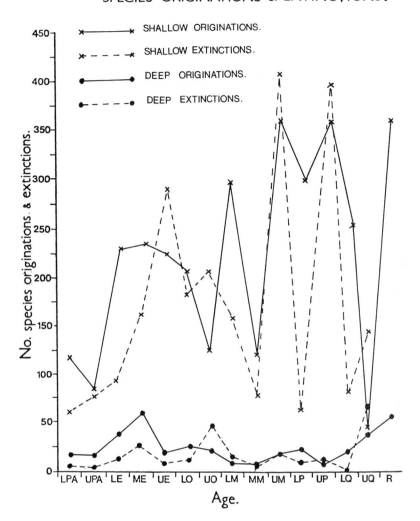

Fig. 6. Species originations and extinctions in the shallow and deeep water regions.

are the lowest of the Eocene-Oligocene interval. A prominent extinction peak occurs in the Upper Oligocene when extinctions greatly exceed originations, marking a significant changeover from Palaeogene to Neogene deep water faunas. This contrasts with the findings of Benson *et al.* (1984) who documented an important taxonomic turnover at approximately 40 MY before the Eocene-Oligocene boundary for global deep-sea Ostracoda (See also Whatley & Coles, in press).

Extinctions just exceed originations in the Lower Miocene, indicating the disappearance of relict Palaeogene species and the low rate of replacement of new species in the early Neogene. Origination rates fall from the Lower Oligocene to the Middle Miocene, rise in the Upper Miocene and Lower Pliocene, fall in the Upper Pliocene, then rise steeply in the Quaternary to Recent. Extinction rates are low from the Lower Miocene to Lower Quaternary, but reach a peak in the Upper Quaternary. This latter phenomenon is an artefact, mainly due to inadequate research on Recent

North Atlantic faunas; further study should lead to the discovery of most Upper Quaternary species in the Recent (see also Van Harten, this volume).

## DISCUSSION: CONTROLS AND FACTORS IN OSTRACOD EVOLUTION AND DIVERSITY

The observed patterns of ostracod evolution and diversity are influenced by numerous factors, which can be classified as being 'artificial' or 'natural'. The 'artificial' controls include such factors as the differential preservation of Ostracoda from the various ages and taxonomic groups, inconsistent taxonomic practices, biases in favour of large and/or ornate species, biases caused by poor stratigraphical control and intensity and quality of research activity. (Schopf *et al.*, 1975; Raup, 1976; Sheehan, 1977; Whatley & Coles, 1987). All of these familiar problems have some effect, and should be borne in mind in any study of this nature.

The 'natural' factors comprise environmental controls (principally temperature, depth, salinity, substrate and ecological niches), sea level (regressions and transgressions) and various palaeoceanographic events. The likely and possible effects of these various factors on evolution and diversity are intimately related and are briefly discussed:

### Environmental factors

**a) Temperature**. As a general rule, ostracod diversity and evolutionary activity increase with rising temperatures, there being far more thermophilic than cryophilic taxa (Fischer, 1960). Numerous thermophilic genera have high rates of evolution in warm, shallow waters, such as *Cytherel-loidea*, *Hermanites*, *Leguminocythereis* and *Quadracythere* mainly in the Palaeogene of North America and Europe, *Actinocythereis* (Palaeogene of the United States), *Cytheretta* (mainly in the Palaeogene and early Neogene of Europe) and *Callistocythere* (mainly in the Miocene to Recent of the Mediterranean and Central Europe). Other diverse and actively evolving genera are

mostly thermophilic with a few cold water species, such as *Aurila* in the Miocene to Recent of Europe, *Cytherura* in the Pliocene to Recent of the United States and *Semicytherura* in the Middle Miocene to Recent of Europe. However, some predominantly cryophilic genera also attain high diversities, frequently extending from the shelf to the deep sea, e.g., *Argilloecia* (Pliocene to Recent), *Cytheropteron* and *Krithe* (Eocene to Recent), while others are exclusively shallow marine to brackish, e.g., *Leptocythere* (Upper Miocene to Recent). Climatic cooling phases cause extinctions in thermophilic genera, e.g., the severe reductions in diversity of *Leguminocythereis* in the Lower Oligocene of northwest Europe, *Cytherelloidea* and *Cytheretta* in the Middle Miocene of Europe and *Aurila* in the Quaternary of the Mediterranean. However, reduced temperatures caused considerable speciation in cold water genera such as *Cytheropteron* and numerous hemicytherids in the Quaternary. Conversely, warming periods lead to increasing diversity and speciation in thermophilic genera such as *Cytheretta*, *Leguminocythereis* and *Quadracythere* in the Eocene worldwide, *Semicytherura* and *Aurila* in the Pliocene of Europe, and *Cytherura* and *Bensonocythere* in the Pliocene of the United States.

However, despite the examples cited above and many others like them, the overall increase in diversity in the Cainozoic contrasts with the overall decrease in temperature over the same interval. It is clear that numerous other factors obscure the pattern, particularly the more intensive studies and better preservation of ostracod faunas in successively younger Cainozoic epochs.

**b) Depth**. As a general rule for marine environments, both diversity and evolutionary rates decrease with increasing depth when the whole ostracod fauna is considered. All other natural and artificial factors being equal, species, generic and suprageneric diversity is higher in shallow shelf environments than in the deep sea, except for some high latitude shelf faunas (Benson, 1975; Mckinney, 1987). Most genera (and suprageneric taxa) are more diverse in, or are confined to, shallow waters. Of the 273 genera in the entire area studied, only 8 are confined to the deep water North Atlantic, none of which include over 6 species. A few genera such as *Argilloecia*, *Eucythere*, *Krithe* and *Pedicythere* are more diverse in the deep sea than in shallow water (Whatley, 1983a; Whatley & Coles, 1987), while some morphologically variable and adaptable genera such as *Bairdia* s.l., *Cytherella*, *Cytheropteron* and *Eucytherura* may be of approximately equal diversity in both shallow and deep environments.

**c) Salinity**. In general, the highest diversity ostracod faunas occur in fully marine environments (30-40°/oo salinity), while those from brackish (0.5-30°/oo) and hypersaline waters (>40°/oo) are much less diverse (Whatley, 1983b; Van Morkhoven, 1972). Most of the species included in this study are stenohaline marine forms, with smaller numbers of euryhaline marine, stenohaline brackish and hypersaline species. While most genera are usually most diverse in waters of normal salinity, an exceptional instance of very high species diversity, rapid evolution and speciation occurred in the brackish Paratethyan seas in the Middle Miocene (Sarmatian) and Upper Miocene (Pontian and Pannonian) of Austria and southeastern Europe. The spread of reduced salinity seas over large areas, combined with rapid palaeogeographical changes and isolation of smaller water bodies resulted in the evolution of numerous endemic species, particularly in the genera *Aurila*, *Cytheridea*, *Leptocythere* (see Whatley & Maybury, 1981), *Loxoconcha* and *Xestoleberis*.

**d) Substrate**. The effects on diversity and evolution of substrate is difficult to assess, although it is probable that finer grained (clay-silt grade) substrates support higher species diversities than sandy and coarser grained sediment substrates (Hulings & Puri, 1964). Evolutionary rates are also high in soft-bottom dwellers such as *Argilloecia*, *Cytheropteron* and *Krithe* (see below).

**e) Ecological niches**. The greater the variety of ecological niches, the higher will be species diversity and, presumably, evolutionary rates. The variety of niches for Ostracoda has been suggested by Maybury & Whatley (1988) as being partially responsible for the extremely high diversity in the Upper Pliocene fauna of Cornwall and North

West France. Shallow water regions obviously have a much greater variety of ecological niches (various sedimentary substrates, rock surfaces, plant and animal hosts etc.) than the deep sea. The relatively low diversity of the North Atlantic fauna (even taking its large areal extent into consideration) is partially due to the limited variety of habitats present, mostly fine-grained carbonate sediments, *Globigerina* ooze and red clay. A few genera, particularly infaunal forms such as *Argilloecia* and *Krithe* are, nevertheless, very diverse in the deep sea. However, some variety of niches is provided there by submerged wood and vegetation in the deep sea, which support a unique ostracod fauna (Maddocks & Steineck, 1987) and ranges back from the Recent to at least the Upper Oligocene (Coles, in preparation; Steineck *et al.*, this volume).

**Sea level: transgressions and regressions**

Marine transgressions generally lead to an increase in ostracod species diversity and evolution as more marine and brackish water habitats become available (Whatley, 1988), a reflection of the 'species-area relationship' (Wise & Schopf, 1981). Transgressions in the Lower and Middle Eocene and Lower Miocene, among others in the Cainozoic result in adaptive radiation of new species and higher taxa. The spread of shallow marine environments, brackish water bodies and similar habitats may lead to increased speciation, particularly if isolation of faunas occurs, for example in the Middle and Upper Miocene of southern and central Europe, or the parapatric speciation of *Bradleya* and *Poseidonamicus* in the S.W. Pacific (Whatley, 1985). In contrast, deep faunas do not seem to be affected by either transgressions or regressions; the oceanic fauna simply advances or retreats down the continental slope and no habitat isolation leads to genetic isolation and speciation. In shallow shelf environments, regressions lead to reduction of available habitat areas and consequent extinctions where the drop in sea level is too rapid for the shelf faunas to adapt or migrate (Newell, 1952). Such regression induced extinctions and drops in species diversity occur at the end of the Oligocene and the Miocene epochs (see species and generic evolution). The stratigraphical resolution of this study and the time periods employed do not allow the recognition of the consequences of smaller scale fluctuations (e.g., Milankovitch cycles) in sea level on ostracod evolution, particularly in the Pliocene and Quaternary.

**Palaeoceanographical events**

In shallow water areas, the principal palaeoceanographical events are due to transgressions and regressions, discussed above, as well as the development of brackish water seas (mainly the Paratethys in central and southeastern Europe; see discussion on salinity above). One obvious palaeoceanographical event in the shelf environments was the desiccation of the Mediterranean in the Messinian Salinity Crisis (Adams *et al.*, 1977). This rapid regression caused mass extinctions of both shallow and deeper water species, but also led to the spread of a new Caspi-brackish fauna (Carbonnel, 1978) and speciation of such genera as *Cyprideis* in hypersaline lagoons (Bassiouni, 1979).

In the deep water North Atlantic, the most significant events are related to water mass changes and the pattern of deep water circulation. Whether the influx of colder, more nutrient-rich bottom waters results in endemic speciation, or merely allows the northward migration of Indo-Pacific species into the N. Atlantic *via* the S. Atlantic is not clear. The most significant influx of psychrospheric species occurs in the Middle Eocene (Coles, Ayress & Whatley, this volume; Whatley & Coles, in press) associated with more vigorous cold water circulation, an event also detected in benthonic foraminifera (Miller, 1983; Wood *et al.*, 1985). However, no detectable faunal turnover at the Eocene-Oligocene boundary such as that described by Benson (1975) or the 40 MY event described by Benson *et al.* (1984) can be identified in the North Atlantic. The Upper Eocene deep-sea fauna closely resembles that of the Lower Oligocene, and the comparatively few originations and extinctions over the interval are of rare species. The enhanced diversity of the deep-sea

faunas in the Oligocene may be the result of increased deep water circulation due to the opening of the Drake Passage between South America and Antarctica, although the timing of this event is controversial (Schnitker, 1980). The Lower Miocene low diversity may possibly be attributed to sluggish circulation in the North Atlantic due to the closure of Tethys (Thomas, 1987; Whatley & Coles, 1987). The 'modern' psychrospheric fauna was essentially established by the Middle Miocene, both for Ostracoda and benthonic Foraminiferida. A global decrease in temperature and spillage of cold North Polar bottom waters into the North Atlantic over the subsided Greenland-Iceland-Faroes-Scotland Ridge produced the modern North Atlantic Deep Water (NADW) and its characteristic faunas (Schnitker, 1980). The subsequent history of the deep North Atlantic fauna has been one of gradual increments of new species, some of which migrated from the Indo-Pacific (Whatley & Ayress, 1988; Coles, Ayress & Whatley, this volume) while others probably evolved *in situ* (e.g., species of *Cytheropteron* and *Krithe*) and a few from the northwest European shelf (e.g., *Muellerina abyssicola*). The Pleistocene glaciations have apparently had no detectable effect on the deep water North Atlantic fauna.

**CONCLUSIONS**

The principal findings of this study are the following:

1. Almost all Cainozoic species are new and do not occur in the Cretaceous. However, up to one third of Cainozoic genera are present in the Mesozoic.

2. Overall species diversity rises in an irregular manner through the Cainozoic. Upper Miocene to Recent diversity exceeds that at any time in the Palaeocene to Middle Miocene.

3. There is a steady, almost linear rise in generic diversity through the Cainozoic, which shows less marked fluctuations than the trend for species diversity.

4. Species and generic diversity in the shallow water environments of Europe and North America is always much greater than that of the deep water North Atlantic, and this difference becomes

increasingly marked through the Cainozoic.

5. In the shallow water regions, the highest species origination peaks occur in the Lower and Middle Eocene, Lower and Middle Miocene, Pliocene and Recent. The times of highest extinction are the Upper Eocene, Upper Oligocene and Upper Miocene. In the deep sea, both origination and extinction rates are much lower; peak originations occur in the Lower Miocene and peak extinctions in the Upper Oligocene, marking the Palaeogene-Neogene faunal changeover.

6. There is a general but irregular trend to a decreasing percentage of new species through the Cainozoic. This trend is interrupted by a notable peak of 60% new species in the Lower Miocene, indicating a Neogene faunal renewal.

7. In general, the most diverse and actively evolving genera inhabit warm-temperate, shallow marine to brackish environments. In contrast, only 6 genera have over 10 known Cainozoic deep water North Atlantic species.

8. When the overall faunal composition is considered, a clear difference is seen between Palaeogene and Miocene to Recent deep-sea faunas. The basic composition of the North Atlantic fauna was established in the Middle Miocene.

9. The suprageneric composition of deep water faunas is much more stable in the Cainozoic compared to shallow water faunas. However, the Cainozoic faunal changes in the entire area studied are much less dramatic than those of the Mesozoic (Whatley, 1988).

10. Numerous factors, both artificial and natural, control the apparent diversity and evolution of Ostracoda. For the Cainozoic of the area studied, the most important natural factors are considered to be variations in sea level, temperature and certain palaeoceanographic events. The first two factors principally affect shallow water Ostracoda, while the latter may influence Ostracoda in both shallow and deep environments.

## ACKNOWLEDGEMENTS

The author wishes to thank Professor R. C. Whatley, Dr C. A. Maybury, Dr M. A. Ayress and two anonymous referees for useful discussions and criticism. Dr D. Loydell and Mrs C. Loydell typed the manuscript. This study was carried out during the tenureship of a N.E.R.C. studentship.

## REFERENCES

Adams, C. G., Benson, R. H., Kidd, R. B., Ryan, W. B. F., Wright, R. C. 1977. The Messinian salinity crisis and evidence of late Miocene eustatic changes in the world ocean. *Nature*, London, **269**, 383-386.

Bassiouni, M. A. A. 1979. Brackische und marine Ostracoden (Cytherideinae, Hemicytherinae, Trachyleberidinae) aus dem Oligozän und Neogen der Turkei. *Geol. Jb.*, Hannover, (B), **31**, 3-195.

Benson, R. H. 1975. The origin of the psychrosphere as recorded in changes of deep sea ostracode assemblages. *Lethaia*, Oslo, **8**(1), 69-83.

Benson, R. H., Chapman, R. E. & Deck, L. T. 1984. Paleoceanographic events and deep-sea ostracodes. *Science*, **224**, 1334-1336.

Carbonnel, G. 1978. La zone a *Loxoconcha djaffarovi* Schneider (Ostracoda, Miocène Supérieur) ou le Messinien de la Vallée du Rhône. *Revta esp. Micropaleont.*, Madrid, **21**(3), 106-118.

Coles, G. P. In preparation. *Cainozoic evolution of Ostracoda from deep waters of the North Atlantic and adjacent shallow water regions.* Unpub. Ph.D. thesis, University of Wales, Aberystwyth.

Coles, G. P. & Whatley, R. C. In press. New Palaeocene to Miocene genera and species of Ostracoda from DSDP Sites in the North Atlantic. *Revta esp. Micropaleont.*, Madrid.

Fischer, A. G. 1960. Latitudinal variation in organic diversity. *Evolution*, Lancaster, Pa., **14**, 64-81.

Harland, W. B., Cox, A. V., Llewellyn, P. G., Pickton, C. A. G., Smith, A. G. & Walters, R. 1982. *A geologic time scale.* 131 pp. Cambridge University Press.

Hulings, N. C. & Puri, H. S. 1964. The ecology of shallow water ostracods of the west coast of Florida. *Pubbl. Staz. zool. Napoli*, Milano & Napoli, **33** suppl., 308-344.

Maddocks, R. F. & Steineck, P. L. 1987. Ostracoda from experimental wood-island habitats in the deep sea. *Micropaleontology*, New York, **33**(4), 318-355.

Maybury, C. & Whatley, R. C. 1988. The evolution of high diversity in the ostracod communities of the Upper Pliocene faunas of St. Erth (Cornwall, England) and North West France. *In* Hanai, T., Ikeya, N. & Ishizaki, K. (Eds), *Evolutionary biology of Ostracoda, its fundamentals and applications*, proceedings of the Ninth International Symposium on Ostracoda, held in Shizuoka, Japan, 29 July - 2 August 1985, Developments in palaeontology and stratigraphy, **11**, 569-596, Kodansha Ltd., Tokyo and Elsevier, Amsterdam, Oxford, New York, Tokyo.

Mckinney, M. L. 1987. Taxonomic selectivity and continuous variation in mass and background extinctions of marine taxa. *Nature*, London, **325**(6100), 143-145.

Miller, K. G. 1983. Eocene-Oligocene paleoceanography of the deep Bay of Biscay: benthic foraminiferal evidence. *Mar.*

*Micropaleontol.*, Amsterdam, 7, 403-440.

Moore, R. C. (Ed.) 1961. *Treatise on Invertebrate Palaeontology, Part Q, Arthropoda*, 3, xxiii + 442 pp., 334 figs. Univ. Kansas Press.

Newell, N. D. 1952. Periodicity in invertebrate evolution. *J. Paleont.*, Ithaca, New York, 26, 371-385.

Raup, D. M. 1976. Species diversity in the Phanerozoic: an interpretation. *Paleobiol.*, Ithaca, New York, 2(4), 289-297.

Schnitker, D. 1980. Global palaeoceanography and its deep water linkage to the Antarctic glaciation. *Earth Sci. Rev.*, Amsterdam, 16, 1-20.

Schopf, T. J. M., Raup, D. M, Gould, S. J., Simberloff, D. S. 1975. Genomic versus morphologic rates of evolution: influence of morphological complexity. *Paleobiol.*, Ithaca, New York, 1, 63-70.

Sheehan, P. M. 1977. Species diversity in the Phanerozoic: A reflection of labor by systematists? *Paleobiol.*, Ithaca, New York, 3(3), 325-328.

Thomas, E. 1987. Late Oligocene to Recent benthic foraminifers from Deep Sea Drilling Project Sites 608 and 610, northeastern North Atlantic. *In* Ruddiman, W. F. *et al.* (Eds), *Init. Reports DSDP*, 94(2), 997-1031. U.S. Govt Printing Office.

Van Morkhoven, F. P. C. M. 1972. Bathymetry of Recent marine Ostracoda in the North-West Gulf of Mexico. *Trans. Gulf Cst Ass. geol. Socs*, 22, 241-252.

Whatley, R. C. 1983a. Some aspects of the Palaeobiology of Tertiary Deep-Sea Ostracoda from the S.W. Pacific. *J. Micropalaeontol.*, London, 2, 83-104.

Whatley, R. C. 1983b. The application of Ostracoda to palaeoenvironmental analysis. *In* Maddocks, R. F. (Ed.), *Applications of Ostracoda*, proceedings of the Eighth International Symposium on Ostracoda, July 26-29, 1982, 51-77. Univ. Houston Geos., Houston, Texas.

Whatley, R. C. 1985. Evolution of the ostracods *Bradleya* and *Poseidonamicus* in the deep-sea Cainozoic of the South-West Pacific. *Spec. Pap. Palaeont.*, Palaeontological Association, London, 33, 103-116.

Whatley, R. C. 1988. Patterns and rates of evolution among Mesozoic Ostracoda. *In* Hanai, T., Ikeya, N. & Ishizaki, K. (Eds), *Evolutionary biology of Ostracoda, its fundamentals and applications*, proceedings of the Ninth International Symposium on Ostracoda, held in Shizuoka, Japan, 29 July - 2 August 1985, Developments in palaeontology and stratigraphy, 11, 1021-1040, Kodansha Ltd., Tokyo and Elsevier, Amsterdam, Oxford, New York, Tokyo.

Whatley, R. C. & Ayress, M. A. 1988. Pandemic and endemic distribution patterns in Quaternary deep-sea Ostracoda. *In* Hanai, T., Ikeya, N., & Ishizaki, K. (Eds), *Evolutionary biology of Ostracoda, its fundamentals and applications*, proceedings of the Ninth International Symposium on Ostracoda, held in Shizuoka, Japan, 29 July - 2 August 1985, Developments in palaeontology and stratigraphy, 11, 739-755, Kodansha Ltd., Tokyo and Elsevier, Amsterdam, Oxford, New York, Tokyo.

Whatley, R. C. & Coles, G. P. 1987. The late Miocene to Quaternary Ostracoda of Leg 94, Deep Sea Drilling Project. *Revta esp. Micropaleont.*, Madrid, 19(1), 33-97.

Whatley, R. C. & Coles, G. P. In press. Global change and the biostratigraphy of Cainozoic deep-sea Ostracoda in the North Atlantic. *J. micropalaeontol.*, London.

Whatley, R. C. & Maybury, C. 1981. The evolution and distribution of the ostracod genus *Leptocythere* Sars, 1925 from the Miocene to Recent in Europe. *Revta esp. Micropaleont.*, Madrid, 13(1), 25-42.

Wise, K. P. & Schopf, T. J. M. 1981. Was marine faunal diversity in the Pleistocene affected by changes in sea level? *Paleobiol.*, Ithaca, New York, 7(3), 394-399.

Wood, K. C., Miller, K. G. & Lohmann, G. P. 1985. Middle Eocene to Oligocene benthic foraminifera from the Oceanic Formation, Barbados. *Micropaleontology*, New York, 31(2), 187-197.

## DISCUSSION

Richard Benson: Your discovery of the change in the North Atlantic from warmer deep faunas to psychrospheric faunas during the earliest Neogene is significant. The North Atlantic was dominated by the outflow of Tethys until this time when the eastern Tethys access to the Indian Ocean was effectively closed. Your data will help reconstruct the change in this dominance.

Graham Coles: The principal faunal changes at the Palaeogene-Neogene boundary are species extinctions; few new species appear in the Neogene until the Upper Miocene.

Martin Angel: You showed that present day Recent ostracod faunas are much more species rich in shallow water habitats than in abyssal depths. In other crustacean groups, notably certain families of Amphipoda and Isopoda the abyssal/shallow water relationship is reversed. There are two theories put forward to account for abyssal species: a) *in situ* speciation and b) down-slope migration. Could you comment on how the ostracods differ from these other crustacean groups?

Graham Coles: Lower ostracod species diversity at abyssal depths relative to shallow waters is the result of several factors. A much greater variety of habitats are available in shallow waters compared to the relatively uniform, fine grained oozes of the deep sea, thus allowing greater niche partitioning, specialization and speciation in shallow waters. However, recorded abyssal ostracod diversity will undoubtedly increase with further research on these relatively poorly sampled habitats, particularly with detailed studies on small (under 0.5mm) species and 'splitting' of genera such as *Krithe*. Some genera, notably *Argilloecia*, *Cytheropteron* and *Krithe* have apparently undergone considerable *in situ* speciation. Some outer shelf and slope (200 to 1000m) species extend into abyssal waters, but many important shelf families, e.g., the Hemicytheridae, are essentially absent from the deep sea.

# 6

# The Carnian salinity crisis: ostracods and palynomorphs as indicators of palaeoenvironment

**Ephraim Gerry[1], Avraham Honigstein[2], Amnon Rosenfeld[3], Francis Hirsch[3] & Yoram Eshet[3]**

[1]The Israel Institute of Petroleum and Energy, Tel Aviv, Israel.
[2]Department of Geophysics and Planetary Sciences, Tel-Aviv University, Israel.
[3] Geological Survey of Israel, Jerusalem, Israel.

## ABSTRACT

Ostracod assemblages in the Middle East (Israel, Jordan, Syria) and Europe (Austria, Italy, Hungary, Spain) indicate an Upper Absaroka A-3 global sea-level lowstand. This regression resulted in a major salinity crisis in the western Tethys during the Carnian and is characterized by abundant gypsum, anhydrite and halite deposits. The marine ostracod *Reubenella avnimelechi* (Anisian -Ladinian) and the brackish-hypersaline *Simeonella brotzenorum* ostracod assemblage zones (late Ladinian-early Carnian) are overlain by the herein established *Renngartenella sanctaecrucis*

Zone of early Carnian (Julian) age. This zone has been encountered in Israel only in the Mohilla Formation of the Devora-2A and Ramallah-1 boreholes and correlates with the *Patinasporites densus* Palynomorph Zone. The abundant occurrence of the euryhaline ostracod *Simeonella brotzenorum* indicates brackish-hypersaline conditions. Species of typically marine genera, such as *Reubenella, Judahella, Mostlerella, Renngartenella, Leviella* and *Mockella*, are found together with *Simeonella brotzenorum* and suggest mixing with or the close adjacence of marine waters.

Regressive-transgressive oscillations were traced by estimating the percentages of anhydrite, reworked palynomorphs and nonmarine ostracods. The resulting regional sea-level changes are in good agreement with the eustatic curve. The common ostracod species of the western Tethys, restricted to the Sephardic 'Muschelkalk' facies belt during the Middle Triassic, extended to the Tethys dolomite-limestone facies during the Cordevolian and the Julian-Tuvalian.

## INTRODUCTION

The Triassic Tethys expanded in a wide belt from Spain in the west to Australia in the southeast. Thick successions of evaporites were deposited during the Triassic. In the Mediterranean region, Triassic sedimentation took place in an arid to subtropical climatic zone. Clastic materials and carbonates were deposited during periods of sea-level highstands as well as in areas with continental fluvial runoff (Druckman *et al.*, 1975).

Triassic sediments in the Middle East crop out in several areas of Jordan, Sinai and southern Israel. About 25 deep boreholes in Israel penetrated the Triassic, among them the Devora-2A and Ramallah-1 boreholes (Fig. 1). The thickness of Triassic sediments in southern Israel is approximately 1000m (e.g., in the Makhtesh Ramon outcrop), but may reach three times as much in the northern part of the country (Devora-2A borehole; Druckman & Kashai, 1981). The sections are composed mainly of carbonates and shales, and evaporites dominate the upper part. The sediments are well dated by megafossils (ammonites and bivalves), and microfauna (foraminifera, conodonts and ostracods), as well as by palynomorphs.

The sequence was deposited in near shore and shallow marine environments, very shallow lagoons and shoaling tidal flats (Druckman, 1974). The transgressive acme was reached within the marine carbonates of the Saharonim Formation of the Middle Triassic. This was followed by a sharp regression in the late Triassic, as witnessed by the evaporites of the Mohilla Formation (Fig. 2).

The Lower Member of the Saharonim Formation (Zak, 1963) consists of cyclical

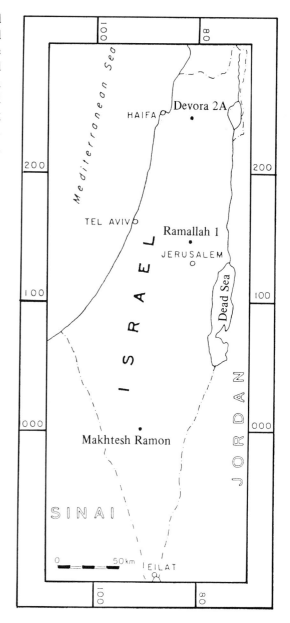

Fig. 1. Map of the localities referred to in Israel.

alternations of fossiliferous limestones and shales, containing cephalopods, pelecypods, brachiopods, echinoid fragments, miliolid foraminifera, ostracods, conodonts and reptile remains (Druckman *et al.*, 1975). This unit was deposited in a low energy, shallow marine environment, with a supply of

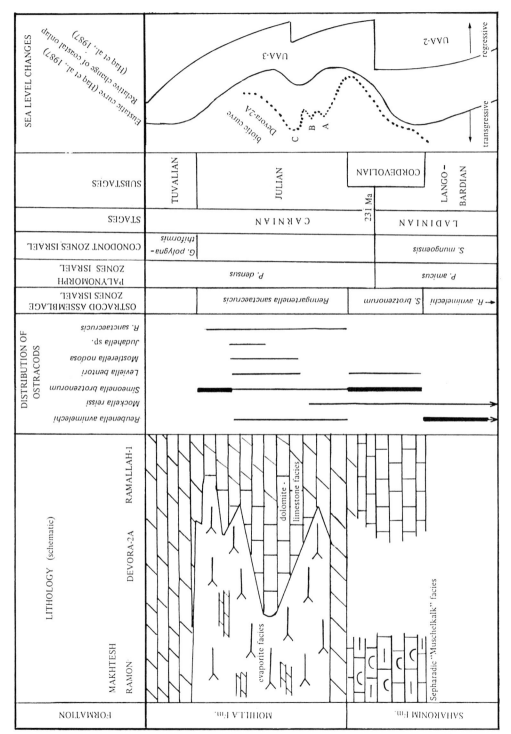

Fig. 2. Middle to late Triassic lithology in Israel, distribution of ostracods, fossil zonations and eustatic sea-level changes. Stage and substage definition modified after Zapfe, 1983; ostracod zones after Derin and Gerry, 1981, and this study; palynomorph zones after Eshet, 1987; conodont zones after Hirsch, 1987; for legend see Fig. 4.

continental derived clays, which rhythmically interrupted the deposition of carbonates. The Middle and Upper members of this formation are characterized by a variety of limestones, dolomites, shales and evaporites (gypsum and anhydrite). They represent gradual shallowing of the sea and salinity increase in a subtidal environment. This can be deduced, among other features, from the great reduction in the abundance of the fauna and its diversity, the presence of algal stromatolites as well as the appearance of tidal channels. The thickness of the Saharonim Formation increases gradually to the northwest, reaching 165m at Makhtesh Ramon, about 400m in Ramallah-1 and approximately 1000m in Devora-2A (Druckman & Kashai, 1981). The sediments of this formation belong to the Sephardic Muschelkalk facies (Hirsch, 1977) and their age is late Anisian to early Carnian (Parnes, 1986).

The overlying sediments of the Mohilla Formation (Zak, 1963) are composed mainly of gypsum, dolomite, limestone and minor amounts of marls. The rare fossils found in these sediments indicate a Carnian age (Eicher & Mosher, 1974). Two main facies were identified within the Mohilla Formation (Fig. 2), the evaporite facies (e.g., Makhtesh Ramon, 210m thick) and the dolomite-limestone facies (e.g., Ramallah-1, about 1000m thick). The approximately 1000m thick succession in the Devora-2A borehole mainly comprises evaporites interfingering with carbonates and may represent a transitional facies type. The deposits of this formation indicate alternations of shallow marine and hypersaline lagoonal environments with laminated anhydrite. Sabkha-like conditions are represented by algal stromatolites and nodular anhydrite (Druckman *et al.*, 1975). The uppermost part of the Mohilla Formation is generally composed of marine carbonates.

## OSTRACODS

Ostracods within the Mohilla Formation were observed in Israel only in cuttings from the Devora-2A (Derin & Gerry, 1979) and the Ramallah-1 (Derin *et al.*, 1980) boreholes and represent the uppermost Triassic ostracod assemblage in the country. This assemblage overlies the brackish-hypersaline water *Simeonella brotzenorum* Ostracod Zone in the late Ladinian-early Carnian, Middle and Upper Members of the Saharonim Formation and the marine *Reubenella avnimelechi* ostracod zone in the late Anisian-early Ladinian Lower Member of the Saharonim Formation. Both of these ostracod zones are well distributed and occur in various boreholes and outcrops in Israel (Gerry, 1967; Sohn, 1968; Hirsch and Gerry, 1974; Derin & Gerry, 1981). Triassic ostracods are also described from the Hisban Formation in Jordan (Basha, 1982), which correlates with the Saharonim Formation.

The ostracod fauna of the Mohilla Formation consists mainly of adult specimens of articulated carapaces. The samples from Devora-2A yield, in general, common to frequent populations; while those of Ramallah-1 are rare, probably due to the dolomitized sediments in this borehole. The assemblages are characterized by the joint occurrence of the euryhaline species *Simeonella brotzenorum*, together with six stenohaline marine species in variable percentages (see Fig. 3). All these ostracods are grouped herein within the *Renngartenella sanctaecrucis* Assemblage Zone. *Renngartenella sanctaecrucis*, *Mostlerella nodosa* and the probably new species of *Judahella* are restricted to this zone. It is assigned to the Julian (Carnian) on the basis of biostratigraphical evidence from other fossil groups in Israel and on ostracod ranges elsewhere (Figs 4 and 5). Lieberman (1979) described a very similar ostracod assemblage of Julian age, containing *Simeonella brotzenorum*, *Reubenella avnimelechi*, *Renngartenella sanctaecrucis* and species of the genera *Mockella*, *Judahella* and *Leviella*, from well-dated sections of the Raibl Formation in northern Italy. In Israel, correlative sediments both below and above the *Renngartenella sanctaecrucis* Ostracod Zone are dated by condonts (*Sephardiella mungoensis* (Langobardian) and *Gondolella polygnathiformis* (Tuvalian); Hirsch, 1987). This ostracod zone is contemporaneous with the Carnian *Patinasporites densus* Palynomorph Zone as indicated below.

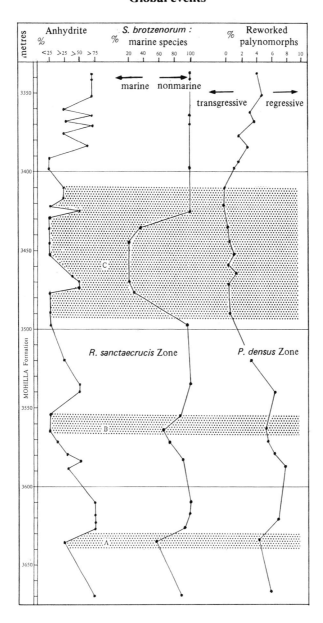

Fig. 3. The Mohilla Formation in the Devora-2A borehole. Percentages of anhydrite, nonmarine ostracods and reworked palynomorphs. The dotted areas define marine transgressive phases.

## PALYNOMORPHS

Palynomorphs were studied in the interval of the *Renngartenella sanctaecrucis* Ostracod Zone in the Devora-2A borehole. The floral assemblage is composed mostly of circumsulcate palynomorphs,

such as *Duplicisporites granulatus* (the most prominant species), together with *Patinasporites densus, Paracirculina scurrilis, Praeciculina granifer* and *Camerosporites secatus* (Palynozone IX; Eshet & Cousminer, 1986). Bisaccate pollen, such as *Lunatisporites acutus, Ovalipollis ovalis* and

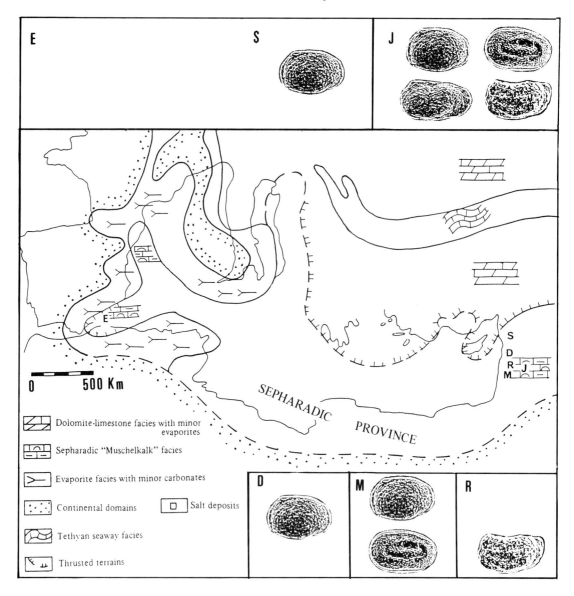

Fig. 4. Palaeogeographical sketch map of the Cordevolian substage, showing the distribution of the facies types (modified after Márquez-Aliaga & Hirsch, 1988) and the ostracod species common to Israel (E = Spain: *L. bentori*; S = Syria: *S. brotzenorum*, age determination uncertain; J = Jordan; *S. brotzenorum, L. bentori, R. sanctaecrucis, M. reissi*; D = Devora-2A: *S. brotzenorum*; M = Makhtesh Ramon: *S. brotzenorum, L. bentori*; R = Ramallah-1: *S. brotzenorum, M. reissi*).

*Striatoabietes aytugii* are present, but rare. All of this flora belongs to the Carnian *Patinasporites densus* Zone (Eshet, 1987). This zone in Devora-2A is underlain by the palynomorphs of the Ladinian *Podosporites amicus* Zone.

Other palynomorphs were determined as being reworked, according to their much older ages (Palaeozoic) and very black, opaque appearance. Since the Permo-Triassic succession in Israel had never been found to be thermally overmature (Bein et al., 1984; Eshet, 1987), black palynomorphs in our samples must be considered as reworked. This

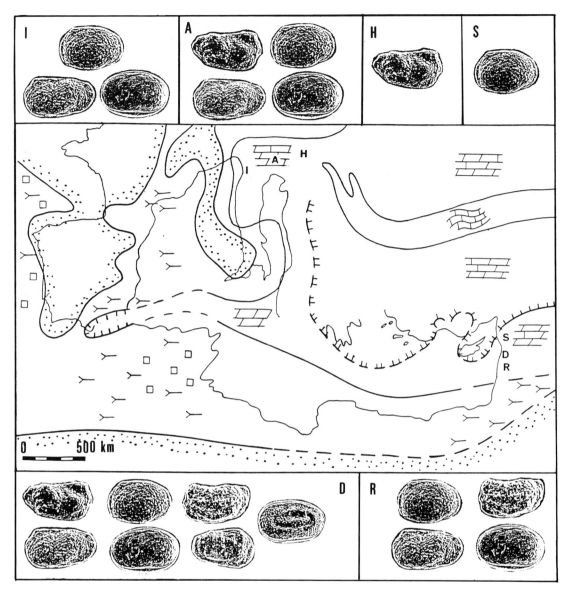

Fig. 5. Palaeogeographical sketch map of the Julian and Tuvalian substages, showing the distribution of the facies types and the ostracod species, common to Israel (I = northern Italy: *S. brotzenorum*, *R. sanctaecrucis*, *R. avnimelechi*; A = Austria: *M. nodosa*, *S. brotzenorum*, *R. sanctaecrucis*, *R. avnimelechi*; H = Hungary: *M. nodosa*; S = Syria: *S. brotzenorum*, age determination uncertain; D = Devora-2A: *M. nodosa*, *S. brotzenorum*, *M. reissi*, *L. bentori*, *R. sanctaecrucis*, *R. avnimelechi*, *Judahella* sp.; R = Ramallah-1: *S. brotzenorum*, *M. reissi*, *R. sanctaecrucis*, *R. avnimelechi*).

organic matter is mostly coalified woody or inertinitic material.

## ENVIRONMENT OF DEPOSITION

An attempt was made to reconstruct regional short-lived environmental changes, due to eustatic sea-level fluctuations, following the Upper Absoraka A-3 global sea-level lowstand (Fig. 2; Haq *et al.*, 1987). Over thirty samples from the interval 3330-3670m within the Mohilla Formation of the Devora-2A borehole were examined for their anhydrite content, the marine:nonmarine ostracod ratio and the percentage of reworked palynomorphs (Fig. 3).

The percentage of anhydrite was determined semiquantitatively for all the samples which yielded fossils. The marine:nonmarine ostracod ratio was calculated on the basis of carapaces and valves of *Simeonella brotzenorum* versus marine species. The percentage of reworked palynomorphs is based on counts of at least ten random traverses of each slide. This technique allows identification of transgressions and regressions in shallow seas, in proximity to land (Eshet *et al.*, 1988).

Fig. 3 shows a good correlation between the three independent criteria, which allows us to delineate a composite sea-level curve from the Devora-2A borehole (Fig. 2). In the studied interval, three pronounced peaks of influx of normal marine waters could be distinguished. The first two occurrences (A, B) are rather short in duration and show a moderate amplitude (up to 50% marine ostracods; 4% reworked palynomorphs). The upper peak is relatively long-lived and has values, which indicate transition to a fully marine environment (up to 80% marine ostracods and almost no reworked palynomorphs). The peaks are correlative with high percentages of carbonates (limestones and dolomites) in the samples. Regressions are marked by the abundance of evaporites, high values of reworked palynomorphs (up to 8%) and monospecific occurrences of *S. brotzenorum*, probably indicating hypersaline lagoons. The alternations of brackish-hypersaline and normal marine environments suggest oscillations of the sea level. Moreover, the carbon isotope depletion in the Mohilla Formation is thought to be related to an influx of continental water floating over a dense evaporite brine (Magaritz & Druckman, 1984).

The resulting biotic curve of Devora-2A is in good agreement with the eustatic curve and the relative coastal onlap (Fig. 2; Haq *et al.*, 1987). The assignment of a normal marine environment to the *Reubenella avnimelechi* Zone, fits the Langobardian global transgressional phase. The major regression between the Ladinian and Carnian (Cordevolian) is expressed in Israel by an almost exclusive monotypical occurrence of *Simeonella*

*brotzenorum*. Three successive highstands of the sea (A, B, C) in the Carnian (Julian) *Renngartenella sanctaecrucis* Zone can be correlated with pulses of a synchronous worldwide transgression.

## PALAEOGEOGRAPHICAL IMPLICATIONS

Figs 4 and 5 depict the distribution of the different facies types in the eastern and western Mediterranean basin during the Cordevolian (Fig. 4) and Julian-Tuvalian substages (Fig. 5). The occurrence of ostracod species in other countries, that are common to those of the *S. brotzenorum* and *R. sanctaecrucis* zones in Israel, is also shown in these figures.

During the late Ladinian-early Carnian (Cordevolian; Fig. 4; Márquez-Aliaga & Hirsch, 1988), the Sephardic 'Muschelkalk' facies still extends from Spain to the Levant, bounded by the 'alpinotype' dolomite-limestone facies in the north and east, and by the evaporite facies in the west. In Jordan (Basha, 1982), marine species are more common than in Israel, but may belong to the uppermost part of the Cordevolian at the transition to the early Julian marine intercalation (see Figs 2 and 3, zone A).

At the end of the late Ladinian, shallowing increased in the Sephardic realm, followed by the worldwide Carnian salinity crisis (Busson, 1982; sea-level lowstand Upper Absaroka A-3, Haq *et al.*, 1987), putting an end to the Sephardic 'Muschelkalk' facies. In its place, the evaporite facies expanded during the Julian-Tuvalian both eastwards and northwards, and the dolomite-limestone facies expanded both southwards and westwards (Fig. 5), Everywhere, the evaporite facies are barren of the mid-Triassic characteristic fauna. The distribution of the frequently occurring western Tethys ostracod species common with those of Israel, is connected to the distribution of dolomite-limestone facies and where they occur in association with the euryhaline *S. brotzenorum* species, they are indicative of marginal marine environments. These assemblages extend in this widely distributed facies belt from the Julian (Israel; this paper; northern Italy, Raibl Formation, Lieberman, 1979) to the Tuvalian (Austria,

Opponitz Formation, Kristan-Tollmann & Hamedani, 1973).

## APPENDIX

### Taxonomic notes on the Ostracoda

**Remarks**. The specimens described and figured in this study are deposited in the ostracod collections of the Geological Survey of Israel, Stratigraphy and Oil Division, Jerusalem, catalogued under 'Devora-2A, Ramallah-1, late Triassic'. All dimensions quoted below are of carapaces, unless otherwise indicated.

Genus *Reubenella* Sohn, 1968
*Reubenella avnimelechi* Sohn, 1968
(Plate 1, figs 1-2)

1968   *Reubenella avnimelechi* Sohn: 18, pl. 1, figs 34-46.
1973   *Reubenella avnimelechi* Sohn; Kristan-Tollmann & Hamedani: 203, pl. 8, figs 7-9, pl. 9, figs 8, 10.
1974   *Reubenella avnimelechi* Sohn; Hirsch & Gerry: pl. 2, figs 3-4.
1979   *Reubenella avnimelechi* Sohn; Lieberman: 103-104.
1982   *Reubenella avnimelechi* Sohn; Basha: pl. 2, figs 7-9.

**Dimensions** (mm).

| Length | Height | Width |
|--------|--------|-------|
| 0.61 | 0.36 | 0.22 |
| 0.51 | 0.33 | 0.21 |

**Remarks**. This species has a long stratigraphical range. It was reported from Anisian-Ladinian strata in Israel (Sohn, 1968; Hirsch & Gerry, 1974), the Ladinian of Jordan (Basha, 1982), the Julian of northern Italy (Lieberman, 1979) and the Tuvalian of Austria (Kristan-Tollmann & Hamedani, 1973).

**Material and distribution**. About sixty carapaces and valves from the Devora-2A and Ramallah-1 boreholes.

Genus *Leviella* Sohn, 1968
*Leviella bentori* Sohn, 1968
(Plate 1, fig. 6)

1968   *Leviella bentori* Sohn: 22, pl. 1, figs 15-18, 23, 25, 26.
1974   *Leviella bentori* Sohn; Kozur *et al.*: 42, fig. 21a.
?1982   *Leviella bentori* Sohn; Basha: pl. 1, fig. 10.

**Dimensions** (mm).

| Length | Height | Width |
|--------|--------|-------|
| 0.47 | 0.29 | 0.13 |

**Remarks**. The outline and ornamentation of our specimens are identical with those of the holotype (Sohn, 1968, pl. 1, fig. 18), but they are smaller. *L. bentori* was described from the Cordevolian of Israel (Sohn, 1968), and Spain (Kozur *et al.*, 1974). The species determination of the specimen, figured from the Cordevolian of Jordan (Basha, 1982), is questionable.

**Material and distribution**. Two carapaces from the Devora-2A borehole.

Genus *Simeonella* Sohn, 1968
*Simeonella brotzenorum* Sohn, 1968
(Plate 1, figs 3-5)

1968   *Simeonella brotzenorum* Sohn: 23, pl. 2, figs 1-4, 6-8, 12-22.
1970   *Lutkevichinella brotzenorum* (Sohn); Wienholz & Kozur: 588, pl. 1, figs 4, 6-8.
1971   *Simeonella brotzenorum brotzenorum* Kozur (in Bunza & Kozur): 3, pl. 1, figs 8-11.
1971   *Simeonella brotzenorum alpina* (in Bunza & Kozur): 4, pl. 1, figs 5-7, 13.
1973   *Simeonella brotzenorum* Sohn; Kristan-Tollmann & Hamedani: 213, pl. 6, figs 13-14, pl. 13, fig 2.
1974   *Simeonella brotzenorum* Sohn; Hirsch & Gerry: pl. 2, figs 1-2.
1979   *Simeonella brotzenorum* Sohn; Lieberman: 103, pl. 5, figs 8-9.
1982   *Speluncella*? *karnica* Kozur (in Bunza & Kozur); Basha: pl. 2, fig. 2.
1982   *Simeonella brotzenorum* Sohn; Basha: pl. 1, fig. 11.

**Dimensions** (mm).

| Length | Height | Width |
|--------|--------|-------|
| 0.56 | 0.38 | 0.32 |
| 0.55 | 0.38 | 0.36 |
| 0.44 | 0.28 | 0.28 |

**Remarks**. This species is widely distributed in the Middle East and Europe (Cordevolian of Israel: Sohn, 1968; Hirsch & Gerry, 1974; Cordevolian of Jordan: Basha, 1982; lower part of late Triassic: Bach-Imam & Sigal, 1985; Carnian of Germany: Wienholz & Kozur, 1970; Julian-Norian of Austria: Bunza & Kozur, 1971; Kristan-Tollmann & Hamedani, 1973; Julian of Italy: Lieberman, 1979).

**Material and distribution**. About two hundred carapaces and valves from the Devora-2A and Ramallah-1 boreholes.

Genus *Mockella* Kozur (in Bunza & Kozur), 1971
*Mockella reissi* (Sohn), 1968
(Plate 1, fig. 8)

1968   *Simeonella reissi* Sohn: 24, pl. 1, figs 1-5, 8-12.
1982   *Simeonella reissi* Sohn; Basha: pl. 2, figs 3-5.

**Dimensions** (mm).

Length Height Width
 0.50    0.26            LV

**Remarks.** The rectangular outline and the three characteristic ribs of *M. reissi* assign this species to the genus *Mockella* Kozur, 1971 (see Kozur, 1973, 9). The range of *M. reissi* (Ladinian-Cordevolian of Israel: Sohn, 1968; Derin *et al.*, 1980; Cordevolian of Jordan: Basha, 1982) is extended herein to the Julian.

**Material and distribution.** Two valves from the Devora-2A and Ramallah-1 boreholes.

Genus *Renngartenella* Schneider, 1957
*Renngartenella sanctaecrucis* Kristan-Tollmann, 1973 in Kristan-Tollmann & Hamedani
(Plate 1, figs 11-13)

1973   *Renngartenella sanctaecrucis* Kristan-Tollmann (in Kristan-Tollmann & Hamedani): 215, pl. 8, figs 1-6, pl. 11, figs 1, 3, 5, 6, pl. 12, fig. 10.
1979   *Renngartenella sanctaecrucis* Kristan-Tollmann (in Kristan-Tollmann & Hamedani); Lieberman: 102, pl. 5, fig. 3.
1982   *Renngartenella sanctaecrucis* Kristan-Tollmann (in Kristan-Tollmann & Hamedani); Basha: pl. 1, fig. 15.

**Dimensions** (mm).

Length  Height  Width
 0.60    0.33    0.26
 0.58    0.31    0.26
 0.56    0.28    0.24

**Remarks.** *R. sanctaecrucis* has been previously recorded from the Cordevolian of Jordan (Basha, 1982), the Julian of Italy (Kristan-Tollmann and Hamedani, 1973; Lieberman, 1979) and the Tuvalian of Austria (Kristan-Tollmann & Hamedani, 1973).

**Material and distribution.** Fifteen carapaces from the Devora-2A and Ramallah-1 boreholes.

Genus *Mostlerella* Kozur (in Bunza & Kozur), 1971
*Mostlerella nodosa* Kozur (in Bunza & Kozur), 1971
(Plate 1, fig. 7)

1971   *Mostlerella nodosa parva* Kozur (in Bunza & Kozur): 39, pl. 4, fig.2
1973   *Mostlerella nodosa* Kozur (in Bunza & Kozur); Kristan-Tollmann & Hamedani: pl. 8, figs 10-11, pl. 13, fig. 3.
1982   *Mostlerella? nodosa* Kozur (in Bunza & Kozur), Basha: pl. 1, fig. 17.

**Dimensions** (mm).

Length Height Width
 0.50    0.26    0.25

**Remarks.** Our specimens are less reticulated than the type material, possibly due to bad preservation. *M. nodosa* also occurs in the Ladinian of Jordan (Basha, 1982), the Julian of Hungary (Bunza & Kozur; 1971) and the Tuvalian of Austria (Kristan-Tollmann & Hamedani, 1973).

**Material and distribution.** Two carapaces from the Devora-2A borehole.

Genus *Judahella* Sohn, 1968
*Judahella* sp.
(Plate 1, figs 9-10)

**Dimensions** (mm).

Length Heigth Width
 0.48    0.29    0.16
 0.49    0.25    0.15

**Remarks.** This probably new species of *Judahella* with narrowly rounded posterior end and straight ventral rib also has three moderately strong nodes dorsally, an elongate sulcus centrodorsally and is weakly reticulate. Our specimens differ from *Judahella gerryi* Sohn, 1968 (16, pl. 3, figs 18-19); Ladinian-early Carnian of Israel) by their larger size, the less prominent nodes and the weaker ventral rib.

**Material and distribution.** Three carapaces from the Devora-2A borehole.

## ACKNOWLEDGEMENTS

The authors wish to thank Drs Y. Druckman and D. Wachs for helpful comments and suggestions, M. Dvorachek and Y. Levy, all of the Geological Survey of Israel, Jerusalem, for the SEM micrographs.

# REFERENCES

Bach Imam, I. & Sigal, J. 1985. Précisions nouvelles sur l'âge Triassique, et non Jurassique, de la majeure partie des formations évaporitiques et dolomitiques des forages de l'Est Syrien. *Rev. Paléobiologie*, Genève, **4**(1), 35-42.

Basha, S. H. S. 1982. Microfauna from the Triassic rocks of Jordan. *Rev. Micropaleont.*, Paris, **25**(1), 3-11.

Bein, A., Feinstein, S., Aizenshtat, Z. & Weiler, Y. 1984. Potential source rocks in Israel: A geochemical evaluation. *Internal Report Oil Explor. Ltd., Tel Aviv*, 1-38.

Bunza, G. & Kozur, H. 1971. Beitraege zur Ostracodenfauna der tethyalen Trias. *Geol. Paläont. Mitt. Ibk*, Innsbruck, **1**(2), 1-76.

Busson, G. 1982. Le Trias comme période salifère. *Geol. Rdsch.*, Stuttgart, **71**(3), 857-880.

Derin, B. & Gerry, E. 1979. Devorah 2A. Stratigraphic log. Scale 1:2000. *Isr. Inst. Petr. Energy, Rep. 2/79*, Tel Aviv, 1 chart.

Derin, B. & Gerry, E. 1981. Late Permian - Late Triassic stratigrapahy in Israel and its signficance for oil exploration. *Geol. Soc. Israel, Symposium Oil Exploration*, Jerusalem, 9-11.

Derin, B., Gerry, E. & Lipson, S. 1980. Ramalla-1. Stratigraphic Log (Segment), 4510 - TD. 6355 m, scale 1:2000. *Isr. Inst. Petrol. Energy, Rep. 1/80*, Tel Aviv, 1 chart.

Druckman, Y. 1974. The stratigraphy of the Triassic sequence in southern Israel. *Geol. Surv. Israel Bull.*, Jerusalem, **64**, 1-94.

Druckman, Y., Gvirtzman, G. & Kashai, E. 1975. Distribution and environments of deposition of Upper Triassic on the northern margins of the Arabian shield and around the Mediterranean Sea. *Proc. 9. Intern. Congr. Sedimentology, Nice*, **5**(1), 183-192. J. P. Mangin, Nice.

Druckman, Y. & Kashai, E. 1981. The Helez Deep and Devora 2A boreholes and their implication to oil prospects in pre-Jurassic strata in Israel. *Geol. Surv. Israel Rep. OD/1/81*, Jerusalem, 1-23.

Eicher, D. B. & Mosher, L. C. 1974. Triassic conodonts from Sinai and Palestine, *J. Paleont.*, **48**, 729-739.

Eshet, Y. 1987. *Palynological aspects of the Permo-Triassic sequence in the subsurface of Israel.* Unpub. Ph.D. thesis, City University, New York, 193 pp.

Eshet, Y. & Cousminer, H. L. 1986. Palynozonation and correlation of the Permo-Triassic succession in the Negev, Israel. *Micropaleontology*, New York, **32**(3), 193-214.

Eshet, Y., Cousminer, H. L., Habib, D., Druckman, Y. & Drugg, W. S. 1988. Reworked palynomorphs and their use in the determination of sedimentary cycles. *Geology*, Boulder, Co., **16**, 662-665.

Gerry, E. 1967. Paleozoic and Triassic ostracodes from outcrops and wells in southern Israel. *Isr. Inst. Petr. Energy, Rep. 1/67*, Tel Aviv, 1-9.

Haq, B. U., Hardenbol, J. & Vail, P. R. 1987. Chronology of fluctuating sea levels since the Triassic. *Science*, **235**, 1156-1167.

Hirsch, F. 1977. Essai de correlation biostratigraphique des niveaux Méso- et Néotriasiques de faciès "Muschelkalk" du domaine Sépharade. *Cuadernos Geol. Ibérica*, Madrid, **4**, 511-526.

Hirsch, F. 1987. Biostratigraphy and correlation of the marine Triassic of the Sepharadic Province. *Cuadernos Geol. Ibérica*, Madrid, **11**, 815-826.

Hirsch, F. & Gerry, E. 1974. Conodont- and Ostracode-Biostratigraphy of the Triassic in Israel. *SchrRreihe erwdiss. Komm. österr. Akad. Wiss.*, Vienna, **2**, 107-112.

Kozur, H. 1973. Beitraege zur Ostracodenfauna der Trias. *Geol. Paläont. Mitt. Ibk*, Innsbruck, **3**(5), 1-41.

Kozur, H., Kampschuur, W., Mulder-Blanken, C. W. H. & Simon, O. J. 1974. Contribution to the Triassic ostracode faunas of the Betic Zone (southern Spain). *Scr. geol.*, Leiden, **23**, 1-56.

Kristan-Tollmann, E. & Hamedani, A. 1973. Eine spezifische Mikrofaunen-Vergesellschaftung aus den Opponitzer Schichten des Oberkarn der niederoesterreichischen Kalkvoralpen. *Neues Jb. Geol. Paläont., Abh.*, Stuttgart, **143**(2), 193-222.

Lieberman, H. M. 1979. Die Bivalven- und Ostracodenfauna von Raibl und ihr stratigraphischer Wert. *Verh. Geol. B.-A., 1979* (2), Vienna, 85-131.

Magaritz, M. & Druckman, Y. 1984. Carbon isotope composition of an Upper Triassic evaporite section in Israel: Evidence for meteoric water influx. *Amer. Assoc. Petrol. Geol., Bull.*, Tulsa, Okla., **68**(4), 502.

Márquez-Aliaga, A. & Hirsch, F. 1988. Migration of the Middle Triassic bivalves in the Sephardic realm. *Congr. Geol. España, 1988, Comm,.* Granada, **1**, 301-304.

Parnes, A. 1986. Middle Triassic cephalopods from the Negev (Israel) and Sinai (Egypt). *Geol. Surv. Israel Bull.* 79, Jerusalem, 9-59.

Sohn, I. G. 1968. Triassic ostracodes from Makhtesh Ramon, Israel. *Geol. Surv. Israel Bull.* 44, Jerusalem, 1-71.

Wienholz, E. & Kozur, H. 1970. Drei interessante Ostracodenarten aus dem Keuper im Norden der DDR. *Geologie*, Berlin, **19**(5), 588-592.

Zak, I. 1963. Remarks on the stratigraphy and tectonics of the Triassic of Makhtesh Ramon. *Isr. J. Earth-Sci.*, Jerusalem, **12**, 87-89.

Zapfe, H. 1983. Das Forschungsprojekt "Triassic of the Tethys Realm", I.G.C.P. Project No. 4, Abschlussbericht. *Schriftenr. Erdwiss. Komm.*, Vienna, **5**, 7-16.

**Plate 1**

Figs 1-2.    *Reubenella avnimelechi* Sohn.

Fig. 1.      Right valve, carapace, Devora-2A, 3467-3473m.
Fig. 2.      Left valve, carapace, Devora-2A, 3467-3473m.

Figs 3-5.    *Simeonella brotzenorum* Sohn.

Fig. 3.      Right valve, carapace, Devora-2A, 3394-3403m.
Fig. 4.      Dorsal view, carapace, Devora-2A, 3394-3403m.
Fig. 5.      Left valve, carapace, Devora-2A, 3394-3403m.

Fig. 6.      *Leviella bentori* Sohn, left valve, carapace, Devora-2A, 3467-3474m.

Fig. 7.      *Mostlerella nodosa* Kozur, right valve, carapace, Devora-2A, 3440-3449m.

Fig. 8.      *Mockella reissi* (Sohn), left valve, Devora-2A, 3632-3638m.

Figs 9-10.   *Judahella* sp.

Fig. 9.      Left valve, carapace, Devora-2A, 3467-3473m.
Fig. 10.     Right valve, carapace, Devora-2A, 3440-3449m.

Figs 11-13. *Renngartenella sanctaecrucis* Kristan-Tollmann.

Fig. 11.     Right valve, carapace, Devora-2A, 3422-3431m.
Fig. 12.     Left valve, carapace, Devora-2A, 3467-3473m.
Fig. 13.     Left valve, carapace, Devora-2A, 3422-3431m.

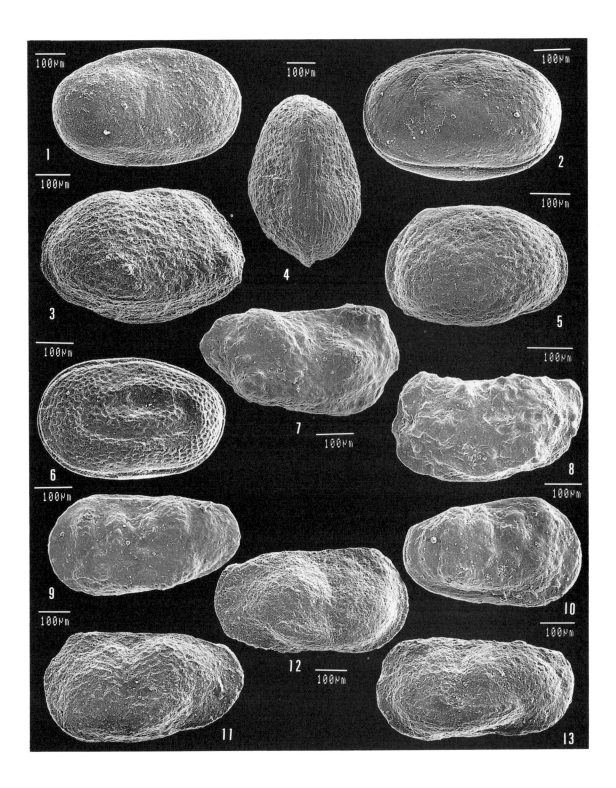

# 7

# The effect of global events on entomozoacean Ostracoda

**Helga Groos-Uffenorde & Eberhard Schindler**

Institute and Museum for Geology and
Palaeontology, University of Göttingen,
Goldschmidt-Str. 3, D-3400 Göttingen, F.R.G.

## ABSTRACT

The problem of whether the mode of life of
entomozoacean ostracods was benthonic or planktonic remains unresolved. The purpose of this
paper is to show that benthonic marine animals
have been affected by global events (mostly
transgressions or 'anoxic' events) within the
Devonian. Entomozoacean ostracods have not
been affected by, for instance, the acme of transgression in the Upper Devonian. Because of their
abundance (juvenile and adult carapaces) in the
black Upper Kellwasser Horizon which lacks
benthos, the benthonic mode of life of entomozoacean ostracods cannot be assumed.

## INTRODUCTION

Several fundamental global biological changes are
known to have taken place during the Palaeozoic. Besides short term extraterrestrial events
(e.g., bolides, asteroid impacts, comet showers),
longer term climatic changes (e.g., glaciation caused
the extinction event in the uppermost Ordovician), major sea level fluctuations and water
chemistry changes (e.g., 'anoxic' events or black
shale events) have been discussed as possible
causes. Combined factors and interactions with
plate tectonics have also been considered. Although a periodicity of extinction events in Mesozoic
and Cainozoic times is assumed by some authors,
'it reveals no evidence of periodicity' in the
Palaeozoic according to Sepkoski (1986, 57). The
effects of these criteria on the distribution of
ostracods have mostly been neglected or have
only been cited according to the *Treatise* (Moore,
1961).

Mass extinctions are clearly visible in the diversity curve of marine families (vertebrates and
invertebrates) numbered 1 to 5 by Raup & Sepkoski
(1982, 1502). Sheehan (1985, 47) correlated these
events with the presence or absence of bioherms.

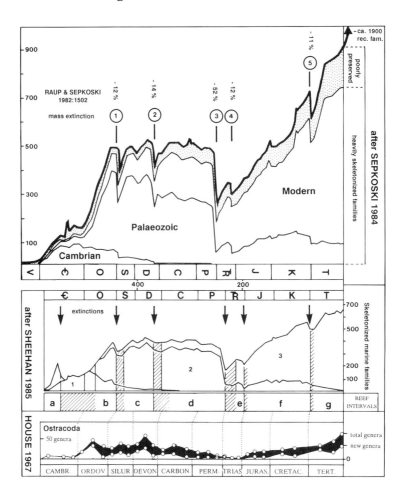

Fig. 1. Diversity of marine animal families compared with reef intervals (a - g = times with distinct reefs) and of ostracod genera.

After each mass extinction a period lacking complex reefs with framework organisms occurred worldwide (see Fig. 1). During intervals with complex reefs (stable community structure in level-bottom communities *sensu* Sheehan (1985, 48), characterized by high diversity) favourable biotopes for benthonic ostracods should have developed, whereas the lack of biohermal reefs may be taken as the cause for the reduced diversity of the ostracods.

It is difficult to find diversity curves for ostracods in the literature. House (1967, 44) compiled the *Treatise* data including ostracod genera. His graph is compared with the diversity of families in Fig. 1 even though the number of recognized

genera (e.g., in the Ordovician) has increased tremendously since the publication of the *Treatise*. In the Fig. 1 of House (1967, 44) peaks are visible within mid-Ordovician, mid-Silurian and Middle Devonian times. The last one coincides with widespread Middle Devonian reef growth; the preceding low diversity at the end of the Ordovician coincides with the glaciation; and the late Devonian low diversity can be related to the reefless interval following the extinction of biohermal reefs in the late Frasnian.

The latter time interval was studied in detail by Copper (1986), who also included ostracod data. His Fig. 3 shows the extinction of four ostracod families at the end of the Givetian (late

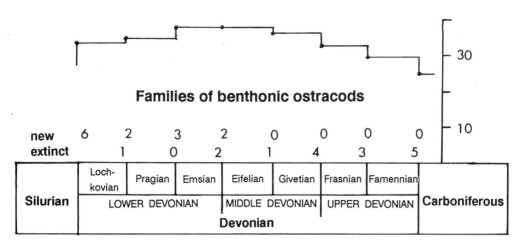

Fig. 2. Distribution of families of Devonian benthonic ostracods (after Gooday & Becker, 1979) excluding Entomozoacea.

Middle Devonian) and three families at the end of the Frasnian (early Upper Devonian). Copper took the data from Gooday & Becker (1979), which are summarized in Fig. 2. The peak in the occurrence of benthonic ostracod families lies within the uppermost Lower Devonian and early Middle Devonian. No new families can be seen in the higher Middle and Upper Devonian, but the extinction rate is clearly visible, perhaps connected with the decrease in fully marine, shallow water carbonate shelf areas. The two entomozoacean families are not included in Fig. 2. Both families are known from the Silurian to the lowermost Carboniferous.

## ENTOMOZOACEAN OSTRACODS

The Entomozoacea, the relatively thin-shelled (poorly calcified) so-called fingerprint ostracods, are well known from the Silurian to the base of the Upper Carboniferous. Their value for stratigraphy and correlation between Europe, North Africa, USSR and China is undisputed (summarized in Gooday, 1983 and Groos-Uffenorde & Wang, 1989; for North Africa see Casier, 1986). The term Entomozoacea is still used although the type species of the type genus is not a fingerprint ostracod, but belongs to the Bolbozoidae within the Myodocopina, the fingerprint ostracods are mostly included in the Halocypridida (?Cladocopina).

They are found in pelagic limestones as well as in shales. We do not agree with Lethiers et al. (1986, 152) that the 'distribution of the presumably pelagic Entomozoidae is not controlled by ecology but rather by selective fossilization in restricted, argillaceous sediments'. Quite clearly, entomozoacean ostracods can only very rarely be found in conodont sample residues of limestones, but they are well preserved in e.g., the black Kellwasser limestones and in weathered light grey limestones as internal and external moulds. They can even be seen in thin sections of limestones, figured e.g., by Buggisch et al. (1986, 26). Nevertheless, selective fossilization (predominantly by diagenesis) is possible. In shale sequences entomozoacean ostracods may be well preserved and the ornamentation including the ribs (in some genera extremely wide extending out of the valve) can easily be demonstrated by latex casts. Entomozoacean ostracods can be found together with dacryoconarids, conodonts, goniatites and thin-shelled small brachiopods or bivalves. Nevertheless, the faunas do not show high diversity. The Devonian 'Cypridinen-Schiefer' of Germany may contain, besides abundant entomozoaceans, occasionally benthonic ostracods, a few benthonic trilobites and rarely different trace fossils. Since Rabien (1956) they have been interpreted as typical sediments of the 'Stillwasser' facies deposited in a relatively deep marine basin, the bottom    water of which may have been poorly oxygenated and dark. Because of their worldwide distribution and rapid evolution, the abundant entomozoacean ostracods of the 'Cypridinen-Schiefer' had been interpreted as pelagic or (nekto)planktonic. However, according to Gooday (1983, 760), there is no convincing morphological basis for rejecting the hypothesis that at least some entomozoaceans were benthonic. On the other hand Casier (1987, 200) believes in the nektobenthonic mode of life of the Entomozoacea, even in shallow water. If these fingerprint ostracods had a benthonic mode of life, they should have been affected by global events caused e.g., by sea level changes or 'anoxic' events. However, they do not show remarkable crises or extinctions comparable to those of other benthonic faunal groups. We agree with Bless (1983, 34) that the Entomozoacea 'were pelagic inhabitants of the marine basin and only occasionally they have been swept into the shelf area'. Alternatively, the basinal environments may have shifted into the shelf (e.g., inter-reef basins).

Entomozoacean ostracods may also be abundant in red shales (e.g., in the Famennian *intercostata* Zone and *hemisphaerica-dichotoma* Zone). These Famennian redbeds are restricted to basinal sites of the Variscan geosyncline of Europe. According to Franke & Paul (1980, 249-250) oligotrophic conditions and oxygen supply are main factors for these pelagic-bathyal sediments. Casier neither discussed the occurrences of entomozoacean ostracods in these beds nor in (cephalopod)limestones. Therefore his statement (1987, 28) that the entomozoaceans could have been nektobenthonic in an environment 'very poor in oxygen' cannot be supported by the present authors. At present the position of the anoxic layer (within the sediment or water column) cannot be proved. The lack of benthonic shelly faunas as well as lack of bioturbation and the presence of undisturbed laminated sediments are thought to be indicators of anaerobic biofacies (evidenced in most parts of the Kellwasser Horizon). Dominance of soft-bodied infauna within a low diversity community, characteristic of dysaerobic biofacies, has not been found.

## GLOBAL EVENTS IN THE DEVONIAN

### Introduction

Because entomozoacean ostracods were widespread in the Devonian, their relationship to global events in the Devonian is discussed. Walliser (1985, 403; right-hand column of Fig. 4) characterized 14 global events within the Devonian, some of which are black shale events and/or extinction or radiation events. The extinctions and radiations are best seen in the evolution of goniatites (see House 1985) and dacryoconarids, and their names have consequently been applied to many of the events.

According to Johnson *et al.* (1985, 567) the Devonian of 'Euramerica contains at least 14 transgressive-regressive (T-R) cycles of eustatic origin'. The post-Lochkovian commonly appears to result from abrupt deepening events followed by prolonged upward shallowing.

The greatest transgression in the Devonian (T-R Cycle II d *sensu* Johnson *et al.* 1985, 578) took place in the late Frasnian and comprises a pair of widely recognized transgressions. The lower one in the Lower *gigas* Conodont Zone and the other at the base of the Lower *Palmatolepis triangularis* Conodont Zone (above the uppermost *gigas* or *Palmatolepis linguiformis* Zone) now being proposed as the Frasnian-Famennian boundary. This Cycle II d includes the time of the two Kellwasser horizons as well as the death of the Devonian reefs.

### The Kellwasser Event

Two horizons of black limestones intercalated in dark calcareous shales have been called 'Kellwasser-Kalk' (Lower and Upper KW horizons in this text) after Roemer (1850). They are best seen in between grey cephalopod limestones settled on submarine rises representing 'normal' pelagic facies. The two KW horizons can easily be distinguished by their conodonts and entomozoacean ostracods.

Many authors do not distinguish between the Kellwasser Event, the two different KW horizons, the Frasnian-Famennian (= F-F) boundary, the German Adorf-Nehden boundary (within the *Palmatolepis triangularis* Zone) respectively the *Cheiloceras* Event. Sometimes, all these terms representing discernible stratigraphical levels are taken as being one single mass extinction; one event in the late Devonian.

The mass extinction 2 in Fig. 1 is not understood as a short global event. According to Raup & Sepkoski (1982, 1502), this late Devonian extinction may have taken place over millions of years.

The late Frasnian Kellwasser Event (Upper Kellwasser Horizon) is considered to be an example of a short-term global event. Calculating the duration of this event is difficult. McLaren (1982, 477) supposes the duration of the F-F extinction 'to be less than a single subzone - 0.5 to 1.0 MY or less' and (1986) 'at one bedding plane'. According to Sandberg *et al.* (1988, 297) the actual extinction (corresponding to level $D_1$ in Fig. 3) 'occurred in far less than 20,000 years and more likely within a few years or days'. Walliser *et al.* (1988, 191) claim for the 'anoxic' event of the Upper Kellwasser Horizon a duration of at least several tens of thousands of years.

**The Kellwasser Horizons**

The KW Event is often equated with the Frasnian-Famennian (F-F) mass extinction event. Below we consider the KW Horizons in greater detail (see also Fig. 3).

The two KW Horizons consist of black or dark grey sediments within light grey micritic cephalopod limestones. They may vary in thickness from some tens of centimetres to a single layer of approximately 10cm. In most cases both horizons yield bituminous, black, micritic, often thinly laminated limestone beds that are intercalated in calcareous dark shales. Microfossils such as conodonts, entomozoacean ostracods and homoctenids (Tentaculoidea) may be abundant and well preserved.

The Lower KW Horizon is situated at the top of the Lower *Palmatolepis gigas* Conodont Zone, respectively upper doI beta/gamma above the *cicatricosa* Entomozoacean zone. It is characterized by the entomozoacean ostracods *Richterina (V.) zimmermanni* (Volk, 1939), *Entomoprimitia nitida* (Roemer, 1850), *Nehdentomis pseudophthalma* (Volk, 1939) and homoctenids. *R. zimmermanni* and *N. pseudophthalma* may occur in high numbers (up to 16 specimens per $cm^2$ of one bedding plane), but mostly single valves. The subdivision with species of *Rabienella* Gründel, 1962 (Entomozoidae) in the Upper Frasnian *sartenaeri* Zone is known only from shale sequences. No detailed work has been done on cephalopod limestones because of the difficulty in finding or preparing the entomozoacean ostracods.

The Upper KW Horizon terminates the Uppermost *Palmatolepis gigas* Conodont Zone (recently named *Palmatolepis linguiformis* Zone by Sandberg *et al.*, 1988), and *splendens* Entomozoacean Zone (uppermost doI near the recently proposed Frasnian-Famennian boundary). The most abundant entomozoacean species is *Entomoprimitia kayseri* (Waldschmidt, 1885). In bed 81 of the Aeketal section (Fig. 3) we found up to 40 specimens per $cm^2$, but mostly carapaces in different orientations including vertical. The Upper KW Horizon shows an extremely detailed succession. This can be observed in many German sections of pelagic cephalopod limestone facies, even with varying thickness and lithology. The best section is the Steinbruch Schmidt (situated 750m northeast of the small village of Braunau near the town of Bad Wildungen in the Ense area of the Kellerwald, eastern Rheinisches Schiefergebirge, West Germany), recently proposed as stratotype for the F-F boundary. At this locality the horizon spans 40cm and, as is to be seen in Fig. 3, single beds can be traced to the much thicker Aeketal section of the Harz Mountains (West Oker reservoir close to the type locality of the 'Kellwasser-Kalk') over a distance of about 100km. The lines A - E indicate correlations based on litho- and biofacies, biostratigraphy and event stratigraphy. The detailed description given below focuses solely on the Upper KW Horizon.

**The causes**

Many different causes for the KW Event have been

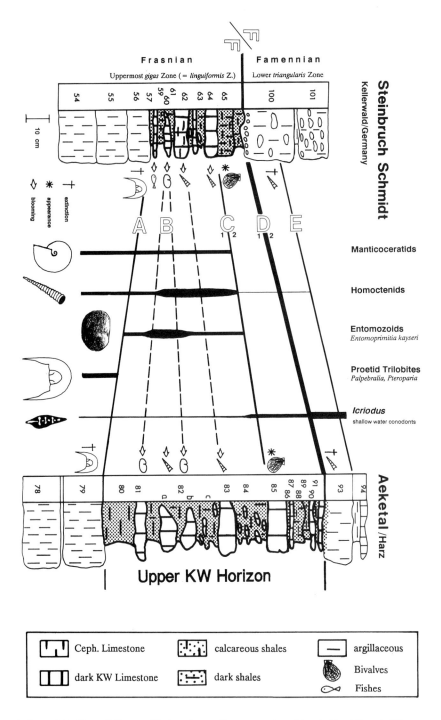

Fig. 3. The main fauna of the Upper Kellwasser Horizon in two German sections. Line A, C and D mark different steps of the KW Event. The dashed lines show the correlation of beds with comparable abundant fauna. At line E the last homoctenid species disappears. For further explanation see text. The samples taken by Schindler are shown with numbers at the left-hand side of the sections.

discussed in the literature:

**1) Bolides or oceanic impact**. Astronomical event *sensu* McLaren (1982, 482), but no geochemical evidence for a large-body impact was found by McGhee *et al.* (1986).

**2) Rapid sea-level fluctuations**. Deepening events and drowning of the reefs e.g., Johnson *et al.* (1986, 141); acme of global Devonian marine transgression *sensu* House (1985, 20); but negation of a major sea-level change in Walliser (1984, 19) and Walliser *et al.* (1988, 192).

**3) Climatic oscillation**. E.g., cooling due to global tectonics in Copper (1986, 838); climatic warming linked with El Niño Southern Oscillation type processes in Thompson & Newton (1987, 225).

**4) Water chemistry changes and stratification.**
a) Stagnation and sapropelitic sedimentation in Rabien (1956, 57) and Buggisch (1972; 41,47).

b) Anoxic event or black shale event e.g., in Walliser (1985, 405), Engel *et al.* (1983, 24), Walliser *et al.* (1988; 191, 192).

c) Oceanic overturn (by upwelled anoxic waters) in Wilde & Berrry (1984, 159), possibly climatically induced (Wilde & Berry 1988, 86); oceanic overturn due to the closure of the 'Frasnian Ocean' (Copper 1986, 838); ocean turnover near the end of the KW Event (Walliser *et al.* 1988, 192).

d) Denitrification of water masses (Berry *et al.* 1987, 32).

**The effects and possible explanations**

Considering the effects within the KW Horizons, we are well aware of the selective preservation of different faunal groups. The fauna is mostly found in the calcareous part due to early diagenetic processes (cementation) in combination with a low sedimentation rate, whereas solution has to be considered in the less calcareous parts. Another difficulty is the correlation of the last occurrences of *Entomoprimitia splendens* (Waldschmidt, 1885) in the Upper KW Horizon with those in shale sections of different areas.

The two Horizons represent 'anoxic' sediments with pelagic faunal elements. The conditions producing the 'anoxic' sediments perturb the whole ecosystem of 'normal' cephalopod limestones. This is indicated by the lack of benthonic organisms and of bioturbation and the presence of sedimentological features such as undisturbed fine lamination.

The succession within the Upper KW Horizon in the two figured sections (Fig. 3) is as follows:

1) At the base of the Upper KW Horizon (line A of Fig. 3) there is a change from well oxygenated conditions, reflected in light grey cephalopod limestones, to 'anoxic' conditions, represented by the dark KW sediments. Benthonic organisms, such as proetid trilobites disappear.

2) Between line A and C, only pelagic faunas (goniatites, entomozoacean ostracods, homoctenids and partly arthrodires?) flourish. No benthonic or endobenthonic life has been recognized.

3) Line B (layer 60 in the Steinbruch Schmidt section and layers 81 + 82 b in the Aeketal section) show excellent preservation and 'blooming' of the two charcteristic entomozoans of the Upper KW Horizon (*Entomoprimitia kayseri* and *E. splendens*). The term 'blooming' is used here in the sense of unusually high numbers of individuals (sometimes of only one species). In addition to the outstanding preservation of valves and carapaces, it is remarkable that both juveniles and adults are present.

4) Line C marks the disappearance of manticoceratid goniatites (Gephuroceratidae), *Entomoprimitia kayseri* and *E. splendens* as well as the strong decrease in homoctenids.

In layer 64 of the Steinbruch Schmidt sequence ($C_1$) there is the first increase in icriodontid conodonts. At level $C_2$ large pelecypods have been found.

5) Line D is drawn at the base of a pronounced lithological change (return to 'normal' cephalopod limestone conditions) together with a strong increase in icriodontids.

At level $D_2$ a brecciated lower part of the first layer above the Upper KW Horizon is developed in many sections.

6) Line E marks the extinction of the last homoctenids that struggled through to the top of the KW Horizon.

The above mentioned succession in the Upper

KW Horizon (= line A - D of Fig. 3) implies that the KW Event was not an instantaneous one, but clearly showed a stepped character. These steps demonstrate that the KW Event, although representing a geologically short time span, was not due to one single catastrophe, e.g., caused by a large-body ocean impact. Each step indicates that within this already perturbed ecosystem each further step has a cumulative effect on the situation as a whole. In areas of black shale deposition, the benthonic fauna was killed (Line A of Fig. 3); only pelagic forms flourish and even bloom.

Within the lower part of the stratigraphically highest limestone layer of Steinbruch Schmidt (bed 64) a change in the conodont biofacies occurs, recognizable by a shift from palma-tolepid-polygnathid to polygnathid-icriodid assemblages (for detailed data see Sandberg *et al.*, 1988). Icriodontid conodonts are often taken as shallow water indicators. Therefore, these authors refer this increase to a first regressive phase leading towards mass extinction. However, at least in layer 64, pelagic homoctenids are still abundant. At line C, the pelagic forms disappear and large pelecypods begin to come in (bed 65). Sandberg *et al.* (1988) identify another drastic shallowing at this position, whereas Walliser *et al.*, (1988, 192) favour a slight shift towards more oxygenated conditions, but still within the dysaerobic zone. The presence of small, thin-shelled pelecypods of the *Buchiola*-group, abundant in the KW horizons, ranging through, and unaffected by, the extinction event indicates a mode of life in a higher part of the water column than the other pelagic elements. According to Schmidt (1935, 83) they were probably attached to drifting plants.

The return to 'normal' oxygen conditions above the F-F-boundary (= line D) is accompanied by another strong increase in numbers of icriodontid conodonts. Sandberg *et al.* (1988) interpret the occurrence of a brecciated base of this layer in many sections as the result of tsunamis or heavy storms. We do not support this idea because the sedimentological features of this breccia suggest an *in situ* formation (crackled breccia) within a continuous section.

The extinction of the homoctenids one layer above the top of the Upper KW Horizon (line E) represents an additional step of the event during which the last survivors disappear. In a less calcareous section (Schaumberg, north of Östrich near Hohenlimburg, Rheinisches Schiefergebirge, West Germany) we found the last *Homoctenus* above the *splendens* Zone together with *Ungerella sigmoidale* (Müller-Steffen, 1964). Rabien (1970, 153) dated the last *Homoctenus* in the Dill Syncline as lower *sigmoidale* Zone ('Nehden-Basis-Horizont', 'Adorf/Nehden Grenzschichten').

We consider the causes of this tremendous bio-event to be a combination and accumulation of effects. An already perturbed ecosystem suffers a stepped increase in changing conditions. This might be controlled by water-depth or oxygen content. In addition, it seems most probable that an oceanic overturn near the end of the Upper KW Horizon (line C) may contribute to the extinction (or drastic reduction, e.g., homoctenids) of planktonic faunal groups.

The effect of the KW Event on entomozoacean ostracods seems to be very distinctive in Fig. 3. This is partly due to the good preservation in the dark limestones and the bad preservation and very difficult preparation in the grey cephalopod limestones. The lack of good conodont faunas within the rare, stratigraphically continuous and entomozoacean-bearing shale sequences (e.g., Schaumberg section), means that the correlation of the conodont zones with the entomozoacean zonation is very difficult above the Upper KW Horizon at the F-F boundary. Recently the Lower *triangularis* Conodont Zone has been reported in several limestone sequences. The correlation, however, into entomozoan-bearing shale sequences, without limestones, is not established. The position of the dashed line between the *splendens* and the *sigmoidale* Zone in Fig. 4 does not correspond to published correlations. However, we found a limestone layer with conodonts of the uppermost *gigas* Zone (= *linguiformis* Zone of Sandberg *et al.*, 1988) in between shales of the *splendens* Zone and shales with homoctenids and *Ungerella sigmoidale* (Schaumberg section). Therefore, the top of the *splendens* Zone may coincide with the top of the Upper Kellwasser Horizon at the recently proposed

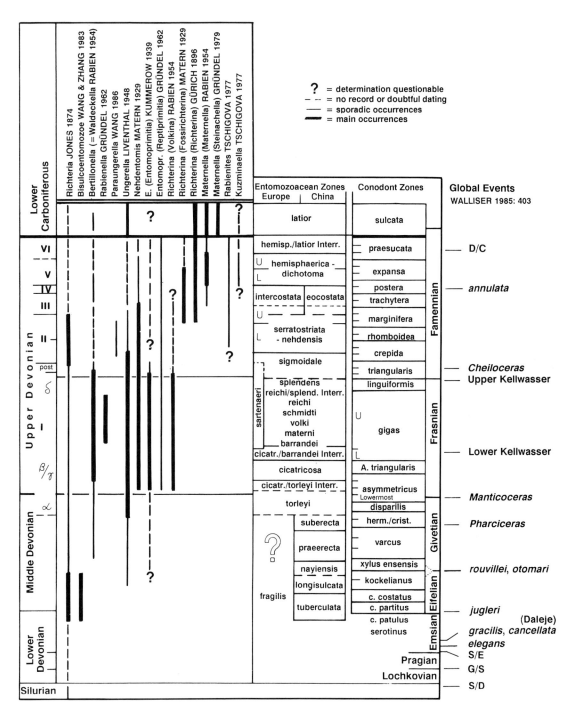

Fig. 4. Occurrences of entomozoacean genera in the Devonian and their relation to global events and worldwide accepted conodont zonation. The correlation of the top of the *serratostriata-nehdensis* Zone with the conodont zonation is questionable. The short dashes mark the previous correlation, the longer dashes show that recently proposed by Buggisch *et al.* (1986, 28).

F-F boundary at the base of the Lower *triangularis* Zone. Only if this correlation is accepted, the entomozoacean ostracods would have been significantly affected by the last step of the KW Event: the possible ocean overturn.

## DEVONIAN EVENTS AND ENTOMOZOACEAN OSTRACODS

The current state of knowledge of a zonation with entomozoacean ostracods is shown in Fig. 4. The correlation after Rabien (1970) and Groos-Uffenorde & Wang (1989) is only slightly modified. There is a gap in the record between the Ludlovian stage (Silurian) and Emsian stage (Lower Devonian) in Europe. No zonation has been proposed for the Middle Devonian of Europe because no complete entomozoacean-bearing sequences have been published, but several isolated occurrences are known. However, the Middle Devonian zonation in South China is more complete than it is in Europe.

The global events in the Devonian are drawn on the right-hand side of Fig. 4. Using published correlations, no clear relationship can be seen between those events and the range of entomozoacean ostracod genera. The distribution of genera and blooming of species can be summarized as follows:

1) The first blooming (abundance of specimens not species) of entomozoacean ostracods occurs after the transgression of the Daleje Event in the uppermost Lower Devonian. *Bisulcoentomozoe* and *Richteria* are very closely related and simultaneously abundant.

2) The next blooming (abundance of several *Ungerella* species) took place after the *Pharciceras* Event. This Taghanic Event (extinction event at the top of the *Maenioceras* Stufe) *sensu* House (1985, 19) was 'followed by a rise in biomass of planktic Ostracoda'.

3) The most important radiation of genera is to be seen one conodont zone after the *Manticoceras* Event.

4) The evolution of the genus and blooming of the index species of *Rabienella* took place between the two Kellwasser horizons within the Upper

*gigas* Zone.

5) The late Frasnian Kellwasser Event is shown in detail in the previous section of this paper.

6) The regression in the late Famennian caused another remarkable extinction event that affected benthonic organisms such as phacopid trilobites as well as the clymenids, which had been controlled by the conditions on the sea floor (Walliser 1984, 18). The species diversity of entomozoacean ostracods in the late Famennian only shows a slight decrease after the *hemisphaerica-dichotoma* Zone.

7) At the Devonian/Carboniferous boundary there is no great change in the entomozoacean fauna. Typical Upper Devonian species do not occur in the Lower Carboniferous but the new species in the Lower Carboniferous are very closely related to Upper Devonian species.

8) The greatest species diversification took place gradually in the cd I (*Gattendorfia* Stufe, Lower Tournaisian) after the latest Famennian mass extinction and just before the transgression e.g., of the 'Liegende Alaunschiefer' of the lowermost *Pericyclus* Stufe (cd II alpha, Upper Tournaisian). The occurrence of entomozoacean ostracods in the Upper Tournaisian and Visean is restricted to isolated localities and the last record is known from the early Namurian.

The number of entomozoacean species in one sample is generally 10 to 15 species in a rich Devonian assemblage, but does not exceed 24 species during the greatest species diversification in the lowermost Carboniferous.

## CONCLUSION

Global biological events in the Devonian are mainly related to transgressions or varying positions of the anoxic zone evidenced by black shales. The effects of global events on entomozoacean ostracods do not result in major crises (extinctions). The global events in the Devonian cannot be demonstrated by simultaneously occurring extinctions or radiations of entomozoacean genera, only by the blooming of some species after an event. In contrast, many marine benthonic animals have been affected by global events and this supports our rejection of the belief that entomozoacean ostracods had a benthonic mode of life.

## ACKNOWLEDGEMENTS

We thank Prof. Dr O. H. Walliser (Göttingen) for stimulating discussions and encouragement and Mrs G. Meyer and Mrs C. Kaubisch (Göttingen) for technical help. The linguistic corrections of the editors and helpful suggestions of unknown referees are greatfully acknowledged.

## REFERENCES

Becker, G. & Bless, M. J. M. 1987. Cypridinellidae (Ostracoda) aus dem Oberdevon Hessens (Unterer Kellwasser-Kalk; Lahn-Dill-Gebiet und östliches Sauerland, Rechtsrheinisches Schiefergebirge). *Geol. Jb.*, Hessen, **115**, 29-56.

Berry, W. B. N., Wilde, P. & Hunt, M. S. Q. 1987. The role of denitrification zone oceanic waters in bio-events. *2nd Internat. Symp. Devonian Syst., Abstr.*, 32.

Bless, M. J. M. 1983. Late Devonian and Carboniferous ostracode assemblages and their relationship to the depositional environment. *Bull. Soc. belge Géol.*, **92**, 31-53.

Buggisch, W. 1972. Zur Geologie und Geochemie der Kellwasserkalke und ihrer begleitenden Sedimente (Unteres Oberdevon). *Abh. hess. Landesamt. Bodenforsch.*, Wiesbaden, **62**, 1-67.

Buggisch, W., Rabien, A. & Hühner, G. 1986. Stratigraphie und Fazies von Oberdevon/Unterkarbon-Profilen im Steinbruch 'Beuerbach' bei Oberscheld. (Conodonten- und Ostracoden-Biostratigraphie, Dillmulde, Rheinisches Schiefergebirge, Blatt 5216 Oberscheld). *Geol. Jb.*, Hessen, **114**, 5-60.

Casier, J.-G. 1986. Présence d'Entomozoacea (Ostracodes) dans la partie Famennienne de la coupe du Djebel Heche (Gourara, Sahara Algerien). *Géobios*, Lyon, **19**, 261-264.

Casier, J.-G. 1987. Étude biostratigraphique et paléoécologique des ostracodes du sommet du Givétien et de la base du Frasnien à Ave-et-Auffe (Bord sud du Bassin de Dinant, Belgique). *Bull. Soc. belg. Géol.*, **96**, 23-34.

Casier, J.-G. 1987. Étude biostratigraphique des ostracodes du récif de Marbre Rouge du Hautmont à Vodelée (Partie supérieure du Frasnien, Bassin de Dinant, Belgique). *Revue Paléobiol.*, **6**, 193-204.

Copper, P. 1986. Frasnian/Famennian mass extinction and cold-water oceans. *Geology*, **14**, 835-839.

Engel, W., Franke, W. & Langenstrassen, F. 1983. Palaeozoic sedimentation in the Northern branch of the Mid-European Variscides. Essay of an interpretation. *In* Martin, H. & Eder, F. W. (Eds), *Intracontinental fold belts, case studies in the Variscan Belt of Europe and the Damara Belt in Namibia*, 9-41. Springer-Verlag.

Franke, W. &. Paul, J. 1980. Pelagic redbeds in the Devonian of Germany - deposition and diagenesis. *Sedim. Geol.*, **25**, 231-256.

Gooday, A. J. 1983. Entomozoacean ostracods from the Lower Carboniferous of South-Western England. *Palaeontology*, London, **26**, 755-788.

Gooday, A. J. & Becker, G. 1979. Ostracodes in Devonian biostratigraphy. *Spec. Pap. Paleont.*, London, **23**, 193-197.

Groos-Uffenorde, H. & Wang, S.-Q. 1989. In press. The entomozoacean succession of South China and Germany (Ostracoda, Devonian). *Cour. Forsch.-Inst. Senckenberg*, Frankfurt a. M.

House, M. R. 1967. Fluctuations in the evolution of Palaeozoic invertebrates. *In* Harland, W. B. *et al.* (Eds), The fossil record. *Geol. Soc. London Spec. Publ.*, **2**, 41-54.

House, M. R. 1985. Correlation of mid-Palaeozoic ammonoid evolutionary events with global sedimentary perturbations. *Nature*, London, **313**, 17-22.

Johnson, J. G., Klapper, G. & Sandberg, C. A. 1985. Devonian eustatic fluctuations in Euramerica. *Geol. Soc. Amer. Bull.*, **96**, 567-587.

Johnson, J. G., Klapper, G. & Sandberg, C. A. 1986. Late Devonian eustatic cycles around margin of Old Red Continent. *Annls. Soc. géol. Belg.*, Liège, **109**, 141-147.

Lethiers, F., Braun, W. K., Crasquin, S. & Mansy, J.-L. 1986. The Strunian Event in Western Canada with reference to ostracode assemblages. *Annls. Soc. géol. Belg.*, Liège, **109**, 149-157.

McGhee, G. R., Jr., Orth, C. J., Quintana, L. R., Gilmore, J. S. & Olsen, E. J. 1986. Geochemical analyses of the Late Devonian "Kellwasser Event" stratigraphic horizon at Steinbruch Schmidt (F.R.G.). *Lecture Notes in Earth Sci.*, **8**, 219-224.

McLaren, D. J. 1982. Frasnian-Famennian extinctions. *Geol. Soc. Amer. Spec. Pap.*, **190**, 477-484.

McLaren, D. J. 1986. Abrupt extinctions. *In* Elliott, D. K. (Ed.), *Dynamics of extinction*, 37-46. John Wiley & Sons.

Moore, R. C. (Ed.) 1961. *Treatise on Invertebrate Paleontology, Pt Q, Arthropoda*, 3, xxiii + 442 pp., 334 figs. Univ. Kansas Press.

Rabien, A. 1956. Zur Stratigraphie und Fazies des Oberdevons in der Waldecker Hauptmulde. *Abh. hess. Landesamt. Bodenforsch.*, Wiesbaden, **16**, 1-83.

Rabien, A. 1970. Oberdevon. *In* Lippert, H. J. (Ed.), *Erl. geol. Kte Hessen, 1:25000, Bl. 5215 Dillenburg*, 78-83, 103-235.

Raup, D. M. & Sepkoski, J. J., Jr. 1982. Mass extinctions in the marine fossil record. *Science*, New York, **215**, 1501-1503.

Roemer, F. A. 1850. Beiträge zur geologischen Kenntnis des nordwestlichen Harzgebirges. 1. Abt. *Palaeontographica*, **3**, 1-67.

Sandberg, C. A., Ziegler, W., Dreesen, R. & Butler, J. L. 1988. Late Frasnian mass extinction: conodont event stratigraphy, global changes, and possible causes. *Cour. Forsch.-Inst. Senckenberg*, Frankfurt a. M., **102**, 263-307.

Schmidt, H. 1935. Die bionomische Einteilung der fossilen Meeresböden. *Fortschr. Geol. Paläont.*, **12**, 1-154.

Sepkoski, J. J., Jr. 1984. A kinetic model of Phanerozoic taxonomic diversity. III. Post-Paleozoic families and mass extinctions. *Paleobiol.*, **10**, 246-267.

Sepkoski, J. J., Jr. 1986. Global bioevents and the question of periodicity. *Lecture Notes in Earth History*, **8**, 47-61.

Sheehan, P. M. 1985. Reefs are not so different. They follow the evolutionary pattern of level-bottom communities. *Geology*, **13**, 46-49.

Thompson, J. B. & Newton, C. R. 1987. Episodic climatic warming and the late Devonian mass extinction. *2nd Internat. Symp. Devonian Syst., Abstr.*, 225.

Walliser, O. H. 1984. Geologic processes and global events. *Terra cognita*, **4**, 17-20.

Walliser, O. H. 1985. Natural boundaries and commission boundaries in the Devonian. *Cour. Forsch.-Inst. Senckenberg*, Frankfurt a. M., **75**, 401-408.

Walliser, O. H., Lottmann, J. & Schindler, E. 1988. Global events in the Devonian of the Kellerwald and Harz Mountains. *Cour. Forsch.-Inst. Senckenberg*, Frankfurt a. M., **102**, 190-193.

Wilde, P. & Berry, W. B. N. 1984. Destabilization of the oceanic density structure and its significance to marine "extinction" events. *Palaeogeogr. Palaeoclimat. Palaeoecol.*, Amsterdam, **48**, 143-162.

Wilde, P. & Berry, W. B. N. 1988. Comment on "Sulfur-isotope anomaly associated with the Frasnian-Famennian extinction, Medicine Lake, Alberta, Canada". *Geology*, **16**, 86.

## DISCUSSION

Robin Whatley: You clearly indicate that the Kellwasser 'event' was stepped. Would you not agree that this is likely to be nearer to the true representation of such events than a uniform extinction at one horizon? The latter one produced either by considering extinction at too high a taxonomic (hierarchical) level or too wide a stratigraphical interval.

Helga Groos-Uffenorde: The stepped character of an event cannot be assumed without very detailed biostratigraphical subdivisions and correlations. Only worldwide studies on Upper Devonian conodonts and recent research on the F-F boundary combined with very detailed studies on favourable sections enabled us to show, that there is no great instantaneously abrupt change. I can imagine that after comparable work, the other bio-events will be considered to be less abrupt. However, this also depends on the importance and nature of the causes for the events.

Roger Kaesler: How are you using the term 'event'? I am especially curious about the Carboniferous-Permian event you have shown.

Helga Groos-Uffenorde: Global events have been subdivided into bio- and litho-events. In general they are a combination of both. Global bio-events are extraordinary, short-term, geologically instantaneous and worldwide occurring abrupt changes in the biosphere. On a small scale, it may be difficult to differentiate global bio-events from long-term mass extinctions documented by the decrease in the diversity curve of marine animals. Bio-events may be extinction events and/or innovation and radiation events. The effects on different taxonomic groups may be the same or different at one time. I do not include short-term sedimentary cycles (Milankovic cycles) or local sedimentary changes (e.g., storm deposits). Litho-events must be recognizable worldwide or at least widely distributed (e.g., time specific facies Walliser, 1984). My graph only showed the mass extinctions of Sepkoski (1984) near the Permo-Triassic Boundary and near the Devonian-Carboniferous Boundary. According to House (1967) the ostracod diversity decreases during the Carboniferous-Permian (see Fig. 1).

Gerhard Becker: In the Lower Kellwasser Limestone (Frasnian) of the Rheinische Schiefergebirge, there is a surprising appearance of new cypridenellid ostracods, later only known from the Carboniferous. How does this fit into the theory of an event?

Helga Groos-Uffenorde: Your recent report on the first occurrence of Cypridinellidae in the Lower Kellwasser Limestone of the Dill Syncline can be confirmed by a few specimens of the same age in our collections from the type locality of the Kellwasser Limestone in the Harz Mountains. However, we do not have any record either from the Upper Kellwasser Horizon or from Famennian strata. At present their occurrence is limited to one pelagic black limestone layer in the late Middle Frasnian. We do not have further results to change the interpretation of Becker & Bless (1987).

# 8

# Major changes in Eocene and Oligocene Gulf Coast ostracod assemblages: relationship to global events

**Joseph E. Hazel**

Department of Geology and Geophysics,
Louisiana State University, U.S.A.

## ABSTRACT

In the middle Tertiary of the southeastern United States there are three stratigraphical intervals in which major changes take place in the taxonomic composition of ostracod assemblages. These, as should be expected, are associated with provincial stage boundaries. These extinction and origination events are compared to known global events.

Many species characteristic of the Claibornian Stage (Lower and Middle Eocene) disappear from the rock record, and many new species typical of the Jacksonian Stage (Middle Eocene to lowest Oligocene) appear in a time interval between about 41.50 Ma and 40.68 Ma. This faunal turnover is associated with a eustatic lowering of sea level and a subsequent transgression during which the Jacksonian fauna developed. Once the Jacksonian fauna was established at about 40.71 Ma, there was little evolutionary change until late in the Jacksonian. During this period of evolutionary stasis there were several major impacts on the earth by extraterrestrial bodies; the ostracods, however, seem not to have been affected. Characteristic Jacksonian forms began to die out shortly after the deposition of the Pachuta Marl and its equivalents. Several disappear at about 37.31 Ma and after this there was a steady depletion of the Jacksonian stock with very few additions until just before the end of the Eocene. Then, beginning at about 36.71 Ma, there was a rapid faunal turnover in a period of about 0.35 MY during which the Vicksburgian (Lower Oligocene) fauna was established. This occurred during a late Eocene-early Oligocene eustatic sea level high stand, which was also a time of global climatic deterioration. The end of the Vicksburgian is coincident with a major eustatic drop in sea level that caused the 30 Ma unconformity of Vail. Only about twelve of the ostracod species that lived in late Vicksburgian time are found in the richly fossiliferous beds of the Chickasawhayan Stage (Upper Oligocene) of the next transgressive cycle.

## INTRODUCTION

The Middle Eocene to Oligocene provincial stages of the Gulf part (Fig. 1) of the Atlantic and Gulf Geological Province of the United States (Murray, 1961) include some of the most fossiliferous deposits in the world. Ostracods that were living in sublittoral habitats are abundant and diverse in these rocks and the stages and their informal subdivisions are easily delineated by ostracod faunal changes. Published and unpublished studies on the Claibornian, Jacksonian, Vicksburgian, and Chickasawhayan stages indicate that together they contain over 500 species of ostracods.

The interval represented by the late Claibornian to Chickasawhayan (Bartonian to Chattian in European terms) witnessed several global or near global changes in the world ocean that include eustatic sea level changes (Haq *et al.*, 1987), development of the psychrosphere and cooling of surface waters as evidenced by decreasing values of $\delta^{18}O$ (Shackleton, 1986; Miller *et al.*, 1987). The late Eocene was also a time of extraterrestrial impacts on the Earth as evidenced by the presence of microtektites and related microspherules at several biochronostratigraphical levels (Keller *et al.*, 1987; D'Hondt *et al.*, 1987; Glass *et al.*, 1985; Byerly *et al.*, 1988; Hazel, 1988; in press). Such impacts have been related to comet showers, which have been suggested as causing stepped mass extinctions (Hut *et al.*, 1987). This report investigates the position in time of major ostracod faunal turnovers relative to these events to see if there are any relationships.

## OSTRACOD ZONES

Ostracod zones useful in the Middle Eocene to Upper Oligocene deposits of the Atlantic and Gulf Coastal Province are discussed below. Four new zones for the Eocene and one for the Oligocene are defined. The relationship of the zones to planktonic zonations, stages and time are shown in Fig. 2.

## *Hazelina couleycreekensis* Zone

*Definition*: Interval from the appearance of *Hazelina couleycreekensis* (Gooch, 1939) to the first appearance of *Cocoaia grigsbyi* (Howe & Chambers, 1935).

*Characterization*: Some other ostracod species that seem to make their first appearance in this zone include *Alatacythere ivani* Howe, 1951; *Actinocythereis davidwhitei* (Stadnichenko, 1927); *A. gosportensis* (Blake, 1950); *Acanthocythereis araneosa*; *Hermanites collei* (Gooch, 1939) and *Hermanites rukasi* (Gooch, 1939). Many typical Claibornian forms have their last appearance in this zone. These include *Haplocytheridea habropapillosa* (Sutton & Williams, 1939); *Phractocytheridea compressa* (Sutton & Williams, 1939) and *Acanthocythereis stenzeli* (Stephenson, 1944).

*Remarks*: Rocks of the upper part (but not uppermost part) of the Claibornian Stage can be placed in this new zone or its biochronozone. This includes the Cook Mountain Formation and the upper part of the Lisbon Formation and their equivalents. It is correlative with the upper *Cubitostrea sellaeformis* Zone of Toulmin (1977), foraminiferal zones upper P12 and P13 of Blow (1979) and nannoplankton zones: upper part of NP16 and NP17 of Martini (1971).

## *Cocoaia grigsbyi* Zone

*Definition*: Interval from the first appearance of *Cocoaia grigsbyi* (Howe & Chambers, 1935) to the first occurrence of *Actinocythereis montgomeryensis* (Howe & Chambers, 1935).

*Characterization*: This zone is characterized by the overlap in range of several typical Claibornian species, such as *Opimocythere martini* (Murray & Hussey, 1942), *Actinocythereis gosportensis*, *Ouachitaia semireticulata* (Stephenson, 1942) and *O. gosportensis* (Stephenson, 1942), with typical Jacksonian forms, such as *Cocoaia grigsbyi* and *Ouachitaia caldwellensis* (Howe & Chambers, 1935).

*Remarks*: This new zone is latest Claibornian age. The Gosport Formation and its equivalents can be placed in this zone or its biochronozone. The zone correlates with foraminiferal zone P14, the lower

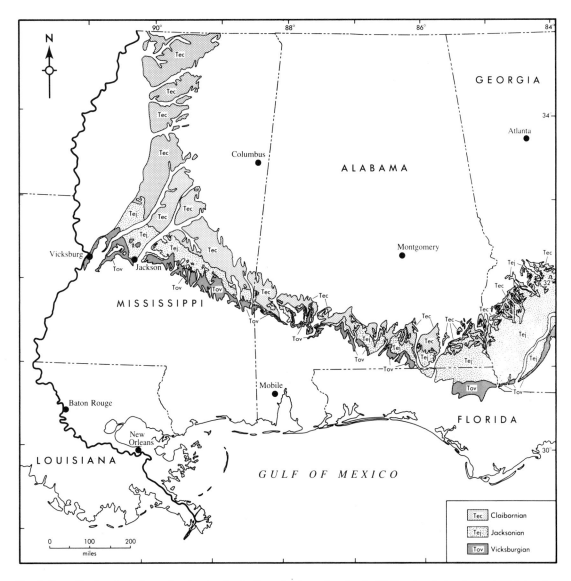

Fig. 1. Generalized geological map showing the distribution of rocks assigned to the Claibornian, Jacksonian and Vicksburgian Provincial stages (latest early Eocene to early Oligocene) in the southeastern United States. Modified from Bennison (1975).

part of nannoplankton zone NP18 and the *Vener-icardia alticostata* Zone of Toulmin (1977).

### *Actinocythereis montgomeryensis* Zone

*Definition*: Interval from the first appearance of *Actinocythereis montgomeryensis* (Howe & Chambers, 1935) to the last appearance of *Clithrocytheridea garretti* (Howe & Chambers,

1935).

*Characterization*: This concurrent range interval zone is characterized by typical Jacksonian assemblages (see Huff, 1970; Howe & Chambers, 1935), which are present in the Moodys Branch Formation and lower and middle Yazoo Formation and their equivalents. The top of the zone is characterized not only by the last appearance of *Clithrocytheridea garretti*, but also the last

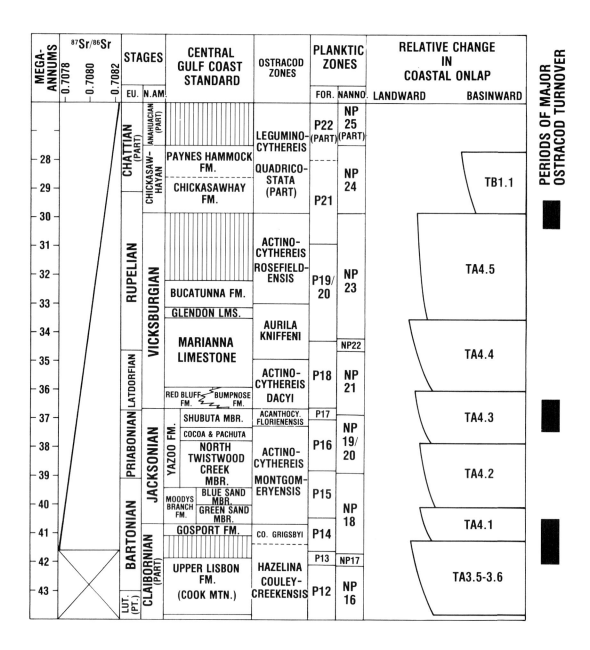

Fig. 2. Chart showing the relationship of periods of major ostracod faunal turnover to cycles of coastal onlap (Haq *et al.*, 1987; Baum & Vail, 1987; Zullo & Harris, 1987), planktonic zonations (Blow, 1979; Martini, 1971), ostracod zones (see text), lithostratigraphical units, the changing ratio of $^{87}Sr/^{86}Sr$ (Hazel, in press, and references therein) and time (Hazel, in press; Hazel *et al.*, 1984). The Gulf Coast Standard section is based on eastern Mississippi and western Alabama (see Toulmin, 1977 and Hazel *et al.*, 1980). It should be pointed out that the considerable hiatus shown separating the Bucatunna and Chicksawhay formations is based on the St. Stephens, Alabama section. To the northwest as the centre of the Mississippi Embayment is approached, there are sediments (Byram Formation) which may be correlated with this interval.

appearance of *Hermanites hysonensis* (Howe & Chambers, 1935), *Ouachitaia caldwellensis, Cyamocytheridea watervalleyensis* (Stephenson, 1937) and others (Huff, 1970; Deboo, 1965; and unpublished data). The last appearance of *Opimocythere mississippiensis* (Meyer, 1887) is also near the top of the zone. In the middle of the zone some species occur that show promise for dividing this relatively long zone. These are '*Acuticythereis*' *cocoaensis* and '*Hulingsina*' *serangodes*, which were described by Krutak (1961). They have only been found to date in the Cocoa Member of the Yazoo Formation in eastern Mississippi and western Alabama and in the shallow subsurface of South Carolina.

*Remarks*: This new zone correlates with foraminiferal zones P14 to P16 and with nannoplankton zones NP18 and NP19/20.

### *Acanthocythereis florienensis* Zone

*Definition*: Interval from the last appearance of *Clithrocytheridea garretti* (Howe & Chambers, 1935) to the first appearance of *Actinocythereis dacyi* (Howe & Law, 1936).

*Characterization*: The last appearances of *Actinocythereis purii* Huff, 1970; *A. gibsonensas*; *A. boldi* Huff, 1970; *Occultocythereis broussardi* (Howe & Chambers, 1935); *Hazelina couleycreekensis* and *Pterygocythere murrayi* Hill, 1954 occur in the lower and middle part of this zone. The last of the typical Jacksonian species, such as *Actinocythereis montgomeryenis* and *Haplocytheridea montgomeryensis*, disappear at the top of the zone. The distinctive species *Actinocythereis quadrata*, described by Howe & Howe (1973), has been found only in the upper few feet of the zone.

*Remarks*: This new partial range interval zone correlates with the upper part of foraminiferal zone P16 and lower P17 and with the upper part of nannoplankton zone NP19/20 and lower part of NP21. Rocks of the middle and upper part of the Shubata Member of the Yazoo Formation and equivalents can be placed in the zone or its chronozone. The extinction of the formaminiferal marker *Turborotalia cerroazulensis* (Cole, 1928) is within the range of *Actinocythereis quadrata*.

### *Actinocythereis dacyi* Zone

*Definition*: Interval from the first appearance of *Actinocythereis dacyi* (Howe & Law, 1936) to the first appearance of *Aurila kniffeni* (Howe & Law, 1936).

*Remarks*: This zone was proposed and discussed in Hazel *et al.* (1980). It correlates approximately with most of Martini's (1971) nannoplankton zone NP21, the foraminiferal zone P18 of Blow (1979) and the *Cassigerinella chipolensis-Pseudohastigerina micra* formaniniferal zone of Bolli & Saunders (1985). It correlates closely with the *Pecten perplanus perplanus* Zone of Glawe (1969). The extinction of the important foraminiferal genus *Hantkenina* occurs in the lowermost part of this zone. The ostracod genus *Hemicyprideis* first appears in the Gulf Basin at the base of this zone (Hazel *et al.*, 1980) and seems to correlate with its first occurrence in Western Europe (Keen, 1978). The oldest known occurrence of the thaerocytherine genus *Puriana* is also in this zone (Howe, 1977).

### *Aurila kniffeni* Zone

*Definition*: Interval from the first appearance of *Aurila kniffeni* (Howe & Law, 1936) to the first appearance of *Actinocythereis rosefieldensis* (Howe & Law, 1936).

*Remarks*: This zone was proposed and discussed in Hazel *et al.* (1980). It correlates closely with the *Pecten perplanus poulsoni* Zone of Glawe (1969) and it approximately correlates with the *Globigerina ampliapertura* Zone of Bolli & Saunders (1985) and the upper part of P18 and the lower part of P19/20 of Blow (1979).

### *Actinocythereis rosefieldensis* Zone

*Definition*: Interval from the first appearance of *Actinocythereis rosefieldensis* (Howe & Law, 1936) to the first appearance of *Leguminocythereis quadricostata* Mumma & Hazel, 1980.

*Remarks*: This zone was proposed and discussed in Hazel *et al.* (1980). It correlates with the *Pecten perplanus byramensis* Zone of Glawe (1969). It also correlates with the middle and upper parts of the nannoplankton zone NP23, the upper P19/20 to the lower part of P21 of Blow (1979) and with the lower and middle *Globigerina opima opima* Zone of Bolli & Saunders (1985). The hemicytherine *Orionina-Caudites* lineage has its earliest occurrence in this zone (Hazel *et al.*, 1980).

### *Leguminocythereis quadricostata* Zone

*Definition*: Interval of the total range of *Leguminocythereis quadricostata* Mumma & Hazel, 1980.

*Characterization*: The ostracod assemblages found within the range of the zonal marker are quite distinct from those of the Vicksburgian and have been descibed by Poag (1972, 1974) and Pooser (1965). The last *Haplocytheridea*, last *Alatacythere*, first *Pterygocythereis* and first *Peratocytheridea* are found in this zone.

*Remarks*: The distribution of the zonal marker was discussed in Hazel *et al.* (1980). This new zone includes rocks of the Chickasawhayan and Anahuacian Stages of Murray (1961) in the Gulf Basin. Along the Atlantic Coast, rocks of the upper part of the Ashley Member of the Cooper Formation and the overlying Edisto Formation and the equivalents of these units contain the zonal marker. The zone correlates approximately with nannoplankton zones NP24 and NP25 and the upper part of Blow's (1979) P21 and P22.

## PERIODS OF MAJOR OSTRACOD FAUNAL CHANGE

### Period 1

The first of three periods of major faunal change is associated with and marks the ClAibornian/Jacksonian stage boundary (Fig. 2). Many species have been described from Claibornian stratigraphical units, although no major comprehensive monographs exist. Of particular importance for the present study is the work of Blake (1950) and a large number of ostracod assemblage slides in the H. V. Howe Collection at Louisiana State University. The Claibornian is

a long stage (about 12 MY) and correlates with the late Ypresian to early Bartonian of Europe. Only the upper part of the Claibornian, that is the upper Lisbon Formation and the Gosport Formation (*Hazelina couleycreekensis* and *Cocoaia grigsbyi* zones) and their equivalents, form part of this study.

By definition, the top of the Claibornian is placed at the top of the richly macrofossiliferous Gosport Formation. In western Alabama, at the well-known Little Stave Creek locality (Bandy, 1949; Toulmin, 1962; Deboo, 1965), the Gosport is conformable with the overlying Moodys Branch Formation of the Jacksonian, but has an unconformable relationship with the underlying Lisbon Formation. The Lisbon-Gosport unconformity would appear to correlate with the sequence boundary between Vail coastal onlap cycles TA3.6 and 4.1 (Baum & Vail, 1987; Zullo & Harris, 1987; Fig. 2). Many typical Claibornian species found in the Lisbon and Cook Mountain formations and their equivalents do not reappear with the succeeding Gosport-Moodys Branch transgression, which began at about 41.50 Ma. These include forms such as *Acanthocythereis stenzeli*, small *Actinocythereis*-like forms common in the Claibornian, some species of *Cyamocytheridea*, the small form of *Clithrocytheridea garretti*, the small form of *Haplocytheridea montgomeryensis* and certain thaerocytherines.

There is a rapid change in faunal composition during the Gosport-Moodys Branch transgression. The Gosport and the lower greensand member of the Moodys Branch are part of the coastal onlap cycle TA4.1 of Haq *et al.*, (1987). The Gosport is characterized by a mixture of typical Claibornian species and species that become typical of the Jacksonian. The Moodys Branch has many new additions and several Claibornian holdovers. Among the new additions to the Coastal Province proper are genera such as *Hirsutocythere* that evolved in Florida (Howe, 1951) in the Middle Eocene. This and the occurrence of large foraminifera (*Camerina*) in the Moodys Branch suggests that the climate was somewhat warmer. However, the generic composition of the ostracod assemblages does not indicate that the early Jacksonian was a much warmer time. Frederiksen (1980a), on the basis of palynological evidence, concludes that the late Claibornian and early Jacksonian climates of the Gulf Coast were similar (humid subtropical to winter-dry tropical). The end of the Claibornian and beginning of the Jacksonian took place during an interval of increasing values of $\delta^{18}O$ in the bottom and surface waters of the open Atlantic (Miller *et al.*, 1987; Shackleton, 1986), which signals cooling and perhaps polar ice buildup. Several species of the spinose planktonic foraminiferal genera *Morozovella*, *Truncorotaloides* and *Acarinina* became extinct in this interval at about 40.46 Ma (Toumarkine & Luterbacher, 1985; Hazel, in press, Table 1).

After the Claibornian to Jacksonian turnover, there was little further change in Jacksonian assemblages (Huff, 1970; Howe & Chambers, 1935; Howe & Howe, 1975) until quite late in the Jacksonian. This period of evolutionary quiescence coincides in part with the formation of the Upper Eocene microspherule layers (Hazel, in press). The bolide impacts that were responsible for the spherule layers seem to have had no effect on the sublittoral ostracods of the Gulf Coast.

## Period 2

At the top of the *Actinocythereis montgomeryensis* Zone at about 37.31 Ma and continuing into the lower and middle parts of the *Acanthocythereis florienensis* Zone, many typical Jacksonian species disappear from the rock record (Huff, 1970; Howe & Howe, 1973 and unpublished data). Most of the rest of the typical Jacksonian forms then disappear during a period of more rapid faunal turnover between 36.71 and 36.36 Ma that marks the Jacksonian-Vicksburgian boundary (Howe & Howe, 1973; Howe, 1977). Vicksburgian marker species begin to appear in the middle of this interval at about 36.66 Ma. This period of faunal change occurred during a period of rising sea level (Haq *et al.*, 1987) and increasing values of $\delta^{18}O$ in sea water (Miller *et al.*, 1987) signalling the development of the psychrosphere and cooler surface waters in the open ocean. Terrestrial climates in the region also apparently became

cooler and probably drier (Frederiksen, 1980a; 1980b). However, the tropical-subtropical genus *Jugosocythereis* is a common constituent of early Vicksburgian facies of sublittoral origin (Hazel *et al.*, 1980). The Jacksonian-Vicksburgian boundary occurs at a high stand of late Eocene-early Oligocene sea level in the middle of coastal onlap cycle TA4.3 of Haq *et al.* (1987).

After the major addition of new taxa and the extinction of a few Jacksonian holdovers, such as *Acanthocythereis florienensis* (Howe & Chambers, 1935), in the early part of the *Actinocythereis dacyi* Zone (Hazel *et al.*, 1980), there was a slower addition of species through the rest of the Vicksburgian (Hazel *et al.*, 1980; Howe & Law, 1936; Howe, 1976). However, there is a reduction in diversity in the *Actinocythereis rosefieldensis* Zone that is related to falling sea level and changing environments. Carbonate deposition ceases in the central Gulf Coast ending a period of warm climate and a southeast to northwest transgression of carbonates over clastics (Hazel *et al.*, 1980, Figs 3, 5). This faunal change, however, is less severe than that of the late Jacksonian and early Vicksburgian or that which ends the Vicksburgian.

## Period 3

The end of the Vicksburgian is coincident with a major eustatic sea level fall that caused the 30 MY unconformity of Haq et al. (1987). Only about twelve of the ostracod species that lived in late Vicksburgian (Hazel *et al.*, 1980) are part of the fauna of 80 species of the succeeding Chickasawhayan Stage (Poag, 1972; 1974). This stage boundary correlates with the TA4/TB1 sequence boundary of Haq *et al.* (1987). Sea level is thought to have fallen below its present level and this seems to have severely affected the sublittoral ostracods. With the return of shallow seas a distinctly different fauna is found.

## SUMMARY

In the middle Tertiary of the southeastern United States three periods of major ostracod extinction and origination can be identified. Not surprisingly, these are coincident with the Claibornian-Jacksonian, Jacksonian-Vicksburgian and Vicksburgian-Chickasawhayan stage boundaries (Fig. 2), which were originally recognized by macrofossil workers. The macrofossils, therefore, demonstrate the same evolutionary pattern as the ostracods (Dockery, 1986; 1988). The first period of major change in the study interval occurs near the Middle Eocene-Upper Eocene boundary and is associated with an unconformity related to a eustatic sea level fall that, in the Coastal Province, marks the sequence boundary between coastal onlap cycles TA3.6 and TA4.1, and the subsequent transgression. Thus, this faunal change is associated with eustatic changes in sea level. Similarly, the faunal differences between the Vicksburgian and Chickasawhayan are also associated with a major eustatic drop in sea level and the subsequent transgression during sea level rise. Conversely, there is no drop in sea level associated with the rapid faunal turnover in the late Jacksonian and early Vicksburgian, which took place during a high stand of sea level at a time when marine and terrestrial climates were deteriorating and the psychrosphere developed.

The observed major changes in ostracod assemblages are most clearly associated with sea level changes, but also occur both at low stands and high stands. Ostracod taxa are apparently stressed during periods of regression and low stands of sea level causing extinctions and accelerating speciation. It is possible that the ostracods are affected during regressive phases in a manner similar to that postulated as a casual mechanism for mollusc extinctions. That is, delta progradation and increased turbidity of shelf waters leading to a loss of stable clear-water, sandy-bottom habitats (Dockery, 1986; 1988). Just how individual taxa are affected during such events is not clear.

During high stands of sea level in the middle part of the Jacksonian and middle part of the Vicksburgian there was only moderate faunal turnover. However, during the late Jacksonian and early Vicksburgian sea level high stand there was a major turnover of species, but in this case the ostracods were responding to stresses caused by climate change.

## ACKNOWLEDGEMENTS

Thanks are due P. R. Krutak, Basin Research Institute, Louisiana State University and the editors of this volume for helpful comments on the manuscript. This research was supported by National Science Foundation Grant EAR-8805268.

## REFERENCES

Bandy, O. L. 1949. Eocene and Oligocene Foraminifera from Little Stave Creek, Clarke County, Alabama. *Bull. Am. Paleont.*, Ithaca, **32**(131), 1-211.

Baum, G. R. & Vail, P. R. 1987. *Sequence stratigraphy, allostratigraphy, isotope stratigraphy and biostratigraphy: putting it all together in the Atlantic and Gulf Paleogene.* Selected Papers and Abstracts, Innovative biostratigraphic approaches to sequence analysis: new exploration opportunities, Eighth Annual Research Conference, Gulf Coast Section SEPM, 15-23. Earth Science Enterprises, Austin.

Bennison, A. P. Compiler. 1975. *Geological Highway Map of the Southeastern Region.* American Association Petroleum Geologists, U. S. Geological Highway Map Series, no. **9**.

Blake, D. B. 1950. Gosport Eocene Ostracoda from Little Stave Creek, Alabama. *J. Paleont.*, Lawrence, **24**(2), 174-184.

Blow, W. H. 1979. *The Cainozoic Globigerinida: a study of the morphology, taxonomy, evolutionary relationships and the stratigraphical distribution of some Globigerinida (mainly Globigeracea)*, vols 1-3, 1413 pp. E. J. Brill, Leiden.

Bolli, H. M. & Saunders, J. B. 1985. Oligocene to Holocene low latitude planktic foraminifera. *In* Bolli, H. M., Saunders, J. B., & Perch-Nielsen, K. (Eds), *Plankton Stratigraphy*, 155-262. Cambridge Univ. Press, Cambridge.

Byerly, G. R., Hazel, J. E. & McCabe, C. 1988. A new late Eocene microspherule layer in central Mississippi. *Mississippi Geology*, Jackson, **8**(4), 1-5.

Deboo, P. B. 1965. Biostratigraphic correlation of the type Shubuta Member of the Yazoo Clay and Red Bluff Clay with their equivalents in southwestern Alabama. *Alabama Geological Survey Bull.*, University of Alabama, **80**, 84 pp.

D'Hondt, S. L., Keller, G. & Stallard, R. F. 1987. Major element compositional variation within and between different late Eocene microtektite strewnfields. *Meteoritics*, Albuquerque (N. Mex.), **22**, 61-69.

Dockery, D. T., III. 1986. Punctuated succession of Paleogene molluscs in the Northern Gulf Coastal Plain. *Palaios*, Ann Arbor, **1**(6), 582-589.

Dockery, D. T., III. 1988. Molluscan extinction rates in question. *Nature*, London, **331**, 123.

Frederiksen, N. O. 1980a. Mid-Tertiary climate of southeastern United States: the sporomorph evidence. *J. Paleont.*, Lawrence, **54**(4), 728-739.

Frederiksen, N. O. 1980b. Sporomorphs from the Jackson Group (upper Eocene) and adjacent strata of Mississippi and western Alabama. *Prof. Pap. U.S. geol. Surv.*, Washington, no. **1084**, 75 pp.

Glass, B. P., Burns, C. A., Crosbie, J. R. & DuBois, D. L., 1985. Late Eocene North American microtektites and clinopyroxene bearing spherules. Lunar Planetary Science Conf. 16, *J. geophys. Res.*, Richmond, (Va.), **90**, D175-D196.

Glawe, L. N. 1969. *Pecten perplanus* stock (Oligocene) of the Southeastern United States. *Alabama Geological Survey Bull.*, University of Alabama, **91**, 179 pp.

Haq, B. U., Hardenbol, J. & Vail, P. R. 1987. Chronology of fluctuating sea levels since the Triassic. *Science*, New York, **235**, 1156-1167.

Hazel, J. E., Edward, L. E. & Bybell, L. M. 1984. Significant unconformities and the hiatuses represented by them in the Paleogene of the Atlantic and Gulf Coastal Province: *In* Schlee, J. (Ed.), Interregional Unconformities. *Am. Assoc. Pet. Geol. Memoir*, Tulsa, **36**, 59-66.

Hazel, J. E. 1988. *How many upper Eocene microspherule layers? More than we thought!* Abstracts presented to the Topical Conference, Global Catastrophes in Earth History, Snowbird, Utah. Lunar and Planetary Institute and The National Academy of Sciences, 72-73.

Hazel, J. E. In press. Chronostratigraphy of Upper Eocene microspherules. *Palaios*, Ann Arbor.

Hazel, J. E., Mumma, M. D. & Huff, W. J. 1980. Ostracode biostratigraphy of the lower Oligocene (Vicksburgian) of Mississippi and Alabama. *Gulf Coast Assoc. Geol. Soc. Trans.*, **30**, 233-240.

Howe, H. J. 1976. Diagnostic central Gulf Coast Vicksburgian ostracods in the H. V. Howe Collection. *Gulf Coast Assoc. Geol. Soc. Trans.*, **26**, 164-177.

Howe, H. J. 1977. A review of the Jacksonian-Vicksburgian boundary in the east-central Gulf Coast by means of Ostracoda. *Gulf Coast Assoc. Geol. Soc. Trans.*, **27**, 291-298.

Howe, H. J. & Howe, R. C., 1975. Central Gulf Coast Jacksonian ostracodes in the H. V. Howe Collection. *Gulf Coast Assoc. Geol. Soc. Trans.*, **25**, 282-295.

Howe, H. V. 1951. New Tertiary ostracode fauna from Levy County, Florida. *Florida Geological Survey Bull.*, Tallahassee, **34**, 82 pp.

Howe, H. V. & Chambers, J. 1935. Louisiana Jackson Eocene Ostracoda. *Louisiana Geological Survey Bull.*, Baton Rouge, **5**, 65 pp.

Howe, H. V. & Law, J. 1936. Louisiana Vicksburg Oligocene Ostracoda. *Louisiana Geological Survey Bull*, Baton Rouge, **7**, 96 pp.

Howe, R. C. & Howe, H. J. 1973. Ostracodes from the Shubata Clay (Tertiary) of Mississippi. *J. Paleont.*, Lawrence, **47**, 629-656.

Huff, W. J. 1970. The Jackson Eocene Ostracoda of Mississippi: *Mississippi Bureau of Geology Bull.*, Jackson, **114**, 389 pp.

Hut, P., Alvarez, W., Elder, W. P., Hansen, T., Kauffman, E. G., Keller, G., Shoemaker, E. M. & Weismann, P. 1987. Comet showers as a cause of mass extinctions. *Nature*, London, **329**(6135), 118-126.

Keen, M. 1978, The Tertiary-Palaeogene, *In* Bate, R. & Robinson, E. (Eds), *A Stratigraphical Index of British Ostracoda. Geol. Journ. Spec. Issue*, No. **8**, 385-450. Seel

House Press, Liverpool.

Keller, G., D'Hondt, S. L., Orth, C. J., Gilmore, J. S., Oliver, P. Q., Shoemaker, E. M. & Molina, E. 1987. Late Eocene impact microspherules: stratigraphy, age, and geochemistry. *Meteoritics*, Albuquerque (N. Mex.), **22**(1) 25-60.

Krutak, P. R. 1961. Jackson Eocene Ostracoda from the Cocoa Sand of Alabama. *J. Paleont.*, Chicago, **35**(4), 769-788.

Martini, E. 1971. Standard Tertiary and Quaternary calcareous nannoplankton zonation. *Proc. II Planktonic Conf., Rome 1970*, 739-783. Edizioni Tecnoscienza, Rome.

Miller, K. G, Fairbanks, R. G. & Mountain, G. S. 1987. Tertiary oxygen isotope synthesis, sea level history, and continental margin erosion. *Paleoceanography*, Washington, **2**(1) 1-19.

Murray, G. E. 1961. *Geology of the Atlantic and Gulf Coastal Province of North America*, 692 pp. Harper and Brothers, New York.

Poag, C. W. 1972. New ostracode species from the Chickasawhay Formation (Oligocene) of Alabama and Mississippi. *Revta esp. Micropaleont.*, Madrid, **4**(1) 65-96.

Poag, C. W. 1974. Late Oligocene ostracodes from the United States Gulf Coastal Plain. *Revta esp. Micropaleont.*, Madrid, **6**(1), 25-38.

Pooser, W. K. 1965. Biostratigraphy of Cenozoic Ostracoda from South Carolina. Arthropoda, Article **8**, *Paleont. Contr. Univ. Kans*, University of Kansas, Lawrence, 80 pp.

Shackleton, N. J. 1986. Paleogene stable isotope events. *Palaeogeog. Palaeoclimatol. Palaeoecol.*, Amsterdam, **57**, 91-102.

Toulmin, L. D. 1962. *Geology of the Hatchetigbee Anticline area, southwestern Alabama*. Gulf Coast Association of Geological Societies, Guide Book, 12th Annual Meeting, New Orleans, La., 1-47.

Toulmin, L. D. 1977. Stratigraphic distribution of Paleocene and Eocene fossils in the eastern Gulf Coast region. *Geological Survey Alabama Monograph*, University of Alabama, **13**, 2 vols, 602 pp.

Toumarkine, M. & Luterbacher, H. 1985. Paleocene and Eocene planktic foraminifera. *In* Bolli, H. M., Saunders, J. B. & Perch-Nielsen, K. (Eds), *Plankton Stratigraphy*, 87-154. Cambridge Univ. Press, Cambridge.

Zullo, V. A. & Harris, W. B. 1987. Sequence stratigraphy, biostratigraphy, and correlation of the Eocene through lower Miocene strata in North Carolina. *Cushman Fdn Foramin. Res. Special Publication*, Washington, **24**, 197-214.

# 9

# Recovering from the effects of an Oceanic Anoxic Event: Turonian Ostracoda from S.E. England

David Horne[1], Ian Jarvis[2] &
Amnon Rosenfeld[3]

[1]Thames Polytechnic, London, U.K.
[2]Kingston Polytechnic, Surrey, U.K.

## ABSTRACT

During the Cenomanian-Turonian Oceanic Anoxic Event (OAE), the bottom waters of the Chalk
Sea in N.W. Europe became increasingly dysaerobic, resulting in the extinction of many taxa. Preliminary results of a study of ostracod assemblages
from the Turonian Chalk at Dover, S.E. England,
show that the post-OAE recolonization of the
Chalk Sea at that locality was a long and gradual
process. The OAE caused the extinction of virtually all Cenomanian podocopid ostracod species,
and the earliest Turonian assemblages consist
almost exclusively of platycopids. Recolonization
was inhibited by the persistence of low seawater
oxygen levels for most of the Turonian, but diversity increased gradually as vacant niches were
filled by new, immigrant species, probably from
shallow-water refuges unaffected by the OAE. As
yet, no such refuges have been identified; the
taxonomic affinities of the fauna suggest that they
might be located in the Boreal rather than the
Tethyan Realm. Many of the species appearing
for the first time are long-ranging, and by the end of
the Turonian the general character of Coniacian-
Maastrichtian ostracod faunas was well established.
The Cenomanian-Turonian OAE was thus a major
influence on Upper Cretaceous marine ostracod
faunas in N.W. Europe. Evidence from equivalent
sequences in other parts of the world is reviewed
and the implications of these data are briefly discussed.

## INTRODUCTION

The anomalous lithological, geochemical, faunal
and floral characteristics of high Cenomanian to
low Turonian marine sediments in many parts of
the world have been attributed to an 'Oceanic
Anoxic Event' (OAE) (Schlanger & Jenkyns, 1976).
Some have argued against the suitability of the
term 'anoxic' in this context (e.g., Whatley, this

volume). There is, however, abundant evidence that at times in the Cretaceous World Ocean an expanded oxygen-minimum zone (and, under extreme conditions, even the bottom water) was devoid of oxygen, resulting in the deposition of laminated black shale facies in the deep ocean (see, e.g., Dean *et al.*, 1984). True anoxia may have been confined to the deep ocean, but the occurrence of oxygen-depleted waters in the Chalk Sea at the same time was part of the same phenomenon. Since we are considering the effects on a shelf sea of an event characterized by anoxia in parts of the World Ocean, the word 'anoxic' seems entirely appropriate.

At Dover most microfossil groups (benthonic and planktonic foraminifera, ostracods, dinoflagellate cysts and calcareous nannofossils) show abundance and diversity minima which correspond closely to the peak of a positive carbon stable-isotope excursion, indicating the position of the OAE in the sequence (Jarvis *et al.*, 1988). The close relationship between the diversity minima and the isotope excursion has been interpreted as indicating that the bottom waters of the Chalk Sea in N.W. Europe became increasingly dysaerobic, resulting in the extinction of many taxa. The disappearances were progressive, with an expanding and intensifying oxygen-minimum zone in the water column increasingly affecting the benthos as it impinged on the sea floor, and causing extinctions of first deeper-water and eventually shallower-water planktonic foraminifera as its top rose towards the surface. The peak of the OAE, at the base of the Melbourn Rock Beds (uppermost Cenomanian), is marked by an abundance of calcispheres and a temporary absence of dinoflagellate cysts.

Virtually all the Cenomanian podocopid ostracod species became extinct during the OAE and the earliest Turonian assemblages consist almost exclusively of surviving platycopids. We have previously documented the subsequent gradual increase in podocopid ostracod diversity through the low Turonian; here we present preliminary results and discussion of a continuation of our study to the top of the Turonian. Our samples were all collected in the vicinity of Dover (Fig. 1)

and are from Abbots Cliff (ABC 19-21), Akers Steps (AKS 1-13, C-I) and East Cliff (ECD 1-5). For details of preparation methods see Jarvis *et al.* (1988), in which the results from ABC 19-21 and AKS 1-13 have already been published.

The lithostratigraphy of the Turonian at Dover and the location of our samples are shown in Fig. 2. The base of the Turonian in East Kent is indicated by a major change in the inoceramid bivalve fauna: late Cenomanian *Inoceramus pictus* J. de C. Sowerby occurs in the basal hardgrounds of the Melbourn Rock (lower part of the Shakespeare Cliff Member, Dover Chalk Formation), immediately above which appear Turonian *Mytiloides* spp. The top of the Turonian is placed immediately below South Foreland Hardground 3 (St Margaret's Member, Ramsgate Chalk Formation), within which the Coniacian ammonite *Forresteria* (*Harleites*) *petrocoriensis* (Coquand) has been found at Langdon Stairs, just east of Dover (Bailey *et al.*, 1984). Coniacian inoceramids (?*Cremnoceramus* spp.) and new forms of *Micraster* also appear around this level.

There is, as yet, no internationally agreed subdivision of the Turonian, although new proposals have been presented recently by Robaszynski (1983). In view of the difficulty of reconciling various authors' subdivisions of the Turonian Stage in different parts of Europe, we prefer at present to use the informal terms low, mid and high. Where we have used formal terms such as Lower, Middle and Upper, it may be assumed that we are simply following the usage of the author or authors in question.

Ostracods of the British post-Cenomanian Chalk are poorly known, those of the Turonian least of all. Neale (1978) was unable to include any low to mid Turonian species in his range chart. Jones and Hinde (1890) recorded 26 species from the high Turonian Chalk Rock of Bedfordshire, Buckinghamshire and Oxfordshire (not developed at Dover, but equivalent to the Lighthouse Down Hardgrounds of the St Margaret's Member), some of which were revised by Kaye (1964). To date, the 12 species whose occurrence at Dover we have already documented (Jarvis *et al.*, 1988) constitute the only published records of British low

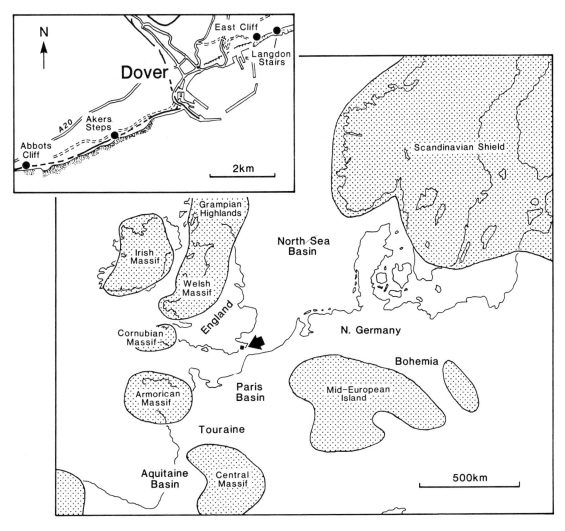

Fig. 1. Palaeogeography of N.W. Europe during the Turonian (based on Hancock, 1986; Tyson & Funnell, 1987). Stippled areas indicate probable landmasses. Inset shows location of sections at Dover.

Turonian ostracods. This dearth of information can be attributed, at least partly, to the facts that Chalk ostracods are much less abundant than foraminifera and that most assemblages are overwhelmingly dominated by species of *Cytherella*. With a little patience, however, diverse faunas may be obtained, particularly from the more easily disaggregated marls (see Weaver, 1982 and Jarvis *et al.*, 1988, for processing techniques). Weaver (1981) pointed out the necessity of picking the finer fractions in order to find small

(less than 300μm long) species of the Cytheruridae.

## THE TURONIAN OSTRACOD FAUNA AT DOVER

Table 1 lists the 28 species we have recorded so far in the Turonian at Dover. At species level, several platycopids survived from the Cenomanian (e.g., *Cytherella ovata* (Roemer, 1840), *Cytherella concava* Weaver, 1982, *Cytherelloidea kayei* Weaver, 1982), but the ostracod fauna established

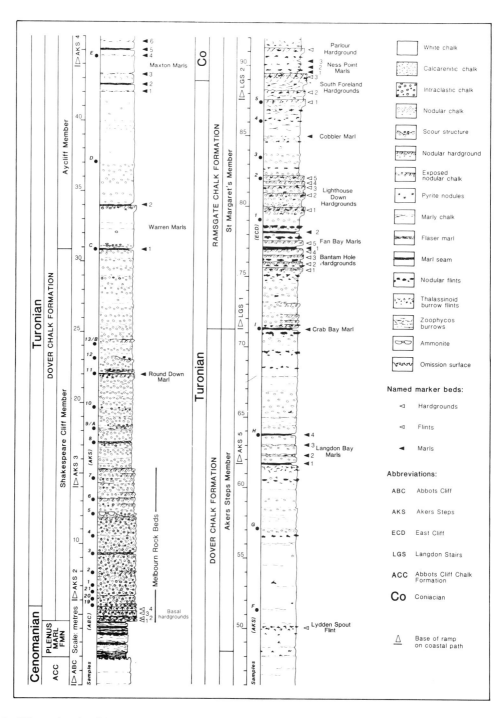

Fig. 2. Lithostratigraphy of the Turonian at Dover, showing sample positions. Composite section logged at Abbots Cliff, Akers Steps and Langdon Stairs (see Fig. 1). Samples were collected at Abbots Cliff (ABC 19-21), Akers Steps (AKS 1-13 & C-I) and East Cliff (ECD 1-5); the East Cliff sequence is slightly expanded compared with that at Langdon Stairs, but these differences do not significantly affect the sample positions. Stratigraphical terminology after Robinson (1986).

Table 1. Species recorded from the Turonian of Dover.

---

*Amphicytherura* sp. A
*Bairdoppilata* sp. A (*sensu* Jarvis *et al.*, 1988)
*Bythoceratina herrigi* Weaver, 1982
*Bythoceratina* cf. *B. montuosa* (Jones & Hinde, 1890)
*Bythoceratina umbonata* (Williamson, 1847)
*Curfsina kafkai kafkai* Pokorný, 1967
*Curfsina senior* Pokorný, 1967
*Cythereis ornatissima* cf. *C. o. altinodosa* Pokorný, 1963
*Cythereis* cf. *C. longaeva* Pokorný, 1963
*Cytherella* cf. *C. chathamensis* Weaver, 1982
*Cytherella concava* Weaver, 1982
*Cytherella* cf. *C. contracta* Van Veen, 1932
*Cytherella ovata* (Roemer, 1840)
*Cytherelloidea granulosa* (Jones, 1849)
*Cytherelloidea hindei* Kaye, 1964
*Cytherelloidea kayei* Weaver, 1982
*Cytherelloidea obliquirugata* (Jones & Hinde, 1890)
*Golcocythere*? *calkeri* (Bonnema, 1941)
*Imhotepia marssoni* (Bonnema, 1941)
*Mosaeleberis* sp. A (*sensu* Jarvis *et al.*, 1988)
*Neocythere* (*Physocythere*) *virginea* (Jones, 1849)
*Parvacythereis subparva* (Pokorný, 1967)
*Pontocyprella* sp. A
*Pterygocythere* cf. *P. diminuta* Weaver, 1982
*Spinoleberis krejcii* Pokorný, 1968
*Trachyleberidea geinitzi* (Reuss, 1874)
*Xestoleberis* sp. A.
*Xestoleberis* sp. B.

---

by the end of the Turonian at Dover is essentially a new one in terms of podocopids. At generic level the Turonian fauna is characterized by a number of new appearances, represented by *Golcocythere*? *calkeri* (Bonnema, 1941), *Parvacythereis subparva* (Pokorný, 1967), *Spinoleberis krejcii* Pokorný, 1968 and *Trachyleberidea geinitzi* (Reuss, 1874), although many species belong to genera that also have Cenomanian species in S.E. England (e.g., *Cythereis, Curfsina, Imhotepia, Neocythere*). The

genus *Spinoleberis* is unknown in Britain before the Turonian, although older species have been described from elsewhere; for example, *S. petrocorica* (Damotte, 1971a) is known from the Upper Cenomanian of the Tethyan margins in southern France: the Dordogne (Damotte, 1971a), the subalpine chains (Donze & Thomel, 1972) and Provence (Babinot, 1980).

The stratigraphical occurrences of *Cytherelloidea* and podocopid species in the Turonian at Dover are shown in Fig. 3. In addition, every sample contained abundant *Cytherella* (not displayed in Fig. 3), the commonest species being *C. ovata* and *C.* cf. *C. contracta* Van Veen, 1932. The earliest Turonian assemblages (ABC 19, 20) at Dover included the last occurrences of the Cenomanian species *Bythoceratina herrigi* Weaver, 1982 and *B. umbonata* (Williamson, 1847), and an isolated record of *Pterygocythere* cf. *P. diminuta* Weaver, 1982. The first additions to the ostracod fauna were *Cythereis* cf. *C. longaeva* Pokorný, 1963 (AKS 1), *Curfsina senior* Pokorný, 1967 and *Mosaeleberis* sp. A (both first appearing in AKS 2), followed near the top of the Shakespeare Cliff Member (AKS 11) by *Cytherelloidea obliquirugata* (Jones & Hinde, 1890) and *Parvacythereis subparva*. *Bairdoppilata* sp. A is a major component of mid to high Turonian assemblages (AKS 13 - ECD 2) and is possibly represented in the low Turonian by juvenile and fragmentary records. High Turonian assemblages are characterized by *Cytherelloidea granulosa* (Jones, 1849) and *Neocythere virginea* (Jones, 1849), both first appearing in AKS E. The genus *Xestoleberis* becomes an important constituent of the fauna near the top of the Turonian, accompanied by several further additions, including *Trachyleberidea geinitzi* and *Imhotepia marssoni* (Bonnema, 1941).

In summary, the trend towards increasing diversity already noted in the low Turonian at Dover (Jarvis *et al.*, 1988) is continued, and by the end of the Turonian the diversity is similar to that of the pre-OAE Cenomanian (15-20 species). Furthermore, many of the new species are long-ranging and are known from Coniacian and younger deposits in N.W. Europe. It required, therefore, most of the Turonian (about 3 Ma; Haq

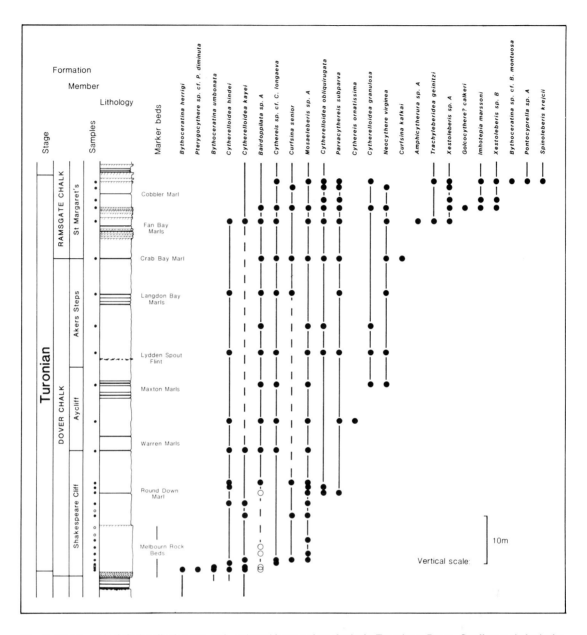

Fig. 3. Distribution of *Cytherelloidea* spp. and podocopid ostracod species in the Turonian at Dover. Small open circles in the sample column (see Fig. 2 for details) indicate material that contained no *Cytherelloidea* or podocopids; all samples yielded *Cytherella* spp. Large open circles indicate identifications based solely on juvenile or fragmentary material. Continuous vertical lines indicate the stratigraphical ranges of species; continuation of ranges into the Cenomanian based on Jarvis *et al.* (1988). Extension of ranges beyond the topmost sample position indicates that these species have been recorded by us in a preliminary examination of Coniacian material at Dover.

*et al.*, 1987) for the ostracod fauna of the Chalk Sea to recover from the effects of the OAE in S.E. England. By the end of the Turonian the basic character of post-OAE ostracod assemblages in Chalk facies was well established.

## COMPARISON WITH OTHER AREAS

In attempting to compare the Turonian ostracods of Dover with faunas of equivalent age from elsewhere in N.W. Europe, we have encountered both taxonomic and stratigraphical difficulties. In the interests of clarity, therefore, the taxonomic notes included in this section are necessarily detailed, although by no means exhaustive.

At Cap Blanc-Nez, on the French coast opposite Dover, Magné & Polvêche (1961) have demonstrated a marked decrease in ostracod diversity at the Cenomanian-Turonian boundary. Their samples from the Plenus Marl yielded, in ascending order, 11, 6 and 4 species respectively, and above these the earliest Turonian faunas consisted only of *Cytherella* species. Their five samples from the Plenus Marl at nearby Sangate showed a similar decrease from 7 species at the base to 3 at the top, with only *Cytherella* species being common to all samples. Unfortunately, few of their identifications were to specific level, none was illustrated, and their sections lack sufficient lithostratigraphical detail to allow more precise comparison with Dover.

Few ostracods are known from the Turonian of the Paris Basin. The five cytheraceans recorded by Damotte (1971b) include *Cythereis longaeva* Pokorný, 1963 and *Imhotepia marssoni multipapillata* Pokorný, 1964 (cited as *Cythereis* sp. 4 *multipapillata*). According to Babinot *et al.* (1985) the faunas are poor, consisting mainly of platycopids, bairdiaceans and a few rare trachyleberids; they interpreted such low diversity as indicating that the depositional environment of the Turonian Chalk was unfavourable to ostracods.

In the S.W. part of the Paris Basin, in the Turonian type-area of Touraine (Fig. 1), a distinct faunal change takes place in the ostracods at the Cenomanian-Turonian boundary, coincident with an abrupt facies change from marginal, sandy, glauconitic marls containing abundant oysters, to typical chalk (Damotte, in Robaszynski *et al.*, 1982). In the Civray-de-Touraine borehole, of 14 podocopid species recorded in the topmost Cenomanian only one, *Spinoleberis petrocorica*, survives into the Turonian. Podocopid diversity is low (4-6 species) in the early Turonian but increases (11 species) by the late Turonian. Although at first sight the type Turonian ostracod fauna has little in common with that of Dover or the rest of the Paris Basin, certain aspects appear to merit further investigation. Firstly, Damotte's record of *Trachyleberidea geinitzi*, restricted to the low Turonian of the type region, is probably not conspecific with (but perhaps ancestral to?) the late Turonian form here illustrated under the same name (Plate 1, Fig. 11). Secondly, the early Turonian *Imhotepia* sp. 81 of Damotte closely resembles the strongly reticulate morph of our *Mosaeleberis* sp. A (Plate 2, Fig. 11). Thirdly, *Cythereis (Rehacythereis) civrayensis* Damotte, 1982, described from the lowermost Turonian of the type area, is remarkably like our *Curfsina senior*; interestingly, Babinot (1973) has illustrated a similar species from the low Turonian of S.E. France as *Curfsina* cf. *senior* Pokorný. Finally, *Spinoleberis petrocorica* is of interest as one of the earliest (Cenomanian -early Turonian) representatives of the genus, and its morphological similarity to the Turonian-Coniacian *S. krejcii* (Plate 1, Fig. 10) suggests an ancestral relationship.

Several Turonian species at Dover were originally described from the Turonian and Coniacian of Bohemia (Fig. 1) by Pokorný (1963-1978), and some of these have also been recorded from Coniacian and younger strata in northern Germany (Clarke, 1983). The first new cytheracean to appear in the Turonian at Dover is *Cythereis* cf. *C. longaeva*, three morphological variants of which are illustrated in Plate 1 (Figs 7-9). Many of our Turonian specimens most closely resemble *C. longaeva longaeva*, found in the Middle to Upper Turonian (and Coniacian?) of Bohemia, and may be referable to the somewhat different, undescribed form noted by Pokorný (1963b) as occurring in the Lower Turonian. We also have specimens (e.g., Plate 1, Fig. 7) similar to *C. longaeva prior* Clarke,

1983, from the Upper Coniacian-Santonian of N.W. Germany, and others resembling *Cythereis zygopleura* Pokorný, 1965, which is known from the Upper Turonian of Bohemia and the Coniacian -Campanian of N.W. Germany.

*Curfsina senior*, another low Turonian appearance at Dover, ranges from Lower to Middle Turonian in Bohemia (Pokorný, 1978b) and from Lower Turonian to Lower Campanian in northern Germany (Clarke, 1983; Gründel, 1969). It has not been recorded in the Turonian of the Paris Basin, although there a species cited as *Curfsina* cf. *nuda* (Jones & Hinde, 1889) by Babinot *et al.* (1982) invites comparison. The reticulate forms of our *Mosaeleberis* sp. A (e.g., Plate 2, Fig. 11) resemble *M. interruptoidea* (van Veen, 1936) *sensu* Pokorný (1978a) from the Lower-Middle Turonian of Bohemia, while the smooth forms (Plate 2, Figs 9-10) may be referable to *Karsteneis (Prosteneis) nodifera* (Kafka, 1886), which ranges through the entire Turonian of Bohemia. *Mosaeleberis* cf. *macrophthalma* (Bosquet, 1847) of Babinot *et al.* (1982) from the Lower Turonian of the Paris Basin also needs comparison with our smooth form. *Parvacythereis subparva*, first appearing in the mid Turonian at Dover, ranges from Lower Turonian to highest Coniacian in Bohemia (Pokorný, 1967) and from Coniacian to Upper Campanian in N.W. Germany (Clarke, 1983).

*Spinoleberis krejcii*, recorded from our topmost Turonian sample at Dover, ranges from Lower Turonian to mid Coniacian in Bohemia. Clarke (1983) recorded it from the Coniacian and lowermost Santonian of N.W. Germany and described a new subspecies, *S. krejcii tenuireticulata*, from Santonian to Upper Campanian deposits. Two other species each occurred in a single sample in the high Turonian at Dover (Fig. 3): in Bohemia, *Curfsina kafkai* Pokorný, 1967 ranges from Middle Turonian to Lower Coniacian, *Golcocythere? calkeri* from Upper Turonian to Middle Coniacian. The latter was also recorded (as *Paracytheretta calkeri*) in the Coniacian and Santonian of N.W. Germany by Clarke (1983).

*Imhotepia marssoni*, appearing towards the top of the Turonian at Dover, possibly includes three subspecies that are not found until the Coniacian in Bohemia (Pokorný, 1964, 1978b) (*I. marssoni marssoni*, *I. marssoni anteglabra* (Pokorný, 1964), and *I. marssoni* subsp. 1 of Pokorný, 1964). Pokorný's (1964) Upper Turonian *I. marsonni multipapillata*, however, is absent. Clarke (1983) recorded the nominate subspecies from Coniacian-Santonian deposits in N.W. Germany. Damotte (1971b) gave the range of the subspecies *multipapillata* in the Paris Basin as mid-Turonian-late Coniacian. She referred it and Pokorný's other subspecies to '*Cythereis* sp. 4', expressing doubts as to whether they were really conspecific with the Maastrichtian *Cythereis marssoni* Bonnema, 1941. Babinot *et al.* (1982) showed '*Imhotepia* gr. *marssoni*' as occurring in both the Lower and Upper (but not the Middle) Turonian of the Paris Basin. The same authors' subsequent (1985) range charts show no pre-Coniacian occurrence of *I. marssoni* in France. *Cythereis ornatissima altinodosa* is confined to the Upper Coniacian in Bohemia; the Turonian form that we have tentatively referred to this subspecies might thus be regarded as an early member of the *C. ornatissima* group.

The stratigraphical and geographical distribution of many of these species may be largely facies-controlled (see, for example, Ohmert, 1971; Pokorný, 1978b), but constructive consideration of this problem cannot proceed until taxonomic difficulties such as those outlined above are resolved. Careful study of comparative specimens of all the above-mentioned species is indicated.

## THE NATURE OF THE CENOMANIAN-TURONIAN OCEANIC ANOXIC EVENT

In the model proposed by Jarvis *et al.* (1988), a phase of increased upwelling led to the expansion and intensification of the oxygen-minimum zone in the World Ocean. As a result, widespread deposition of black shales occurred in many ocean basins, with increased burial of organic matter causing a major positive shift in seawater carbon stable-isotope ratios. DSDP data show that black shale facies extended between palaeodepths of 0.5 and 2.5km in the eastern North Atlantic (Schlanger *et al.*, 1987), and possibly as deep as 3.5km in the eastern South Atlantic (Stow & Dean, 1984; Dean et al., 1984), indicating anoxic or near-anoxic bottom waters at those depths. Microfossil data suggest that at the peak of the OAE oxygen-depleted conditions may have extended to within 50-100m of the surface (Jarvis et al., 1988).

The available information suggests that the Cenomanian-Turonian OAE was global in extent (Schlanger et al., 1987); it has been recognized in S.W. Australia, the central Pacific, the Americas from Peru to Alaska, Africa, the North and South Atlantic and Europe. Distinct changes can be recognized between Cenomanian and Turonian-Coniacian ostracod faunas in many parts of the world, for example France (Babinot, 1980), Spain (Colin et al., 1982), Tunisia (Bismuth et al., 1981), Israel (Rosenfeld & Raab, 1974), off eastern Canada (Ascoli, 1988) and South Africa (Dingle, 1985), although detailed studies of boundary sections are lacking.

At Dover the effect of the OAE, as defined by the positive carbon stable-isotope excursion, was relatively short-lived, probably less than 500,000 years. Macrofaunal and other evidence indicate that although the bottom waters may have become severely oxygen-depleted, they were never truly anoxic. However, carbon stable-isotope values do not return to pre-OAE levels within the low Turonian at Dover, and data from several sites in southern England (Scholle & Arthur, 1980, Fig. 2) indicate that they declined progressively through the remainder of the Turonian, reaching a minimum close to the Turonian-Coniacian boundary. This trend may reflect a gradual increase in seawater oxygen levels, suggesting that throughout much of the Turonian conditions remained difficult for many ostracod groups, retarding their recolonization of the Chalk Sea. It took most of the Turonian for the ostracod fauna to regain its pre-OAE diversity (15-20 species in the late Cenomanian). Benthonic foraminifera show a similar rate of recovery in the low Turonian but then reach a diversity plateau of around 10 species and had still not returned to pre-OAE diversity (15-20 species) by the end of the Turonian (Hart et al., 1981). Planktonic groups (foraminifera and nannofossils), on the other hand, recovered relatively

rapidly, returning to pre-OAE diversity levels within the basal Turonian.

## ORIGINS OF THE TURONIAN OSTRACOD FAUNA IN S.E. ENGLAND

The Turonian podocopid fauna at Dover is completely new, replacing the Cenomanian fauna extinguished by the OAE. However, many of the Turonian species have closely related, although morphologically distinct, counterparts in the Cenomanian of S.E. England. For example, the high Turonian Xestoleberis sp. A is remarkably similar to X. planus Weaver, 1982, recorded from the Cenomanian of other sites in S.E. England although not found by us at Dover; the Turonian species differs only in details of shape and in possessing small postero-marginal tubercles in the female carapace (Plate 2, Figs 7-8). The mid to late Cenomanian Pontocyprella robusta is replaced in the Turonian by the more elongate Pontocyprella sp. A. Other late Cenomanian-Turonian congeneric pairs are Imhotepia euglyphea - Imhotepia marssoni, Curfsina derooi - Curfsina senior, Neocythere kayei - Neocythere virginea and Bairdoppilata pseudoseptentrionalis - Bairdoppilata sp. A (Table 2). A less convincing correspondence is apparent between the Turonian Cythereis longaeva group (Plate 1, Figs 7-9) and the Albian-Cenomanian Cythereis reticulata - hirsuta - folkstonensis group (cf. Van der Wiel, 1978), while Cythereis ornatissima cf. C. ornatissima altinodosa (Plate 1, Fig. 4) is similar to two Cenomanian species: C. humilis humilis Weaver, 1982, from S.E. England, and Cythereis condemiensis Breman, 1976, from Spain (Breman, 1976) and southern France (Babinot et al., 1978). Further afield, several other British Cenomanian species appear to have counterparts (descendants?) in the post-Cenomanian Cretaceous of Europe. For example, Bythoceratina herrigi Weaver, 1982 closely resembles the Campanian and Maastrichtian subspecies of B. pedatoides (Bonnema, 1941) illustrated by Clarke (1983) from N.W. Germany. Variants of Bythoceratina umbonatoides (Kaye, 1964) have been illustrated from the Cenomanian (e.g., by Weaver, 1982) and the Coniacian to Campanian (e.g., by Clarke, 1983)

Table 2. Some Cenomanian species and their Turonian equivalents.

| Cenomanian | Turonian |
| --- | --- |
| *Imhotepia euglyphea* Weaver, 1982 | *Imhotepia marssoni* (Bonnema, 1941) |
| *Curfsina derooi* Weaver, 1982 | *Curfsina senior* Pokorný, 1967 |
| *Bairdoppilata pseudoseptentrionalis* Mertens, 1956 | *Bairdoppilata* sp. A |
| *Neocythere kayei* Weaver, 1982 | *Neocythere virginea* (Jones, 1849) |
| *Xestoleberis planus* Weaver, 1982 | *Xestoleberis* sp. A |
| *Pontocyprella robusta* Weaver, 1982 | *Pontocyprella* sp. A |
| *Cythereis humilis humilis* Weaver, 1982 | *Cythereis ornatissima* cf. *C. ornatissima altinodosa* Pokorný, 1963 |
| group | group |
| *Cythereis reticulata* Jones & Hinde, 1890 | *Cythereis longaeva* Pokorný, 1963 |
| *Cythereis hirsuta* Damotte & Grosdidier, 1963 | (various subspecies) |
| *Cythereis folkstonensis* Kaye, 1964 | |

of N.W. Europe, although we believe these forms to be separable at least at the subspecies level.

The above observations rest on the assumption that our taxonomic judgements are correct. We are well aware that other authors' concepts of species might be broader than ours, and that they might argue for fewer extinctions in the OAE, suggesting instead that many of the Turonian forms are conspecific with Cenomanian taxa. Even accepting the latter view, however, there is still the problem of absence of these species from S.E. England and elsewhere throughout a considerable stratigraphical interval.

The global nature of the OAE and the close affinities between many European Cenomanian and Turonian ostracods both indicate that the Turonian species are of local (i.e., N.W. European) origin. Whether we consider the Turonian species to be the same as, or closely related descendants of, the Cenomanian species, the implications are much the same. A group of N.W. European Cenomanian species must have survived in a refuge (or refuges) unaffected by the OAE, to subsequently recolonize the Chalk Sea in the Turonian.

The enlarged bathymetrical extent of the oxygen-minimum zone at the height of the OAE would seem to preclude a deep-water refuge. Consequently, the root stock for the Turonian fauna must be sought among Cenomanian species that survived somewhere in shallow, well-oxygenated waters on the margins of the basins, such as in parts of Saxony, northern Germany (Hilbrecht & Hoefs, 1986).

The OAE has been linked to a phase of maximum eustatic transgression during the late Cenomanian-earliest Turonian (Arthur *et al.*, 1987). The bulk of the Turonian, in contrast, is characterized by a regression culminating close to the Turonian-Coniacian boundary (Hancock, 1986). Falling sea levels may thus have been the driving force behind the contraction of the oxygen-minimum zone, leading to better oxygenation within epicontinental seas and a migration of shallow-water facies and faunas towards the centres of basins.

## CONCLUSION

The Turonian ostracod fauna at Dover was probably derived from Cenomanian stock that survived the OAE in well-oxygenated, shallow-water refuges on the basin margins. Unfortunately, such shallow-water facies are preserved in only a few areas in N.W. Europe and their ostracod faunas are poorly known. These problems, combined with a general lack of taxonomic and stratigraphical precision, make the identification of probable source areas virtually impossible at present. The predominantly boreal affinities of the Turonian fauna suggest that its origins should be sought in northern rather than southern Europe.

## ACKNOWLEDGEMENTS

We wish to thank Andy Gale and Kym Jarvis for their help in collecting the samples. Technical assistance was provided by Rose Gent. Robin Whatley and two anonymous referees are thanked for their many helpful criticisms and suggestions.

## REFERENCES

Arthur, M. A., Schlanger, S. O. & Jenkyns, H. C. 1987. The Cenomanian-Turonian Oceanic Anoxic Event, II: palaeoceanographic controls on organic matter production and preservation. *In* Brooks, J. & Fleet, A. (Eds), *Marine Petroleum Source Rocks, Geol. Soc. spec. Publ.*, **26**, 401-420. Blackwell Scientific Publ. Ltd, Oxford

Ascoli, P. 1988. Mesozoic-Cenozoic foraminiferal, ostracod and calpionellid zonation of the North Atlantic margin of the North America: Georges Bank-Scotian basins and northeastern Grand Banks (Jeanne d'Arc, Carson and Flemish Pass basins). Biostratigraphic correlation of 51 wells. *Geol. Surv. Can.*, Ottawa, No. 1791, 1-41.

Babinot, J.-F. 1973. Ostracodes turoniens de la région de Cassis-La Bédoule (Bouches-du-Rhône, Var): associations et affinités paléogéographiques. *Géobios*, Lyon, **6**, 27-48.

Babinot, J.-F. 1980. Les ostracodes du Crétacé Supérieur de Provence. *Trav. Lab. Géol. hist. Pal.*, Université de Provence, Marseille, **10**, 1-640.

Babinot, J.-F., Colin, J.-P., Damotte, R. & Donze, P. 1978. Les ostracodes du Cénomanien français. Mise au point biostratigraphique et paléogéographique. *Géol. Mediter.*, Université de Provence, Marseille, **5**, 19-26.

Babinot, J.-F., Colin, J.-P. & Damotte, R. 1982. Les ostracodes du Turonien français. *Mém. Mus. natn. Hist. nat. Paris*, N.S., C, **49**, 189-196.

Babinot, J.-F., Colin, J.-P. & Damotte, R. 1985. Ostracodes du Crétacé supérieur. *In* Oertli, H. J. (Ed.), *Atlas des ostracodes de France*, 211-255. *Bull. Centres Rech. Explor.-Prod., Elf-Aquitaine*, Pau, Mém. **9**.

Bailey, H. W., Gale, A. S., Mortimore, R. N., Swiecicki, A. & Wood, C. J. 1984. Biostratigraphical criteria for the recognition of the Coniacian to Maastrichtian stage boundaries in the Chalk of north-west Europe, with particular reference to southern England. *Bull. geol. Soc. Denm.*, Copenhagen, **33**, 31-39.

Bismuth, H., Boltenhagen, C., Donze, P., Le Fevre, J. & Saint-Marc, P. 1981. Le Crétacé moyen et supérieur du Djebel Semmama (Tunisie du Centre-Nord); microstratigraphie

et évolution sédimentologique. *Bull. Centres Rech. Explor.-Prod., Elf-Aquitaine*, Pau, **5**, 193-269.

Breman, E. 1976. Paleoecology and systematics of Cenomanian and Turonian Ostracoda from Guadalajara and Soria (central Spain). *Revta esp. Micropaleont.*, Madrid, **8**, 71-122.

Clarke, B. 1983. Die Cytheracea (Ostracoda) im Schreibekriede - Richtprofil von Lägerdorf-Kronsmoor-Hemmoor(Coniac bis Maastricht; Norddeutschland). *Mitt. geol.-paläont. Inst. Univ. Hamburg*, **54**, 65-168.

Colin, J.-P., Lamolda, M. A. & Rodríguez-Lázaro, J. M. 1982. Los ostrácodos del Cenomaniense superior y Turoniense de la Cuenca Vasco-Cantábrica. *Revta esp. Micropaleont.*, Madrid, **14**, 187-220.

Damotte, R. 1971a. Quelques ostracodes du Cénomanien de Dordogne et de Touraine. *Rev. Micropaléont.*, Paris, **14**, 3-20.

Damotte, R. 1971b. Les ostracodes marins dans le Crétacé du Bassin de Paris. *Mém. Soc. géol. Fr.*, Paris, N.S., No. 113, 147 pp., 8 pls.

Dean, W. E., Arthur, M. A. & Stow, D. A. V. 1984. Origin and geochemistry of Cretaceous deep-sea black shales and multi-coloured claystones, with emphasis on Deep Sea Drilling Project Site 530, southern Angola Basin. *In* Hay, W. W., Sibuet, C.-C. *et al.*, *Init. Reports DSDP*, **75**, 819-844. U.S. Govt. Printing Office, Washington.

Dingle, R. V. 1985. Turonian, Coniacian and Santonian Ostracoda from south-east Africa. *Ann. S. Afr. Mus.*, Cape Town, **96**, 123-239.

Donze, P. & Thomel, G. 1972. Le Cénomanien de la Foux (Alpes de Haute-Provence). Biostratigraphie et faunes nouvelles d'ostracodes. *Eclogue geol. Helv.*, Lausanne, Basel, **65**, 369-389.

Gründel. J. 1969. Ostracoden aus der plenus-Zone (Oberkreide) Sachsens. *Freiberger ForschHft*, Berlin, C245, 83-89.

Hancock, J. M. 1986. Cretaceous. *In* Glennie, K. W. (Ed.), *Introduction to the Petroleum Geology of the North Sea*, 2nd edition, 161-178. Blackwell, Oxford.

Haq, B. U., Hardenbol, J. & Vail, P. R. 1987. Chronology of fluctuating sea levels since the Triassic. *Science*, New York, **235**, 1156-1167.

Hart, M. B., Bailey, H. W., Fletcher, B. Price, R. & Sweicicki, A. 1981. Cretaceous. *In* Jenkins, D. G. & Murray, J. W. (Eds), *Stratigraphical Atlas of Fossil Foraminifera*, 121-137. Ellis Horwood, Chichester, for British Micropalaeontological Society.

Hilbrecht, H. & Hoefs, J. 1986. Geochemical and palaeontological studies of the δ13C anomaly in Boreal and North Tethyan Cenomanian-Turonian sediments in Germany and adjacent areas. *Palaeogeog. Palaeoclimat. Palaeoecol.*, Amsterdam, **53**, 169-189.

Jarvis, I., Carson, G. A., Cooper, M. K. E., Hart, M. B., Leary, P. N., Tocher, B. A., Horne, D. & Rosenfeld, A. 1988. Microfossil assemblages and the Cenomanian-Turonian (late Cretaceous) Oceanic Anoxic Event. *Cret. Res.*, Academic Press, London, **9**, 3-103.

Jones, T. R. & Hinde, G. J. 1890. A supplementary monograph of the Cretaceous Entomostraca of England and Ireland. *Monogr. Palaeontogr. Soc.*, London, i-viii, 1-70, 4 pls.

**Plate 1**

The following abbreviations are used: RV = right valve, LV = left valve, car. = carapace. All specimens deposited in the British Museum (Natural History) (oS numbers). All external lateral views. Approx. magnifications: Figs 1-3 X85, Figs 4-9 X60, Figs 10-11 X85.

Figs 1-2. *Curfsina kafkai kafkai* Pokorný, 1967.
Fig. 1.    Male LV (oS 13283, 660μm long), AKS I.
Fig. 2.    Female LV (oS 13284, 620μm long), AKS I.

Fig. 3.    *Parvacythereis subparva* (Pokorný, 1967). LV (oS 13285, 520μm long), ECD1.

Fig. 4.    *Cythereis* cf. *C. ornatissima altinodosa* Pokorný, 1963. LV (oS 13286, 830μm long), AKS D.

Fig. 5.    *Golcocythere*? *calkeri* (Bonnema, 1941). Car., right side (oS 13287, 700μm long), ECD 2.

Fig. 6.    *Imhotepia marssoni* (Bonnema, 1941). Car., right side (oS 13288, 600μm long), ECD 2.

Figs 7-9. *Cythereis* cf. C. *longaeva* Pokorný 1963.
Fig 7.    Car., left side (oS 13289, 760μm long), ECD 5.
Fig. 8.    Car., left side (oS 13290, 790μm long), AKS D.
Fig. 9.    LV (oS 13291, 760μm long), AKS F.

Fig. 10.    *Spinoleberis krejcii* Pokorný 1968. RV (oS 13292, 560μm long), ECD 5.

Fig. 11.    *Trachyleberidea geinitzi* (Reuss, 1874). RV (oS 13293, 530μm long), ECD 5.

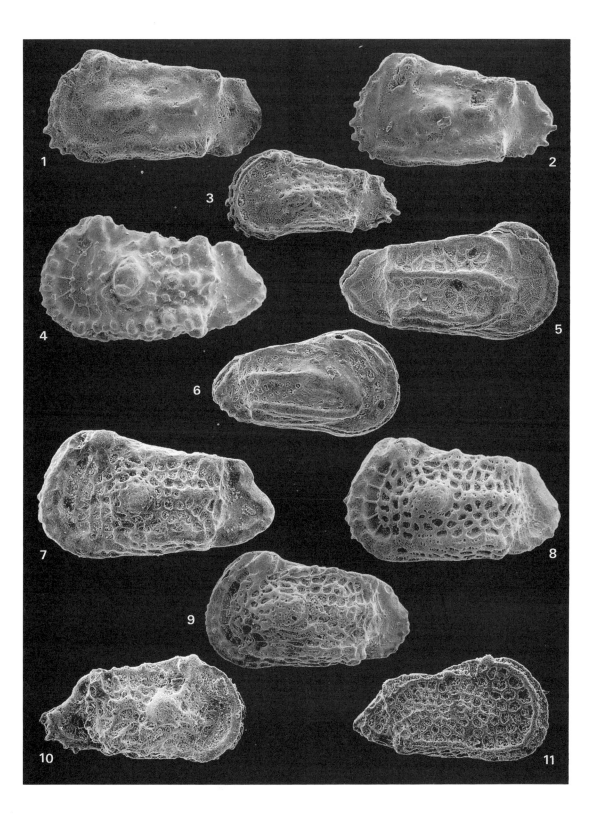

**Plate 2**

The following abbreviations are used: RV = right valve, LV = left valve, car. = carapace. All specimens deposited in the British Museum (Natural History) (oS numbers). All external lateral views. Approx. magnifications: Figs 1-5 x90, Fig. 6 X45, Figs 7-8, X100, Figs 9-11 X75.

Fig.  1.     *Cytherelloidea kayei* Kaye, 1964. RV (oS 13294, 590µm long), AKS C.

Fig.  2.     *Cytherelloidea hindei* Kaye, 1964. Car., right side (oS 13295,  620µm long), AKS F.

Fig.  3.     *Cytherelloidea granulosa* (Jones, 1849). RV (oS 13296, 720µm long),  AKS F.

Fig.  4.     *Cytherelloidea obliquirugata* (Jones & Hinde, 1890). Car., right side  (specimen lost after photography; 580µm long), AKS F.

Fig.  5.     *Neocythere* (*Physocythere*) *virginea* (Jones, 1849). LV (oS 13297,  570µm long), ECD 1.

Fig.  6.     *Bairdoppilata* sp. A (*sensu* Jarvis *et al.*, 1988). LV (oS 13298, 1,100µm long), AKS F.

Figs 7-8.  *Xestoleberis* sp. A.
Fig.  7.     Female RV (oS 13300, 450µm long), ECD 2.
Fig.  8.     Male RV (oS 13299, 430µm long), ECD 2.

Figs 9-11. *Mosaeleberis* sp. A (*sensu* Jarvis *et al.*, 1988).
Fig.  9.     Smooth form, female car., left side (oS  13301,  650µm long), AKS E.
Fig.  10.    Smooth form, male LV (oS 13302, 750µm long), ECD 2.
Fig.  11.    Reticulate form, female car., left side (oS 13303, 720µm long), AKS C.

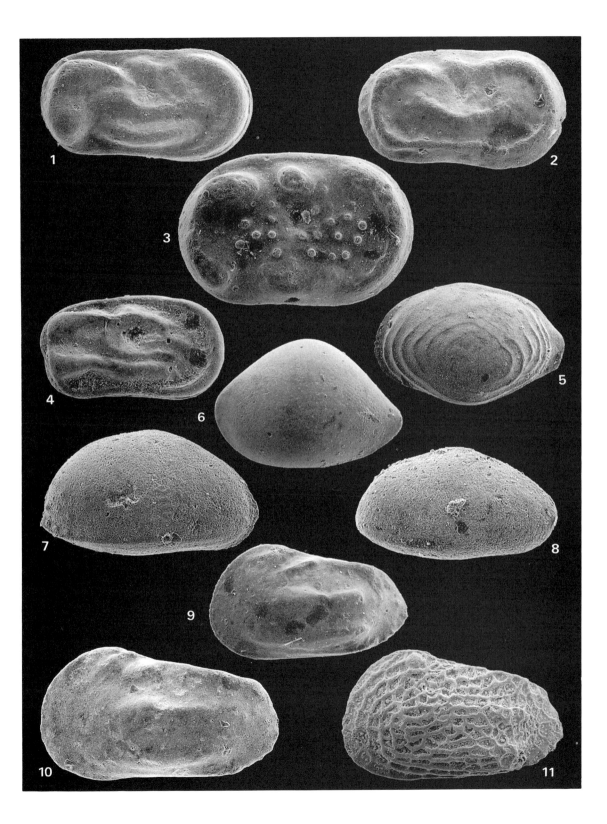

Kaye, P. 1964. Revision of British marine Cretaceous Ostracoda with notes on some additional forms. *Bull. Br. Mus. nat. Hist. (Geol.)*, London, **10**, 37-79, 9 pls.

Magné, J. & Polvêche, J. 1961. Sur le niveau à *Actinocamax plenus* (Blainville) du Boulonnais. *Annls Soc. géol. N.*, Lille, **81**, 47-62.

Neale, J. W. 1978. The Cretaceous. *In* Bate, R. & Robinson, E. (Eds), *A stratigraphical index of British Ostracoda. Geol. Journ. Spec. Issue* No. **8**, 325-284. Seel House Press, Liverpool.

Ohmert, W. 1971. Ecology of some Trachyleberididae (Ostracoda) from the Bavarian Upper Cretaceous. *In* Oertli, H. J. (Ed.), *Paléoécologie Ostracodes, Bull. Centre Rech. Pau-SNPA*, **5** suppl., 601-614.

Pokorný, V. 1963a. *Karsteneis* gen. n. (Ostracoda, Crustacea) from the Upper Cretaceous of Bohemia. *Cas. Miner. Geol.*, Praha, **8**, 39-44.

Pokorný, V. 1963b. The revision of *Cythereis ornatissima* (Reuss, 1846) (Ostracoda, Crustacea). *Rozpr. cesl. Akad. Ved*, Praha, **73**, 1-59.

Pokorný, V. 1964. The phylogenetic lines of *Cythereis marssoni* Bonnema, 1941 (Ostracoda, Crustacea) in the Upper Cretaceous of Bohemia, Czechoslovakia. *Acta Univ. Carol., Geol.*, Praha, **3**, 255-274.

Pokorný, V. 1965. New species of *Cythereis* (Ostracoda, Crustacea) from the Turonian of Bohemia. *Acta Univ. Carol., Geol.*, Praha, **1**, 75-89.

Pokorný, V. 1967. The genus *Curfsina* (Ostracoda, Crustacea) from the Upper Cretaceous of Bohemia, Czechoslovakia. *Acta Univ. Carol., Geol.*, Praha, **4**, 345-361.

Pokorný, V., 1968. *Spinoleberis krejcii* sp. n. (Ostracoda, Crustacea) from the Upper Cretaceous of Bohemia, Czechoslovakia. *Acta Univ. Carol., Geol.*, Praha, **4**, 375-389.

Pokorný, V. 1978a. The genus *Mosaeleberis* (Ostracoda, Crust.) in the Upper Cretaceous of Bohemia, Czechoslovakia. *Acta Univ. Carol., Geol.*, Praha, No. 1-2, 145-161.

Pokorný, V. 1978b. *Ostracode biostratigraphy of the Turonian and Coniacian of Bohemia, Czechoslovakia.* Paleontologicka Konference '77, 243-251. Univerzita Karlova Praha 1978.

Robaszynski, F. 1983. Conclusions to the colloquium on the Turonian Stage: integrated biostratigraphic charts and facies maps (France and adjacent areas). *Zitteliana*, München, **10**, 585-594.

Robaszynski, F., Alcaydé, G., Amédro, F., Badillet, G., Damotte, R., Foucher, J.-C., Jardiné, S., Legoux, O., Manivit, H., Monciardini, C. & Sornay, J. 1982. Le Turonian de la région-type: Saumurois et Touraine: Stratigraphie, biozonations, sédimentologie. *Bull. Centres Rech. Explor.-Prod., Elf-Aquitaine*, Pau, **6**, 119-225.

Robinson, N. D. 1986. Lithostratigraphy of the Chalk Group of the North Downs. *Proc. Geol. Ass.*, London, **97**, 141-170.

Rosenfeld, A. & Raab, M. 1974. Cenomanian-Turonian ostracodes from the Judea Group in Israel. *Geol. Surv. Israel Bull.*, Jerusalem, **62**, 1-64.

Schlanger, S. O. & Jenkyns, H. C. 1976. Cretaceous oceanic anoxic events: causes and consequences. *Geologie Mijnb.*, Den Haag, **55**, 179-184.

Schlanger, S. O., Arthur, M. A., Jenkyns, H. C. & Scholle, P. A. 1987. The Cenomanian-Turonian Oceanic Anoxic Event, I: Stratigraphy and distribution of organic carbon-rich beds and the marine δ13C excursion. *In* Brooks, J. & Fleet, A. (Eds), *Marine Petroleum Source Rocks, Geol. Soc. spec. Publ.* **26**, 371-399. Blackwell Scientific, Oxford.

Scholle, P. A. & Arthur, M. A. 1980. Carbon isotope fluctuations in Cretaceous pelagic limestones: potential stratigraphic and petroleum exploration tool. *Bull. Am. Ass. Petrol. Geol.*, Tulsa, **64**, 67-87.

Stow, D. A. V. & Dean, W. E. 1984. Middle Cretaceous black shales at site 530 in the southeastern Angola Basin. *In* Hay, W. W., Sibuet, J.-C. *et al.* (Eds), *Init. Repts DSDP*, **75**, 809-817. U.S. Govt Printing Office, Washington.

Tyson, R. V. & Funnell, B. M. 1987. European Cretaceous shorelines, stage by stage. *Palaeogeog. Palaeoclimat. Palaeoecol.*, Amsterdam, **59**, 69-91.

Van der Wiel, A. M. 1978. Ostracoda from the Albian of Petit Blanc-Nez (NW France) with emphasis on evolution in the *Cythereis reticulata - hirsuta - folkstonensis -* group. *Proc. K. ned. Akad. Wet.*, Amsterdam, **81**, 248-262.

Weaver, P. P. E. 1981. The distribution of Ostracoda in the British Cenomanian. *In* Neale, J. W. & Brasier, M. D. (Eds), *Microfossils from Recent and fossil shelf seas*, 156-162. Ellis Horwood, Chichester, for British Micropalaeontological Society.

Weaver, P. P. E. 1982. *Ostracoda from the British Lower Chalk and Plenus Marls. Palaeontogr. Soc. (Monogr.)*, London, **135**, 1-127.

# 10

# A setback for the genus *Sinocytheridea* in the Japanese mid-Pleistocene and its implications for a vicariance event

Kunihiro Ishizaki

Institute of Geology and Paleontology, Tohoku University, Japan.

## ABSTRACT

Twenty-eight sediment samples from a core drilled in Osaka Bay off the western coast of the Kii Peninsula, Southwest Japan, yielded 24 ostracod species belonging to 19 genera. Among these is *Sinocytheridea impressa* (Brady), found during an interval of 0.12 MY or more, from pre-0.44 to 0.32 Ma, as dated by calcareous nannofossils. Fossil and living records show this species to have lived in Chinese waters since the Pliocene and it is dominant in shallow coastal seas from the northern South China Sea to the Yellow Sea.

The mid-Pleistocene occurrence of *S. impressa* in Osaka Bay suggests that the contemporary shelf of the East China Sea was, at least in part, covered by a very shallow sea extending to the Japanese Islands, through which this species could cross the East China Sea.

Two lines of evidence show that the contemporary Sea of Japan developed an oceanic connection, and concurrently, the East China Sea deepened causing a fragmented distribution of *S. impressa* (manifestation of a vicariance event) 0.3 to 0.35 Ma. This habitat fragmentation might have triggered the abrupt disappearance of *S. impressa* from Japan.

## INTRODUCTION

It has been known for many years that shallow water benthonic podocopid ostracods exhibit high levels of endemism, particularly latitudinally (Pokorný, 1978; Whatley, 1983, 1986, 1988; Titterton & Whatley, 1988). Since this group lacks pelagic larvae, they can only be distributed passively, or actively in a very limited way along areas of continuous continental shelf (Teeter, 1973; Titterton & Whatley, 1988). They can, therefore, be used to elucidate past palaeogeographical events (Schallreuter & Siveter, 1985; Whatley, 1988).

The Japanese Islands are today separated from mainland China-Korea by the East China Sea, the shallowest part of which is less than 150m in the Korea Strait (Mammerickx, *et al.*, 1976) (Fig. 1).

It is generally considered, however, that during the early mid-Pleistocene (Günz-Mindel Glaciation), mainland China was connected to the Japanese Islands by a land bridge, over which the *Stegodon-Ailuropoda* fauna crossed from South China to Japan, where it evolved into the *Stegodon-Paleoloxodon* fauna peculiar to the Japanese Islands (e.g., Kamei, 1962).

*Sinocytheridea latiovata* and *S. longa*, Hou & Cheng (1982), are considered by Zhao & Whatley (1987) to be conspecific with *Cytheridea impressa* G. W. Brady. This form occurs in the Chinese Pliocene (Leizhou Peninsula; Gou *et al.*, 1983) to Holocene interval (Jiangsu Province; Hou & Chen, 1982), and is dominant in Recent shallow waters (less than 20m or so (Wang *et al.*, 1980; Wang *et al.*, 1985c; Zhao *et al.*, 1985) along the coast from the northern South China Sea to the Yellow Sea (Fig. 1). This species, however, is absent from comparable modern environments in Japan and existed there for only a very limited period during the mid-Pleistocene (pre-0.44 to 0.32 Ma) in Osaka Bay (Fig. 2; Plate 1; Table 1). This discrepancy in the modern occurrence of this species between Japan and China may be related to an extensive transgression which took place from 0.3 to 0.35 Ma in the N.W. Pacific, particularly around the Japanese Islands (Wang *et al.*, 1985b; Takayanagi *et al.*, 1987). This took place after an extended period of lower sea level, when mainland China was connected to the Japanese Islands by a land bridge, which may have, at least in part, been inundated by waters much shallower than the present East China Sea, providing a route by which *Sinocytheridea impressa* could pass.

## MATERIAL

The study is based on an offshore core (56-9) from Osaka Bay (Fig. 1). The core was 380m in length and penetrated both the Sennan-oki and the overlying Koku-juma formations. The possibly unconformable junction between the two being encountered at 172.6m, above which is a 10m thick conglomerate. The lower formation comprises alternating freshwater sands and muds, into which above 250m, are intercalated four marine clay horizons. Above its basal conglomerate, the Koku-jima Formation is mainly argillaceous with intercalated levels of pebbles and cobbles (Nakaseko *et al.*, 1984).

Twenty-eight of the 32 samples studied from marine horizons yielded ostracods (Table 1).

The upper part of the Sennan-oki Formation belongs to the Matsuyama reversed polarity epoch and the Koku-jima Formation to the Brunhes normal polarity epoch (Nakaseko *et al.*, 1984). A lignite at 20m was dated as 22.99+/-2.07 thousand years and Yamauchi & Okamura (1984) and Okamura & Yamauchi (1984), on nannofossil data, assigned the ages as shown on the left of Fig. 2.

## OCCURRENCE OF OSTRACODS AND DEPOSITIONAL ENVIRONMENTS

The 28 ostracod-bearing samples yielded 4394 ostracods, belonging to 24 species and 20 genera (Table 1). Most samples yielded more than one specimen per 1ml sediment; a maximum of 169 individuals occurred in T145 and few individuals between T141 and T9 with the exception of T107. Sparsity of ostracods at certain horizons may be attributed to the intermittent occurrence of non-marine environments, as indicated by frequent intercalations of strata which lack marine diatoms (Koizumi & Sakou, 1984).

The structure of the ostracod assemblages was analysed by using the Shannon-Wiener's information function as amended by Buzas & Gibson (1969). The examined assemblages show a very simple structure, having a diversity of less than 2.10 and a number of species less than 12 (Table 1). In three samples (T144, 37 & 19), equitability is extremely low: T19 lies immediately above a sand-gravel layer and lacks marine diatoms; the other two samples yield marine diatoms, but each includes a distinct dominant species, *Sinocytheridea impressa* in T144 and *Bicornucythere bisanensis* (Okubo) in T37. These two species, occurring in such abundance and dominance in these two samples, indicate a restricted marine or even brackish water environment (Whatley, 1983).

Table 1. The occurrence of Ostracoda in core 56-9 from the eastern margin of Osaka Bay.

| Specific names \ Samples | T6 | T9 | T12 | T15 | T19 | T31 | T33 | T35 | T37 | T39 | T76 | T77 | T78 | T79 | T80 | T82 | T94 | T95 | T96 | T97 | T99 | T101 | T102 | T107 | T141 | T141' | T144' | T145' |
|---|---|---|---|---|---|---|---|---|---|---|---|---|---|---|---|---|---|---|---|---|---|---|---|---|---|---|---|---|
| Ambtonia obai (Ishizaki, 1971) | 2 | 3 | - | - | - | 15 | 1 | 1 | - | - | - | - | - | - | - | - | - | - | - | - | - | - | - | - | - | - | 2 | 3 |
| Amphileberis gibbera Guan, 1978 | - | - | - | - | - | 1 | 1 | - | - | - | - | - | - | - | - | - | - | - | - | - | 3 | - | - | 1 | - | 10 | 2 | - |
| Aurila cymba (Brady, 1869) | - | - | - | - | - | - | 3 | - | 6 | - | - | - | - | - | - | - | - | - | - | 8 | - | 3 | 3 | - | 1 | - | - | - |
| Bicornucythere bisanensis (Okubo, 1975) | 21 | 39 | 71 | 44 | 7 | 74 | 95 | 117 | 174 | 78 | 11 | 15 | 68 | 83 | 60 | 80 | 4 | 5 | 9 | 18 | 21 | 8 | 3 | 1 | 1 | 63 | 15 | 28 |
| Callistocythere alata Hanai, 1957 | - | - | - | 2 | - | - | - | - | 10 | 4 | - | - | - | - | - | - | 117 | 53 | 24 | 9 | 8 | - | 4 | - | - | 7 | - | 9 |
| Callistocythere reticulata Hanai, 1957 | - | - | - | - | - | - | - | - | - | - | - | - | - | - | - | - | - | - | - | - | - | - | - | - | - | 12 | - | - |
| Cornucoquimba tosaensis (Ishizaki, 1968) | - | - | - | - | - | - | - | - | - | - | - | - | - | - | - | - | - | - | - | - | - | - | - | 1 | - | 10 | - | 1 |
| Cytherois nakanoumiensis Ishizaki, 1969 | - | - | - | - | - | - | - | - | - | - | - | 1 | 1 | 3 | - | - | - | - | - | - | - | - | - | - | - | - | - | - |
| Cytherois uranouchiensis Ishizaki, 1968 | - | - | - | - | - | - | - | - | - | - | 4 | 4 | 4 | 7 | 5 | - | - | - | - | - | - | - | - | - | - | - | - | - |
| Cytheromorpha acupunctata (Brady, 1880) | 3 | 4 | 8 | 6 | 27 | 10 | 10 | 20 | 9 | 58 | - | - | - | - | - | - | 4 | 4 | 11 | 8 | 2 | - | - | 26 | - | 18 | 2 | 9 |
| Krithe japonica Ishizaki, 1971 | 1 | 1 | - | - | - | - | 7 | 2 | - | - | - | - | - | - | - | - | - | - | - | - | - | - | - | - | - | - | - | - |
| Loxoconcha pulchra Ishizaki, 1968 | - | - | 1 | 15 | - | - | - | - | - | - | - | - | - | - | - | - | - | - | - | - | - | - | - | - | - | - | - | - |
| Loxoconcha viva Ishizaki, 1968 | 3 | 4 | 2 | 2 | - | 2 | 15 | 2 | 1 | - | - | - | - | - | - | - | - | - | - | - | - | - | - | - | - | - | - | - |
| Loxoconcha sp. | 4 | - | - | - | - | - | - | - | - | - | - | - | - | - | - | - | - | - | - | - | - | - | - | - | - | - | - | - |
| Neomonoceratina delicata Ishizaki and Kato, 1976 | - | - | - | 19 | 19 | 29 | 6 | 1 | 20 | 10 | 38 | 45 | 74 | 95 | 8 | 1 | - | - | - | 1 | 1 | - | - | - | - | - | - | - |
| Neopellucistoma inflatum Ikeya and Hanai, 1982 | - | - | - | - | - | - | - | - | - | - | - | - | - | - | 1 | 1 | - | - | 1 | - | 1 | - | - | - | - | - | - | - |
| Nipponocythere bicarinata (Brady, 1880) | - | - | 3 | - | 1 | 1 | - | 3 | 6 | 3 | 7 | 17 | 37 | 13 | 6 | 7 | 3 | 3 | 2 | - | - | - | - | - | - | - | - | - |
| Pacambocythere humilitorus Malz, 1982 | - | - | - | - | - | - | - | - | - | - | - | - | - | - | - | - | - | - | - | - | 4 | - | 13 | 4 | - | - | - | - |
| Parakrithella pseudadonta (Hanai, 1959) | - | - | - | - | - | - | - | - | - | 1 | - | - | - | - | - | - | - | - | - | - | 7 | - | 16 | 7 | - | - | - | - |
| Pistocythereis bradyformis (Ishizaki, 1968) | - | - | - | - | - | - | - | - | - | - | 7 | - | - | - | - | - | - | - | - | - | - | - | 1 | - | - | - | - | - |
| Pistocythereis bradyi (Ishizaki, 1968) | 3 | 4 | 3 | - | - | 6 | 3 | 23 | 12 | 20 | 9 | 2 | 5 | 13 | 31 | 38 | 17 | 34 | 9 | 7 | 3 | 8 | 3 | 2 | 4 | 3 | - | 5 |
| Sinocytheridea impressa (G. S. Brady, 1869) | 9 | 7 | 5 | 17 | 207 | 7 | 6 | 8 | 31 | 101 | 37 | 32 | 57 | 95 | 64 | - | - | 3 | 8 | 1 | - | - | - | 22 | - | 124 | 131 | 82 |
| Spinileberis quadriaculeata (Brady, 1880) | 7 | - | 12 | 7 | 6 | 5 | 8 | 31 | 37 | 32 | 57 | 95 | 19 | 3 | 8 | 1 | 3 | 7 | 1 | - | 5 | - | - | 22 | - | 32 | 13 | 76 |
| Trachyleberis scabrocuneata (Brady, 1880) | 3 | 6 | 7 | 9 | 1 | 35 | 14 | 18 | 6 | 38 | 36 | 31 | 14 | 19 | 7 | 26 | 14 | 25 | 8 | 1 | 4 | 1 | 1 | 49 | 5 | 11 | 1 | 11 |
| Gen. et sp. indet. | - | - | - | 1 | - | - | - | 1 | - | - | - | - | - | 3 | 1 | 7 | - | - | - | - | - | - | - | - | - | - | - | - |
| **Total** | 45 | 73 | 96 | 83 | 257 | 195 | 184 | 216 | 222 | 189 | 193 | 121 | 220 | 262 | 313 | 305 | 196 | 98 | 110 | 69 | 28 | 13 | 9 | 198 | 28 | 287 | 164 | 220 |
| Diversity | 1.6 | 1.6 | .96 | 1.4 | .70 | 1.9 | 1.7 | 1.5 | .9 | 1.3 | 1.5 | 1.7 | 1.8 | 1.7 | 1.7 | 1.7 | 1.4 | 1.5 | 2.0 | 1.8 | 1.8 | 1.5 | 1.2 | 2.1 | 1.9 | 1.7 | .7 | 1.5 |
| Equitability | .63 | .52 | .44 | .57 | .40 | .55 | .46 | .4 | .3 | .6 | .5 | .6 | .6 | .6 | .7 | .4 | .5 | .8 | .5 | .7 | .5 | .9 | .8 | .7 | .8 | .6 | .4 | .5 |
| Number of species | 8 | 10 | 6 | 7 | 5 | 12 | 12 | 10 | 10 | 7 | 8 | 9 | 10 | 9 | 9 | 9 | 9 | 9 | 10 | 11 | 8 | 5 | 4 | 11 | 9 | 9 | 6 | 9 |
| Individuals/sediments (ml) | 1 | 2 | 3 | 4 | 5 | 8 | 6 | 7 | 6 | 13 | 4 | 4 | 10 | 10 | 10 | 4 | 3 | 14 | 3 | 2 | 0 | 1 | 0 | 5 | 0 | 8 | 3 | 169 |
| Sediment character | m | m | m | m | m | m | m | m | m | m | m | m | m | m | m | m | m | m | m | m | m | m | m | m | msh | m | m | msn |

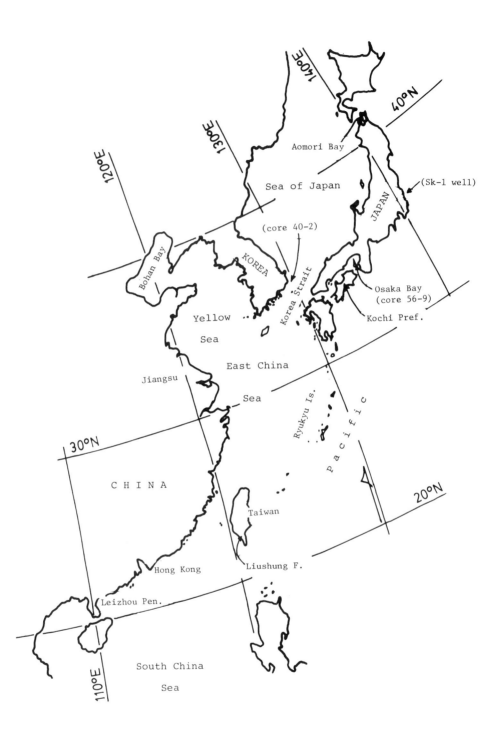

Fig. 1. A map of the Northeast Asian littoral giving the areas studied by previous workers and location of cores referred to in the text.

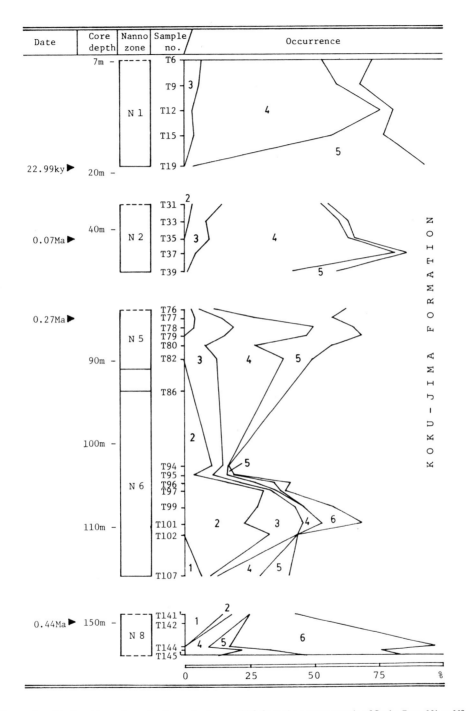

Fig. 2. The stratigraphical positions of examined samples in core 56-9 from the eastern margin of Osaka Bay. N1 to N8 represent calcareous nannofossil zones. The 'occurrence' column shows the percentage frequencies of selected index ostracod species or species groups: 1 = *Pacambocythere humilitorus*; 2 = *Aurila cymba - Pistocythereis bradyformis*; 3 = *Cytherois uranouchiensis - Pistocythereis bradyi - Loxoconcha viva*; 4 = *Bicornucythere bisanensis*; 5 = *Spinileberis quadriaculeata*; 6 = *Sinocytheridea impressa*. Dates include the disappearance of *Pseudoemiliania lacunosa* (0.44 Ma), the appearance of *Emiliania huxleyi* (0.27 Ma), the onset of the acme zone of *Emiliania huxleyi* (0.07 Ma) and ¹⁴C age of a lignite bed (22.99 kY).

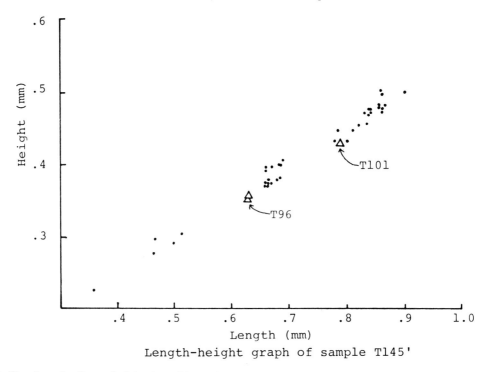

Length-height graph of sample T145'

Fig. 3. Size dispersion diagram for left valves of female *Sinocytheridea impressa* from sample T145' and comparison with those from samples T101 and T96.

The present day ecology of the fossil species associations is also indicative of restricted marine environments such as occur within bays. *Bicornucythere bisanensis*, *Cytheromorpha acupunctata* (Brady), *Pistocythereis bradyi* (Ishizaki), *Spinileberis quadriaculeata* (Brady), *Trachyleberis scabrocuneata* (Brady) are all common or abundant in such environments today, and the absence of *Neonesidea* is further evidence (Table 1). The majority of ostracod-bearing samples are largely clay with shell fragments in T145' and 141', and represent conditions relatively free of terrigenous material or disturbance due to wave action. These characteristics indicate that the sediment samples were deposited in a bay backed by a low hinterland and protected from the open sea.

In an attempt to find evidence of temporal fluctuations of the environment, the following six species or species groups were selected. Most of them occur living in Japanese bays at the present day such as: Uranouchi Bay in Kochi Prefecture (10.5km long and less than 1km in average width;

Ishizaki, 1968) and Aomori Bay in Aomori Prefecture (Ishizaki, 1971):

1) *Pacambocythere humilitorus* Malz (=*Buntonia parascorta* Ishizaki, 1983): This form is rare in the Pliocene Ananai Formation which occurs sporadically along the southeastern coast of Kochi Prefecture, Shikoku (Ishizaki, 1983; Ishizaki & Tanimura, 1985). The latter study revealed this species to have been distributed under the influence of the then prevailing Kuroshio Current or in an open shelf area. The occurrence of this form in the mid-Pleistocene of Osaka Bay may, therefore, imply that the area was subject to the influence of open sea waters.

2) *Aurila cymba* (Brady) - *Pistocythereis bradyformis* (Ishizaki): This species association was found in the area extending from the centre to the mouth of Uranouchi Bay.

3) *Cytherois uranouchiensis* Ishizaki - *Pistocythereis bradyi* - *Loxoconcha viva* Ishizaki: In contrast to the above species association, this was found in the area extending from the centre of

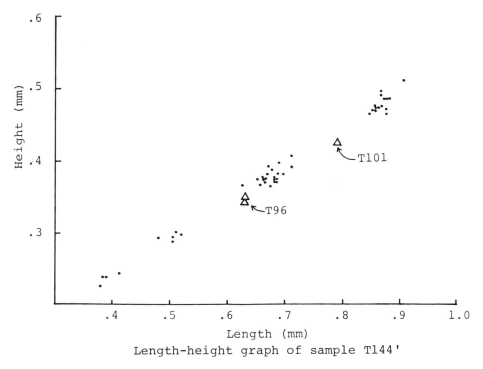

Length-height graph of sample T144'

Fig. 4. Size dispersion diagram for left valves of female *Sinocytheridea impressa* from sample T144' and comparison with those from samples T101 and T96.

Uranouchi Bay to slightly landwards.

4) *Bicornucythere bisanensis*: This species is generally distributed in close association with *Spinileberis quadriaculeata* in Uranouchi Bay.

5) *Spinileberis quadriaculeata*: this species is distributed widely in Uranouchi Bay in association with *Bicornucythere bisanensis*, but its dominance is, more or less, confined to the inner part of the bay. In Aomori Bay, this species is not as abundant as *B. bisanensis* at water depths of 20m or more. It is commoner at shallower depths, showing no definite relationship with substrate, implying that the relative frequencies of both species are dependent on water depth (Ishizaki, in press).

6) *Sinocytheridea impressa*: Previously the occurrence of this species in Japan was based on one immature valve from the Pleistocene of the Atsumi Peninsula (Yajima, 1987). Common individuals occur in many samples betwen T145 and T96 in core 56-9 from Osaka Bay (Table 1; Fig. 2). Despite its rarity in Japan, this species is widely distributed along the coast of mainland China from

the northern part of the South China Sea to the Yellow Sea, where it particularly dominates shallow seas at depths of less than 20 to 30m (Wang *et al.*, 1985c; Zhao *et al.*, 1985).

Temporal environmental vicissitudes in the Pleistocene of Osaka Bay can be predicted by means of up-core fluctuations in frequencies of the six selected indicators (Fig. 2). This figure clearly shows that the interval immediately below zone N5 is dominated by three of the indicators in turn, in the order 1-2-3. Such changes in faunal dominance suggest the waxing and waning of open sea marine influence during the interval. Thereafter, the area gradually became a more protected embayment, as suggested by the disappearance of the indicators 1 and then 2 and the increasing frequency of 4 and 5. *Bicornucythere bisanensis* possibly became dominant as the bay deepened and *Spinileberis quadriaculeata* as it became shallower, resulting in fluctuating frequencies of these two species in the later part of the time interval.

## OSTRACOD CARAPACE SIZE

It is known that the size of podocopid ostracod carapaces can vary according to such environmental factors as salinity (Barker, 1963), carbonate content (Urlichs, 1971), temperature expressed as water depth (Benson & Sylvester-Bradley, 1971) or latitude (McKenzie, 1969), food supply or population density (Puri, 1968; Keen, 1971), seasonal variability (Szczechura, 1971) and grain size of sediments (Puri, 1968).

With the above in mind, the length and height of several selected ostracod species were measured. Plots were made using a Hewlett-Packard 7470A plotter.

Length-height plots show no distinct differences in valve size for corresponding size groups from sample to sample, with the single exception of *Sinocytheridea impressa*.

Substantial numbers of specimens (female left valves) of *S. impressa* are available from T145' and T144' (Figs 3 & 4; intervals prior to 0.44 Ma), but most specimens had suffered dissolution in T141'. Higher in the core, until the level of the last appearance of that species, all samples yielded a few specimens for measurement. In Figs 3 & 4 one small adult instar from T101 is plotted and two A-1 instars from T96 also lie at the smaller end of the size range.

Specimens from T101 and T96 were insufficient for statistical analysis. The probability that any one of the valves would plot at such an aberrant point is small, however, being less than 10% by visual inspection. The probability that all these three points taken together show such deviation should be significantly small. Such dwarfing of *S. impressa* could be caused by environmental constraints severe enough to influence valve size in these two instars and eventually to lead to its extinction in Japan.

## OCCURRENCE OF *SINOCYTHERIDEA IMPRESSA*

This species has been recorded from the Pliocene and Quaternary of the Chinese coast (Hou & Chen,

1982; Gou *et al.*, 1983; Wang *et al.*, 1985b; Wang *et al.*, 1985d;) and from the Recent of Malaysia (Zhao & Whatley, in press), Hong Kong (Brady, 1869; Whatley & Zhao, 1988; Zhao & Whatley, 1987, 1988a, b) and along the coast of mainland China (Wang & Bian, 1985; Wang & Zhao, 1985; Wang *et al.*, 1985a, d, e; Zhao, 1985).

Although Wang & Zhao (1985) show that this form is distributed in shallow waters at depths of less than 30-50m in the South China Sea and less than 90-100m in the East China Sea, their data imply that it is dominant in coastal waters at depths of less than 50m, extending into estuaries, marshes and tidal pools. In their study of Pleistocene sediment cores from the East China Sea and the Yellow Sea, Wang *et al.* (1985d) also mentioned that *Sinocytheridea latiovata*, 'together with *Sinocythere sinensis* Hou, was an inhabitant of polyhaline shallow waters at depths of less than 20m'.

Apart from mainland China, Hu & Yeh (1978) reported *S. impressa* (as '*Cyprideis yehi*') from the Pleistocene Liushuang Formation in the Tainan district of southwestern Taiwan (Fig. 1); and Cheong *et al.* (1986) found infrequent occurrences of *Sinocytheridea* sp. in submarine cores off the east coast of Korea. It is possible that this form is conspecific with *S. impressa* as judged from their figure and it still lives in coastal areas of Korea.

From this data, it is evident that the distribution of *S. impressa* along the continental margin of mainland Asia is very different from that of the Japanese Archipelago, from which the species disappeared in the Pleistocene.

## OCEANOGRAPHICAL IMPLICATIONS

Principal components factor analysis of planktonic foraminifera from the surface of bottom sediments in the northwestern coastal Pacific shows the relative importance of the Tsugaru Warm Current (F 1), the Oyashio Current (F 2), the Kuroshio Current (F 3), and oceanic water (F 4) (Oda *et al.*, 1983). An exploratory oil and gas well, drilled off the coast of Kashima, Ibaraki Prefecture, central Japan (Kashima-oki Sk-1 well) (Takayanagi *et al.*, 1987) (Fig. 1) provides very important information regarding the then prevailing oceanography (Fig. 5). The most remarkable of the temporal fluctuations in foraminiferal assemblage composition occur in intervals from near the top part of the examined sequence to a well depth of 330m, where the first and third varimax assemblages are in reciprocal relationship with each other.

Such fluctuations indicate oceanographical conditions subject to countervailing effects of the Tsugaru Warm Current (F 1) and the Kuroshio Current (F 3) at the site of the Kashima-oki Sk-1 well: the stronger the Kuroshio Current became, the less the Tsugaru Warm Current extended to the south; and the weaker the Kuroshio Current became, the farther the Tsugaru Warm Current extended southwards. The fact that no such countervailing effects are observed below that well depth, may suggest a substantial shift in oceanographical conditions at and around the Kashima-oki Sk-1 well. This horizon is dated as 0.35 Ma on the basis of calcareous nannofossils (Takayanagi *et al.*, 1987). These lines of evidence may imply that the Sea of Japan came to be connected in some way to the open sea and that concurrently the East China Sea and adjacent shelf areas may have deepened at about 0.35 Ma.

In addition, Wang *et al.* (1985b) studied foraminifera from cores on the inner shelf of the East China Sea and recognized five extensive transgressions since the beginning of the Pleistocene. They mentioned an extensive *Spirillina* transgression (dated as 0.30 Ma on the basis of magnetic events), possibly equivalent to the Mindel/Riss interglaciation, following the *Paromalina* transgression of 2.26 Ma after a considerable period during which shelf areas were mostly emergent. During the greater part of this period (between 2.26 Ma and 0.30 Ma) the Japanese Islands may have been connected to mainland Asia.

Ichikura & Ujiié (1976) studied the lithology and planktonic foraminifera from cores in the Sea of Japan and concluded that this may have been an enclosed marginal sea during almost the entire Pleistocene.

Such sources of information are very important in understanding the history of Pleistocene eustatic sea-level changes. After a long period of emergence of the shelf areas, an extensive transgression took place between 0.30-0.35 Ma, inundating the land bridge between Japan and the continent and leading to fragmentation of the previously continuous land mass.

## EVIDENCE FROM THE DISTRIBUTION OF *SINOCYTHERIDEA IMPRESSA*

Surrounding the land bridge were shallow seas, through which *S. impressa* extended its range from the littoral of the Asian continent to the Japan-Taiwan area. This extended distribution may have been established pre-0.44 Ma but lasted only until the following extensive transgression, which took place 0.30-0.35 Ma. During times of lower sea level, Japanese shallow water ostracod assemblages were enriched by the appearance of *Spinileberis quadriaculeata*, *Pistocythereis bradyi* and *P. bradyformis* in association with *S. impressa* (Ishizaki, in press).

The subsequent transgression, by inundating the shelf, would have tended to isolate shallow water assemblages, especially those of the Japanese Islands (manifestation of a vicariance event). The theory of island biogeography by MacArthur & Wilson (1967) has theoretically widespread application (Raup & Stanley, 1978). After the isolation of its habitat, therefore, extinction due to further fragmentation or insularization of that habitat, and positive relationships between the number of species and extent of island size (Hope, 1973) can be implicated in the disappearance of *S. impressa* from Japan.

The effect of habitat fragmentation on different species is highly variable, possibly depending upon various factors such as abundance, body size, food preference and habitat (Diamond, 1984). *S. impressa* may be confined to and be able to dominate only in shallow waters, which could be one of the factors that led to its extinction in Japan. A possible reflection of such critical conditions may be the dwarfism observed in the populations of *S. impressa* from samples T101 and T96, which represent periods immediately before its disappearance.

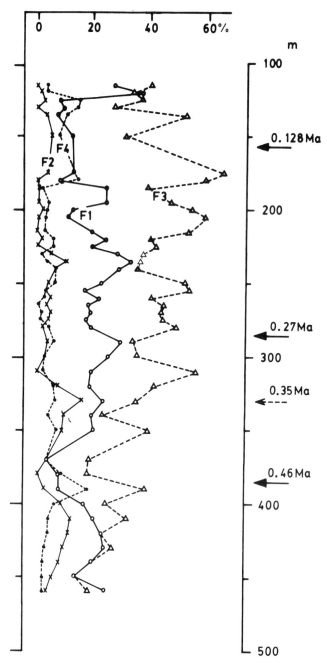

Fig. 5. The percentage frequencies of the first four varimax assemblages of planktonic foraminifera (F1 to F4) along well depth of Kashima-oki Sk-1 well. Indicators of the Tsugaru Warm Current (F1), the Oyashio Current (F2), the Kuroshio Current (F3) and oceanic water (F4) were determined by Oda, Ishizaki & Takayanagi (1983) through examination of calculated factor scores. Dates include the disappearance of *Emiliania annula/E. ovata* (0.46 Ma), the appearance of *Emiliania huxleyi* (0.27 Ma) and the boundary between oxygen isotope stages 5 and 6 (0.128 Ma), as determined by the oxygen isotopic record of the planktonic foraminifer *Globorotalia inflata* (d'Orbigny) (Takayanagi *et al.*, 1987). The sediments at a well depth of 330m are dated as 0.35 Ma on the basis of interpolation between 0.46 Ma and 0.27 Ma datum planes of calcareous nannofossils.

The ostracods of Osaka Bay show that after the 0.30-0.35 Ma transgression, a shallow water inhabitant, *Spinileberis quadriaculeata*, replaced *Sinocytheridea impressa*. Wang *et al.* (1985a) mentioned that the conspicuous absence of the worldwide euryhaline ostracod genus *Cyprideis* along the greater part of the Chinese coast is presumably due to its replacement by '*Sinocytheridea latiovata*'. By the same token, *S. impressa* may have been replaced by *Spinileberis quadriaculeata* inhabiting similar habitats, which could account for its disappearance from Japan 0.30-0.35 Ma.

## CONCLUSIONS

The Japanese Islands are separated today from mainland China by the East China Sea. These two areas may have been connected by a land bridge by the emergence of shelf areas during lowered sea level episodes in the early Middle Pleistocene. This land bridge seems to have, at least in part, given way to shallow seas, where the shallow marine ostracod *Sinocytheridea impressa* was able to extend its range into the China-Korea-Japan-Taiwan area. This species then disappeared from Osaka Bay, Japan, 0.32 Ma, in marked contrast to its presence in the littoral of mainland Asia, where it is widespread throughout a considerable geographical and ecological range.

Such a contrast may have been driven by two factors: the splitting of the *S. impressa* populations (a vicariance event) because of deepened shelf areas between the opposite emergent areas, which resulted from the transgression occurring 0.30 to 0.35 Ma; the higher extinction proneness of this species, possibly due to its habitat speciality to dominate only shallow coastal waters.

Hence, it is evident that the waxing and waning of some fossil ostracods can be related to sea level changes, occurring mostly on a regional or even wider, rather than local scale.

## ACKNOWLEDGEMENTS

I am grateful to Professor Y. Takayanagi, Institute of Geology and Paleontology, Faculty of Science, Tohoku University for reading the manuscript. I am also indebted to Professor Emeritus K. Nakaseko, Osaka University, who kindly placed at my disposal one of the sediment cores drilled in Osaka Bay. Particular thanks are due to the editors of the Proceedings and two anonymous reviewers for comments and criticisms which improved the quality of the final version of this paper.

## REFERENCES

Barker, D. 1963. Size in relation to salinity in fossil and Recent euryhaline ostracods. *J. mar. biol. Ass. U.K.*, Plymouth, **43**, 785-795.

Benson, R. H. & Sylvester-Bradley, P. C. 1971. Deep-sea ostracodes and the transformation of ocean to sea in the Tethys. *In* Oertli, H. J. (Ed.), *Paléoécologie des Ostracodes, Pau 1970, Bull. Centre Rech. Pau-SNPA*, **5** suppl., 63-91, 15 figs, 1 tab., 1 pl.

Brady, G. S. 1869. Les entomostracés des Hong Kong. *In* Folin, L. De & Periér, L. (Eds), *Les Fonds de la Mer*, 1(1), 115-159. Savy, Paris.

Buzas, M. A. & Gibson, T. G. 1969. Species diversity: Benthic foraminifera in western North Atlantic. *Science, N.Y.*, New York, **163**, 72-75.

Cheng Hae-Kyung, Lee Eui-Kyeong, Paik Kwang-Ho & Chang Soon-Keun. 1986. Recent ostracodes from the southwestern slope of the Ulleung basin, east sea, Korea. *J. Paleont. Soc. Korea*, Seoul, **2** (1986), 38-53.

Diamond, J. M. 1984. "Normal" extinctions of isolated populations. *In* Nitecki, M. H. (Ed.), *Extinctions*, 191-246. Univ. Chicago Press, Chicago.

Gou Yunsian, Zheng Shuying & Huang Baoren. 1983. Pliocene ostracode fauna of Leizhou Peninsula and northern Hainan Islands, Guangdong Province. *Palaeont. Sin.*, Geological Survey of China, Peking, No. **163**, N. S. B, No. 18, 1-134. (In Chinese.)

Hope, J. H. 1973. Mammals of the Bass Strait Islands. *Roy. Soc. Victoria, Proc.*, Melbourne, **85**(2), 163-195.

Hou Yu-tang & Chen Te-chiung. 1982. *Sinocytheridea latiovata* Hou et Chen gen. et sp. nov. & *Sinocytheridea longa* Hou et Chen gen. et sp. nov. *In* Hou You-tang, Chen Te-chiung, Yang Heng-ren, Ho Jun-de, Zhou Quan-chun & Tian, Muqu. *Cretaceous-Quaternary Ostracode Fauna from Jiangsu Province*, 164-166. Geol. Publ. House, Peking. (In Chinese.)

Hu Chung-Hung & Yeh Kuei-Yu. 1978. Ostracod faunas from the Pleistocene Liushung Formation in the Tainan area, Taiwan. *Proc. geol. Soc. China*, Taipei, No. 21, 151-162.

Ichikura, M. & Ujiié, H. 1976. Lithology and planktonic foraminifera of the Sea of Japan piston cores. *Nat. Sci. Mus. Bull.*, Tokyo, Ser. C, **2**, 151-178.

Ishizaki, K. 1968. Ostracodes from the Uranouchi Bay, Kochi Prefecture, Japan. *Sci. Rep. Tohoku Univ.*, Sendai, 2nd Ser., **40**(1), 1-45.

Ishizaki, K. 1971. Ostracodes from Aomori Bay, Aomori Prefecture, northeast Honshu, Japan. *Sci. Rep. Tohoku Univ.*, Sendai, 2nd Ser., **43**(1), 59-97.

Ishizaki, K. 1983. Ostracoda from the Pliocene Ananai Formation, Shikoku, Japan -Description-. *Trans. Proc. palaeont. Soc. Japan*, Tokyo, N. S., No. **131**, 135-158.

Ishizaki, K. In press. Sea level change in mid-Pleistocene time and effects on Japanese ostracode faunas. *Bull. mar. Sci.*, Coral Gables.

Ishizaki, K. & Tanimura, Y. 1985. Ostracoda from the Pliocene Ananai Formation, Shikoku, Japan, Faunal analysis. *Trans. Proc. palaeont. Soc. Japan*, Tokyo, N.S., No. **137**, 50-63.

Kamei, T. 1962. Some problems on the succession of the Quaternary mammalian faunas in Japan. *Earth Sci. Ass. Geol. Collab.*, Tokyo, Japan, No. **60/61**, 23-24. (In Japanese.)

Keen, M. C. 1971. A palaeoecological study of the ostracod *Hemicyprideis montosa* (Jones & Sherborn) from the Sannoisian of NW Europe. *In* Oertli, H. J. (Ed.) *Paléoécologie des Ostracodes, Pau 1970, Bull. Centre Rech. Pau-SNPA*, **5** suppl., 523-543.

Koizumi, I. & Sakou, T. 1984. Detailed survey on diatoms in the drilling core samples at the Kansai International Airport in Osaka Bay off Senshu, Central Japan. *In* Nakaseko, K. (Ed.), *Geological Survey of the Submarine Strata at the Kansai International Airport in Osaka Bay, Central Japan*, 57-68. Rep. Calamity Sci. Inst., Osaka. (In Japanese.)

MacArthur, R. H. & Wilson, E. O. 1967. *The Theory of Island Biogeography*. 203 pp. Princeton Univ. Press, Princeton.

McKenzie, K. G. 1969. Notes on the Paradoxostomatids. *In* Neale, J. W. (Ed.), *The Taxonomy, Morphology and Ecology of Recent Ostracoda*, 48-66, Oliver & Boyd, Edinburgh.

Mammerickx, J., Fisher, R. L., Emmel, F. J. & Smith, S. M. 1976. *Bathymetry of the East Asian Seas*, Scripps Inst. Oceanogr., San Diego.

Nakaseko, K., Takemura, K., Nishiwaki, N., Nakagawa, Y., Fusutani, M. & Yamauchi, M. 1984. Stratigraphy of the submarine strata at the Kansai International Airport in Osaka Bay off Senshu, Central Japan. *In* Nakaseko, K. (Ed.), *Geological Survey of the Submarine Strata at the Kansai International Airport on Osaka Bay, Central Japan*, 191-198. Rep. Calamity Sci. Inst., Osaka. (In Japanese.)

Oda, M., Ishizaki, K. & Takayanagi, Y. 1983. Analysis of planktonic foraminifera in the surface marine sediments of east of Honshu. Spec. Proj. Res. *"The Ocean Characteristics and their Changes" Newsletter*, No. **11**, 3-9. (In Japanese.)

Okamura, M. & Yamauchi M. 1984. Detailed survey of nannofossils at the Kansai International Airport in Osaka Bay, Central Japan. *In* Nakaseko, K. (Ed.), *Geological Survey of the Sub-marine Strata at the Kansai International Airport in Osaka Bay, Central Japan*, 19-28. Rep. Calamity Sci. Inst., Osaka. (In Japanese.)

Pokorný, V. 1978. Ostracodes. *In* Haq, B. U. & Boersma, A. (Eds), *Introduction to Marine Micropaleontology*, 109-149. Elsevier, New York, Oxford.

Puri, H. S. 1968. Ecologic distribution of Recent Ostracoda. *In* Proc. Symp. Crustacea, pt. 1, *Mar. Biol. Ass. India (1966)*, 457-495.

Raup, D. M. & Stanley, S. M. 1978. *Principles of Paleontology*, 2nd ed., 481 pp. Freeman & Co., San Francisco.

Schallreuter, R. E. L. & Siveter, D. J. 1985. Ostracodes across the Iapetus ocean. *Palaeontology*, London, **28**(3), 577-598.

Szczehura, J. 1971. Seasonal changes in a reared fresh-water species, *Cyprinotus (Heterocypris) incongruens* (Ostracoda), and their importance in the interpretation of variability in fossil ostracodes. *In* Oertli, H. J. (Ed.), *Paléoécologie des Ostracodes, Pau 1970, Bull. Centre Rech. Pau-SNPA*, **5** suppl., 191-205.

**Plate 1**

Figs 1-4. *Sinocytheridea impressa* (Brady) from core 56-9 (sample T144) from the eastern margin of Osaka Bay. Length of upper scale bar for Figs 1-4a = 500μm and length of lower scale bar for Fig. 4b = 100μm.

Fig. 1. External lateral view of right valve, IGPS coll. cat. no. 98844.
Fig. 2. External lateral view of left valve, IGPS coll. cat. no. 98845.
Fig. 3. Internal lateral view of left valve, IGPS coll. cat. no. 98846 .
Fig. 4a. Internal lateral view of right valve, IGPS coll. cat. no. 98847.
Fig. 4b. Right valve, detail of muscle scars, IGPS coll. cat. no. 98847.

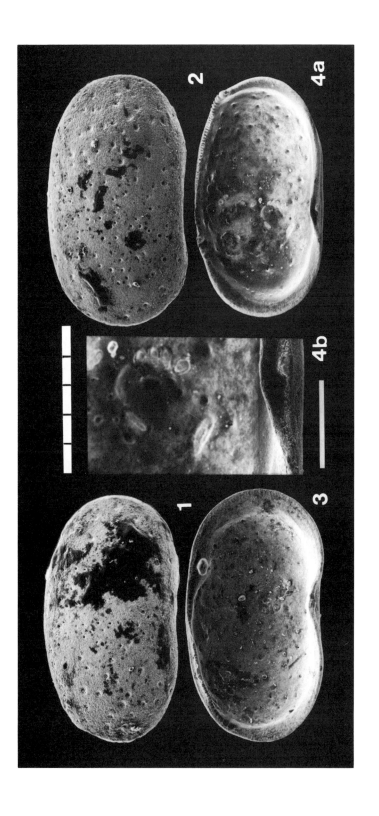

Takayanagi, Y., Saito, T., Okada, H., Ishizaki, K., Oda, M., Hasegawa, S., Okada, H. & Manickam, S. 1987. Mid-Quaternary paleoceanographic trend in near-shore waters of the Northwest Pacific - A case study based on an offshore well. *Sci. Rep. Tohoku Univ.*, Sendai, 2nd Ser., **57**(2), 105-137.

Teeter, J. W. 1973. Geographic distribution and dispersal of some Recent shallow-water marine Ostracoda. *Ohio J. Sci.*, Columbus, **73**(1), 46-54.

Titterton, R. & Whatley, R. C. 1988. The provincial distribution of shallow water Indo-Pacific marine Ostracoda: Origin, antiquity, dispersal routes and mechanisms. *In* Hanai, T., Ikeya, N. & Ishizaki, K. (Eds), *Evolutionary biology of Ostracoda, its fundamentals and applications*, proceedings of the Ninth International Symposium on Ostracoda, held in Shizuoka, Japan, 29 July - 2 August 1985, Developments in palaeontology and stratigraphy, **11**, 759-786, Kodansha Ltd., Tokyo and Elsevier, Amsterdam, Oxford, New York, Tokyo.

Urlichs, M. 1971. Variability of some ostracods from the Cassian Beds (Alpine Triassic) depending on the ecology. *In* Oertli, H. J. (Ed.), *Paléoécologie des Ostracodes, Pau 1970, Bull. Centre Rech. Pau-SNPA*, **5** suppl., 695-715.

Wang Pinxian & Bian Yunhua. 1985. Foraminifera and Ostracoda in bottom sediments of the Bohai Gulf and their bearing on Quaternary paleoenvironments. *In* Wang Pinxian *et al.* (Eds), *Marine Micropaleontology of China*, 133-150. China Ocean Press Beijing, Springer-Verlag, Berlin, Heidelberg, New York, Tokyo.

Wang Pinxian, Hong Xueqing & Zhao Quanhong. 1985a. Living foraminifera and Ostracoda: Distribution in the coastal area of the East China Sea and the Huanghai Sea. *In* Wang Pinxian *et al.* (Eds), *Marine Micropaleontology of China*, 243-255. China Ocean Press Beijing, Springer-Verlag, Berlin, Heidelberg, New York, Tokyo.

Wang Pinxian, Min Qiubao & Bian Yunhua. 1980. Distribution of foraminifera and Ostracoda in bottom sediments of the north-western part of the southern Yellow Sea and its geological significance. *Papers on Marine Micropaleontology*, 61-83. China Ocean Press Beijing.

Wang Pinxian, Min Qiubao, Bian Yunhua & Cheng Xinrong. 1985b. On Micropaleontology and stratigraphy of Quaternary marine transgressions in East China. *In* Wang Pinxian *et al.* Eds), *Marine Micropaleontology of China*, 265-284. China Ocean Press Beijing, Springer-Verlag, Berlin, Heidelberg, New York, Tokyo.

Wang Pinxian, Min Qiubao & Gao Jianxi. 1985c. A preliminary study of foraminiferal and ostracod assemblages of the Huanghai Sea. *In* Wang Pinxian *et al.* (Eds), *Marine Micropaleontology of China*, 115-132. China Ocean Press Beijing, Springer-Verlag, Berlin, Heidelberg, New York, Tokyo.

Wang Pinxian, Zhang Jijun & Gao Jianxi. 1985d. A close-up view of lowered sea-level microfauna from the East China and Huanghai Seas in the Late Pleistocene. *In* Wang Pinxian *et al.* (Eds), *Marine Micropaleontology of China*, 256-246. China Ocean Press Beijing, Springer-Verlag, Berlin, Heidelberg, New York, Tokyo.

Wang Pinxian, Zhang Jijun & Min Qiubao. 1985e.

Distribution of foraminifera in surface sediments of the East China Sea. *In* Wang Pinxian *et al.* (Eds), *Marine Micropaleontology of China*, 34-69. China Ocean Press Beijing, Springer-Verlag, Berlin, Heidelberg, New York, Tokyo.

Wang Pinxian & Zhao, Quanhong. 1985. Ostracod distribution in bottom sediments of the East China Sea. *In* Wang Pinxian *et al.* (Eds), *Marine Micropaleontology of China*, 70-92. China Ocean Press Beijing, Springer-Verlag, Berlin, Heidelberg, New York, Tokyo.

Whatley, R. C. 1983. The application of Ostracoda to palaeoenvironmental analysis. *In* Maddocks, R.F. (Ed.), *Applications of Ostracoda*, proceedings of the Eighth International Symposium on Ostracoda, July 26-29, 1982, 51-77. Univ. Houston Geos., Houston, Texas.

Whatley, R. C. 1986. The southern end of Tethys: An important locus for the origin and evolution of both deep and shallow water Ostracoda. *In* McKenzie, K. G. (Ed.), *Shallow Tethys 2*, 461-474. A. A. Balkema, Rotterdam.

Whatley, R. C. 1988. Ostracoda and palaeogeography. *In* De Deckker, P., Colin, J.-P. & Peypouquet, J.-P. (Eds), *Ostracoda in the Earth Sciences*, 103-123. Elsevier, Amsterdam, Oxford, New York, Tokyo.

Whatley, R. C. & Zhao Quanhong. 1988. A revision of Brady's 1869 study of the Ostracoda of Hong Kong. *J. micropalaeontol.*, London, **7**(1), 21-29.

Yajima, M. 1987. Pleistocene Ostracoda from the Atsumi Peninsula, Central Japan. *Trans. Proc. palaeont. Soc. Japan*, Tokyo, N. S., No. **146**, 49-76.

Yamauchi, M. & Okamura, M. 1984. Synopsis on the occurrence of nannofossils of the submarine strata at the Kansai International Airport in Osaka Bay, Central Japan. *In* Nakaseko, K. (Ed.), *Geological Survey of the Submarine Strata at the Kansai International Airport in Osaka Bay, Central Japan*, 13-17. Rep. Calamity Sci. Inst., Osaka. (In Japanese.)

Zhao Quanhong. 1985. Distribution of Recent ostracodes in coastal zones of East China Sea and Yellow Sea. *Acta Oceanologica Sinica*, Peking, **7**(2), 193-204.

Zhao Quanhong, Wang Pinxian & Zhang Qinglan. 1985. Ostracoda in bottom sediments of the South China Sea off Guangdong Province, China. Their taxonomy and distribution. *In* Wang Pinxian *et al.* (Eds), *Marine Micropaleontology of China*, 196-218. China Ocean Press Beijing, Springer-Verlag, Berlin, Heidelberg, New York, Tokyo.

Zhao Quanhong & Whatley, R. C. 1987. On *Sinocytheridea impressa* (Brady). *Stereo-Atlas Ostracod Shells*, London, **14**(1), 13-16.

Zhao Quanhong & Whatley, R. C. 1988a. Distribution of ostracod assemblages in bottom sediments of Malaysian coastal waters. *J. Tongji Univ. Shanghai*, **16**(2), 159-168. (In Chinese.)

Zhao Quanhong & Whatley, R. C. 1988b. A revision of Brady's 1869 Ostracoda of Hong Kong. *Acta Micropalaeontol. Sinica*, Beijing, **5**(1), 15-23. (In Chinese.)

Zhao Quanhong & Whatley, R. C. In press. The Ostracoda of the Sedili River estuary and Jason Bay, SW Malaysia. *Micropaleontology*, New York.

# 11

# Ostracoda, sea-level changes and the Eocene-Oligocene boundary

Michael Keen

Department of Geology and Applied Geology,
University of Glasgow, Glasgow G12 8QQ, U.K.

## ABSTRACT

Biostratigraphy has so far proved inadequate for resolving the problem of correlation at the Eocene-Oligocene boundary, between oceanic sediments and marginal marine sediments, and between high latitude and low latitude deposits. This is especially the case with the European stratotypes and the international biozones. The ostracod faunas of late Eocene and early Oligocene strata on the Isle of Wight, southern England, have been utilized to construct salinity profiles from which it is possible to recognize four distinct eustatic events. These correlate with recently published global eustatic curves, and help to tie-in the English sequence with the international scale. The Eocene and the Oligocene ostracod faunas exhibit important differences at the species level, but not at higher taxonomic levels. There is no evidence for a mass extinction.

## INTRODUCTION

The Oligocene was not included in Lyell's original division of the Tertiary (1833), but was introduced by Beyrich in 1854 as the name of a new era in earth history marked by a marine transgression thought to be of worldwide extent. During the past thirty years considerable controversy has existed over the exact definition of the Oligocene; at one time it seemed that removal of its lower and upper stages to the Eocene and Miocene respectively, would leave it as no more than a single stage, i.e. the Rupelian. However, more recent studies have shown that it was an important period during which fundamental changes occurred. Most debate has centred around the definition of the base of the epoch. It is now very apparent that major biological and physical changes can be recognized in many parts of the world during the time interval of 34-40 million years ago. This 'Terminal Eocene

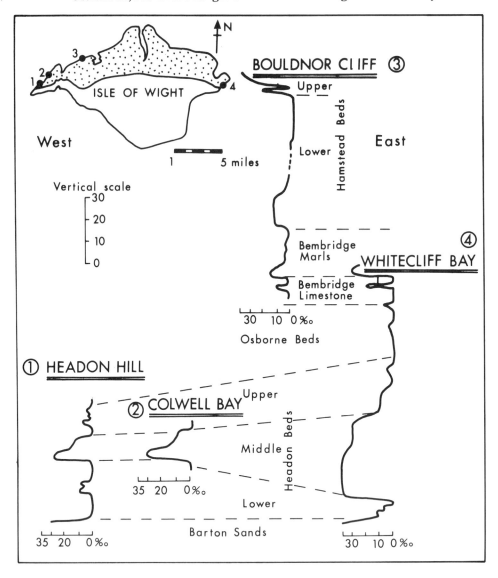

Fig. 1. Salinity curves derived from ostracod distributions for the four principle localities on the Isle of Wight. Note that the transgressive phase (increased salinity values) is sharply defined, the regressive phase more gradual.

Event' includes marked climatic cooling, perhaps related to the formation of the first significant sea ice around Antarctica, deepening of the CCD and an associated expansion of the psychrosphere, significant sea-level changes (Haq *et al.*, 1987), widespread oceanic hiatus and onshore unconformities, an important extinction event, and a possible extraterrestrial impact indicated by tektite and microtektite strewnfields, iridium abundance, and an impact crater. These events are fully described in Cavalier *et al.* (1981), and further details can be seen in the symposium volume on terminal Eocene events edited by Pomerol & Premoli-Silva (1986). However, correlation of these events has proved difficult and it is not clear how synchronous they really are and whether any are causally related. The extinction event has attracted some attention because it is one of the clearest peaks on the curves of Raup & Sepkoski (1984), and appears to support their ideas on the 26 MY periodicity of mass extinctions.

Correlation of strata deposited during this period is difficult because none of the proposed biostratigraphical zonal schemes is satisfactory. There is considerable debate over the recognition of nannofossil and planktonic foraminiferal zones in deep *versus* shallow water, and in low latitudes compared with high latitudes. It is not easy to tie in the classical European stratotypes with the oceanic record of biostratigraphy and magnetostratigraphy. Radiometric dating has not helped either; works published during the 1980s date the boundary at various ages between 32.5 MY and 39 MY. In summary, the Eocene-Oligocene boundary lies somewhere within the planktonic foraminiferal zones P15-P18 and nannoplankton zones NP18-NP22. This problem cannot be satisfactorily resolved until a boundary stratotype is chosen and correlation is better understood. In the meantime, the classification proposed by Haq *et al.* (1987) is followed in this paper.

## SEA-LEVEL CHANGES IN THE LATE EOCENE AND EARLY OLIGOCENE OF THE HAMPSHIRE BASIN, ENGLAND

Biostratigraphy has so far proved incapable of resolving the problem of correlation around the Eocene-Oligocene boundary. A possible alternative is 'event stratigraphy', of which the two most obvious candidates are magnetostratigraphy and sequence stratigraphy. In the oceanic realm the Eocene-Oligocene boundary has been placed in polarity chronozone 13, but this has not yet been recognized with certainty within any of the European stratotypes. Townsend & Hailwood (1985) have described the sequence of magnetic polarity reversals for the Eocene of southern England, but their work does not extend to the late Eocene and Oligocene.

The study of cycles of deposition has a long and distinguished history in Tertiary stratigraphy. Stamp (1921) was one of the earliest workers to use this method of event stratigraphy for correlation; more recent proponents of the method in England have

been King (1981) and Plint (1983, 1988) who have recognized cycles of sedimentation related to transgressive and regressive events. Plint (1988) has been able to match the sequences of the Eocene of the Hampshire Basin with the recently published coastal onlap chart of Haq *et al.* (1987). He did this primarily on the basis of sedimentary facies, recognizing estuarine, marine and alluvial facies. The late Eocene and Oligocene sediments cannot be classified so easily without reference to their fauna, and amongst the fauna the ostracods are of prime importance. The conditions of deposition were never those of the open shelf, but of a coastal margin of lakes, lagoons and shallow marine embayments. In such a situation transgressive and regressive events cannot be recognized by varying water depth, but only on the basis of fluctuating salinity, provided that changes due to localized coastal changes can be separated from those due to global eustatic change.

The ostracods are ideal for palaeosalinity studies, and their palaeoecology has been comprehensively studied for these strata by Keen (1971, 1972, 1977). The strata studied range from the Barton Sands at the base, to the Upper Hamstead Beds at the top. The stratigraphical terminology adopted is the traditional one of Bristow *et al.* (1889), although more recent nomenclature is available (Insole & Daley, 1985). The late Eocene and early Oligocene sediments of the Hampshire Basin give a fairly complete record through an interval of some 5 million years, making it highly probable that any major eustatic events will be preserved. Salinity curves have been constructed for the major sections on the Isle of Wight, using ostracod assemblages described in the works of Keen cited above. This allows the recognition of freshwater, oligohaline, lower mesohaline, upper mesohaline, polyhaline and euhaline salinities shown in Fig. 1.

Eustatic events can be separated from more localized events by their occurrence at more than one locality and by the magnitude of the change. Examples of salinity fluctuations caused by minor and localized coastal changes can be seen in the Headon Beds (see Fig. 1 for stratigraphy). In the Lower Headon Beds at Headon Hill, the *Cyrena pulchra* Bed records a flood plain lake which became a brackish lagoon when the sea breached a nearby barrier; the lagoon was short-lived (*circa* 100 years?) and soon reverted to a freshwater lake. The coastal breaching was perhaps due to a storm, and does not indicate global eustatic change. In the Upper Headon beds the two mile long section between Headon Hill and Colwell Bay shows how minor fluctuations of a barrier between a fresh-water lake and a brackish lagoon allowed fluctuating freshwater and brackish sediments, fauna and flora to be produced. The deposits produced by such changes are extremely localized and cannot be traced over more than one or two square miles.

The salinity curves from the various localities have been amalgamated to give a composite curve for the sequence, and from this a eustatic curve constructed (Fig. 2). This has been compared with the coastal onlap and eustatic curves of Haq *et al.* (1987) and see Fig. 3. Obviously some biostrati-graphical or magnetostratigraphical control is needed in order to recognize tie-in points. Plint (1988) has been able to do this for lower parts of the Eocene succession where the biostratigraphy and magne-tostratigraphy are more reliable. The important regression at 39.5 MY is regarded as one of two principal lowstand events by Plint and this agrees very well with the results of the present study. If this correlation is accepted, the curves produced here for the late Eocene-early Oligocene show a reasonable match with the curves of Haq *et al.* The transgression of the Middle Headon Beds can be correlated with that of their Cycle TA4.1 and that of the Bembridge Oyster Marl with their Cycle TA4.2. A problem then arises as to the importance of the minor transgressive event recorded in the Lower Hamstead Beds. If this is disregarded, then the transgression of the Upper Hamstead Beds would correlate with Cycle TA4.3 and not with the major Rupelian transgression of Cycle TA4.4. Conventional biostratigraphy offers a solution to this predicament (see below), by suggesting that the Upper Hamstead Beds should be placed in TA4.4, and that the higher salinities suggested for the Lower Hamstead Beds are indeed a weak record of a eustatic transgressive event.

If these correlations are correct the four

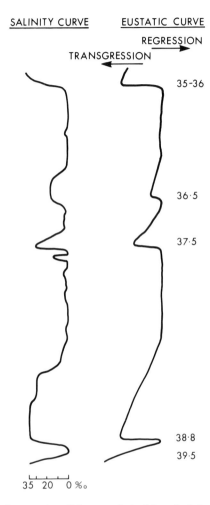

Fig. 2. A composite salinity curve derived from the information given in Fig. 1, with a corresponding interpretative eustatic curve. The four transgressive events are dated using the time scale of Haq *et al.*, 1987.

marine transgressions can be correlated with those occurring at 38.8 MY (Middle Headon Beds), 37.5 MY (Bembridge Oyster Marl), 36.5 MY (*Nematura* Bed and White Band) and 35-36 MY (Upper Hamstead Beds). If the Bartonian is now included in these results (after Plint, 1988 and see Fig. 3), the match is good and all the global eustatic events can be recognized, but just as importantly the overall picture of a regressive Priabonian sand-wiched between two periods of higher sea levels is very clear and lends strength to the correlations.

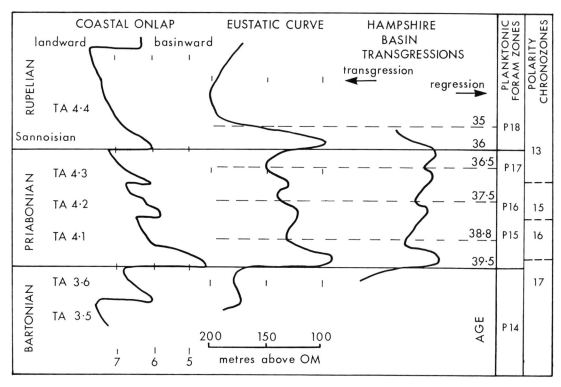

Fig. 3. Sequence stratigraphy across the Eocene-Oligocene boundary. The classification is that of Haq *et al.*, 1987. TA3.5-TA3.6 and TA4.1-TA4.4 are cycles, part of the supercycles TA3 and TA4, which are subdivisions of the Tejas Megacycle.

## CHANGES IN OSTRACOD FAUNAS AT THE EOCENE-OLIGOCENE BOUNDARY

The marine horizons described in the previous section have all been considered at one time or another to represent the basal Oligocene. The ostracods give a quite unequivocal answer (Keen, 1968, 1972, 1978) in correlating the Upper Hamstead Beds with the lower Oligocene of the Paris Basin and Belgium. Furthermore, the difference between this fauna and that of older beds is the most conspicuous in the Palaeogene of western Europe. Considering the Hampshire Basin succession, of the 13 brackish and marine species recorded in the Upper Hamstead Beds only two occur in older horizons (the Bembridge Beds) and none occurs in the Headon Beds. Of these 13 species, 12 are known in the lower Oligocene of the Paris Basin (Sannoisian and lower Stampian) and 10 are present in the Tongrian and Rupelian of Belgium. The Sannoisian of the Paris Basin has

yielded 27 species, none of which is found in older deposits in France, although two species occur in the Bembridge Beds of England. The Bembridge Oyster Marl contains only seven brackish and marine species, four are known from older deposits, two from younger and one is indigenous. The Middle Headon Beds have yielded some 40 species; four are found in the Bembridge Oyster Marl, 14 in the Barton Beds, while the remainder are indigenous. These figures emphasize the great faunal break occurring in the English succession, and its coincidence with the Eocene-Oligocene boundary. Considering the English, Paris and Belgian faunas together, from a combined Lower Oligocene ostracod fauna of some 33 species and subspecies only 2 are known in older rocks. Of the remainder approximately half can be phylogenetically related to local late Eocene species, but the other half have no known ancestors in the area. The difference between the faunas cannot be explained by facies differences because similar environments had existed in immediately preceding periods. Such a great faunal break can only be explained by the arrival of new immigrants which rapidly colonized the area, together with accelerated evolution of indigenous species.

The correlations suggested here on the basis of global eustatic events have some implications for Palaeogene stratigraphy. If the base of the Rupelian defines the base of the Oligocene, then in England this lies at the base of the Upper Hamstead Beds; if this is so, magnetic chronozone 13 must occur in the Upper Hamstead Beds. The Headon, Osborne, Bembridge and Lower Hamstead Beds are of Priabonian age, allowing correlation beyond the northwest European area.

What of ostracod faunas in other parts of the world? In Aquitaine, southwestern France, Ducasse & Peypouquet (1986) state that the Eocene-Oligocene boundary shows an important renewal of the ostracod fauna and that this can be easily distinguished in both 'margino-littoral' and very deep realms, but that the same faunal type persists through time in the more stable epi-mesobathyal realm. Barbin & Guernet (1988) have studied the ostracods of the type area of the Priabonian in northern Italy and record no significant difference across the boundary, but also give an alternative interpretation, i.e. the top of the Priabonian is older than the base of the Oligocene in northern Europe which, they argue, accounts for the faunal continuity. In Germany the Oligocene fauna is so much better documented than the Eocene, that any statistics would be misleading. The faunas from other parts of Europe are too poorly known to allow any firm conclusions to be reached. Van den Bold (1986) has considered the ostracods across the Eocene-Oligocene boundary in the Gulf of Mexico Caribbean region. In Alabama and Mississippi he records 30 late Eocene species of which 17 cross the boundary and are joined by 21 new species, although the latter do not all appear immediately above the boundary.

This discussion has so far centred upon species changes. At the generic level the Eocene-Oligocene boundary is barely detectable, with only one or two genera becoming extinct and a similar number appearing. There are no extinctions at the family level, nor do any new families appear.

## THE *HEMICYPRIDEIS* DATUM

In large areas of Europe north of the Tethyan realm, from southern England to Turkey and southwest Russia, shallow marine or brackish water sediments occur near the base of the Rupelian. This is the 'Sannoisian facies' of the Anglo-Gallic area. Elements of this fauna, including some freshwater species, together with the succeeding Rupelian ostracod fauna can be recognized throughout the region. A *Hemicyprideis* datum can also be recognized by the first appearance of this brackish-water, shallow marine genus. The exact age of this event at any one locality is obviously facies controlled. There are difficulties in determining the age of the datum; it is found in the Middle Priabonian Cluj Formation of Roumania (Bombita *et al.*, 1975) where it clearly antedates the beginning of the Oligocene, and also occurs in the Bembridge Marls of England. Its earliest appearance can probably be dated to planktonic foraminiferal zone P16, but in most regions this probably happens in P17 (NP20). Van den Bold (1986) also records the appearance of *Hemicyprideis* at the base of the Oligocene in the Gulf coast of the U.S.A., where it is said to replace the Eocene genus *Haplocytheridea*. There is, therefore, the possibility of this datum being recognized outside Europe.

## THE TERMINAL EOCENE EXTINCTION EVENT

As indicated above, there is an important change in ostracod faunas at the Eocene-Oligocene boundary at the species level, but not at higher taxonomic levels. The extinction appears to have been a fairly short-lived event, but the new fauna may have appeared over a rather longer time span. This appears to have been the case in North America, while the abruptness of the change in northwest Europe was probably due to migration. This is not the place for a detailed discussion on the faunal changes associated with the boundary, but in passing it can be noted that other groups of invertebrates show important changes, while the 'Grande Coupure' of Stehlin (1909) indicates the most important change in land mammals of the European Palaeogene. However against this it should be noted that planktonic organisms, other vertebrates and mammals outside Europe do not show abrupt changes. Descriptions of faunal and floral events around the Eocene-Oligocene boundary can be read in Cavalier *et al.*, (1981), and in Pomerol & Premoli-Silva (1986). The oceanic record has assumed a leading role in the discussion of the terminal Eocene events, and it seems clear (e.g., Corliss *et al.*, 1984) that there is an absence of catastrophic change at any time between the Middle Eocene and Middle Oligocene. Some authors (e.g., Keller, 1986) have argued for stepped extinction in the late Eocene, but as each of their 'steps', or periods of more rapid change, is usually associated with a submarine hiatus, it is likely that the actual faunal change was more gradual. In summary, it seems that the latest Eocene changes merely reflect an acceleration of important biotic changes having their origin in the late Middle Eocene (Pomerol & Premoli-Silva, 1986). There is no real evidence for a mass extinction at this time, and even in areas such as northern Europe where there was a significant faunal turnover, this can be adequately explained by migration.

## REFERENCES

Barbin, V. & Guernet, C. 1988. Contribution à l'étude du Priabonien de la région-type (Italie du nord): les ostracodes. *Revue de Micropaleont.*, Paris, **30**, 209-231

Beyrich, H. E. 1854. Uber die Stellung der Hessischen Tertiärbildungen. *Ber. Verh. preuss. Akad. Wiss. Berlin.*, 640-666.

Bold, W. A., van den. 1986. Distribution of Ostracoda at the Eocene-Oligocene boundary in deep (Barbados) and shallow marine environment (Gulf of Mexico), *In* Pomerol, C. & Premoli-Silva. I. (Eds), *Terminal Eocene events*, Developments in palaeontology and stratigraphy, **9**, 259-263. Elsevier, Amsterdam, Oxford, New York, Tokyo.

Bombita, G., Gheta, N., Iva, M. & Olteanu, R. 1975. Éocène moyen-supérieur et Oligocène inférieur des environs de Cluj, *In* Micropaleontolgical guide to the Mesozoic and Tertiary of the Romanian Carpathians, 14th. European Micropaleontological Colloquium, Institute of Geology and Geophysics, Bucharest, Excursion P, 163-174.

Bristow, H. W., Reid, C. & Strachan, A. 1889. The Geology of the Isle of Wight (2nd. ed.). *Mem. geol. Surv.* U.K., xiv + 349 pp. H.M.S.O., London.

Cavalier, C., Chateauneuf, J.-J., Pomerol, C., Rabussier, D., Renard, M. & Vergnaud-Grazzini, C. 1981. The Geological events at the Eocene/Oligocene boundary. *Palaeogeogr. Palaeoclimat. Palaeocol.*, Amsterdam, **36**, 223-248.

Corliss, B. H., Aubry, M. P., Berggren, W. A., Fenner, J. M., Keigwin, L. D., jr. & Keller, G. 1984. The Eocene-Oligocene boundary event in the Deep Sea. *Science*, N.Y., New York, **226**, 806-810.

Ducasse, O. & Peypouquet, J.-P., 1986. Ostracods at the Eocene-Oligocene boundary in the Aquitaine Basin. Stratigraphy, phylogeny, palaeonenvironments. *In* Pomerol, C. & Premoli-Silva, I. (Eds), *Terminal Eocene events*, Developments in palaeontology and stratigraphy, **9**, 265-274. Elsevier, Amsterdam, Oxford, New York, Tokyo.

Haq, B. U., Hardenbol, J. & Vail, P. R. 1987. Chronology of fluctuating sea levels since the Triassic. *Science*, New York, **235**, 1156-1167.

Insole, A. & Daley, B. 1985. A revision of the lithostratigraphical nomenclature of the Late Eocene and Early Oligocene strata of the Hampshire Basin, southern England. *Tertiary Res.*, **7**, 67-100.

Keen, M. C. 1968. Ostracodes de l'Éocène supérieur et l'Oligocène inférieur dans les basins de Paris, Hampshire, et de la Belgique, et leur contribution à l'échelle stratigraphique. *In* Colloque sur l'Éocène, *Mém. Bur. Rech. géol. Minièr.*, Paris, no. 58, 137-145.

Keen, M. C. 1971. A palaeoecological study of the ostracod *Hemicyprideis montosa* (Jones and Sherborn) from the Sannoisian of North-West Europe. *In* Oertli, H. J. (Ed.), *Paleoecologie des ostracodes, Pau 1970, Bull. Centre Rech. Pau-SNPA*, **5** suppl., 523-543, 4 figs, 2 tabs, 2 pls.

Keen, M. C. 1972. The Sannoisian and some other Upper Palaeogene Ostracoda from north west Europe. *Palaeontology*, London, **15**, 267-325.

Keen, M. C. 1977. Ostracod assemblages and the depositional environments of the Headon, Osborne, and Bembridge Beds (Upper Eocene) of the Hampshire Basin. *Palaeontology*, London, **20**, 405-445.

Keen, M. C. 1978. The Tertiary - Palaeogene, *In* Bate, R. H., & Robinson, E. (Eds), *A Stratigraphical Index of British Ostracoda*, 384-450. Seel House Press, Liverpool.

Keller, G., 1986. Late Eocene impact events and stepwise mass extinctions, *In* Pomerol, C. & Premoli, I. (Eds), *Terminal Eocene events*, Developments in palaeontology and stratigraphy, **9**, 403-412. Elsevier, Amsterdam, Oxford, New York, Tokyo.

King, C. 1981. The stratigraphy of the London Clay and associated deposits. *Tertiary Res. Special Papers*, Leiden, **6**, 158 pp.

Plint, A. G. 1983. Facies, environments and sedimentary cycles in the Middle Eocene, Bracklesham Formation of the Hampshire Basin: evidence for global sea-level changes? *Sedimentology*, Amsterdam and New York, **30**, 625-653.

Plint, A. G. 1988. Global eustacy and the Eocene sequence in the Hampshire Basin, England. *Basin Research*, **1**, 11-22.

Pomerol, C. & Premoli-Silva, I., (Eds), 1986. *Terminal Eocene events*, Developments in palaeontology and stratigraphy, 414 pp. Elsevier, Amsterdam, Oxford, New York, Tokyo.

Raup, D. & Sepkoski, J. 1984. Periodicity of extinctions in the Geologic past. *Proc. natn. Acad. Sci. U.S.A.*, Washington, **81**, 801-805.

Stamp, L. D. 1921. On cycles of sedimentation in the Eocene strata of the Anglo-Franco-Belgian Basin. *Geol. Mag.*, London, **58**, 108-114, 146-157, 194-200.

Stehlin, H. G., 1909. Remarques sur les faunules de Mammifères des couches éocènes et oligocènes du Basin de Paris. *Bull. Soc. géol. Fr.*, Paris, 4(18), 488-520.

Townsend, H. A. & Hailwood, E. A. 1985. Magnetostratigraphic correlation of Palaeogene sediments in the Hampshire and London Basins, southern UK. *J. Geol. Soc. Lond.*, **142**, 957-982.

## DISCUSSION

Joe Hazel: The Eocene-Oligocene boundary as based on the extinction of the *Turborotalia cerroazulensis* gp. actually would be in the middle of the Cycle Ta4.3 in the American Gulf Coast. This may not have anything to do with the classical boundary of Europe. What do you think?

Mike Keen: As you are aware, it really comes down to a matter of definition. Until a boundary stratotype is chosen, it is impossible to say where the boundary is, and it may be that if the stratotype is chosen outside Europe (as seems likely), then the base of the Oligocene could be older than the 'traditional' base in Europe. However, as you well know, there is no agreement within Europe as to where the base should be taken. The important point is surely to concentrate on correlation, and to try to get this correct.

Franz Gramman: Where do you place the Eocene-Oligocene boundary in the Isle of Wight sequence?

Mike Keen: A good question, which I have carefully avoided in my diagrams! The results of this study suggest that if the transgression marking the beginning of Cycle TA4.4 is accepted as defining the base of the Oligocene, in the Isle of Wight this coincides with the base of the Upper Hamstead Beds. This is the Rupelian transgression, which is heralded by the Sannoisian, and possibly the Lattorfian. However, there is no accepted definition for the base of the Oligocene; please also see the reply to Hazel.

Roger Kaesler: Haq's model shows slow transgression and rapid regression; your diagrams indicate rapid transgression and slow regression. We see a similar pattern to yours in the late Carboniferous and early Permian of the U.S.A., with very thin transgressive limestones overlain by deep water black shales, followed by thicker regressive limestones. These are believed to be Glacio-eustatic; could your transgressions be glacially controlled?

Mike Keen: The change from freshwater or low salinity sediments and fauna to high salinity is usually sharply defined, and sometimes erosive, while the regressive phase is invariably quite gradual in these sequences on the Isle of Wight. The eustatic curve has been derived directly from the salinity curve, so reflects this fact. In my experience this is quite typical for transgressive-regressive events. However, when this is translated into time, this asymmetry may be artificial, i.e. transgressions tend to be erosive with little sedimentation, while the regressive phase allows the accumulation of much thicker sediments. Indeed I have heard it argued that the plane of erosion (or unconformity, if prominent enough) represents the real transgression, and all the succeeding sediments belong to the regression. I am not sure whether this pattern can be used as evidence for glaciation, although there is strong evidence for Antarctic ice from the late Eocene onwards.

Robin Whatley: Will you accept that the unusual type of asymmetry exhibited by your transgression-regression curves is due to the rapid inundation and subsequent much slower silting up of a small embayment?

Mike Keen: Yes. Please see the reply to Kaesler.

# 12

# Namurian entomozoacean Ostracoda and eustatic events

I. P. Wilkinson & N. J. Riley

British Geological Survey, Keyworth, Nottingham, NG12 5GG, U.K.

## ABSTRACT

Namurian entomozoacean ostracods occur in at least five marine horizons in the Namurian *Eumorphoceras* Ammonoid Biozone of Northern England and form an integral part of the associated dysaerobic faunal community. Previously published records from the *Eumorphoceras* Ammonoid Biozone are restricted to those of the genus *Truyolsina* Becker & Bless in Spain. The assemblages from northern England, discussed herein, include new species and the youngest unequivocal record of entomozoacean ostracods. The range of *Maternella* is extended into the Arnsbergian. Entomozoacean ostracods, in common with so many other members of the early Namurian dysaerobic community, failed to survive the mid-Carboniferous faunal crisis which took place during the latest part of the *Eumorphoceras* Ammonoid Biozone, when a eustatic regression occurred.

## INTRODUCTION

This paper is the first account of Namurian (Silesian) entomozoaceans outside Spain and records the youngest known representatives of the superfamily. Their occurrence in the British Namurian is documented and their extinction is related to the global eustatic event during the mid-Carboniferous boundary interval. The detailed taxonomy of the fauna will be the subject of a subsequent publication.

The Entomozoacea is a somewhat enigmatic group of Palaeozoic ostracods. The superfamily has traditionally been placed in the order Myodocopida and various suborders e.g., Cladocopina (Polenova & Zanina, 1960, 330), Myodocopina (Sylvester-Bradley, 1961) and Entomozocopina (Gründel, 1969). However, as Siveter *et al.* (1987) point out the lack of features such as a rostrum and rostral incisure must preclude the

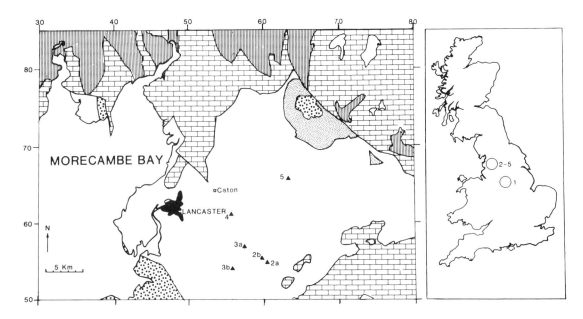

Fig. 1. Namurian entomozoacean ostracod localities in northern England. Main map: localities 2-5, Forest of Bowland; geology adapted from British Geological Survey, 1:250,000 UTM Map Series, Sheet 53N 04W, Liverpool Bay, Solid Geology (1978). Vertical hatching, pre-Carboniferous; brick ornament, Dinantian; fine stipple, Westphalian; coarse stipple, Permo-Triassic; the Namurian is left blank.

entomozoacea from Myodocopida and any subordinal relationships must also be in doubt. There has been further confusion at the family and subfamily level, stemming from the fact that the type genus, *Entomozoe*, with the type species *E. tuberosa* (Jones, 1861), should not be included with those taxa commonly called 'fingerprint' ostracods, the entomozoids of most authors. In this paper, the 'fingerprint' Ostracoda are considered to fall within the Superfamily Entomozoacea Pribyl, 1950.

Considerable information is available for the late Devonian and earliest Carboniferous (summarized by Groos-Uffenorde, 1984), but very little is known about the group in the younger parts of the Dinantian and early Namurian. Of the genera that thrived during the Famennian and Courceyan, only *Richterina* Gürich, *Maternella* Rabien, *Entomoprimitia* Kummerow and *Ungerella* Livental survived the Courceyan, although greatly reduced in specific diversity; by the Asbian only *Maternella* was left. Hitherto, the youngest known species was *Maternella (Steinachella) geniceraensis* Jordan & Bless, 1970; which first appears during the late

Courceyan-early Chadian *Scaliognathus anchoralis* Conodont Biozone, and ranges up into the late Asbian or early Brigantian part of the *Gnathodus bilineatus* Conodont Biozone in the Rheinisches Schiefergebirge (Groos-Uffenorde, 1984). The material described herein extends the range of the genus up into the Arnsbergian.

Information on Silesian entomozoaceans is sparse and the only genus described is *Truyolsina* Becker & Bless 1975, from the *Eumorphoceras* Ammonoid Biozone of the Cantabrian Mountains of northern Spain (Becker & Bless, 1975; Becker, 1976) where three species are known: *T. truyolsi* (two subspecies), *T.* cf. *truyolsi* and *T. necopinata*. According to Becker & Bless (1975) and Becker (1976), these occurrences are within the E2a Chronozone, although the Pendleian (E1) horizon is probable in the case of *T. ?necopinata*.

Little work has been done on British entomozoaceans, despite being recorded in the Devonian of southwest England over 130 years ago (Sandberger, 1852; Roemer, 1853). T. R. Jones described several late Silurian (Jones, 1861, 137) and early Carboniferous (Jones, 1873) species from

| STAGE | GENUS-ZONE | CHRONO-ZONE | HORIZON | INDEX |
|-------|------------|-------------|---------|-------|
| ARNSBERGIAN | EUMORPHOCERAS | E2c | Nuculoceras nuculum | E2c4 |
| | | | Nuculoceras nuculum | E2c3 |
| | | | Nuculoceras nuculum | E2c2 |
| | | | Nuculoceras stellarum | E2c1 |
| | | E2b | Cravenoceratoides nititoides | E2b3 |
| | | | Cravenoceratoides nitidus | E2b2 |
| | | | Cravenoceratoides edalensis | E2b1 |
| | | E2a | Eumorphoceras yatesae | E2a3 |
| | | | Eumorphoceras ferrimontanum | E2a2 |
| | | | Cravenoceras cowlingense | E2a1 |
| PENDLEIAN | | E1c | Cravenoceras malhamense | E1c1 |
| | | E1b | Tumulites pseudobilinguis | E1b2 |
| | | | Cravenoceras brandoni | E1b1 |
| | | E1a | Cravenoceras leion | E1a1 |

Fig. 2. Subdivision of the *Eumorphoceras* Ammonoid Biozone in the British Isles.

southern Scotland, but few advances were made until Gooday (1973, 1974, 1978, 1983) carried out extensive research on late Devonian and early Carboniferous taxa from South West England. Post-Courceyan entomozoaceans, however, have no published record.

## STRATIGRAPHICAL AND GEOGRAPHICAL LOCATION OF BRITISH NAMURIAN ENTOMOZOACEAN OSTRACODS

**Lithostratigraphy.** In northern England, basinal Namurian sediments comprise pelagic marine and non-marine mudstones interbedded with deltaic deposits. Repeated phases of deltaic advance and abandonment resulted in a cyclical sequence. Thin, widespread marine bands are bounded by thicker non-marine strata, which vary laterally in facies and thickness according to delta type and proximity, and the palaeotopography of the basin floor. Reviews of Namurian sedimentation can be found in Besly & Kelling (1988) and a eustatic model

has been proposed by Ramsbottom (1977).

Entomozoacean ostracods have been found in the Namurian deposits of the Forest of Bowland, Lancashire, and Derbyshire (Fig. 1). Those from the latter (locality 1) are from a condensed sequence distal to deltaic influence. In contrast those from the Forest of Bowland (localities 2 to 5) are in a proximal setting of alternating mudstones and sandstones over 500m thick (Roeburndale Formation of Arthurton *et al.*, 1988, and Caton Shales of Moseley, 1954). This sequence thins to about 50m of mudstone in the more distal setting only 25km to the South East, where the full complement of early Arnsbergian marine bands in North West England was first documented (Riley, 1985).

**Bio-chronostratigraphy.** Each of the Namurian marine bands contains a distinctive combination of nekto-pelagic and epibenthonic fauna dominated by epibyssate bivalves and ammonoids, and it is the ammonoids that form the main basis for correlation and subdivision. The index taxa are, with few exceptions (e.g., *Nuculoceras nuculum* Bisat) unique to particular marine bands. The subdivision of the *Eumorphoceras* Ammonoid Biozone in the British Isles (Fig. 2) is an extension and adaptation of schemes proposed by earlier workers including Yates (1962), Ramsbottom (1977), Ramsbottom *et al.* (1978), Ramsbottom & Saunders (1985) and Riley *et al.* (1987).

The faunas associated with the Namurian Entomozoacea in northern England are listed and classified in the Appendix under their respective ecological groupings. These faunas represent low-diversity high-abundance assemblages typical of late Palaeozoic dysaerobic marine environments (see Kammer *et al.*, 1986). The entomozoacean ostracods can be considered an integral opportunistic vagrant nekto-benthonic component of the dysaerobic assemblage, colonizing during brief spells when the oxygen level was suitably high.

**Locality details.** Entomozoacean ostracods have been recovered from several marine bands in the Namurian in northern England:
1) The *Cravenoceras leion* Marine Band (E1a1), disused railway cutting (SK 1953 7113), 1km west of Longstone Station, Derbyshire, locality and stratigraphy described in Aitkenhead *et al.* (1985).
2) The *Cravenoceras cowlingense* Marine Band (E2a1), two sections;
a) Screes End (SD 6038 5517), 1.5km southeast of the Mountain Rescue Post, Tarnbrook Wyre, near Lancaster, Lancashire.
b) Screes End (SD 5985 5540), 0.5km east southeast of the Mountain Rescue Post, Tarnbrook Wyre, near Lancaster, Lancashire.
3) *Eumorphoceras ferrimontanum* Marine Band (E2a2) two sections;
a) Unnamed stream on Lee Fell (SD 5729 5690), 30m north northwest of nearby aqueduct crossing, 2.05km southwest of the westernmost triangulation point on Ward's Stone, Tarnbrook Wyre, near Lancaster, Lancashire.
b) River Marshaw Wyre (SD 5682 5430), 0.75km south southwest of Lee Bridge, Abbeystead, near Lancaster, Lancashire.
4) *Eumorphoceras yatesae* Marine Band (E2a3),

Sweet Beck (SD 5507 6107), 0.7km south southeast of The Crag, Littledale, near Caton, Lancashire.
5) *Cravenoceratoides nitidus* Marine Band (E2b2), 118m depth, Wray Geothermal Borehole (SD 6320 6570), 0.45km northeast of Leyland House, Wray, near Caton, Lancashire.

All material and details of sections are curated with the Biostratigraphy Research Group, British Geological Survey, Keyworth. Localities 2-4 occur in an area of restricted access and it is imperative that permission is sought from the Abbeystead Estate Office.

**Taphonomy.** The ostracods occur in dark-grey (*circa* N3-4, Geological Society of America colour chart), calcareous, fissile mudstone, in the form of crushed carapaces or disarticulated valves, some of which are fragmentary. They are preserved as flattened moulds and in the case of overlapping or articulated valves, superposition of ornament results. Up to five valves or valve fragments per square centimetre are found on one or a limited number of closely-spaced bedding planes within a marine band. Orientation is apparently random, but the material is too limited for statistical analysis. The consistent mode of preservation at different localities and horizons, and the extremely fragile nature of the valves suggest it is reasonable to conclude that each assemblage has undergone little or no post-mortem transportation.

**Taxonomic relationships of the entomozoaceans encountered**

**E1a1.** The oldest entomozoaceans found during the present study are those from the *Cravenoceras leion* Marine Band (E1a1) near Longstone (locality 1, Fig. 1). Their taxonomic relationships are uncertain, however, due to their fragmentary preservation. Several specimens show an ovate, near circular, outline with concentrically disposed ribbing, but limited anastomosis, reminiscent of *Maternella*. The ornament is characterized by numerous cross ribs giving an almost reticulate appearance (Fig. 3a) and, in this respect, resembles the Lower Carboniferous species *Maternella clathrata* (Kummerow). There is no evidence

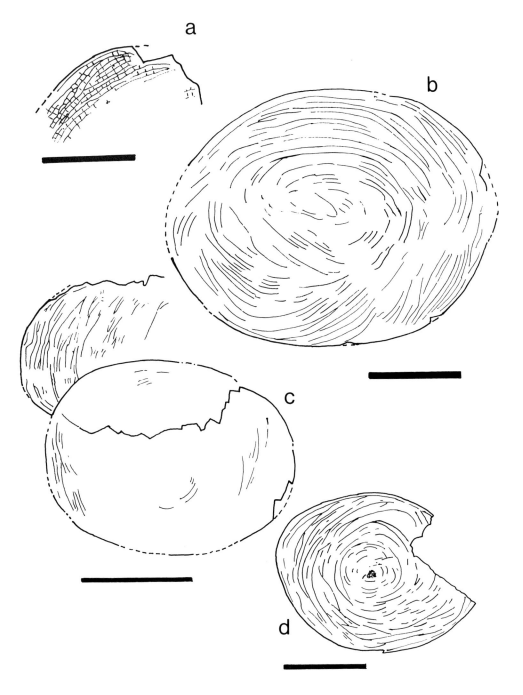

Fig. 3. Namurian entomozoacean Ostracoda from northern England (scale bars=1mm, specimen numbers prefixed FW and Ro refer to BGS registration series).

Fig. 3a. ?*Maternella* sp. FW 8585 from the *Cravenoceras leion* Marine Band (E1a1), locality 1, showing the anastomosing and intercalated ribs.

Fig. 3b. *Maternella (Steinachella)* sp. Ro 8319 from the *Eumorphoceras ferrimontanum* Marine Band (E2a2) of locality 3b.

Figs 3c - 3d. *Maternella (Steinachella)* sp. Ro 9293 from the *Eumorphoceras yatesae* Marine Band (E2a3) of locality 4.

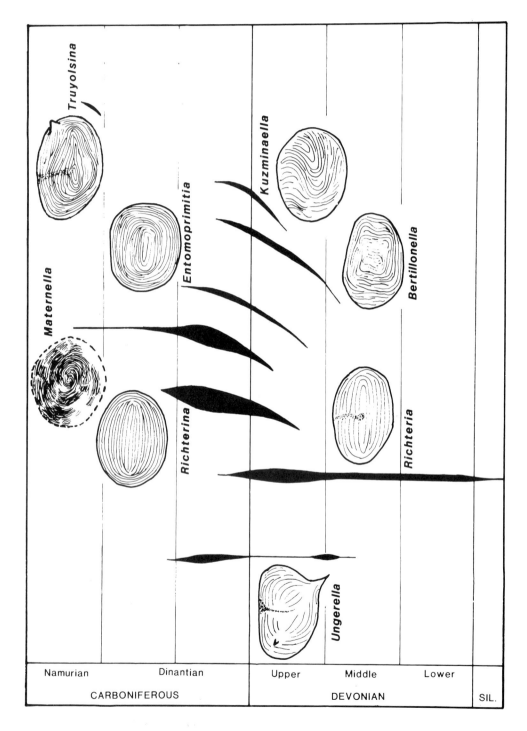

Fig. 4. Stratigraphical distribution of the main entomozoacean ostracod genera in the Upper Palaeozoic. Cartoons redrawn after Kummerow (1939); Jordan & Bless (1970); Becker & Bless (1975); and Gooday (1983). Bar widths indicate relative radiations of the genera.

of a sulcus and what may be interpreted as a weakly developed muscle patch as seen in some specimens, may be a preservational artefact.

**E2a1.** Rare specimens occur in the *Cravenoceras cowlingense* Marine Band at Tarnbrook Wyre (localities 2a and b). Unfortunately preservation is too poor to recognize any details in shape or ornament, except for a few short ribs. Nevertheless, considering the rarity of the superfamily in the Namurian, it is considered important to record their presence.

**E2a2.** The ostracod fauna from the *Eumorphoceras ferrimontanum* Marine Band (E2a2) at Tarnbrook and Marshaw (localities 3a and b) is monospecific, comprising an undescribed species of *Maternella (Steinachella)* (to be described elsewhere) that is approximately contemporaneous with *Truyolsina truyolsi* from Spain. This large, ovate species (Fig. 3b) is characterized by concentric ribbing, spiralled in the peripheral areas, but rarely branched or anastomosing, and slightly drawn out in the direction of the length of the valve. Sulci and spines are absent. The largest specimen is 3.39mm long and 2.63mm high. This taxon differs from *Truyolsina* in being much larger, less rhombic and more ovate in shape, and in lacking the anastomosing ribs and 'stagonoid' (*sensu* Becker & Bless, 1975) ornamentation. In addition, adults of *Truyolsina* have a variably developed sulcus, which extends to a distinct subcentral muscle pit, and a postero-dosal spine. *Maternella (Steinachella) geniceraensis* Jordan & Bless, previously the stratigraphically youngest published species, exhibits more anastomosis and additional intercalated ribs.

**E2a3.** Specimens from the *Eumorphoceras yatesae* Marine Band (E2a3) at Sweet Beck (locality 4) can be assigned to *Maternella (Steinachella)* (Figs 3c and 3d). They have an ovate outline and finely ribbed ornamentation which is concentrically disposed about the centre of the valve. A peripheral rib parallel to the margin, can be traced around most of the valve and anastaomosing and intercalated ribs are frequent. Although a sulcus is absent, a weak muscle patch is seen in the better preserved specimens. It differs from the type material of *M. (S.) geniceraensis* in having more widely spaced, less anastomosing, ribs (Jordan &

Bless, 1970) and is more ovate than the specimens figured by Groos-Uffenorde (1984).

**E2b2.** Specimens from the *Cravenoceratoides nitidus* Marine Band (E2b2) of the Wray Borehole (locality 5) are now stratigraphically the youngest known members of the superfamily. Although small patches of very fine ribbing may sometimes be seen, preservation is so poor that their relationship with other members of the superfamily is unknown. The record is significant, however, in that this species represents the last member of Entomozoacea before their apparent extinction in the middle part of the Arnsbergian.

## EUSTASY AND ENTOMOZOACEAN DISTRIBUTION

One of the most striking features in the distribution of entomozoacean ostracods is their sudden diversification during the late Devonian (Fig. 4). During the mid- to late Frasnian diversification was rapid and, despite a decline during the early part of the Famennian, the superfamily reached its acme during the latest Famennian and early Courceyan. However, an abrupt decline took place during the late Courceyan and early Chadian after which the superfamily did not recover with only a few species continuing through into the Namurian.

Comparison of these trends with the global eustatic sea level changes shows remarkable correlation (Fig. 5). The early to mid-Frasnian was a time of rapid transgression followed by a short regressive phase during the early Famennian. A further, more significant, global transgressive event took place towards the close of the Devonian (late Famennian) and continued into the Carboniferous (Courceyan). The diversity of the Entomozoacea, at least during the Devonian and basal Carboniferous, therefore, appears to be related to global eustatic events; high peaks in diversity corresponding to times of maximum transgression.

The decline in diversity following the major eustatic regression during the late Courceyan, was apparently caused by the reduction of the biospace available to the entomozoaceans and the introduction of major elements of stress related to such aspects as environmental and resource instability

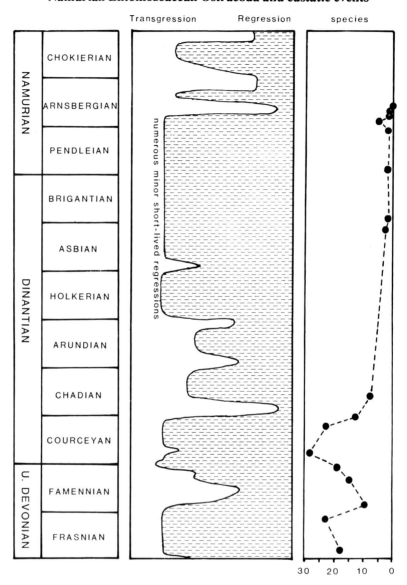

Fig. 5. A graphical comparison between the number of entomozoacean ostracod species and generalized eustatic sea level fluctuations during the late Devonian to Silesian. Eustatic curve modified from data by House (1975) and Ramsbottom (1977).

and biological controls such as competition and changes in the trophic structure and food web. So profound were these, that the entomozoaceans were unable to recover their former diversities despite the long period of relatively high global sea levels during the Holkerian, Asbian and Brigantian. There is some evidence that by the earliest Namurian a recovery started with several new taxa appearing approximately contemporaneously. However, this proved short-lived and the final demise of the group coincides with the major regression that persisted from the late Arnsbergian until the Kinderscoutian. This latter event had a dramatic effect upon the dysaerobic community as a whole and resulted in the rapid species turnover seen in this and other unrelated communities around the mid-Carboniferous Boundary (Ramsbottom *et al.*, 1982; Lane & Ziegler, 1985; Riley 1987).

## CONCLUSIONS

Namurian entomozoacean ostracods were part of the marine dysaerobic community, with at least two genera represented, including *Maternella* from Britain, and *Truyolsina* from Spain. They appear to have some biostratigraphical potential in the early Namurian, but it would be premature to propose a formal zonal scheme at the moment. The final demise of the entomozoaceans can be linked to the eustatic regression towards the end of the *Eumorphoceras* Ammonoid Biozone. This event affected dysaerobic and other marine communities on a global scale and forms the basis for the recognition of the mid-Carboniferous Boundary.

## ACKNOWLEDGEMENTS

The authors wish to thank the Abbeystead Estate for granting access permission, and our colleagues: A. Brandon, E. W. Johnson and A. A. Wilson for providing unpublished lithostratigraphical information. The faunas from locality 1 were collected by M. J. Reynolds and J. Pattison, the rest by N. J. Riley. This paper is published with the permission of the Director, British Geological Survey (NERC).

## APPENDIX

**Associated macrofaunas:** The associated faunas are listed and classified below in their respective ecological groupings:

**a) Nekto-pelagic**
Cephalopods including orthoconic nautiloids, localities 1, 2a,b, 4 & 5
*Anthracoceras* sp., locality 5
*Cravenoceras cowlingense* (Bisat), localities 2a & b
*Cravenoceras gairense* Currie, locality 4
*Cravenoceras* spp., localities 1, 3a & b
*Eumorphoceras ferrimontanum* (Yates), localities 3a & b
*Eumorhoceras grassingtonense* (Dunham & Stubblefield), localities 2a & b
*Eumorphoceras* spp., localities 1 & 4
*Metadimorphoceras* sp., locality 2a

Conodonts, localities 2a, b, 3a, 4 & 5
Fish debris, localities 1, 2a & 5
**b) Fixo-sessile**
Hexactinellid calcisponge spicules, locality 5
Crinoid debris, localities 3a, b & 4
Inarticulate brachiopod *Orbiculoidea*, locality 5
Smooth spiriferoid brachiopod *Martinia* sp. locality 1
Myalinid bivalve *Selenimyalina variabilis* (Hind), locality 5
Pectinoid bivalves including *Dunbarella yatesae* Brandon, localities 3a & 4
*Obliquipecten costatus* Yates, localities 2a & 3a
*Posidonia corrugata* Etheridge, localities 1-5
*Posidonia lamellosa* (de Koninck), localities 2a & b
Rostraconch *Chaenocardiola* sp., localities 2a & 3b
**c) Vagrant epifaunal-shallow infaunal benthos** (locality 5 only)
Arenaceous foraminifera
Bellerophontid archaeogastropod *Euphemites* sp.
Indeterminate microscopic turreted mesogastropods.

## REFERENCES

Aitkenhead, N., Chisholm, J. I. & Stevenson, I. P. 1985. Geology of the country around Buxton, Leek and Bakewell: Memoir for 1:50 000 geological sheet 111 (England and Wales). *Mem. Br. Geol. Surv.*, 168 pp. H.M.S.O., London.

Arthurton, R. S., Johnson, E. W. & Munday, D. J. C. 1988. Geology of the Country around Settle: Memoir for the 1:50,000 geological sheet 60 (England and Wales). *Mem. Br. Geol. Surv.*, 147 pp. H.M.S.O., London.

Becker, G. 1976. Oberkarbonische Entomozoidae (Ostracoda) im Kantabrischen Gebirge (N. Spanien). *Senckenberg. leth.*, Frankfurt a.M., **57**, 201-223.

Becker, G. & Bless M. J. M. 1975. *In* Becker, G., Bless, M. J. M. & Kullmann, J. Oberkarbonische Entomozoen-Schiefer im Kantabrischen Gebirge (Nordspanien). *Neues Jb. Geol. Paläont. Abh.*, Stuttgart, **150**, 92-110.

Besly, B. M. & Kelling, G. (Eds), 1988. *Sedimentation in a synorogenic basin complex: the Upper Carboniferous of northwest Europe*, 276 pp. Blackie, Glasgow and London.

Gooday, A. J. 1973. *Taxonomic and stratigraphic studies on the Upper Devonian and Lower Carboniferous Entomozoidae and Rhomboentomozoidae (Ostracoda, ?Myodocopida) from south west England.* Unpub. Ph.D. Thesis, University of Exeter, 260 pp., 37 pls.

Gooday, A. J. 1974. Ostracod ages from the Upper Devonian purple and green slates around Plymouth. *Proc. Ussher Soc.*, Camborne, **3**, 51-62.

Gooday, A. J. 1978. The Devonian. *In* Bate, R. & Robinson, E. (Eds), A stratigraphical index of British Ostracoda. *Geol.*

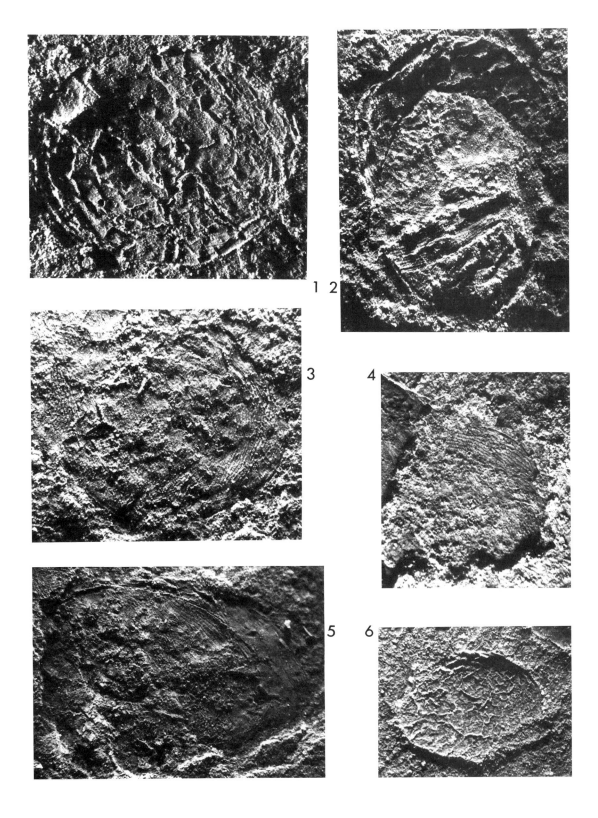

*Jl Special Issue*, **8**, 101-120, pls 1-3. Seel House Press, Liverpool.

Gooday, A. J. 1983. Entomozoacean ostracods from the Lower Carboniferous of south-western England. *Palaeontology*, London, **26**(4), 755-788.

Groos-Uffenorde, H. 1984. Review of the stratigraphy with entomozoid Ostracodes. *IX International Carboniferous Conference, Urbana 1979*, 2 (Biostratigraphy), Carbondale, 212-222.

Gründel, J. 1969. Neue taxonomische Einheiten der Unterklasse Ostracoda (Crustacea). *Neues Jb. Geol. Paläont. Mh.*, Stuttgart, **6**, 353-361.

House, M. R. 1975. Faunas and time in the marine Devonian. *Proc. Yorks. geol. Soc.*, Leeds, **40**, 459-490.

Jones, T. R. 1861. *In* Howell, H. H. & Geikie, A. The geology of the neighbourhood of Edinburgh. *Mem. Geol. Surv. G.B.*, 151 pp. H.M.S.O., London.

Jones, T. R. 1873. Notes on the Palaeozoic bivalved Entomostraca No. X. *Entomis* and *Entomidella. Ann. Mag. nat. Hist.*, London, Series 4, **11**, 413-417.

Jordan, H.-P. & Bless, M. J. M. 1970. Nota preliminar sobre los ostrácodos de la Formación Vegamian. *Breviora geol. astúr.*, Oviedo, **14**, 37-44.

Kammer, T. W., Brett, C. E., Boardman II, D. R. & Mapes, R. H. 1986. Ecologic stability of the dysaerobic biofacies during the Late Paleozoic. *Lethaia*, Oslo, **19**, 109-121.

Kummerow, E. H. E. 1939. Die Ostrakoden und Phyllopoden des deutschen Unterkarbon. *Abh. preuss. geol. Landesanst.*, N. F., **194**, 1-107, pls 1-7.

Lane, H. R. & Ziegler, W. (Eds), 1985. Toward a Boundary in the Middle of the Carboniferous: Stratigraphy and Paleontology. *Cour. Forsch-Inst. Senckenberg*, Frankfurt a.M., **74**, 192 pp.

Moseley, F. 1954. The Namurian of the Lancaster Fells. *Q. Jl. geol. Soc. Lond.*, London, **109**, 423-454.

Polenova, E. N. & Zanina, I. E. 1960. Cladocopa. *In* Orlov, Y. A. (Ed.), *Basic Palaeontology*, 330-332. State Scientific-Technological Publishing House, Moscow. (In Russian.)

Ramsbottom, W. H. C. 1977. Major cycles of transgression and regression (mesothems) in the Namurian. *Proc. Yorks. geol. Soc.*, Leeds, **41**, 261-91.

Ramsbottom, W. H. C., Calver, M. A., Eagar, R. M. C., Hodson, F., Holliday, D. W., Stubblefield, C. J. and Wilson, R. B. 1978. A Correlation of the Silesian Rocks in the British Isles. *Spec. Rep. geol. Soc., Lond.*, **10**, 81 pp.

Ramsbottom, W. H. C. & Saunders, W. B. 1985. Evolution and evolutionary biostratigraphy of Carboniferous ammonoids. *J. Paleont.*, **59**, 123-139.

Ramsbottom, W. H. C., Saunders, W. B. & Owens, B. (Eds), 1982. *Biostratigraphic data for a Mid-Carboniferous Boundary*, 156 pp. Subcommission on Carboniferous Stratigraphy, Leeds.

Riley, N. J. 1985. *Asturoceras* and other dimorphoceratid ammonoids from the Namurian (E2) of Lancashire. *Proc. Yorks. geol. Soc.*, Leeds, **45**, 219-224.

Riley, N. J. 1987. Type ammonoids from the Mid-Carboniferous Boundary interval in Britain. *Cour. Forsch-Inst. Senckenberg*, Frankfurt a.M., **98**, 25-38.

Riley, N. J., Varker, W. J., Owens, B., Higgins, A. C. &

Ramsbottom, W. H. C. 1987. Stonehead Beck, Cowling, North Yorkshire, England: A British Proposal for the Mid-Carboniferous Boundary Stratotype. *Cour. Forsch-Inst. Senckenberg*, Frankfurt a.M., **98**, 159-178.

Roemer, F. A. 1853. Mitteilung an Professor Bronn gerichtet. *Neues Jb. Miner. Geol. Paläont.*, Stuttgart, 810-818.

Sandberger, F. 1852. Mitteilungen an Geheimen-Rath von Leonhard. *Neues Jb. Miner. Geol. Palaont.*, Stuttgart, 56-57.

Siveter, D. J., Vannier, J. M. C. & Palmer, D. 1987. Silurian myodocopid ostracodes: their depositional environments and the origin of their shell microstructures. *Palaeontology*, London, **30**, 783-813.

Sylvester-Bradley, P. C. 1961. *In* Moore, R. C. (Ed.), *Treatise on Invertebrate Paleontology, Part Q, Arthropoda 3*, 387-406. Univ. Kansas Press.

Trewin, N. 1970. A dimorphic goniatite from the Namurian of Cheshire. *Palaeontology*, London, **13**, 40-46.

Yates, P. J. 1962. The palaeontology of the Namurian rocks of Slieve Anieran, Co. Leitrim, Eire. *Palaeontology*, London, **5**, 355-443.

## DISCUSSION

Gerhard Becker: I have found *Truyolsina* in northern Spain, mostly in the lower part of the E2 Zone with *Eumorphoceras bisulcatum*. Only *T. necopinata* occurs in the uppermost part of E1. There is no relationship to your material which very much resembles *Maternella*. How many specimens do you have from Great Britain?

Ian Wilkinson: About 120 at the moment, but mostly fragments. The faunas are in a bad state of preservation.

# BIOLOGY AND GENETICS

# 13

# What the sex ratio tells us: a case from marine ostracods

**Katsumi Abe**

Geological Institute, University of Tokyo, Japan

## ABSTRACT

The sex ratio of ostracods changes extensively in a living population, but a fairly constant value, the tertiary sex ratio, can be expected in some fossil samples. In a living population where the tertiary sex ratio differs significantly from 1:1, sexual dimorphism is often observed in the penultimate instar and it is always assumed that mortality rates differ between the sexes before maturation with the result that the ratio will be biased in new adults, in favour of females in most cases. Consequently, the sex ratio in a fossil sample suggests when in ontogeny mortality had begun to differ between the sexes. Furthermore the sex ratio could be an indicator of both reproductive strategy and (palaeo)environment. This is so because the sex ratio of any species may reflect certain aspects of its reproductive strategy which they it has acquired in the process of adaptation.

## INTRODUCTION

'Many species of fresh-water ostracodes lack males, as indicated both from observations in nature and from cultures in aquaria. *Cyprinotus incongruens* is parthenogenetic in one geographic range and syngamic in another. Furthermore, laboratory cultures of this species can be changed from syngamic to parthenogenetic by isolation of the females, and from parthenogenetic to syngamic by placing the females on a near-starvation diet' (Kesling, 1951). The literature on marine ostracods shows a general predominance of females over males (cf. Van Morkhoven (1962) and Whatley and Stephens (1977), for examples to the contrary). The sex ratio may be an important factor when we examine the evolution of sex, including the cause of change between syngamic and parthenogenetic reproduction. However, few workers have discussed the sex ratio of marine ostracods as a major subject (Van Harten, 1983; Kamiya, 1988b; for Palaeozoic ostracods, see Martinsson, 1956).

Recently some species in a small cove of Japan have been studied in detail, with respect to their sex ratio and population dynamics (Abe, 1983, on a single species from a muddy substrate and Kamiya (1988b) on two species from *Zostera* beds). These studies lead to the recognition that, in syngamic species of marine ostracods, not only sex ratio but also population dynamics vary greatly from species to species and that each species has its own particular pattern in the seasonal change of sex ratio. It was also suggested that the sex ratio could be understood more precisely in connection with the adaptive strategy of the species.

In this study, firstly it is demonstrated, referring to a well examined species, *Keijella bisanensis* (Okubo), how the value of the sex ratio will vary less in the fossil record than it does in a living population. Secondly, several possible patterns of population dynamics of ostracods are illustrated in order to make the subsequent discussion more comprehensive. Thirdly, the point in ontogeny at which the mortality begins to make a difference between males and females is discussed, together with what the result of this will be on field observations. Lastly, it is suggested that the sex ratio is useful in estimating reproductive strategy and the preferred micro-environment of the species.

This case study is based on a limited number of syngamic ostracod species, which live in the shallow sea of the Temperate Zone around Japan. Indeed the number of species investigated here is not sufficient for generalization, but the conclusions are consistent for all of the studied species.

## THE SEX RATIO IN FOSSIL OSTRACODA

Most studies on marine ostracods have been caried out by palaeontologists who have usually ignored the sex ratio, except in some cases where it has been recorded for adults. However, it is an advantage of ostracods, that in many species the sex of adult specimens can be determined from the carapace morphology alone, and such an advantage is rarely found in other marine invertebrates. Sexual dimorphism can also be detected in the

carapaces of juveniles (Whatley & Stephens, 1977). Furthermore, in discussing the sex ratio the problem of predation can be ignored in some species, since a predator, if any, is unlikely to prefer one sex to the other. Indeed, in some ostracods, males are more active, move around more often and thus are found more easily by a predator, but the difference of activity does not always result in selective predation. In *K. bisanensis*, for example, the walking speed of adult males is about one and a half times that of adult females. Possible predators are gobioid fish, which are often observed in the habitat of *K. bisanensis*. However, a study of the food habits of gobioid fish revealed them to feed mainly on Copepoda, Amphipoda, Cumacea and myodocopid Ostracoda, in descending order of frequency (Hayashi & Goto, 1979). The fact that both the tertiary sex ratio (i.e. the ratio in new adults; see Pianka, 1974, 111) of living adults and the sex ratio in the thanatocoenosis are 50%, also suggests that the gobioid fish probably do not feed on the benthonic ostracods, including *K. bisanensis*, and are, therefore, not responsible for changing the sex ratio of their adult populations.

Abe (1983) showed that male mortality was approximately twice as high as that of females in an adult population of *K. bisanensis*, and thus the sex ratio of the living population is female-biased throughout a year (Fig. 1). During the second half of the reproductive period, when fresh adults of the new generation increase greatly in number, the ratio approaches 50%, but it is not exactly 50:50. This slight difference can be ascribed to the fact that the population at this time still includes a small number of old females from the generation produced in the previous reproductive season (Abe, 1983, 482). In the present study it has been verified by culturing experiments, that the tertiary sex ratio is actually 1:1 in *K. bisanensis*. Among 121 penultimate instars, 88 (73%) succeeded in moulting in a petri dish, of which 41 were males, and 47 females. (In the same sample, the new adult population was composed of 171 males and 163 females.)

Because Abe (1983) approached the study from a palaeontological viewpoint, he developed the

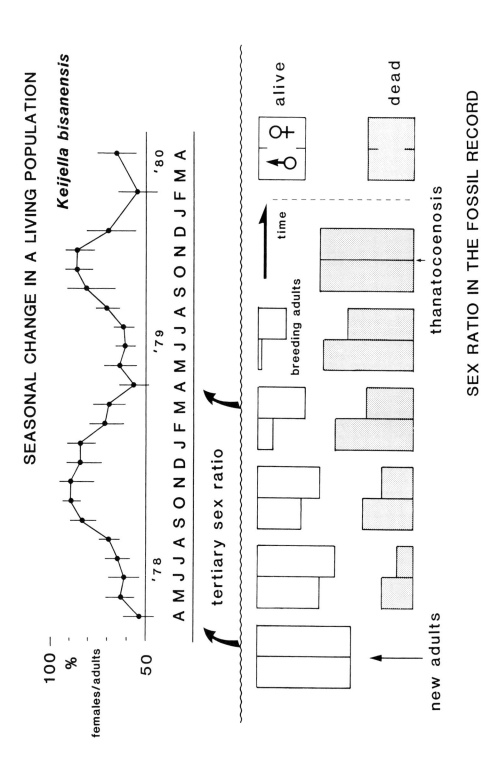

Fig. 1. Seasonal changes in the sex ratio in a living adult population of *Keijella bisanensis*. The ratio around April is an approximate value of the tertiary sex ratio. All the carapaces of newly grown adults have the same potential to eventually accumulate in the sediments, irrespective of the time of death. Therefore, the tertiary sex ratio, regardless of whether it is 50% or not, is expected to be preserved in the fossil record, if the predation is not significant and dead carapaces are accumulated near the original habitat.

subject into an inquiry as to how far the seasonal change of sex ratio would be preserved or biased in the thanatocoenosis, and therefore, in the population structure of fossil species. In spite of the strongly female-biased sex ratio of the living adults, fossils of *K. bisanensis* nearly always keep the tertiary sex ratio of approximately 50%. This is because all the individuals surviving the maturation moult have the same probability of leaving their carapaces to acccumulate in the sediments regardless of the time when they die (the probability of post-maturation moulting is extremely low, cf. Kornicker, 1975; Baker, 1977), and because there is no significant difference in shape and size between male and female carapaces of this species, in terms of their sorting properties during transportation. Basically it should also apply to other ostracod species, which do not have significant differences of carapace morphology between the sexes, that the sex ratio in the thanatocoenosis maintains the tertiary sex ratio. Therefore, the expectation that the sex ratio in the thanatocoenosis of most ostracod species will be around 50% seems to be inconsistent with the fact that so many populations of supposedly syngamic ostracod species seem to have unequal, and mostly female-biased, sex ratios in fossil samples (cf. Van Harten, 1983), even if we consider only those thanatocoenoses deposited near the oiriginal habitat. Indeed in some species the tertiary sex ratio deviates from 50%, and thus it is necessary in such species to assume an unequal sex ratio of pre-adult instars in a living population, even though their sex is not detected.

## POPULATION DYNAMIC PATTERNS

Ecological and palaeoecological studies on marine invetebrates, including ostracods, have been focused mainly on the geographical distribution of a certain species and/or species assemblage. Some of these studies have elucidated in which environment and at which horizon which faunas are dominant. Other studies have been carried out with a view to correlation between the morphology of carapace and environmental parameters such as water temperature, salinity, substrate, depth etc. (see the short review of Hanai *et al.*, 1985). However, the population dynamics of marine or brackish water ostracods does not seem to have attracted many researchers (with the exceptions of Elofson, 1941; Theisen, 1966 and Heip, 1976a; 1976b on *Cyprideis torosa* (Jones); Abe, 1983 on *K. bisanensis*; Kamiya, 1988a on two species of *Loxoconcha*; and also see Horne (1983) for a review of 14 studies by 11 researchers on 54 species). One reason may be that the elucidation of population dynamics requires enormous efforts in the periodical sampling of a large number of specimens. This may be proved by the fact that the ostracods so far investigated are all shallow marine species.

Analysis of population dynamics is one of the most fundamental aspects of population ecology, and it has been carried out widely on both terrestrial and marine animal species. Mathematical ecology has recently also been greatly developed. Marine ostracod research, however, lags behind that on other groups in the development of such studies. Therefore, at present it is difficult to collect and classify all the patterns of ostracod population dynamics which would exist in nature. All that can be done is to indicate the possible patterns based on the limited amount of previous work.

Simple computer graphics can be used to illustrate several possible patterns of life cycle (Fig. 2). One closed pulse represents one generation of a population. The upper figure of each set is designed to represent the cumulative effect of several generations which are superimposed in the lower figure (except for type B-2). Thus, the upper figure will be closer to the field observation in most cases, in which it is not easy to tell with certainty to which generation an individual belongs. Here a 'generation' includes a variety of developing stages from eggs through to adult forms. The height (or the number of individuals) and the bottom length (or the life span) of the pulse is not important. What we should pay attention to is the way in which generations overlap one another. Needless to say, there must be a period when two adjacent generations overlap, otherwise the population would become extinct.

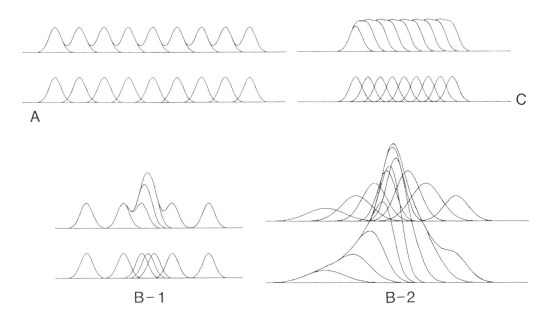

Fig. 2. Possible patterns of population dynamics of ostracods. The upper figure of each set represents the cumulative effect of several generations (except for type B-2). Each pulse stands for one generation, including every developing stage from eggs to adult forms. See text.

In a population whose pattern of population dynamics falls into type A, the alternation of generations is uncomplicated. The reproductive season is confined to a distinct period in a year, when a new generation is produced by a very small number of adults of the previous generation. *K. bisanensis* is a representative species of this type.

Conversely, species of type B do not have a restricted period of reproduction in the year (Fig. 2, type B-2 illustrates a more probable pattern, in which changing of the span and the density of generations is taken into consideration. Such changes may be ascribed to water temperature). In type B-2, the ostracods reproduce almost all the year round with a relatively short period of especially high productivity; many generations overlap one another, and the population size fluctuates extensively during the year. One of the representatives of this type is *Loxoconcha japonica* Ishizaki studied by Kamiya (1988b, 1988c). *Cyprideis torosa* living in brackish water may be another (cf. Heip, 1976b). Intermediate between the two extremes is type C, in which the total density maintains a rather constant value. The author does not

know any example of this type, but some ostracod species might follow this pattern of life cycle.

## WHEN IN ONTOGENY DOES THE MORTALITY RATE BEGIN TO DIFFER BETWEEN THE SEXES?

There are few studies in which both sex ratio and measurements of pre-adult carapaces are provided which would, therefore, be useful for discussing the relationship of the sex ratio in adults and the size distribution of the penultimate instar. The following are those examples which make such discussion possible.

Populations of *K. bisanensis* were studied which live on muddy bottoms in a rather calm embayment. The tertiary sex ratio shows equal numbers of both sexes and sexual dimorphism is not recognized among any pre-adult instars (Fig. 3a).

Two *Loxoconcha* species, living in *Zostera* beds, were studied by Kamiya (1988a, 1988b, 1988c) with special reference to their adaptive morphology and strategy. He showed that differences in carapace shape between two species should be ascribed to the microhabitat and to the way they copulate. He also investigated differences in their population dynamics and provided the basic data to consider the relationship between their sex ratio and reproductive strategy.

*Loxoconcha japonica* lives on the fronds of *Zostera marina* (sea grass), which grows rapidly in the spring and dies back in the autumn. The fronds almost disappear during the winter, and the species thus loses a large part of its habitat. In such an environment, *L. japonica* has acquired a female-biased tertiary sex ratio (62%, Fig. 3e). In contrast *L. uranouchiensis* Ishizaki, which lives on a sandy bottom, just below the *Zostera* fronds and in a relatively stable environment, has almost a 1:1 tertiary sex ratio (Fig. 3b), as does *K. bisanensis* on its quite stable muddy bottom environment.

A female-biased tertiary sex ratio means that mortality is already higher in males than in females in the juvenile forms or during the last moult, if we make the likely assumption of an equal ratio at fertilization or hatching.

*Cyprideis* species have often been investigated by European researchers from various viewpoints (Heip, 1976a, 1976b; Van Harten, 1983). Subjects of their studies include population dynamics and spatial distribution pattern (Vesper, 1975; Heip, 1976a, b). It has been pointed out that in *Cyprideis* species 'bimodality is nearly always present in the penultimate instar, but it may well appear earlier in larval development' (Van Harten, 1983, p. 571). 'From the shape of the length histograms it appears that the sexes occur in subequal proportion when bimodality arises' (Figs 3c, d, g).

A living population and its thanatocoenosis of *Spinileberis quadriaculeata* (Brady), which occurs dominantly on sandy mud bottoms, was studied by Oishi (1978) in the brackish waters of Hamana-ko Bay (for general geographic features and ostracod assemblages of the Bay see Ikeya & Hanai (1982) and Ikeya *et al.*, 1986). In this species, sexual dimorphism in the carapaces can be recognized only in adults. The tertiary sex ratio of this species is almost 50% (52 males and 56 females).

*Cytheromorpha acupunctata* (Brady) was studied in the same bay (Ikeya & Ueda, 1988). The penultimate instars are strongly dimorphic but maintain a subequal sex ratio until the last moult. The tertiary sex ratio is biased for females (65%, Fig. 3f).

These examples and all the other species which the author studied may be classified, as in Fig. 4, in terms of presence-absence of sexual dimorphism and ontogenetic change of sex ratio. These two criteria may be independent of each other, but classification as in Fig. 4 is useful from a practical point of view for the reason explained below. In the species where no sexual dimorphism arises before maturation, the sex ratio of pre-adult instars is unknown but is presumed to be 50:50. Selection pressure or mortality must be significantly different somewhere during the period indicated by hatching in Fig. 4. Every species can be represented by a horizontal line (alpha, beta, gamma and delta). *K. bisanensis*, for example, will find its position near line beta. In cases where the difference of mortality rate between the sexes becomes significantly large for the first time during the last moulting or in the later period of the penultimate stage, field observation on the thanatocoenosis will show that 1) the sex ratio of the penultimate instar is subequal but 2) the tertiary sex ratio of adults is biased. This is the case in those *Cyprideis* species so far studied.

An important finding is that the combination of a biased tertiary ratio in the adult and no dimorphism in the penultimate instar has not been found in any population the author has examined in the field (Fig. 4). However, an equal tertiary ratio and strong dimorphism in the penultimate instar is typical in the case of *L. uranouchiensis*. Therefore, an unequal tertiary sex ratio seems to be associated with sexual dimorphism in the penultimate instar. Imagine the hypothetical case of a thanatocoenosis in which the sex ratio is biased in the adults but equal in the penultimate instar. In such a case, it cannot be proved whether at the last moult or earlier, during the later period of existence of the penultimate instar, the ratio had been biased. Only the seasonal observation of a living population can determine whether A or B in Fig. 4 is the pathway of species gamma. However, case A (bias only at the last moult) will be very rare, because it is

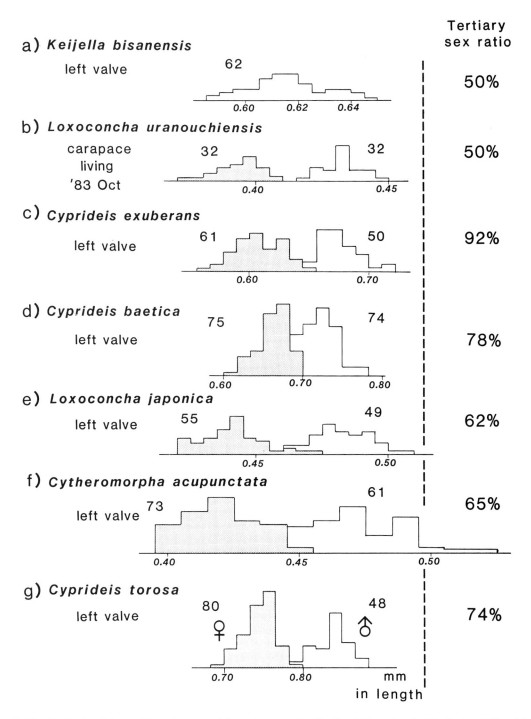

Fig. 3.  Size distribution of the penultimate instars and the tertiary sex ratios (for *Cyprideis* species after Van Harten, 1983 and for *Loxoconcha* species after Kamiya, 1988b). The figures above the open histograms stand for the individual number of males and the solid ones for females, except for *K. bisanensis*, in which the sex is unknown in the penultimate instar. The material is from thanatocoenoses except for that of *L. uranouchiensis*.

reasonable to assume the difference in the quality and intensity of selection pressure between males and females where strong dimorphism exists in the carapace morphology. A population which has a biased ratio and exhibits dimorphism in the earlier stage (shown by the lower horizontal line in the figure) may be the more r-selective (see MacArthur & Wilson, 1967; Pianka, 1974). The most extreme case might be interpreted as parthenogenesis, but the role of chromosomal evolution should not be overlooked.

## SEX RATIO AS A POSSIBLE INDICATOR OF (PALAEO)ENVIRONMENT

Several approaches have been developed so far to estimate the palaeoenvironment based on the fossil record. They include those in which the occurrence of a particular species or assemblage plays a great role (Whatley, 1983, 1988) and others in which the chemical composition of the carapace is used (Forester, 1986). Hanai *et al.* (1985) suggested another approach, in which they partly succeeded in establishing some correspondence between the environment of the microhabitat and sensory organ morphology of marine ostracods as shown by the carapace pore canals. This could be called the 'organ level approach', while the traditional way is at the species or assemblage level. The present author believes that the sex ratio provides yet another approach.

Some species have acquired and developed their own reproductive strategy in the process of adaptation to the microenvironment of their habitat. This strategy may be reflected in various ecological aspects of the species. Mating behaviour may also be one of these aspects, together with the sex ratio and pattern of population dynamics. Mating behaviour is closely related to the spatial distribution pattern of individuals. For example, in *K. bisanensis*, adult females are always aggregated, while adult males are distributed at random (see Abe, 1983, for details). In the reproductive season, the sex ratio is highly biased in favour of females, to the extent of nine females to one male. Thus a kind of polygamy is suggested for the species. The advantage of the ratio being biased in

favour of females should be understood in the context of reproductive strategy, which is beyond the scope of this study.

Van Harten (1983) discussed resource competition as a possible cause of the sex ratio in *Cyprideis* species. In *K. bisanensis*, also, some interference including competition is suggested among the individuals of the penultimate instar, because their spatial distribution pattern has been found to be uniform (Abe, 1983, 451). Resource competition, however, may be significant only in a species which has too high a population density for the available resources to support, as in *Cyprideis* species, and may not apply to low density species, such as *K. bisanensis*.

Although reproductive strategy could be reflected in several ecological aspects (sex ratio, mating behaviour, distribution pattern, pattern of population dynamics etc.), sex ratio alone among them is capable of preservation in the fossil record. Generally a one to one correspondence between sex ratio and reproductive strategy, or between sex ratio and the microenvironment will not be expected in all syngamic animal species. However, as far as marine ostracods are concerned, every observed example seems to have a good correspondence and, in fact, is well interpreted by means of a simple law. This states that species which have basically equal tertiary sex ratios (*K. bisanensis, L. uranouchiensis, S. quadriaculeata*) are more *K*-selective and inhabit relatively stable environments. On the contrary, species which have biased tertiary sex ratios (*L. japonica* and *Cyprideis* species) are more *r*-selective and inhabit relatively unstable environments. While the number of examples are not sufficient to suggest that this should apply to all ostracods, it is significant that it is applicable to all those studied to date.

## CONCLUSIONS

It may be concluded that the sex ratio in some fossil populations of species tells us (1) when mortality started to differ between males and females. It also suggests (2) what type of reproductive strategy the species had acquired and developed, and (3) whether the palaeoenvironment which the

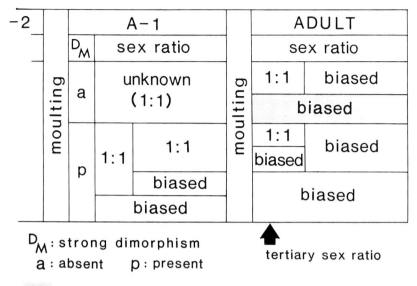

$D_M$ : strong dimorphism
a : absent   p : present

tertiary sex ratio

: Those species which would pass this category
are unknown in the field observation.

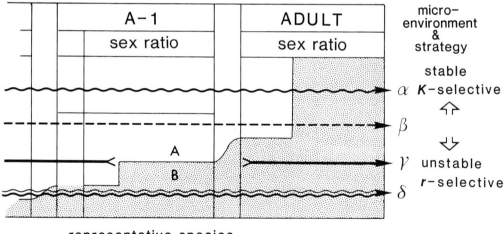

micro-
environment
&
strategy

stable
$\alpha$ **K**-selective

$\beta$

$\gamma$ unstable
$\delta$ **r**-selective

representative species
$\alpha$: *Keijella bisanensis*   $\beta$: *Loxoconcha uranouchiensis*
$\gamma$: *Cyprideis exuberans*   $\delta$: *Cyprideis torosa*

Fig. 4. What the sex ratio tells us: (1) the difference in mortality between the sexes, (2) type of reproductive strategy, (3) environment of the habitat. Note that the combination of a biased tertiary sex ratio in the adult and no strong dimorphism in the penultimate instar has not been found in any population the author has examined.

species inhabited was stable or variable (at least as regards Ostracoda). The author hopes that other workers will work in this field to realize the potential of the sex ratio as an indicator of palaeoenvironment.

## ACKNOWLEDGEMENTS

I wish to thank Professor J. W. Neale for his critical reading of the first version of the manuscript. I am also grateful to Professor N. Ikeya who kindly permitted me to examine the ostracod collection of Shizuoka University and to Professor R. C. Whatley, whose advice I followed to improve the manuscript.

## REFERENCES

Abe, K. 1983. Population structure of *Keijella bisanensis* (Okubo) (Ostracoda, Crustacea) - An inquiry into how far the population structure will be preserved in the fossil record. *J. Fac. Sci. Tokyo Univ.*, Tokyo, **20**, 443-488.

Baker, J. H. 1977. Life history patterns of the myodocopid ostracod *Euphilomedes producta* Poulsen, 1962. *In* Löffler, H. & Danielopol, D. (Eds), *Aspects of ecology and zoogeography of Recent and fossil Ostracoda*, 245-254. Dr. W. Junk b.v. Publishers, The Hague.

Elofson, O. 1941. Zur Kenntnis der marinen Ostracoden Schwedens. *Zool. Bidr. Upps.*, Stockholm, **19**, 215-534.

Forester, R. M. 1986. Determination of the dissolved anion composition of ancient lakes from fossil ostracodes. *Geology*, Boulder, Colorado, **14**, 796-799.

Hanai, T., Abe, K., Kamiya, T. & Tabuki, R. 1985. Sensory organs of Ostracoda and palaeoenvironment. *In* Kajiura, K. (Ed.), *Ocean Characteristics and their Changes*, 424-429. Kosei-sha Kosei-kaku. (In Japanese).

Hayashi, M. & Goto, Y. 1979. On the seasonal succession and the food habits of the gobioid fishes in the Odawa Bay, Yokosuka, Japan. *Sci. Rep. Yokosuka Cy Mus.*, Kurihama, Yokosuka, Japan, **26**, 35-56. (In Japanese with English abstract).

Heip, C. 1976a. The spatial pattern of *Cyprideis torosa* (Jones, 1850) (Crustacea: Ostracoda). *J. mar. biol. Ass. U.K.*, Plymouth, **56**, 179-189.

Heip, C. 1976b. The Life-Cycle of *Cyprideis torosa* (Crustacea, Ostracoda). *Oecologia*, Berlin, **24**, 229-245.

Horne, D. J. 1983. Life-cycles of podocopid Ostracoda - A review (with particular reference to marine and brackish-water species). *In* Maddocks, R. F. (Ed.), *Applications of Ostracoda*, proceedings of the Eighth International Symposium on Ostracoda, July 26-29, 1982, 581-590. Univ. Houston Geos., Houston, Texas.

Ikeya, N. & Hanai, T. 1982. Ecology of Recent Ostracods in the Hanana-ko Region, the Pacific coast of Japan. *In* Hanai, T. (Ed.), *Studies on Japanese Ostracoda*, 15-59. University of Tokyo Press.

Ikeya, N., Oishi, M. & Ueda, H. 1986. Seasonal distribution pattern of benthic ostracod faunas of a station in Hamana-ko Bay, Pacific coast of central Japan. *Rep. Fac. Sci. Shizuoka Univ.*, Shizuoka, **20**, 189-204.

Ikeya, N. & Ueda, H. 1988. Morphological variations of *Cytheromorpha acupunctata* (Brady) in continuous populations at Hamana-ko-Bay, Japan. *In* Hanai, T., Ikeya, N. & Ishizaki, K. (Eds), *Evolutionary biology of Ostracoda, its fundamentals and applications*, proceedings of the Ninth International Symposium on Ostracoda, held in Shizuoka, Japan, 29 July - 2 August 1985, Developments in palaeontology and stratigraphy, **11**, 319-340, Kodansha Ltd., Tokyo and Elsevier, Amsterdam, Oxford, New York, Tokyo.

Kamiya, T. 1988a. Morphological and ethological adaptations of Ostracoda to microhabitats in *Zostera* beds. *In* Hanai, T., Ikeya, N., & Ishizaki, K. (Eds), *Evolutionary biology of Ostracoda, its fundamentals and applications*, proceedings of the Ninth International Symposium on Ostracoda, held in Shizuoka, Japan, 29 July - 2 August 1985, Developments in palaeontology and stratigraphy, **11**, 301-316, Kodansha Ltd., Tokyo and Elsevier, Amsterdam, Oxford, New York, Tokyo.

Kamiya, T. 1988b. The meaning of sex-ratios obtained from fossil Ostracoda. *Senckenberg. leth.*, Frankfurt a.M., **68**(5/6), 337-345.

Kamiya, T. 1988c. Contrasting population ecology of two species of *Loxoconcha* (Ostracoda, Crustacea) in Recent *Zostera* (eelgrass) beds: adaptive differences between phytal and bottom-dwelling species. *Micropaleontology*, New York, **34**, 316-331.

Kesling, R. V. 1951. The Morphology of Ostracod Molt stages. *Illinois biol. Monogr.*, Urbana, **21**, 126 pp.

Kornicker, L. S. 1975. Antarctic Ostracoda (Myodocopina). *Smithson. Contr. Zool.*, Washington, D.C., **163**, 1-720.

MacArthur, R. H. & Wilson, E. O. 1967. *The theory of island biogeography*. 203 pp. Princeton University Press.

Martinsson, A. 1956. Ontogeny and development of dimorphism in some Silurian ostracods. *Bull. geol. Instn Univ. Uppsala*, **37**, 2-42.

Oishi, M. 1978. *Seasonal change of ostracod species and assemblages from Hamana-ko, the Pacific coast of Japan*. Unpub. thesis, Shizuoka University, 36 pp. (In Japanese).

Pianka, E. R. 1974. *Evolutionary ecology*, 356 pp. Harper & Row.

Theisen, B. T. 1966. The Life History of Seven Species of Ostracods from a Danish Brackish-water Locality. *Meddr Danm. Fisk. og Havunders.*, Kobenhavn, **4**, 215-270.

Van Harten, D. 1983. Resource competition as a possible cause of sex ratio in benthic ostracodes. *In* Maddocks, R. F. (Ed.), *Applications of Ostracoda*, proceedings of the Eighth International Symposium on Ostracoda, July 26-29, 1982, 568-580. Univ. Houston Geos., Houston, Texas.

Van Morkhoven, F. P. C. M. 1962. *Post-Palaeozoic Ostracoda. Their Morphology, Taxonomy, and Economic Use*, Volume I, 204 pp. Elsevier Publishing Company, Amsterdam, London, New York.

Vesper, B. 1975. To the problem of noding on *Cyprideis torosa* (Jones 1850). *Bull. Am. Paleont.* , Ithaca, **65**, 205-216.

Whatley, R. C. 1983. The application of Ostracoda to palaeoenvironmental analysis. *In* Maddocks, R.F. (Ed.), *Applications of Ostracoda*, proceedings of the Eighth International Symposium on Ostracoda, July 26-29, 1982, 51-77. Univ. Houston Geos., Houston, Texas.

Whatley, R. C. 1988. Population structure of ostracods: some general principles for the recognition of palaeoenvironments. *In* De Deckker, P., Colin, J.-P. & Peypouquet, J.-P. (Eds) *Ostracoda in the Earth Sciences*, 245-256. Elsevier, Amsterdam, Oxford, New York, Tokyo.

Whatley, R. C. & Stephens, J. M. 1977. Precocious sexual dimorphism in fossil and Recent Ostracoda. *In* Löffler, H. & Danielopol, D. (Eds), *Aspects of Ecology and Zoogeography of Recent and Fossil Ostracoda*, 69-91. Dr. W. Junk b.v. Publishers, The Hague.

## DISCUSSION

Dan Danielopol: High mortalities occur in females having a high fecundity. In both males and females high mortality seems to be the cost the animals pay when they invest a high amount of energy in reproductive activities. Could you comment on this?

Katsumi Abe: In *Keijella bisanensis*, field observations concerning population density, spatial distribution pattern and sex ratio suggest that 1) sex ratio is highly biased in favour of females around the beginning of the reproductive period, and that 2) mating is a kind of polygamy in which one male mates with several females and one female mates with several males. Females of the last generation, which are recognized on the basis of their dark colour and the dirtiness of their carapace, lay eggs in the first half of the reproductive period and then die.

Robin Whatley: You seem to make a whole series of very positive assertions on the basis of studies which, however detailed, are of very few taxa. Are you confident that your conclusions are applicable to the majority of ostracods in marine environments?

Katsumi Abe: I do not know if my conclusions are applicable to the majority of marine ostracods. This is because there are few works on this kind of research. As far as the species I examined are concerned, the conclusions are right. I only hope that other workers will work in this field.

Robin Whatley: You need only to study the population structure and sex ratio of very few fossil species to demonstrate that your comments on the parity of the sexes in fossil faunas are completely misleading. I would welcome your comments on this.

Katsumi Abe: Your criticism seems to come from midsunderstanding what I meant. The parity of the sexes in fossils is not observed for all the ostracod species. What I would like to demonstrate is that the tertiary sex ratio of the living populations I studied are preserved in the thanatocoenosis which accumulated near their habitat, and even in transported ones, there is not a significant difference with respect to sorting properties between the sexes. Please note that the tertiary sex ratio differs from species to species. Selective predation was not observed in *K. bisanensis*.

# 14

# Cytogenetic studies on marine myodocopid Ostracoda: the karyotypes of some species of *Vargula* Skogsberg, 1920

**Alicia Moguilevsky**

Institute of Earth Studies, University College of Wales, Aberystwyth, U.K.

## ABSTRACT

The number and morphology of the chromosomes of 8 shallow water species of *Vargula* Skogsberg are analysed and compared. Of these species one, *V. hilgendorfi* (G.W. Müller), was recovered from Misaki, Japan, while the remaining 7 were collected in the Caribbean. Six of these Caribbean species are as yet unnamed and, together with *V. graminicola* Cohen & Morin, they form part of a major study of the taxonomic relationships and bioluminescent characteristics of species of *Vargula* currently being carried out by Cohen and Morin. The karyotypes of at least 5 of these species, for which it was possible to ascertain the number, consists of 2n = 16 (14A+XX) for the female, and 2n = 15 (14A+XO) for the male. As well as sharing the same number, the chromosomes of these species exhibit considerable uniformity in their morphology and characteristics which include the presence of only one chiasma per bivalent as shown in the metaphase I plates studied. The possible evolutionary implications of this single chiasma are discussed. The karyotypes obtained are consistent with the results shown in previously studied myodocopid ostracods and which clearly distinguish them from the Podocopida. One of the aims of this cytotaxonomic study of the available species of *Vargula*, aside from establishing their karyotypes, was to analyse to what extent these karyotypes reflected the varied bioluminescent patterns in addition to the ecological preferences of the species as observed by Cohen and Morin.

## INTRODUCTION

Species of *Vargula* are widespread in all oceans between latitudes 80°N and 74°S and depths of 0-3431m (Kornicker, 1984). The adult female (always larger than the male) of the majority of the species which occur in the western Atlantic, has a carapace length of less than 3mm, except *V. magna* Kornicker, which is 3.52-4.06mm long.

Many species of *Vargula* are known to be bioluminescent. Bioluminescent nocturnal displays, presumed to be associated with sexual communication, have been observed in species of this genus from the U.S. Virgin Islands by Morin & Bermingham (1980). Similar displays were observed during nocturnal surveys of the reefs and surrounding habitats of the San Blas Islands (Panama) carried out by Morin (1986). He discovered a wide variety of distinctive bioluminescent display patterns produced by ostracods that were closely associated with particular habitats. Cohen & Morin (1986) and Morin & Cohen (1988), subsequently demonstrated that each species is associated with a particular habitat and that possibly a distinct niche occupation could be recognized for each one. Some of these *Vargula* species, such as *V. graminicola* and *V. shulmanae* Cohen & Morin, are sibling species differing in few morphological characters but distinctive in diet, habitat and bioluminescent pattern.

Given the above, the present author decided to test the possibility that these species could be distinguished on the basis of their respective karyotypes.

## MATERIAL AND METHODS

From previous studies of myodocopid species (Moguilevsky, 1985; Moguilevsky & Whatley, 1988), it was found that tissue taken from young embryos extracted from gravid females and spermatogonia from testis tissue squash, provided the best preparations for chromosome analysis. This type of material was not available for all the species studied herein and the results reflect this.

The complement of at least 5 of these species of *Vargula*, for which it was possible to ascertain the number, consists of $2n = 16$ (14A + XX) for the female and $2n = 15$ (14A + XO) for the male. As well as sharing the same number, the chromosomes of all these species exhibit considerable uniformity in their morphology and characteristics, which include the presence of only one chiasma per bivalent, as shown in the metaphase I plates of meiosis. The species studied are *Vargula hilgendorfi* (Misaki, Japan) and *V. graminicola* (Maca-

roon Reef, California).

The names used herein for the following six, as yet unnamed, species are the 'working initials' or 'code names' given to them by Cohen and Morin (pers. comm.) based on, and describing their bioluminescent patterns:

**'SWD'** (Barkers Point, San Salvador, Bahamas): 'Streaking Wide Downers; a widely spaced downward display with the pulses often produced as a streak rather than a spot'.

**'SU'** (Playa Kalki, Curacao): 'Shallow Upper, for the upward luminescent displays made in very shallow water'.

**'WHU'** (Cayo Enrique, Isla Marqueyes, Puerto Rico): 'Wide, High Upper; for the widely spaced upward display'.

**'OLD'** (Rocky Point, San Salvador, Bahamas): 'Other Low Downer; for downward display that may be different from a previously described downward display'.

**'BLU'** (Cayo Enrique, Isla Marqueyes, Puerto Rico): 'Bullae - Like Upper; they have an upward display similar to the display seen in *V. 'bullae'* Poulsen from St. Croix'.

**'LD'** (Playa Kalki and Carmabi Reef, Curacao): 'Low Downer, for the downward displays they make, low over the bottom'.

Cohen and Morin (pers. comm., this volume), in their cladistic analysis of the genus *Vargula*, have asigned the species they studied into different groups according to, among other characters, similarities and differences in morphology, ecological preferences and bioluminescent display patterns:

| 'F' - Group | 'T' - Group |
|---|---|
| *V. graminicola* | **'SU'** |
| **'SWD'** | **'LD'** |
| | **'BLU'** |
| | **'WHU'** |
| | **'OLD'** |

*V. hilgendorfi* is placed in a group of its own, although seemingly intermediate between *V. magna* Kornicker and the 'F'-Group, it is more distant from the 'T'-Group. (Cohen & Morin, this volume).

## THE SPECIES

### *Vargula hilgendorfi* (G.W. Müller, 1890)

*V. hilgendorfi* is a bioluminescent species recorded from various localities in Indonesia, from coastal waters off Thailand and Singapore and also from the Kei Islands in the Celebes Sea (Poulsen, 1962). It is also found living along the coast of South Korea and around Japan, where the bioluminescence of this species has been intensively studied (Hanai, 1974). To most biologists it is still more familiar as '*Cypridina hilgendorfi*' since the majority of studies on this species were carried out prior to 1961 when Sylvester-Bradley raised *Vargula* to generic rank. Skogsberg (1920) considered *Vargula* to be a subgenus of *Cypridina*.

The depth of occurrence of the species in those areas from which it has been recorded, is in the range 1-25m (Poulsen, 1962). The size of the female carapace is 3.4mm in length and 2.3mm in height and that of the male 2.8mm in length and 1.9mm in height (Hiruta, 1980). The specimens used in this study were collected by Karen Watson at Misaki, Japan in July 1985. They were recovered from littoral sediments in about 30cm water depth by means of baited bottle traps left overnight. The specimens were taken to the Marine Biological Station of Misaki where they were kept and cultured in a 0.05% solution of colchicine and sea water for about 24 hours, and subsequently fixed and preserved in 1:3 acetic alcohol. No males were collected in the sample and only 2 gravid females with developing embryos were available for this study. The low numbers of specimens recovered could be accounted for by the effects of a typhoon which affected the area in June 1985.

Since testis tissue was not available and the female ovaries did not yield good preparations for the analysis of meiosis, the counts and description of the chromosome morphology was based only on the study of mitotic metaphase plates obtained from embryo tissue. A number of preparations were made in the usual manner using embryos removed from each of the gravid females (Moguilevsky, 1985). Unfortunately, in some of the preparations, the chromosomes appeared highly condensed and in others some degree of overlap rendered the counts, measurements and analysis of the morphology, somewhat difficult (see Plate 1, Figs 2-4). However, by combining the observations extracted from a number of these preparations, a karyotype composed of 16 chromosomes (2n) was obtained. The chromosome complement obtained from female embryos was 14A + XX and, 14A + XO from male embryos. The chromosomes are all metacentric or very slightly submetacentric and they range in size from 5.71 to 4.20μm. Their overall similarity renders them individually unidentifiable.

### *Vargula graminicola* Cohen and Morin, 1986

*V. graminicola* is a benthonic species which occurs in the Caribbean and is abundant in shallow (3-10m) sea grass beds (*Thalassia testudinum* and *Syringodium filiformis*). In Panama, it lives in current-swept sea grass beds within the 'turf', rubble and top few millimetres of sand, away from coral reefs and sand 'blow-outs'. Individuals are infaunal by day and epibenthonic and demersal by night. The size range for the length of the female carapace is 1.8-2.25mm and 1.52-1.85mm for that of the male. Details of the morphology and ecology of this species are given by Cohen & Morin (1986).

The specimens used in this study were recovered from sea-grass bed samples collected from the Macaroon Reef, San Blas Islands (Panama), by Morin in December 1985, during the bioluminescent period of the species (about 1 or 2 hours after sunset). The specimens were cultured in 0.001g/ml of 'demecolcine' in sea water for about 38 hours. They were then stored for incubation at room temperature (about 32°C) in the dark, after which they were fixed and preserved in 1:3 acetic alcohol. The samples were subsequently stored for most of the time in a refrigerator before being posted to the U.K.

The sample studied, the only one for this species, consisted of a single gravid female, two adult males and three juveniles. All specimens were dissected and analysed. Whereas the tissue obtained from the juveniles (gut and immature gonads) did not show cells in division, the preparations made from testes tissue of the two males, showed only the last stages of spermiogenesis and, therefore, only spermatozoa were visible. Although, no dividing cells were observed either in the ovaries of the gravid female, some good preparations were obtained in the usual manner using tissue of the developing embryos extracted from this female.

The analysis of the karyotype was, therefore, carried out on the limited number of mitotic metaphases encountered. Although most of the chromosomes were clearly outlined, a certain amount of overlap tended to obscure some of them. In all of the metaphase plates which were clear enough for chromosome counts, the chromosome number was rather high; most cells had 32 chromosomes and others 64.

The fact that 16 is the chromosome number encountered in the only other species of *Vargula* (*V. hilgendorfi*) for which the chromosome number is clearly known and that the most frequent numbers found in this species (either 32 or 64) are both multiples of 16, leads the present author to suspect that these high numbers are an indication of polyploidy.

Polyploidy in animals is relatively rare. It has been reported in ostracods but only in some freshwater, parthenogenetic species (Tetart, 1978). On the other hand, although both colchicine and colcemid ('demecolcine') are alkaloids used to arrest cell division at metaphase based on their spindle inhibiting properties, colchicine has also been used to induce polyploidy in plant tissue (Macgregor and Varley, 1983). In some cases, as a result of overculturing, polyploidy in animal tissues has also been observed. It is possible that colchicine may also have induced polyploidy in some cells in the present material. Therefore, although not previously encountered as a problem in the culturing of any of the other species of myodocopid ostracods studied, it is suggested in this work that the high number of chromosomes found for this species may be an artefact of the culturing method.

The 32 chromosomes seen in Figs 1 and 2 of Plate 2 can, in both cases, be interpreted as representing 2 diploid sets each of 16 chromosomes (Figs 1a, b, and 2a, b). Of these 16 chromosomes, 14 are metacentric or very slightly submetacentric and, 2 acrocentric. The length of the metacentric individuals varies very gradually. The size range is 11.5 to 8.8µm for the largest and 7.8 to 5.8µm for the smallest. The size range of the acrocentric chromosomes is between 6.4 and 4.8µm.

Until experiments can be carried out on more samples, testing different culturing times and colchicine concentrations, it is here suggested that the karyotype of *V. graminicola* is composed of a diploid number of 16 for the female and 15 for the male. The chromosome complement is made up of 14 autosomes and 2 'X' and, 14 autosomes and one unpaired 'X' chromosome (XO) respectively.

For each of the remaining as yet unnamed species: **'SWD'**, **'WHU'**, **'BLU'**, **'OLD'**, **'LD'** and **'SU'**, the following culturing treatments were carried out:

(a) Control: the specimens were left in sea water, without cochicine, for 12 hours after which an equal volume of 0.36M $MgCl_2$ (isotonic to sea water) was added to anaesthetize them. About 2ml of freshwater was then dropped slowly into 6ml of the solution and the specimens were then fixed in fresh, cold 1:3 acetic alcohol for 20 minutes. The solution was changed twice in the following 2 days.

(b) Cultured in 0.05% colchicine solution (0.5g/l, colchicine in sea water) for 12 hours. The colchicine stock solution was made up as 1g of colchicine per litre of sea water which was then added slowly (over 1 hour) to the ostracods, kept in 5ml of sea water, until the final volume was 10ml. The specimens were fixed and preserved as explained in (a).

(c) Cultured for 12 hours as in (b), after which the volume of the culturing solution was reduced to about 5ml, and freshwater added slowly over about 6 minutes, until the volume was again 10ml. The specimens were fixed and preserved as in (a).

(d) Cultured as in (b) but for 24 hours, fixed and preserved as in (a).

(e) Cultured as in (c) but for 24 hours, fixed and preserved as in (a).

## Vargula 'SWD'

The specimens used in this study were collected from Barkers Point (about 1km from Rocky Point) on the northwest side of San Salvador, Bahamas, on 24 July, 1986. They were recovered from high above mixed coral patches and reef pavement with gorgonians, and occasionally sand, at depths of 2 to at least 6m. They co-occur with 'OLD' in shallow water. Only males (14) were present in the available samples. The concentration of the colchicine solution used in culturing these specimens was estimated since no balance was available at the time of the experiments. Of the 14 specimens, 7 were cultured for 12 hours and the remaining 7, for 24 hours. All specimens were fixed and preserved as described in (a).

A number of slides were made from specimens of both groups. Whereas the preparations obtained from dissecting the specimens cultured for 24 hours showed hardly any dividing tissue (only sperms were visible), those obtained from the 12 hour group yielded some cells at metaphase I and II of meiosis (Plate 3, Figs 1-4). Although the number and morphology of chromosomes are not normally described from meiotic plates, as explained above, no mitotic tissue was available for this species (nor for any of the remaining Caribbean species). Therefore, a suggested karyotype will be offered for these species by interpreting the configuration of the bivalents at metaphase I of meiosis.

Plate 3, Fig. 1 shows 14 bivalents and 2 univalent 'X' (arrowed), which are interpreted as representing two cells with 7 bivalents and 1 univalent each. The interpretation and diagram of this figure, show the metacentric or submetacentric nature of the chromosomes and the location of the single chiasma observed in each bivalent. Plate 3, Figs 3 and 4, show chromosomes at metaphase II and, although the boundaries of the cells are rather uncertain in this material, the morphology of these chromosomes confirms the interpretation of those in Figs 1 and 2.

On the basis of the above, it is suggested that the karyotype for this species is composed (for the male) of 14 autosomes (metacentric/submetacentric) plus an unpaired 'X', [15 = 14A+XO], and for the female: 14A+XX.

The relative size of chromosomes is usually obtained from numerous measurements taken on mitotic metaphase photographs. Although these were not available for the Caribbean species and despite the fact that these measurements are less reliable, since the degree of contraction and the position of the chiasma slightly distorts the actual size of the chromosomes, an approximate size range is given based on the photographs of metaphase I and II of meiosis. The length of the autosomes varies between 3 and 8.5µm, and that of the 'X' chromosomes between 3.8 and 6.3µm, approximately.

## Vargula 'SU'

The specimens available for this study (only males) were collected in June 1986 at Playa Kalki, in the northwest end of the island of Curacao. They were recovered from very shallow water (<3m), usually among and above the coral *Acropora palmata* and the hydrocoral *Millepora complanata* . Twenty-five specimens were divided into groups of 5 individuals. Each group was cultured according to one of the 5 culturing treatments as described above. All individuals were dissected and testes tissue analysed.

Only in the preparations obtained from the groups treated according to methods (a) - control group with no colchicine added and (c) - cultured in colchicine for 12 hours plus hypotonic solution, were chromosomes observed. No mitotic plates were found in any of the slides; only chromosomes at metaphase I stage of meiosis were detected in preparations of either groups.

Plate 4, Fig. 1 is a photomicrograph obtained from the control group (specimens fixed and preserved but not cultured in colchicine). This is a side view in which the bivalents (12) appear congressed on the spindle. They are co-oriented and centromere pairs of each bivalent lie equidistant above and below the spindle equator (sister centromeres lie adjacent to one another and so face towards the same spindle pole). The univalent 'X' (arrowed), is orientated to one pole only and lies off the spindle equator. The effect of colchicine on meiotic plates is a disarray of this type of formation, as shown in Figs 2-4. The diagram in Fig. 1a, shows the interpretation of the location of the single chiasma present in each bivalent (2 in the distal and 4 in the interstitial position) as well as the morphology of the chromosomes involved (metacentric/submetacentric). The photomicrograph in Fig. 2 shows 7 bivalents, not aligned on the equator, and a univalent 'X' (arrowed). The chromosomes show similar configurations to those in Fig. 1 due to the presence of a single chiasma per pair (top view). This specimen was cultured in colchicine for 12 hours.

The scattering of meiotic chromosomes, due to the effect of the colchicine, can also be seen in Figs 3 and 4.

The cell boundaries are uncertain in all the photomicrograhs obtained for this species. This uncertainty renders even the suggestion of a possible karyotype, very difficult; a complement of either 15 = (14A+XO) or 13 = (12A+XO) for the male, is equally possible. The size range of these meiotic chromosomes is between 5.35 and 7.20μm.

### Vargula 'WHU'

Six male specimens were collected by sweep net from Cayo Enrique in Puerto Rico. They were found mostly above mixed coral (especially *Monastrea annularis*) shoreward of, or near the edge of drop-offs, at depths of 2-8m and most often on promontories. They can co-occur with 'BLU' in these areas. Half of them were cultured in colchicine for 12 hours and the remaining 3, for 24 hours. A number of preparations were made from the testes tissue extracted from all individuals. No chromosomes were seen in any of the slides obtained from the second group (only sperms).

Plate 5 (Figs 1-3) shows chromosomes at the metaphase I stage of meiosis, observed in the preparations obtained from the first group. The cell boundaries are uncertain in all of these photographs although, in Figs 1 and 2, there appears to be two cells involved with 7 bivalents and one univalent 'X' in each. The diagrams in Figs 1a and 2a are an interpretation of the morphology

of the chromosomes concerned (all metacentric), and the location of the single chiasma observed in each bivalent: 5 in the distal and 2 in the interstitial position (1a) and 4 in the distal and 3 in the interstitial position (2a). Fig. 3 shows a scattering of chromosomes similar in morphology (all metacentric) to those shown in Figs 1 and 2. There is also, a single chiasma per bivalent as observed previously. The approximate size range of these chromosomes is estimated at between 4.1 and 6.3μm.

A suggested possible karyotype for this species (male only) could be 2n = 15 (14A+XO).

### Vargula 'OLD'

The 20 available male specimens of this species were collected from Rocky Point in San Salvador, Bahamas. They were found near shore above shallow patch reefs (with scattered coral growth - <50% cover), reef sides and beach rock, usually in 0.5-3m of water. They occasionally display over sand. Half of the specimens were cultured in colchicine for 24 hours and the remaining 10, for 12 hours and treated in a hypotonic solution prior to fixation. All individuals were dissected and testes tissue analysed. In the preparations obtained from the first group neither mitotic nor meiotic chromosomes were found, only sperms were present in them.

The results obtained from the analysis of the second group were also disappointing, since only a few scattered chromosomes at metaphase stages I and II of meiosis were identifiable (Plate 5, Figs 4 and 5).

Plate 5, Fig. 4 shows 8 bivalents and a single univalent 'X' (arrowed). The chromosomes all appear to be metacentric and of similar size. The location of the single chiasma per bivalent has been interpreted in Fig. 4a.

The uniformity of these chromosomes, in terms of their morphology (all metacentric) and size, seems to be confirmed by those seen at metaphase II (Plate 5, Figs 5 and 5a). The approximate size range of these meiotic chromosomes is estimated as between 4 and 5μm.

The uncertainty of the cell boundaries in the

material studied, renders even a 'suggested' chromosome complement impossible.

## *Vargula* 'LD'

Twenty-five male specimens were recovered from two localities in Curacao: Playa Kalki in the northwest of the island, and Carmabi Reef on the west-central part. They were found among gorgonians, especially *Pseudopterogorgia* , and corals surrounded by sand on the shallow, flat shelf that surrounds much of the island and on slopes beyond the shelf (at about 35 degrees or less). These individuals were divided in 5 groups and cultured and preserved as for *Vargula* 'SU' above. All specimens were dissected and testes tissue analysed.

The results were rather disappointing since only the preparation obtained from a specimen cultured for 25 hours, yielded some chromosomes, and these were scattered and their cells boundaries uncertain. Fig. 1 of Plate 6 shows these chromosomes at metaphase I of meiosis, while Fig. 1a offers an interpretation of the morphology of the chromosomes and the location of the single chiasma present per bivalent.

Although the majority of the chromosomes appear to be metacentric or very slightly submetacentric, this is the only species of *Vargula* of this study, except for *V. graminicola* , in which acrocentric chromosomes have also been observed. The size range of these meiotic chromosomes is estimated as between 3.75 and 5.3µm.

The information gained from the preparations of this species, is obviously insufficient to offer a certain karyotype.

## *Vargula* 'BLU'

Twenty male specimens were collected by sweep net from Cayo Enrique in Puerto Rico. They were broadly distributed over mixed scleractinian corals (especially *Acropora cervicornis*), often with gorgonians and the nearby (within 2m) sand at shallow depths of 3-6m. Ten of these individuals were cultured in colchicine for 12 hours (see Method (b)), and the remaining 10 for 24 hours (see

Method (d)). All of them were dissected and the testes tissue analysed.

Only one of the preparations obtained, yielded chromosomes; this was from one of the specimens cultured for 12 hours. Plate 6, Fig. 2 shows these chromosomes at metaphase II of meiosis. Although the cell boundaries are uncertain, this photograph seems to show the presence of two groups of 7 chromosomes (2 cells?) of similar morphology (all metacentric). Note that the chromosomes, each consisting of two chromatids, are not very dissimilar in terms of size (see interpretative diagram in Plate 6, Fig. 1). The estimated size range of these meiotic chromosomes is 2.2-5µm.

The insufficient data obtained for this species precludes the author from suggesting a karyotype until more material can be analysed.

## DISCUSSION

With only one possible exception (*V. kuna*), Cohen and Morin (1986) have found that in all the luminescent species of *Vargula* they have studied to date, 'each unique type of luminescent display consistently correlates with a particular and distinct suite of morphological characters in specimens producing that display'. Based on the high correlation between luminescent behaviour and morphology, they were able to conclude that each display type, which involves spatial differencies, display period and specific microhabitat selection, represents the signals produced by a distinct species. Given the above, the present author decided to test the possibility that these species could also be distinguished on the basis of their karyotype, or at least to analyse to what extent that karyotype reflected their similarities and differences.

Of the eight shallow water species of *Vargula* available for this study, for only two (*V. hilgendorfi*, *V. graminicola*) was it possible to establish their karyotype with certainty, based on mitotic plates obtained from embryo tissue squash.

Although the chromosome complement is the same for both species: 16 = 14A+XX (female) and 15 = 14A+XO (male), the morphology and size of these chromosomes is slightly different, enabling their karyotypes to be easily distinguished and recognized. While the karyotype of *V. hilgendorfi* is very uniform (all chromosomes are metacentric/submetacentric and similar in size, ranging between 4.2 and 5.71μm). In *V. graminicola* 14 of the 16 chromosomes are metacentric/submetacentric and the remaining 2 are acrocentric and slightly smaller. The size of the larger metacentric chromosomes varies gradually between 6.8 and 10.5μm and that of the acrocentric between 4.8 and 6.4μm.

Since for the remaining 6 and as yet unnamed species mitotic tissue was not available, the karyotypes suggested herein have been interpreted by analysing the configurations of bivalents seen in photomicrographs of metaphase I and II of meiosis (the only stages obtained) taken of testes tissue squash preparations.

With only one exception (*Vargula* '**LD**'), in which acrocentrics were also present, the morphology of the chromosomes observed in these 6 species is similar in all of them: metacentric or slightly submetacentric.

The suggested karyotype of all of these species (males only) for which a chromosome complement was possible to ascertain ('**SWD**', '**SU**'?, '**WHU**') is composed of 15 = (14A+XO) and is similar to that found for the male of both *V. hilgendorfi* and *V. graminicola* .

The uncertainty of the cell boundaries in the material available for the remaining 3 species ('**OLD**', '**LD**', '**BLU**') renders it impossible to propose even a suggested karyotype.

The size range of the chromosomes of all the species studied herein is summarized below:

| | | |
|---|---|---|
| *V. hilgendorfi* | 4.2- 5.7μm | (metacentric) |
| *V. graminicola* | 6.8- 10.2μm | (metacentric); |
| | 4.8- 6.4μm | (acrocentric) |
| '**SWD**' | 3.0- 8.5μm | (metacentric) |
| '**SU**' | 5.4- 7.2μm | (metacentric) |
| '**WHU**' | 4.1- 6.3μm | (metacentric) |
| '**OLD**' | 4.0- 5.0μm | (metacentric) |
| '**LD**' | 3.0- 4.8μm | (metacentric); |
| | 3.6- 4.2μm | (acrocentric) |
| '**BLU**' | 2.2- 5.0μm | (metacentric). |

Although, as seen above, some differences and similarities between the size ranges seem to be apparent, it must be noted that the range given for the 6 unnamed species was 'estimated' from the meiotic plates obtained, and, therefore, until further, more suitable material becomes available, one must proceed with caution in drawing definite conclusions.

Notwithstanding the fact that not all the material on which this study was based was ideally suited for karyotype analysis, the results obtained and the similarities and differences observed between the karyotypes, are noticeable and encouraging. The study of the karyotypes, together with that of morphological and ecological characteristics of the species, aids in the better understanding of their relationships and evolution.

The analysis of the karyotypes obtained for the species of *Vargula* studied herein shows a considerable homogeny for the group in terms of chromosome numbers and morphology, which is also consistent with the results obtained for three other myodocopid species: *Gigantocypris dracontovalis* Cannon, 1940, *G. muelleri* Skogsberg, 1920 and *Macrocypridina castanea* (Brady, 1897) (Moguilevsky, 1985; Moguilevsky & Whatley, 1988) and which also distinguishes them clearly from the Podocopida.

The chromosome number for *Gigantocypris dracontovalis*, *G. muelleri* and *Macrocypridina castanea* is the same (2n = 18) and only slightly higher than that found for the *Vargula* species (2n = 16). These numbers fall within the lower part of the range found by Tetart (1978) for the Podocopida (14-35).

Whereas only metacentric and submetacentric chromosomes are found in the 2 *Gigantocypris* species, the karyotype of *M. castanea* shows also the presence of a pair of acrocentric chromosomes. Of the *Vargula* species, only two (*V. graminicola* and *Vargula* '**LD**') have acrocentric elements (1 pair) together with the metacentric chromosomes which are the majority and the only type found in the remaining 6 species.

Acrocentric chromosomes are the most typical morphological feature of a great number of podocopid karyotypes (Tetart, 1978) and are the only type present in the Darwinulacea. Tetart has suggested that the more 'primitive' karyotypes are those composed of only 'acrocentric' elements while more evolved karyotypes have less 'acrocentric' and more metacentric chromosomes. The Darwinulacea range from the Devonian to Recent!

The analysis of further species of Myodocopida is needed to test and confirm that Robertsonian fusions, which are one of the possible evolutionary processes involved in the reduction of chromosome numbers and morphological change from acrocentric to metacentric and claimed by Tetart to have occurred within the Podocopida, have also taken place within the Myodocopida.

The largest chromosomes are found in the 2 species of *Gigantocypris* (*G. muelleri* : 19-24μm; *G. dracontovalis* : 16-22μm). The size range for *M. castanea* is 9-19μm, which is slightly higher than the largest of the *Vargula* chromosome range: 6.8-10.2μm (*V. graminicola*). On average, the size range of the chromosomes of all the *Vargula* species is higher than that found by Tetart (1978) for the Podocopida [Cytheracea (1 species): 3μm; Darwinulacea (1 species): 0.5-1μm; Cypridacea (22 species): 0.5-6μm].

Although the significance of the relationship is not, as yet clearly understood, from the results obtained so far there seems to be a direct correlation between the size of the chromosomes and the overall body size of the adults. Both *Gigantocypris* species and *M. castanea* are considerably larger than the *Vargula* species which in their turn are larger than the Podocopida studied by Tetart. A more extensive database is needed to be able to confirm Hinegardner's (1976) suggestion that a positive correlation exists between adult body size and DNA contents in certain animal species.

As described above, the chromosome complement of the 5 species of *Vargula* for which the karyotype was possible to ascertain, is the same (2n = 16). It is not unusual for different species of a genus to have the same number of chromosomes and also for all chromosomes to be of similar overall length (Lazzaretto-Colombera, 1979, 1981, 1983).

Chromosome banding is a useful resource in the identification of individual chromosomes and understanding the processes of speciation (Moguilevsky in prep, a). It is intended to apply this

technique to future material of this genus to enhance our understanding of its processes of speciation. This is of great importance given the suggestion that luminescent displays by *Vargula* species are 'probably sexually isolating behavioural characters' (Morin, 1986).

As shown by Stebbins and Ayala (1985), reproductive isolation is well recognized in both animals and plants. A good example of this is found in many species of *Drosophila* in the United States Mainland. Here, many species are morphologically alike but reproductively isolated, whereas Hawaiian species of this genus exhibit the opposite, i.e. clear morphological differences are found between species that are only little differentiated genetically. This seems to be also the case even between geographically separated, interfertile populations.

As Stebbins and Ayala (1985) point out, morphological change and the development of reproductive isolation are genetically distinct phenomena that can occur either together or separately.

One striking feature found in 6 of the 8 species of *Vargula* studied in this work (those for which meiosis was available) is the presence of a single chiasma per bivalent. Although it is far too early to speculate on its evolutionary implications, this feature poses some interesting questions. Is this single chiasma per bivalent common to all *Vargula* species? What is the relationship between the presence, in these shallow water dwellers, of a single chiasma per pair of chromosomes and the demands of the environment, taking into account that the significance of chiasmata is two fold? The answers are that a) they perform an essential mechanical function, by holding together the homologue of which the bivalents are composed, and b) they contribute to genetic recombination since the number of chiasmata within each bivalent is what determines the level and pattern of recombination between genes on the same chromosome (White, 1973).

It must be remembered, however, that chiasma formation is heritable (as is pairing, centromere orientation and dysjunction). Some degree of localization of chiasmata is probably universal although they are probably never found entirely at random along the length of the chromosome. Localization of chiasmata obviously affects the overall level of genetic recombination since the closer a chiasma is to a chromosome end, the less recombination it causes.

In contrast to these shallow water *Vargula* species, both species of *Gigantocypris* (*G. muelleri* Skogsberg and *G. dracontovalis* Cannon) as well as *Macrocypridina castanea* (Brady), which live in meso to bathypelagic environments, exhibit between 2 and 3 chiasmata per bivalent (Moguilevsky, in prep., b).

The author considers it premature at this stage, to draw any firm conclusions from the comparison of these two groups, since an insufficient number of preparations (and those only males) were available for 6 of the *Vargula* species and instances are known where the distribution of chiasmata in males and females of the same species is noticeably different (Rees & Jones, 1977).

## ACKNOWLEDGEMENTS

The author would like to thank Dr Karen Watson for collecting the Misaki material and Dr Anne Cohen and Prof. James Morin for all their help in collecting and culturing the Caribbean species and personal communications regarding them. I am grateful to Robin Whatley for useful discussions and advice and critically reading the manuscript. My most grateful thanks are owed to Dr Neil Jones of the Agricultural Sciences Department, University College of Wales, Aberystwyth for his constant help and guidance in the fascinating and complex world of genetics.

**Plate 1**

*Vargula hilgendorfi* (G.W. Müller, 1890) female, Misaki, Japan.
(Specimens cultured in 0.05% colchicine in sea water for 24 hours.)

Fig. 1.    Low power showing 3 c-metaphases in young embryo tissue.
Fig. 2.    Enlargement of one of the cells in Fig. 1 showing 15 highly contracted metacentric chromosomes (male embryos).
Fig. 3.    C-metaphase in a female embryo squash showing 16 metacentric chromosomes.
Fig. 4.    C-metaphase in young embryo tissue showing chromosomes more contracted than in Fig. 3.
Fig. 5.    C-mitosis in young female embryo tissue showing 16 metacentric chromosomes.
Fig. 5a.   Karyotype from Fig. 5 showing uniformity of morphology and the gradual variation of size between the largest (*circa* 7.14µm) and the smallest (*circa* 4.28µm) chromosomes.

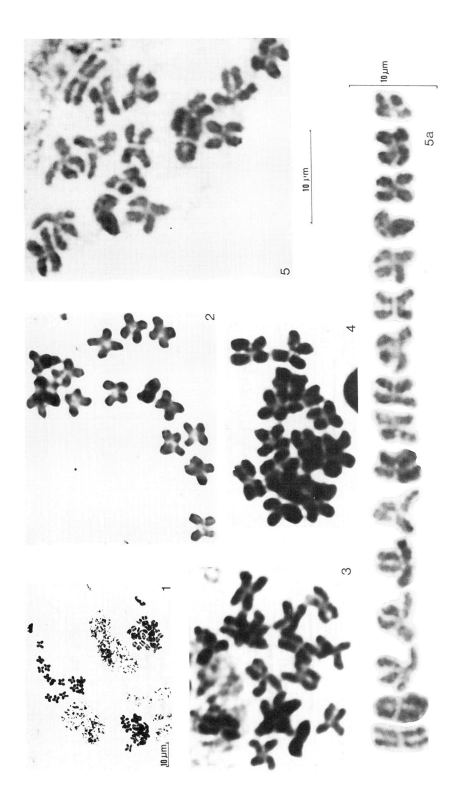

**Plate 2**

*Vargula graminicola* Cohen & Morin, 1986, female, Macaroon Reef, San Blas, Panama.
(Specimens cultured in 0.001g/ml of 'demecolcine' in sea water for 38 hours.)

Fig. 1.    Photomicrographs of c-mitosis in young embryo tissue squash, showing 32 chromosomes at metaphase. These are interpreted as representing two cells of 16 chromosomes each, as shown in Figs 1a and b.

Figs 1a, b. Karyotypes from Fig. 1 showing respectively, 14 metacentric chromosomes which vary gradually in size between the largest (9.09-10μm) and the smallest (6.06-6.36μm), and two smaller acrocentric chromosomes of about the same size (5.71μm and 5.75μm).

Fig. 2.    Photomicrograph of c-mitosis in young embryo tissue squash showing 32 chromosomes at metaphase. These are interpreted as representing two cells of 16 chromosomes each, as shown in Figs 2a and b.

Figs 2a, b. Karyotypes from Fig. 2 showing respectively, 14 metacentric chromosomes which vary gradually in size between the largest (10.6-10.9μm) and the smallest (6.06-8.18μm), and two smaller acrocentric (6.66 and 6.36μm).

Figs 3, 4. Photomicrographs of c-mitosis in two young embryo tissue squashes showing 32 chromosomes at metaphase in each.

Fig. 5.    Photomicrograph of c-mitosis in a young embryo squash: late prophase showing 16 chromosomes.

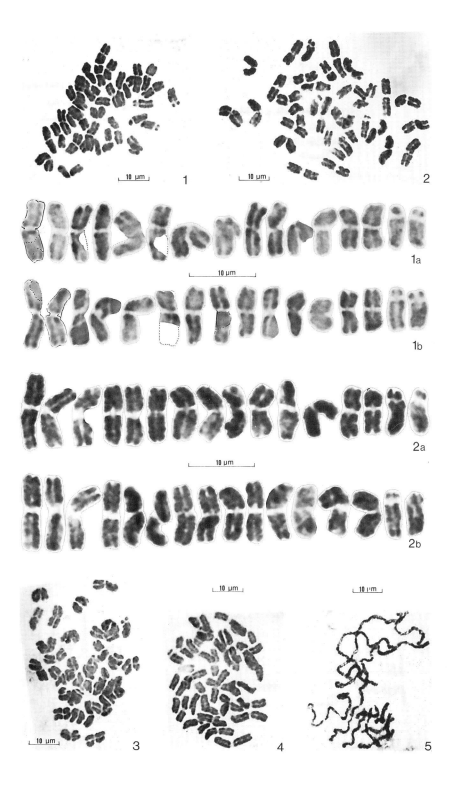

10 μm    1

10 μm    2

1a

10 μm

1b

2a

10 μm

2b

10 μm

10 μm

10 μm

10 μm    3

4

5

**Plate 3**

*Vargula* 'SWD', male, San Salvador, Bahamas.
(Specimen cultured in colchicine for 12 hours and treated in a hypotonic solution prior to fixation.)

| | |
|---|---|
| Figs 1, 2. | Photomicrographs of meiosis from testes tissue squash. |
| Fig. 1. | Fourteen bivalents and 2 univalent 'X' (arrowed) which are interpreted as representing 2 cells with 7 bivalent and 1 univalent each. |
| Figs 1a, a', b, b'. | Interpretation and diagrams of Fig. 1 showing the metacentric or submetacentric chromosomes and the location of the single chiasma observed in each bivalent. |
| Fig. 2. | Another photomicrograph of the same specimen as in Fig. 1 showing 6 bivalents (one missing?) and one univalent 'X' (arrowed). |
| Figs 2a, a'. | Interpretation and diagram of Fig. 1 showing the metacentric or submetacentric chromosomes and the location of the single chiasma observed in each bivalent. |
| Figs 3, 4. | Photomicrographs of the same specimen as in Fig. 1 showing chromosomes at metaphase II of meiosis. The boundaries of the cells involved are uncertain in this material, but the morphology of the chromosomes observed confirms the interpretation of those in Figs 1 and 2 (metacentric or very slightly submetacentric). The 'X' chromosomes are arrowed. |

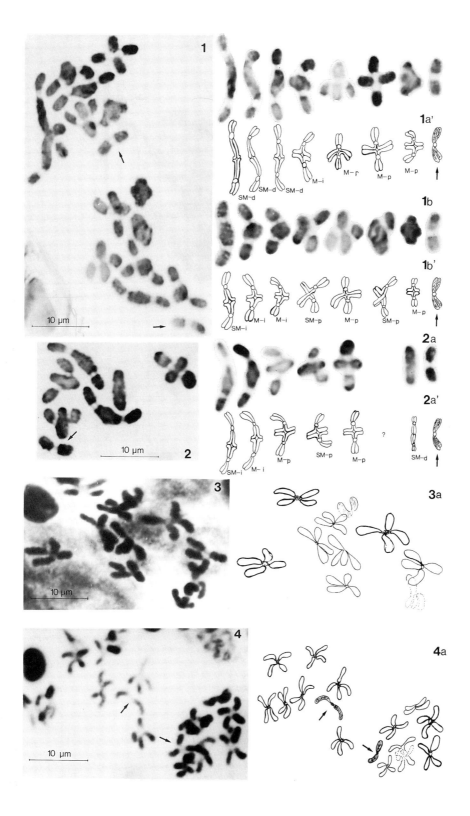

**1**

**1a'**

SM-d    SM-d SM-d    M-i    M-r    M-p    M-p

**1b**

**1b'**

SM-i    M-i    M-i    SM-p    M-p    SM-p    M-p

**2**

**2a**

**2a'**

SM-i M-i    M-p    SM-p    M-p    ?    SM-d

**3**

**3a**

**4**

**4a**

10 μm

*Vargula* 'SU' male, Curacao.
Figs 1-4. Metaphase I of meiosis obtained from testes tissue squash.

Fig. 1.    Twelve bivalents congressed on the spindle, side view. Bivalents are co-orientated and centromere pairs of each bivalent lie equidistant above and below the spindle equator (sister centromeres lie adjacent to one another and so face towards the same spindle pole). The univalent 'X' (arrowed), is orientated to one pole only and lies off the spindle equator. (Specimen in Fig. 1 was not cultured in colchicine.)

Fig. 1a.   Diagram of Fig. 1 showing the interpretation of the location of the single chiasma observed in each bivalent: 2 in distal and 4 in interstitial position.

Fig. 2.    Seven bivalents plus univalent 'X' (arrowed) not aligned on the equator but showing chromosomes in similar configurations as in Fig. 1 due to the presence of a single chiasma per pair (top view). (Specimen cultured in cochicine for 12 hours and treated in a hypotonic solution prior to fixation).

Fig. 2a.   Diagram of Fig. 2.

Fig. 3.    Eleven bivalents plus univalent 'X' (arrowed) showing similar configurations to those in Fig. 2. (Same specimen as in Fig. 2.)

Fig. 4.    Six bivalents plus univalent 'X' (arrowed) showing similar configurations to those in Fig. 2. (Same specimen as in Fig. 2.)

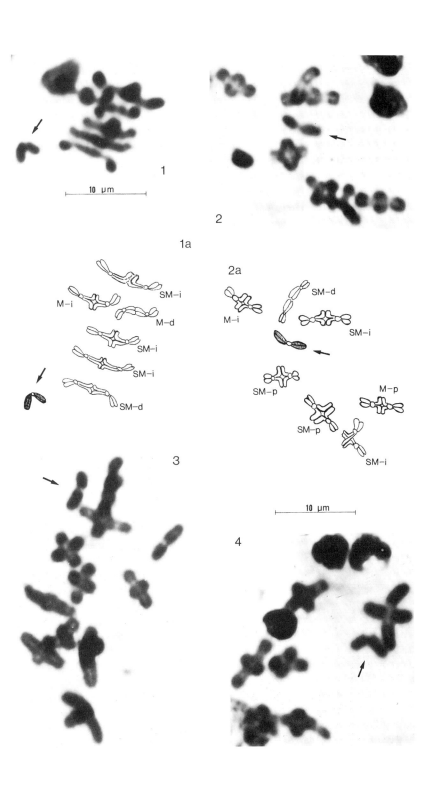

1

10 μm

2

1a

SM-i
M-i
M-d
SM-i
SM-i
SM-d

2a

M-i
SM-d
SM-i
SM-p
M-p
SM-p
SM-i

3

10 μm

4

**Plate 5**

Figs 1- 3. *Vargula* **'WHU'**, male, Isla Marqueyes, Puerto Rico.

Fig. 1.     Photomicrograph of metaphase I of meiosis obtained from testes tissue squash showing 7 bivalents and 1 univalent 'X' (arrowed). (Specimen cultured in colchicine for 12 hours and treated in a hypotonic solution prior to fixation.)

Fig. 1a.   Diagram of Fig. 1 showing the interpretation of the morphology of the chromosomes (metacentric) and the location of the single chiasma observed in each bivalent: 5 in distal and 2 in interstitial position.

Fig. 2.     Another photomicrograph of metaphase I of meiosis from testes tissue squash, showing 7 bivalents and 1 univalent 'X' (arrowed). (Same specimen as Fig. 1.)

Fig. 2a.   Diagram of Fig. 2 showing the interpretation of the morphology of the chromosomes (metacentric) and the location of the single chiasma observed in each bivalent: 4 in distal and 3 in interstitial position.

Fig. 3.     Photomicrograph of metaphase I of meiosis from testes tissue squash. The number of cells and hence the number of chromosomes per cell, is uncertain in this material. (Specimen cultured in colchicine for 24 hours and treated in a hypotonic solution prior to fixation.)

Fig. 3a.   Diagram of Fig. 3 showing the interpretation of the morphology of the chromosomes (metacentric) and the location of the single chiasma observed in each bivalent. (Morphology and configurations similar to those shown in Figs 1 and 2.)

Figs 4, 5. *Vargula* **'OLD'**, male, San Salvador, Bahamas.
(Specimen cultured in colchicine for 12 hours.)

Figs 4, 4a. Photomicrograph and diagram respectively, of metaphase I of meiosis obtained from testes tissue squash. The number of cells and hence, the number of chromosomes per cell is uncertain in this material. The diagram shows the uniformity of the morphology of the chromosomes (all metacentric or very slightly submetacentric) and the location of the single chiasma observed in each bivalent.

Fig. 5.     Chromosomes at metaphase II of meiosis obtained from testes tissue.

Fig. 5a.   Diagram of Fig. 5. Note the chromosomes, each consisting of two cromatids, exhibiting uniformity in terms of morphology (all metacentric) and size (as observed in Fig. 4).

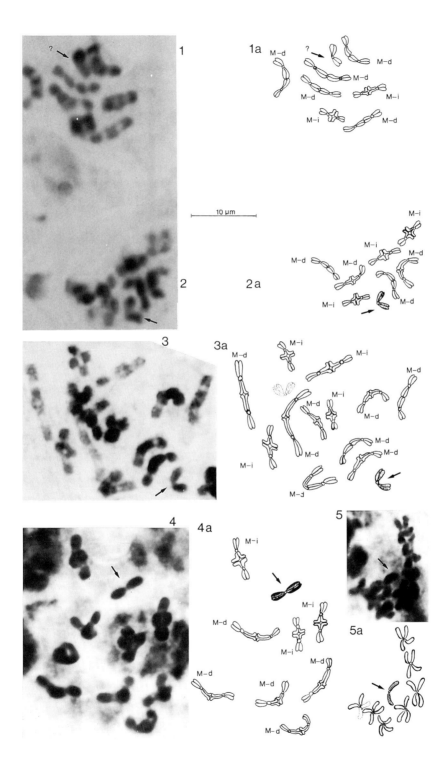

1    1a    M-d    ?    M-d
                  M-d
                  M-i
        M-i    M-d

10 µm

                  M-i
            M-i    M-d
    M-d    M-d
2    2a    M-i    M-d

3    3a    M-d    M-i    M-i    M-d
        M-d    M-i
        M-i    M-d    M-d
            M-d
        M-i    M-d

4    4a    M-i
        M-d    M-i
            M-i
    M-d    M-d    M-d
        M-d

5    5a

**Plate 6**
*Vargula* '**LD**', male, Curacao.
(Specimen cultured in colchicine for 25 hours and treated in a hypotonic solution prior to fixation.)

Fig. 1.     Chromosomes in metaphase I of meiosis obtained from testes tissue showing the location of the single chiasma present in each bivalent. Univalent 'X', arrowed. The number of cells involved and, therefore, the number of chromosomes per cell is uncertain in this material.

Fig. 1a.   Interpretative diagram of Fig. 1.

*Vargula* '**BLU**' male, Isla Marqueyes, Puerto Rico.
(Specimen cultured in colchicine for 12 hours and treated in a hypotonic solution prior to fixation.)

Fig. 2.     Chromosomes at metaphase II of meiosis obtained from testes tissue. Note the chromosomes, each consisting of two cromatids, exhibiting uniformity in terms of morphology (all metacentric) and size. Possibly two cells involved, with 7 chromosomes each.

Fig. 2a.   Interpretative diagram of Fig. 2.

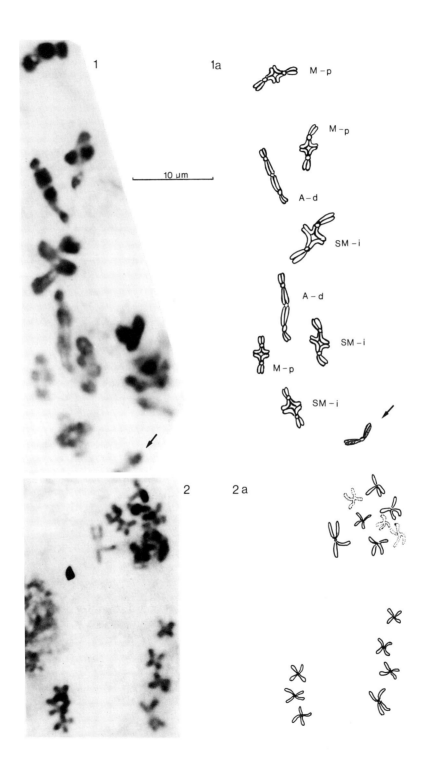

1  1a

M – p

M – p

10 μm

A – d

SM – i

A – d

SM – i

M – p

SM – i

2  2 a

## REFERENCES

Cohen, A. C. & Morin, J. G. 1986. Three new luminescent ostracodes of the genus *Vargula* (Myodocopida, Cypridinidae) from the San Blas region of Panama. *Contrib. in Science, Nat. Hist. Mus.*, Los Angeles Co., **373**, 1-23.

Hanai, T. 1974. Notes on the taxonomy of Japanese cypridinids. *Geosci . Man* , Baton Rouge, **VI**, 117-126.

Hinegardner, R. 1976. Evolution of genome size. *In* Ayala, F. (Ed.), *Molecular Evolution* , 179-199. Sinauer Association, Sunderland, Massachusetts.

Hiruta, S. 1980. Morphology of the larval stages of *Vargula hilgendorfi* (G.W. Müller) and *Euphilomedes nipponica* Hiruta from Japan (Ostracoda: Myodocopina). *J. Hokkaido Univ. Educ.*, Sapporo, Section IIB, **30**(2), 145-167.

Kornicker, L. S. 1984. Cypridinidae of the continental shelves of southeastern North America, the northern Gulf of Mexico, and the West Indies (Ostracoda: Myodocopina). *Smithson. Contr. Zool .*, Washington, D.C., **401**, 37 pp.

Lazzaretto-Colombera, I. 1979. Karyological differences between the sibling species *Tisbe reluctans* and *Tisbe persimilis* (Copepoda, Harpaticoida). *Acc. Lincei-Rend. d. cl. di Sc. fis., mat. e nat.*, Roma, **XVI**, 587-591.

Lazzaretto-Colombera, I. 1981. Karyological comparison between three sibling species of the *Tisbe reticulata* group (Copepoda, Harpaticoida). *Zool. Scr.*, Stockholm, **10**, 33-36.

Lazzaretto-Colombera, I. 1983. Karyology and chromosome evolution in the genus *Tisbe* (Copepoda). *Crustaceana*, Leiden, **45**(1), 85 - 95.

Macgregor, H. C. & Varley, J. M. 1983. *Working with Animal Chromosomes* , 250 pp. John Wiley & Sons.

Moguilevsky, A. 1985. Cytogenetic studies on marine ostracods: the karyotype of *Gigantocypris muelleri* Skogsberg, 1920 (Ostracoda, Myodocopida). *J. micropalaeontol.*, London, **4**(2), 159-164.

Moguilevsky, A. In prep., a. Chromosome banding in three species of pelagic myodocopid Ostracoda.

Moguilevsky, A. In prep., b. Analysis of chiasma frequency and distribution in some species of Cypridinacea (Myodocopina, Ostracoda).

Moguilevsky, A. & Whatley, R. C. 1988. Cytogenetic studies on marine myodocopid Ostracoda: the karyotypes of *Gigantocypris dracontovalis* Cannon, 1940 and *Macrocypridina castanea* (Brady, 1897). *In* Hanai, T., Ikeya, N. & Ishizaki, K. (Eds), *Evolutionary biology of Ostracoda, its fundamentals and applications*, proceedings of the Ninth International Symposium on Ostracoda, held in Shizuoka, Japan, 29 July - 2 August 1985, Developments in palaeontology and stratigraphy, **11**, 293-300, Kodansha Ltd., Tokyo and Elsevier, Amsterdam, Oxford, New York, Tokyo.

Morin, J. G. 1986. "Firefleas" of the sea: Luminescent signaling in marine ostracode crustaceans. *Fla. Ent.*, Gainesville, **69**(1), 105-121.

Morin, J. G. & Bermingham, E. L. 1980. Bioluminescent patterns in a tropical ostracod. *Am. Zool.*, Utica, N.Y., **20**(4), 851. (Abstract).

Morin, J. G. & Cohen, A. C. 1988. Two new luminescent ostracodes of the genus *Vargula* (Myodocopida: Cypridinidae) from the San Blas region of Panama. *J. Crustacean Biol.*,

Massachusetts, **8**, 620-638.

Poulsen, E. 1962. Ostracoda - Myodocopa, Part 1: Cypridiniformes-Cypridinidae. *Dana Rep.*, Copenhagen, **57**, 1-414.

Rees, H. & Jones, R.N. 1977. *Chromosome Genetics*, 151 pp. Edward Arnold.

Skogsberg, T. 1920. Studies on marine ostracodes, 1: Cypridinids, halocyprids, and polycopids. *Zool. Bidr. Upps.*, Stockholm, Suppl., **1**, 1-784.

Stebbins, G. L. & Ayala, F. J. 1985. The Evolution of Darwinism. *Scient. Am.*, New York, **253**(1), 54-64.

Sylvester-Bradley, P. C. 1961. *In* Moore, R. C. (Ed.), *Treatise on Invertebrate Palaeontology*, *Pt . Q*, Arthropoda, **3**, xxiii + 422 pp., 334 figs, Univ. Kansas Press.

Tetart, T. 1978. Les garnitures chromosomiques des ostracodes d'eau douce. *Trav. Lab. Hydrobiol.*, Grenoble, **69-70**, 113-140.

White, M. J. D. 1973. *Animal Cytology and Evolution* , 3rd edition, viii + 961 pp. Cambridge University Press, London and New York.

## DISCUSSION

Dan Danielopol: Is there some evidence of polyploidy in the deep sea myodocopids, which live in a harsh environment? This occurs in some freshwater ostracods living in shallow, astatic habitats (see Tetart's publications).

Alicia Moguilevsky: There is no evidence of polyploidy in any of the deep sea myodocopid species that I have studied to date. All these species are syngamic, unlike the freshwater species reported by Tetart. All the reported cases of polyploidy are of parthenogenetic species.

Ken McKenzie: Do myodocopids have multiple sex chromosomes? These have been reported in freshwater ostracods.

Alicia Moguilevsky: The chromosome complement for the female of all the myodocopid species that I have studied to date is composed of a number of autosomes and 2 'X' and the complement for the male is of autosomes plus a single unpaired 'X' (XO).

Ken McKenzie: What is the size of chromosomes in myodocopids?

Alicia Moguilevsky: Refer to the text.

Richard Reyment: I seem to remember polyploidy in T. Waterman's (1960) *Physiology of the Crustacea* . How does this connect to the results you have presented?

Alicia Moguilevsky: Polyploidy in animals is relatively rare and by and large, restricted to the plant kingdom. Complications with the sex-determining chromosomes and some kind of physiological disturbance are some of the possible reasons which might explain the disparity between plants and animals. It has been reported in ostracods but only in some freshwater, parthenogenetic species (see Tetart, 1978).

# BIOSTRATIGRAPHY

# 15

# Lower and Middle Jurassic Ostracoda from Argentina

Sara C. Ballent

Division of Invertebrate Palaeozoology,
Museum of Natural Sciences, La Plata,
Argentina.

## ABSTRACT

The ostracod microfaunas of Lower and Middle Jurassic age recognized up to the present in central western Argentina are analyzed. The ammonite zones are well known and have facilitated the dating of the microfaunas. In the Upper Pliensbachian the cytheraceans are mainly represented by Bythocytheridae and Cytheruridae; cypridaceans, bairdiids, cytherellids, darwinulids and healdiids are secondary. All genera are pandemic and they occur widely mainly in the Northern Hemisphere. Neither the Protocytheridae nor the Progonocytheridae have yet been found in the Lower Jurassic of central west Argentina.

At the Aalenian-Bajocian boundary, only cytheraceans are recognized. The Cytheruridae are most abundant; the Protocytheridae are represented by the typically boreal genus *Ektyphocythere* Bate, and the Progonocytheridae by some new taxa.

In the Lower-Middle Bajocian the ostracods are represented by Protocytheridae and Cytherellidae.

In the Middle-Upper Callovian the cytheraceans are represented by Bythocytheridae, Cytheruridae and Progonocytheridae. Cypridaceans and cytherellids are also present. Most of the genera are known in the Northern Hemisphere.

The genus *Sondagella* Dingle, which is restricted in its occurrence to southern Gondwanaland from the Upper Jurassic to the Lower Cretaceous, is present at the Callovian-Oxfordian boundary.

Considering the present knowledge of Jurassic ostracods in Argentina, it is too early to talk about an 'austral characteristic' of the microfaunas, at least for the Lower and Middle Jurassic.

## INTRODUCTION

Neuquén Province, in central western Argentina (Fig. 1) has provided most of the Jurassic ostracod assemblages known at present in Argentina.

The microfaunas are associated with well-known ammonite zones and have been described and figured in several papers (Musacchio, 1978,

1979b, 1979c; Ballent, 1985, 1986a, 1987).

The present paper deals with the ostracod microfaunas of the Lower and Middle Jursassic from Neuquén Province. Its purpose is to contribute to the somewhat sparse micropalaeontological information on Jurassic ostracods in the Southern Hemisphere.

## MICROFAUNAS

**Lower Jurassic** (see Table 1)

In the Lower Jurassic, the ostracods are stratigraphically associated with the *Fanninoceras* Assemblage Zone which indicates a late Pliensbachian age (Riccardi, 1984, 563). The Cytheracea are represented by Bythocytheridae and Cytheruridae which constitute 40% of the total of species of the ostracod fauna. The Cypridacea, bairdiids, cytherellids, darwinulids and healdiids are secondary. All genera are pandemic and occur widely mainly in the Northern Hemisphere.

The cytherellids are represented by the cosmopolitan genera *Cytherella* and *Cytherelloidea*. Neither the Protocytheridae, which are very important in the Lower Jurassic of the European Province (see Bate, 1977), nor the Progonocytheridae, which first occur in the Sinemurian of the European Province (see Bate, 1977, p. 234), have yet been found in the Lower Jurassic of central western Argentina.

Some selected Argentinian ostracod taxa have the potential to be useful for age determination and stratigraphical correlation:

*Eucytherura*? *isabelensis* Ballent (Plate 1, Fig. 1) is easy to identify because it is ornamented with 5 tubercles on a regularly reticulate surface with papillae.

*Liasina*? sp. (Plate 1, Fig. 2), is very close to *Liasina* Gramann externally, but is unknown internally. The genus *Liasina* has been recorded from the Sinemurian to Lower Toarcian in central western Europe (Apostolescu, 1959; Gramann, 1963; Bate & Colman, 1975; Michelsen, 1975; Lord, 1978; Sivhed, 1980; Donze, 1985; Riegraf, 1985; Ainsworth, 1987).

*Isobythocypris* sp. (Plate 1, Fig. 3). The genus

*Isobythocypris* Apostolescu has been recorded mainly from the Sinemurian-Lower Toarcian of central western Europe (Drexler, 1958; Apostolescu, 1959; Herrig, 1969; Michelsen, 1975; Lord, 1978; Sivhed, 1980; Donze, 1985; Riegraf, 1985; Ainsworth, 1987); it has been also cited from the Sinemurian and Pliensbachian of DSDP Site 547, off Morocco (Bate *et al.*, 1984) and from the Toarcian of Sinai, Egypt (Rosenfeld *et al.*, 1987).

*Ogmoconcha* sp. (Plate 1, Fig. 4). The genus *Ogmoconcha* Triebel is particularly characteristic of the earliest part of the Jurassic and it is only present up to the basal Toarcian. It has been recorded from Hettangian to basal Toarcian in central western Europe (Triebel, 1941; Drexler, 1958; Apostolescu, 1959; Barbieri, 1964; Lord, 1971; Bate & Coleman, 1975; Michelsen, 1975; Lord, 1978; Sivhed, 1980; Donze, 1985; Riegraf, 1985; Ainsworth, 1987) and from the Pliensbachian of DSDP Site 547, off Morocco (Bate *et al.*, 1984). It also occurs in the Rhaetian of Great Britain (Bate, 1978).

**Middle Jurassic**

In the Middle Jurassic, ostracods have been recovered from the Aalenian-Bajocian boundary and from early-middle Bajocian and Callovian strata.

**Aalenian-Bajocian boundary** (see Table 1). The ostracods here are associated with the *Puchenquia malarguensis* Assemblage Zone which indicates an Aalenian-Bajocian age (Riccardi, 1984, 567).

To date, only cytheraceans have been recognized. They are mainly represented by Cytheruridae (almost 40% of the total species); Protocytheridae and Progonocytheridae have also been recorded. Among the Cytheruridae, the genus *Procytherura* Whatley has beeen recognized with three species (Plate 1, Figs 5-7). *Procytherura* is a long-ranging genus known mainly from Toarcian to Oxfordian in central western Europe (Whatley, 1970; Bate & Coleman, 1975; Bate, 1978; Sivhed, 1980; Ware & Whatley, 1980; Riegraf, 1985; Dépêche, 1985; Ainsworth, 1986); it has also been recognized in the Sinemurian of DSDP Site 547, off

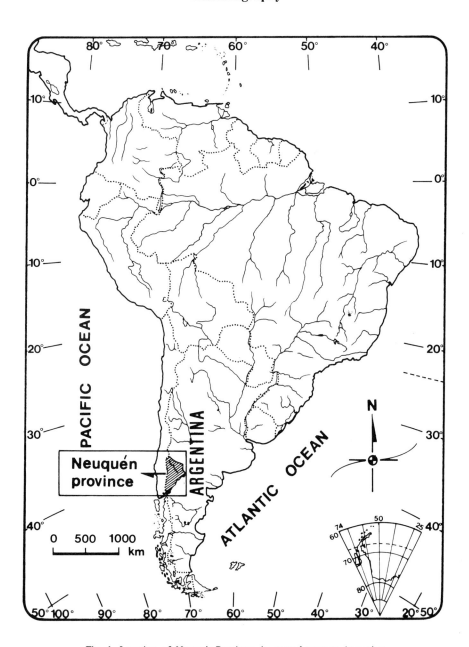

Fig. 1. Location of Neuquén Province, in central western Argentina.

Morocco (Bate *et al.*, 1984); from the Upper Jurassic of Tanzania (Bate, 1975); the Kimmeridgian of DSDP Site 260, off Western Australia (see Indet. sp. E Oertli, 1974, 949 = *Procytherura aerodynamica* Bate, 1975, 204); the Neocomian of the Neuquén Basin, Argentina (Musacchio, 1979a) and of South Africa (Brenner & Oertli, 1976; McLachlan, *et al.*, 1976a; Dingle, 1984; Valicenti & Stephens, 1984) and Middle Albian of DSDP Site 327 on the Malvinas Plateau (Dingle, 1984).

Among the Cytheruridae, *Rutlandella* sp. A (Plate 1, Fig. 8) is close to the type species

Table 1.  Stratigraphical ranges of ostracod species recovered in central western Argentina.

| | Lower Jurassic | | Middle Jurassic | | | | |
|---|---|---|---|---|---|---|---|
| Series | | | | | | | |
| Stage | Pliensbachian | Toarc. | Aalenian | Bajocian | | Bath. | Callovian |
| | | | | Lower | Upper | | Low. / Mi. / Upp. |
| Ammonite Zones — Europe | U. jamesoni; T. ibex; P. davoei; A. margaritatus; P. spinatum | | L. opalinum; L. murchisonae; G. concavum | H. discites; S. ovales; W. laeviuscula; O. sauzei; S. humphriesianum | S. subfurcatum; G. garantiana; P. parkinsoni | | M. macrocephalus; M. gracilis; K. jason; E. coronatum; P. athleta; Q. lamberti |
| Ammonite Zones — Argentina | Miltoceras F.; Uptonia F.; Fanninoceras Z. | | Bredya Z.; Z. groeberi Z.; Puchenquia malarguensis Z. | Pseudotoites singularis Z.; Emileia giebeli Z.; S. humphriesianum Z. | M. ? rotundum Z. | | Eurycephalites Z.; Reineckeia Z. |

OSTRACODS

(in Ballent, 1987)
- Cytherella sp. A
- C. sp. B
- Cytherelloidea sp. C
- Bairdia sp.
- Bythocypris ? sp. A
- B. ? sp. B
- Isobythocypris sp.
- Paracypris sp. A
- P. sp. B
- P. ? sp.
- Liasina ? sp.
- Darwinula sp.
- Monoceratina sp. B
- M. ? sp. C
- M. ? sp. D
- Eucytherura ? isabelensis Ballent
- Procytherura ? sp.
- Rutlandella ? sp.
- Cytheracea indet.
- Ogmoconcha sp. (Ballent, in prep.)
- Rutlandella sp. A
- Monoceratina ? nov. sp. A
- Eucytherura sp.
- Procytherura nov. sp. A
- P. nov. sp. B
- P. nov. sp. C
- Progonocytheridae nov. sp.
- Palaeocytheridea ? sp.
- Ektyphocythere australis Ball. (in Ball., 1985)
- Cytherella sp. A
- Cytherelloidea sp. A
- C. sp. B
- Cytherella sp. (in Musacchio, 1979 b)
- Cytherelloidea sp.
- Paracypris sp. 1
- P. sp. 2
- P. sp.
- Monoceratina sp. 1
- M. sp. 2
- M. ? sp. 3
- Eucytherura ? leufuensis Musacchio
- Polycope sp.
- Gen. et. sp. indet. (in Ball., 1985)
- Progonocythere neuquenensis Mus.
- Sondagella sp.

*R. transversiplicata* Bate & Coleman. The Argentinian species has been also found in strata of late Pliensbachian age in Neuquén province. *Rutlandella* has been cited from the Toarcian and Aalenian of central western Europe (Bate & Coleman, 1975; Bate, 1978; Penn *et al.*, 1980; Knitter, 1983; Exton & Gradstein, 1984; Ainsworth, 1986); it has also been recorded from the Pliensbachian of DSDP, Site 547, off Morocco (Bate *et al.*, 1984) and from the Bajocian of Sinai, Egypt (Rosenfeld *et al.*, 1987).

The Progonocytheridae are represented by some new taxa (Plate 1, Figs 11-12).

The Protocytheridae are represented sparsely by *Ektyphocythere australis* Ballent (Plate 1, Fig. 9). This is close to *Procytheridea exampla* Peterson, 1954, of which the Argentinian species can be distinguished by a distinctive ornament of ridges arranged in a triangular pattern and intercostal surface with cross partitions. *Ektyphocythere* Bate has been recorded widely in the Northern Hemisphere: Lower and Middle Jurassic of central western Europe (Bate, 1963; Bate & Coleman, 1975; Lord, 1974; Michelsen, 1975; Lord, 1978; Penn *et al.*, 1980; Sivhed, 1980; Herrig, 1982; Morris, 1983; Knitter, 1983; Exton & Gradstein, 1984; Donze, 1985; Riegraf, 1985; Ainsworth, 1986); Bajocian and Bathonian of Jordan (Basha, 1980); Toarcian of Morocco (Boutakiout *et al.*, 1982); Sinemurian of DSDP Site 547, off Morocco (Bate *et al.*, 1984) and Toarcian and Bajocian to Callovian of Sinai, Egypt (Rosenfeld *et al.*, 1987) (see Fig. 2). The record of *E. australis* in Argentina is, up to the present, the only occurrence of the genus in the Southern Hemisphere. In Neuquén province *E. australis* also has been found in early-middle Bajocian and early-middle Callovian strata (Ballent, 1986a, 40). It is mainly recovered in calcareous sandstones and arenaceous marls of shallow, rather high energy environments which suggests that its presence is facies-controlled.

**Lower-Middle Bajocian** (see Table 1). The ostracod microfauna is associated with the *Emileia giebeli* Assemblage Zone which indicates an early-middle Bajocian age (Riccardi, 1984, 567).

This fauna is poor and is represented by Cytherellidae with the cosmopolitan genera *Cytherella* and *Cytherelloidea*, and in addition Protocytheridae with *Ektyphocythere australis* Ballent (Plate 1, Fig. 10) (see remarks above).

**Middle-Upper Callovian** (see Table 1). This ostracod is associated with the *Reineckeia* Assemblage Zone which indicates a middle-late Callovian age (Riccardi, 1984, 568).

The cytheraceans are mainly represented by Bythocytheridae, Cytheruridae and Progonocytheridae which amount to nearly 50% of the total species. Cypridaceans and cytherellids are fewer in number. Most of the genera are known from the Northern Hemisphere.

Dingle (1988) analysed the marine ostracod distributions during the breakup of southern Gondwanaland, which began during mid-Jurassic times. This author recognized the South Gondwana Fauna A (SGFA - Bajocian to Aptian) which is characterized by the presence of various taxa (see Dingle, 1988, 842).

Neither *Amicytheridea* Bate and *Afrocytheridea* Bate, two characteristic genera of SGFA (in the east), nor *Majungaella* Grekoff, which is exclusive to the South Gondwana Fauna with a wide distribution in the Middle-Upper Jurassic and Cretaceous, have been found in Callovian sediments of Neuquén Province. Up to the present, this last genus has been recorded from the Neocomian of the Austral Basin (Volkheimer & Musacchio, 1980) and of the Neuquén Basin (Ballent, 1988), in Argentina.

The genus *Sondagella* Dingle (Plate 1, Fig. 14), which is restricted to the SGF (in the west) and found mainly in the Upper Jurassic to Lower Cretaceous of Argentina (Musacchio, 1978; 1979a, 1979c, 1980; Ballent, 1986b, 1988) and South Africa (Dingle, 1969; Brenner & Oertli, 1976; McLachlan *et al.*, 1976a, 1976b; Dingle, 1984; Valicenti & Stephens, 1984), has been recognized in sediments near the Callovian-Oxfordian boundary in Neuquén Province.

## CONCLUSIONS

The cytheracean Cytheruridae and Bythocytheridae are dominant in the Lower Jurassic. All genera are pandemic and they occur widely, mainly in the Northern Hemisphere.

At the Aalenian-Bajocian boundary, the cytherurids are still important while the protocytherids and prognocytherids first appear at this level in Argentina, later than in the Northern Hemisphere, probably indicating a North-South migration.

The record of *Sondagella* on strata near the Callovian-Oxfordian boundary could be the oldest known of the genus.

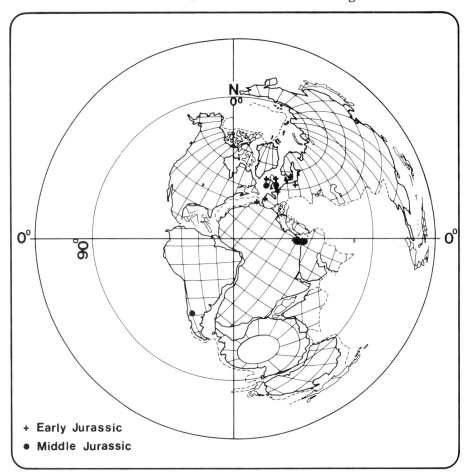

Fig. 2. Palaeogeographical distribution of the genus *Ektyphocythere* Bate during the Jurassic. Palaeocontinental reconstruction from Briden *et al.* (1974) for the 'Jurassic' (about 170 ± 15 MY).

The absence of *Amicytheridea* and *Afrocytheridea*, two characteristic genera of the South Gondwana Fauna (in the east) and the presence of *Sondagella*, a typical genus of the South Gondwana Fauna (in the west) confirms the incipient breakup of southern Gondwanaland, which began during mid-Jurassic times.

The definition of austral characteristics of the microfaunas is not yet possible, at least for the Lower and Middle Jurassic, in view of the present state of our knowledge of Jurassic ostracods from Argentina.

## REFERENCES

Ainsworth, N. R. 1986. Toarcian and Aalenian Ostracoda from the Fastnet Basin, offshore Southwest Ireland. *Bull. geol. Surv. Ir.*, Dublin, **3**, 277-336.

Ainsworth, N. R. 1987. Pliensbachian Ostracoda from the Fastnet Basin, offshore Southwest Ireland. *Bull. geol. Surv. Ir.*, Dublin, **4**(1), 41-62.

Apostolescu, V. 1959. Ostracodes du Lias du Bassin de Paris. *Revue Inst. Fr. Pétrole*, Paris, **14**(16), 795-826.

Ballent, S. C. 1985. *Taxonomía y Bioestratigrafía de los microfósiles calcáreos del Jurásico inferior y medio de la Republica Argentina.* Fac. Cienc. Nat. y Mus., Univ. Nac. La Plata. Tesis 443, 1-272. (Unpublished.)

Ballent, S. C. 1986a. Una nueva especie del género *Ektyphocythere* Bate (Ostracoda) en el Jurásico medio de la provincia del Neuquén, Argentina. *Notas Mus. La Plata*, Buenos Aires, **21**, Pal., (105), 40-46.

Ballent, S. C. 1986b. *Estudio micropaleontológico de tres pozos de la Cuenca neuquina.* Fac. Cienc. Nat. y Mus., Univ. Nac. La Plata-Petrolera Pérez Companc. (Unpublished.)

Ballent, S. C. 1987. Foraminíferos y ostrácodos del Jurásico inferior de Argentina. *Rev. Mus. La Plata*, Buenos Aires, **9**, Pal., (53), 43-118.

Ballent, S. C. 1988. *Estudio micropaleontológico de dos pozos de la Cuenca Neuquina.* Fac. Cienc. Nat. y Mus., Univ. Nac. La Plata - Petrolera Pérez Companc. (Unpublished.)

Barbieri, F. 1964. Micropaleontología del Lias e Dogger del Pozzo Ragusa I (Sicilia). *Riv. ital. Paleont.*, **70**(4), 709-830.

Basha, S. 1980. Ostracoda from the Jurassic system of Jordan, including a stratigraphical outline. *Revta esp. Micropaleont.*, Madrid, **12**(2), 231-254.

Bate, R. 1963. Middle Jurassic Ostracoda from North Lincolnshire. *Bull. Br. Mus. nat. Hist. (Geol.)*, London, **8**(4), 173-219.

Bate, R. 1975. Ostracods from callovian to Tithonian sediments of Tanzania, East Africa. *Bull. Br. Mus. nat. Hist. (Geol.)*, London, **26**(5), 163-223.

Bate, R. 1977. Jurassic Ostracoda of the Atlantic Basin. *In* Swain, F. (Ed.), *Stratigraphic Micropaleontology of Atlantic Basin and Borderlands*, 231-244. Elsevier, Amsterdam.

Bate, R. 1978. The Jurassic. Part II Aalenian to Bathonian. *In* Bate, R. & Robinson, E. (Eds), *A Stratigraphical Index of British Ostracoda. Geol. Journ. Spec. Issue*, No. **8**, 213-258. Seel House Press, Liverpool.

Bate, R. & Colman, B. M. 1975. Upper Lias Ostracoda from Rutland and Huntingdonshire. *Bull. geol. Surv. Gt Br.*, London, **54**, 1-42.

Bate R., Lord, A. & Riegraf, W. 1984. Jurassic Ostracoda from Leg 79, Site 547. *In Init. Reports DSDP*, 79, 703-710. U.S. Govt Printing Office, Washington.

Boutakiout, M., Donze, P. & Oumalch, F. 1982. Nouvelles espèces d'ostracodes du Lias moyen et supérieur du Jbel Dhar En Nsour (Rides Sud-Rifaines, maroc septentrional). *Revue Micropaléont.*, Paris, **25**(2), 94-104.

Brenner, P. & Oertli, H. 1976. Lower Cretaceous Ostracodes (Valanginian to Hauterivian) from the Sundays River Formation, Algoa Basin, South Africa. *Bull. Cent. Rech. SNPA*, Pau, **10**(2), 471-533.

Briden, J., Drewry, G. & Smith, G. 1974. Phanerozoic equal-area world maps. *J. Geol.*, Chicago, **82**, 555-574.

Dépêche, F. 1985. Lias supérieur, Dogger, Malm. *In* Oertli, H. J. (Ed.), *Atlas des ostracodes de France*, 118-145. *Bull. Centres Rech. Explor.-Prod. Elf-Aquitaine*, Pau, Mém. **9**.

Dingle, R. 1969. Marine Neocomian Ostracoda from South Africa. *Trans. R. Soc. S. Afr.*, Cape Town, **38**(2), 139-163.

Dingle, R. 1984. Mid-Cretaceous Ostracoda from Southern Africa and the Falkland Plateau. *Ann. S. Afr. Mus.*, Cape Town, **93**(3), 97-211.

Dingle, R. 1988. Marine Ostracod Distributions during the Early Breakup of Southern Gondwanaland. *In* Hanai, T., Ikeya, N. & Ishizaki, K. (Eds), *Evolutionary biology of Ostracoda, its fundamentals and applications*, proceedings of the Ninth International Symposium on Ostracoda, held in Shizuoka, Japan, 29 July - 2 August 1985, Developments in palaeontology and stratigraphy, **11**, 841-854, Kodansha Ltd., Tokyo and Elsevier, Amsterdam, Oxford,

New York, Tokyo.

Donze, P. 1985. Lias inférieur et moyen. *In* Oertli, H. J. (Ed.), *Atlas des ostracodes de France*, 101-107. *Bull. Centres Rech. Explor.-Prod. Elf-Aquitaine*, Pau, Mém. **9**.

Drexler, E. 1958. Foraminiferen und Ostracoden aus dem Lias alpha von Siebeldingen/Pfalz. *Geol. Jb.*, Hannover, **75**, 475-554.

Exton, J. & Gradstein, F. 1984. Early Jurassic Stratigraphy and Micropaleontology of the Grand Banks and Portugal. *In* Westermann, G. (Ed.). *Jurassic-Cretaceous Biochronology and Paleogeography of North America. Geol. Assoc. Canada Spec. Paper*, **127**, 13-30.

Gramann, F. 1963. *Liasina* n. gen. (Ostracoda) aus dem deutschen Lias. *Geol. Jb.*, Hannover, **82**, 65-74.

Herrig, E. 1969. Ostracoden aus dem Ober-Domerien von Grimmen westlich von Greifswald. *Geologie*, **18**(9), 999-1128.

Herrig, E. 1982. Ostrakoden aus dem Lias von Thüringen. Die Familien Progonocytheridae, Cytherethidae und Brachycytheridae. *Z. geol. Wiss*, Berlin, **10**(11), 1449-1461.

Knitter, H. 1983. Biostratigraphische Untersuchungen mit Ostracoden in Toarcien Süddeutschlands. *Facies*, 8, 2, 13-262.

Lord, A. 1971. Revision of some Lower Lias ostracoda from Yorkshire. *Palaeontology*, London, **14**(4), 642-665.

Lord, A. 1974. Ostracods from the Domerian and Toarcian of England. *Palaeontology*, London, **17**(3), 599-622.

Lord, A. 1978. The Jurassic. Part I Hettangian-Toarcian. *In* Bate, R. & Robinson, E. (Eds) *A Stratigraphical Index of British Ostracoda. Geol. Journ. Spec. Issue*, No. **8**, 189-212. Seel House Press, Liverpool.

McLachan, I., McMillan, I. & Brenner, P. 1976a. Micropalaeontological Study of the Cretaceous beds at Mbotyi and Mngazana, Transkei, South Africa. *Trans. geol. Soc. S. Afr.*, Johannesburg, **79**(3), 321-340.

McLachan, I., Brenner, P. & McMillan, I. 1976b. The Stratigraphy and Micropaleontology of the Cretaceous Brenton Formation and the PB-A/1 Well, near Knysna, Cape Province. *Trans. geol. Soc. S. Afr.*, Johannesburg, **79**(3), 341-370.

Michelsen, O. 1975. Lower Jurassic biostratigraphy and ostracods of the Danish Embayment. *Danm. geol. Unders.*, Kobenhavn, II **104**, 1-287.

Morris, P. 1983. Palaeoecology and Stratigraphic distribution of Middle Jurassic Ostracods from the Lower Inferior Oolite of the Cotswolds, England. *Palaeogeogr. Palaeoclimat. Palaeoecol.*, Amsterdam, **41**, 289-324.

Musacchio, E. 1978. Microfauna del Jurásico y Cretácico inferior. *VII Congreso Geológico Argentino. Relatorio*, 147-161.

Musacchio, E. 1979a. Ostrácodos del Cretácico inferior en el Grupo Mendoza, Cuenca del Neuquén, Argentina. *VII Congreso Geológico Argentino, Actas II*, 459-473.

Musacchio, E. 1979b. Foraminíferos y Ostrácodos del Jurásico en las inmediaciones del arroyo Picún Leufú y la Ruta 40 (Provincia del Neuquén, Argentina), con algunas consideraciones sobre la estratigrafía de la Formación Lotena. *In* Dellapé, D., Pando, G., Uliana, M. & Musacchio, E. (Eds), *VII Congreso Geológico Argentino, Actas II*,

**Plate 1**

Repository: MLP-Mi = Museo de Ciencias Naturales de La Plata-Micropaleontología. Scale bar = 0.1mm.

Fig. 1. *Eucytherura*? *isabelensis* Ballent (MLP-Mi 576/1), female carapace, left lateral view. Upper Pliensbachian.

Fig. 2. *Liasina*? sp. (MLP-Mi 565), juv. male? carapace, right lateral view. Upper Pliensbachian.

Fig. 3. *Isobythocypris* sp. (MLP-Mi 559), carapace, right lateral view. Upper Pliensbachian.

Fig. 4. *Ogmoconcha* sp. (MLP-Mi 584), carapace, left lateral view. Upper Pliensbachian.

Fig. 5. *Procytherura* sp. A (MLP-Mi 623), carapace, right lateral view, Aalenian-Bajocian boundary.

Fig. 6. *Procytherura* sp. B (MLP-Mi 624), carapacecaparace, right lateral view. Aalenian-Bajocian boundary.

Fig. 7. *Procytherura* sp. C (MLP-Mi 625), carapace, left lateral view, Aalenian-Bajocian boundary.

Fig. 8. *Rutlandella* sp. A (MLP-Mi 622), carapace, right lateral view, Aalenian-Bajocian boundary.

Fig. 9. *Ektyphocythere australis* Ballent (MLP-Mi 628), juv., right lateral view, Aalenian-Bajocian boundary.

Fig. 10. *Ektyphocythere australis* Ballent (MLP-Mi 500/1), female carapace, left lateral view. Early-Middle Bajocian.

Fig. 11. *Palaeocytheridea*? sp. (MLP-Mi 626), right valve, external view. Aalenian-Bajocian boundary.

Fig. 12. Progonocytherid sp. (MLP-Mi 627), carapace, right lateral view. Aalenian-Bajocian boundary.

Fig. 13. *Progonocythere neuqenensis* Musacchio (MLP-Mi 620), male carapace, right lateral view. Middle-Upper Callovian.

Fig. 14. *Sondagella* sp. (MLP-Mi 621), carapace, right lateral view. Callovian-Oxfordian boundary.

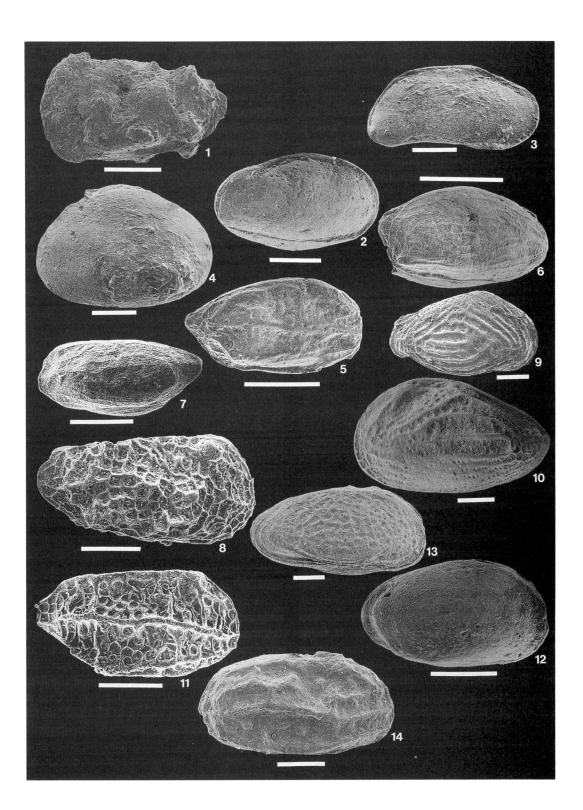

489-507.

Musacchio, E. 1979c. Datos paleobiogeográficos de algunas asociaciones de foraminíferos, ostrácodos y carofitas del Jurásico medio y el Cretácico inferior de Argentina. *Ameghiniana*, Buenos Aires, **14**(3-4), 247-271.

Musacchio, E. 1980. South American Jurassic and Cretaceous Foraminifera, Ostracoda and Charophyta of Andean and Sub-andean regions. *In* Volkheimer, W. & Musacchio, E. (Eds), *Cuencas Sedimentarias del Jurásico y Cretácico de América del Sur*, **2**, 461-698. Comité Sudamericano del Jurásico y Cretácico. Buenos Aires.

Oertli, H. 1974. Lower Cretaceous and Jurassic ostracods from DSDP Leg 27. A preliminary account. *In Init. Reports DSDP*, **27**, 947-965. U.S. Govt Printing Office, Washington.

Penn, I., Dingwall, R. & Knox O. B. R. 1980. The inferior Oolite (Bajocian) sequence from borehole in Lyme Bay, Dorset. *Rep. Inst. geol. Sci.*, London, **79**(3), 1-27.

Peterson, J. 1954. Jurassic Ostracoda from the ''Lower Sundance'' and Rierdon Formations, western interior United States. *J. Paleontol.*, Chicago, **28**(2), 153-176.

Riccardi, A. 1984. Las asociaciones de amonitas del Jurásico y Cretácico de la Argentina. *IX Congreso Geológico Argentino, Actas 4*, 559-595.

Riegraf, W. 1985. Mikrofauna, Biostratigraphie und Fazies im Unteren Toarcium Südwestdeutschlands und Vergleiche mit benach barten Gebieten. *Tübinger Mikropaläontologische Mitteilungen*, **3**, 1-232.

Rosenfeld, A., Gerry, E. & Honigstein, A. 1987. Jurassic Ostracodes from Gebel Maghara, Sinai, Egypt. *Revta esp. Micropaleont.*, Madrid, **19**(20), 251-280.

Sivhed, U. 1980. Lower Jurassic ostracods and Stratigraphy of Western Skane Southern Sweden. *Sver. geol. Unders.*, Ser. C a, **50**, 3-85.

Triebel, E. 1941. Zur Morphologie und Oekologie des fossilen Ostracoden mit Beschreibung eineger neuer Gattungen und Arten. *Senckenberg. leth.*, Frankfurt a.M., **23**(4-6), 249-400.

Valicenti, H. & Stephens, J. 1984. Ostracods from the Upper Valanginian and Upper Hauterivian of the Sundays River Formation, Algoa Basin, South Africa. *Revta esp. Micropaleont.*, Madrid, **16**(1-2-3) 171-239.

Volkheimer, W. & Musacchio, E. 1980. The continental margin of Gondwana, principally in central western Argentina: Jurassic and Lower Cretaceous palynomorphs and calcareous microfossils. *In* Creswell, M. & Vella, P. (Ed.), Gondwana Five. *Proc. Fifth International Gondwana Symposium*. Balkema A.A. Rotterdam.

Ware, M. & Whatley, R. 1980. New genera and species of Ostracoda from the Bathonian of Oxfordshire, England. *Revta esp. Micropaleont.*, Madrid, **12**(2), 199-230.

Whatley, R. 1970. Scottish Callovian and Oxfordian Ostracoda. *Bull. Br. Mus. nat. Hist. (Geol.)*, London, **19**(6), 299-358.

# Stratigraphical distribution of fresh and brackish water Ostracoda in the late Neogene of Hispaniola

Willem A. van den Bold

Department of Geology and Geophysics,
Louisiana State University, U.S.A.

## ABSTRACT

Brackish water ostracods are used in combination with fresh water and marine species to establish a biostratigraphical zonation of formations in the Plio-Pleistocene of the Dominican Republic. Ornamentation of some *Cyprideis* species may provide clues to hydrological conditions during deposition.

## INTRODUCTION

Brackish water environments are found in several Neogene formations of the Dominican Republic. Especially in the Pliocene and Quaternary, brackish and freshwater Ostracoda occur in strata alternating with marine deposits. These brackish (and hypersaline) beds occur in the Arroyo Blanco Formation of the Azua Basin and its gypsiferous member on the south side of the Sierra de Neiba, and the Las Salinas and Jimaní formations of the Enriquillo Basin (Fig. 1).

The Jimaní Formation was named by Arick, a geologist with the Dominican Seaboard Oil Company in 1941 (Bermúdez, 1949), to include the strata above the Las Salinas Formation and below the reef limestone later shown by Mann *et al.* (1984) to be Holocene. Its type locality is 1km north of Jimaní and the sequence from which the ostracod fauna is recorded in Table 1 can be regarded as the type section (Fig. 2 **AR**). Only a few marine Ostracoda are recorded here: *Bairdia, Loxoconcha* (*Loxocorniculum*), *Xestoleberis*; while freshwater ostracods include *Cyclocypris, Cypridopsis, Cytheridella, Darwinula, Dolerocypris?, Hemicypris, Limnocythere* and *Strandesia.* These are generally not abundant (except *Cyclocypris?* sp. 1, *Cytheridella* and *Limnocythere*, locally) and many of them extend their range into brackish water, where *Cyprideis* and *Perissocytheridea* dominate. In the probably contemporaneous beds of East Barbarita (Table 2, Fig. 2 **EB**), north of Lake Enriquillo, more marine species are found, including *Basslerites minutus* Bold, 'Campylocythere' perieri (Brady), *Caribella yoni* (Puri), *Cativella navis* Coryell & Fields, *Cytherella* spp, *Jugosocythereis pannosa* (Brady) (which is the dominant ostracod in the Holocene reef), *Neocaudites* sp., *Occultocythereis angusta* Bold, *Orionina serrulata* (Brady), *Paracytheridea tschoppi* Bold and *Radimella confragosa* (Edwards). In

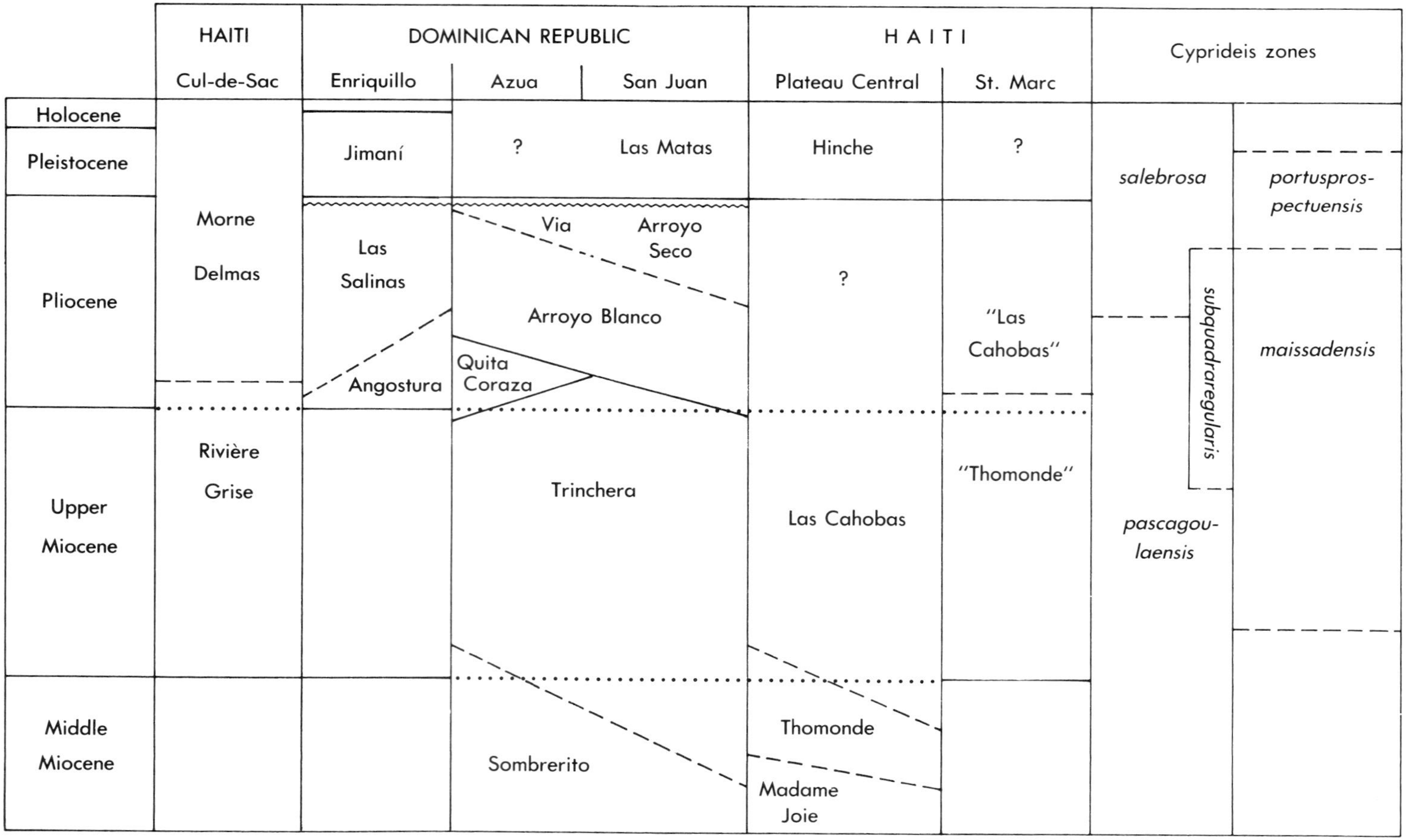

Fig. 1. Stratigraphy of the Southwest Dominican Republic and Central and southern Haiti, with approximate position of *Cyprideis*-zones.

Table 1. Ostracod distribution in the Jimaní Formation at 'Artifact Ridge', north of Jimaní. Samples collected by P. Mann and Charlotte Glenn (Fig. 2 **AR**).

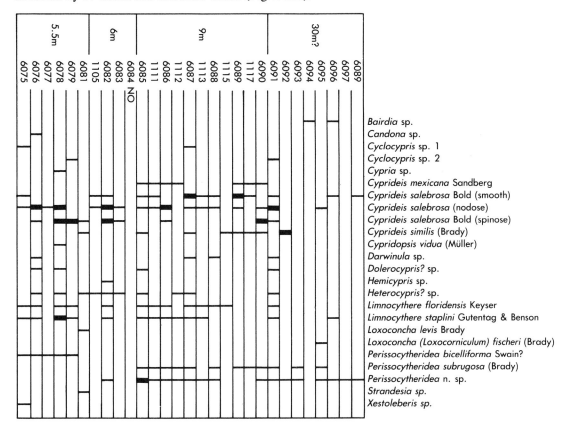

both sections, *Cyprideis mexicana* Sandberg, *C. salebrosa* Bold and *C. similis* (Brady) occur together with several species of *Perissocytheridea*.

A very similar brackish water association (Bold, 1975a) is found in the Las Salinas Formation (Cooke in Vaughan *et al.*, 1920 emended by Bermúdez, 1949). Near the top of the formation (Arroyo Aculadero section, Fig. 2 **AA**, Table 3), this is replaced by *Cyprideis portusprospectuensis* Bold and *Peratocytheridea karlana* (Stephenson), probably indicating a change from lower to higher salinity, perhaps even hypersalinity.

The Loma de Yeso Member of the Arroyo Blanco Formation (Cooper, 1983) has a similar, but much impoverished fauna. The Arroyo Blanco Formation (Bermúdez, 1949) has practically no fresh water species and marine species are much more common than in the Enriquillo Basin and

include *Loxoconcha (Touroconcha) lapidiscola* Hartmann, *Proteoconcha? evai* Bold, *Quadracythere producta* (Brady), *Radimella confragosa* and *Uroleberis torquata* Bold. Here, *Cyprideis maissadensis* Bold is the dominant brackish water species, but *C. mexicana*, *C. salebrosa* and *C. similis* are also present. In most of the Azua Basin the Arroyo Blanco overlies the Trinchera Formation, which here extends into the Pliocene (N 18).

## BIOSTRATIGRAPHY

As Ostracoda appear to be the only means of establishing a biostratigraphical zonation in brackish water environments, it is essential to determine the exact top and bottom of a total range zone based on the distribution of species of *Cyprideis*, without interference from changing ecological

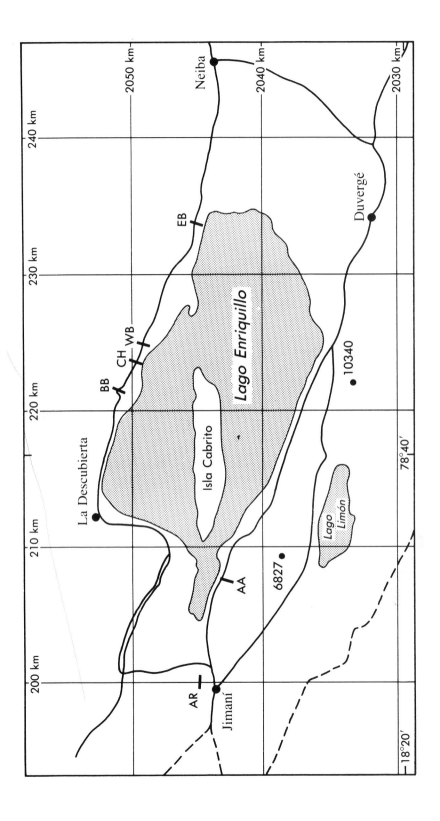

Fig. 2. Location of sections around Lake Enriquillo. AA: Arroyo Aculadero (Las Salinas Formation); AR: 'Artifact Ridge', type area of the Jimaní Formation; CH: Cañada Honda (Pre-Holocene reef sediments, probably Jimaní Formation); EB: East Barbarita Gully (Pre-Holocene reef sediments, probably Jimaní Formation); BB: 'Big Bend' (Post-Holocene reef sediments); WB: 'Wooden Bridge' (Post-Holocene reef sediments). 6827 R. Beall location, 10340 J. W. Hunter location (Dominican Seaboard Oil Company), both Jimaní Formation.

Table 2. Distribution of Ostracoda in East Barbarita Gully, in southward dipping Pre-Holocene reef sediments (Jimaní Formation?). Samples collected by P. Mann and Charlotte Glenn (Fig. 2 **EB**).

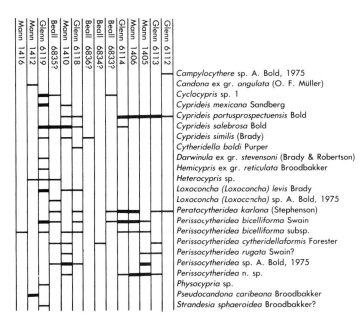

Campylocythere sp. A. Bold, 1975
Candona ex gr. angulata (O. F. Müller)
Cyclocypris sp. 1
Cyprideis mexicana Sandberg
Cyprideis portusprospectuensis Bold
Cyprideis salebrosa Bold
Cyprideis similis (Brady)
Cytheridella boldi Purper
Darwinula ex gr. stevensoni (Brady & Robertson)
Hemicypris ex gr. reticulata Broodbakker
Heterocypris sp.
Loxoconcha (Loxoconcha) levis Brady
Loxoconcha (Loxoconcha) sp. A. Bold, 1975
Peratocytheridea karlana (Stephenson)
Perissocytheridea bicelliforma Swain
Perissocytheridea bicelliforma subsp.
Perissocytheridea cytheridellaformis Forester
Perissocytheridea rugata Swain?
Perissocytheridea sp. A. Bold, 1975
Perissocytheridea n. sp.
Physocypria sp.
Pseudocandona caribeana Broodbakker
Strandesia sphaeroidea Broodbakker?

Table 3. Distribution of Ostracoda in the Las Salinas Formation, underlying Holocene reef sediments in Arroyo Aculadero (Fig. 2 **AA**).

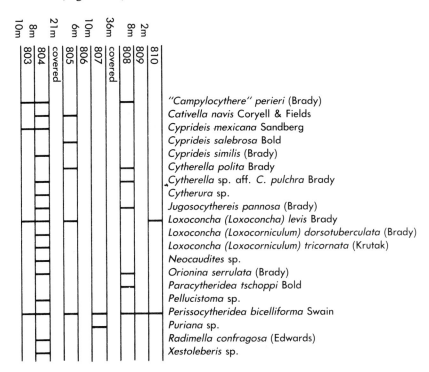

"Campylocythere" perieri (Brady)
Cativella navis Coryell & Fields
Cyprideis mexicana Sandberg
Cyprideis salebrosa Bold
Cyprideis similis (Brady)
Cytherella polita Brady
Cytherella sp. aff. C. pulchra Brady
Cytherura sp.
Jugosocythereis pannosa (Brady)
Loxoconcha (Loxoconcha) levis Brady
Loxoconcha (Loxocorniculum) dorsotuberculata (Brady)
Loxoconcha (Loxocorniculum) tricornata (Krutak)
Neocaudites sp.
Orionina serrulata (Brady)
Paracytheridea tschoppi Bold
Pellucistoma sp.
Perissocytheridea bicelliforma Swain
Puriana sp.
Radimella confragosa (Edwards)
Xestoleberis sp.

conditions. Hispaniola is one of the places where there are, in the late Neogene, extensive lagoonal deposits, partly interfingering with lacustrine and marine deposits, where such a zonation might be verified. The zonation, originally proposed (Bold, 1976) has only been slightly emended (Bold, 1983), and a simplified version is included in Fig. 1. It should be emphasized, however, that no direct comparison can be made with the well established planktonic foraminiferal and nanno-fossil zonations.

In the Greater Antilles *Cyprideis pascagoulaensis* (Mincher) has not been found earlier than the late Middle Miocene (Bold, 1981); its latest documented occurrence is in the Upper Miocene Cercado Formation of the Cibao Basin (Bold, 1988), but, as brackish water conditions gave way to normal marine ones, its real range could very well extend into the Lower Pliocene. Its latest occurrence is well above the base of the *Radimella confragosa* Zone (N 17).

*Cyprideis maissadensis* occurs from a little above the Middle-Upper Miocene boundary (N15) in Haiti and extends in the Azua (Dominican Republic) and La Cruz (Cuba) basins well into the Pliocene (co-occurrence with *C. salebrosa*).

*Cyprideis salebrosa* (and probably *C. mexicana* and *C. similis*) are first found well above the base of the *Radimella confragosa* Zone and some 2000m above the Mio-Pliocene boundary as established on planktonic foraminifera in the Azua Basin. All three species are extant. This distance above the Mio-Pliocene boundary suggests that the base of the occurrence of *C. salebrosa* lies above the N18-N19 boundary. The range of *C. subquadraregularis* (Brady) overlaps with that of *C. pascagoulaensis* and *C. salebrosa* and, therefore, is Upper Miocene to Lower Pliocene.

*Cyprideis portusprospectuensis* and *C.* sp. aff. *C. portusprospectuensis* occur well above the first occurrence of *C. salebrosa* and above the last occurrences of *C. subquadraregularis* and *C. maissadensis*. They have not been found in Holocene deposits above the reef of the Enriquillo Basin. *Cyprideis portusprospectuensis* is the 'reversed' overlap form and occurs in the Las Salinas and Jimaní formations of the Enriquillo Basin and the

Harbour View Beds of Jamaica. *Cyprideis* sp. aff. *C. portusprospectuensis* is the 'normal' form which occurs in the Las Cahobas Formation of the St. Marc area of Haiti and the Arroyo Blanco Formation of the Azua Basin. It is believed that these species occur in age-equivalent beds, but apart from their similar position in the sequence, no proof of their true ages exists. *Cyprideis portusprospectuensis* occurs together with '*Campylocythere*' *perieri* in the Jimaní Formation (Fig. 2, localities 6827 and 10340; Table 2) which may indicate a Pleistocene age for part of its range, as the latter species has not been found in Pre-Quaternary deposits.

## PALAEOHYDROLOGY

In several studies, (e.g., Carbonel, 1982) the increase in punctation of *Cyprideis* species has been linked to increased $Mg^{2+}/Ca^{2+}$ ratio of the environment. *Cyprideis maissadensis* in the Maissade Beds of the Las Cahobas Formation in Central Haiti is generally heavily punctate to reticulate, which is in accordance with the presence of high organic content (lignites) in the environment. In the Arroyo Blanco Formation this species is always lightly punctate and lived there in a more calcareous *milieu*. The Arroyo Blanco Formation often contains thin beds of reef limestone and/or reef rubble at its base. Organic content is always low.

In the Las Salinas Formation of the Arroyo Aculadero section (Fig. 2 **AA**, Table 3) *C. salebrosa* varies from smooth to punctate; in contrast *C. portusprospectuensis* varies from smooth through punctate to reticulate, which may indicate a loss of $Ca^{2+}$ due to deposition of limestone and the approach of conditions in which gypsum was deposited near the top of the section (cf. Carbonel, 1982).

In the area north of Jimaní, about a hundred metres of section of the Jimaní Formation is exposed below the Holocene reef (Fig. 2 **AR**). Here, *C. salebrosa* is nearly always punctate and the punctate forms dominate over smooth or reticulate ones. In this section nodose variants (the number of nodes varying generally from 1 to 5) always dominate over non-nodose forms, and strongly spinose forms occur in the lower 10m and again

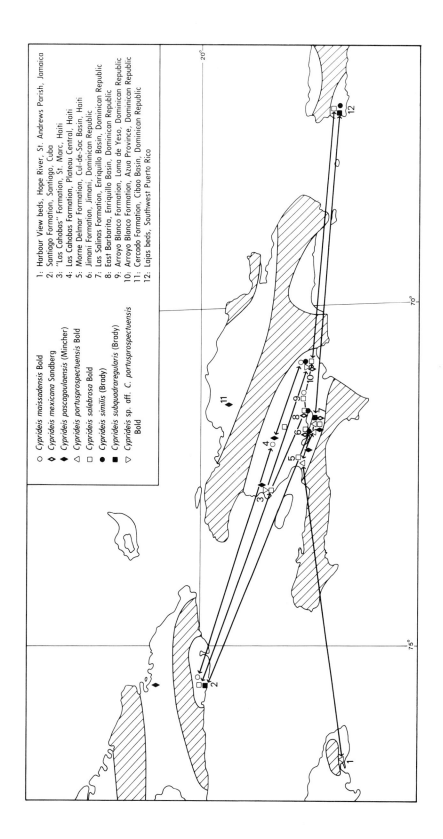

Fig. 3. Geographical distribution of *Cyprideis* species in the late Miocene and Pliocene of the Greater Antilles. The arrows connect occurrences of the same species and indicate possible migration routes.

about 10m higher in the section (Table 1). To date it is not possible to identify the environment in which these hollow spined forms occur. However, there is a negative correlation with the occurrence of *C. mexicana* and *C. similis*. So far the spinose form has never been found in Recent deposits.

Post-Holocene reef sediments in the Enriquillo Basin (Fig. 2 **CH**, **WB** and **BB**) show an overall succession from a marine through a lagoonal to a lacustrine *milieu* with *Cytheridella* and *Limnocythere*. A local repetition of the last two environments may occur. As documentation of this is not yet complete, no further comment will be given here.

## PALAEOBIOGEOGRAPHY

The brackish water genus *Cyprideis* has an interesting distribution in the Greater Antilles. Whereas *C. salebrosa* and *C. pascagoulaensis* appear to occur wherever conditions were favourable (the absence of *C. salebrosa* in the Cibao Basin, northern Dominican Republic (Bold, 1988) is due to deeper water conditions in the later Pliocene) several other species have a more limited distribution.

*Cyprideis maissadensis*, which is the common *Cyprideis* species from the late Miocene to late Pliocene in the Plateau Central of Haiti (Bold, 1981) and the Azua Basin in the Dominican Republic, has not been found in the Enriquillo-Cul de Sac basins of southern Hispaniola (Fig. 3). It was misidentified in the La Cruz and Santiago formations in southwest Cuba (Bold, 1975b) as *C. bensoni* Sandberg.

*Cyprideis subquadraregularis*, on the other hand, occurs from Cuba through southern Hispaniola to southwest Puerto Rico. Its co-occurrence with *C. pascagoulaensis* and *C. salebrosa* indicates, together with the occurrence of *Radimella confragosa*, that it was present there in the late Miocene and early Pliocene.

*Cyprideis portusprospectuensis* is known from the Harbour View beds in Jamaica (from where it was originally described) to southern Hispaniola. It rarely occurs together with *C. salebrosa* and

represents slightly different (more saline) conditions than those under which *C. salebrosa* lived. Its stratigraphical position is as yet unclear; it is possibly late Pliocene and early Pleistocene. *C. portusprospectuensis* is a 'reversed' overlap *Cyprideis* with the right valve larger and overlapping the left valve. *C.* sp. aff. *C. portusprospectuensis* (= *Cyprideis* n. sp?, Bold, 1981), on the other hand, is the mirror-image with 'normal' overlap. It has been found so far in the St. Marc area of Haiti and the Azua Basin in the Dominican Republic, presumably in beds of approximately the same age as those in which *C. portusprospectuensis* occurs (Fig. 3).

Some species of *Perissocytheridea* also show a peculiar distribution: *P. compressa* occurs from southern Cuba through Central Haiti and the northern Dominican Republic (= *P. cahobensis* Bold, 1981, 1988) to Azua (Bold, 1983) and is absent in the Enriquillo Basin. Other species (e.g., *P. plauta* Forester, in Bold, 1975a) appear endemic to southern Hispaniola whereas others occur throughout the Caribbean. The latter are probably species that could survive normal marine salinities (e.g., *P. bicelliforma* Swain, *P. subrugosa* Brady).

## CONCLUSIONS

Ostracod studies in the Dominican Republic show that in the partially connected Azua and Enriquillo basins the environment changed from shallow marine to lagoonal (brackish and hypersaline) in Pliocene to early Holocene times. The Holocene reef of Lake Enriquillo represents a return to shallow marine conditions and is followed by a renewed reduction in salinity.

# REFERENCES

Bermúdez, P. J. 1949. Tertiary smaller Foraminifera of the Dominican Republic. *Cushman Laboratory Foraminiferal Research, Special Publication*, **25**, iv + 322 pp., 26 pls.

Bold, W. A. van den, 1975a. Neogene biostratigraphy (Ostracoda) of southern Hispaniola. *Bull. Am. Paleont.*, Ithaca, **66**(286), 549-625, pls. 58-62, 19 text-figs., 15 tables.

Bold, W. A. van den. 1975b. Ostracodes from the late Neogene of Cuba. *Bull. Am. Paleont.*, Ithaca, **68**(289), 121-167, pls. 14-19, 4 text-figs., 5 tables.

Bold, W. A. van den. 1976. Distribution of species of the tribe Cyprideidini (Ostracoda, Cytherideidae) in the Neogene of the Caribbean. *Micropaleontology*, New York, **22**(1), 1-43, pls. 1-5, 18 text-figs., 10 tables.

Bold, W. A. van den. 1981. Distribution of Ostracoda in the Neogene of Central Haiti. *Bull. Am. Paleont.*, Ithaca, **79**(312), 1-136, pls. 1-6, 18 text-figs., 17 tables.

Bold, W. A. van den. 1983. Shallow marine biostratigraphic zonation in the Caribbean Post-Eocene. *In* Maddocks, R. F. (Ed.), *Applications of Ostracoda*, proceedings of the Eighth International Symposium on Ostracoda, July 26-29, 1982, 400-416, 9 text-figs. Univ. Houston Geos., Houston, Texas.

Bold, W. A. van den. 1988. Neogene Paleontology in the northern Dominican Republic. 7. The Subclass Ostracoda. *Bull. Am. Paleont.*, Ithaca, **94**(329), 1-105, pls. 1-13, 15 text-figs., 5 tables.

Carbonel, P. 1982. Les ostracodes traceurs des variations hydrologiques dans des systèmes de transition eaux douces - eaux salées. *Mém. Soc. géol. Fr. Paléont.*, Paris, **144**, 17-128, 8 text-figs., 2 tables.

Cooper, J. C., 1983. *Geology of the Fondo Negro Region, Dominican Republic*. Unpub. Magister thesis, State University of New York at Albany., 145 pp.

Mann, P., Taylor, P. W., Burke, K. and Kulstad, R., 1984. Subaerially exposed Holocene coral reef, Enriquillo Valley, Dominican Republic. *Bull. geol. Soc. Am.*, New York, **95**(9), 1084-1092, 9 text-figs.

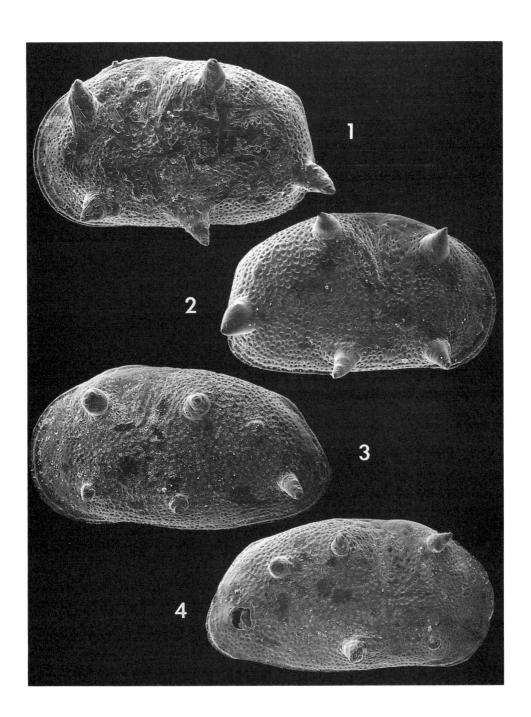

## DISCUSSION

Franciszek Adamczak: Have you any explanation for the reversed overlap (right over left) in your fauna?

Willem van den Bold: At first sight the distribution of 'reversed' forms appears to be more geographically than environmentally controlled. The distribution of the 'normal' form (*Cyprideis* sp. aff. *C. portusprospectuensis*) coincides with the presence of *C. maissadensis* and *Perissocytheridea compressa*, which both indicate lower than normal marine salinites; whereas *C. protusprospectuensis* and *P. bicelliforma* (= *P. cribrosa* (Klie)) are indicators of marine salinites. Overlap reversal has been reported previously in *Haplocytheridea*, *Hemicyprideis* and *Cyprideis*. *Hemicyprideis cubensis chicoyensis*, the 'reversed' form, is always nodose and occurs in lower salinities than the 'normal' form (*C. cubenisis cubensis*). Krstic found a 'reversed' species of *Cyprideis* in the Balkans and another species that is sometimes reversed, but all are always much rarer than the 'normal' form (*Cyprideis major* Kollmann). So far no-one, to my knowledge, has provided a satisfactory explanation of this phenomenon. It might be a response to marginal conditions for those particular species.

Pierre Carbonel: In your third slide, you showed in the the upper part of the graph, the development of a *Cyprideis* indicative of hypersaline conditions. Have you an explanation for its presence with other species (generally the diversity is very low in such conditions; 1-2 species, for example, in Lake Tunis)? Is it, in your opinion, the result of cyclical seasonal changes, that causes phases of hypersalinity during the year?

Willem van den Bold: In the upper part of the section (Table 2) *Cyprideis portusprospectuensis* is accompanied by several species of *Perissocytheridea* and *Paratocytheridea*. The approach of hypersaline conditions is only suggested by the increased reticulation of *Cyprideis*. *Perissocytheridea* is known to occur in hypersaline environments, but I have yet to evaluate the *Perissocytheridea* species as environmental markers. Salinity levels must have been definitely lower than those presently obtaining in Lake Enriquillo (*Cyprideis edentata* and *Perissocytheridea cribriformis*). There is no evidence for seasonal change.

# 17

# Palaeocopid Ostracoda across the Siluro-Devonian international stratotype boundary in Czechoslovakia

**Miroslav Kruta**

Thalmannova 4, Prague 6, Czechoslovakia.

## ABSTRACT

Over the Siluro-Devonian international strato-type boundary sections at Klonk and Budnany Rock, Czechoslovakia, it is possible to observe a change in the ostracod fauna. The Prídolian fauna is represented by *Klonkina cornigera, Vania perdita, 'Ctenobolbina' bohemica* and *Parahippa rediviva*, whereas the Lochkovian fauna is represented by *Tricornina navicula, Berounella rostrata* and *'Ectoprimitia' krausei*.

## INTRODUCTION

One of the most important tectonic features of the early Palaeozoic sediments of the Barrandian Basin of central Bohemia is the lack of distur-bance due to Caledonian movements. The sediments of the Upper Silurian Prídolian Stage pass into sediments of the Devonian Lochkovian Stage without interruption. Equally important, at the Siluro-Devonian boundary sections at Klonk and Budnany Rock, this transgression occurs without any change in facies. In the stratotype section at Klonk, both Silurian and Devonian strata are developed as alternating layers of fine grained limestones and calcareous shales. The boundary is currently defined only on palaeontological criteria, in particular from graptolites, trilobites and conodonts. The auxiliary section at Bundnany Rock is lithologically similar but with the addition of a thick bed of organodetrital limestone, the so-called *Scyphocrinites* bed, just above the bound-ary. This paper assesses, for the first time, the nature and significance of the sequence of Ostra-coda across the stratotype boundaries at Klonk, Budnany Rock and adjacent localities.

Do ostracod faunas change across the Siluro-Devonian boundary? Ostracods can be found in all sections, but only in the limestone layers. Because their carapaces are composed of calcite, until now no chemical treatment has been successful in separating them from the limestone matrix. This study was carried out on material isolated from limestone entirely by mechanical preparation. Fortunately, several of the limestone layers produced many carapaces and broken valves. In the stratotype at Klonk, limestone layers 9 and 13 produced most of the ostracod fauna. Layer 13 is characterized by the last occurrence of the graptolite *Monograptus transgrediens* and is situated 2m below the boundary. The first Devonian Lochkovian ostracod fauna are found at the Budnany Rock stratotype in the bed 52, approximately 1m above the Siluro-Devonian boundary.

## HISTORICAL SURVEY

Prídolian and Lochkovian ostracod faunas were described by Boucek as early as 1936. However, at that time the whole Lochkovian sequence was considered to belong to the Silurian, so both the late Silurian and early Devonian faunas were mixed and documented together as Upper Silurian ostracods. Boucek (1936) described most of the important taxa but, unfortunately, his drawings were poor and they falsely depicted some important morphological features. As a result, many of the taxa which he described did not receive international recognition. The main part of Boucek's Prídolian fauna was probably collected at the base of the series, at the Koledník locality and his Lochkovian fauna came from bed 52 at Budnany Rock.

## THE PRESENT STUDY

### Prídolian fauna

The Prídolian ostracods in the sections studied were all recovered from the limestone layers. They mostly form lumachelles, a few millimetres thick, situated on the upper bedding plane of individual beds. They were studied from four localities in the Barrandian Basin: At the base of the Prídolian at Koledník and Kosov, from the Lower Prídolian at Kosov, from the Upper Prídolian

at Klonk and from the entire sequence of the Prídolian at Pozáry. The same species were found in the entire sequence of the Prídolian and the most important of these are:

1) *Microchilina jarovensis* Boucek, 1936 (Plate 1, Fig. 3). This is the commonest species in the entire Prídolian sequence from Bohemia. However, possibly because Boucek's original drawings show very little resemblance to the true shape of the carapace, it has not been recognized in other regions. Sethi (pers. comm.) reports that the genus occurs in the Silurian of Sweden.

2) *Klonkina cornigera* Kruta, 1986 (Plate 1, Fig. 4). A small distinct ostracod which has been found in all the localities studied.

3) '*Ctenobolbina*' *bohemica* Boucek, 1936 (Plate 1, Figs 1, 2). This species was recognized by Boucek (1936), but only on the basis of a very young instar. The present material, from Koledník and Klonk, also yielded mostly instars. However, two adult females were obtained from Klonk, which demonstrate that this species belongs to the beyrichiacean subfamily Amphitoxotidinae Martinsson, 1962.

4) *Vania perdita* Kruta & Siveter, 1989 (Plate 1, Fig. 5). This species has a very distinct lateral outline but its taxonomic position is uncertain. It is more common in the upper parts of the Prídolian, but, nevertheless, is also found at its base.

5) *Parahippa rediviva* Barrande, 1872. This species is rare but present throughout the Prídolian. This genus is also known from the Silurian of Sweden Sethi, 1979 and is reported to occur in Silurian erratics from Germany (Schaffer, pers. comm.).

### Lochkovian fauna

The mode of preservation of ostracods in the Lower Lochkovian is the same as in the underlying Prídolian. Ostracods occur in 1-2mm thick lumachelles. These lumachelles contain more fragments and fewer whole carapaces than the Prídolian. Elongated fragments (mainly long spines) on the bedding planes show parallel orientation.

In the limestone layers overlying bed 20 at

Klonk, only poorly developed ostracod lumachelles were found, in beds 21 and 52. They contained mainly debris and just a few specimens of *Microchilina jarovensis* and one specimen of *Berounella rostrata* in bed 21.

At the auxiliary stratotype at Budnany Rock, the ostracod faunas are better preserved. In bed 52 a rich ostracod lumachelle containing complete carapaces occurs. The most important species found in the Lochkovian are:

1) *Microchilina jarovensis* Boucek, 1936 (Plate 1, Fig. 3). This species occurs in both the Prídolian and Lochkovian. According to Boucek (1936) the specimens from the Lochkovian belonged to a species which he called *M. acuta*. However, the present material does not support Boucek's original opinion that there are two stratigraphically separated species.

2) *Berounella rostrata* Boucek, 1936 (Plate 1, Fig. 6). According to present taxonomic opinion the genus *Berounella* is not a palaeocope (Sohn & Berdan, 1960). However, as it is one of the most distinctive new species to appear at the base of the Lochkovian it is included here.

3) *Tricornina navicula* Boucek, 1936 (Plate 1, Fig. 8). This is another species, which first appears at the base of Devonian. Lateral spines are broken in specimens with valves preserved parallel to the bedding plane, but a few specimens were found with carapaces orientated vertically and these show that the length of the spines reaches at least half the maximum length of the carapace.

4) '*Ectoprimitia*' *krausei* Boucek, 1936 (Plate 1, Fig. 7). An ostracod species with a rather small carapace, common in the Lower Lochkovian.

## CONCLUSION

Across the international Siluro-Devonian boundary in Central Bohemia it is possible to observe a distinct change in the ostracod fauna. Some important Prídolian species disappear (*Klonkina cornigera*, '*Ctenobolbina*' *bohemica*, *Vania perdita*, *Parahippa rediviva*) and new species appear (*Tricornina navicula*, *Berounella rostrata*, '*Ectoprimitia*' *krausei*). Also of possible significance is a change in the general shape of carapaces. The Lower

Devonian ostracod fauna is characterized by the occurrence of valves with rather long spines (*B. rostrata*, *T. navicula*) not present in the Upper Silurian, which could reflect a change in prevailing life styles. This trend can be observed in the Pragian, where the spines of *Berounella* and *Tricornina* species reach even greater lengths, exceeding the length of the carapace.

## REFERENCES

Barrande, J. 1872. *Systême Silurien du centre de la Bôheme.* Supplément au Vol.I., Trilobites, crustacés divers et poissons. Prague, Paris.

Boucek, B. 1936. Die Ostrakoden des Böhmischen Ludlows. (Stufe eß). *N. Jb. Mineral. Geol. Paläont.*, Stuttgart, Beil.-Bd., **76**, Abt. B., 31-98.

Boucek, B. & Pribyl, A. 1955. O silurských ostrakodech a stratigrafii vrstev budnanských e a nejblizsího okolí Kosova a Koledníku u Berouna. *Sbor. Ústr. Úst. geol.*, Praha, Odd. paleont., 577-662.

Kruta, M. 1968. The new ostracode genus *Klonkina* from the Upper Silurian of Bohemia. *N. Jb. Geol. Palaont. Mh.*, Stuttgart, 444-448.

Kruta, M. & Siveter, D. J. 1989. In press. On *Vania perdita* gen. et sp. nov. *Stereo-Atlas Ostracod Shells*, London.

Martinsson, A. 1962. Ostracodes of the family Beyrichiidae from the Silurian of Gotland. *Bull. geol. Inst. Univ. Uppsala*, Uppsala, **41**, 1- 369.

Sethi, D. K. 1979. Paleocope and eridostracan ostracodes. *In* Jaanusson, V., Laufeld, S. & Skoglund, E. (Eds), *Lower Wenlock faunal and floral dynamics - Vattenfallet section, Gotland. Sver. geol. Unders.*, Ser. C, Uppsala, **762**, 142-166.

Sohn, I. G. & Berdan, J. M. 1960. The Ostracode family Berounellidae, new. *J. Paleont.*, Tulsa, Okla, **34**(3), 479-482.

## DISCUSSION

Gerhart Becker: In both Dr Boucek's and my opinion, there is a facies change in the Lochkovian. There are different faunas (Thuringian Assemblage) and signs of transgression.

Miroslav Kruta: There is no distinct change in lithology, but the ostracod fauna is different and shows signs of belonging to another assemblage.

Gerhart Becker: Have you found any silicified ostracod faunas; have you any conodonts?

Miroslav Kruta: There are conodonts, but we had no success in finding silicified ostracods earlier than the Pragian and in the Pragian we succeeded only in one layer at the top of the sequence.

**Plate 1**

Fig. 1.  '*Ctenobolbina*' *bohemica* Boucek, 1936. RV, adult technomorph. Prídolian, Klonk. X45.

Fig. 2.  '*Ctenobolbina*' *bohemica* Boucek, 1936. RV, adult heteromorph. Prídolian, Klonk. X45.

Fig. 3.  *Microchilina jarovensis* Boucek, 1936. LV, Prídolian, Klonk. X75.

Fig. 4.  *Klonkina cornigera* Kruta, 1986. LV, Prídolian, Kosov. X120.

Fig. 5.  *Vania perdita* Kruta & Siveter, 1989. LV, Prídolian, Klonk. X70.

Fig. 6.  *Berounella rostrata* Boucek, 1936. LV, Lochkovian, Budnany Rock. X60.

Fig. 7.  '*Ectoprimitia*' *krausei* Boucek, 1936. LV, Lochkovian, Budnany Rock. X160.

Fig. 8  *Tricornina navicula* Boucek, 1936. LV, Lochkovian, Budnany Rock. X60.

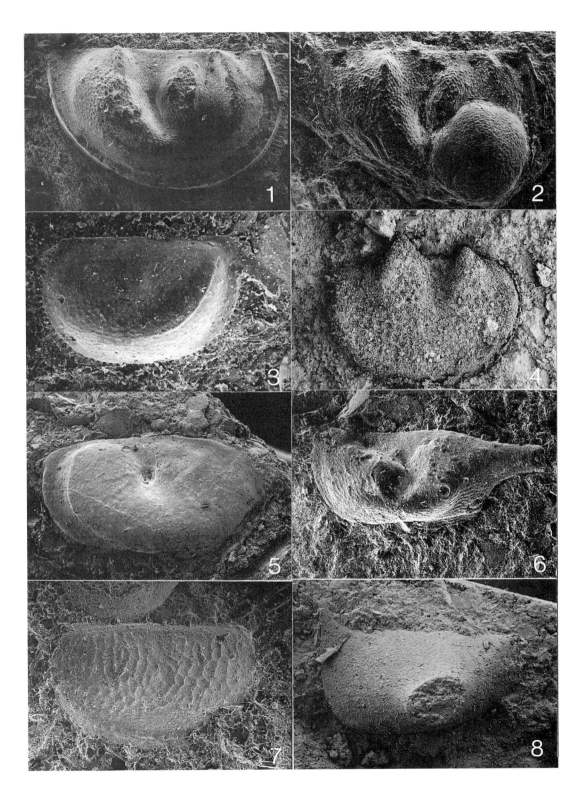

# 18

# The biostratigraphical sequence of Mesozoic non-marine ostracod assemblages in northern China

Pang Qiqing[1] & Robin Whatley[2]

[1]Hebei College of Geology, Hebei, China.
[2]Institute of Earth Studies, University College of Wales, Aberystwyth, Dyfed, U.K.

## ABSTRACT

The non-marine Triassic to Upper Cretaceous strata of the intermontane and piedmont basins of northern China are characterized by six distinct ostracod faunas and twelve assemblages. The Triassic is characterized by the *Darwinula-Lutkevichinella-Tungchuania* Fauna with three distinct faunal assemblages. The Middle Jurassic yields the *Darwinula sarytirmenensis-D. impudica* Fauna; the Upper Jurassic, the *Djungarica-Mantelliana* Fauna; the terminal Jurassic and earliest Cretaceous, the *Eoparacypris-Luanpingella-Yanshanina* Fauna with two assemblages, while the remainder of the early Cretaceous is characterized by the *Cypridea* Fauna followed by the Upper Cretaceous *Cypridea-Talicypridea* Fauna.

## INTRODUCTION

Non-marine Mesozoic strata are widely and well developed in the intermontane and piedmont basins of northern China, especially in Hebei, Shanxi, Inner Mongolia, Beijing, western Henan and the Eerduosi (Shaanxi-Gansu-Ningxia) Basin (Table 1, Fig. 1). In these inland basins, fluvio-lacustrine argillaceous and arenaceous sediments were deposited during the entire Mesozoic. Deposits of the Lower and Middle Triassic are characterized by their red colour, Upper Triassic and early Middle Jurassic strata by their coal seams and the mid-Upper Jurassic by its contained volcanics. Lower Cretaceous strata comprise red fluvio-lacustrine rocks intercalated with volcanics. Outcrops of Upper Cretaceous strata are restricted to northern Hebei,

## Table 1. The stratigraphical division and correlation of Mesozoic erathem in North China.

| Age | Eerduosi basin | Inner Mongolian Autonomous Region — Yinshan | Inner Mongolian Autonomous Region — Daxinganling | Shanxi | Beijing | Tianjin | Hebei — Plain of N.C. | Hebei — Jibei | West Henan | Liaoning |
|---|---|---|---|---|---|---|---|---|---|---|
| Over Strata | Oligocene ($E_3$) / Quaternary (Q) | Quaternary (Q) / Guyang Fm | Eocene ($E_2$) / Erliandabusu Fm | Hannuoba Fm ($N_2$) / Zhumapu Fm / Zuoyun Fm | Changxindian Fm ($E_2$) | Kongdian Fm ($E_2$) | Tertiary | Hannuoba Fm (N) / Hulaguu | Jiyuen Group ($E_{2-3}$) | Pleistocene (Q) |
| $K_2$ | | | | | | | | | | Sunjiawan Fm |
| $K_1$ | Zhidan Group | Lisangou Fm / Bainuyangpan Fm / Daqinghshan Fm / Changhangou Fm / Zhaugou Fm / Wudanggou Fm / Nansuleitu Fm | Xilinguole Fm / Chaganlimennuoer / Bayanhua Fm / Bulagenhada Fm / Sandaogou Fm / Duolun Fm / Daotenuoer Fm / Shuiquancun Fm / Chagannuoer Fm / Zhangjiakou Fm (Xinganling Group) | Zhongzhuangpu Group / Hunyuan Group | Xiazhuang Fm / Lushangfen Fm / Tuoli Fm / Dahuichang Fm / Donglanggou Fm | Lower Cretaceous | Liujiadong Fm / Lincheng Fm | Tujingzi Fm / Huangjiapu Fm / Jiufutang Fm / Jianchang Fm / Jingangshan Fm / Jixian Fm / Dabeigou Fm / Zhangjiakou Fm | J – K | Fuxin Fm / Jiufutang Fm / Jianchang Fm / Jingangshan Fm / Yixian Fm |
| $J_3$ | Anding Fm / Zhiluo Fm | | | Tianchihe Fm | Donglingtai Fm / Zhangjiakou Fm / Houcheng Fm / Tiaojishan Fm / Jiulongshan Fm | | | Houcheng Fm / Tiaojishan Fm / Jiulongshan Fm | Hanzhuang Fm | Tuchengzi Fm / Lanqi Fm / Haifangou Fm |
| $J_2$ | Yanan Fm | | Alatanheli Fm | Yungang Fm / Datung Heifeng Fm / Yongdingzhuang Fm | Mentougou Fm / Nandaling Fm / Xingshikou Fm | | | Mentougou Fm / Xiahua-yuan Fm / Nandaling Fm / Xingshikou Fm | Mawa Fm / Yima Fm | Beipao Fm |
| $J_1$ | Yanchang Fm | Laowopu Fm | | Yanchang Fm | | | | | Anyao Fm | Xinglonggou Fm / Kuntouboluo Fm |
| $T_3$ | Tongchuan Fm / Ermaying Fm | | | Tongchuan Fm / Ermaying Fm | | | Liuquan Fm | Huzhangzi Fm | Tanzhuang Fm / Chunshuyao Fm / Youfangzhuang Fm / Ermaying Fm | Laohugou Fm |
| $T_2$ | Heshanggou Fm | | | Heshanggou Fm | | | Heshanggou Fm | Xiabancheng Fm | Heshanggou Fm | |
| $T_1$ | Liujiagou Fm | | | Liujiagou Fm | | | Liujiagou Fm | Tingjiagou Fm | Liujiagou Fm | |
| Understrata | Sunjiagou Fm | Naobaogou Fm ($P_2$) | Baocraobao Fm ($P_2$) | Sunjiagou Fm | Shuanquan Fm ($P_2$) | | Sunjiagou Fm | Sunjiagou Fm | Sunjiagou Fm | Sunjiagou Fm |

Shaanxi and Inner Mongolia and are mainly coarse to medium fluvio-lacustrine sands and clays.

All these strata contain abundant non-marine Ostracoda and diversity is particularly high in the Upper Jurassic and Cretaceous. The Ostracoda are of fundamental importance in the dating and correlation of Mesozoic strata in northern China and also have a wider potential for international and even intercontinental correlation.

## THE SEQUENCE OF OSTRACOD FAUNAS AND ASSEMBLAGES (Table 2)

### I. The Triassic *Darwinula-Lutkevichinella-Tungchuania* Fauna

This fauna is mainly found in western Henan and northern Shaanxi and is divisible into three assemblages in stratigraphical order:

1) **Lower Triassic. The *Darwinula triassiana-D. fengfengensis* Assemblage**. This assemblage comprises 13 species of *Darwinula* and is found mainly in the red mudstones and grey-green silty mudstones of the Heshangou Formation around the town of Yima, Dengfeng County in the western part of Henan and the counties of Yoaxian, Tongchuan, Hangchen, Wupu and Fugu in northern Shaanxi.

The 13 species are: *Darwinula triassiana* Belousova, *D. fengfengensis* Pang, *D. rotundata* Lubimova, *D. recondita* Mandelstam, *D. pseudoinornata* Belousova, *D. subaccepta* Pang, *D. zouzhigouensis* Pang, *D. subpromissa* Pang, *D. ingrata* Lubimova, *D. parva*, *D. oblonga*, *D. fragilis* Schneider and *D. cuspicata* var. *dilucida* Jiang.

Of the ostracods cited above, some are of particular biostratigraphical significance: *D. triassica* and *D. pseudoinornata*, *D. ingrata*, *D. rotundata*, *D. recondita* also occur in the Lower Triassic of the USSR, some as far afield as the Donetz Basin. All of these species also occur in western China, in the Lower Triassic of the Zhungeer Basin, Xinjiang. Also, some of the endemic species, such as *D. subaccepta*, *D. subpromissa*, and *D. zouzhigouensis* have very close morphological counterparts in the Triassic and particularly Lower Triassic of the USSR (i.e.

*D. accepta* Lubimova, *D. promissa* Lubimova, *D. pseudoinornata* Belousova). *D. fragilis* is very widely distributed. It ranges back to the Upper Permian in the USSR, but it is also Triassic there and in the Bunstsanstein of Germany, the Rhaetian of Britain and the Keuper of Poland.

Supporting palaeontological evidence for the Lower triassic age of this assemblage is the occurrence in the Heshangou Formation of the vertebrate *Xilousuchus*, the charophyte *Stellatochara-Stenochara* flora and other plants, such as the *Pleuromeia sternbergi-P. rossia* flora which is late Lower Triassic.

2) **Middle Triassic. *Darwinula-Lutkevichinella* Assemblage**. This assemblage is from purple, red and grey-green mudstones of the Ermaying Formation and, in Yaoxian and Hangcheng Counties and around the town of Tongchuan in northern Shaanxi, it comprises 4 genera and the following are the most important 15 species: *Darwinula gerdae* Glebovskaja, *D. oblonga*, *D. fragilis* Schneider, *D. subovaliformis* Su & Pang, *D. accuminata* Belousova, *D. liulingchuanensis* Zhong, *D. alta* Jiang, *Lutkevichinella ornata*, *L. ornatula*, *L. brachicostata*, *L. minuta*, *L. ansulca* Su & Pang, *Shensinella praecipua*, *S. gaoyadiensis* Su & Pang and *Tungchuannia quadratiformis* Su & Pang.

In the western part of Henan Province, the diversity of this assemblage is very much reduced and comprises only the following 2 genera and 7 species not all of which are common between the two areas: *Darwinula tersiensis* Mandelstam, *D. oblonga* Schneider, *D. subovatiformis* Su & Pang, *D. subarta* Pang, *L. minuta* Su & Pang, *L. sublongovata* and *L. yimaensis* Pang.

Many of the ostracods in this assemblage either range up from the Lower Triassic or up into the Upper Triassic, imparting to the fauna a transitional aspect between these two divisions of the system. Most of the species are endemic to northern China while others, such as *D. gerdae*, *D. oblonga* and *D. accuminata* are also known from contemporary strata in the USSR.

The formation also yields a rich flora of Middle Triassic charophytes (*Stellatochara, Cuneatochara, Porochara, Altochara, Vladiminella*) and

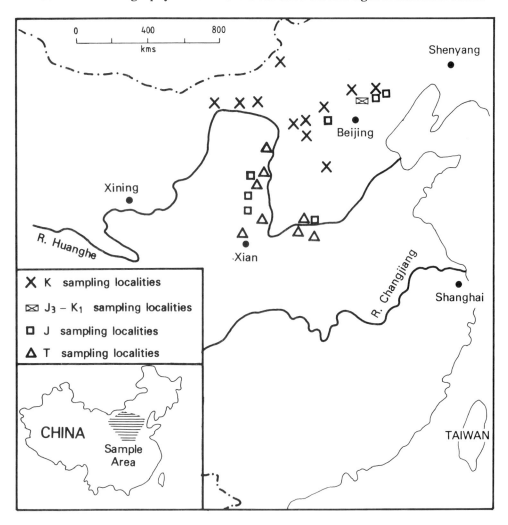

Fig. 1. The important Mesozoic, non-marine ostracod localities in North China.

abundant vertebrates (*Sinokannemeyeria*, *Parakannemeyeria*, *Shansiodon*, *Chasmatosaurus?*, *Sinognathus*, *Neoprocolophon* and *Shansisuchus*) all of which are well-known indices of the Middle Triassic.

**3)  Middle and Upper Triassic.  *Darwinula-Tungchuania* Assemblage.** This assemblage has been recognized in the Middle Triassic Youfanzhuang Formation and the Upper Triassic Chuanshuyao and Tanzhuang formations in Jiyuan County in western Henan; also from the Middle Triassic Tongchuan and the Upper Triassic Yanchang

formations of northern Shaanxi. The ostracod faunas which typify the various formations are given below:

**Youfanzhuang Formation:** *D. accuminata, D. gerdae, D. medialis* Zhong, *S. gaovadiensis, T. quadratiformis, T. agrestate* Zhong, and *L. minuta.*

**Tongchuan Formation:** *T. houae, T. aurita* Zhong, *D. accuminata, D. alta, D. gerdae, D. liulingchuanensis, D. oblonga* and *D. subovatiformis.*

**Yanchang Formation:** *D. gerdae, D. accuminata, D. liulingchuanensis, D. opinabilis* Zhong, *D. oblonga, D. zichangensis* Su & Pang, *D. medialis, T. aurita, T. houae, T. agrestata,*

Table 2. The stratigraphical distribution of Mesozoic non-marine ostracod assemblages.

| Age | | Strata | | Fauna | | Assemblages |
|---|---|---|---|---|---|---|
| Cretaceous | Late | Zhumapu Fm. | | VI | 12 | Cypridea-Talicypridea-Candona assemblage |
| | Early | Zhuoyun Fm. / Xianzhuang Fm. | | V | 11 | Cypridea-Candona-Rhinocypris assemblage |
| | | Huangjiapu Fm. / Guyang Fm. | | | 10 | Cypridea-Lycopterocypris assemblage |
| | | Jiufutang / Lisangou Fm. | | | 9 | Cypridea-Djungarica-Rhinocypris assemblage |
| | | Jianchang Fm. | | | | (2)  Cypridea unicostata-C.koskulensis subassemblage |
| | | Jinganshan Fm. | | | | (1)  Cypridea vitimensis-C.luanpingensis subassemblage |
| Jurassic | Late | Yixian Fm. | | IV | 8 | Rhinocypris-Yanshanina-Cypridea assemblage |
| | | Dabeigou Fm. | | | 7 | Eoparacypris-Luanpingella-Pseudoparacypridopsis assemblage |
| | | Zhangjiakou Fm. | | III | 6 | Djungarica-Mantelliana-Damonella assemblage |
| | | Houcheng Fm. | | | | |
| | Middle | Tiaojishan Fm. | | II | 5 | Darwinula sarytirmenensis-D.impudica-D.magna assemblage |
| | | Jiulongshan Fm. | | | | |
| | | Mentougou Fm. / Xiahuayuan Fm. | | | 4 | Darwinula sarytirmenensis-Timiriasevia catenularia assemblage |
| | Early | Nandaling Fm. | | | | |
| | | Xingshikou Fm. | | | | |
| Triassic | Late | Yanchang Fm. | Tanzhuang Fm. | I | 3 | Darwinula-Tongchuania assemblage |
| | | | Chuanshuyao Fm. | | | |
| | Middle | Tongchuan Fm. / Youfanzhuang Fm. | | | 2 | Darwinula-Lutkevichinella assemblage |
| | | Ermaying Fm. | | | | |
| | Early | Heshangou Fm. | | | 1 | Darwinula triassiana-D.fengfengensis assemblage |
| | | Liujiagou Fm. | | | | |

VI  Cypridea-Talicypridea Fauna

V  Cypridea Fauna

IV  Eoparacypris-Luanpingella-Yanshanina Fauna

III  Djungarica-Mantelliana Fauna

II  Darwinula sarytirmenensis-D.impudica Fauna

I  Darwinula triassiana-Lutkevichinella-Tongchuania Fauna

*T. perelengana* Zhong, *L. costata* (Zhong), *L. ornatula* and *Gomphocythere*? *pulchra* Zhong.

**Chuanshuyao and Tanzhuang formations**: *D. liulingchuanensis*, *D. gerdae*, *D. medialis*, *D. opinabilis*, *T. agrestata* and *T. houae*.

The most important ostracod species for the recognition of this assemblage are: *D. medialis*, *D. opinabilis*, *D. zichangensis*, *T. perelegana* and *G.? pulchra*.

On the basis of the Ostracoda, the abundant Yangchan flora and the bivalve fauna (*Shaanxiconcha clinovata* Group, *S. triangulata* Group and *Sibireconcha*) the Yangchan Formation is clearly Upper Triassic. However, while the ostracods of the Tongchuan Formation are rather similar to those of the Yangchan Formation, the flora (*Pleuromeia*, *Glossopteris*, *Tongchuanophyllum*, *Danaeopsis magnifolia*, *D. plana*, *Neocalamites carcinoides*, *N. carrerei*, *Equisetites brevidentatus*, *E. sthenodon*, *Taenicradopsis rhizomoides* and *Bernoullia zeilleri*) is considered to represent the late Middle Triassic.

## II. The Middle Jurassic *Darwinula sarytirmenensis-D. impudica* Fauna

This fauna is found in northwestern Hebei, western Henan and northern Shaanxi and comprises two assemblages.

### 1) Early Middle Jurassic. The *Darwinula sarytirmenensis-Timiriasevia catenularia* Assemblage.
This is mainly known from the dark mudstones and calcareous mudstones of the Xiahuayuan Formation in northwestern Hebei and comprises: *D. sarytirmenensis*, *D. impudica* Sharapova, *D. dongchengensis* Pang, *Timiriasevia catenularia* Mandelstam, *T. bella* Su & Li and *T. yangyuanensis* Pang.

The ostracods in this assemblage described by Soviet authors were first recorded from the coal-bearing Jurassic strata of Mount Sartirmen on the eastern shore of the Caspian Sea and subsequently they became widely known in the Middle Jurassic rocks of the northwest and south of China.

### 2) Late Middle Jurassic. *D. sarytirmenensis -D. impudica -D. magna* Assemblage.
This assemblage is mainly encountered in the Jiulongshan Formation around Xiaofanzhngzi in Chengde County, Yanshan region, Hebei: *D. sarytirmenensis*, *D. impudica*, *D. magna*, *D. incurva* Bate, *D. changxinensis* Ye, *D. lufengensis* Wang & Ye, *D. paracontracta*, *D. magna yunlongensis* Ye, *D. yibinensis* Su & Li, *D. submagna*, *D. chengdeensis*, *D. stenimpudica*, *D. xiaofanzhangziensis*, *D. paracrassiformis* Pang, *T. catenularia*, *T. armeniacumiformis* Zhong and *T. humilis* Zhong.

This assemblage is also found in the late Middle Jurassic Mawa Formation in Jiyuan County, western Henan but here the diversity is lower: *D. sarytirmenensis*, *D. impudica*, *T. epidermiformis* Mandelstam, *T. catenularia* and *T. mackerrowi* Bate.

The contemporary Anding Formation of northern Shaanxi yields the following species: *D. sarytirmenensis*, *D. impudica*, *D. magna*, *Timiriasevia shensiensis* Zhong, *T. mackerrowi*, *T. bella* Su & Li, *T. armeniacumiformis*, and *T. humilis* Zhong.

Although many of the ostracods of this assemblage also occur in the underlying one, it also includes species which range up into the Upper Jurassic. On this, and other grounds, the assemblage is thought to typify the late Middle Jurassic.

## III. The Upper Jurassic. *Djungarica-Mantelliana* Fauna

This fauna, which comprises 6 genera and 19 species, is typical of the Houcheng Formation of the Yanshan region, Hebei: *Djungarica yunnaensis* Ye, *D. pingquanensis*, *D. yangshulingensis* Pang, *Mantelliana reniformis*, *M. subreniformis*, *M. yangshulingensis*, *M. obliquovata* Pang, *Damonella ovata*, *D. depressa* Gou, *D. orbiculata*, *D. suborbiculata*, *D.? subquadrata* Pang, *Eoparacypris pingquanensis* Pang, *Stenestroemia pingquanensis* Pang, *S. yangshulingensis* and *Darwinula yangshulingensis* Pang.

The most marked difference between this fauna and Fauna II is the decrease in the importance of *Darwinula* and the appearance of *Djungarica*,

*Mantelliana* and *Damonella*. The fauna contains a number of species very similar to those Purbeck-Wealden beds of England and the genus *Stenestroemia* is also found there, in the Lower Serpulit bed in Germany and in the Rabekke Formation of Denmark. The fauna is clearly younger than Fauna II but appears older than Fauna IV and is probably very late Jurassic.

## IV. The uppermost Jurassic-lowermost Cretaceous *Eoparacypris-Luanpingella-Yanshanina* Fauna

This fauna spans the Jurassic-Cretaceous boundary and is only found in the Yanshan region of northern Hebei. It comprises two assemblages:

**1)** *Eoparacypris-Luanpingella-Pseudoparacypridopsis* **Assemblage.** This is made up of 7 genera and 14 species and is typified by the absence of *Cypridea*. It occurs mainly in the terminal Jurassic - lowermost Cretaceous Dabeigou Formation in Luanping County, Yanshan region of northern Hebei: *Eoparacypris macroselina, E. surriensis* Anderson, *E. jingshangensis* Yang, *E. obesa, E. dadianziensis* Pang, *Luanpigella postacuta* Yang, *L. dorsincurva, L. postacuminata* Pang, *Pseudoparacypridopsis mountfieldensis* Anderson, *P. luanpingensis, P. dorsalta* Pang, *Limnocypridea subplana, L.* aff. *abscondida* Lubimova, *Yanshanina dabeigouensis* (Yang) together with *Rhinocypris* sp. and *Darwinula* sp.

**2)** *Rhinocypris-Yanshanina-Cypridea* **Assemblage.** This assemblage is known mainly from the Upper Jurassic-Lower Cretaceous Yixian Formation of Luanping County, Hebei and comprises 6 genera and 11 species: *Rhinocypris echinata* (Lubimova), *R. foveata* (Jiang), *R. subechinata, R. dadianziensis* Pang, *Yanshanina subovata, Y. elongata, Y. postitruncata* Pang, *Y. dabeigouensis* (Yang), *Cypridea altidorsangulata, C. xitaiyangpoensis* Pang, *Timiriasevia opinabilis* Mandelstam and *Darwinula* sp.

This fauna contains both late Jurassic and early Cretaceous elements and is potentially important for the resolution of the position of the Jurassic-Cretaceous boundary in China, a problem which has remained unresolved for 50 years.

The first of the two assemblages contains many species first described from the Purbeck beds of England, such as *Eoparacypris macroselina* (Lower and Upper Purbeck), *E. surriensis* (*Cypridea vidrana* Zone), *Pseudoparacypridopsis mountfieldensis* (*C. granulosa granulosa* Zone) Other elements are known from Mongolia. Most significant, however, is the absence of *Cypridea* from this assemblage. This absence could be attributed to stratigraphy (the strata pre-date the early Cretaceous arrival of *Cypridea* in China (Whatley, in press) or to environmental factors, such as relatively high salinity levels (Anderson & Bazley, 1971). This assemblage could, therefore, be of either uppermost Jurassic or lowermost Cretaceous age.

The second assemblage is noted for the fact that *Cypridea* begins to appear but is never abundant or diverse. *C. xitaiyangpoensis* and *C. altidorsangulata* are very similar, respectively, to *C. dunkeri dunkeri* and *C. dunkeri carinata* from the Middle Purbeck beds of England, which are at the Jurassic-Cretaceous boundary. *Yanshanina* is similar to *Mongolianella* and *Jinggunella* from the Upper Jurassic of Yunnan and Sichuan in southwestern China, while *Timiriasevia opinabilis* occurs in the Neocomian (Hauterivian-Barremian) of the West Siberian lowlands of the USSR. All this evidence suggests that the assemblage probably spans the boundary between the two systems.

## V. The *Cypridea* Fauna

This fauna is widely distributed throughout the area in question and comprises three ostracod Assemblages:

**1) Early Cretaceous.** *Cypridea-Djungarica-Rhinocypris* **Assemblage.** This assemblage was found principally in Hebei and Inner Mongolia and is divisible into 3 subassemblages:

i) *Cypridea vitimensis-C. luanpingensis* **Subassemblage**. This is an abundant and diverse fauna of 8 genera and 42 species mainly known from the Jingangshan Formation in Luanping County and the Qingquan basin of Weichang County, Yanshan region, Hebei: *Cypridea vitimensis, C. sulcata* Mandelstam, *C. peltoides peltoides, C. tumescens acrobeles, C. dunkeri inversa* Anderson, *C. granulosa granulosa* (J. de C. Sowerby), *C. granulosa protogranulosa, C. coelnothi, C. segena, C. varians* Anderson, *C. dadianziensis, C. granulosa subgranulosa, C. granuliformis, C. (C.) jiandeensis tubercularis* Pang, *C. (C.) huoshanensis* Gou, *C. dabeigouensis* Yang, *C. stenolonga, C. lahailiangensis, C. obliquoblonga, C. chengdeensis, C. altidorsangulata, C. gaiziliangensis, C. subsulcata* Pang, *Djungarica stolida* Jiang, *D. saidovi* Galeeva, *D. elongata* Pang, *Timiriasevia polymorpha* Mandelstam, *T. subopinabilis, T. pusilloretirugata, T. araneosa* Pang, *Rhinocypris echinata, R. dadianziensis, Lycopterocypris circulata* Lubimova, *Yanshanina subovata, Ziziphocypris linchengensis* Su & Li, *Darwinula contracta, D. barabinskensis* Mandelstam, *D. leguminella* (Forbes), *D. oblonga* (Roemer), *D. dadianziensis* and *D. lahailiangensis* Pang.

ii) *Cypridea unicostata-C. koskulensis* **Subassemblage**

This is widely distributed in Juifutang Formation in Pingquan and Fengning Counties of the Yanshan region of Hebei and in the Lincheng Formation of Linchen County in the same province. It also occurs in the Dahuichang Formation in the Fengtai and Chaoyang districts of Beijing and the Lisangou Formation of Guyang County, Inner Mongolia. The faunas of the various formations of each region are given below:

**Jiufutang Formation**: *Cypridea unicostata* Galeeva, *C. koskulensis* Mandelstam, *C. luozhangziensis, C. dayingensis, C. subfoveolata, C. pingquanensis* Pang, *C. (C.) shouchangensis* Yang & Ye, *C. (C.)* cf. *angusticaudata* Cao & Yang, *C. (C.)* aff. *concisa* Jiang, *Limnocypridea abscondida, L. reticulata, Lycopterocypris infantilis* Lubimova, *L. debilis, L. louzhangziensis, Damonella pingquanensis*

Pang, *Timiriasevia obesiconvexa* Pang, *Darwinula subparallela* Ye, *D. tubiformis* Lubimova, *D. barabinskensis* and *D. jonesi* Anderson.

**Dahuichang Formation**: *Cypridea unicostata, C. paraunicostata* Zhang, *C. dahuichangensis, C. acutidorsangulata* Pang, *C. (C.) multispinosa* Hou, *C. (C.)* aff. *multispinosa* Hou, *C. (C.) shouchangensis* Yang & Ye, *Rhinocypris strumospinata, R. substrumospinata* Zhang, *Lycopterocypris infantilis, L. debilis* Lubimova, *L. taiyanggongensis* Zhang, *L. dorsofornicata* Zhang.

**Lincheng Formation**: *Cypridea yumenensis* Hou, *C. taihangshanensis* Su & Li, *Cypridea* spp. *Rhinocypris echinata, Ziziphocypris linchengensis* Su & Li, *Theriosynoecum zhubiensis* Su & Li and *Darwinula contracta*.

**Lisangou Formation**: *Cypridea unicostata, C.* aff. *unicostata, C. (C.) multispinosa* Hou, *C.* aff. *justa, C. trita* Lubimova, *Djungarica saidovi* Galeeva, *D. stolida, D.* aff. *stolida* Jiang, *D. triangulata* Pang, *R. echinata, R. foveata* (Jiang), *L. infantilis, L. eggeri* Mandelstam, *L. multifera* Lubimova, *Z. linchengensis, D. contracta, D. custella* Pang, *Clinocypris scolia* Mandelstam and *Timiriasevia* sp.

Also, from boreholes in the Beijing area the following are recorded: *Cypridea propunctata* Sylvester-Bradley, *C. foveolata* (Egger), *C. obvellatispinata, C. spinacliniata* Zhang, *C. (Pseudocyporidina) ellipseloides* Hou, *C. (Ulwellia) subproducta* Hou, Ye & Cao and a few specimens of *Timiriasevia* and *Darwinula*. These may, however, belong to the succeeding subassemblage.

2) *Cypridea-Lycopterocypsis* **Assemblage**

This assemblage has been recovered from the Guyang Formation, Guyang County, Inner Mongolia: *Cypridea (P.) parallela, C. (C.) concina, C. (C.) multispinosa, C. (C.)* aff. *uninoda* Hou, *C. curtorostrata* Hao, *C. hayehudongensis* Zhang, *C. doliobovata* Pang, *C. parahayehudongensis* Zhang, *C. guyangensis, C. subpolita, C. subvalida* Pang, *L. eggeri, L. infantilis* and *D. contracta*. The ostracods are preserved in dark mudstones and marls within the coal-bearing strata.

**3) Late early Cretaceous *Cypridea-Candona-Rhinocypris* Assemblage**

This assemblage has been found in the Xiazhuang Formation of West Mountain, Beijing, the Zuoyun Formation in Zuoyun County, Shaanxi and the Tujingzi Formation in Wanquan County, northwest Hebei. The typical ostracod faunas of each formation are given below:

**Xiazhuang Formation**: *Cypridea (Y.) arca* Hou, *C.* aff. *gigantea* Ye, *C. (Ulwellia) fengtaiensis*, *C. (U.) xiazhuangensis*, *C. (U.) wangzuoensis* Pang, *C.* sp. 1 & 2, *L. infantilis*, *Triangulicypris fertilis* Ye, *T. dorsacuta*, *T. postacuta*, *T. subpostacuta* Pang, *Candona rectangulata*, *C. nitidaformis* Pang, *Candoniella candida* Hao, *C. minuta* Ye, *C. subminuta* Pang, *Clinocypris dorsoconvexa*, *Cyclocypris subcalculaformis* Pang, *C. minipisiformis* Pang & Zhang, *Cyprois latirotunda*, *Mongolianella? latireniformis* Pang, *Ziziphocypris simakovi* (Mandelstam), *Timiriasevia principalis* Lubimova, *T. kaitunensis* Liu, *T. miniscula* (Ye), *T. pulchra* (Ho), *T. xiazhuangensis*, *T. dorsarca* Pang, *Darwinula leguminella* (Forbes), *D. longyouensis* Yu, *D. xishanensis*, *D. obliquovata* and *D. longovata* Pang.

**Zuoyun Formation**: *C. (P.) ningxiaensis*, *C. (P.) opima* Qi, *C. zhamagouensis* Pang, *C. elliptica*, *C. postangusta*, *C. zuoyunensis*, *C. occultirostrata*, *C. rectangulata* Pang, *Candona disjuncta* Hao, *C. prona* Su, *Candoniella minuta*, *C. zhanmagouensis* Pang, *Rhinocypris yumenensis* Qi, *R. zuoyunensis* Pang, *Z. simakovi*, *Z. costata* (Galeeva), *Lycopterocypris debilis* Lubimova, *L. placida* Pang, *Clinocypris altidorsalis*, *C. sublongula* Pang, *Typhlocypris arrecta* Chen & Ho and *Paracypretta? zuoyunensis* Pang.

The ostracod assemblage from the Tujingzi Formation of Inner Mongolia is very similar to that of the Zuoyun Formation but is less abundant and diverse.

This entire fauna is characterized by a dominant, diverse and abundant fauna of *Cypridea* and the absence of the Upper Cretaceous *Talicypridea* and is clearly of Lower Cretaceous age. Many of the taxa characterizing the assemblages and subassemblages of the fauna are

of considerable biostratigraphical importance and also occur outside northern China.

The first subassemblage of Assemblage 1 contains 23 species/subspecies of *Cypridea* of which, no less than 8 occur in the Purbeck-Wealden beds of England. Other species, described by Soviet authors, are from the Hauterivian and Barremian of western Siberia and still others are known from the Lower Cretaceous of Mongolia and Gansu Province in China. Although the subassemblage contains some species also known in the uppermost Jurassic, the faunal evidence is overwhelmingly in favour of its being early Lower Cretaceous.

The second subassemblage contains less species common to the English Purbeck beds and more in common with Lower Cretacous faunas from elsewhere in China, Mongolia and Soviet Asia. It is thought to be somewhat younger in age than the first subassemblage.

The second assemblage is much less diverse than the first and contains few of the species of the latter. It contains only three genera (*Cypridea, Lycopterocypris, Darwinula*), of which *Cypridea* is overwhelmingly dominant and *Darwinula*, relatively rare. Some of the dominant species, *C. (P.) parallela*, *C. (C.) concina*, *C. (C.)* aff. *uninoda*) and *C. curtorostrata* are also found in the Xinminpu, Liupanshan and Songhuajiang groups of the early Cretaceous elsewhere in China and there are also some species (*C. guyangensis*, *C. subpolita*, *C. subvalida*) which are very similar to species in the Djunbain Formation of Mongolia (*C. polita*, *C. valida*). Its age is clearly Lower Cretaceous and younger than assemblage 1.

The composition of the 3rd assemblage of this fauna is very different from earlier ones. Although some inherited species belonging to *Cypridea, Lycopterocypris, Triangulicypris, Timiriasevia* and *Mongolialella* occur, they are no longer dominant. Increasingly, through the remainder of the Cretaceous and into the Cainozoic, their place is taken by species of *Candona, Candoniella, Cyprois, Cyclocypris* and *Typhlocypris*. Many of the species of this assemblage occur elswhere in China in strata which is considered to be late Cretaceous. For example, *C. (Y.) arca* was first found ranging from the Fulongquan Formation to the Sifangtai Formation in the Cretaceous of the Songliao Plain, and *C. gigantea* is common in the Sifangtai and Mingshui formations of the same area. There are many other similar examples. Overall the ostracods of this assemblage contain a mixture of both early and late Cretaceous species. However, since it lacks the genus *Talicypridea*, it is considered to be late Lower Cretaceous in age.

## VI. The early Upper Cretaceous *Cypridea-Talicypridea* Fauna

This fauna comes from northern Shaanxi, northwestern Hebei and Inner Mongolia. It is characterized by the presence of the endemic asiatic genus *Talicypridea*. It comprises a single assemblage, the *Cypridea-Talicypridea-Candona* Assemblage which occurs in the Zhumapu Formation of Shaanxi, the Hulaigou Formation of Hebei and the Erliandabusu Formation of Inner Mongolia. The most important species of each of the formations is given below:

**Zhumapu Formation**: *C. (P.) lenta* Hou, *C. (P.) vulgaria* Yang, *C. infedelis* Ye, *C. youyunensis* Zhang, *C. concinaformis* Su, *C. zhumapuensis, C. rectangulata, C. shanxiensis, C. datongensis, C. sublonga, C. unornata, C. shibagouensis, C. tarsorbita, C. lijianpuziensis, C. paucispinata, C. paratereti s* Pang, *Talicypridea gibbera* (Yuan), *T. obesa* Li, *T. tumens, T. exilicristata* Pang, *Candona nitida, C. disjuncta* Hao, *C. declivis* Ruan, *C. subrectangulata, C. subincognata, C. shuiquangouensis* Pang, *Candoniella mordvilkoi* Mandelstam, *C. candida, C.* cf. *minuta, Rhinocypris pulchra* Pang, *R. intermedia* Zhang, *R. hangetanensis, R. unispinata, R. tenuis* Pang, *Z. simakovi* Mandelstam, *Z. costata* (Galeeva), *Mongolianella perlucida* Su, *M. ordinata* Lubimova, *M. panda* Zhang, *M. shanxiensis* Zhang, *Cyprois hangetaensis* Pang, *Cyprinotus* cf. *mohuanensis* Yu, *Hemicyprinotus* aff. *trapeziodes* Ho, *Clinocypris scolia* Mandelstam, *C. subscoliosa* Pang, *Pinnocypridea? orbicularis* Pang, *D. contracta* Mandelstam and *D. zuoyunensis* Pang.

**Hulaigou Formation**: This is a much more impoverished assemblage: *Cypridea youyuensis* Zhang, *C. (P.)* aff. *corpulenta* Yu, *Talicypridea tumens, L. contrita, Rhinocypris* sp. and *D. contracta.*

**Erliandabusu Formation**: *Cypridea tera* Su, *Talicypridea obesa* Li, *Lycopterocypris cuneata* (Tsao), *L. profunda* and *Theriosynoecum? erlianensis* Zhang. The fauna of this formation is virtually the same as that of the Zumapu Formation.

Most of the *Cypridea* species which dominate this fauna are smooth; reticulate or spinose forms and are much less common than in earlier assemblages. Many of these *Cypridea* species are known from Upper Cretaceous deposits elswhere in China, notably in the Songliao and Jianghan plains and in Jiangsu Province. Biostratigraphically, however, the single most important occurrence in this assemblage is *Talicypridea*. This genus is typical of Upper Cretaceous non-marine ostracod communities in Mongolia and China and occurs at the very inception of the fauna, and clearly establishes its age.

## CONCLUSIONS

While it can be seen from the paper that our present knowledge allows the clear recognition of a succession of faunas, assemblages and subassemblages throughout all those parts of the Mesozoic which are represented by suitable ostracod-bearing non-marine facies, it is also obvious that further research is required to refine this basic biostratigraphical scheme. Of particular importance is the precise definition of the Jurassic-Cretaceous boundary on the basis of Ostracoda. Also demanding further attention are problems of more detailed correlation between this area of northern China and basins of Mesozoic non-marine deposition in North West Europe and both European and Asiatic regions of the USSR. Of particular importance in this respect is to establish a more precise chronostratigraphical relationship between the Jurassic and Cretaceous faunas discussed above and those from the classic sections of England.

There is also, within China, considerable room for improvement in correlation between the various regions and also between closely adjacent basins of deposition.

## ACKNOWLEDGEMENTS

The authors wish to thank two anonymous referees for their helpful comments.

## REFERENCES

[Editorial comment: Although the majority of the following references are not cited in the text; they are included as a valuable list of literature relevant to the area.]

Anderson, F. W. 1973. The Jurassic-Cretaceous transition: the nonmarine ostracod faunas. *In* Casey, R. & Rawson, P. F. (Eds), *The Boreal Lower Cretaceous. Geol. Jour.* Special Issue 5, 101-110. Seel House Press, Liverpool.

Anderson, F. W. & Bazley, R. A. B. 1971. The Purbeck Beds of the Weald (England). *Bull. geol. Surv. Gt. Br.*, London, No. 34, 1-173, pls 1-23.

Bate, R. H. 1956. Freshwater ostracods from the Bathonian of Oxfordshire. *Palaeontology*, London, 8(4), 794-756, pls 109-111.

Galeeva, L. I. 1955. *Ostracoda from Cretaceous deposits of the Peoples' Republic of Mongolia.* VNIGNI, Gostoptechizdat, Moscow, 95 pp., 15 pls. (In Russian.)

Gou Yunxian. 1983. Cretaceous ostracods from the Yanbeian area, Jilin Province. *Acta palaeont. sin.*, Beijing, 22(1), 42-45, pls 1-3. (In Chinese.)

Gou Yunxian, Wang Zongzhe, Yang Jiedong & Wan Wensheng. 1986. Cretaceous Ostracoda from the Eren Basin of Inner Mongolia along with sedimentary environments. *In* Nanjing Instit. Geol. and Palaeontol., Academia Sinica and the First Exploration Company, North China Oil Field, Ministry of Oil Industry. (Editoral Board), *Cretaceous ostracod and sporo-pollen fossils from the Eren Basin, Inner Mongolia. Cainozoic-Mesozoic Palaeontology and Stratigraphy of East China*, 2, 1-104, pls 1-34. Anhui Science and Technology Publishing House, Hefei. (In Chinese.)

Hao Yichun, Su Deying, Li Yougui, Ruan Peihua & Yuan Fengtain. 1974. *Cretaceous-Tertiary Ostracoda*, 93 pp. 30 pls. Geological Publishing House, Beijing. (In Chinese.)

Li Yunen. 1983. On the nonmarine Jurassic-Cretaceous boundary in the Sichuan Basin by Ostracoda. *Bull. Chengdu Inst. Geol. M.R., Chinese Acad. Geol. Sci.*, Geological Publishing House, Beijing, 4, 77-89. (In Chinese.)

Lubimova, P. S. 1956. Ostracoda from Cretaceous depoosits of the eastern part of the Peoples' Republic of Mongolia. *Trudy VNIGRI*, Leningrad, 93, 1-174, pls 1-25. (In Russian.)

Lubimova, P. S. 1965. Ostracoda from Lower Cretaceous deposits of the Caspian Depression. *Trudy VNIGRI*, Leningrad, 244, 1-199. (In Russian.)

Lubimova, P. S., Kazmina, T. A., & Reshetnikova, M. A. 1960. Ostracoda from Mesozoic and Cainozoic deposits of the west Siberian Lowland. *Trudy VNIGRI*, Leningrad, 160, 27-125. (In Russian.)

Pang Qiqing. 1982a. Ostracoda. *In* Bureau of Geology of Inner Mongolian Autonomous Region. (Editorial Board), *The Mesozoic stratigraphy and palaeontology of the Guyang coal-bearing basin, Inner Mongolian Autonomous Region, China*, 57-84, pls 7-16. Geological Publishing House, Beijing. (In Chinese.)

Pang Qiqing. 1982b. Middle-Upper Jurassic Ostracoda from the Yanshan region, Hebei. *Journal Hebei College of Geology*, Hebei, 1(2) (summary 17-18), 89(110), pls 1-3. (In Chinese.)

Pang Qiqing. 1984. Fossils and the boundary of the terrestrial Jurassic-Cretaceous system in the Yanshan area, Hebei province. *Journal Hebei College of Geology*, Hebei, 3 (summary 27), 1-16, pls 1-5. (In Chinese.)

Pang Qiqing. 1985. Preliminary discussion on the continental Permian-Triassic boundary in the northern foothills of the Tienshan Mountains, Xinjiang. *Xinjiang Geology*, Xingjiang People's Publishing House, Urumqi, 3(4), 93-98. (In Chinese.)

Pang Qiqing, Zhang Lixian & Wang Qiang. 1984. Ostracoda. *In* Tianjin Institute of Geology and Mineral Resources. (Editorial Board), *Palaeontological Atlas of North China, III Micropalaeontological Volume*, 59-199, pls 13-75. Geological Publishing House, Beijing. (In Chinese.)

Su Deying & Li Yougui. 1981. Ostracoda from the eastern foothills of the Taihang Mountains and the geological age. *Bull. Inst. Geol., Chinese Acad. Geol. Sciences*, Geological Publishing House, Beijing, 3, 110-127, pls 1-2. (In Chinese.)

Su Deying, Li Yougui, Pang Qiqing & Chen Sue. 1980. Ostracoda. *In* Institute of Geology Chinese Academy of Geological Sciences. (Editorial Board), *The Mesozoic stratigraphy and palaeontology of the Shaanxi-Gansu-Ningxia Basin*, II, 48-83, pls 118-119, Geological Publishing House, Beijing.

Whatley, R. C. In press. The reproductive and dispersal strategies of Cretaceous non-marine Ostracoda: the key to pandemism. *Proceedings of the 1st International Symposium on Cretaceous non-marine correlations*, IGCP 245, Urumqi, China (1987).

Yang Renquan. 1981. The fossil ostracod assemblage from the Dabeigon Formation of the Luanping Group, northern Hebei and its chronological significance. *In* Palaeontological Society of China. (Editorial Board), *Selected papers from the 1st Convention of the Micropalaeontological Society of China (1979)*, 76-84, pls 1-2. Science Publishing House, Beijing. (In Chinese.)

Ye Chun-hui. 1984. The sequence of Jurassic-Cretaceous nonmarine ostracod assemblages in China. *Bull. Nanjing Inst. Geol. and Palaeontol., Academia Sinica*, Jiangsu Science and Technology Publishing House, Nanjing, 9, 219-235, pls 1-7. (In Chinese.)

Ye Chun-hui, Gou Yunxian, Hou Youtang & Cao Meizhen. 1977. Mesozoic-Cainozoic ostracod faunas in Yunnan. *In* Nanjing Inst. Geol. and Palaeontol., Academia Sinica. (Editorial Board), *Mesozoic fossils of Yunnan*, 2, 153-330, pls 1-24. Jiangsu Science and Technology Publishing House, Nanjing. (In Chinese.)

Ye Chun-hui, Gou Yunxian, Hou Youtang & Cao Meizhen.

1980. Jurassic and Cretaceous ostracods. *In* Nanjing Inst. Geol. and Palaeontol., Academia Sinica. (Editorial Board),*The classification and correlation of Mesozoic volcanic sedimentary strata in Zhejiang and Anhui Provinces.* 173-220, pls 1-4. Science Publishing House, Beijing. (In Chinese.)

Zhang Lijun. 1985. Nonmarine ostracod faunas of the late Mesozoic in western Liaoning. *In* Zhang Lijun, Pu Ronggan & Wu Hongzhang. (Eds), *Mesozoic stratigraphy and palaeontology of western Liaoning*, 2, 1-90, pls 1-2. Geological Publishing House, Beijing, (In Chinese.)

Zhang Lijun & Zhang Yingju. 1982. Ostracods from the Fuxing Formation (Lower Cretaceous) in the Fuxing Basin, Liaoning. *Acta palaeont. sin.*, Beijing, **21**(3), 362-369, pls 1-2. (In Chinese.)

Zhong Xiachun. 1964. Upper Triassic and Middle Jurassic Ostracoda from the Ordos Basin. *Acta palaeont. sin.*, Beijing, **12**(3), 426-465, pls 1-3. (In Chinese.)

# 19

# The geological and exploration significance of Cretaceous non-marine Ostracoda from the Hailaer Basin, northwestern China

Ye Dequan

Daqing Petroleum Administrative Bureau,
Heilongjiang, China.

## ABSTRACT

Non-marine Ostracoda have been recovered from two levels in the Cretaceous of the Hailaer Basin in northwestern China. The lower level is in the mudstone facies of the Damoguaihe Formation and its ostracod fauna is of rather low diversity and is dominated by *Limnocypridea*, with subordinate *Cypridea* and *Djungarica*. This fauna can be correlated with other Lower Cretaceous faunas in China and Inner Mongolia. The higher level, in the Qingyuongang Formation, also in mudstone facies, yields a fauna dominated by the Upper Cretaceous endemic genus *Talicypridea* with subordinate *Cypridea*. The two faunas are analysed biostratigraphically and palaeoenvironmentally and it is also suggested that the presence of large numbers of Ostracoda can be used to identify source rocks. This hypothesis is supported by evidence from organic geochemistry.

## INTRODUCTION

The Hailaer Basin is situated in Heillongjiang Province in northwestern China, near the border with the USSR. The southern part of the basin is in the Mongolian People's Republic, to the east it is bounded by the River Yiming and its western flank extends just to the west of Lake Hulun. Major geographical features and the position of sampling localities are indicated in Fig. 1.

In early geological surveys of the basin, very few fossils were found. However, since 1984, with increased impetus in hydrocarbon exploration, some 20 wells have been drilled. From these, 2376 samples have been analysed which have yielded an abundant fauna and flora which includes non-marine ostracods, gastropods, angiosperms, charophytes and algae.

Palaeontological evidence has been used to demonstrate that Cretaceous deposits in the Hailaer

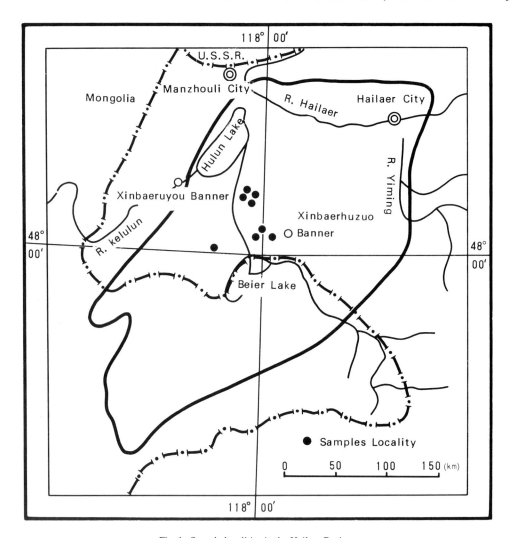

Fig. 1. Sample localities in the Hailaer Basin.

Basin are some 4000m thick and entirely continental in facies. The greater part of this thickness is made up by the early Cretaceous which, however, is very limited in its outcrop. The late Cretaceous is much less well developed, thinner, and largely confined to the southern and western parts of the basin. Because the overlying Cainozoic is thin and poorly developed, the Cretaceous in the basin is not deeply buried.

## STRATIGRAPHY

The Cretaceous rocks of the Hailaer Basin are represented by 5 formations, as shown in Tables 1 and 2. The Beier, Damoguaihe, Yiming and Hulun formations are all early Cretaceous; the Qingyuangang Formation is late Cretaceous. The detailed stratigraphy of the >4000m section is as follows, in stratigraphical order:

### Beier Formation ($K_1d$)

This is a lithologically variable unit, comprising variegated conglomerates and grey sandstones, intercalated with thin-bedded arenaceous conglomerates, silty sandstones and dark mudstones in

Table 1. The ecology and vertical distribution of the biotae from the Hailaer Basin.

| Environ / Strata | Fresh Water — Ostracoda — Cypridea | Limnocypridea | Lycopterocypris | Talicypridea | Gastr. | Sporo-Pollen | Choro. | Land — Plants | Fresh, Brackish — Dinof. | Acrit. |
|---|---|---|---|---|---|---|---|---|---|---|
| Qingyuangang Fm. | — | | | — | | | — | | | |
| Hulun Fm. | | | | | | — | — | | | |
| Yiming Fm. 3rd | | | | | | — | | | | |
| Yiming Fm. 2nd | | | | | | — | | | | |
| Yiming Fm. 1st | | | | | | — | | | — | — |
| Damoguaihe Fm. 4th | | | | | | — | | — | — | — |
| Damoguaihe Fm. 3rd | | | | | — | | | | — | — |
| Damoguaihe Fm. 2nd | — | — | — | | | | | | — | — |
| Damoguaihe Fm. 1st | | | | | | — | | | | |
| Beier Fm. | | | | | | — | | | | |

Table 2. Cretaceous stratigraphy of the Hailaer Basin.

| System | Series | Formation | Member | Basin evolutional stages |
|---|---|---|---|---|
| K | K₂ | Qingyuangang Fm. | | Uplifting |
| | K₁ | Hulun Fm. | | Shrinking |
| | | Yiming Fm. | 3rd | Subsiding |
| | | | 2nd | |
| | | | 1st | |
| | | Damoguaihe Fm. | 4th | Subsiding |
| | | | 3rd | |
| | | | 2nd | |
| | | | 1st | |
| | | Beier Fm. | | Fault-subsiding |

the Wuersun Depression; degraded andesites and red mudstones in the Beier Depresssion and basalt in the Hongqi Depression. Everywhere it rests unconformably on the metamorphosed Palaeozoic basement. Its thickness ranges between 596 and 1233m and it has yielded spores and pollen.

**Damoguaihe Formation (K₁d)**

This rests unconformably on the underlying Beier Formation. It is 630-1700m thick and is divisible, from bottom to top, into the following members:
**Member 1. (K₁d₁).** Dark mudstones, oil shales and argillaceous silty sandstones intercalated with medium to thick-bedded silty sandstones and yielding spores and pollen. 0-472.5m thick.
**Member 2. (K₁d₂).** The lower part comprises grey-white fine to medium grained sandstones intercalated with dark mudstones and thin-bedded coals and mottled arenaceous conglomerates. Above this are grey silty sandstones which are

interbedded and intercalated within both thick and thin-bedded mudstones. The latter yield: *Limnocypridea grammi, L. abscondida, Cypridea* sp., *Lycopterocypris* sp. and *Djungarica* sp. as well as spores, pollen, dinoflagellates and 'acritarchs'. 286-526m thick.
**Member 3. (K₁d₃).** This member comprises dark grey mudstones, silty mudstones and various silty sandstones which are both oil and coal-bearing. Because of the lateral stability and vertical cyclicity of the facies, this member is an excellent source and resevoir unit. It has yielded gastropods (Viviparidae), spores, pollen, dinoflagellates and 'acritarchs'. 158-562m thick.
**Member 4. (K₁d₄).** Dark grey and grey silty to sandy mudstones, muddy siltstones and silty sandstones with intercalated coals. Numerous macro plant fossils (*Pagiophyllum, Taenipteris, Cladophlebis*), spores, pollen, dinoflagellates and 'acritarchs'. 150-431m thick.

Table 3. Ostracod abundance.

| System | Series | Formation | Fossil name and quantity | |
|--------|--------|-----------|---------------------------|---|
| K | Upper K₂ | Qingyuangang Fm. | Talicypridea amoena | ■ |
| | | | T. elevata | ▦ |
| | | | T. gibbera | ▨ |
| | | | T. parvita | ▦ |
| | Lower K₁ | Hulun Fm. | T. sp. | ▢ |
| | | | Cypridea gigantea | ▨ |
| | | | C. apiculata | ▦ |
| | | | C. eximia | ▦ |
| | | Yiming Fm. | C. aff. cavernosa | ▩ |
| | | | Candona glaber | ▦ |
| | | | C. longa | ▦ |
| | | | Mongolianella perlucida | ▨ |
| | | | Harbinia hapla | ▨ |
| | | Damoguaihe Fm. | Ziziphocypris sp. | ▦ |
| | | | Limnocypridea grammi | ▦ |
| | | Beier Fm. | L. obscondida | ▦ |
| | | | Cypridea sp. | ▦ |
| | | | Djungarica sp. | ▦ |

Fossil quantity

| ▢ < 5 | ▦ 5 - 10 | ▨ 10 - 20 | ▩ 20 - 35 | ■ > 35 |
|-------|----------|-----------|-----------|--------|

## Yiming Formation (K₁y)

Depending on the locality, this may be conformable or unconformable on the underlying Damoguahaihe Formation. It is divided into the following three members, in stratigraphical order:

**Member 1. (K₁y₁).** This member comprises dark grey mudstones, muddy silts and sandy mudstones and grey-green silty sandstones and intercalated coal seams. It yields spores, pollen, dinoflagellates and 'acritarchs'. 251-558m thick.

**Member 2. (K₁y₂).** Similar to the underlying member in lithology but also with grey-white sandstones and arenaceous conglomerates. Pollen and spores. 257-419m thick.

**Member 3. (K₁y₃).** Comprising grey and green-grey mudstones, muddy silts and silty sandstones with spores and pollen. 261-470m thick.

## Hulun Formation (K₁h)

This everywhere rests unconformably on the underlying Yiming Formation. It consists of variegated conglomerates which are intercalated within purple-red and grey-green mudstones. The formation yields charophytes (*Eunclistochara mundula*, *Atopochara* sp.), pollen and spores. Ranging up to 275m in thickness.

## Qingyuangang Formation (K₂q)

This rests either disconformably or unconformably on the Hulun Formation and also has a disconformable relationship with the overlying Tertiary. It is composed mainly of pink, red and grey silty and sandy mudstones and muddy silty sandstones which occur in a rythmic order and which are interbedded with sandstones and conglomerates. The argillaceous units are very calcareous and yield very abundant Ostracoda such as: *Talicypridea amoena*, *T. elevata*, *Harbinia* aff. *hapla*, *Cypridea cavernosa*, *C. bianzhaoensis*, *C. gigantea* and also the charophytes *Atropochara ulaensis*, *Otusochara altamulaensis*, *Aclistochara jilinensis*. Ranging up to 334m in thickness.

## THE OSTRACOD ASSEMBLAGES

Previous studies of the Ostracoda of the Hailaer Basin found only a relatively small fauna from the Qingyuangang Formation. Recently, however, very large numbers of ostracods have been recovered from not only the Qingyuangang Formation but also the Damoquaihe Formation (Table 3). The ostracods occur as two morphologically distinct assemblages:

### 1) *Limnocypridea* Assemblage

This assemblage occurs in dark coloured mudstones in the 2nd member of the Damoquaihe Formation. The fauna is dominated by the genus *Limnocypridea*, especially the species *L. grammi* and *L. abscondida*. Several species of the genus *Cypridea* and *Djungarica* sp. complete the fauna. These species have mainly smooth carapaces and are well preserved. They include both large and smaller species and the Ostracoda occur together with an abundant associated fauna and flora of bivalves (*Unio* sp., *Ferganoconcha* sp.), gastropods (*Viviparus*), dinoflagellates, 'acritarchs',

spores and pollen.

### 2) *Talicypridea* Assemblage

This assemblage occurs in the pink, purplish-red and grey-green mudstones. It differs from the *Limnocypridea* Assemblage in its much higher diversity of Ostracoda (10 genera, 28 species) and in that the small genus *Talicypridea* is the dominant taxon. Another difference is that most of the species are ornamented rather than smooth, many bearing protuberances, nodes or with a honeycomb, reticulate or 'fingerprint' ornament. The dominant species are: *Talicypridea amoena*, *T. elevata*, *T. angusta*, *T. gibbera*, *T. parvita* and *T.* sp. coexisting with numerous species of *Cypridea*, such as *C. apiculata*, *C. eximia* and *C.* aff. *cavernosas*. Most of these are large species with honeycomb or reticulate ornament. The remainder of the fauna comprise a mumber of well-preserved, mainly smooth and smaller species, such as: *Candona glober*, *C. longa*, *Candoniella* sp., *Mongolianella perlucida*, *Ziziphocypris* sp, and *Harbinia hapla*.

## THE APPLICATION OF THE OSTRACODA

The relatively recent discovery of large faunas of ostracods is due to the large scale nature of hydrocarbon exploration in the Hailaer Basin. This discovery has important biostratigraphical and palaeoenvironmental implications:

**Biostratigraphy**. The principal species of the *Limnocypridea* Assemblage from the Damoguaihe Formation, *L. grammi* and *L. abscondida* also occur in dark-coloured mudstones of the middle part of the Zhunayin Formation in the Kelulun and Tugolige regions of Mongolia and also from the upper parts of the Jiufotang Formation in western Liaoning Province. They are also recorded from the upper part of the Saihantala Formation of the Bayehua Group in the Erlian Basin and *L. grammi* occurs in the early Cretaceous of the Zhungeer Basin, Xinjiang Province in western China. Many of the other species in this assemblage, the gastropods and the associated micro and macro flora are also of common occurrence in the early Cretaceous of China.

Table 4. The correlation of the Hailaer Basin with its adjacent areas.

| | | Hailaer Basin | Songliao Basin | Erlian Basin | Liaoxi Basin |
|---|---|---|---|---|---|
| K | K$_2$ | | | Minshui Fm. | Erliandabusu Fm. | |
| | | Qingyuangang Fm. | Sifangtai Fm. | | |
| | | | Nenjiang Fm. | | |
| | | | Yaojia Fm. | | |
| | | | Qingshangou Fm. | | |
| | | | Quantou Fm. | | Shunjiawan Fm. |
| | | Hulun Fm. | Denglouku Fm. | | |
| | K$_1$ | Yiming Fm. | Yingcheng Fm. | Saihantala Fm. | Fuxin Fm. |
| | | Damoguaihe Fm. | Shahezi Fm. | Duhonmo Fm. | Shahai Fm. |
| | | | | Tenggeer Fm. | Jiufotang Fm. |
| | | Beier Fm. | Huoshiling Fm. | Aershan Fm. | Yixian Fm. |

The genus *Talicypridea* is endemic to China and Mongolia in the Upper Cretaceous. The principal species of the *Talicypridea* Assemblage from the Qingyuangang Formation (*T. amoena, T. elevata, T. gibbera, T. parvita, Cypridea gigantea, C.* aff. *cavernosa, C. bianzhaoensis, Candona glaber, C. longa, Mongolianella perlucida*) are all common to the late Cretaceous of the Songliao Basin (Table 4). The assemblage is also found in other major late Cretaceous basins in China, notably the Zhungeer Basin, the Jianghan Basin, Nanxiong Basin, the Erliandabusu Formation in the Erlian Basin and the Shanginshade Formation in the eastern Mongolian People's Republic. *Harbinia hapla* is confined to the Upper Cretaceous of China. Similarly, the charophytes of the Qingyuangang Formation are widespread throughout the late Cretaceous of China.

**Palaeoenvironmental reconstruction.** The palaeoecology of the ostracods of the Damoguaihe Formation suggests that at the time of its deposition, the area was a very large permanent lake. The ostracods are encountered in dark-coloured, fine-grained mudstones deposited in quiet, low energy conditions. These mudstones are both thick and widely distributed throughout the basin; attesting both to the large size and the stability of the lake.

Above the mudstones, the facies changes rapidly and the grain size of the sediments increases and their colour changes progressively from dark to light. Evidently, within the same lake the environment changed from deep to littoral, either due to its silting up or to the elevation of its bed.

Most of the ostracods are smooth and are thought to be adapted to quiet, low energy conditions in relatively deep water on a soft substrate which typified the depositional environment of the 2nd member of the Damoduaihe Formation. Their absence from the overlying 3rd and 4th members of the formation was probably due to the increase in energy levels and in the grain size of the sediments which went hand in hand with the shallowing of the lake. The presence in the two upper members of dinoflagellates and 'acritarchs' indicates that conditions were not inimical to all life.

The more ornate ostracods of the *Talicypridea* Assemblage in the Qingyuangang Formation, while also occurring in fine grained mudstones, probably lived under somewhat higher energy and shallower water conditions than those of the preceding assemblage.

**Exploration significance.** Large accumulations of ostracods can be indicative of concentrations of organic oil-generating material. Those beds in which ostracods are concentrated in the Cretaceous rocks of the Hailaer Basin may, therefore, be high in organic carbon. According to the geochemical index, the mean level for organic carbon is 1.77%, Chloroform 'A' 0.05% and residual

Table 5. The organic richness of the Damoguaihe Formation in the Hailaer Basin.

| Member | Organic Carbon (%) | Chloroform 'A' (%) | Hydrocarbons (ppm) |
|---|---|---|---|
| 4th | 1.96 | 0.0634 | 336 |
| 3rd | 2.86 | 0.0646 | 319 |
| 2nd | 2.50 | 0.1010 | 495 |
| 1st | 2.49 | 0.1576 | 1106 |

Table 6. The various types of organic matter found in the Damoguaihe Formation of the Hailaer Basin.

| Member | H/C ratio | O/C ratio | Hydrogen index mgHC/gc | Oxygen index mgCO$_2$/gc | Kerogen Identif. |
|---|---|---|---|---|---|
| 4th | 0.74 | 0.12 | 85 | 56 | II - - III |
| 3rd | 0.72 | 0.10 | 81 | 26 | III - - II |
| 2nd | 0.94 | 0.09 | 280 | 25 | II - - II |
| 1st | 0.92 | 0.06 | 251 | 34 | II - - II |

carbon is 1.77%, Chloroform 'A' 0.05% and residual hydrocarbon 0.06-1.35%. By these standards, the Damoguaihe Formation is a good source rock (Table 5).

The type of kerogen (saprohumolite or humosaprolite) is also important in assessing the potential of a source rock. The ostracod-bearing 2nd member of the Damoguaihe Formation on the basis of its dark organic rich deep lacustrine mudstone facies and favourable kerogen type is an ideal source rock. The 3rd member, however, made up of mudstones of a swampy fluvial facies, although it has a higher organic content, contains a poorer type of kerogen and has less potential as a source rock (Table 6).

## CONCLUSIONS

The importance of the discovery of large numbers of non-marine Cretaceous ostracods in two levels in the Hailaer Basin is not confined to biostratigraphical considerations. The faunas are also of great importance in reconstructing the environment of deposition and in assessing the likely source rock potential of these beds.

## ACKNOWLEDGEMENTS

The author wishes to thank the two anonymous referees for their very many useful comments and also the editors who have greatly improved the paper.

## REFERENCES

[Editorial comment: The following literature is relevant to this study, although it is not cited in the text.]

Editorial and author group of areal stratigraphy scale of Inner Mongolia. 1978. *Areal stratigraphy scale of northern China, Part for Inner Mongolia*, Geological Publishing House, Beijing. (In Chinese.)

Editorial and author group of areal stratigraphy scale of Helongjiang Province. 1979. *Areal stratigraphy scale of the northeastern district of China, Part for Helongjiang Province*. Geological Publishing House, Beijing. (In Chinese.)

Institute of Development, Daqing Oilfield. 1976. *Ostracod fossils of the Cretaceous in the Sonliao Basin.* Science Press. (In Chinese.)

Ye Dequan. 1983. Cretaceous ostracods from the Sonliao Basin and their significance. *Petroleum Geology and Development in Daqing,* **2**(1). (In Chinese.)

Ye Dequan, Zhao Zhuanben & Zhang Ying. 1980. Palaeontological features of the Cretaceous continental facies in the Sonliao Basin. *Acta Petroleum (Supplement).* (In Chinese.)

**Plate 1**

(All specimens from the Qingyuangang Formation, Upper Cretaceous.)

Figs 1-2.   *Cypridea* sp.
Fig. 1.     Right lateral view, X60.
Fig. 2.     Dorsal view, X60.

Figs 3-4.   *Cypridea eximia* Ye.
Fig. 3.     Left lateral view, X50.
Fig. 4.     Dorsal view, X50.

Figs 5-6.   *Talicypridea* sp.
Fig. 5.     Right lateral view, X50.
Fig. 6.     Dorsal view, X50.

Figs 7-8.   *Talicypridea bianzhaoensis* Ye.
Fig. 7.     Right lateral view, X60.
Fig. 8.     Dorsal view, X60.

Figs 9-10.  *Harbinia* aff. *hapla* Tsao.
Fig. 9.     Right lateral view, X50.
Fig. 10.    Dorsal view, X50.

Figs 11-12. *Talicypridea* aff. *amoena* Liu.
Fig. 11.    Right lateral view, X60.
Fig. 12.    Dorsal view, X60.

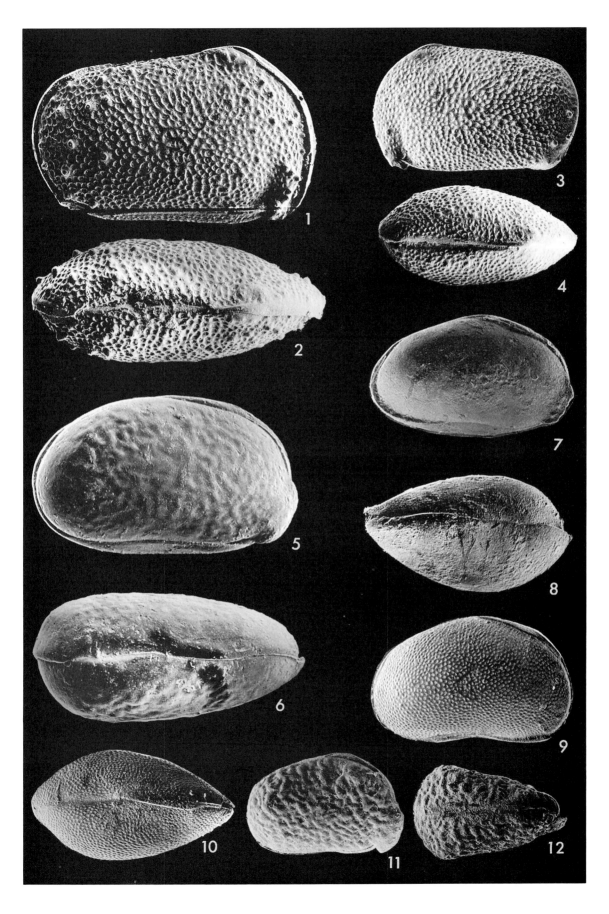

# 20

# A review of the Sarmatian Ostracoda of the Vienna Basin

Jaromír Zelenka

Czechoslovakian Geological Survey, Prague.

## ABSTRACT

Based on the ostracod fauna, the Sarmatian is subdivided into the *Cytheridea hungarica -Aurila mehesi* Assemblage Zone and the *Aurila notata* Total Range Zone. The latter zone, can be subdivided into the *Aurila notata-Cyamocytheridea leptostigma leptostigma* Acme Zone and the *Hemicytheria hungarica-Leptocythere cejcensis* Subzone. A list of the typical ostracod fauna for each biozone is presented. The ostracod zonal division correlates well with zones based on foraminiferal fauna and can be distinguished throughout the entire control area of the Paratethys.

## INTRODUCTION

The Vienna Basin is a part of the central Paratethys, which represents marine areas of Neogene sedimentation in the wider region of the Carpathians, Eastern Alps and Dinarides. The Sarmatian stage is one of the most typical stages of the Paratethys and it coincides with the beginning of the isolation of Paratethys from other marine influences and its evolution within enclosed basins with the consequent typical endemic fauna.

The original standard division of the Sarmatian was made on the basis of Mollusca and Foraminiferida. Subsequently, the Ostracoda have been shown to be of great importance in the biostratigraphy of this stage. The new ostracod zonation of the Sarmatian of the Vienna Basin is discussed and a possible application to the entire central Paratethys is outlined in the present paper.

## PREVIOUS WORK

The first palaeontologist to study Sarmatian ostracods was Reuss in his classic work (1850) on Neogene ostracods of the Austro-Hungarian Empire, in which he described the following species: *Aurila notata, Aurila hispidula, Hemicytheria omphalodes omphalodes, Cyamocytheridea leptostigma leptostigma, Leptocythere tenuis, Xestoleberis pilosella* and *Xestoleberis tumida*. The next work was that of Héjjas (1894) in which *Hemicyprideis dacica dacica* was first described.

Méhes (1908) also made an important contribution to the knowledge of the ostracod fauna, and described *Cyprideis pannonica, Callistocythere egregia, Leptocythere naca* and *Loxoconcha kochi*. Zalányi (1913) described the following important species: *Aurila mehesi, Aurila merita, Cytheridea hungarica, Miocyprideis sarmatica, Senesia vadaszi* and *Cytheridea gigantea* together with others. These ostracods described by Héjjas, Méhes and Zalányi from the Pannonian Basin are also abundant in the Sarmatian deposits of the Vienna Basin.

Kollman (1960) lists the Sarmatian localities in Austria and described two new species, *Miocyprideis janoscheki* and *Cyamocytheridea leptostigma foveolata*. Cernajsek (1971; 1972; 1974) in his study of Austrian Sarmatian described *Aurila kollmanni, Loxoconcha schmidi* and *?Bythocypris pappi* (= *Phlyctenophora farkasi*).

Sarmatian ostracods recorded from the Vienna Basin are mentioned in Dornic & Kheil (1963) from Czechoslovakia. Jirícek (197;, 1972; 1974; 1975a; 1975b; 1983) utilized the ostracods in biostratigraphical studies of the Sarmatian. Some of his taxonomic determinations, however, now require revision. Based on the ostracod fauna, Jirícek divided the Sarmatian into several biozones. This author, in his various papers, frequently changed the range, number and the fauna content of these biozones. The present author (Zelenka, 1989; in press a) investigated the ostracod fauna of the Sarmatian. In this extensive work the author describes in detail all species occurring at many outcrops and in boreholes of this stage from the Czechoslovakian and Austrian parts of the Vienna and Danube basins. A list of all these species is presented for each locality. Based on the ostracod fauna, the new stratigraphical division of the Sarmatian was proposed and new biozones erected. These biozones correlate well with the zones based on foraminifera. The main results are presented in this paper.

**BIOSTRATIGRAPHY**

The Sarmatian ostracod faunas of the Vienna Basin are quite distinct from those of the marine Badenian and the lower brackish to freshwater Pannonian. The brackish Sarmatian sea was characterized by endemic species, low species diversity and high incidence of individuals.

The fauna of Sarmatian sediments (mainly Mollusca, Foraminiferida) indicates a relatively shallow sea, which probably did not exceed one hundred metres in depth. This is confirmed by the known ecology of the ostracod genera *Aurila, Hemicytheria, Leptocythere, Callistocythere, Cytheridea, Cyamocytheridea* and *Cyprideis*.

The majority of Sarmatian mesohaline to pliohaline ostracods are usually thick-shelled, weakly ornamented, often punctate. There are exceptions, however, such as *Hemicytheria omphalodes omphalodes, Loxoconcha kochi* and *Leptocythere naca*, which are species with stronger ornamentation. Species occurring in the 'lower part of the Sarmatian are more strongly calcified than those in the upper part'.

The Sarmatian of the Vienna Basin was divided by Grill (1943) into three biozones on the basis of foraminifera, from the base to the top these are:
1) Zone of large elphiids (the *Elphidium reginium* Zone) of 'lower' Sarmatian *age.*
2) *Elphidium hauerinum* Zone of 'middle' Sarmatian age.
3) *Nonion granosum,* now *Porosononion subgranosum* Zone of 'upper' Sarmatian age.

Based on Ostracoda a new stratigraphical division of the Sarmatian is presented.
1) *Cytheridea hungarica-Aurila mehesi* Assemblage Zone (equivalent to the *Elphidium reginum* Zone). The characteristic ostracod species of this zone are *Aurila mehesi* and *Cytheridea hungarica.* Also occurring are *Aurila jollmanni, Aurila merita, Sensia vadaszi, Cytheridea gigantea* and *Cyamocytheridea leptostigma foveolata.*
2) *Aurila notata* Total Range Zone which is equivalent to the *Elphidium hauerinum* Zone and *Porosonion subgranosum* Zone. The *Aurila notata* Total Range Zone is informally divided into two parts. The lower part of the *Aurila notata* Total Range Zone is equivalent to the *Elphidium hauerinum* Zone. Ostracods are relatively scarce in this lower part. Species of *Leptocythere, Callistocythere* and *Xestoleberis* occur and *Aurila notata, Cymocytheridea leptostigmaleptostigma, Loxoconcha porosa* appear

for the first time. The upper part is equivalent to the *Porosononion subgranosum* Zone and within it the *Aurila notata-Cyamocytheridea leptostigma leptostigma* Acme Zone and the *Hemicytheridea hungarica-Leptocythere cejcensis* Subzone can be distinguished. The *Aurila notata-Cyamocytheridea leptostigma leptostigma* Acme Zone is equivalent to the lower part of *Porosononion subgranosum* Zone.

The *Aurila notata* Range Zone is rich in ostracods and it is recognized by the abundant occurrence of *Aurila notata* and *Cyamocytheridea leptostigma leptostigma*. Other prominent species are *Aurila hispidula, Cyprideis pannonica, Cyprideis pokornyi, Miocyprideis janoscheki, Miocyprideis sarmatica, Loxoconcha kochi, Loxoconcha porosa, Leptocythere naca, Phlyctenophora farkasi* and *Cytheretta dadayi*.

The *Hemicytheria hungarica-Leptocythere cejcensis* Subzone is equivalent to the upper part of the *Porosononion subgranosum* Zone. *Hemicytheria hungarica* and *Leptocythere cejcensis* characterize this subzone and *Miocyprideis janoscheki, Cyprideis pannonica, Cyprideis pokornyi* and *Leptocythere naca* are also typical. The following species range throughout the Sarmatian: *Hemicytheria omphalodes omphalodes, Hemicypridea dacica dacica, Leptocythere tenuis, Callistocythere egregia, Xestoleberis pilosella* and *Xestoleberis sera*.

By comparison with the works of other authors, such as Choczewski (1956), Sheremeta (1961), Krstic (1963), Széles (1963), Buryndina (1974), Galic (1968), Krstic & Obradovic (1980), all of whom are concerned with Sarmatian ostracod faunas from other parts of the central Paratethys, it appears that the proposed ostracod biostratigraphy is applicable throughout the entire central Paratethys area.

## ACKNOWLEDGEMENTS

The author wishes to thank Professor Robin Whatley and Dr Caroline Maybury and two anonymous referees for reading the manuscript and for making valuable suggestions and corrections. The author also wishes to acknowledge Dr Fred Rögl of the Natural History Museum, Vienna, who made it possible to study Reuss and Kollmanns' original specimens, and Dr Andrea Korecz of Hungarian Geological Survey, Budapest, who allowed him to study Zalányi and Méhes' classical collections. The author is also grateful to Dr Jan Kulich of Charles University, Prague, for his help in producing the S.E.M. photographs. The author also wishes to thank the organizing committee of the 10th International Symposium for the assistance that enabled him to attend the Symposium.

## REFERENCES

Buryndina, L. V. 1974. Nekotoryje novyje vidy ostrakod iz sarmatskich otlozenij Zakarpat´ja. (On some new species of ostracods from Sarmatian deposits of the Transcarpathians.) *Paleont. sbor.*, Lvov, **11**(2), 67-70.

Cernajsek, T. 1971. Die Entwicklung und Abgrenzung der Gattung *Aurila* Pokorný (1955) im Neogen Österreichs. *Verh. Geol. Bundesanst.*, Wien, **3**, 571-575.

Cernajsek, T. 1972. Zur Palöekologie der Ostrakodenfaunen am Westrand des Wiener Beckens. *Verh. Geol. Bundesanst.*, Wien, **2**, 237-246.

Cernajsek, T. 1974. Die Ostracodenfaunen der Sarmatischen Schichten in Österreich. *In* Papp, A., Marinescu, F., Senes, J. *et al.* (Eds), *Chronostratigraphie und Neostratotypen, Bd. IV, Miozän M₅, Sarmatien. VEDA, vydavatelstvo SAV,* Bratislava, 458-491.

Choczewski, J. 1956. Malzoraczki sarmatu dolnego w Dwikozach kolo Sandomierza. (Ostracoda of Lower Sarmat at Dwikozy near Sandomierz.) *Rocz. Pol. Tow. geol.*, Kraków, **25**, 55-87.

Dornic, J. & Kheil, J. 1963. Príspevek k mikrobiostratigrafii a tektonice severozápadních okrajových cásti Vídenské pánve a tzv. hradistského príkopu. (Ein Beitrag zur Mikrobiostratigraphie und Tektonik der NW-Randteile des Wiener Beckens und des sog. Uherské Hradiste - Grabens.) *Sbor. geol. Ved.*, Geol., Praha, **3**, 85-107.

Galic, N. 1968. Tortonska i donjosarmatska mikrofauna sire okoline Koceljeva (zapadna Srbija). (La microfaune Tortonienne et Sarmatienne inférieure des environs plus larges de Koceljevo (Serbie occidentale.) Vesnik, Ser. A, *Zav. geol. geofiz. Istraz.*, Beograd, **26**, 229-242.

Grill, R. 1943. Über mikropaläontologische Gliederungsmöglichkeiten im Miozän des Wiener Beckens. *Mitt. Reichsamts Bodenforsch.*, Zweigst., Wien, **6**, 33-44.

Héjjas, E. 1894. Neue Beiträge zur fossilen Ostracodenfauna Siebenbürgens. *Értesitö erdél. muz.-egyl. orvos-természettudom. Szakostalyabol*, Koloszvar, **19**, 99-112.

Jirícek, R. 1970. Stratigraciceskie i tektoniceskie korreljacii na granice torton-sarmat Paratetidy. *In Buglovskie sloi miocena.* (Materially Vsesojuznogo simpoziuma. Lvov, 6 - 16 sentjabrja 1966 g.), Naukova dumka, Kiev, 152-176.

Jirícek, R. 1972. Problém hranice sarmat/panon ve Vídenské,

**Plate 1**

Fig. 1.     *Aurila mehesi* (Zalányi), right valve, external view, X56. Borehole Cunin - 1 (250.7 - 258.3m).

Fig. 2.     *Aurila hispidula* (Reuss), left valve, external view, X60. Borehole Mikulcice - Tessice Cf - 20 (132.4 - 134.0m).

Figs 3-4.   *Hemicytheria omphalodes omphalodes* (Reuss).
Fig. 3.     Left valve, internal view, X64. Borehole Bucany - 97 (396.0 - 400.0m).
Fig. 4.     Left valve, external view, X64. Borehole Bucany - 97 (396.0 - 400.0m).

Fig. 5.     *Aurila merita* (Zalányi), left valve, external view, X64. Locality Podivín.

Figs 6-7.   *Senesia vadaszi* (Zalányi).
Fig. 6.     Carapace, ventral view, X70. Borehole Holíc H III - K8 (without determination of depth).
Fig. 7.     Carapace, left lateral view, X70. Borehole Holíc H III - K 8 (without determination of depth).

Figs 8-9.   *Loxoconcha porosa* Méhes.
Fig. 8.     Right valve, internal view, X90. Locality Cejc.
Fig. 9.     Left valve, external view, X86. Locality Cejc.

Figs 10-11. *Leptocythere naca* (Méhes).
Fig. 10.    Carapace, left lateral view, X132. Borehole Spacince-4 (374.0 - 377.0m).
Fig. 11.    Carapace, dorsal view, X132. Borehole Spacince-4 (374.0 - 377.0m).

Fig. 12.    *Aurila notata* (Reuss), left valve, external view, X66. Borehole Alt Lichtenwarth - 9 (645.0 - 650.0m).

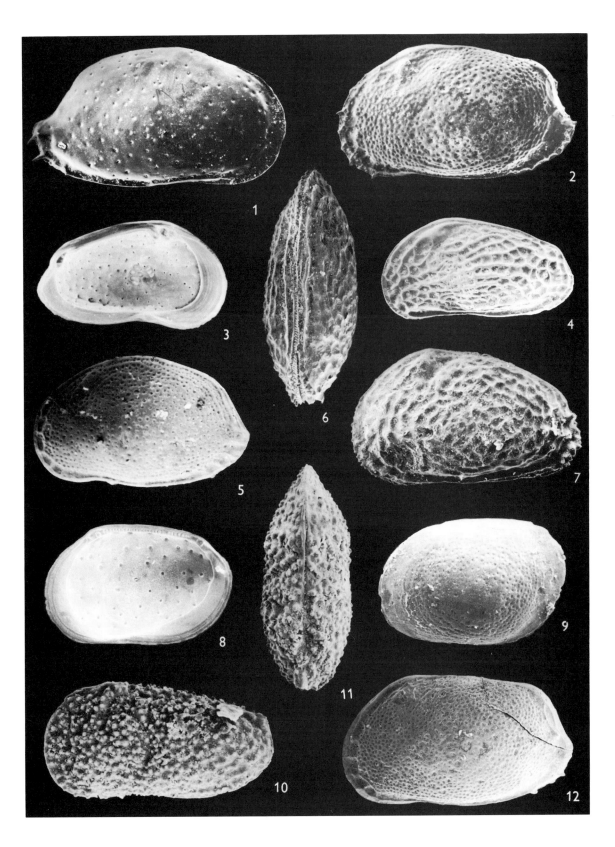

**Plate 2**

Figs 1-3.    *Cyprideis pannonica* (Méhes).
Fig. 1.      Left valve, external view, X67. Borehole Cunín Cf - C  6 (55.0 - 58.0m).
Fig. 2.      Right valve, external view, X67. Borehole Cunin Cf - C6 (55.0 - 58.0m).
Fig. 3.      Left valve, internal view, X67. Borehole Alt Lichtenwarth - 9 (645.0 - 650.0m).

Fig. 4.      *Leptocythere cejcensis* Zelenka, left valve, external view, X106. Locality Cejc.

Figs 5-6.    *Hemicyprideis dacica dacica* (Héjjas).
Fig. 5.      Right valve, external view, X68. Borehole Bucany 60 (407.0 - 411.0m).
Fig. 6.      Left valve, internal view, X68. Borehole Bucany 60 (407.0 - 411.0m).

Figs 7-8.    *Miocyprideis sarmatica* (Zalányi).
Fig. 7.      Right valve, external view, X70. Borehole Hodonín - Nesyt N - 10 (170.0m).
Fig. 8.      Left valve, external view, X74. Borehole Hodonín - Nesyt N - 10 (170.0m).

Figs 9-10.   *Miocyprideis janoscheki* Kollmann.
Fig. 9.      Left valve, external view, X80. Borehole Hodonín Cf - 4 (60.0 - 61.0m).
Fig. 10.     Right valve, internal view, X80.  Borehole  Alt Lichtenwarth - 9 (645.0 - 650.0m).

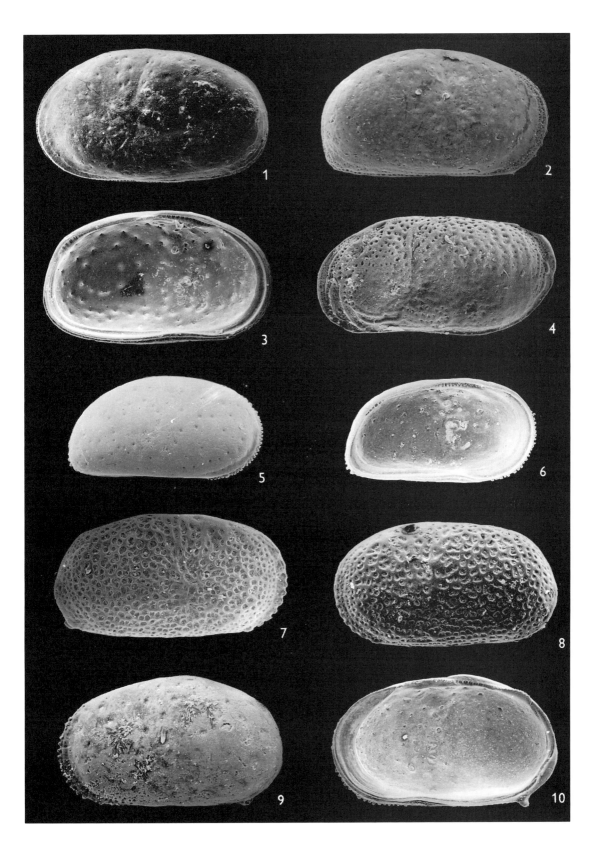

Podunajské a východoslovenské pánvi. (Das Problem der Grenze Sarmat/Panon in dem Wiener Becken, dem Doanubecken und dem ostslowakischen Bechen). *Miner. slov.*, Spisská Nová Ves., **4**(14), 39-81.

Jiríček, R. 1974. Biostratigraphische Bedeutung der Ostracoden des Sarmats s. str. *In* Papp, A., Marinescu, F., Senes, J. *et al.* (Eds), *Chronostratigraphie und Neostratotypen, Bd. IV. Miozän M₅, Sarmatien*, 434-457. VEDA, vydavatelstvo SAV, Bratislava.

Jiríček, R. 1975a. Biozonen der Zentralen Paratethys. *VI. Congr. RCMNS, Bratislava 1975*, Nafta - Gbely, 1-16.

Jiríček, R. 1975b. Ostracod zones in the Neogene of the Central Paratethys. *In* Cicha, I. (Ed.), *Biozonal division of the Upper Tertiary basins of the Eastern Alps and West Carpathians.* Proc. 6. Congr. RCMNS, Bratislava 1975, 57-69. ÚÚG Praha.

Jiríček, R. 1983. Redefinition of the Oligocene and Neogene ostracod zonation of the Paratethys. *Miscellanea micropalaeont.*, Knihovnicka Zem. Plynu Nafty, Hodonín, **4**, 195-236.

Kollmann, K. 1960. Cytherideinae und Schulerideinae n. subfam. Ostracoda aus dem Neogen des östlichen Österreich. *Mitt. Geol. Gesell.*, Wien, **51**, 1958, 89-195.

Krstic, N. 1963. Prethodno saopstenje o vertikalnom resprostranjenju ostrkoda u neogenu Sribje. (Vorläufige Mitteilung über die Verteilung der Ostrakoden im Neogen Serbiens.) *Zap. Srp. geol. Drus.*, za 1960 i 1961 g., Beograd, 169-171.

Krstic, N. & Obradovic, J. 1980. Osladjeni Miocenski sedimenti Stare Karaburme u Beogradu. (Freshwater Miocene sediments of Stara Karuburma a part of Belgrade.) *Symposium de Géol. Régionale et Paléont.*, 399-413. Rudarsko-geoloski fak. Beograd.

Méhes, G. 1908. Beiträge zur Kenntnis der pliozänen Ostracoden Ungarns. II. Die Darwinulidaeen und Cytheridaeen der unterpannonischen Stufe. *Földt. közl.*, Budapest, **38**, 7-10, 601-635.

Reuss, A. E. 1850. Die fossilen Entomostraceen des österreichischen Tertiärbeckens. *Haidinger's Naturwiss. Abh.*, Wien, **3**, 41-92.

Sheremeta, V. G. 1961. Nekotoryc novye vidy ostrakod iz sarmatskich i pannonskich otlozenij zakarpat´ja. *Paleont. Sbor.*, Lvov, **1**, 113-120.

Széles, M. 1963. Szarmáciai és pannóniai korú kagylósrákfauna a Duna-Tisza közi sékely- és mélyfúrásokból. (Sarmatische und pannonische Ostrakodenfaunen aus Bohrungen zwischen Donau und Theiss.) *Földt. Közl.*, Budapest, **93**(1), 106-116.

Zalányi, B. 1913. Miocäne Ostrakoden aus Ungarn. Mitt. Jb. kgl. Ung. geol. Reichanst., Budapest, **21**(4), 85-152.

Zelenka, J. 1989. Ostrakodová fauna sarmatu vídenské a podunajské pánve. *MS Geological Survey*, Praha, 1-186, 47 pls.

Zelenka, J. In press a. Príspevek k poznání ostrakodové fauny sarmatu vídenské pánve. *Zpr. geol. Vyzk.*, v Roce 1988, Praha.

Zelenka, J. In press b. On *Leptocythere cejcensis* n. sp. (Crustacea, Ostracoda) from the Upper Sarmatian of the Vienna Basin. *Vest. Ústr. Úst. geol.*, Praha.

# DEEP SEA

# 21

# Food in the deep Ocean

**Martin V. Angel**

Institute of Oceanographic Sciences Deacon
Laboratory, Wormley, Godalming, Surrey, GU8
5UB, U.K.

## ABSTRACT

The deep ocean environment is very different from most others because of its remoteness from most primary producers. How food is supplied to it is a key factor in determining the composition, structure and functioning of its communities. Interpretations of the geological record of deep-sea faunas must take account of the mechanisms whereby food is delivered and also how these mechanisms may vary in time and space within the overlying water column. The deposition of allochthonous forms will be *via* the same mechanisms. This paper reviews the status of knowledge about food supply in the deep ocean, in the context of the spate of recent publications.

## INTRODUCTION

Organisms must achieve three main objectives if they are to be successful. In this context, an individual's success is based on it contributing to the continuance of the species. These objectives are: 1) to survive, 2) to feed, and 3) to reproduce. The attainment of these three objectives will result in the partitioning of resources and risk limitation, both of which will vary during the organism's life-cycle. Each species will have evolved a pattern of responses and physiological adaptations, which will be tuned to the biotic and abiotic characteristics of the environment which it inhabits (i.e. its ecological niche). These patterns will show different degrees of variation, which will be influenced by the variability of the habitat both in time and space. They will also be influenced by the ecological history of the habitat. The longer the habitat has persisted and the greater its predictability, the more finely will the life-cycles of the organisms be tuned to the characteristics of the environment. One of the major factors involved in this tuning will be the quantity, quality and predictability of the food supply. Patchiness in supply at all time/space scales will have a strong influence on the selective pressures modifying the adaptations of the species. In this brief paper, I will review some of the recent developments in the theories involving the fluxes of organic material to the deep ocean, and assess their relevance to benthonic and benthopelagic ecology. There will be little direct reference to Ostracoda because there is very little information about them in this context.

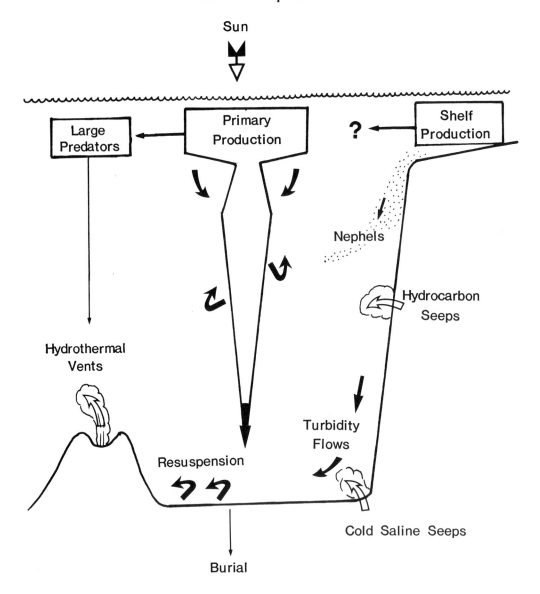

Fig. 1. Sources of energy for deep ocean ecosystems.

## SOURCES

The sun's energy is the primary source. Trapped by the photosynthetic pigments, it is used to convert simple molecules into more complex compounds. Until recently, contemporary photosynthesis was considered to be the one ultimate source, but now the existence of chemosynthetic activity around hydrothermal vents and cold saline hydrocarbon seeps has been recognized, the role of photoautotrophs in the global ecosystem can no longer be considered unique. In global terms chemosynthesis is probably trivial, but within the deep ocean it can be locally dominant. The reported short-lived activity of the vents (i.e. a few decades at most) has implications for the functional ecology of the vent communities which are outside the scope of this review.

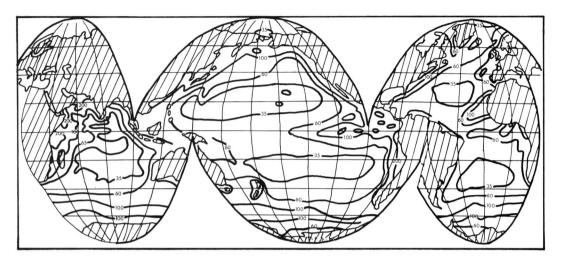

Fig. 2. Distribution of mean primary production (gC m⁻²) in the world's oceans (after Berger *et al.*, 1987).

The products of photosynthesis which reach the deep ocean come from two main sources, *in situ* oceanic production and off-shore transport of shelf production and terrigenous material. Fig. 1 summarizes how the material from some of these sources may reach the deep ocean, and it also provides a structural basis for the ensuing discussion.

## PELAGIC PRODUCTION

There is surprisingly little known about global patterns of primary production in the open ocean. The classical picture based on early $^{14}C$ data has only recently been updated by Berger *et al.* (1987) and their new figure is shown (Fig. 2). The central oligotrophic gyres show up as centres of low productivity, and the zones of high production associated with the areas of coastal and equatorial upwelling are clearly seen. Another feature is the poleward increase in productivity that is associated with the subtropical convergences, which, in most oceans, are situated at around the 40° latitudes. Productivity is controlled by the level of solar insolation that reaches and penetrates the sea surface, the stability of the water column and the availability of nutrients. In some circumstances grazing, as classical theory suggested, may have a limiting effect on the level of production, by reducing the populations of autotrophs. Solar

radiation input is determined by astronomical factors and the reflectivity of the ocean surface. The marked seasonality in both radiation and the heating cycle at polar to temperate latitudes (i.e. latitudes >40°) generates a pulsed production cycle, which is further enhanced by the nutrient supply in the euphotic zone, being linked with the turn-over in the water column during winter. There are also marked regional differences; for example, the nitrate and phosphate concentrations in the deep water of the Pacific are higher than in the Atlantic, because there the deep water is older and has had more time to become enriched by water column remineralization (Tsunogai & Noriki, 1987). Hence the comparable waters in the Pacific seem to be twice as productive as in the Atlantic (Hinga, 1985).

An important concept that has emerged in recent years is that of partitioning primary production into two components. 'Old production' is the component which is supported by nutrients recycled within the euphotic zone, and 'new production' is that which is supported by nutrients supplied by turbulent mixing and diffusion from deep water (Eppley & Peterson, 1979). It is the proportion represented by the level of new production which is available for export to the deep ocean.

In waters of the central and eastern Pacific, estimates for new production amount to 18-56% of total production (Chavez & Barber, 1987). Hayward (1987) found that there was no correlation between the shape of the nutricline and primary production. He argues that there must be other mixing processes which supplement the supply of nutrients to the photic zone provided by turbulent diffusion. Longhurst & Harrison (1988) and Angel (1989) have recently suggested that diel vertical migration by pelagic organisms may be the mechanism whereby nutrients are either added to or lost from the wind mixed layer, thus disturbing the predicted levels of productivity. Mixing is also subject to interannual variation and, even in regions where the water column appears to be highly stable, clear evidence of interannual variations in secondary production exists. Lawes et al. (1987) showed that primary production in one area in the Pacific rose from $57gC\ m^{-2}\ yr^{-1}$ to $100gC\ m^{-2}\ yr^{-1}$ between 1968 and 1985. In the North Atlantic there may be significant longitudinal changes in production related to changes in the depth of winter mixing, which may mimic the latitudinal changes induced by climatic variations in other oceans. Hinga (1985) summarized the available data for sediment community respiration in the Pacific and Atlantic Oceans and showed that, whereas new production was 1.8 times higher in the Pacific, profiles of sediment respiration showed even greater enhancement in activity. Hinga (1985) concluded that in the Pacific the dynamics of the biological communities differ so that a greater percentage of the primary production is passed down to and utilized by the benthonic infauna.

## TRANSFER TO THE DEEP OCEAN

Broecker & Peng (1982) wrote, 'While much of the plant material is consumed by animals living in the surface ocean, some insoluble indigestibles (i.e. fecal matter) move into the deep-sea under the influence of gravity'. There is now a rapidly growing literature on the sedimentation of organic material into the deep ocean and the factors which modify the flux (see the reviews by Fowler & Knauer (1986) and Alldredge & Silver (1988)). Approximately 10% of the primary production is collected in sediment traps at the base of the euphotic zone (e.g., Small et al., 1987), but only 1-3% reaches the abyssal sea-bed.

Some of the available data on particle fluxes in deep water derived from sediment trap observations have been summarized recently by Wefer (1989). The observations range from polar to tropical latitudes, and Table 1 summarizes some of the data given by Wefer. Total fluxes range from $7.2$-$124.7 g\ m^{-2}\ yr^{-1}$. There are clear regional differences in the proportions of carbonate and opal in the flux, but the mean percentage of organic carbon is $5.94 \pm 2.63\%$ (ignoring the exceptionally high value observed at Bear Island). Another feature that should be borne in mind is the seasonal variability of the fluxes. Wefer et al. (1988) reported that at a depth of 1588m in the Bransfield Strait, 97% of the total annual flux of $110 g\ m^{-2}$ occurred during the two most productive months. This was mostly in the form of krill faecal pellets, and just 9% was in more usual particulate forms (equivalent to $3\ gC\ m^{-2}$).

Deployment of sediment traps at depths shallower than about 500m will not sample material in the guts of vertically migrating organisms and will, therefore, underestimate the amount of material leaving the euphotic zone. Angel (1989) conservatively estimated this element of the flux to be equivalent to about 1-2% of the primary production. Profiles of the vertical flux based on sediment trap data almost invariably show an increase in the fluxes at mesopelagic depths at around 500-700m. Initially these maxima were interpreted as indicating lateral advection into the area, and in some localities this will continue to be considered the main reason for the peak. However, Walsh et al. (1988) have stated recently that 'packaging and transport by zooplankton and nekton is the most likely cause of the widely observed mid-water column flux maximum'. There is now preliminary evidence that freely suspended bacteria play a major role in the reprocessing of organic material within the water column (Cho & Azam, 1988). Previous suggestions that 'snow' and other aggregates are centres of greatly enhanced biological activity (e.g., Alldredge & Youngbluth, 1985) have not always been confirmed by sediment trap observations (e.g., Karl et al., 1988). Two factors may

Table 1 Sediment flux rates and relative composition of sedimenting material observed at various long-term stations (modified from Wefer, 1989).

| Locality | Depth m | Flux in g m$^{-2}$ yr$^{-1}$ Total | Organic C | % composition Carbonate | Opal |
|---|---|---|---|---|---|
| Fram Strait 78°52'N 1°22'E | 2442 | 7.2 | 0.41 | 43.8 | 18.8 |
| Bear Island 75°51'N 11°28'E | 1700 | 28.0 | 2.85 | 48.6 | 14.4 |
| Greenland Basin 74°35'N 6°43'W | 2823 | 10.2 | 0.4 | 46.3 | 36.8 |
| Norwegian Sea 70°N 1°58'W | 2749 | 16.8 | 0.5 | 73.6 | 11.9 |
| Lofoten Basin 69°39'N 10°E | 2761 | 22.8 | 1.37 | 71.9 | 6.9 |
| Norway Abyss 65°31'N 0°64'E | 2630 | 17.4 | 0.59 | 70.1 | 12.8 |
| Bransfield Strait 62°15'S 57°31'W | 1588 | 107.7 | 4.3 | 9.6 | 71.6 |
| Weddell Sea 62°26'S 34°45'W | 863 | 0.37 | 0.025 | 2.73 | 79.2 |
| N.E. Pacific 50°N 145°W | 3800 | 45.0 | 1.1 | 49.0 | 42.5 |
| Sargasso Sea | 3200 | 13.2 | 0.66 | 62.1 | circa 20.0 |
| Tropical Pacific 11°N 140°W | 3400 | 13.2 | 0.59 | 55.3 | 29.5 |
| Panama Bight 5°22'N 85°35'W | 3560 | 124.7 | 4.31 | 48.8 | circa 40.0 |
| Pacific      year 1 | 3495 | 21.8 | 1.19 | 56.6 | 32.1 |
| 1°N 139°W year 2 | 2908 | 41.74 | 1.89 | 55.0 | 37.4 |

lead to a slowing of microbial degradation with increasing depth down the water column: firstly the inhibitory effects of reducing temperature and increasing hydrostatic pressure on the bacteria which originally colonized the particles at shallow depths (although this will not apply to the hyperbaric bacteria which colonize them at depth), and secondly the relatively short residence time of the large, and hence faster falling, particles within the water column. Furthermore, the reductions in standing crop of pelagic organisms with depth will reduce the likelihood of particles being intercepted, eaten and repackaged the deeper they sink.

There is also a trend for the mean size of pelagic organisms to increase with depth (Angel, 1989). Consequently, because of the empirical relationship between the size of faecal pellets, the size of the organisms generating them and their sinking rates (Fowler & Knauer, 1986) these increases in body size will generate variations in the flux patterns with both depth and latitude.

One of the more startling recent observations has been the rapidity with which aggregates of phytodetritus begin to sediment out onto the abyssal sea floor following the onset of the spring bloom at temperate latitudes in the northeastern Atlantic

(Billett *et al.*, 1983; Lampitt, 1985). This peak in detrital flux stimulates responses in the benthonic meiofaunal organisms (Gooday, 1988), so that the newly-deposited floc layer is soon inhabited by a very different foraminiferan community from that occupying the underlying sediment. The macro- and megabenthos also respond behaviourally, and in many species seasonal reproduction is stimulated. Once on the bottom, the flocs are resuspended and redistributed by tidal oscillations, especially after the structural fabric of the aggregates has been degraded by bacteria. Thus the arrival of these massive inputs of floc greatly enhance the food supply not only at the sediment/water interface, but also in the mixed water layer of the benthonic boundary zone, which is inhabited by the benthopelagic community.

## MOULTED CARAPACES

Crustaceans have to moult in order to grow. Some, like the euphausiids, continue to moult throughout their life-cycle, whereas others like copepods and ostracods, cease to moult once they have reached maturity. Lasker (1976) estimated that 9% of the secondary production of *Euphausia pacifica* is lost in the moults which each adult animals sloughs every 5-10 days. Chitin is rela- tively refractory but it contains 51% carbon and 8% nitrogen by weight. Terazaki & Wada (1988) estimated that the carapaces of the copepod *Calanus cristatus* collected from the Japan Sea may have been drifting for more than a year. Mutilated copepods and carapaces begin to be obvious in samples at depths >2000m (e.g., Farran, 1926). In deep abyssal benthopelagic zones cope- pod carapaces can constitute >80% of plankton samples (Roe *et al.*, 1987). Some pelagic ostra- cods have been reported to have gut contents con- sisting purely of folded chitinous remains (Angel, 1972), but it is not clear whether they are able to break chitin down or whether their food source is the attached bacteria. Locally the rain of this material may be immense, such as beneath mega- swarms of krill (*Euphausia superba*) as recorded off Elephant Island in 1981 (Macaulay *et al.*,1984; Brinton & Antezana, 1984). Chitin does not

accumulate in oxic sediments, and so in the deep ocean it must be degraded and, therefore, enter the food chain.

## LARGE FOOD FALLS

Some of the early experiments utilizing baited cameras (e.g., Dayton & Hessler, 1972) indicated the existence of large populations of mobile benthopelagic scavengers in the deep ocean. Haedrich & Rowe (1977) estimated that the standing crop of these mobile scavengers was roughly equivalent to that of the sedentary epifauna and infauna. Stockton & DeLaca (1982) while acknowledging the local importance of large food falls, did not consider they made a major direct contribution to the global input of organic material to the deep ocean. They pointed out that the influence of any fall would be to create concentric areas affected by the spread of water-borne Particulate Organic Matter (POM) and Dissolved Organic Matter (DOM), and also faecal material transported away by the departing scavengers. They also showed that, whereas small items were quickly consumed, very large falls, such as those created by the corpse of a large tuna or whale, persist for hundreds of days or even years because they overwhelm the ability of the local populations of scavengers to deal with such a large amount of material. Recently the skeleton of a whale has been located off the coast of California. Its age has been estimated to be around fifty years old, but the bones are still leaking oil and are populated by a specialized microbial and metazoan epifauna (Smith, pers. comm.). Rowe & Staresinic (1979) attempted to draw up budgets of the relative importance of the various inputs into the Sargasso Sea and concluded that inputs of POM amounted to 4gC m$^{-2}$ yr$^{-1}$. In addition, the input of *Sargassum* weed amounted to 0.4gC m$^{-2}$ yr$^{-1}$, the large food falls to 0.05gC m$^{-2}$ yr$^{-1}$, and chemoautotrophic production to 1-10 x 10$^{-5}$gC m$^{-2}$ yr$^{-1}$. Smith (1985) analysed photographic series of fauna attracted to baits in the Santa Catalina Basin off California. He showed that a succession of species was attracted and that several species rested near the bait after feeding and possibly returned to feed again, a behaviour described as 'tank-topping'. Smith further concluded that large food falls were much more important than other workers had thought, and that they might provide on average as much as 11% of the annual needs of benthonic respiration. Large food falls have also been observed in highly oligotrophic areas. Roe *et al.* (1987) describe how a corpse of a *Pyrosoma*, a salp, was photographed on the sea-bed at a depth of 5540m on the Madeiran Abyssal Plain. It remained in the field of view of a time-lapse camera over a period of 16 days, during which time it was fed upon twice by a squat lobster, (*Munidopsis* sp.) and once by a starfish (*Hyphalaster inermis*). Eventually it disappeared from the field of view. Three other possible *Pyrosoma* corpses were observed in 1191 photographs of the sea-bed taken along transects which covered an area of around 2380m².

Analogous to the falls of animal remains, which are undegraded and highly labile, are the falls of plant material derived either from land or shelf regions. Turner (1973) first described the role of obligate wood-boring bivalves together with bacteria and fungi, in converting refractory woody material into a more utilizable organic form in the deep ocean. Wolff (1979) further extended these observations to remains of sea grasses and *Sargassum* weed, which are utilized by a highly specialized fauna. More recently, Maddocks & Steineck (1987) have described a unique association of podocopid ostracods from wooden blocks experimentally deployed by submersibles at depths in the range 1800-4000m. These faunas were unlike any known deep-sea podocopid assemblage and were more closely related to coastal species. Turner (1981) considered natural 'wood-islands' together with hydrothermal vent communities to be centres of diverse communities in the deep sea.

## CHEMOSYNTHETIC SOURCES

The discovery of large concentrations of organisms in the vicinity of hydrothermal vents and hydrocarbon seeps which are either feeding directly on the high concentrations of suspended chemosynthetic bacteria, or 'gardening' them by maintaining them symbiotically within specialized organs, poses a number of ecological problems.

The vents are ephemeral features, associated with spreading centres and back-arc systems. They have been observed to be populated by apparently flourishing communities, which three years later, have totally vanished. At the vents, the chemosynthetic production is based on the oxidation of dissolved ions to sulphate. The vents which occur along the mid-ocean ridges are linked between oceans, and hence faunal continuity between many of the communities is to be expected, but still awaits confirmation. However, the ridge systems of the northeastern Pacific and of the back-arc systems are isolated and these have proved to show a much greater degree of local endemicity.

The hydrocarbon seeps are predominantly slope features associated mainly with sedimentary fans, such as the Mississippi Fan in the Gulf of Mexico. In this case, the chemosynthetic production is based on the oxidation of methane, and the seeps are likely to persist for hundreds of years. The greatly enhanced food supply at these centres leads to the build up of localized high concentrations of biomass. As in terrestrial island faunas, these islands of richness might be expected to result in the selection of forms with reduced dispersion. While such an adaptation will continue to be favourable around the persistent hydrocarbon seeps, the ephemeral character of the hydrothermal vents may lead to the selection of species which can switch their life-cycle tactics from a low dispersive to a high dispersive regime. The rate of spreading and the local seismicity will influence the longevity of each vent system, and it may prove difficult to predict the likely persistence of each individual system. So far, studies of these features have concentrated on the communities inhabiting the immediate vicinity of the vents. There will be gradients of enrichment radiating out around each centre of high production, which may support specialized communities of smaller organisms including ostracods (e.g., Wiebe *et al.*, 1988). Studies on the colonization of new systems should provide important insights into mechanisms of gene-flow within deep ocean communities. Another factor requiring critical re-examination is the accuracy of Rowe & Staresinic's (1979) estimate that chemoautotrophy provides only $10^{-2}$-$10^{-3}$% of the benthonic input of organic material to typical abyssal communities.

Recently cold, deep-sea brine seeps of hyper-saline sulphide-rich fluid have been discovered. These have been found to support similar populations to those associated with hot water vents. Some of the cold water seeps are associated with subduction zones (Juniper & Sibuet, 1987; Sibuet et al., 1988) and others with tectoni-cally inactive regions (Hecker, 1985).

## EXPORT FROM SHELF SEAS

The shelf-break appears to function as a partial barrier to faunal distribution and to the exchange of water and materials. Generally it has been thought that there is very little export of POM from the shelf to the deep ocean. However, Walsh (1988) has recently estimated that on average as much as 30-47gC m$^{-2}$ yr$^{-1}$ may exit the shelf-break to storage centres along the slope. Much of the carbon probably accumulates as buried debris within the sediments. The results of the Shelf Edge Exchange Processes Program (SEEP) which was conducted off the east coast of the U.S.A. in 1983-1984, showed that the production of the Spring Bloom was neither quickly decomposed nor exported. The quantity eventually exported proved to be a function of shelf width, slope depth and variations in the nutrient supply. The form of the carbon exported differed from that generated by oceanic systems and could also be assessed by its isotopic characteristics (e.g., $^{210}$Pb and $^{13}$C). The export is also likely to be influenced by topogra-phical features. For example, off the west coast of Ireland during winter the shelf waters are cooled and become denser than the oceanic waters offshore, and so periodically they cascade down the slope. These cascades generate high turbidity layers (nephels) which extend to depths of some 600m down the slope (Dickson & McCave, 1986). The nephels are enriched with POM and enhance the local food supply for suspension-feeders on the slope. Biscaye et al. (1988) have published profiles showing that although down-slope movement of particulates accounts for <10% of the biogenic

carbon fixed on the continental shelf, $^{210}$Pb data imply that some 20% of biogenic and abiogenic particles from the shelf are exported to the slope.

Canyons also play an important role, for not only do they act as traps for suspended material being advected along-shelf, but also they act as channels for intermittent turbidity flows. Can-yons were little studied biologically until the devel-opment of submersibles and are still difficult environments. Some descriptions are beginning to be published, and of interest here are the observa-tions by Cacchione et al. (1978) who described how a rain of corpses from a bloom of salps over the shelf of the East coast of the U.S.A. was funnelled via a canyon out onto the abyssal plain. The location of Dumpsite 106 for New York's wastes is at the head of a canyon, down which much of the suspended material is advected.

## VARIABLILITY IN TIME AND SPACE

The availability of food varies extensively in time, as seen in the seasonal and interannual fluctuations of sediment trap fluxes observed by Deuser (1986). Spatial heterogeneity is well exemplified by the change from impoverished deep-sea conditions to the richness of the hydrothermal vents and cold-water seeps. These time/space variations occur at all scales and strongly influence the composi-tion, the functioning and the distribution patterns of the sea-bed communities. Therefore, they also strongly influence the precision with which we are able to sample and quantify the communities.

At small time/space scales there are likely to be quite wide fluctuations in the rain of particles sedimenting onto the sea floor, although this is yet to be confirmed by closely set sediment traps. The surface production of biogenic material is very patchy, and this temporal and spatial patchiness is transmitted down into the body of the ocean. The slower the sedimentation rate, the more evenly spread will the rain become, partly as a result of the shears and other mixing processes which occur within the water column, and partly because of the varying sinking rates of differently sized particles. Once the material arrives on the sea-bed, so long as it remains mobile it will be subject to resuspension and redistribution by tidal currents and 'benthic storms' (Lampitt, 1985). Microtopography generated both by physical forces (e.g., ripple marks and scour patterns) and by biological activity creating mounds and hollows will result in marked localized patchiness. Such patchiness leads to variability even at the small scale of a box-corer.

Large scale topographical features will also generate substantial variations. For example Genin & Boehlert (1985) have shown how a sea-mount may influence the temperature structure of the overlying water and enhance the productivity of the euphotic zone both above it and in the downstream plume. The flow patterns over the top of sea mounts influence the distributions of suspension-feeding organisms such as corals (Genin, *et al.*, 1988). Even in slope environments the interactions between water mass distribution current flow and topography can result in marked distribution patterns. When Lampitt *et al.* (1986) published their biomass profiles for the Porcupine Seabight region in the Northeast Atlantic, they had to explain the large peak in biomass that occurred at about 1300m. There was a twenty-fold increase in the ash-free dry weight of the megabenthos which was caused by a zone of the large suspension-feeding hexactinellid sponge *Pheronema carpenteri*. This sponge reached peak abundances of 475/1000m³ at the centre of its 50m wide bathymetric zone. Further round the rim of the Seabight the sponge is replaced by dense concentrations of the coral *Lophelia pertusa* in one direction and the sea-pen *Kophobelemnon stelliferum* in the other. Each species has associated with it, a specialized community of commensals which are almost entirely restricted to these zones. The band of *Pheronema* coincides with the lower boundary of the Mediterranean Outflow Water in the region. Some of these associated species will be linked biologically to the dominant species, while others rely on the structural habitat the dominant species provides. Hence if the coral bank dies because of a shift in the current patterns, the former will die out rapidly but the latter may persist until the coral skeletons either collapse or become buried.

Major oceanographical features are likely to determine many of the zoogeographical shifts in deep community structure through changes in the sedimentation regime. For example, upwelling regions, where productivity is greatly enhanced, are underlain by communities which are richer in both species and in biomass than those which occur beneath the surrounding regions. Another feature which is of considerable influence is the subtropical convergence. This marks the limit to which wintertime cooling at low-latitude is sufficient to erode the seasonal thermocline so that it breaks down and nutrients are resupplied to the euphotic zone by vertical mixing. On the low latitude side of the convergence, the water is permanently stratified and primary production shows relatively little seasonal variation. On the high latitude side, not only is the average annual production higher, but it is also highly pulsed seasonally. Merrett (1987) has recently identified this boundary as being coupled with major shifts in the faunal assemblages of demersal fish, and probably most other groups of benthonic taxa will be similarly affected. The CLIMAP program (Cline & Hayes, 1976) showed that in the North Atlantic the Subtropical Convergence has shifted very little latitudinally between 15,000 BP, during the height of the last glaciation, and the present time. The Polar Front, however, has shifted dramatically, reducing the areal extent of the temperate zones almost to nothing. In other oceans, these features shifted latitudinally and the size of the temperate regions remained little changed. To what extent the ecological memory of these past events is still retained within the communities we study at present in unclear.

## THE ECOLOGICAL IMPLICATIONS OF THE TIME/SPACE CHARACTERISTICS IN FOOD SUPPLY

In the deep ocean the four main sources of food supply are: 1) finely suspended material, 2) sedimenting particulate organic matter (POM), 3) large packages, and 4) chemoautotrophy. Each source provides material of very different quality and with very different characteristics of predictability. Fine suspended material tends to be refractory, or as in the case of freely-suspended bacteria, difficult to extract from their highly diluted concentration in the water. Where concentrations are increased as a result of the hydrological regime and where localized current regimes favour the use of low-energy collection mechanisms, suspension-feeders can become the dominant forms. These organisms may be totally sedentary and, when adult, have no dispersal ability. They are probably relatively long-lived and slow to respond to environmental change. Any behavioural responses they show will be limited to activity cycles and changes in their orientation to currents.

Those organisms which rely on the sedimentation of POM need either to have a limited ability to move or the ability to sweep a relatively large area of the sediment interface where there is the most abundant supply. The material available will tend to be more labile and more copious than the suspended material but more variable in occurrence, particularly at latitudes where surface production is seasonally pulsed. In the smaller meiofaunal species, response to heavy deposition of flocs may be behavioural (i.e. by migrations up out of the underlying sediment) or reproductive (Gooday, 1988). In many macrobenthonic and megabenthonic groups the seasonal deposition of flocs at higher latitudes stimulates seasonal breeding. Moreover, the ability to move slowly over the sea bed enables organisms such as holothurians constantly to sweep up new material and to move between patches of accumulated floc. Infaunal species may either exploit the relatively small quantities of organic material that become buried, or sweep the surface area around their burrows, or, by creating hollows, induce a localized accumulation of material.

Large packages tend to occur unpredictably. The plant material which arrives is refractory, but the organic material in large animal corpses tends to be very highly labile. The persistence of smaller packages is related to their degree of refractoriness. Small fish corpses up to a few kilograms in weight tend to be consumed in a matter of a few hours or days, whereas plant material or *Pyrosoma* corpses (see above), in which the organic material

is less readily degraded, persist for much longer. However, the persistence of very large carcases may not be determined by the labile nature of their content, because they may overload the ability of the benthonic community to exploit so much material quickly. Benthopelagic scavengers which exploit these large high quality food falls have to be highly mobile in order to locate their food. They are believed to move up olfactory gradients and seem to be most active at certain phases of the tidal oscillations (Lampitt *et al.*, 1983). Few trap experiments have been conducted in a manner whereby the smaller organisms which are attracted to such baits will be sampled, but some myodocopids have been taken in baited traps set out from *Discovery*.

Likewise the rich, but highly localized and rare sources of chemoautotrophic input occurring around hydrothermal vents and hydrocarbon seeps are likely to be exploited by as many members of the micro- and macrobenthos as the megabenthos which, hitherto, have received most attention. The ecological conflict between the need to minimize mobility while the source is still fully operating and the need to disperse once it begins to fail has been discussed above.

## SUMMARY

Food supply is generally limited in the deep ocean and decreases as the depth increases. There is a range of sources which vary greatly in quality, quantity and predictability. The exploitation of each of these different sources will tend to select for quite different life styles and life cycle characteristics. Deep ocean food-webs are fundamentally different from those of most terrestrial and shallow water environments because their source of primary production is remote. They are based primarily on inputs of detrital material, and so the options of feeding regimes available to the inhabitants are limited to exploitation of the detrital sources or to carnivory/parasitism. Present day communities are structured and exist according to the present environmental conditions but they also retain an ecological memory as a result of the ability of organisms to persist in regions which they could no longer recolonize if locally exterminated. Also the ability of each species to adapt fully to a highly stable and predictable environment may take many generations, even for those which are genetically flexible. The fine tuning of life cycles and community function to predictable environments which can be seen in some deep-sea communities, may take time scales that are long compared to the cycle of change generated by glaciations in terrestrial and shallow-water environments. This creates problems in sampling Recent communities, because the relevant time/space scales may neither be obvious nor accessible to our sampling methodology. Hence there are always going to be difficulties in interpreting geological records in the context of what is known about Recent faunas. Furthermore, the role of the Ostracoda within these deep ocean communities has received so little attention that the interpretations of faunistic studies may contain fundamental misconceptions.

# REFERENCES

Alldredge, A. L. & Silver M. W. 1988. Characteristics, dynamics and significance of marine snow. *Progress in Oceanography*, **20**, 41-82.

Alldredge, A. L. & Youngbluth M. J. 1985. The significance of macroscopic aggregates (marine snow) as sites for heterotrophic bacterial production in the mesopelagic zone of the subtropical Atlantic. *Deep-Sea Res.*, **32**, 1445-1456.

Angel, M. V. 1972. Planktonic oceanic ostracods: Historical, present and future. *Proc. R. Soc. Edinb.*, Edinburgh, **B73**, 213-228.

Angel, M. V. 1989. Does mesopelagic biology affect the vertical flux. *In* Berger, W. H. (Ed.), *Productivity of the Ocean: Past and Present*, Dahlem Workshop Reports, **37**, 155-173. John Wiley.

Berger, W. H., Fischer, K., Lai C. & Wu G. 1987. Ocean carbon flux: Global maps of primary production and export production. *In* Agegian, C. (Ed.), *Biogeochemical Cycling and Fluxes between the Deep Euphotic Zone and other Oceanic Realms*, NOAA Symposium Series for Undersea Research. NOAA Undersea Research Progress **3**, University of California.

Billett, D. S. M., Lampitt, R. S., Rice, A. L. & Mantoura, R. F. C. 1983. Seasonal sedimentation of phytoplankton to the deep-sea benthos. *Nature*, London, **302**(5908), 520-522.

Biscaye, P. E., Anderson, R. F. & Deck, B. L. 1988. Fluxes of particles and constituents to the eastern United States continental slope and rise: SEEP-1. *Continental Shelf Research*, **8**, 855-904.

Brinton, E. & Antezana, T. 1984. Structures of swarming and dispersed populations of krill (*Euphausia superba*) in Scotia Sea and South Shetland waters during January-

March 1981, determined by Bongo Nets. *J. Crustacean Biol.*, Lawrence, Kansas, **4** (Spec. No 1), 45-66.

Broecker, W. S. & Peng, T.-H. 1982. *Tracers in the Sea*, 690 pp. Palisades, NY Columbia University.

Cacchione, D. A., Rowe, G. T. & Malahoff, A. 1978. Submersible investigation of Outer Hudson Submarine Canyon. *In* Stanley, D. J. & Kelling, G. *Sedimentation in Submarine Canyons, Fans and Trenches*, 422-450. Dowden, Hutchinson and Ross, Stroudsberg, Penn.

Chavez, F. P. & Barber, R. T. 1987. An estimate of new production in the equatorial Pacific. *Deep-Sea Res.*, **34**, 1229-1243.

Cho, B. C. & Azam F., 1988. Major role of bacteria in biogeochemical fluxes in the ocean's interior. *Nature*, London, **332**(6163), 441-443.

Cline, R. M. & Hayes J. D. (Eds), 1976. *Investigation of Late Quaternary Paleoceanography and Paleoclimatology*. Geological Society of America Memoirs, **145**, 1-464.

Dayton, P. K. & Hessler, R. R. 1972. Role of biological disturbance in maintaining diversity in the deep sea. *Deep-Sea Res.*, **19**, 199-208.

Deuser, W. G. 1986. Seasonal and interannual variations in deep-water particle fluxes in the Sargasso Sea and their relation to surface hydrography. *Deep-Sea Res.*, **33**, 225-246.

Dickson, R. R. & McCave I. N. 1986. Nepheloid layers on the continental slope west of Porcupine Bank. *Deep-Sea Res.*, **33**, 791-818.

Eppley, R. W. & Peterson, B. J. 1979. Particulate organic matter flux and planktonic new production in the deep ocean. *Nature*, London, **282**(5006), 677-680.

Farran, G. P. 1926. Biscayan plankton collected during a cruise of H.M.S. Researcher, 1900 Part 14 The Copepoda. *J. Linn. Soc. Zoology*, London, **36**, 219-310.

Fowler, S. W. & Knauer, G. A. 1986. Role of large particles in the transport of elements and organic compounds through the oceanic water column. *Progress in Oceanography*, **16**, 147-194.

Genin, A. & Boehlert, G. W. 1985. Dynamics of temperature and chlorophyll structures above a seamount: An oceanic experiment. *J. mar. Res.*, New Haven, **43**, 907-924.

Genin, A., Dayton, P. K., Lonsdale, R. F. & Spiess F. N. 1988. Corals on seamount peaks provide evidence of current acceleration over deep-sea topography. *Nature*, London, **322**(6074), 59-61.

Gooday, A. J. 1988. A response by benthic Foraminifera to the deposition of phytodetritus in the deep sea. *Nature*, London, **322**(6074), 70-73.

Haedrich, R. L. & Rowe G. T. 1977. Megafaunal biomass in the deep sea. *Nature*, London, **269**(4823), 141-142.

Hayward, T. 1987. The nutrient distribution and primary production in the central North Pacific. *Deep-Sea Res.*, **34**, 1593-1628.

Hecker, B. 1985. Fauna from a cold sulfur-seep in the Gulf of Mexico: comparison with hydrothermal vent communities and evolutionary implications. *In The Hydrothermal Vents of the Eastern Pacific: An overview. Bull. biol. Soc. Wash.*, **6**, 465-473.

Hinga, K. R. 1985. Evidence for a higher average primary productivity in the Pacific than in the Atlantic Ocean.

Deep-Sea Res., 32, 117-126.

Juniper, S. K. & Sibuet, M. 1987. Cold seep benthic communities in Japan subduction zones: spatial organisation, trophic strategies and evidence for temporal evolution. Marine Ecology - Progress Series, 40, 115-126.

Karl, D. M., Knauer, G. A. & Martin, J. H. 1988. Downward flux of particulate organic matter in the ocean: a particle decomposition paradox. Nature, London, 332(6078), 438-441.

Lampitt, R. S. 1985. Evidence for the seasonal deposition of detritus to the deep-sea floor (Porcupine Bight, N.E. Atlantic) and its subsequent resuspension. Deep-Sea Res., 32, 885-897.

Lampitt, R. S., Billett, D. S. M. & Rice, A. L. 1986. Biomass of the invertebrate megabenthos from 500 to 4100m in the northeast Atlantic Ocean. Mar. Biol., 93, 69-81.

Lampitt, R. S., Merrett, N. R. & Thurston, M. H. 1983. Interrelations of necrophagous amphipods, a fish predator and tidal currents in the deep sea. Mar. Biol., 74, 73-78.

Lasker, R. 1976. Feeding, growth, respiration and carbon utilization of a euphausiid shrimp. J. Fish. Res. Bd Can., Ottawa, 23, 1291-1323.

Lawes, E. A., DiTullio, G. R. & Redalje, D. G. 1987. High phytoplankton growth and production rates in the North Pacific subtropical gyre. Limnol. Oceanogr., Baltimore, 32, 905-918.

Longhurst, A. R. & Harrison W. G. 1988. Vertical nitrogen flux from the ocean photic zone by diel migrant zooplankton and nekton. Deep-Sea Res., 35, 881-889.

Macaulay, M. C., English, T. S. & Mathisen, O. A. 1984. Acoustic characterisation of swarms of Antarctic krill (Euphausia superba) from Elephant Island and Bransfield Strait. J. Crustacean Biol., Lawrence, Kansas, 4 (Spec. No. 1), 16-44.

Maddocks, R. F. & Steineck, P. L. 1987. Ostracoda from experimental wood-island habitats in the deep sea. Micropaleontology, New York, 33, 318-355.

Merrett, N. R. 1987. A zone of faunal change in assemblage of abyssal demersal fish in the eastern Northern Atlantic Ocean: A response to seasonality in production? Biological Oceanography, New York, 5, 137-151.

Roe, H. S. J., Badcock, J., Billett, D. S. M., Chidgey, K. C., Domanski, P. A., Ellis, C. J., Fasham, M. J. R., Gooday, A. J., Hargreaves, P. M. D., Huggett, Q. J., James, P. T., Kirkpatrick, P. A., Lampitt, R. S., Merrett N. R., Muirhead, A., Pugh, P. R., Rice, A. L., Thurston, M. H. & Tyler, P. A. 1987. Great Meteor East: A biological characterisation. Institute of Oceanographic Sciences Report, 248, 1-260, + 18 figs.

Rowe, G. T. & Staresinic, N. 1979. Sources of organic matter to the deep-sea benthos. Ambio spec. Rep., Royal Swedish Academy of Sciences, Stokholm, 6, 19-23.

Sibuet, M., Juniper, S. K. & Pautot, G. 1988. Cold-seep benthic communities in the Japan subduction zones: Geological control of community development. J. mar. Res., New Haven, 46, 333-348.

Small, L. F., Knauer, G. A. & Tuel, M. D. 1987. The role of sinking fecal pellets in stratified euphotic zones. Deep-Sea Res., 34, 1705-1712.

Smith, C. R. 1985. Food for the deep sea: utilization, and flux of nekton falls at the Santa Catalina Basin Floor. Deep-Sea Res., 32, 417-442.

Stockton, W. L. & DeLaca, T. E. 1982. Food falls in the deep sea: occurrence, quality and significance. Deep-Sea Res., 29, 157-169.

Terazaki, M. & Wada, W. 1988. Occurrence of large numbers of carcases of large, grazing copepods Calanus cristatus from the Japan Sea. Mar. Biol., 97, 177-183.

Tsunogai, S. & Noriki, S. 1987. Organic matter fluxes and the sites of oxygen consumption in deep water. Deep-Sea Res., 34 755-767.

Turner, R. D. 1973. Wood-boring bivalves, opportunistic species in the deep sea. Science, New York, 180, 1377-1379.

Turner, R. D. 1981. Wood island and thermal vents as centers of diverse communities in the deep sea. The Soviet Journal of Marine Biology, 7, 1-10.

Walsh, J. J. (Ed.), 1988. Shelf edge exchange processes of the mid-Atlantic Bight. Continental Shelf Research, 8, 433-946.

Walsh, J. J., Dymond, J. & Colllier, R. 1988. Rates of recycling of biogenic components of settling particles in the ocean derived from sediment trap experiments. Deep-Sea Res., 35, 43-58.

Wefer, G. 1989. Particle flux in the ocean: Effects of episodic production. In Berger, W. H. (Ed.), Productivity of the Oceans: present and past. Dahlem Workshop Reports, 37, 139-154, John Wiley.

Wefer, G., Fischer, G., Fueterer, D. & Gersonde, R. 1988. Seasonal particle flux in the Bransfield Strait, Antarctica. Deep-Sea Res., 35, 891-898.

Wiebe, P. H., Copley, N., Van Dover, C., Tamse, A. & Manrique, F. 1988. Deep-water zooplankton of the Guaymas Basin hydrothermal vent field. Deep-Sea Res., 35, 985-1013.

Wolff, T. 1979. Macrofaunal utilization of plant remains in the deep-sea. Sarsia, Universitetet i Bergen, 64, 117-136.

## DISCUSSION

Roger Kaesler: Hydrothermal vents (not events!) have been much in the news and are biologically fascinating. How important are they quantitatively?

Martin Angel: Recent estimates imply that in global terms vent production and faunas make a trivial contribution to deep-sea production. However, geochemically vents are extremely important and the modification of the vent waters by biological processes may prove to be the most important aspect of these specialized faunas.

Dan Danielopol: The patchy distribution of the food resources and subsequently of the fauna and later the necessity for active movement of the animals between various patches is very similar to what one finds in intensive freshwater habitats.

Martin Angel: The scale, both in time and space, of the patchiness is very different between the abyssal oceanic environments and freshwater interstitial habitats Movement between patches only has survival value if there is a good chance of finding a rich patch. Foraging strategies will be dependent on many factors of which the scale of the patchiness is but one. Many of the mechanisms of disturbance in these two types of habitat are driven by very different physical events, and hence their differences in scale. However, the apparent long-term persistence in geological time scales of both types of habitats and their remoteness from sources of photo-autotrophic production may be why the communities inhabiting these environments show some striking similarities such as the presence of relict faunal elements.

# 22

# A comparison of North Atlantic and Pacific Cainozoic deep-sea Ostracoda

**Graham Coles[1], Michael Ayress[1] & Robin Whatley[2]**

[1]Geochem Laboratories Ltd, Chester St., Saltney, Chester CH4 8RD, U.K.
[2]Institute of Earth Studies, University College of Wales, Aberystwyth, Dyfed, U.K.

## ABSTRACT

The diversity of deep-sea Cainozoic Ostracoda from the North Atlantic and Pacific Oceans is documented. Species diversity increases in an overall, but non-uniform manner throughout the Cainozoic in both oceans. The largest increment of new species is in the Middle Eocene, coincident with the global development of the psychrospheric fauna. During the Cainozoic at least 95 species are common to both oceans. Our present knowledge of the stratigraphical ranges of these common species suggests that more species entered the deep sea in the North Atlantic than in the Pacific. The number of contemporary common species increases steadily from 2 in the Palaeocene to 85 in the Quaternary and Recent. Most deep-sea genera can be traced back to the Cretaceous. At least 38 of the genera from both oceans are recorded from Upper Cretaceous shelf environments, particularly in North West Europe, and, to a lesser extent the Gulf Coast-Caribbean area as well as the Southern Hemisphere (South Africa and West Australia).

## INTRODUCTION

Whatley & Ayress (1988) demonstrated that many more Ostracoda are pan-abyssal than had been previously thought (Brady, 1880; Benson, 1969, 1975; Whatley, 1983). They documented 65 cosmopolitan species which were common to the Quaternary of the North Atlantic, Indian and South West Pacific oceans and suggested that most species entered the deep sea in the Neogene of the South West Pacific (especially in the region between Australia and New Zealand) and subsequently migrated to other areas of the world's oceans. However, their study was confined largely to the Neogene and Quaternary and did not

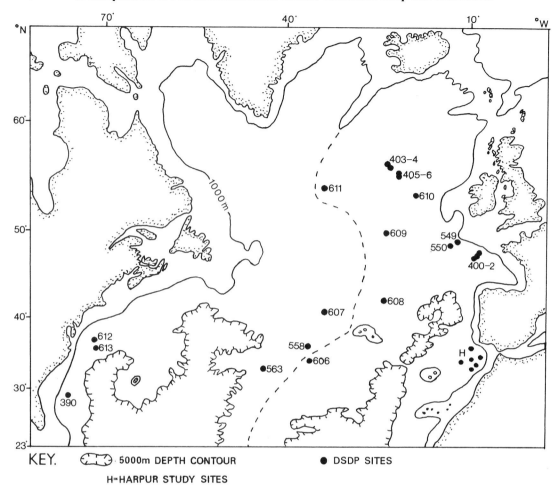

Fig. 1. The North Atlantic Ocean showing the DSDP sites studied by Coles (in prep.), Whatley & Coles (1987), Ducasse & Peypouquet (1979), Guernet (1982) and Cronin & Compton-Gooding (1987) and the study sites of Harpur (1985).

consider Palaeogene faunas. Recent studies on the Palaeogene Ostracoda of the South West Pacific (Millson, 1987) and of the North Atlantic (Coles & Whatley, in press; Coles, in prep. and Whatley & Coles, in press) have recorded diverse faunas which provide detailed information on the composition and origins of Cainozoic deep-sea faunas and which, to an extent, modify the conclusions of Whatley & Ayress (1988). More detailed biostratigraphical evidence, particularly from the Palaeogene of the deep oceans in the Southern Hemisphere and especially from the Indian Ocean, is required to substantiate what can here only be the suggested origins and migration

routes of many ostracod taxa.

## SAMPLES AND OSTRACODA STUDIED

The samples studied originate from two areas:

1) The North Atlantic Ocean below 1000m present day water depth (PDWD) and north of the Tropic of Cancer (Latitude 23°N). The locations of the samples are shown in Fig. 1.

2) The western Pacific Ocean below PDWD 1000m, between Latitudes 32°N and 52°S, with most samples from the South West Pacific. The locations of the samples are shown in Fig. 2. Most of the samples studied are from DSDP sites; the

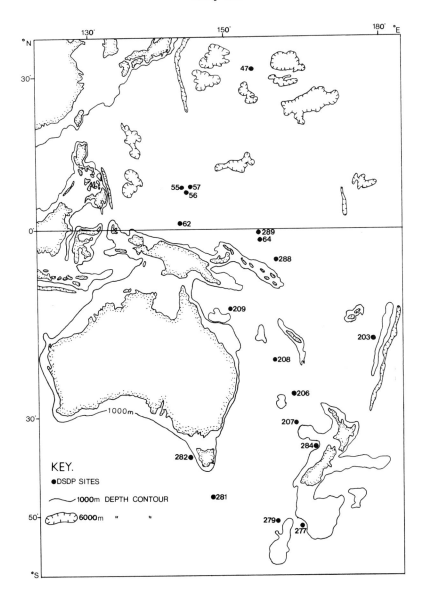

Fig. 2. The western Pacific Ocean showing the DSDP sites studied by Whatley (1983), Downing (1985), Millson (1987) and Ayress (1988).

locations, water depths and stratigraphical ages of these are detailed in Appendix 1 and include the samples studied by Whatley (1983) from the Miocene to Quaternary of the South West Pacific and Whatley & Coles (1987) from the Upper Miocene to Quaternary of Leg 94 in the North Atlantic. Details of the Quaternary and Recent samples from the North Atlantic studies of Porter (1984) and Davies (1981) respectively are given in Whatley & Ayress (1988). The Palaeogene samples from the South West Pacific (Millson, 1987) and North Atlantic (Coles, in prep.) are detailed in Appendix 1, as are additional Quaternary samples studied by Harpur (1985) and Ayress

(1988). The statigraphical ranges of Ostracoda in the two study areas have been compiled from all the available published and unpublished sources given in the reference list. Only verifiable records (by examination of type specimens or illustrations) are included in order to eliminate synonyms. The number of species considered has been reduced by the omission of the following:

1) All Krithinae (*Krithe* and *Parakrithe*) species; although *Krithe* is the most diverse genus in the deep sea (38 species occur in the Cainozoic of the North Atlantic alone) and both constituent genera include some cosmopolitan species, their taxonomy currently remains very problematical.

2) Very rare species represented by single specimens, since their stratigraphical ranges are almost certainly incomplete.

3) All suspected shallow water contaminant species (identified by the possession of an eye tubercle, details of preservation and known ecology).

4) Species endemic to the Ita Mai Tai Guyot (DSDP Site 200 in the West Pacific) because of the unique and unusual fauna of this site.

## DIVERSITY PATTERNS THROUGH THE CAINOZOIC

A total of 363 species are recorded from the Cainozoic of the Pacific, compared with 284 from the same interval in the North Atlantic. The total simple species diversity (number of species) of both oceans by stage and epoch and the number and percentage of species common to both oceans by stage and epoch, and the number of originations and extinctions of shared species are given in Table 1, together with the number of samples studied from both oceans of each stage and epoch. For the Miocene, only Middle Miocene samples were available from the Pacific, and no attempt was made to subdivide the Pliocene because of the relatively short duration of that epoch. Due to the paucity of studies on Recent faunas, especially from the Pacific, the Quaternary and Recent have been combined. Simple species diversity in both oceans and the total number of common species are plotted for each stage in

Fig. 3. These data indicate the following:

1) Total species diversity is 22% higher in the Pacific (363 species), than in the North Atlantic (284 species), despite the greater number of samples studied from the latter region (435) compared to that of the Pacific (211). However, the difference in diversity between the Pacific and North Atlantic is less than that indicated by Whatley & Coles (1987), mainly because of further research on North Atlantic faunas (Coles, in prep.), the elimination of synonyms, and the fact that Guyot endemic species (Site 200) are not considered in this study.

2) Species diversity increased through the Cainozoic in both oceans, reaching peaks of 169 and 154 species in the Quaternary and Recent of the Pacific and North Atlantic respectively. However, some differences in the pattern of species diversity in the two oceans are evident from Fig. 3. Palaeocene diversity is low in both oceans; this is only partially due to the low number of Palaeocene samples. Mean species abundance in Palaeocene samples is low; averaging 8.1 species per productive sample in the North Atlantic (Coles, in prep.) and 10.8 species in the Pacific (Millson, 1987). Palaeocene faunas most closely resemble those of Upper Cretaceous chalks in composition (see 'Origins' Section below) and lack many typical deep-sea genera such as *Bradleya*, *Henryhowella*, *Parakrithe*, *Pedicythere*, *Pennyella*, *Poseidonamicus*, and '*Thalassocythere*', which we do not encounter in either ocean until the Eocene (see Table 3). Eocene faunas are much more diverse than those of the Palaeocene in both oceans, with more than three or four times the number of species present. Although this is partly explained by virtue of the increased number of samples studied, the Eocene (especially Middle Eocene) increase in diversity reflects the establishment of a diverse, deep-sea fauna in both oceans, including many Tethyan cosmopolitan taxa (Benson, 1975; Steineck *et al.*, 1984, 1988). Mean species diversity in North Atlantic Eocene samples is 26, increasing from 18 in the Lower Eocene to 51 in the Upper Eocene (Coles, in prep.). Overall Oligocene diversity is slightly lower than Eocene diversity in both oceans. This is probably

Table 1. Species diversity and shared species in the North Atlantic and Pacific oceans.

| Age | K (U) | Palaeocene (L, U) | Eocene (L, M, U) | Oligocene (L, U) | Miocene (L, M, U) | Pliocene | Quaternary & Recent |
|---|---|---|---|---|---|---|---|
| Total N. Atlantic species | 15 | 23 (25) | 56 98 82 (113) | 99 104 (106) | 59 52 61 (66) | 66 | 154 |
| Total Pacific species | 12 | 38 (41) | 53 118 74 (140) | 37 79 (104) | (118) | 159 | 169 |
| Total shared species | 1 | 2 (2) | 10 15 13 (17) | 19 24 (24) | 22 31 32 (38) | 45 | 85 |
| N. Atlantic % shared species | 7 | 9 (8) | 18 15 16 (15) | 19 23 (23) | 37 59 52 (58) | 68 | 55 |
| Pacific % shared species | 8 | 5 (5) | 19 13 18 (12) | 51 30 (23) | (32) | 28 | 50 |
| Origination of shared species | 5 | 3 6 (9) | 15 16 7 (38) | 12 7 (19) | 2 6 3 (11) | 7 | 6 |
| Extinction of shared species | 0 | 0 0 (0) | 0 1 0 (1) | 2 0 (2) | 1 4 (5) | 2 | |
| Total N. Atlantic samples | 5 | 5 (10) | 21 13 12 (46) | 10 19 (29) | 10 30 47 (87) | 89 | 174 |
| Total Pacific samples | 9 | 3 (12) | 4 21 9 (34) | 11 12 (23) | (44) | 58 | 40 |

| | |
|---|---|
| Total number of Pacific Species | = 363 (26% shared) |
| Total number of N. Atlantic species | = 284 (33% shared) |
| Total number of shared species | = 95 |
| Shared species with first record in the Pacific | = 26 |
| Shared species with first record in the N. Atlantic | = 47 |
| Shared species which appear in the same epoch in both oceans | = 22 |

because of the fewer samples studied (especially in the Lower Oligocene of the Pacific) since Oligocene mean species diversity per sample exceeds that of the Eocene in the North Atlantic, where peak diversity actually occurs in the Upper Oligocene.

In the North Atlantic, species diversity decreases rapidly from the Oligocene to the Miocene, although three times as many Miocene as Oligocene samples were studied. This reflects over 40 species extinctions at the end of the Oligocene and the relatively few species originating (10) in the Lower Miocene, possibly because of the closure of Tethys (Thomas, 1987). In contrast, Miocene diversity in the Pacific is higher than Oligocene, probably due to the larger number of Miocene samples. Whatley & Coles (in press) also suggest that the diversity increase to its Upper Oligocene acme followed by a decline into the Miocene could, in the context of the North Atlantic, be also a function of evolution towards a climax community followed by a diversity crash due to a complex of competitive and other causes.

Diversity in the Miocene and Pliocene of the Pacific is notably higher than that of the North Atlantic during the same interval, despite the greater number of North Atlantic Neogene samples compared to the Pacific. The difference partly reflects the greater latitudinal and PDWD range of the Neogene Pacific samples (32° N - 52° S and 1389m - 3300m) compared with the North Atlantic samples (33° N - 53° N and 2417m - 3883m). Thus, the Neogene North Atlantic faunas documented herein and by Whatley & Coles (1987) undoubtedly lack many species confined to bathyal (1000m - 2000m) water depths. However, high diversity in the Pacific is also due to many species having originated in the S.W. Pacific region and which were subsequently introduced into the deep sea (to abyssal depths in the late Neogene) perhaps by crustal foundering (Whatley et al., 1983).

Pliocene diversity in the North Atlantic is very similar to that of the Miocene, whereas over 40 more species are recorded from the Pliocene of the Pacific than from the Miocene. This contrasts with the data of Whatley (1983) and is probably due to the elimination of homonyms and contaminant species in this study. Quaternary and Recent diversity in the Pacific is similar to that of the Pliocene, while the much higher Quaternary and Recent diversity in the North Atlantic relative to the Pliocene, reflects the higher number and greater latitudinal and PDWD range of the Quaternary and Recent samples compared with those from the Pliocene.

## COMMON SPECIES

At least 95 species are common to the Cainozoic of both the North Atlantic and the Pacific. The stratigraphical ranges of these in both oceans are given in Table 2. Of these, at least 65 also occur in the Indian Ocean (Whatley & Ayress, 1988) and others are also present in the South Atlantic (Benson, 1977; Benson & Peypouquet, 1983) or the Caribbean (Steineck et al., 1984). These oceans are not considered here because of the relatively incomplete sample coverage through the Cainozoic.

The number of species common to both the North Atlantic and Pacific increases steadily through the Cainozoic, with small decreases in the Upper Eocene and Lower Miocene (Table 1, Fig. 3). One third of all North Atlantic Cainozoic species also occur in the Pacific, whereas one quarter of all Pacific species also occur in the North Atlantic. A notable feature is the much higher percentage of shared species in the North Atlantic Miocene and Pliocene compared to the Pacific; the relatively high percentage of shared species in the Lower Oligocene of the Pacific reflects the low diversity of Lower Oligocene faunas in the Pacific. Only 2 Palaeocene species (Aversovalva alveiformis (Deltel) and Eucythere circumcostata Whatley & Coles) are common to both oceans, but the number of shared species increases rapidly in the Eocene. A maximum of 85 of the 95 shared species occur in the Quaternary and Recent of both oceans, comprising 50% and 55% of Pacific and North Atlantic species respectively, representing a marked increase over the 68 shared species reported by Whatley & Ayress (1988). The percentage of shared species in the North Atlantic reaches a peak of 68% in the Pliocene and then decreases due to an increase in the number of apparently endemic bathyal Quaternary and Recent species (Porter, 1984; Davies, 1981). Since all the North Atlantic Pliocene samples are from abyssal (>2000m) water depths, this suggests that abyssal species are more likely to be widely distributed than bathyal species and may reflect more homogeneous environmental conditions encountered at those depths or

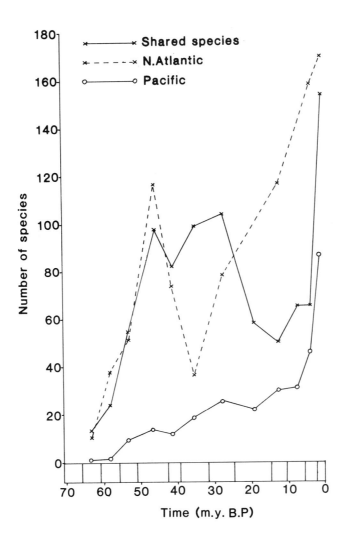

Fig. 3. Simple species diversity in the North Atlantic and western Pacific oceans and the number of shared species through the Cainozoic by stage.

the greater area of the abyssal plains and their virtue as a migratory pathway. Most (38) of the common species first appear in the Eocene and progressively fewer common species appear in subsequent epochs. This indicates that the longest-lived species are the most widely distributed in the world's oceans and that a diverse, cosmopolitan fauna was well established by the Middle Miocene. This fauna probably achieved circumequatorial distribution in the Tethys (Steineck *et al.*, 1984; 1988). However, not all typical deep-sea

genera are present in both oceans by the Eocene (see 'Origins' Section below). Younger true deep-sea species achieved worldwide distribution *via* the psychrosphere which allowed migration between oceans with increasing ease through the Neogene with the breakdown of barriers to global circulation, i.e. the opening of the Drake Passage. This is reflected in the high number of common species in the Pliocene and Quaternary. Also notable is the very low extinction rates of common species; only 10 become extinct in the entire

Table 2. Stratigraphical ranges of shared species in the North Atlantic and Pacific Oceans.

| Species | North Atlantic | Pacific |
| --- | --- | --- |
| *Abyssocypris atlantica* (Maddocks, 1977) | Recent | M. Eoc.-Quat. |
| *Abyssocythere trinidadensis* (van den Bold, 1957) | M. Eoc.-U. Mio. | L. Olig.-Plio. |
| *Abyssocythereis sulcatoperforata* (Brady, 1880) | U. Mio.-Quat. | L. Olig.-Quat. |
| *Agrenocythere hazelae* (van den Bold, 1946) | L. Olig.-Rec. | U. Olig.-Rec. |
| *Ambocythere caudata* (van den Bold 1965) | M. Mio.-Rec. | Plio.-Quat. |
| *Ambocythere* sp. Downing, 1985 | Rec. | Plio. |
| *Aratrocypris praealta* Whatley *et al.*, 1985 | Olig. | U. Pal.-M. Mio. |
| *Aratrocypris* n. sp. Whatley *et al.*, in press | U. Pal.-Plio. | L. Eoc.-Quat. |
| *Argilloecia* sp. 3 Coles, in preparation | L.-M. Eoc. | Eoc. |
| *Argilloecia* sp. 6 Coles, in preparation | M. Eoc.-M. Mio. | U. Pal.-Quat. |
| *Argilloecia* sp. 8 Coles, in preparation | M. Eoc.-L Mio. | Plio.-Quat. |
| *Argilloecia* sp. 10 Coles, in preparation | U. Eoc.-U. Olig. | Plio.-Quat. |
| *Australoecia micra* (Bonaduce *et al.*, 1975) | U. Eoc.-Rec. | M. Eoc.-Quat. |
| *Australoecia* n. sp. Coles & Whatley, in press | M. Eoc.-Olig. | Quat. |
| *Aversovalva alveiformis* (Deltel, 1964) | L. Pal.-U. Olig. | U. Pal.-U. Olig. |
| *Aversovalva atlantica* Whatley & Coles, 1987 | L. Mio.-Rec. | Plio.-Quat. |
| *Aversovalva hydrodynamica* Whatley & Coles, 1987 | L. Olig.-Rec. | M. Mio.-Quat. |
| *Aversovalva* cf. *pinarense* (van den Bold, 1946) | L. Eoc.-Rec. | L. Pal.-Quat. |
| *Aversovalva* n. sp. Coles & Whatley, in press | L. Olig.-L. Mio. | M. Eoc.-Quat. |
| *Aversovalva* sp. Ayress, 1988 | Recent | Plio.-Quat. |
| *Bairdia* gr. *subcircinata* Brady & Norman, 1869 | L. Eoc.-Rec. | U. Olig.-Quat. |
| *Bairdoppilata cassida* (van den Bold, 1946) | M. Eoc.-L. Mio. | U. Cret.-Quat. |
| *Bathycythere audax* (Brady & Norman, 1869) | Plio.-Rec. | M. Mio.-Quat. |
| *Bradleya dictyon* (Brady, 1880) | U. Olig.-Rec. | L. Olig.-Rec. |
| *Bythoceratina scaberrima* (Brady, 1886) | Plio.-Rec. | M. Mio.-Quat. |
| *Bythocypris affinis* (Brady, 1886) | Rec. | Quat. |
| *Bythocypris aturica* (Deltel, 1964) | L. Eoc.-M. Mio. | L. Eoc.-U. Olig. |
| *Bythocypris mozambiquensis* (Maddocks, 1969) | Plio.-Quat. | Plio.-Quat. |
| 'Bythocythere' *bathytatos* Whatley & Coles, 1987 | L. Eoc.-Plio. | Quat. |
| 'Bythocythere' n. sp. Coles & Whatley, in press | M. Eoc.-Rec. | Plio.-Quat. |
| *Cardobairdia asymmetrica* (van den Bold 1946) | M. Eoc.-Rec. | U. Cret.-M. Mio. |
| *Chejudocythere* sp. 1 Coles, in preparation | U. Olig.-Rec. | Quat. |
| 'Cluthia' sp. Downing, in preparation | Plio.-Rec. | Plio.-Quat. |
| *Cytherella gamardensis* Deltel, 1964 | L. Eoc.-U. Olig. | M. Eoc. |
| *Cytheropteron branchium* Whatley *et al.*, 1986 | M. Eoc.-Quat. | Plio.-Quat. |
| *Cytheropteron garganicum* Bonaduce *et al.*, 1975 | Quat.-Rec. | Plio.-Quat. |
| *Cytheropteron lineoporosa* Whatley & Coles, 1987 | U. Eoc.-Rec. | Plio.-Quat. |
| *Cytheropteron paucipunctatum* Whatley & Coles, 1987 | L. Olig.-Quat. | Quat. |
| *Cytheropteron pherozigzag* Whatley *et al.*, 1986 | L. Olig.-Quat. | Plio.-Quat. |
| *Cytheropteron retrosulcatum* Colalongo & Pasini, 1980 | Quat.-Rec. | M. Mio.-Quat. |
| *Cytheropteron testudo* Sars, 1869 | U. Mio.-Rec. | M. Eoc.-Quat. |
| *Cytheropteron tressleri* Whatley & Coles, 1987 | Plio.-Quat. | Plio.-Quat. |

Table 2. Stratigraphical ranges of shared species in the North Atlantic and Pacific Oceans - (continued)

| Species | North Atlantic | Pacific |
|---|---|---|
| *Cytheropteron trifossata* Whatley & Coles, 1987 | L. Olig.-Quat. | Plio.-Quat. |
| *Cytheropteron* sp. 1 Coles, in preparation | Olig. | Plio.-Quat. |
| *Cytheropteron* sp. 6 Coles, in preparation | L. Eocene | U. Pal.-Plio. |
| *Dutoitella suhmi* (Brady, 1880) | U. Mio.-Rec. | U. Olig.-Quat. |
| *Echinocythereis echinata* (Sars, 1865) | L. Olig.-Rec. | M. Mio. |
| *Eopaijenborchella cymbula* (Ruggieri, 1950) | Quat.-Rec. | Quat. |
| *Eucythere circumcostata* Whatley & Coles, 1987 | L. Pal.-Quat. | U. Cret.-Quat. |
| *Eucythere concinna* Ciampo, 1981 | U. Olig.-U. Mio. | M. Mio. |
| *Eucythere hyboma* Whatley & Coles, 1987 | M. Mio.-Quat. | U. Eoc.-Quat. |
| *Eucythere multipunctata* Whatley & Coles, 1987 | U. Pal.-Rec. | L. Eoc.-Quat. |
| *Eucythere parapubera* Whatley & Downing, 1983 | M. Eoc.-Rec. | U. Cret.-Quat. |
| *Eucythere calabra* (Colalongo & Pasini, 1980) | L. Eoc.-Rec. | L. Eoc.-Quat. |
| *Eucytherura* n. sp. Coles & Whatley, in press | M. Eoc.-U. Olig. | M. Mio.-Quat. |
| *Eucytherura* sp. 2 Whatley & Coles, 1987 | U. Olig.-Rec. | U. Cret.-Quat. |
| *Eucytherura* sp. 1 Whatley & Coles, 1987 | L. Olig.-Quat. | M. Eoc.-Quat. |
| *Eucytherura* sp. 1 Ayress, 1988 | Rec. | Quat. |
| *Eucytherura* sp. 2 Ayress, 1988 | Rec. | Quat. |
| *Heinia dryppa* Whatley & Coles, 1987 | L. Mio.-Quat. | L. Eoc.-Quat. |
| *Henryhowella* gr. *asperrima* (Reuss, 1850) | M. Eoc.-Rec. | M. Eoc.-Rec. |
| *Henryhowella dasyderma* (Brady, 1886) | M. Mio.-Rec. | M. Mio.-Quat. |
| *'Hyphalocythere' presequenta* (Benson, 1977) | L. Eoc.-Rec. | Plio.-Quat. |
| *Palmoconcha* sp. B (Cronin, 1983) | Rec. | L. Eoc.-Quat. |
| *Palmoconcha?* sp. 2 (Whatley & Coles, 1987) | M. Eoc.-Plio. | M. Mio.-Quat. |
| *Parahemingwayella* n. sp. Coles & Whatley, in press | U. Eoc.-U. Olig. | Quat. |
| *Pariceratina* n. sp. Millson, 1987 | U. Pal.-M. Eoc. | L. Eoc.-M. Mio. |
| *Pedicythere* cf. *phryne* Bonaduce *et al.*, 1975 | L. Eoc.-Rec. | Plio.-Quat. |
| *Pedicythere polita* Colalongo & Pasini, 1980 | Quat.-Rec. | L. Eoc.-Quat. |
| *Pelecocythere sylvesterbradleyi* Athersuch, 1979 | U. Eoc.-Rec | Plio.-Quat. |
| *Pelecocythere trinidadensis* (van den Bold, 1960) | U. Eoc.-U. Mio. | U. Olig.-Quat. |
| *Pennyella dorsoserrata* (Brady, 1880) | U. Olig.-Rec. | M. Mio. |
| *Pennyella horridus* (Whatley & Coles, 1987) | U. Mio.-Quat. | Plio.-Quat. |
| *Pennyella* sp. Coles, in preparation | U. Mio. | M. Mio.-Quat. |
| *Phacorhabdotus* n. sp. Coles & Whatley, in press | M. Eoc.-L. Mio. | L. Olig.-Quat. |
| *Posacythere* cf. *undata* (Colalongo & Pasini, 1980) | L. Olig.-Rec. | Quat. |
| *Poseidonamicus anteropunctatus* Whatley *et al.*, 1986 | L. Olig.-U. Mio. | M. Mio.-Quat. |
| *Poseidonamicus* gr. *major* Whatley *et al.*, 1986 | U. Mio.-Rec. | M. Mio.-Quat. |
| *Poseidonamicus minor* Benson, 1972 | U. Olig.-U. Mio. | M. Mio.-Quat. |
| *Poseidonamicus praenudus* Whatley *et al.*, 1986 | U. Mio.-Quat. | Plio.-Quat. |
| *Propontocypris* sp. 1 Coles, in preparation | L. Eoc.-U. Olig. | Plio.-Quat. |
| *Propontocypris* cf. *trigonella* (Sars, 1865) | Quat. | Quat. |
| *Pterygocythere mucronalatum* (Brady, 1880) | L. Pal.-Rec. | L. Olig.-Rec. |
| *Rimacytheropteron longipunctata* (Breman, 1975) | Plio.-Quat. | Plio.-Quat. |

Table 2. Stratigraphical ranges of shared species in the North Atlantic and Pacific Oceans - (continued)

| Species | North Atlantic | Pacific |
| --- | --- | --- |
| *Rimacytheropteron* n. sp. Coles & Whatley, in press | L. Eoc.-U. Mio. | Quat. |
| *Rockallia enigmatica* Whatley *et al.*, 1978 | U. Mio.-Rec. | Quat. |
| *'Rostrocythere'* n. sp. Coles & Whatley, in press | L. Olig.-Rec. | Quat. |
| *Ruggieriella* sp. A Cronin, 1983 | Rec. | L. Eoc.-Quat. |
| *Saida* n. sp. Coles & Whatley, in press | L. Eoc.-U. Olig. | M. Mio. |
| *Saida* sp. A Cronin, 1983 | Rec. | Quat. |
| *Semicytherura coeca* Ciampo, 1986 | U. Olig. | Plio.-Quat. |
| *'Thalassocythere'* acanthoderma (Brady, 1880) | U. Olig.-Rec. | M. Mio.-Quat. |
| *Xestoleberis abyssoris* Whatley & Coles, 1987 | U. Eoc.-Plio. | Plio.-Quat. |
| *Xestoleberis profundis* Whatley & Coles, 1987 | M. Eoc.-Rec. | M. Mio.-Quat. |
| *Xiphichilus*? cf. *gracilis* (Chapman, 1915) | M. Eoc.-U. Olig. | Quat. |

Note *Bathycythere audax* : *Cythere audax* Brady & Norman 1869 from the Recent of the North Atlantic is a senior subjective homonym of the type species of *Bathycythere*, *B. vanstraateni* Sissingh 1971.

Cainozoic (Table 1).

## ORIGINS

Benson (1969) was impressed with 'the possible Cretaceous aspect of the deep-sea ostracod fauna' and subsequently it has been suggested that many taxa entered the deep-sea from ancestral stocks from the Cretaceous continental shelves when oceanic thermal gradients were much less marked than after the formation of the psychrosphere and the isolation of the deep sea fauna in the Middle Eocene (Benson, 1975; 1979; Benson *et al.*, 1984; Steineck *et al.*, 1984). This is demonstrated by the strong similarity, at least at the generic level, between the Cainozoic (especially Palaeogene) deep-sea faunas of the North Atlantic and Pacific Oceans and the Upper Cretaceous chalk faunas, especially in North West Europe. Table 3 gives the stratigraphical ranges of 38 Cainozoic deep-sea genera which are undoubtedly recorded from the Upper Cretaceous. The stratigraphical range of each genus in the North Atlantic and Pacific oceans (if present in both) and their geographical occurrence in the Upper Cretaceous of six well-studied regions are

shown. These regions are: southern England (Neale, 1978; Weaver, 1982), East Germany: Rügen (Herrig, 1966), The Netherlands (Bonnema, 1941), the Gulf Coast of the United States (Alexander, 1929; Smith, 1978; Maddocks, 1985), South Africa (Dingle, 1981; 1984) and western Australia (Bate, 1972; Neale, 1975). In addition, the presence of these genera in other, less well studied areas, or areas with few genera in common with Cainozoic deep-sea faunas, is indicated, i.e. in France (Babinot *et al.*, 1985), the Caribbean (van den Bold, 1946), Argentina (Bertels, 1975) and the Falklands Plateau (Dingle, 1981). The stratigraphical ranges of the additional 23 genera which occur in the North Atlantic and Pacific, but are not definitely recorded from the Cretaceous are also given, in order to investigate their origins.

From the data in Table 3, it is clear that the Cretaceous chalk faunas of East Germany and The Netherlands have most genera in common with Cainozoic deep-sea faunas, with 29 and 27 genera in common respectively. Particularly notable are the first records of *Aratrocypris* as *Paradoxostoma*? *cretacea* Bonnema 1941 (Whatley, Witte & Coles, in press), *Chejudocythere* as *Loxocythere subtrigonalis* Herrig, 1963 and new

Table 3: Stratigraphical ranges of shared genera in the Cainozoic of the North Atlantic and Pacific, and the occurrence of deep-sea genera in the Upper Cretaceous of selected regions; England (1), East Germany (2), Netherlands (3), United States Gulf Coast (4), South Africa (5), Western Australia (6), other areas (7).

| Cainozoic deep-sea genera | Range in the North Atlantic Ocean | Range in the Pacific Ocean | Occurrence in Cretaceous | | | | | | |
|---|---|---|---|---|---|---|---|---|---|
| | | | 1 | 2 | 3 | 4 | 5 | 6 | 7 |
| *Abyssocypris* | M. Eoc.-Rec. | M. Eoc.-Rec. | | | | | | | |
| *Abyssocythere* | U. Pal.-Rec. | L. Olig.-Rec. | | | | | | | x |
| *Abyssocythereis* | U. Mio.-Quat. | L. Olig.-Rec. | | | | | | | |
| *Agrenocythere* | U. Pal.-Rec. | L. Olig.-Rec. | | | | | | | |
| *Agulhasina* | | L. Pal.-Quat. | | | | | x | x | |
| *Ambocythere* | M. Eoc.-Rec. | L. Olig.-Rec. | | | | | | | |
| *Anchistrocheles* | M. Eoc.-Rec. | U. Quat. | | | | | | | |
| *Aratrocypris* | U. Pal.-Rec. | U. Pal.-Rec. | | x | | | | | |
| *Argilloecia* | L. Pal.-Rec. | U. Pal.-Rec. | | x | x | x | | | x |
| *Australoecia* | L. Eoc.-Rec. | M. Eoc.-Quat. | ? | | | | | | |
| *Aversovalva* | L. Pal.-Rec. | L. Pal.-Quat. | x | x | x | x | x | x | x |
| *Bairdia* | L. Pal.-Rec. | U. Pal.-Rec. | x | x | x | x | x | x | x |
| 'Bairdia' n. gen. | M. Eoc.-L. Plio. | L. Olig.-L. Mio. | | | | | | | |
| *Bairdoppilata* | L. Pal.-Rec. | U. Pal.-Rec. | x | x | x | x | x | | x |
| *Bathycythere* | L. Plio.-Rec. | M. Mio.-Quat. | | | | | | | |
| *Bradleya* | U. Olig.-Rec. | M. Eoc.-Rec. | | | | | | | |
| *Bythoceratina* | L. Eoc.-Rec. | L. Eoc.-Rec. | x | x | x | x | | | x |
| *Bythocypris* | L. Eoc.-Rec. | L. Eoc.-Rec. | x | x | x | x | x | x | x |
| 'Bythocythere' n. gen. | L. Eoc.-Rec. | L. Plio-Quat. | x | ? | | | | ? | |
| *Cardobairdia* | U. Pal.-Rec. | M. Mio. | x | x | x | | | | x |
| *Chejudocythere* | L. Eoc.-Rec. | Quat. | x | | | | | | |
| *Cytheralison* | | L. Eoc.-Quat. | | | | | | x | |
| *Cytherella* | L. Pal.-Rec. | L. Pal.-Rec. | x | x | x | x | x | x | x |
| *Cytheropteron* | L. Pal.-Rec. | U. Pal.-Rec. | x | x | x | x | x | x | x |
| *Dutoitella* | M. Eoc.-Rec. | M. Eoc.-Rec. | | | | | x | | |
| *Echinocythereis* | L. Olig.-Rec. | M. Mio. | | | | | | | |
| *Eopaijenborchella* | U. Quat.-Rec. | U. Plio.-Quat. | | | | | | | |
| *Eucythere* | L. Pal.-Rec. | L. Pal.-Quat. | x | x | x | x | | | |
| *Eucytherura* | L. Pal.-Rec. | L. Pal.-Quat. | x | x | x | x | x | x | x |
| *Heinia* | L. Eoc.-Rec. | L. Eoc.-Quat. | | | | | | | |
| *Hemiparacytheridea* | L.-M. Eoc. | M. Mio. | x | x | x | | | x | x |
| *Henryhowella* | M. Eoc.-Rec. | M. Eoc.-Rec. | | | | | | | x |
| *Krithe* | L. Pal.-Rec. | U. Pal.-Rec. | | x | x | x | x | x | x |
| *Macrocypris* | L. Pal.-Rec. | M. Eoc.-Rec. | x | x | x | x | | x | |
| *Palmoconcha* | U. Pal.-Rec. | L. Eoc.-Quat. | | x | x | x | | | |
| *Parahemingwayella* | L. Pal.-Rec. | M. Mio.-Quat. | | | | | | x | |
| *Parakrithe* | M. Eoc.-Rec. | M. Eoc.-Rec. | | | | | | | ? |

Table 3. - (continued)

| Cainozoic deep-sea genera | Range in the North Atlantic Ocean | Range in the Pacific Ocean | Occurrence in Cretaceous 1 | 2 | 3 | 4 | 5 | 6 | 7 |
|---|---|---|---|---|---|---|---|---|---|
| *Pariceratina* | U. Pal.-M. Eoc. | L. Eoc.-M. Mio. | x | x | x | | x | x | |
| *Pedicythere* | L. Eoc.-Rec. | L. Eoc.-Quat. | x | x | x | | x | x | x |
| *Pelecocythere* | L. Eoc.-Rec. | U. Olig.-Quat. | | | | | | | |
| *Pennyella* | L. Eoc.-Rec. | M. Mio.-Quat. | | | | | | x | |
| *Phacorhabdotus* | L. Pal.-L. Mio. | L. Olig.-Quat. | x | x | x | x | | | x |
| *Polycope* | L. Quat.-Rec. | M. Mio.-Quat. | x | x | x | | | | |
| *Posacythere* | L. Olig.-Rec. | Quat. | x | | | | | | ? |
| *Poseidonamicus* | L. Olig.-Rec. | M. Eoc.-Rec. | | | | | | | |
| *Propontocypris* | L. Eoc.-Rec. | U. Plio.-Quat. | | | | | | | |
| *Pseudocythere* | M. Eoc.-Rec. | L. Eoc.-Rec. | | x | x | | | | |
| *Pterygocythere* | L. Pal.-Rec. | M. Mio.-Rec. | x | x | x | x | x | | x |
| *Rimacytheropteron* | L. Eoc.-Rec. | U. Plio.-Quat. | | | | | | | |
| *Rockallia* | L. Olig.-Rec. | M. Eoc.-Quat. | | | | | | | |
| 'Rostrocythere' n. gen. | L. Olig.-Rec. | Quat. | | x | x | | | | |
| *Ruggieriella* | Rec. | L. Eoc.-Quat. | | | | | | | |
| *Saida* | L. Eoc.-Rec. | M. Eoc.-Quat. | x | x | x | | | x | |
| *Semicytherura* | U. Olig. | Plio.-Quat. | | x | x | x | | x | |
| *Swainocythere* | M. Eoc.-Quat. | Quat. | | | | | | | |
| 'Thalassocythere' | L. Eoc.-Rec. | M. Mio.-Rec. | | | | | | ? | |
| *Toolongella* | | L. Eoc.-Plio. | | | | | | x | |
| *Trachyleberidea* | L. Pal.-Rec. | | x | x | x | x | | | x |
| *Xestoleberis* | L. Pal.-Rec. | L. Pal.-Rec. | x | x | x | x | x | | x |
| *Xylocythere* | U. Olig.-Rec. | L.-M. Mio. | | | | | | | |
| *Zabythocypris* | Rec. | L. Plio.-Quat. | | | | | | | |

Note 1. '*Bairdia*', '*Bythocythere*', '*Rostrocythere*' and '*Thalassocythere*' (='*Hyphalocythere*') are assigned to new genera in Coles & Whatley (in press). 2. *Dutoitella* Dingle 1981 includes '*Neoatlanticythere*' and '*Suhmicythere*' of Benson and *Pennyella* Neale 1974 includes '*Oxycythereis*' of Benson (Coles & Whatley, in press).

genera described in Coles & Whatley (in press) herein referred to as '*Bythocythere*' and '*Rostrocythere*'. Fewer genera are shared with the Upper Cretaceous of Western Australia and South Africa, although certain genera, namely *Agulhasina*, *Cytheralison*, *Pennyella* and *Toolongella* are only recorded from these areas in the Cretaceous. All of these genera, except *Pennyella* Neale, 1975, only occur in the Southern Hemisphere and not in the North Atlantic. The other deep-sea genera which are not recorded from the Cretaceous of the regions in Table 3 occur in the Cretaceous of other areas, i.e. *Henryhowella* in Argentina (Bertels, 1975) and *Parahemingwayella* on the Falkland Plateau (Dingle, 1984).

Most of the 38 Cainozoic genera also present in

the Cretaceous probably range throughout the Cainozoic of both oceans, although some small, rare genera have apparently limited ranges. The recorded extinctions of *Hemiparacytheridea*, *Pariceratina* and *Phacorhabdotus* in the North Atlantic are considered to be reliable, as are those of *Pariceratina* and *Toolongella* in the Pacific. From the stratigraphical ranges of the genera in Table 3, some observations can be made on the geographical origins of certain genera. A total of 34 genera have their earliest records in the North Atlantic, compared to only 13 genera first appearing in the Pacific and 14 genera which appear contemporaneously in both oceans. While some of these records will probably be modified by additional sampling, especially in the case of small, rare genera, other genera clearly appear in the North Atlantic earlier than in the Pacific or *vice versa*, for example:

1) *Abyssocythere* ranges back to the Upper Cretaceous and Palaeocene of the Caribbean (van den Bold, 1957), Danian of Senegal (Colin, 1987) and Upper Palaeocene of the North Atlantic (Upper Palaeocene of DSDP Site 549, Coles, in prep.), but does not reach the Pacific until the Lower Oligocene (Steineck *et al.*, 1988; Millson, 1987, in DSDP Site 289).

2) *Abyssocythereis* is first recorded from the Lower Oligocene of the Pacific (Millson, 1987), but does not appear in the North Atlantic until the Upper Miocene (Zone NN 9) of the North Atlantic in DSDP Site 563 (Coles, in prep.).

3) *Agrenocythere* is common in Palaeocene and Eocene bathyal sediments in the North Atlantic, western Europe and the Caribbean (Benson, 1972; Steineck *et al.*, 1984; Ducasse *et al.*, 1985; Coles, in prep., in DSDP Site 549), but does not appear in the Pacific until the Lower Oligocene (Steineck *et al.*, 1988). Guernet (1985) recorded *Agrenocythere* from the Middle Eocene of the Indian Ocean (DSDP Site 220) suggesting an eastward migration through the Tethys, or *via* the Cape of Good Hope route.

4) *Ambocythere* is present in the Middle Eocene of Barbados (Steineck *et al.*, 1984) and DSDP Site 390 in the Bahama Basin (Guernet 1982, as *Orionina*? sp.) and the Upper Eocene of

Cuba (van den Bold, 1958), but does not occur in the Pacific until the Lower Oligocene (Steineck *et al.*, 1988).

5) *Bradleya* occurs in the Middle Eocene of the South West Pacific (Whatley, 1985), but is only known from the Upper Oligocene (Zone NP 23) of the North Atlantic in DSDP Site 549 (Coles, in prep.).

6) *Echinocythereis* ranges back to the Lower Oligocene (Zone NP 21) of DSDP Site 549 in the North Atlantic (Coles, in prep.) and is moderately common from the Middle Miocene to the Recent of the North Atlantic. However, it is only known in the Middle Miocene of the Pacific (Harlow, lit. comm., 1985).

7) *Pelecocythere* ranges back to the Lower Eocene in the North Atlantic in DSDP Site 550 (Coles, in prep.) and the Middle Eocene of the Caribbean (as *Eocytheropteron*? *trinidadensis* in Steineck *et al.*, 1984), but does not reach the Pacific until the Upper Oligocene (Steineck *et al.*, 1988; Millson, 1987 in DSDP Site 208).

8) *Phacorhabdotus* is widespread in the Palaeocene of the North Atlantic (DSDP sites 549 and 550, Coles, in prep.) and in West Africa (Colin, 1987), but only appears in the Lower Oligocene of the Pacific (Steineck *et al.*, 1988; Millson, 1987 in DSDP Site 206).

9) *Poseidonamicus* first appears in the Middle Eocene of the South West Pacific (Whatley, 1985) and does not reach the North Atlantic until the Lower Oligocene (Zone NP 21) in DSDP Site 549 (Coles & Whatley, in press).

10) *Pterygocythere* ranges back to the Lower Palaeocene of the North Atlantic (Coles, in prep.; in DSDP Site 550), but only appears in the Pacific in the Lower Oligocene, represented by *P. mucronalatum* (Brady, 1880) in Steineck *et al.* (1988).

11) *Rimacytheropteron* is first reported from the Lower Eocene of DSDP Site 549 in the North Atlantic (Coles & Whatley, in press), but does not occur in the Pacific until the Upper Pliocene of DSDP Site 208 (Downing, 1985; Whatley & Coles, 1987).

12) *Trachyleberidea* probably entered the deep sea from Cretaceous stocks in Europe or

North America; it first appears in the Lower Palaeocene of DSDP Site 550 (Coles, in prep.) and is widespread throughout the Palaeogene of the North Atlantic (Ducasse & Peypouquet, 1979; Cronin & Compton-Gooding, 1987: Coles, in prep. in the Upper Palaeocene to Upper Oligocene of DSDP Site 549). Although *Trachyleberidea* is known from the Middle Eocene of the Indian Ocean (Guernet, 1985, in DSDP Site 214), it is unknown in the Pacific.

Other genera only occur either in the Cainozoic of the North Atlantic or the Pacific, and are not recorded from the Cretaceous. Most deep-sea Pacific genera which do not occur in the North Atlantic are rare, as yet undescribed and mainly belong to the Trachyleberididae or Bythocytheridae (Whatley, 1983; Table 2). However, a few genera such as *Tongacythere* (Upper Palaeocene to Quaternary) seem to be confined to the Pacific, while *Typhlocythere* occurs in the Middle Miocene to Quaternary of the Pacific Ocean and Upper Miocene of the Mediterranean Basin, but is unrecorded from the North Atlantic.

Several North Atlantic Cainozoic genera appear to be absent from the Pacific. *Buntonia* ranges from the Middle Eocene (DSDP Site 549, Coles, in prep.) to the Recent of the North Atlantic and probably orginated from stocks on the West African shelf. *Rectobuntonia* and *Loxoconchidea* occur in the Quaternary and Recent of the North Atlantic and may have orginated in the Mediterranean, although *Loxoconchidea* also occurs in the Quaternary of the Timor Sea in DSDP Site 262 (Ayress, 1988). *Muellerina abyssicola* (Sars, 1865) and *Thaerocythere crenulata* (Sars, 1865) both occur at depths in excess of 1000m in the North Atlantic and originated in North West Europe (Whatley & Ayress, 1988).

Of the 95 common species, 47 first appear in the North Atlantic, 26 first appear in the Pacific and 22 appear in the same epoch of both oceans. These figures are in close agreement with those for the origins of deep-sea genera, and demonstrate the varied origins of the Palaeogene to Recent cosmopolitan deep-sea fauna, contrasting with Whatley & Ayress (1988) who suggested that most Quaternary cosmopolitan species originated

in the Pacific and migrated down a diversity gradient to the Atlantic. However, further detailed studies, especially of Palaeogene faunas in the world's oceans could modify our conclusions concerning the origin and migrational history of many of the deep-sea taxa.

## CONCLUSIONS

The results of this study indicate that:

1)   The ostracod faunas of both the North Atlantic and Pacific increase in diversity through the Cainozoic to a maximum in the Quaternary and Recent. High diversity (over 80 species) also occurs in the Middle Eocene to Upper Oligocene of the North Atlantic and in the Middle Eocene, Middle Miocene and Pliocene of the Pacific. Cainozoic Pacific faunas are 22% more diverse than those from the North Atlantic. However, the observed diversity patterns are strongly influenced by the availability, latitudinal and depth range of samples.

2)   At least 95 species are common between the North Atlantic and Pacific, further confirming the cosmopolitan nature of Cainozoic deep-sea faunas. One third of all North Atlantic species also occur in the Pacific, while one quarter of all Pacific species also occur in the North Atlantic. Faunal similarity is very low in the Palaeocene, but greatly increases in the Eocene, coincident with the development of a cosmopolitan psychrospheric fauna. The number of common species increases through the Cainozoic to reach a marked peak in the Quaternary and Recent. The steep increase in pandemism during the Neogene to Quaternary is possibly related to the opening of the Drake Passage.

3)   The Cainozoic deep-sea faunas of the North Atlantic and the Pacific derive mainly from Cretaceous ancestors; at least 38 genera are recorded from Upper Cretaceous shallow water environments, especially the chalk sea of North West Europe. The greater number of both Cainozoic deep-sea genera and species first recorded in the North Atlantic compared to the Pacific suggests that a substantial, if not major proportion of the Cainozoic deep-sea fauna may

have originated in the North Atlantic and Caribbean region, with a lesser proportion originating in the Pacific.

## ACKNOWLEDGEMENTS

The authors wish to thank the two anonymous referees for their helpful comments and criticisms. They also wish to thank Dr C. Maybury for her extensive editorial improvements, her skill and patience. Drs S. E. Downing, K. J. Millson and N. R. Ainsworth are thanked for allowing us access to unpublished data as are P. Smith, W. Harpur, H. Davies and C. Porter. G. Coles wishes to acknowledge a NERC studentship which allowed him to study North Atlantic deep-sea Ostracoda.

## APPENDIX 1

The location, present day water depth (PDWD) and age of samples used in this study.

### A. North Atlantic

DSDP samples (Legs 80, 82 and 94)

| Site | Age of samples | PDWD(m) | Latitude | Longitude |
|------|----------------|---------|----------|-----------|
| 549 | U. Palaeocene - U. Oligocene | 2515 | 49° 05.28'N | 13° 05.88'W |
| 550 | L. Palaeocene - L. Eocene | 4420 | 48° 30.91'N | 13° 26.37'W |
| 558 | L. Oligocene - U. Miocene | 3754 | 37° 46.20'N | 37°20.61'W |
| 563 | M. - U. Miocene | 3786 | 33° 38.50'N | 43° 46.04'W |
| 606 | L. Pliocene - L. Quaternary | 3007 | 37° 20.32'N | 35° 29.99'W |
| 607 | U. Miocene - Quaternary | 3427 | 41° 00.07'N | 32° 57.44'W |
| 608 | U. Miocene - Quaternary | 3526 | 42° 50.21'N | 23° 05.25'W |
| 609 | U. Miocene - Quaternary | 3884 | 49° 52.67'N | 24° 14.29'W |
| 610 | U. Miocene - Quaternary | 2417 | 53° 13.30'N | 18° 53.21'W |
| 611 | U. Miocene - Quaternary | 3201 | 52° 50.47'N | 30° 19.58'W |

Operation Navado (1964-1971) boreholes, Extra-Iberian Portal Region (Harpur, 1985). All samples are of Upper Quaternary age.

| Borehole | PDWD(m) | Latitude | Longitude |
|----------|---------|----------|-----------|
| BEAUMONT | 1200 | 35° 08'N | 07° 24'W |
| NAVADO IIB | 1350 | 34° 35'N | 07° 09'W |
| THEO II | 1440 | 35° 46'N | 07° 38'W |
| BALEN | 2417 | 34° 56'N | 08° 09'W |
| HAIGH | 2421 | 36° 19'N | 08° 38'W |
| DAY | 3700 | 34° 56'N | 09° 08'W |

### B. West Pacific

All DSDP samples:

| Site | Age of samples | PDWD(m) | Latitude | Longitude |
|------|----------------|---------|----------|-----------|
| 47 | Pliocene | 2689 | 32° 26.9 'N | 157° 42.7'E |
| 55 | M. Miocene - Pliocene | 2850 | 9° 18.1 'N | 142° 32.1'E |
| 56 | M. Miocene | 2508 | 8° 22.4 'N | 143° 33.6'E |
| 57 | Pliocene | 3300 | 8° 40.0 'N | 143° 32.0'E |

Appendix 1. The location, present day water depth (PDWD) and age of samples used in this study - (continued)

| Site | Age of samples | PDWD(m) | Latitude | Longitude |
|------|----------------|---------|----------|-----------|
| 62 | M. Miocene - Pliocene | 2591 | 10° 52.2 'N | 141° 56.30'E |
| 64 | M. Miocene - Quaternary | 2052 | 1° 44.4 'S | 158° 36.54'E |
| 203 | U. Pliocene - Quaternary | 2720 | 22° 09.22'S | 177° 32.70'E |
| 206 | L. Palaeocene - Quaternary | 3196 | 32° 00.75'S | 165° 27.15'E |
| 207 | U. Cretaceous - Quaternary | 1389 | 36° 57.75'S | 165° 26.06'E |
| 208 | U. Cretaceous - Quaternary | 1545 | 26° 06.61'S | 161° 13.27'E |
| 209 | M. Eocene - Quaternary | 1428 | 15° 56.19'S | 152° 11.27'E |
| 277 | L. Palaeocene - Quaternary | 1214 | 52° 13.43'S | 166° 11.48'E |
| 279 | Quaternary | 3341 | 51° 20.14'S | 162° 38.10'E |
| 281 | Quaternary | 1591 | 47° 59.84'S | 147° 45.85'E |
| 282 | Quaternary | 4202 | 42° 14.67'S | 143° 29.18'E |
| 284 | Quaternary | 1066 | 40° 30.48'S | 167° 40.81'E |
| 288 | M. Miocene - Pliocene | 3000 | 5° 38.35'S | 161° 41.53'E |

## REFERENCES

Alexander, C. I. 1929. Ostracoda of the Cretaceous of North Texas. *Texas Univ. Bull.*, 2907, 137pp.

Ayress, M. A. 1988. *Late Pliocene to Quaternary Deep-sea Ostracoda from the Eastern Indian Ocean and Southwestern Pacific Oceans.* Unpubl. Ph.D. thesis, University of Wales, 2 vols., 1088 pp.

Babinot, J-F., Colin, J-P. & Damotte, R. 1985. Crétacé supérieur. *In* Oertli, H. J. (Ed.), *Atlas des ostracodes de France*, 211-255. *Bull. Centres Rech. Explor.-Prod. Elf-Aquitaine*, Pau, Mém. 9.

Bate, R. H. 1972. Upper Cretaceous Ostracoda from the Carnarvon Basin, Western Australia. *Spec. Pap. Palaeont.*, London, 10, 85 pp.

Benson, R. H. 1969. Preliminary report on the study of abyssal Ostracoda. *In* Neale, J. W. (Ed.), *The taxonomy, morphology and ecology of Recent Ostracoda*, 475-480. Oliver & Boyd, Edinburgh.

Benson, R. H. 1972. The *Bradleya* problem with descriptions of two new psychrospheric ostracode genera, *Agrenocythere* and *Poseidonamicus* (Ostracoda: Crustacea). *Smithson. Contr. Paleobiol.*, Washington, 12, 138 pp.

Benson, R. H. 1975. The origin of the psychrosphere as recorded in changes of deep-sea ostracode assemblages. *Lethaia*, Oslo, 8, 69-83.

Benson, R. H. 1977. The Cenozoic fauna of the Sao Paulo Plateau and the Rio Grande Rise (DSDP Leg 39, Sites 356 and 357). *In* Supko, P. R., Perch-Nielson, K. *et al.* (Eds), *Init. Repts DSDP*, 39, 860-883. U.S. Govt Printing Office, Washington.

Benson, R. H. 1979. In search of lost oceans: a paradox in discovery. *In* Gray, J. & Boucot, A. J. (Eds), *Historical biogeography, plate tectonics, and the changing environment*, 379-389. Oregon State University Press.

Benson, R. H., Chapman, R. E. & Deck, L. T. 1984. Paleooceanographic events and deep-sea ostracodes. *Science*, New York, 224, 1334-1336.

Benson, R. H. & Peypouquet, J-P. 1983. The upper and mid-bathyal Cenozoic ostracode faunas of the Rio Grande Rise found on Leg 72 Deep Sea Drilling Project. *In* Barker, P. F. *et al.* (Eds), *Init. Repts DSDP*, 72, 805-818. U.S. Govt Printing Office, Washington.

Bertels, A. 1975. Upper Cretaceous (Middle Maastrichtian) ostracodes of Argentina. *Micropaleontology*, New York, 21(1), 97-130.

Bold, W. A. van den. 1946. *Contribution to the study of Ostracoda with special reference to the Tertiary and Cretaceous microfauna of the Caribbean Region*, 167 pp. De Bussy, Amsterdam.

Bold, W. A. van den. 1957. Ostracoda from the Palaeocene of Trinidad. *Micropaleontology*, New York, 3(1), 1-18.

Bold, W. A. van den. 1958. *Ambocythere*, a new genus of ostracodes. *Ann. Mag. nat. Hist.*, London, Series 12, 10, 801-813.

Bonnema, J. H. 1941. Ostracoden aus der Kreide des Untergrundes der nordöstlichen Nierderlande. *Natuurh. Maandblad*, Maastricht, 29-30, 27pp.

Brady, G. S. 1880. Report on the Ostracoda dredged by H.M.S. Challenger during the years 1873-1876. *Report on the voyage of H.M.S. Challenger, Zoology*, 1, 1-184.

Coles, G. P. In preparation. *Cainozoic evolution of Ostracoda from deep waters of the North Atlantic and adjacent shallow water regions.* Unpub. Ph.D. thesis, University of Wales.

Coles, G. P. & Whatley, R. C. In press. New Palaeocene to Miocene genera and species of Ostracoda from DSDP Sites in the North Atlantic. *Revta esp. Micropaleont.*, Madrid.

Colin, J. P. 1987. Étude systematique des ostracodes de la formation des Madeleines (Danien du Senegal). *Cah. Micropaleontol.* **2**(3/4), 113-124.

Cronin, T. M. 1983. Bathyal ostracodes from the Florida-Hatteras slope, the Straits of Florida, and the Blake Plateau. *Marine Micropaleontology*, **8**, 89-119.

Cronin, T. M. & Compton-Gooding, E. E. 1987. Cainozoic Ostracoda from Deep Sea Drilling Project Leg 95 off New Jersey (Sites 612 and 613). *In* Poag, C. W., Watts, A. B. *et al.* (Eds), *Init. Repts DSDP*, **95**, 439-451. U.S. Govt Printing Office, Washington.

Davies, H. C. 1981. *The areal and depth distribution of N. E. Atlantic Ostracods*. Unpub. M.Sc. thesis, University of Wales, 311 pp.

Dingle, R. V. 1981. The Campanian and Maastrichtian Ostracoda of South-East Africa. *Ann. S. Afr. Mus.*, Cape Town, **85**(1), 181pp.

Dingle, R. V. 1984. Mid-Cretaceous Ostracoda from Southern Africa and the Falkland Plateau. *Ann. S. Afr. Mus.*, Cape Town, **93**(3), 97-211.

Downing, S. E. 1985. *The taxonomy, palaeoecology, biostratigraphy and evolution of Pliocene Ostracoda from the W. Pacific*. Unpub. Ph.D. thesis, University of Wales, 2 vols., 1073 pp.

Ducasse, O. Guernet, C. & Tambareau, Y. 1985. Paléogène. *In* Oertli, H. J. (Ed.), *Atlas des ostracodes de France*, 257-311. *Bull. Centres Rech. Explor.-Prod. Elf-Aquitaine*, Pau, Mém. **9**.

Ducasse, O. & Peypouquet, J.-P. 1979. Cenozoic ostracodes: their importance for bathymetry, hydrology and biogeography. *In* Montadert, L. & Roberts, D. G. (Eds), *Init. Repts DSDP*, **48**, 343-363. U.S. Govt Printing Office, Washington.

Guernet, C. 1982. Contribution à l'étude des faunes abyssales: Les ostracodes Paléogènes du Bassin des Bahamas, Atlantique Nord (DSDP, Leg 44). *Revue Micropaléont.*, Paris, **25**(1), 40-56.

Guernet, C. 1985. Ostracodes paléogènes de quelques sites 'D.S.D.P.' de l'ocean Indien (Legs 22 et 23). *Revue Paléobiol.*, **4**(2), 279-295.

Harpur, W. K. 1985. *Late Quaternary deep-sea Ostracoda from the extra - Iberian Portal region*. Unpub. M.Sc. thesis, University of Wales, 185pp.

Herrig, E. 1966. Ostracoden aus der Weissen Schreibkreide (Unter-Maastricht) der Insel Rugen. *Paläont. Abh.*, Berlin, Abt. A, **2**(4), 693-1024.

Maddocks, R. F. 1985. Ostracoda of Cretaceous - Tertiary contact sections in Central Texas. *Trans. Gulf Cst Ass. geol. Socs*, **35**, 445-56.

Millson, K. J. 1987. *The palaeobiology of Palaeogene Ostracoda from Deep Sea Drilling Project cores in the South West Pacific*. Unpub. Ph.D. thesis, University of Wales, 2 vols. 733 pp.

Neale, J. W. 1975. The ostracod fauna from the Santonian Chalk (Upper Cretaceous) of Gingin, Western Australia. *Spec. Pap. Palaeont.*, London, **16**, 1-81.

Neale, J. W. 1978. The Cretaceous. *In* Bate, R. & Robinson, E. (Eds), *A Stratigraphical Index of British Ostracoda. Geol. Journ. Spec. Issue*, No. **8**, 325-384. Seel House Press, Liverpool.

Porter, C. 1984. *Late Quaternary Ostracoda from the N.E. Atlantic*. Unpub. M.Sc. thesis, University of Wales, 293 pp.

Smith, J. K. 1978. Ostracoda of the Prairie Bluff Chalk, Upper Cretaceous (Maastrichtian) and the Pine Barren member of the Clayton Formation, Lower Paleocene (Danian) from exposures along Alabama State Highway 263 in Lowndes County Alabama. *Trans. Gulf Cst Ass. geol. Socs*, **28**(2), 539-579.

Steineck, P. L., Breen, M., Nevins, N. & O'Hara, P. 1984. Middle Eocene and Oligocene deep-sea Ostracoda from the Oceanic Formation, Barbados. *J. Paleont.*, **58**(6), 1463-1469.

Steineck, P. L., Dehler, D. Hoose, E. M. & McCalla, D. 1988. Oligocene to Quaternary ostracods of the Central Equatorial Pacific (Leg 85, DSDP - IPOD). *In* Hanai, T., Ikeya, N. & Ishizaki, K. (Eds), *Evolutionary biology of Ostracoda, its fundamentals and applications*, proceedings of the Ninth International Symposium on Ostracoda, held in Shizuoka, Japan, 29 July - 2 August 1985, Developments in palaeontology and stratigraphy, **11**, 597-617, Kodansha Ltd., Tokyo and Elsevier, Amsterdam, Oxford, New York, Tokyo.

Thomas, E. 1987. Late Oligocene to Recent benthic foraminifers from Deep Sea Drilling Project Sites 608 and 610, northwestern North Atlantic. *In* Ruddiman, W. F. *et al.* (Eds), *Init. Repts DSDP*, **94**(2), 997-1031. U.S. Govt Printing Office, Washington.

Weaver, P. P. E. 1982. Ostracoda from the British Lower Chalk and Plenus Marls. *Palaeontogr. Soc. (Monogr.)*, London, **562**, 1-127.

Whatley, R. C. 1983. Some aspects of the palaeobiology of Tertiary deep-sea Ostracoda from the S.W. Pacific. *J. Micropalaeontol.*, London, **2**, 83-104.

Whatley, R. C. 1985. Evolution of the ostracods *Bradleya* and *Poseidonamicus* in the deep-sea Cenozoic of the South-west Pacific. *Spec. Pap. Palaeont.*, London, **33**, 103-116.

Whatley, R. C. & Ayress, M. A. 1988. Pandemic and endemic distribution patterns in Quaternary deep-sea Ostracoda. *In* Hanai, T., Ikeya, N. & Ishizaki, K. (Eds), *Evolutionary biology of Ostracoda, its fundamentals and applications*, proceedings of the Ninth International Symposium on Ostracoda, held in Shizuoka, Japan, 29 July - 2 August 1985, Developments in palaeontology and stratigraphy, **11**, 739-755, Kodansha Ltd., Tokyo and Elsevier, Amsterdam, Oxford, New York, Tokyo.

Whatley, R. C. & Coles, G. P. 1987. The late Miocene to Quaternary Ostracoda of Leg 94, Deep Sea Drilling Project. *Revta esp. Micropaleont.*, Madrid, **19**, 33-97.

Whatley, R. C. & Coles, G. P. In press. Global change and the biostratigraphy of Cainozoic deep-sea Ostracoda in the North Atlantic. *J. micropalaeontol.*, London.

Whatley, R. C., Harlow, C. J., Downing, S. & Kesler, J. 1983. Some observations on the origin, evolution, dispersion and ecology of the genera *Poseidonamicus* Benson and *Bradleya* Hornibrook. *In* Maddocks, R. F. (Ed.),

*Applications of Ostracoda*, proceedings of the Eighth International Symposium on Ostracoda, July 26-29, 1982, 492-509. Univ. Houston Geos., Houston, Texas.

Whatley, R. C. Witte, L. & Coles, G. P. In press. New data on the ostracod genus *Aratrocypris* Whatley *et al.* 1985, with descriptions of species from the Upper Cretaceous of Europe and the Cainozoic of the North Atlantic. *J. Micropalaeontol.*, London.

# 23

# Xylophile Ostracoda in the deep sea

**P. Lewis Steineck[1], Rosalie F. Maddocks[2], Ruth D. Turner[3], Graham Coles[4] & Robin Whatley[4]**

[1]Division of Natural Sciences, SUNY College at Purchase, New York, U.S.A.
[2]Department of Geosciences, University of Houston, U.S.A.
[3]Museum of Comparative Zoology, Harvard University, U.S.A.
[4]Institute of Earth Studies, University College of Wales, Aberystwyth, U.K.

## ABSTRACT

A taxonomically and ecologically novel association of podocopid Ostracoda has been found living on experimental wood islands emplaced by submersible at 1800 to 4000m on the sea floor, which became riddled by bivalve wood borers (Family Pholadidae, Subfamily Xylophagainae). The association is characterized by species of the ostracod genera *Xylocythere, Paradoxostoma, Cytherois, Propontocypris* and *Parapontoparta*, quite distinct from the normal soft-sediment deep-sea benthos. Affinities with phytal, commensal and other specialized shallow-water inhabitants suggest a submergent origin and possible dependence on larger invertebrates for nutrition. The occurrence of some species at several distant stations demonstrates a capacity for long-distance dispersal. Newly recognized fossil occurrences, in Oligocene to Holocene deep-sea sediments, confirm the antiquity of this hitherto unsuspected xylophile fauna and encourage fuller investigation of the diversity, sources and biotic interactions of deep-sea ostracods.

## THE XYLOPHILE BIOCOENOSIS

In the deep sea, bacterial and fungal decomposers and wood-boring bivalves of the Subfamily Xylophagainae (Family Pholadidae) convert refractory wood parcels of terrestrial origin into more usable food sources (Turner, 1973, 1977, 1981; Wolff, 1976, 1979; Stockton & DeLaca, 1982). This localized, transient energy enrichment creates a eutrophic habitat-island colonized by a diverse, trophically complex community with a high biomass of macro-invertebrates (Turner, 1977, 1981; Wolff, 1979). Interest in this xylophile ('wood-loving') deep-water community has been heightened by recent recognition of ostracods and sipunculids with shallow-water affinities (Maddocks & Steineck, 1987; Rice, 1985) and of

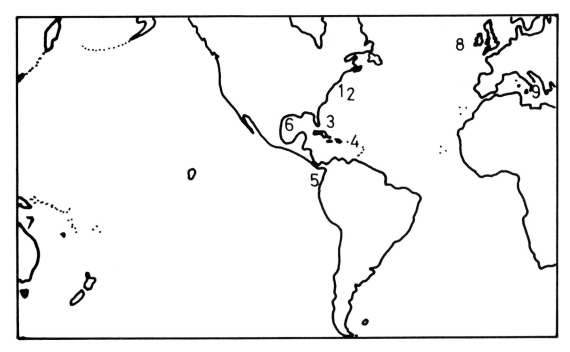

Fig. 1. Locations of experimental wood-island stations (1-5) and fossil occurrences of species of *Xylocythere* (6-0). 1 = DOS-1, 110 miles south of Woods Hole, 39°46'N, 70°41'W, depth 1830m. 2 = DOS-2, 190 miles southeast of Woods Hole, 38°18.4'N, 69°35.6'W, depth 3506m. 3 = Tongue of the Ocean, Bahama Islands, 24°53.2'N, 77°40.2'W, depth 2066m. 4 = off north coast of St. Croix, 17°56.63'N, 64°48.6'W, depth 4000m. 5 = Panama Basin, 05°20.75'N, 81°56.19'W, depth 3900 m. 6 = UH 4129, Northwest Gulf of Mexico, Texas A&M University R. V. *Gyre* cruise 77-G-14, station 10A, 26°34'N, 93°34.6'W, depth 1500m, Smith-MacIntyre box-core, Holocene. 7 = Queensland Plateau, southwest Pacific, Deep Sea Drilling Site 209, 15°56.19'S, 152°11.27'E, present water depth 1428m, core 9, section 3, core catcher, planktonic foraminiferal zone N14, Middle Miocene (see Whatley, 1983, Fig. 1). 8 = Goban Spur, Northeast Atlantic Ocean, Deep Sea Drilling Site 549A, 49°05.9'N, 13°05.89'W, present water depth 2513 m, core 7, section 5, core catcher, planktonic foraminiferal zone P21, Upper Oligocene; and also core 5, section 5, core catcher, planktonic foraminiferal zone N15, Upper Miocene. 9 = Lower Pleistocene of Calabria, reported by Colalongo and Pasini (1980). 0 = Central equatorial Pacific, Deep Sea Drilling Site 575A, 5°51'N, 135°02.16'W, present water depth 4536m, core 18, section 3, depth 65-70cm, planktonic foraminiferal zone N5, Lower Miocene.

endemic higher taxa of molluscs and echinoderms (Marshall, 1985a, 1985b; Baker *et al.*, 1986; Turner, 1977, 1981, 1985).

A unique and hitherto unknown association of podocopid Ostracoda has been found living in exuberant abundance on five 'wood-island' experimental stations positioned on the sea floor by Turner using the deep-diving submersible *Alvin* (Maddocks & Steineck, 1987). At each station (see Fig. 1, localities 1 to 5), these blocks and panels of wood were left on the sea floor for a year or more; emplacement and retrieval techniques were described by Turner (1981) and Maddocks & Steineck (1987). Fourteen species of living ostracods including the new genus *Xylocythere* were described from these five sampling sites by

Maddocks & Steineck (1987) (see Table 1). At a typical station, up to several hundred living ostracods were recovered in the washings from the riddled wood, belonging to 2 to 7 species in 2 or 3 of the following genera: *Xylocythere* Maddocks & Steineck, 1987 (Family Cytheruridae, Subfamily Eucytherurinae), *Paradoxostoma* Fischer, 1855; *Cytherois* Müller, 1884; *Sclerochilus* Sars, 1866; *Propontocypris* Sylvester-Bradley, 1947 and *Parapontoparta* Hartmann, 1955 (see Table 1).

Two attributes of this assemblage are noteworthy. Firstly, live individuals are several orders of magnitude more abundant than is normal for food-limited, abyssal soft-sediment faunas. Secondly, most representatives of the wood-island ostracod faunas are not known to live elsewhere in the deep

Table 1. Known occurrences of xylophile Ostracoda.

| Species name | Locality (see Fig. 1 for location) | | | | | | | | | |
|---|---|---|---|---|---|---|---|---|---|---|
| | 1 | 2 | 3 | 4 | 5 | 6 | 7 | 8 | 9 | 0 |
| *Ambocythere ramosa* Bold, 1965 | | 1 | | | | | | | | |
| *Cytherois lignicola* Maddocks & Steineck, 1987 | | | | 9 | | | | | | |
| *Cytherois paralignicola* Maddocks & Steineck, 1987 | | 2 | | | | | | | | |
| *Paradoxostoma turnerae* Maddocks & Steineck, 1987 | 36 | | | | | | | | | |
| *Paradoxostoma* sp. 1 Maddocks & Steineck, 1987 | 1 | | | | | | | | | |
| *Parapontoparta spicacarens* Maddocks & Steineck, 1987 | | | | | 20 | | | | | |
| *Propontocypris (P.) excussa* Maddocks & Steineck, 1987 | 600 | 148 | | 4 | | | | | | |
| *Propontocypris (P.) repanda* Maddocks & Steineck, 1987 | | | 23 | | | | | | | |
| *Propontocypris (P.) sectilis* Maddocks & Steineck, 1987 | | 47 | | | | | | | | |
| *Sclerochilus contortus* (Norman, 1862) | 1 | | | | | | | | | |
| *Xylocythere pointillissima* Maddocks & Steineck, 1987 | | 88 | 13 | 257 | | | | | | |
| *Xylocythere producta* (Colalongo & Pasini, 1980) | | | | | | | | | * | |
| *Xylocythere rimosa* Maddocks & Steineck, 1987 | | | | | 4 | | | | | |
| *Xylocythere tridentis* Maddocks & Steineck, 1987 | 14 | | | | | | | | | |
| *Xylocythere turnerae* Maddocks & Steineck, 1987 | | 72 | 157 | 15 | | | | | | |
| *Xylocythere* sp. 1 | | | | | | | | | 5 | |
| *Xylocythere* sp. 2 | | | | | | | 1 | | | |
| *Xylocythere* sp. 3 | | | | | | 6 | | | | |
| *Xylocythere* sp. 4 | | | | | | | 1 | | | |
| *Xylocythere* sp. 5 | | | | | | | | 1 | | |
| *Xylocythere* sp. 6 | | | | | | 6 | | | | |
| *Xylocythere* sp. 7 | | | | | | | | | | 1 |

sea. Their closest taxonomic relatives live on algae, sea grasses, corals, bryozoans, mangrove-swamp muds and other organic-rich substrates in the photic zone, or are ectosymbionts on larger invertebrates at varying depths. These findings suggest that this distinctive fauna is submergent in origin and endemic to wood. Among potential sources of nutrition which these deep-sea wood communities provide for ostracods are bacteria, fungi, faecal pellets and perhaps, in the case of *Paradoxostoma* and *Propontocypris*, the body fluids and external mucus secretions of larger invertebrates. Conspicuous anatomical modifications such as those for filter feeding are absent.

*Xylocythere*, the most characteristic genus of this association, is thought to be endemic to wood falls in the deep sea. It may be that other members of the Eucytherurinae, which are rare but persistent in deep-sea assemblages, also occupy localized, ephemeral habitats. The accompanying Paradoxostomatidae, Pontocyprididae and Paracypridinae impart an incongruously shallow-water aspect to this *in situ* bathyal to abyssal association. Although the single specimen of *Ambocythere ramosa* Bold recorded by Maddocks & Steineck (1987) may be a stray from the nearby soft-bottom fauna, the remainder of this expatriate ostracod biocoenosis is entirely unlike the normal deep-sea community.

## THE SOFT-SEDIMENT FAUNA AT DOS-2

At Locality 2 (Fig. 1, see also Table 2), direct comparison of wood-panel faunas with those of

Table 2. Distribution of living Ostracoda at DOS-2 (Locality 2).

| Species name | On wood panels | Alvin 790 box core: 1 | 3 | 4 | Alvin 1699 mud box: 1-83 | 2-83 |
|---|---|---|---|---|---|---|
| *Ambocythere ramosa* Bold, 1965 | 1 | | | | | |
| *Cytherois paralignicola* Maddocks & Steineck, 1987 | 2 | | | | | |
| *Propontocypris (P.) excussa* Maddocks & Steineck, 1987 | 148 | | | | | |
| *Propontocypris (P.) sectilis* Maddocks & Steineck, 1987 | 47 | 6 | 7 | 3 | | |
| *Xylocythere pointillissima* Maddocks & Steineck, 1987 | 88 | | | | | |
| *Xylocythere tridentis* Maddocks & Steineck, 1987 | 14 | | | | | |
| *Xylocythere turnerae* Maddocks & Steineck, 1987 | 72 | | | | | 2 |
| *Abyssocypris atlantica* (Maddocks, 1977) | | 1 | | 1 | | |
| *Bythocypris* cf. *affinis* (Brady, 1886) | | 1 | | | | |
| *Krithe* spp. | | 6 | 5 | 1 | 4 | |
| *Cytheropteron porterae* Whatley & Coles, 1987 | | | | | | 2 |
| *Echinocythereis echinata* (Sars, 1866) | | | | | | 5 |
| *Jonesia?* spp. | | | | | 1 | 5 |
| *Pelecocythere sylvesterbradleyi* Athersuch, 1979 | | | | | 1 | 1 |
| *Polycope* sp. | | | | | 1 | |
| *Poseidonamicus major* Benson, 1972 | | | | | | 3 |
| *Pseudocythere* spp. | | | | | | 3 |
| *Legitimocythere acanthoderma* (Brady, 1880) | | | | | | 1 |
| *Bythoceratina scaberrima* (Brady, 1886) | | | | | | 5 |

neighbouring sediment is possible. Through the courtesy of Dr F. J. Grassle and Ms Linda Morse-Porteous, we were able to examine the living ostracods from three sediment box-cores taken on *Alvin* Dive 790 at the DOS-2 station (Locality 2). Each core was located 15cm from a wood block (emplaced earlier by Turner and not yet retrieved, but assumed to have the same ostracod fauna as the panels that have been retrieved at this site) and sampled a 225cm² area. The cores contained dark muds enriched by the spillover of organic material and faecal pellets fom the adjacent wood-based community (Grassle, 1977; Grassle & Morse-Porteous, 1988; see Turner, 1977, Fig. 1, for photograph of the equivalent phenomenon at Locality 3). Macro-invertebrate diversity and density were higher in these cores than in other cores collected 60cm from the wood panels (Grassle &

Morse-Porteous, 1988); the latter cores yielded no living ostracods.

Each of the four subcores per core yielded as few as 0 to as many as 6 living ostracods, while the total populations for each of the three cores ranged from 4 to 15 individuals per 225cm². These population densities are quite high for deep-sea sediments, while the variability is a familiar result of the small-scale patchiness of meiobenthonic populations at these depths (Coull *et al.*, 1977). The most numerous species is *Propontocypris (P.) sectilis*, a good swimmer and member of the xylophile association. The other species (*Krithe* spp., *Abyssocypris atlantica* and *Bythocypris* cf. *affinis*) belong to the local sediment-dwelling population.

We also examined the live ostracods from two recolonization trays (*Alvin* 1699 Mud Box 1-83 and 2-83) that were deployed on the bottom at this site from July 1983 to June 1986 (Grassle & Morse-Porteous, 1988). Each tray was 0.25m² in area, filled with azoic mud (frozen and thawed) collected previously from the same area, and covered with 2mm Nytex screening to exclude predators. The first tray yielded 7 specimens of 4 species of normal soft-bottom dwellers. The other produced 28 specimens of 13 species, a highly diverse assemblage dominated by characteristic deep-sea dwellers but also including 2 specimens of *Xylocythere turnerae* (see Table 2).

These results show that both swimming (*Propontocypris*) and crawling (*Xylocythere*) xylophile ostracods are capable of at least short-range dispersal. It appears likely that the xylophile ostracod population inhabits the neighbouring trophically enriched mud as well as the wood panels themselves. The two individuals in the recolonization tray may have been waifs passively displaced from the preferred wood habitat, new immigrants approaching the wood island, or emigrants seeking new wood. Alternatively, the halo of trophic enrichment around wood islands may be sufficiently diffuse that elements of the xylophile association intermingle with the normal soft-bottom fauna over an extensive area.

## DISPERSAL MECHANISMS

Long-term propagation of populations of wood-dwelling ostracods, which lack demersal larvae, in the deep sea requires the transfer of live individuals or eggs from older, exhausted sites to fresh wood parcels. Colonization may take place either by 'island hopping' between nearby parcels or by more extensive dissemination. In the wood-island samples the presence of juveniles for most species suggests that the life cycle is completed *in situ*. However, the recurrence of some species at distant stations indicates that effective long-distance dispersal mechanisms also exist. Such mechanisms might include: (1) entrainment in high-energy bottom flows, (2) transport of viable eggs in the gut of fishes, (3) hitch-hiking on fish or mobile invertebrates (or on their demersal larvae) by secreting adhesive threads for attachment (cytheraceans only), (4) active swimming (cypridaceans)

or crawling (cytheraceans), and (5) deployment of a sail-like network of secreted threads to harness ambient current energy more efficiently (cytheraceans only). Although there is no direct evidence for any of these methods, it is perhaps relevant that both Paradoxostomatidae and *Xylocythere* have very large spinnaret glands in both sexes and that both *Propontocypris* and *Parapontoparta* are active swimmers.

## THE XYLOPHILE TAPHOCOENOSIS

Taphonomic processes (bioturbation, current-sorting and selective destruction of fragile carapaces) affecting deep-sea sediments and their microfaunas make it unlikely that the geologically ephemeral wood-island ostracod biocoenosis would survive burial intact (Kontrovitz & Nicolitch, 1979; Kidwell, 1986; Van Harten, 1986). In most depositional environments, mixing and differential preservation obliterate fine-scale ecological zonations (Berger, 1981; Schindel, 1983; Smith & Schafer, 1984a; 1984b; Furisch, 1978; Frydl, 1982; Izuka & Kaesler, 1986). Fossil carapaces of wood-dwelling ostracods would become rare constituents in a time-averaged assemblage dominated by soft-sediment taxa.

Specialists studying Cainozoic deep-sea Ostracoda should be alert for seemingly anomalous or exotic forms, which up to now have been interpreted as representing post-mortem transportation from shallow-water habitats, but which may, in fact, be indigenous wood-dwelling species. Their recognition may mitigate the difficulties in recognizing allochthony and determining lower depth limits for bathyal and abyssal ostracods (Cronin, 1983). Given the abundance of individuals in the experimental wood-island communities, fossil representatives are likely to be patchily distributed, i.e., fairly abundant when encountered at all. It may be significant that all five specimens of *Xylocythere* sp. 1 at DSDP Site 549A occur in a single sample. Because the extant ostracod fauna of deep-sea wood parcels is taxonomically distinctive and, therefore, recognizable in the fossil record, it will help in palaeobiogeographical and palaeoceanographical reconstructions, providing information about proximity to major river systems, mangrove coasts and the paths of surface and bottom currents.

## FOSSIL RECORD OF THE XYLOPHILE FAUNA

Maddocks & Steineck (1987) recognized that biostratinomic processes would cause xylophile ostracods to be, at best, rare as fossils. Nevertheless, they optimistically predicted that fossil xylophile ostracods awaited discovery in the oceanic core record and in deep-water subaerially exposed sections. This has proved to be the case. We now enumerate newly discovered or reinterpreted fossil occurrences of this taxonomically distinctive fauna, which greatly expand its geographical and chronostratigraphical range.

The most compelling evidence is provided by seven additional species of *Xylocythere*, from the Oligocene and Miocene of the Northeast Atlantic, the Miocene of the central and Southwest Pacific and the Holocene of the Gulf of Mexico (Table 1, Fig. 1). *Xylocythere producta* (Colalongo & Pasini, 1980) was previously described from the Pleistocene of Calabria in an assemblage interpreted by Cronin (1983) as upper bathyal.

*Xylocythere* sp. 1 is represented by five specimens in one horizon at Locality 8, DSDP Site 549A, Northeast Atlantic Ocean, Upper Oligocene. The very rich, diverse fauna (47 species) is dominated by *Krithe* (six species), *Poseidonamicus* sp. nov., *Argilloecia* (5 species), *Cytherella serratula* (Brady), *Cytherella harmoniensis* Bold, *Henryhowella* ex gr. *asperrima* (Reuss), *Parakrithe vermunti* (Bold), *Australoecia* sp., *Agrenocythere hazelae* (Bold), *Agrenocythere* sp. and many other rare species, including *Eucytherura calabra* (Colalongo & Pasini).

*Xylocythere* sp. 2 is represented by a single specimen at Locality 7, DSDP Site 209, Middle Miocene, in the Southwest Pacific. It is remarkably similar to *Xylocythere producta* (Colalongo & Pasini). The associated typical deep-sea ooze fauna is dominated by *Krithe* spp., *Bradleya* spp., *Poseidonamicus* spp., *Argilloecia* spp., *Henryhowella* ex gr. *asperrima* (Reuss), etc., with

rarer *Eucytherura* spp.

*Xylocythere* sp. 3 is represented by three specimens in Holocene sediment at Locality 6, UH 4129, Northwest Gulf of Mexico. The ostracod assemblage at this station comprises more than 40 species, dominated by *Bradleya dictyon* (Brady), *Bythocypris affinis* (Brady), *Agrenocythere hazelae* (Bold), *Parakrithe vermunti* (Bold) and *Krithe* spp. Two very rare species of *Cytherois*? may also signal a xylophile component, although *Propontocypris* is not present.

*Xylocythere* sp. 4 is known from a single specimen at the same locality, Holocene, Northwest Gulf of Mexico.

*Xylocythere* sp. 5 designates a single specimen at Locality 8, DSDP Site 549A, Northeast Atlantic Ocean, Upper Miocene. The specimen very closely resembles *Xylocythere* sp. 6, although a heavy crust of coccoliths prevents detailed comparison of arrangements of fossae and cluster pores. The associated fauna includes *Krithe* spp., *Bradleya dictyon* (Brady), *Parakrithe vermunti* (Bold), *Poseidonamicus pintoi* Benson, *Henryhowella dasyderma* (Brady), etc.

*Xylocythere* sp. 6 is represented by six specimens at Locality 6, Holocene, which also yields *Xylocythere* spp. 3 and 4. *Xylocythere* sp. 7 is represented by a single specimen at DSDP Site 575A, core 18, section 3, planktonic foraminiferal zone N5, Lower Miocene, in the central equatorial Pacific. The associated fauna is dominated by *Bradleya johnsoni* Benson, *Poseidonamicus* spp., *Henryhowella* sp., *Abyssocythere trinidadensis* (Bold) and numerous species of *Krithe* and *Parakrithe*.

More circumstantial are the occasional deep-sea records of Paradoxostomatidae and *Propontocypris*, which, because they have numerous species in shallow-water habitats, have commonly been interpreted as down-slope contaminants (e.g., Bonaduce *et al.*, 1983; Van Harten, 1986). Nevertheless, single living specimens of *Propontocypris* at 5320m and *Paradoxostoma* at 2600m in the Indian Ocean were reported by Hartmann (1985). Van Harten (this volume) records a large fauna of 'Paradoxostomatidae', including 'several undescribed taxa which may or may not rightfully belong in this group proper as well as some affirmed paradoxostomatid species', from the Recent sediments collected on the eastern flank of the Mid-Atlantic Ridge in the North Atlantic. Fossil records of *Propontocypris* include the Miocene to Quaternary of the Southwest Pacific (Whatley, 1983), the Oligocene to Quaternary of the Indian, Pacific and North Atlantic oceans (Whatley & Coles, 1987; Whatley & Ayress, 1988) and the Lower to Middle Eocene and Lower Oligocene at Locality 8 (Coles, litt. comm., 1988) the Cainozoic of the South Atlantic (Benson *et al.*, 1985) and the Middle Eocene through Lower Oligocene at Locality 8 (Coles, litt. comm., 1988). Re-examination of previously studied faunas will probably identify additional evidence for a widespread xylophile ostracod presence in the deep sea, especially in areas that have experienced high levels of wood falls in the past.

## ORIGIN AND GEOLOGICAL HISTORY OF THE XYLOPHILE FAUNA

The xylophile ostracod fauna probably evolved from involuntary arrivals on sunken driftwood and may depend in part on larger invertebrates for dispersal and nutrition. Shallow-water Ostracoda living on wood and marine macrophytes, in organic-rich sediments, or as symbionts on the companion macro-invertebrate fauna are routinely displaced from their preferred euphotic setting when clumps of vegetation are detached by storms, floods and rivers and transported seawards (Wolff, 1976, 1979; Stockton & DeLaca, 1982). Once on the sea floor, some of the involuntarily expatriated ostracod species are able to tolerate the new physicochemical *milieu* and successfully seek out an environment most closely resembling their ancestral habitat, sunken wood parcels. Here, they perfect preadapted or newly evolved traits needed for life in a localized and depletable deep-sea ecosystem. This ongoing process may have a long history, having been especially active during times of lowered sea level, when rates of redeposition of shallow-water debris increased, and when rivers debouched more directly into the ocean.

An ancient origin for the ecological association between deep-sea wood parcels and ostracods is suggested by six observations: (1) the occurrence of similar generic and suprageneric elements at distant stations, (2) the distinctive species composition at each station, (3) the co-occurrence of as many as three congeneric species at one station, (4) the loss of colour bands and eyes possessed by shallow-water relatives, (5) the stratigraphical range (Oligocene to present), large number of species, and broad geographical distribution of *Xylocythere*, and (6) the archaic aspect of the pore clusters that are the most distinctive aspect of *Xylocythere*.

Xylophagaine wood-boring molluscs initially radiated in the late Cretaceous (Turner, 1969), which was also a time of rapid expansion for cytheracean and cypridacean ostracods (Whatley & Stephens, 1976; Whatley, 1988) as well as for woody angiosperms. The first xylophile ostracods may have become part of deep-sea wood-based communities at that time or soon after.

## ACKNOWLEDEGMENTS

Ruth Turner's wood-island experiments were funded by the Office of Naval Research Contract 14-76-CO 281, NR 104-687. Dr Jan R. Factor, SUNY-Purchase, assisted with the S.E.M. micrographs. We thank Dr Fred Grassle and Ms Linda Morse-Porteous, Woods Hole Oceanographic Institution, for making available the sediment box-cores at DOS-2.

**Plate 1**

Scale bar = 100µm.

Figs 1-2.   *Xylocythere* sp. 1.
Fig.  1.      2658 RV exterior.
Fig.  2.      2657 LV exterior.

Fig.  3.      *Xylocythere* sp. 2, 2662 LV exterior.

Fig.  4.      *Xylocythere* sp. 3, 2635 LV exterior.

Fig.  5.      *Xylocythere* sp. 4, 2640 RV exterior.

Fig.  6.      *Xylocythere* sp. 5, 2655 LV exterior.

Figs 7-9.   *Xylocythere* sp. 6.
Fig.  7.      3022 RV exterior.
Fig.  8.      3023 LV exterior.
Fig.  9.      2633 RV exterior.

Fig.  10.    *Xylocythere* sp. 7, 2632 LV exterior.

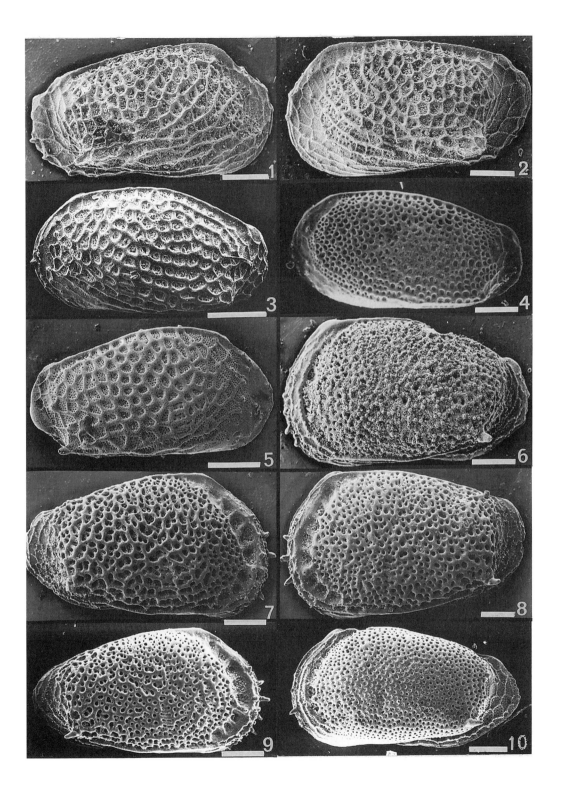

**Plate 2**

Scale bar = 100µm (Figs 1-6, 8, 10) and 10µ (Figs 7, 9).

Figs 1-2. *Xylocythere* sp. 1.
Fig. 1.    2657 LV interior.
Fig. 2.    2658 RV interior.

Fig. 3.    *Xylocythere* sp. 2, 2662 LV interior.

Figs 4, 9. *Xylocythere* sp. 4.
Fig. 4.    2640 RV interior.
Fig. 9.    2640 RV interior detail of anterior hinge and cluster pores.

Fig. 5.    *Xylocythere* sp. 5, 2655 LV interior.

Fig. 6.    *Xylocythere* sp. 6, 2639 RV interior.

Fig. 7.    *Xylocythere* sp. 3, 2635 RV, details of exterior, normal pore canals and cluster pores.

Fig. 8.    *Xylocythere* sp. 1, 2659W, dorsal view of whole carapace.

Fig. 10.   *Propontocypris* (*Propontocypris*) *sectilis* Maddocks & Steineck. 1987:  2520F = USNM 231753, exterior of female LV.

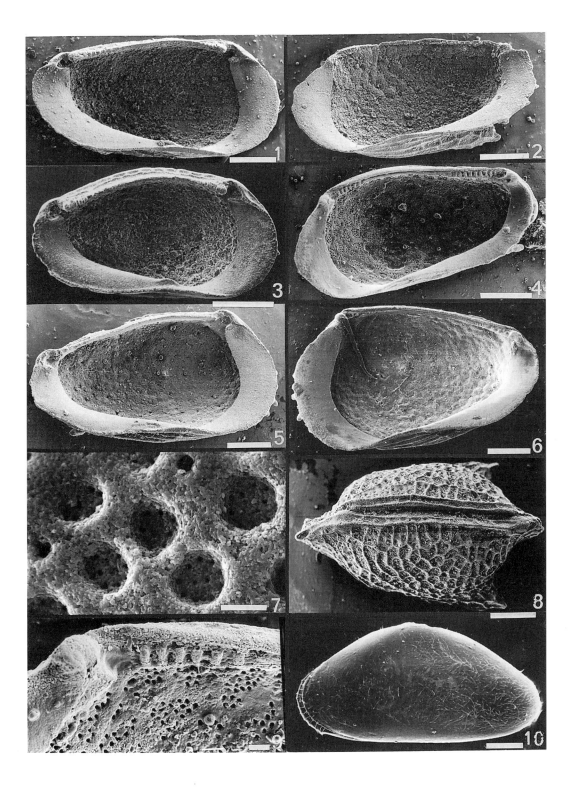

## REFERENCES

Baker, A. N., Rowe, F. W. E. & Clark., H. E. S. 1986. A new class of Echinodermata from New Zealand. *Nature*, London, **321**(6073), 862-864.

Benson, R. H., Chapman, R. E. & Deck, L. T. 1985. Evidence from the Ostracoda of major events in the South Atlantic and worldwide over the past 80 million years. *In* Hsu, K. J. & Weissart, J. H., (Eds), *South Atlantic Paleoceanography*, 325-350, Cambridge University Press.

Berger, W. H. 1981. Paleoceanography: the deep-sea record. *In* Emiliani, C. (Ed.), The Oceanic Lithosphere. *The Sea*, **7**, 1437-1519. John Wiley and Sons, New York.

Bonaduce, G., Ciliberto, B., Minichelli, G., Masoli, M. & Pugliese, N. 1983. The deep-water benthic ostracodes of the Mediterranean. *In* Maddocks, R. F. (Ed.), *Applications of Ostracoda*, proceedings of the Eighth International Symposium on Ostracoda, July 26-29, 1982, 459-471. Univ. Houston Geos., Houston, Texas.

Colalongo, M. L. & Pasini, G. 1980. La Ostracofauna plio-pleistocenica della Sezione Vrica in Calabria (con considerazioni sul limite Neogene/Quaternario). *Boll. Soc. paleont. ital.*, Modena, **19**, 44-126.

Cronin, T. M. 1983. Bathyal ostracodes from the Florida-Hatteras Slope, the Straits of Florida, and the Blake Plateau. *Marine Micropalentology*, Amsterdam, **8**, 89-110.

Coull, B. C., Ellison, R. L., Fleeger, J. W., Higgins, R. P., Hope, W. D., Hummon, W. D., Rieger, R. M., Sterrer, W. E., Thiel, H. & Tietjen, J. H. 1977. Quantitative estimates of the meiofauna from the deep sea off North Carolina. *Mar. Biol.*, New York, **39**, 233-240.

Frydl, P. 1982. Holocene ostracods in the southern Boso Peninsula. *In* Hanai, T. (Ed.), *Studies on Japanese Ostracoda*, University of Tokyo, University Museum Bulletin **20**, Tokyo, 61-140.

Fursich, F. T. 1978. The influence of faunal condensation and mixing on the preservation of fossil benthic communities. *Lethaia*, Oslo, **11**, 185-258.

Grassle, J. F. 1977. Slow recolonization of deep-sea sediment. *Nature*, London, **265** (5595), 618-619.

Grassle, J. F. & Morse-Porteous, L. S. 1988. Macrofaunal colonization of disturbed deep-sea environments and the structure of deep-sea benthic communities. *Deep-Sea Res.*, Oxford, **34**, 1911-1961.

Hartmann, G. 1985. Ostracoden aus der Tiefsee des Indischen Ozeans und der Iberischen See sowie von ostatlantischen sublitoralen Plateaus und Kuppen. *Senckenberg. marit.*, Frankfurt a. M., **17**, 89-146.

Izuka, S. T. & Kaesler, R. L. 1986. Biostratinomy of ostracode assemblages from a small reef flat in Maunalua Bay, Oahu, Hawaii. *J. Paleont.*, Chicago, **60**, 347-360.

Kidwell, S. M. 1986. Models for fossil concentrations: paleobiologic implications. *Paleobiol.*, Ithaca, **12**, 6-24.

Kontrovitz, M. & Nicolitch, M. J. 1979. On the response of ostracode valves and carapaces to water currents. *In* Krstic, N. (Ed.), *Taxonomy, Biostratigraphy and Distribution of Ostracodes, proceedings of the Seventh International Symposium on Ostracodes*, 269-272. Serbian Geological Society, Belgrade.

Maddocks, R. F. & Steineck, P. L. 1987. Ostracoda from experimental wood-island habitats in the deep sea. *Micropaleontology*, New York, **33**, 318-355.

Marshall, B. A. 1985a. Recent and Tertiary Cocculinidae and Pseudococculinidae (Mollusca: Gastropoda) from New Zealand and New South Wales. *N. Z. Jl Zool.*, Wellington, **12**, 505-546.

Marshall, B. A. 1985b. Recent and Tertiary deep-sea limpets of the genus *Pectonodonta* Dall (Mollusca: Gastropoda) from New Zealand and New South Wales. *N. Z. Jl Zool.*, Wellington, **12**, 273-282.

Rice, M. E. 1985. Description of a wood dwelling sipunculan, *Phascolosoma turnerae*, new species. *Proc. biol. Soc. Wash.*, Washington, D.C., **98**, 54-60.

Schindel, D. E. 1983. Resolution analysis: a new approach to the gaps in the fossil record. *Paleobiol.*, Ithaca, **8**, 340-353.

Smith, J. N. & Schafer, C. T. 1984a. Bioturbation processes in continental record. *Paleobiol.*, Ithaca, **8**, 340-353.

Smith, J. N. & Schafer, C. T. 1984b. Bioturbation processes in continental slope and rise sediments delineated by Pb-210, microfossil and textural indicators. *J. mar. Res.*, New Haven, **42**, 1117-1145.

Stockton, W. L. & DeLaca, T. E. 1982. Food falls in the deep sea: occurrence, quality and significance. *Deep-Sea Res.*, Oxford, **29**, 157-169.

Turner, R. D. 1969. Superfamily Pholadacea. *In* Cox, L. R. *et al.* (Eds), *Treatise on Invertebrate Paleontology*, Pt. N, *Mollusca*, **6**, Bivalvia 2, 702-742. Geological Society of America, Boulder.

Turner, R. D. 1973. Wood-boring bivalves, opportunistic species in the deep sea. *Science*, New York, **180**, 1377-1379.

Turner, R. D. 1977. Wood, mollusks and deep-sea food chains. *Bull. Am. malac. Un.*, Seaford, N. Y., **26**, 13-19.

Turner, R. D. 1981. Wood islands and thermal vents as centers of diverse communities in the deep sea. *Sov. J. Mar. Biol.*, Vladivostok, **7**, 1-10.

Turner, R. D. 1985. Notes on mollusks of deep-sea vents and reducing sediments. *In* Prezant, R. S. & Counts, C. L. III. (Eds), *Perspectives in Malacology*, Bull. Amer. Mal. Union, Hattiesburg, Special Ed. 1., 23-24.

Van Harten, D. 1986. Use of ostracodes to recognize downslope contamination in paleobathymetry and a preliminary reappraisal of the paleodepth of the Prasás marls (Pliocene), Crete, Greece. *Geology*, Boulder, **14**, 856-859.

Whatley, R. C. 1983. Some aspects of the palaeobiology of Tertiary deep-sea Ostracoda from the S. W. Pacific. *J. micropaleont.*, London, **2**, 83-104.

Whatley, R. 1988. Patterns and rates of evolution among Mesozoic Ostracoda. *In* Hanai, T., Ikeya, N. & Ishizaki, K. (Eds), *Evolutionary biology of Ostracoda, its fundamentals and applications*, proceedings of the Ninth International Symposium on Ostracoda, held in Shizuoka, Japan, 29 July - 2 August 1985, Developments in palaeontology and stratigraphy, **11**, 1021-1040, Kodansha Ltd., Tokyo and Elsevier, Amsterdam, Oxford, New York, Tokyo.

Whatley, R. C. & Ayress, M. 1988. pandemic and endemic distribution patterns among Cainozoic deep-sea

Ostracoda. *In* Hanai, T., Ikeya, N. & Ishizaki, K. (Eds), *Evolutionary biology of Ostracoda, its fundamentals and applications*, proceedings of the Ninth International Symposium on Ostracoda, held in Shizuoka, Japan, 29 July - 2 August 1985, Developments in palaeontology and stratigraphy, **11**, 739-755, Kodansha Ltd., Tokyo and Elsevier, Amsterdam, Oxford, New York, Tokyo.

Whatley, R. C. & Coles, G. 1987. The late Miocene to Quaternary Ostracoda of Leg 94, Deep-Sea Drilling Project. *Revta esp. Micropaleont.*, Madrid, **19**, 33-97.

Whatley, R. C. & Stephens, J. M. 1976. The Mesozoic explosion of the Cytheracea. *Abh. Verh. naturw. Ver. Hamburg* N. F., 18/19 (Suppl.), 191-200.

Wolff, T. 1976. Utilization of seagrass in the deep sea. *Aq. Bot.*, Amsterdam, **2**, 161-174.

Wolff, T. 1979. Macrofaunal utilization of plant remains in the deep-sea, *Sarsia*, Universitetet i Bergen, **64**, 117-136.

## DISCUSSION

Roger Kaesler: Have closely related ostracods been found in wood in shallow water, especially wood bored by *Teredo*? (I suppose these ostracodes could be said to have 'gone down on the stakes of *Teredo*'.)

Paul Steineck: Data on the microfauna of *Teredo*-infested wood in shallow water are few. Several species of ostracods have been found as commensals on the cosmopolitan wood-boring isopod *Limnoria* spp. in Ireland, San Diego Harbour, Gulf of St. Lawrence and the Netherlands. These forms are usually found attached to the flat exterior surface of the abdomen of the isopod. Three genera have been recognized: *Aspidoconcha* de Vos, 1953; *Redakea* de Vos, 1953 and *Lacoonella* de Vos & Stock, 1956. Dr J. M. C. Holmes (1987, *Ir. Nat. J.*, Belfast, **22**, 317-319) assigned *Aspidoconcha* to the Paradoxostomatidae, a well-represented group in the deep-sea wood-dwelling fauna. *Laocoonella* shows some similarity to *Xylocythere* but description and illustrations of this genus differ *in litt.* and an exacting comparison of the two forms is not possible at present.

Rosalie Maddocks: Yes, a few Pontocyprididae, but I agree that we need to know more about these shallow-water relatives. We have written to researchers studying shallow wood communities in several parts of the world and hope to get more material.

Ian Wilkinson: What is the time period involved in the colonization of the wood islands and what is the abundance of the 'opportunistic' species?

Paul Steineck: There is no evidence about the rate of colonization of the experimental wood-islands by ostracods. This is an interesting question but difficult to approach experimentally, because Turner's wood islands can be sampled and/or observed only by deep-submersible and diving opportunities are infrequent. Among macro-invertebrates, galatheid crabs and capitellid and ampharetid polychaetes appear to be the first to reach recently submerged wood, in one case less than five months after submergence.

David Horne: What are the taxonomic affinities of *Xylocythere*?

Paul Steineck: Maddocks & Steineck (1987, *Micropaleontology*, New York, **33**(4), 320-323) discussed in detail the taxonomic relationships of *Xylocythere*. This genus was placed in the subfamily Eucytherurinae Puri, 1974 *emend.* on the basis of pore clusters occupying sola in the reticulum, which penetrate to the interior of the carapace, producing a pattern of clustered perforations corresponding directly to the exterior reticulate ornament without accompanying setae. Unlike the small, subquadrate to rhomboid species of *Eucytherura* and related genera, *Xylocythere* exhibits an elongate-ovoid shape, sinuous dorsal and ventral margins, a merodont-entomodont hinge with a smooth median element and a prominent small posteroventral spine in each valve. Genera which may be related to *Xylocythere* include an unnamed taxon based on '*Bythoceratina*' *reticulata* Bonaduce, Ciampo & Masoli, 1976, from the Adriatic Sea and *Laocoonella* De Vos, 1953, which interestingly is commensal on the wood-boring isopod *Limnoria ligorum* (Rathke).

# 24

# Modern abyssal ostracod faunas of the eastern Mid-Atlantic Ridge area in the North Atlantic and a comparison with the Mediterranean

**Dick Van Harten**

Institute of Earth Sciences, Free University, Amsterdam, The Netherlands.

## ABSTRACT

Surface sediments from sites at abyssal depths between lat. 53°24' N; long. 27°29' W and lat. 28°35' N; long. 38°42'W on the eastern flank of the Mid-Atlantic Ridge were sampled and their ostracods examined from fractions down to the 150μm grain size limit. More than 100 species belonging to over 40 genera were found, one site alone yielding some 65 species. There is evidence to support a faunal break, possibly related to a bottom water-mass boundary, in the area of the Azores. Several samples contained large numbers of paradoxostomatids which are believed to be autochthonous. It is estimated that overall the North Atlantic abyss houses at least 60 ostracod genera and somewhat over twice as many species. Several species found in the Mid-Atlantic Ridge area are also known from the Mediterranean, where some of them apparently prefer bathyal rather than abyssal conditions.

## INTRODUCTION

During the APNAP-I expedition of the R/V *Tyro* to the North Atlantic (August-September 1986), abyssal sediments on the eastern flank of the Mid-Atlantic Ridge were sampled using box-coring devices (Fig. 1, Table 1). Although primarily intended for the study of water-mass boundaries and their effect on plankton distribution, these samples also contained a wealth of benthonic ostracods. Taxonomic investigations are still in progress but in all, more than 100 ostracod species seem to be present belonging to some 40 genera. Not a few species are new to science and one or two new genera may be erected based on this material.

The taxonomic details of these faunas will be published elsewhere and nomenclature is deliberately left open here. This paper aims to report the broad outlines only.

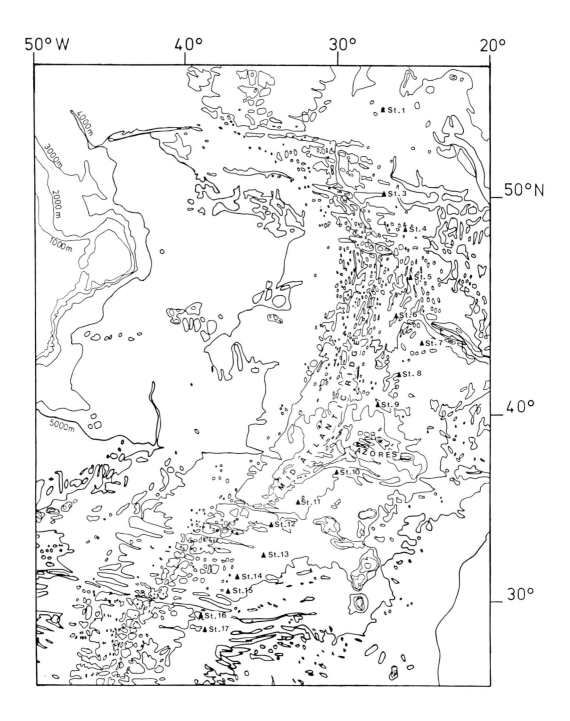

Fig. 1. Stations occupied by the Dutch APNAP-1 expedition in the North Atlantic.

Table 1. Location and water-depths of stations used in the study.

| STATION | LATITUDE N | LONGITUDE W | DEPTH (m) |
|---|---|---|---|
| T86-01 | 53° 24' | 27° 29' | 2580 |
| T86-03 | 50° 09' | 27° 02' | 3113 |
| T86-05 | 46° 53' | 25° 21' | 3121 |
| T86-07 | 43° 55' | 24° 58' | 2664 |
| T86-08 | 42° 15' | 25° 40' | 3232 |
| T86-09 | 40° 48' | 27° 27' | 2026 |
| T86-10 | 37° 07' | 30° 02' | 2610 |
| T86-11 | 35° 34' | 32° 33' | 2220 |
| T86-13 | 32° 46' | 34° 42' | 2992 |
| T86-14 | 31° 26' | 36° 15' | 3071 |
| T86-15 | 30° 29' | 36° 57' | 3271 |
| T86-16 | 29° 06' | 38° 50' | 3504 |
| T86-17 | 28° 35' | 38° 42' | 4015 |

## MATERIAL AND METHODS

The material studied represents the soft top layer (1-2cm) of the cored sediments. Immediately upon recovery aboard ship, the samples were stored in a staining solution of ethanol and Rose Bengal using a modification of the method described by Lutze (1964). After having been left to stand for a saturation period of two months, the samples were washed over a nylon cloth with 70µm aperture width. The residues were gently boiled in a soda solution to ensure better cleaning and finally washed again.

The ostracods were picked from all grain size fractions larger than 150µm. Picking is very tedious in the 250-150µm size grade and it was found useful to strew the washings on the picking tray holding the latter in a strongly tilted rather than horizontal position. This resulted in the rolling away of the mostly spherical pelagic components, leaving flatter elements such as ostracods behind. By tilting the tray in the opposite direction and letting the grains roll towards the other edge and repeating this process several times, a measure of separation was achieved that greatly facilitated speedy and representative picking.

## RESULTS AND DISCUSSION

### Staining

The Rose Bengal staining had little effect. Finds of specimens that had taken up the dye to an extent as to suggest they had been collected alive were rare. Such strongly stained specimens usually contained soft parts, whereas no chitinous remains were seen in unstained shells. There is no reason, therefore, to conclude that the Rose Bengal technique is ineffective in the case of ostracods. Our disappointing results may merely reflect the rarity or patchiness of benthonic life in the deep ocean that has already been commented on by several other workers (see Maddocks & Steineck, 1987). However, in view of the large numbers of ostracods in some samples (e.g., 523 adult valves were found in *circa* 150ml of raw sediment at station T86-10), the near-absence of live specimens seems remarkable. It is worth mentioning that many valves of *Krithe* were slightly stained, suggesting the presence of some kind of proteinaceous matter. No traces of chitin were observed in these specimens and it is believed they were already dead when collected. Whatley & Wall (1969) suggested that staining of empty ostracod shells might be due to micro-organisms having become aggregated in them.

## The ostracods encountered

A qualitative account of the taxa encountered is given in Table 2. This table underestimates true diversity to the extent that many separate species are hidden in open nomenclature as 'spp.'. This is especially true of *Krithe* (with an estimated 10-15 species), *Cytheropteron* (at least 12 species), *Eucytherura* (at least 9 species), *Pseudocythere* (number of species unknown but apparently large), *Argilloecia* (probably some 6 species) and *Polycope* (number of species unknown).

Most of the taxa displayed in Table 2 seem rather regularly distributed in the study area, but diversity is probably higher to the south of the Azores (straddled by stations T86-09 and T86-10; see Fig. 1) than it is to the north: in terms of the taxa discerned here, mean diversity is 19.3 between stations T86-01 to 09 and 27.3 between stations T86-10 to 17 (see Table 2). The Azores also mark the northern limit of *Australoecia micra* (Bonaduce *et al.*), *Bythocythere* spp., Gen. A sp. 1, *Jonesia* spp., *Metapolycope* sp. 1, *Ruggieriella* cf. *decemcostata* Colalongo & Pasini, and *Sclerochilus* sp. Failing further evidence, the cause of this apparent break remains uncertain. The Azores Ridge, however, has been reported to mark an important boundary between bottom watermasses (Murray *et al.*, 1986).

What is grouped in Table 2 as 'paradoxostomatids' comprises several undescribed taxa which may or may not belong in this group proper, as well as some definite paradoxostomatid species. Paradoxostomatids possess modified mouth parts which are thought to be primarily adapted to derive food from vegetable matter such as plants and plant debris. They are, therefore, usually considered to be restricted to photic waters and just underneath. In the present material, however, there is no evidence to indicate that these 'paradoxostomatids' are allochthonous. Whatley (verb. comm., 1987) suggested that they might have fallen off weeds floating on the surface but Sargasso weeds collected during the cruise, although encrusted with foraminifera and bryozoans and containing many small vagile gastropods and even crabs, did not yield a single ostracod. The assumption that the 'paradoxostomatids' are authentic deep-sea dwellers is supported by their preservation which is similar to that of the rest of the ostracod assemblage. They may belong to the 'wood-island habitat' of Maddocks & Steineck (1987) although none of the species described by these authors could be identified in our material.

## Contamination

At station T86-01, the northernmost of the series, single valves of *Cytheropteron nodosoalatum* Neale & Howe and *Eucythere anglica* Brady were found. On the basis of their preservation alone, these are undoubtedly shallow-water contaminants, probably imported through ice rafting (Whatley & Coles, 1987). No further unambiguous contaminants were seen.

## Most frequent genera

Table 3 is a breakdown in terms of most frequently occurring genera. Since picking in the 250-150µm fraction was aimed to be representative rather than complete, and *Eucytherura*, *Pedicythere*, *Pseudocythere*, *Cytheropteron*, *Argilloecia*, *Bythocythere* and the 'paradoxostomatids' are either confined to this size grade or strongly overrepresented in it, the resultant percentages are rough approximations only. This does not detract, however, from the dominance at most stations of *Krithe*, the 'paradoxostomatids', *Cytheropteron* and *Poseidonamicus*.

Again, the faunas in the north appear to differ from those in the south. This finds expression in far lower percentages for *Krithe* and *Poseidonamicus* and markedly high values for *Eucytherura* and *Pedicythere*. Station T86-09 stands out as the only site to yield *Echinocythereis*, *Bradleya*, *Henryhowella* and *Ambocythere* in percentages higher than 5.

## Generic and specific diversity

The generic and specific diversities found in the

Table 2. Taxa found in study area.

| STATION T86- | 01 | 03 | 05 | 07 | 08 | 09 | 10 | 11 | 13 | 14 | 15 | 16 | 17 |
|---|---|---|---|---|---|---|---|---|---|---|---|---|---|
| *Ambocythere* sp. 1 | x | | | | | x | x | x | x | | x | | |
| *Argilloecia* spp. | x | x | x | | | x | x | x | x | x | | x | |
| *Australoecia micra* (Bonaduce et al.) | | | | | | | x | | | x | | | |
| *Aversovalva hydrodynamica* Whatley & Coles | x | | x | x | x | | x | x | x | x | | | |
| *Bairdoppilata* cf. *victrix* (Brady) | | | x | x | | | x | x | x | x | x | | x |
| *Bathycythere vanstraateni* Sissingh | | x | x | | x | | | x | | x | | x | |
| *Bradleya dictyon* (Brady) | x | | x | x | | | x | x | x | | | | |
| *Bythoceratina scaberrima* (Brady) | | | x | | | | x | x | x | x | | x | |
| *Bythocypris* spp. | | | x | x | x | | | | | x | x | x | |
| *Bythocythere* spp. | | | | | | | x | x | | x | x | x | x |
| *Cytherella serratula* (Brady) | | | | x | | | x | x | x | | x | | |
| *Cytheropteron* spp. | x | x | x | x | x | x | x | x | x | x | x | x | x |
| *Echinocythereis echinata* (Sars) | | x | x | x | x | x | x | x | x | x | x | | x |
| *Eucythere* spp. | x | | | x | | | x | x | | | x | | |
| *Eucytherura* spp. | x | x | x | x | | x | x | x | x | x | x | x | x |
| Gen. A sp. 1 | | | | | | | x | x | | | x | x | |
| Gen. B sp. 1 | | | x | x | | | x | x | | x | | | x |
| *Heinia dryppa* Whatley & Coles | | | x | x | x | x | x | x | x | x | | | |
| *Henryhowella asperrima* (Reuss) | | x | x | | | x | x | x | x | x | | x | x |
| *Jonesia* spp. | | | | | | | | x | x | x | | | |
| *Krithe* spp. | x | x | x | x | x | x | | x | x | x | x | x | x |
| *Macrocypris* spp. | | | x | x | | | x | x | x | x | x | x | |
| *Metapolycope* sp. 1 | | | | | | | x | x | | | | | |
| *Monoceratina* sp. 1 | | | | | | | | | | x | | | |
| *Nannocythere* sp. 1 | | | | x | | | x | x | x | | | | |
| "paradoxostomatids" | | | x | x | | | x | x | x | x | x | x | |
| *Pedicythere* spp. | | | x | x | | | x | x | x | x | x | x | x |
| *Pennyella dorsoserrata* (Brady) | | x | | | | | x | | x | x | x | x | x |
| *Pennyella horrida* Whatley & Coles | | | | | | | | x | | | | | |
| *Polycope* spp. | x | | x | x | x | x | x | x | x | x | x | x | x |
| *Pontocypris* sp. | | | x | x | | | | x | x | x | | | x |
| *Poseidonamicus* cf. *minor* Benson | | | x | x | x | x | x | x | | x | x | | |
| *Poseidonamicus pintoi* Benson | | | x | | x | x | | x | x | x | x | | x |
| *Pseudocythere* spp. | | | x | x | | | x | x | x | x | x | x | |
| *Pterygocythere mucronalata* (Brady) | | | x | x | | x | | x | x | x | | x | x |
| *Rimacytheropteron longipunctatum* (Breman) | | | | | | x | | x | | | x | | x |
| *Rimacytheropteron* sp. 1 | | | | | | x | | | | | | | |
| *Rockallia enigmatica* Whatley et al. | | | x | | | x | x | x | | | x | | |
| *Rostrocythere* spp. | | | x | x | | | | | | | x | x | |
| *Ruggieriella* cf. *decemcostata* Colalongo & Pasini | | | | | | | x | x | x | | | | |
| *Sclerochilus* sp. | | | | | | | x | x | | | | | |
| *Semicytherura* spp. | | | | x | | | | x | | | | | |
| "*Suhmicythere*" *suhmi* (Brady) | | | | | x | | | x | | | x | | x |
| *Swainocythere* spp. | | | | x | | | | x | | | x | x | |
| "*Thalassocythere*" *acanthoderma* (Brady) | | | x | x | | x | x | x | x | | x | x | x |
| *Xestoleberis profundis* Whatley & Coles | | | x | x | | | x | x | x | x | | | |
| *Zabythocypris* sp. 1 | x | | x | | | | | | x | x | x | x | |
| NUMBER OF TAXA | 8 | 14 | 26 | 28 | 15 | 25 | 37 | 31 | 30 | 29 | 26 | 21 | 17 |

present study compare well with recent data reported from the abyssal North Atlantic by other authors. Whatley & Coles (1987) using Leg 94 DSDP material quoted 34 as the generic and 61 as the specific diversity for the Quaternary (omitting the Krithinae). Taking several other Quaternary sites into account as well, Whatley & Ayress (1988) mentioned 130 species, again omitting the Krithinae. The present Mid-Atlantic Ridge material contains more than 40 genera and over 100 species (including *Krithe* with an estimated 10-15 species). All these values are definitely of

Table 3. Most frequent genera in samples yielding more than 50 (T86-05) or 100 (remaining stations) adult valves. Figures are percentages of adult valve populations. Percentages lower than 5 are omitted.

| STATION T86- | 05 | 07 | 09 | 10 | 11 | 13 | 14 | 15 | 16 |
|---|---|---|---|---|---|---|---|---|---|
| *Bythocypris* | 11 | | | | | | | | |
| *Eucytherura* | 20 | 10 | | | | | | | |
| *Polycope* | 11 | | | | 9 | | | | |
| *Pedicythere* | 13 | 21 | | | | | 5 | | 9 |
| *Krithe* | 7 | 6 | 30 | 27 | 19 | 31 | 38 | 22 | 29 |
| "*paradoxostomatids*" | 14 | 11 | | 7 | 13 | 15 | 14 | 15 | 18 |
| *Pseudocythere* | 5 | 6 | | | | 7 | 6 | 10 | 16 |
| *Cytheropteron* | | 18 | | 14 | 9 | 15 | | | 5 |
| *Echinocythereis* | | | 6 | | | | | | |
| *Bradleya* | | | 10 | | | | | | |
| *Henryhowella* | | | 10 | | | | | | |
| *Ambocythere* | | | 8 | | | | | | |
| *Argilloecia* | | | 11 | 8 | 8 | | | | |
| *Poseidonamicus* | | | 7 | 11 | 7 | 6 | 9 | 12 | |
| *Bythocythere* | | | | | | | | 6 | |

the same order of magnitude, however, and differences may be partly taken to reflect sample size. True North Atlantic abyssal ostracod diversity is certain to be higher than single available studies show and may, on the basis of current knowledge, be conservatively estimated at some 60 genera and somewhat over twice as many species. Interestingly, Whatley (1983) reported approximately the same generic diversity (58) but a much higher specific figure (365) for the Quaternary of the bathyal and abyssal S.W. Pacific. This implies that S.W. Pacific deep-sea ostracod genera on average comprise about three times as many species as do North Atlantic ones, suggesting a higher overall speciation level in the S.W. Pacific. However, the study area in the S.W. Pacific was much greater in latitudinal and longitudinal range than that in the North Atlantic.

The above values are totals for large ocean basins. Single site diversities will, of course, tend to be lower. The maximum number of species encountered at a single site in the present study is *circa* 65 (station 10). In this context, it is worth noting that Cronin (1983) found an average specific diversity per site of 33.3 between 200 and 1100m on the Florida-Hatteras slope, while quoting 61 as the highest single site value for that area (at 220m depth).

## COMPARISON WITH THE DEEP MEDITERRANEAN

The deep Mediterranean has several ostracod species in common with the bathyal and abyssal North Atlantic (Whatley & Ayress, 1988; Coles, lit. comm., 1988) although diversity is far lower in the former. Bonaduce *et al.* (1983) recognized only 43 deep-sea species as autochthonous in the western Mediterranean. As a consequence of an anoxic event in the early Holocene, the deep waters of the eastern basin have even fewer species (Van Harten, 1987; Van Harten & Droste, 1988).

Table 4 lists the species, which are known to the author from Recent sediments of the Mediterranean, and which were also encountered in the Mid-Atlantic Ridge area covered by this study. Species of *Pedicythere*, *Polycope* and *Pseudocythere* may need to be added once taxonomic scrutiny is completed.

Considering the Messinian Salinity Crisis and the attendant desiccation of the Mediterranean, an oceanic provenance is likely for those deep-water species this sea has in common with the Atlantic. Atlantic deep-sea species were able to enter the Mediterranean in the Pliocene when the circulation was estuarine and there was a direct psychrospheric connection (Van Harten, 1984). The

Table 4. Species common to the sampled area of the Mid-Atlantic Ridge and the Mediterranean. Asterisks indicate species that are no longer extant in the eastern Mediterranean Basin.

---

*Argilloecia acuminata* Müller
*Argilloecia cylindrica* Sars
*Australoecia micra* Bonaduce *et al.*
*\*Bathycythere vanstraateni* Sissingh
*Cytheropteron testudo* Sars
*Echinocythereis echinata* (Sars)
*Henryhowella asperrima* (Reuss)
*\*Macrocypris minna* (Baird)
*Rimacytheropteron longipunctata* (Breman)

---

subsequent reversal of the circulation to anti-estuarine, with an outflow along the bottom prevented any further ingress of bathyal and abyssal Atlantic species. Genetic isolation of the species thus 'imprisoned' in the deep Mediterranean is thought to have been a major factor enabling these one-time psychrospheric immigrants to rapidly adapt to the thermospheric conditions which they experience today.

Far from all the original Atlantic immigrants survived in the Mediterranean. The vast majority probably became extinct in the Pliocene and several others died out in the Pleistocene and early Holocene. The oceanic species *Bythoceratina scaberrima, Eucytherura calabra, Paijenborchella malaiensis cymbula* and '*Hyphalocythere*' sp. 1, quoted by Whatley and Ayress (1988) as recorded in the Mediterranean, belong in these categories. Whether *Cytheropteron testudo*, which is occasionally found on the bottom of the deep Mediterranean, is really still extant there is uncertain.

Remarkably, several species that did survive apparently adjusted their depth range. *Bathycythere vanstraateni, Echinocythereis echinata* and *Henryhowella asperrima*, all species which, in the Atlantic, descend far into the abyss, are rarely if ever found at greater than bathyal depths in the Mediterranean (Van Harten, 1987; Van Harten & Droste, 1988). Adults of these species are noticeably smaller in the Mediterranean than they are in

the North Atlantic.

## FUTURE WORK AND CONCLUSIONS

APNAP-I which collected the samples reported here was the first expedition in an intended series of four to the same area. On this first venture, geochemical work was primarily aimed at establishing parameters relevant to water-mass identification in the upper parts of the water-column. For the remaining cruises, however, it is planned to pay full attention to the chemistry and physics of the bottom water and the sediment itself. The ensuing data are expected to shed light on what causes the apparent faunal break in the region of the Azores and may help to unravel problems connected with *Krithe* and its controversial systematics.

In relation to total yield, very little living material was encountered. The apparent patchiness in space and time of life in the abyss is a problem in itself and worthy of future attention.

Until recently, it was doubted by few that the ocean floors were dark and gloomy deserts of extremely low biological diversity. Benson, for instance, claimed as late as 1984 that in samples from depths greater than 1000m and having 300 ostracod specimens, a maximum of only 10 to 12 species were likely to be found belonging to 9 or 10 genera. Recent research has clearly shown, however, that the floor of the world's ocean is in fact home to a profusion of genera and species (Cronin, 1983; Whatley, 1983; Whatley and Coles, 1987; Whatley and Ayress, 1988) and that single samples may compare in variety with those from far shallower waters. While the present results underscore these findings, this report is only a link in a process that will undoubtedly see the unveiling of still greater riches in the future. There is no doubt that much of what still awaits discovery lies concealed in the smaller size grades of the sediment. If anything, this paper intends to emphasize the importance of examining this fine-grained material, tedious a task as it may be.

## ACKNOWLEDGEMENTS

I thank APNAP-I chief scientist Gerald Ganssen of the Free University, Amsterdam and his crew for collecting the samples and placing them at my disposal. For co-examining and discussing their contents I am indebted to Robin Whatley, Michael Ayress, and Graham Coles of the University College of Wales, Aberystwyth. This is a publication of Werkgroep Mariene Geologie en Technologie, Amsterdam.

## REFERENCES

Benson, R. H. 1984. Estimating greater paleodepths with ostracodes, especially in past thermospheric oceans. *Palaeogeogr. Palaeoclimat. Palaeoecol.*, Amsterdam, **48**, 107-141.

Bonaduce, G., Ciliberto, B., Masoli, M., Minichelli, G. & Pugliese, N. 1983. The deep-water benthic ostracodes of the Mediterranean. *In* Maddocks, R. F. (Ed.), *Applications of Ostracoda*, proceedings of the Eighth International Symposium on Ostracoda, July 26-29, 1982, 459-471. Univ. Houston Geos., Houston, Texas.

Cronin, T. M. 1983. Bathyal ostracodes from the Florida-Hatteras slope, the Straits of Florida, and the Blake Plateau. *Marine Micropaleontology*, Amsterdam, **8**, 89-119.

Lutze, G. F. 1964. Zum Färben rezenter Foraminiferen. *Meyniana*, Kiel, **14**, 43-47.

Maddocks, R. F. & Steineck, P. L. 1987. Ostracoda from experimental wood-island habitats in the deep sea. *Micropaleontology*, New York, **33**, 318-355.

Murray, J. W., Weston, J. F., Haddon, C. A. & Powell, A. D. J. 1986. Miocene to Recent bottom water masses of the north-east Atlantic: an analysis of benthic foraminifera. *In* Summerhayes, C. P. & Shackleton, N. J. (Eds), *North Atlantic Palaeoceanography, Spec. Publ. Geol. Soc.*, London, **21**, 219-230.

Van Harten, D. 1984. A model of estuarine circulation in the Pliocene Mediterranean based on new ostracod evidence. *Nature*, London, **312**(5992), 359-361.

Van Harten, D. 1987. Ostracodes and the early Holocene anoxic event in the Eastern Mediterranean - Evidence and implications. *Mar. Geol.*, Amsterdam, **75**, 263-269.

Van Harten, D. & Droste, H. J. 1988. Mediterranean deep-sea ostracods, the species poorness of the eastern basin as a legacy of an Early Holocene anoxic event. *In* Hanai, T., Ikeya, N. & Ishizaki, K. (Eds), *Evolutionary biology of Ostracoda, its fundamentals and applications*, proceedings of the Ninth International Symposium on Ostracoda, held in Shizuoka, Japan, 29 July - 2 August 1985, Developments in palaeontology and stratigraphy, **11**, 721-737, Kodansha Ltd., Tokyo and Elsevier, Amsterdam, Oxford, New York, Tokyo.

Whatley, R. C. 1983. Some aspects of the palaeobiology of Tertiary deep-sea Ostracoda from the S.W. Pacific. *J. micropalaeontol.*, London, **2**, 83-104.

Whatley, R. C. & Ayress, M. 1988. Pandemic and endemic distribution patterns in Quaternary deep-sea Ostracoda. *In* Hanai, T., Ikeya, N. & Ishizaki, K. (Eds), *Evolutionary biology of Ostracoda, its fundamentals and applications*, proceedings of the Ninth International Symposium on Ostracoda, held in Shizuoka, Japan, 29 July - 2 August 1985, Developments in palaeontology and stratigraphy, **11**, 739-755, Kodansha Ltd., Tokyo and Elsevier, Amsterdam, Oxford, New York, Tokyo.

Whatley, R. C. & Coles, G. 1987. The Late Miocene to Quaternary Ostracoda of leg 94, Deep Sea Drilling Project. *Revta esp. Micropaleont.*, Madrid, **19**, 33-97.

Whatley, R. C. & Wall, D. R. 1969. A preliminary account of the ecology and distribution of Recent Ostracoda in the southern Irish Sea. In Neale, J. W. (Ed.), *The Taxonomy, Morphology and Ecology of Recent Ostracoda*, 268-298. Oliver & Boyd, Edinburgh,

## DISCUSSION

Robin Whatley: I would like to confirm that, having seen the fauna, I must agree with Dr Van Harten that the 'paradoxostomatids' he has found are certainly indigenous.

Dick Van Harten: Thank you. Well, in this case there really seems to be no other explanation but it makes one wonder about the provenance of similar forms elsewhere in the deep sea, the Mediterranean for instance.

# ECOLOGY

# 25

# The external mechanisms responsible for morphological variability in Recent Ostracoda: seasonality and biotope situation: an example from Lake Titicaca

P. Carbonel, Ph. Mourguiart & J.-P. Peypouquet

Department of Geology and Oceanography,
University of Bordeaux, France.

## ABSTRACT

The environmental mechanisms directly responsible for variations recorded in the ornamentation of ostracods are well known. They concern mainly the carbonate equilibrium at the water/sediment interface (reticulation phenomenon *sensu lato*) and the impact of the input of fine-grained and all allochthonous matter on biotopes (nodation and microconation phenomena).

Although these processes can be observed in all sorts of environments, whether continental shelf, coastal, lagoonal or estuarine, it is difficult to determine the factor(s) governing such equilibria.

Studies in Lake Titicaca revealed the existence of two major groups of factors responsible for the morphological variability of ostracods (mainly in the intertropical zone):

1) contrasting seasonality i.e. alternation between a dry and wet season with all the subsequent effects on the circulation, input from the flanks of the basin and chemical equilibria at the water-sediment interface.

2) biotope positioning. A maximum of species and morphs were observed at the borderline between the phytal and 'deep' zones.

With present-day conditions, where the seasonal contrast is relatively important, polymorphism is high. Conversely, when climatic conditions were different, polymorphism changed from being very intense under contrasting seasons (7500 BP), into monomorphism (towards 4500 BP) under almost uniform climatic conditions during all seasons. This is perhaps the first step towards speciation.

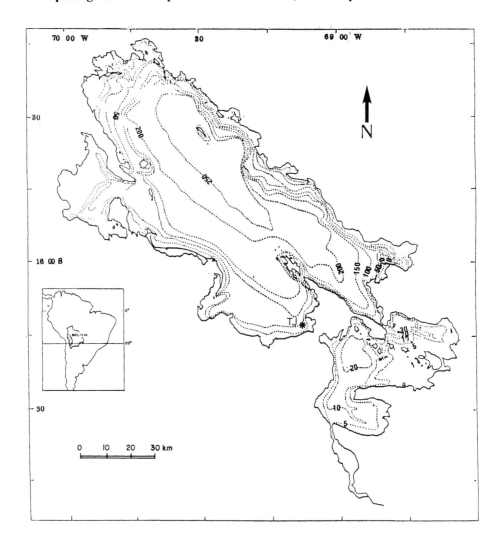

Fig. 1. Lake Titicaca, location map - site of core TJ.

When global climatic changes occurred, e.g., during the Pretiglian glacial and the Tiglian interglacial in North West Europe in inner shelf environments, very similar aspects as those described above must have existed.

## INTRODUCTION

Some aspects of morphological variability in ostracods are more or less governed by variations in the chemical equilibria at the water-sediment interface. Ostracods take all the components of their shells from the water during each moulting stage without storing them in the soft parts (Turpen & Angell, 1971). This morphological variability is mainly expressed as 3 types of variation that can sometimes be observed together on the carapace:

1) reticulation, depending upon carbonate equilibria: this is the 'agradation-degradation' phenomenon defined by Peypouquet *et al.* (1987, 1988),

2) nodation, spinosity and microconation, connected with the supplies of fine-grained sediment and allochthonous organic matter,

3) size (not examined here).

These phenomena are observable in all environments; marine, lagoonal, lacustrine. Our aim is to show that ostracod variability is not haphazard but linked to the fluctuation of global parameters operating at the water-sediment interface where most ostracods live and moult.

First, we discuss the morphological variability which exists only in certain places and later the succession of valves with different morphologies on the bottom, representing each stage of the seasonal cycle. The example chosen is present day ostracod fauna living in Lake Titicaca, particularly two groups of *Limnocythere*.

## VARIABILITY IN *LIMNOCYTHERE* FAUNA IN THE PRESENT DAY ENVIRONMENTS OF LAKE TITICACA

Lake Titicaca is located between the Andean Cordillera at 3800m above sea level (Fig. 1). It has an area of about 8000km$^2$ and its maximum depth is approximately 220m. It is subdivided into the Great Lake and Little Lake (or Lake Huynaymarca) connected by the Tiquina Strait. The Great Lake is considered to be a warm monomictic and eutrophic lake (Hutchinson & Löffler 1956) with a productivity of 500g cm$^{-2}$ yr$^{-1}$. The Little Lake with a maximum depth of 40m (main part < 10m), is oligotrophic and has a productivity of 20g cm$^{-2}$ yr$^{-1}$. Its ecology shows 2 phases (Fig. 2):

1) homothermy and homogenization during the austral winter,

2) stratification during summer and autumn.

Such a contrast entails, in the hypolimnion, an increase of $CO_2$ (caused by the consumption of organic matter by bacteria) and a decrease of pH during the stratification phase and later, saturation of $O_2$, with the recycling of chemical elements during the homogenization phase.

## BATHYMETRIC ZONATION

The distribution of ostracods is the same in both parts of the lake. In the shallower parts, their distribution is controlled by that of phytal communities. Six successive zones are distinguished from the shallowest to the deepest parts of the lake (Fig. 3):

1) 0-2.5m, coastal zone. Low diversity of phytal fauna living on plants with *Elodaea* and *Myriophyllum*, generally less than 1m deep. This zone is physically very unstable and has a high energy level.

2) 2.5-4.5m, *Totora*. Ostracods are absent (oxygen depletion on the bottom): vegetative parts of *Totora* are developed on the water surface.

3) 4.5-7.0m, *Chara*. Weakly calcified ostracods living on *Chara*. No evidence of thanatocoenosis.

4) 7.5-12.5m, lower boundary of vegetation. Maximum number of species, individuals and morphs. This is particularly true for *Limnocythere*.

5) 12.5-20.0m, aphytal zone. Decrease in abundance and diversity of species.

6) more than 20m, deep zone. Ostracods rare.

In zones 5 and 6, which are deeper and less rich in nutrients than those above, ostracods become fewer and less diversified. The plants serve as regulators of food and energy levels and live under the direct influence of annual climatic variations.

This distribution is schematic, local characteristics exist, such as the density of *Totora* and the extension of zones 4, 5 and 6 depending on topography and water circulation. In the Great Lake, zones 4 and 5 are extended between 8-30m and 30-60m respectively, because of steeper slopes.

The abundance and morphological variability of *Limnocythere* are greatest in zone 4. This area is, therefore, the key sector where the impact of seasonal parameters is highest without any evidence of interference because of the filtering action of the water plants.

## THE *LIMNOCYTHERE* FAUNA IN LAKE TITICACA

The *Limnocythere* fauna living in Lake Titicaca is abundant (often the main group) and diverse (Mourguiart, 1987). A taxonomic study of this group is in process and we will give only the principal elements necesssary to the understanding of the present paper. Seven species have been identified: *Limnocythere titicaca* Lerner-Segueev and the *Pampacythere* group Whatley & Cholich,

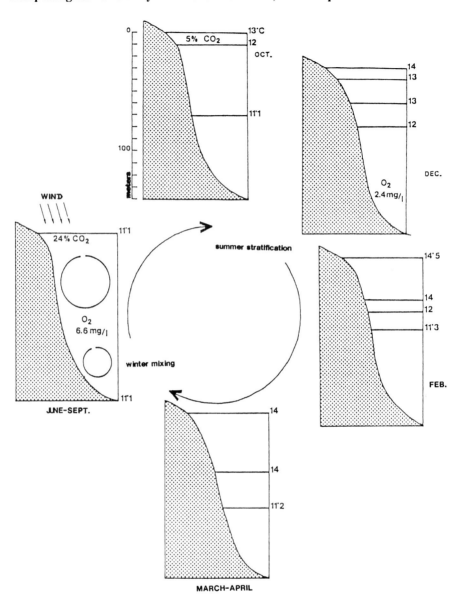

Fig. 2. Seasonal changes of water circulation in Lake Titicaca.

*L. bradburyi* Forester, *L.* sp. X Mourguiart, *L.* sp. Y Mourguiart, *L.* sp. Z Mourguiart, L. gr. 'A' Mourguiart and L. gr. 'B' Mourguiart. These species can be grouped together into 2 or 3 genera, but the confusion existing concerning their recognition (with the exception of *Pampacythere* and the type-genus) as well as the lack of interpretation of the soft parts, are the reasons why we have chosen to retain the denomination *'Limnocythere'* for the whole group. Two specific groups are particularly variable: *Limnocythere* gr. 'A' and 'B'. Their abundance and wide distribution account for their being chosen for this study. The main differences between the two groups are the following:

| | Group 'A' | Group 'B' |
|---|---|---|
| Lateral view | subrectangular to slightly reniform | pseudorectangular |
| Dorsal view | oval | flat anteriorly |
| Calcification | variable, generally weak | variable, weak |
| Ant. margin | rounded, large with relief | rounded and flat |
| Dors. margin | straight | straight |
| Ventr. margin | ant. 1/3 slightly concave | ant. 1/3 slightly concave |
| Size | medium to large (L/H=0.53) | large (L/H=0.46) |
| Overlap | RV=LV | RV=LV |

The morphological variability observed in both groups is similar (Fig. 6). In the group 'A', 4 'morphotypes' are recognized, A1 standard, A2 with one spine per valve, A3 with tubercle and microcones, A4 with some expansions similar to *Neolimnocythere haxaceros* Delachaux. In the group 'B', 8 'morphotypes' occur; these include B1 standard, B2 with one spine, B3 with microcones, B4 with expansions and B7 living on *Chara*. Members of group 'A' occur more frequently than those of group 'B', possibly due to ecological differences.

## SEASONAL VARIATION OF *LIMNOCYTHERE* GROUP 'A' AND 'B'

Seven successive phases can be observed throughout the year with respect to the lake's ecology (Mourguiart, 1987).

1) During the austral winter (Fig. 4a) the water mass is homogeneous, the bottom is anoxic with no possibility for ostracods to live.

2) At the end of winter in September (Fig. 4b) maximum photosynthesis occurs and supersaturation of carbonates at the water-sediment interface gives rise to very large shells.

3) At the end of September (Fig. 4c), detritus from plants yields nutrients and elements of macrophytes are dissolved resulting in the occurrence of spines.

4) In the austral spring (October) (Fig. 4d), with increasing temperature, the water column begins to be stratified. The diatom bloom entails a strong depletion in silica and the death of these organisms (Carmouze *et al.*, 1984) gives rise to a rain of

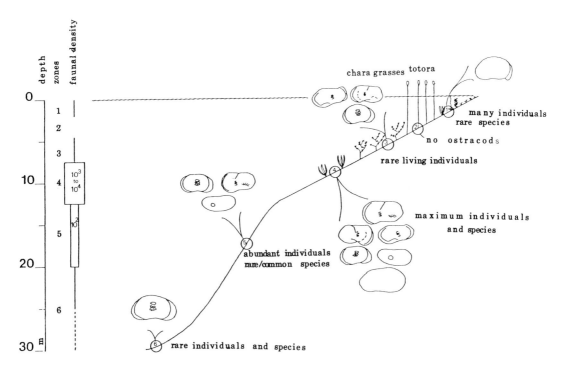

Fig. 3. Bathymetric zonation of the ostracod assemblages.

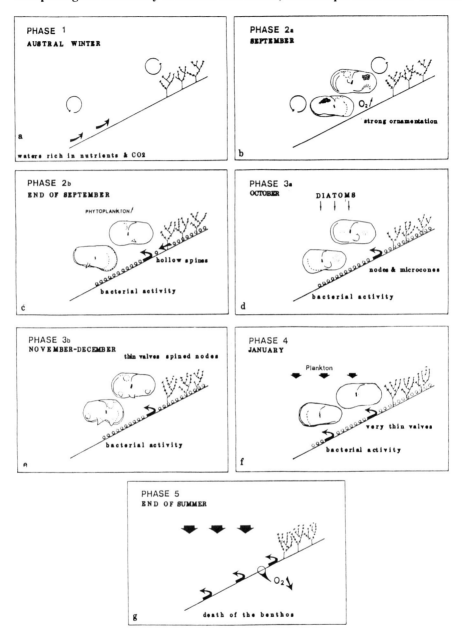

Fig. 4. Seasonal variability of *Limnocythere* gr. 'A' and 'B'.

frustules, comparable to the input of fine detritus (Kirk, 1985). This rain induces the development of microcones and nodes on the surface of the valves (Mourguiart, 1987). This mechanism is similar to that described by Abe & Choe (1988, 371) for the formation of microcones in *Pistocythereis bradyi* (morph D). 'Small sub-conical projections on the muri of *P. bradyi* are well developed around the pores of sensory hairs which are situated only at the junction of the muri and they are naturally assumed to affect the function of the sensory organ, perhaps by protecting the hair or preventing the

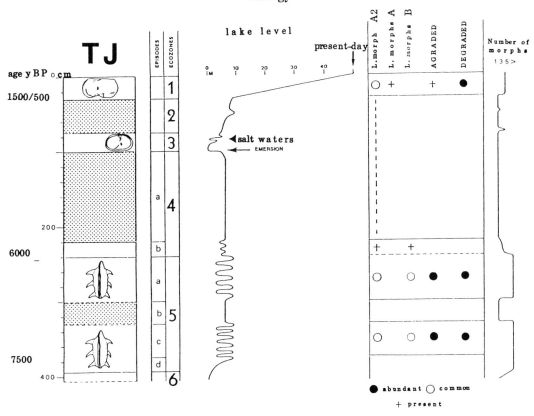

Fig. 5. Core TJ, zonation.

intrusion by foreign matter.'

5) In the austral spring (November-December) (Fig. 4e), the calcification of the shells decreases (with increase of $CO_2$ due to intense bacterial activity). Recycling of dissolved elements causes the development of spined nodes.

6) During the austral summer (Fig. 4f), primary productivity increases. Ostracod carapaces become very thin, as bioprecipitation of $CaCO_3$ is very difficult under these conditions.

7) At the end of summer (Fig. 4g) the water column is completely stratified; the bottom becomes almost depleted of oxygen. Calcite bioprecipitation becomes impossible (Kelts & Hsu, 1978). Most of the benthos dies.

Fig. 4 shows the successive seasonal stages in the morphological variability of *Limnocythere* groups 'A' and 'B' within an average climatic situation, i.e. a situation with a well-marked but not strong seasonal contrast. It is interesting to

examine the response of *Limnocythere* to the environment under palaeoclimatic situations where seasonal contrasts were similar to those prevalent today.

## THE PALAEOFAUNA OF LAKE TITICACA DURING THE HOLOCENE

We have selected a core from the Great Lake in Yunguyo Bay (Fig. 1) (50m deep), i.e. with a present day water depth that is sufficient for the recording of important variations of the lake level. This core is 4.06m long. An overall analysis of the fauna (Carbonel *et al.*, 1988) shows a zonation of 6 ecozones (Fig. 5). From bottom to top these are:

1) Ecozone 6, beach deposits with gypsum and without ostracods.

2) Ecozone 5, (394-240cm), zone with strong fluctuations. Maximum polymorphism occurring

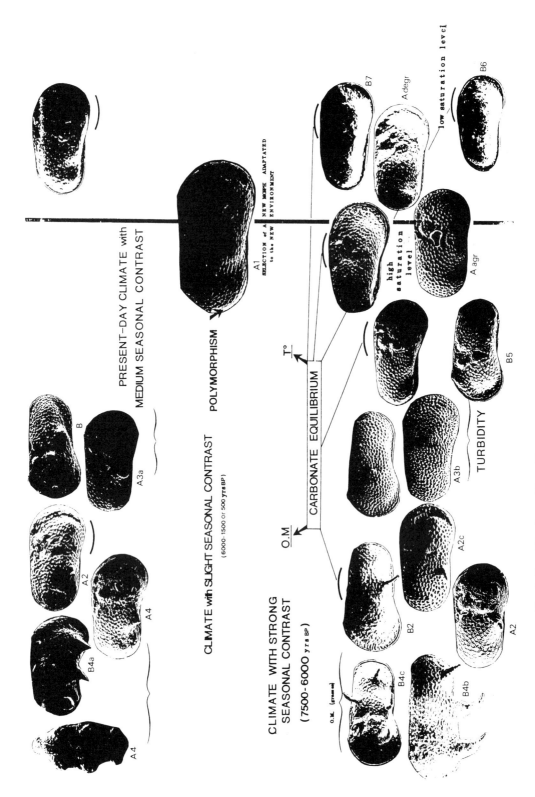

Fig. 6. Core TJ, polymorphism of *Limnocythere* gr. 'A' and 'B' during the last 7500 years.

at the same time as the maximum development of the ostracod fauna (Mourguiart, 1987).

3) Ecozone 4, (240-100cm), proximity of *Chara*. Very slight polymorphism changing into monomorphism.

4) Ecozone 3, (100-75cm), very low lake level with emergent phases giving rise to greater water concentration (occurrence of *Cyprideis*).

5) Ecozone 2, (75-30cm), similar to ecozone 4.

6) Ecozone 1, (30 cm-top), typical present-day fauna in this part of the lake. High polymorphism (less than in ecozone 5).

Three main phases can be observed in the morphological variability of *Limnocythere* (Fig. 6), maximum in ecozone 5 (7500-6000 BP), minimum in ecozones 2 and 4 (6000-4500 and 2000-500) and medium in ecozone 1 (500-present day).

Each type of variability can be related to a type of climate. Between 7500 and 6000 BP, the lake's level increases very slowly in association with a summer monsoon with intense stormy and rainy periods and very strong winter aridity. This phenomenon coincides with the sinking of the Inter Tropical Convergence Zone (ITCZ) towards the Bolivian Altiplano (Servant & Fontes, 1984).

Between 6000 and 4500 BP, and 2000 and 500 BP, the seasonal contrast decreases and the climate is characterized by rain showers, a stabilization of the lake's level and a large extension of the peat-bog in the high valleys. This change results probably from the rise of the ITCZ to the North, a rise that is stronger than that of today. This rise occurs together with a strong jet-stream on the top of the Andes and with an intensification of the El Niño phenomenon on the coasts (Martin *et al.*, 1987). In the final phase, the lake's level is comparable to the present day and the seasonal contrast recurs with the summer precipitations.

Variations in the seasonal contrast associated with the displacement of ITCZ seem to be indirectly related to the morphological variability of *Limnocythere*. This last point raises the following question. Is the selection of morphs by the environment (*sensu* Clark, 1976) the starting point for speciation? In the example given, it is impossible to know with certainty

because the time factor is very short.

The same question can be asked for a longer period of time. Is it possible that a long term climatic situation can lead to speciation through geographical isolation and long term environmental stability? Several examples exist showing the same selection effects and stabilization and expansion of new morphs, and perhaps species, during longer time periods when intense climatic changes occur as, for example, during the Palaeocene (Peypouquet *et al.*, 1988) and during the initial glacial phases around the Plio-Quaternary boundary (Kasimi, 1966; Braccini, 1988).

## CONCLUSIONS

The morphological variability of the *Limnocythere* groups appears in very specific areas of a limnic system (the same observation can be made in a marine coastal system) immediately beneath large phytal asemblages. This variability also depends on the seasonal evolution of the equilibria at the water-sediment interface, controlled by the seasonal contrast. During phases without seasonal contrast, there is only one morph present, while during phases with strong seasonal contrast, many morphs can be observed. The succession of such phases in the past may have probably resulted in the formation of new species. From a general point of view, can polymorphism be a necessary step towards speciation? The discussion is open.

## REFERENCES

Abe, K. & Choe, K.-L. 1988. Variation of *Pistocythereis* and *Keijella* species in Gamagyang Bay, South coast of Korea. *In* Hanai, T., Ikeya, N. & Ishizaki, K. (Eds), *Evolutionary biology of Ostracoda, its fundamentals and applications*, proceedings of the Ninth International Symposium on Ostracoda, held in Shizuoka, Japan, 29 July - 2 August 1985, Developments in palaeontology and stratigraphy, **11**, 367-373, Kodansha Ltd., Tokyo and Elsevier, Amsterdam, Oxford, New York, Tokyo.

Braccini, E. 1988. Étude d'une espèce polymorphe du Plio-Pleistocène normand. Son intérêt dans les reconstitutions paléoclimatiques, paléohydrologiques et paléoenvironnementales. *Mém. DEA, Bordeaux I*, 1-27.

Carbonel, P., Mourguiart, Ph. & Peypouquet, J.-P. 1988. Le polymorphisme induit par l'environnement chez les Ostracodes: Rôle du rythme saisonnier. *Travaux C.R.M.*,

Nice, **8**, 1-12.

Carmouze, J. P., Arze, C. & Quintanilla, J. 1984. Le lac Titicaca stratification physique et métabolisme associé. *Rev. Hydrobiol. Trop.*, **17**(1), 3-11.

Clark, W. C. 1976. The environment and the genotype in polymorphism. *Zool. J. Linn. Soc.*, London, **58**, 255-262.

Hutchinson, G. C. & Löffler, H. 1956. The thermal stratification of Lakes. *Proc. Nat. Acad.*, **1**(42), 84-86.

Kasimi, R. 1966. Les ostracodes et les paléoenvironnements du Plio-Pleistocéne en Normandie. Signification paléogéographie et paléoclimatique. *Thèse 3ème cycle Bordeaux I*, **no 2147**, 255 pp.

Kelts, K. & Hsu, K. J. 1978. Freshwater carbonate sedimentation. *In* Lerman, P. (Ed.), *Lakes chemistry, geology, physics*, 295-323.

Kirk, J. T. O. 1985. Effects of suspensoids (turbidity) on penetration of solar radiation in aquatic ecosystems. *Hydrobiologia*, Den Haag, **125**, 195-208.

Martin, L., Flexor, J. M. & Suguio, K. 1987. Inversion de la direction de la houle dominante au cours des 5000 dernières années dans la région de l'embouchure du Rio Doce (Brésil) en liaison avec une modification de la circulation atmosphérique. *Geodynamique*, **2**(2), 121-123.

Mourguiart, Ph. 1987. Les ostracodes lacustres de l'Altiplano bolivien. Le polymorphisme, son intérêt dans les reconstitutions paléohydrologiques et paléoclimatiques de l'Holocène. *Thèse 3ème cycle Bordeaux I*, n° **2191**, 1-293.

Peypouquet, J.-P., Carbonel, P., Ducasse, O., Tölderer-Farmer, M. & Lété, C. 1987. Environmentally cued polymorphism of ostracods - a theoretical and practical approach. A contribution to geology and to the understanding of ostracod evolution. *In* Hanai, T., Ikeya, N. & Ishizaki, K. (Eds), *Evolutionary biology of Ostracoda, its fundamentals and applications*, proceedings of the Ninth International Symposium on Ostracoda, held in Shizuoka, Japan, 29 July - 2 August 1985, Developments in palaeontology and stratigraphy, **11**, 1003-1019, Kodansha Ltd., Tokyo and Elsevier, Amsterdam, Oxford, New York, Tokyo.

Peypouquet, J.-P., Carbonel, P., Ducasse, O., Tölderer-Farmer, M. & Lété C. 1988. Le polymorphisme induit par l'environnement chez les Ostracodes: son intérêt pour l'évolution. *Travaux C.R.M., GRECO 1988*, Nice, **8**, 13-19.

Servant, M., Fontes, J. C. 1984. Les basses terasses du Quaternaire récent des Andes boliviennes. Datations par le [14]C. Interprétation paléoclimatique. *Cah. Off. Rech. Sci. Tech. Outre-Mer*, Paris, Série Géologie, **14**(1), 15-28.

Turpen, J. B. & Angell, R. W. 1971. Aspects of moulting and calcification in the ostracode *Heterocypris*. *Biol. Bull. mar. biol. Lab. Woods Hole, Mass.*, **140**, 331-338.

# 26

# Ostracoda of the terrigenous continental platform of the southern Gulf of Mexico

María Luisa Machain-Castillo[1],
Ana María Pérez-Guzmán[2] &
Rosalie F. Maddocks[2]

[1]Instituto de Ciencias del Mar y Limnología,
Universidad Nacional Autónoma de México,
Ap. Postal 70-305, México.
[2]Department of Geosciences, University of
Houston, Texas, U.S.A.

## ABSTRACT

Sixty-two sediment samples were collected across the terrigenous continental shelf off Veracruz and Tabasco, Mexico, ranging in depth from 17 to 370m. Their ostracod fauna includes 175 species, of which at least 90 also occur on the terrigenous shelf of the northern Gulf of Mexico, and 20 are also reported from the Caribbean. Total diversity appears to be a little higher than in the northern Gulf of Mexico, and many species are probably new.

Three principal ostracod assemblages are distributed in bathymetric belts parallel to the coast. Assemblage I, dominated by *Loxoconcha moralesi*, *Cytherella vermillionensis*, and species of *Loxoconcha* and *Cytherura* with additional species of *Cytheropteron*, *Pterygocythereis*, *Protocytheretta*, *Neomonoceratina*, and *Macrocyprina*, is found between the 20 and 60m isobaths and has

the highest species diversity. Assemblage II, dominated by *Echinocythereis margaritifera* and species of *Cytheropteron* with less abundant *Munseyella louisianensis*, *Buntonia* n. sp., *Malzella* and *Xestoleberis*, occurs between the 50 and 110m isobaths. Assemblage III is found in depths over 100m and has the lowest diversity. It is dominated by species of *Krithe*, *Parakrithe*, *Cytherella* and *Argilloecia* with less abundant *Ambocythere* and *Echinocythereis spinireticulata*.

Systematic changes in faunal composition take place from west to east within each assemblage, corresponding to increased salinity and carbonate content of sediments. A factor analysis estimates the relative effects of bathymetry, substrate, salinity, influence of local rivers and lagoons, and water-mass circulation on faunal composition and diversity.

## INTRODUCTION

Although the Recent Ostracoda of the Gulf of Mexico have been studied since 1867, most of the numerous papers deal with marginal marine, estuarine, lagoonal, mangrove and deltaic facies. Reviews of previous ostracod studies in the Gulf of Mexico were presented by Hulings (1967), Maddocks (1974), Garbett & Maddocks (1979) and Cronin (1979); Fig. 1 shows the location of additional research published since 1979. The vast majority of these works concern the northern coast of the Gulf of Mexico. The only comprehensive, published, regional investigations of offshore marine faunas in the Gulf are those by Puri (1960) and Benson & Coleman (1963) on the west coast of Florida, by Kontrovitz (1976) on the Louisiana shelf, and by Krutak and co-workers on the Veracruz-Anton Lizardo reefs (Krutak, 1982; Krutak & Rickles, 1979; Krutak *et al.*, 1980). In Mexican waters, lagoonal ostracods have been described from Laguna de Terminos (Morales, 1966), Laguna Mandinga (Krutak, 1971, 1974, 1977), and Laguna Madre (Sandberg, 1964). Consequently, comparatively little is known of the ostracods of the southern Gulf of Mexico, and a comprehensive understanding of their distribution in the Gulf of Mexico is lacking. Improved taxonomic and ecological analyses of living ostracods are needed for application to palaeoenvironmental intrepretations of older rocks and eventual reconstruction of the Cainozoic history of the Gulf of Mexico.

Recently the University of Mexico has undertaken a long-term, multidisciplinary investigation of the biology, physical and chemical oceanography, sedimentology, and micropalaeontology of the southern Gulf ('Estudios Multidisciplinarios de la Zona Económica Exclusiva del Sur del Golfo de México: Caracteristícas Geológicas, Físicas, Químicas y Biológicas'). Two major micropalaeontological research projects are included, one on the Holocene microfauna (ostracods, planktonic and benthonic foraminifera), and the other on the Quaternary history of these microfaunas in the southern Gulf of Mexico and Mexican Caribbean regions. One short-term goal is to define the ostracod assemblages of the southern Gulf and compare them with those found in the Atlantic, Caribbean and northern Gulf. Other immediate contributions are expected to be the bathymetrical zonation of the species or genera, and an evaluation of the oceanographical and sedimentological parameters controlling their distribution. Long-term goals include clarification of the taxonomy and delineation of biogeographical patterns for ostracods in the Gulf of Mexico. Machain-Castillo (1987) presented an overview of ostracod faunas in the southern Gulf of Mexico. This paper presents preliminary results from a study of the Ostracoda of the continental shelf off Veracruz and Tabasco, Mexico.

## HYDROLOGY AND SEDIMENTOLOGY

The study area lies on the continental shelf between latitudes 18°12.5'N and 19°14.7'N and longitudes 92°29'W and 95°36.9'W (see Fig. 1). Rich fisheries and petroleum resources give this area great economic importance, although oceanographical details are still poorly known.

The major factors affecting the hydrology of this area of the continental shelf are discharge from rivers (the Grijalva-Usumacinta, Coatzacoalocos and Papaloapan), lagoonal circulation (the Terminos, Carmen and Machona), evaporation, and the intrusion of Caribbean waters. During the summer, the Gulf of Campeche is the most saline area of the Gulf of Mexico with values up to 36.5°/oo (Nowlin, 1972; Lizarraga-Partida *et al.*, 1984). Haline fronts associated with river discharges (Fig. 2A) in the Bay of Campeche include a low-salinity mass between the Usumacinta-Grijalva Rivers and the mouth of the Coatzacoalcos River (<35.5°/oo), and another low-salinity (<34.0°/oo) mass northeast of the Papaloapan River (Padilla *et al.*, 1986). According to these authors, a high-salinity (>37.0°/oo) water mass is located between the Terminos Lagoon and the Usumacinta-Grijalva rivers.

Temperature and oxygen values average 23°C and 4.5 ml/l, respectively, showing very little change throughout the Bay area. The Yucatan Current dominates circulation in the Gulf of Mexico. It enters from the Caribbean sea as a warm saline

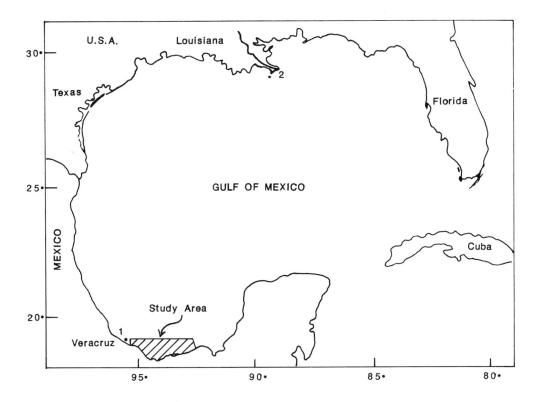

Fig. 1. Location of study area and of studies of Holocene marine Ostracoda published since 1979 or not in previous reviews. 1 = Krutak, 1982; Krutak & Rickles, 1979; Krutak et al., 1980. 2 = Howe and van den Bold, 1975.

current through the Yucatan Channel, flowing westward into the central eastern Gulf (Nowlin, 1972). From this area, the current turns eastward, southward, and again eastward, flowing across the Florida Strait into the Atlantic as the Florida Current (Nowlin, 1972). According to Nowlin, the extension, penetration and local gyral circulation of the Yucatan Current vary seasonally. In winter, waters from the western portion of the Gulf are directed southward into the Gulf of Campeche, forming a western boundary current known as the Mexican Current.

The sediments in the Bay of Campeche (Fig. 2B) are derived from source areas in the distant mountains *via* local rivers and from coral reefs to the east (Creager, 1958). None of the numerous rivers are even remotely equivalent in discharge or clay-size sediment load to the Mississippi River, and there are correspondingly important differences in physiography and sedimentology

between the southern and northern continental shelves of the Gulf of Mexico. The clastic sediments on the Campeche shelf rapidly become finer-grained seaward, from sandy silt or silty sand near the river mouths to fine silts and clays near the edge of the continental shelf and slope. The carbonate content increases eastward toward the Yucatan carbonate platform (Creager, 1958; Lecuanda & Ramos, 1985).

## METHODS

The samples studied are part of the OGMEX multidisciplinary project and were collected by the oceanographic vessel 'Justo Sierra' of the National University of Mexico during the ABACO I (1985), II (1986), and OGMEX I, II, III (1987) cruises. One hundred and five Smith-McIntyre grab samples were analyzed. However, only 62 samples (ranging from 17 to 370m water depth)

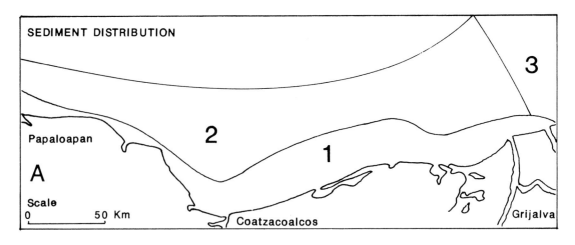

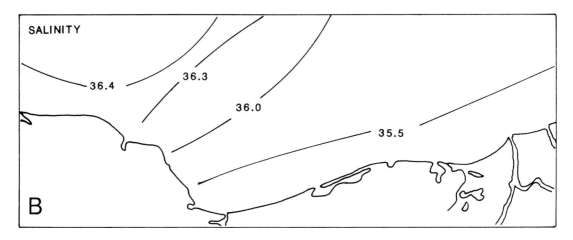

Fig. 2. Bottom-water isohalines (A) and sediment distribution (B) in the Bay of Campeche (based on Lizarraga-Partida et al., 1984; Lecuanda and Ramos, 1985. 1 = sand and sandy silt,  2  = terrigenous mud, 3 = carbonate mud.

that yielded more than 100 ostracod specimens are considered in this report (Fig. 3A, Table 1.)

Subsamples (30ml) were dried, weighed, washed over a 63μm sieve, and all size fractions picked for Ostracoda. Some subsamples were split to a smaller size yielding at least 300 specimens, while for other samples a second 30ml subsample was picked. Those samples with rare ostracods (fewer than 100 specimens after picking two subsamples) were deleted from this study. The specimens recovered were determined taxonomically, and their absolute and relative abundance recorded (whole carapaces counted as two specimens). Very few specimens had soft parts

enclosed, and the population counts do not distinguish between living and dead occurrences.

## OSTRACOD ASSEMBLAGES

At least 175 species were recognized, some of which were identified from published literature, comparison with diverse collections from elsewhere in the Gulf of Mexico and Caribbean regions, and comparison with primary types and other identified specimens in the Henry V. Howe Ostracod Collection at Louisiana State University. Others are new, and many require more detailed taxonomic investigation. The most diverse genera are *Cytherura* with at least 15 species, *Loxoconcha* with 11 species, and *Argilloecia*, *Cytherella*, *Cytheropteron* and *Krithe* with at least 5 species each. Only the 99 most abundant species were used in the factor analysis; a list of these species is given in Table 2. Inspection of the occurrence data shows that three principal assemblages of ostracods are fairly predictably distributed in bathymetric belts parallel to the coast.

Assemblage I is found between the 17 and 60m isobaths and corresponds in its distribution to that of sandy sediments, which increase in carbonate content eastward. The assemblage is dominated by *Loxoconcha moralesi* Kontrovitz, *L.* sp. A, *Cytherella vermillionensis* Kontrovitz, and species of *Pontocythere* and *Cytherura*, all with relative abundances between 10 and 30%. *Macrocyprina skinneri* Kontrovitz, *Neomonoceratina mediterranea* (Ruggieri), *Protocytheretta* sp. A, *Pterygocythereis alophia* Hazel and species of *Cytheropteron* are also important in this assemblage. Near the mouths of major rivers and lagoons, allochthonous specimens of estuarine and lagoonal species are found, such as *Paracytheroma texana* Garbett & Maddocks, *Proteoconcha gigantica* (Edwards), *Limnocythere friabilis* Benson & MacDonald, *Loxoconcha matagordensis* Swain, *Megacythere repexa* Garbett & Maddocks and species of *Cyprideis*. As a result, this thanatocoenosis has the highest species diversity of the three assemblages.

Assemblage II is dominated by *Echinocythereis margaritifera* (Brady) (10 to 50%) and species of *Cytheropteron* (4 to 13%), including *C. morgani* Kontrovitz, *C. yorktownensis* Kontrovitz, and *C.* sp. aff. *C. hamatum* Sars. Also important are *Munseyella louisianensis* Kontrovitz, *Buntonia* n. sp. and species of *Xestoleberis* and *Malzella*. This assemblage is found between the 60 and 110m isobaths on muddy sediments. The total number of species is lower than in Assemblage I, but nearly all are autochthonous.

Assemblage III is found in depths below 100m on fine muds and has the lowest species diversity. It is dominated by species of *Krithe* and *Parakrithe* (totalling 11 to 61%), *Argilloecia* (10 to 36%) and *Cytherella* sp. A (2 to 29%). Abundance of *Krithe*

and *C.* sp. A increases dramatically with depth; below 130m they have relative abundances above 30 and 20%, respectively. Other important companion species in this thanatocoenosis are *Henryhowella* ex. gr. *asperrima* (Reuss), *Pseudopsammocythere* ex. gr. *vicksburgensis* (van den Bold), *Cytherella* sp. B and *Ambocythere* sp.

Within each assemblage, significant changes take place from west to east, perhaps related to the increasing carbonate content and higher salinity (>37°/oo) to the east near Terminos Lagoon. Notably, *Loxoconcha moralesi*, *Cytherella vermillionensis*, *Cytheropteron* sp. aff. *C. hamatum*, *Neomonoceratina mediterranea*, *Paracytheroma texana*, *Protocytheretta* sp. A and *Paradoxostoma ensiforme* Brady become increasingly abundant to the east, whereas species of Bairdiidae (derived from the Veracruz reefs), *Aurila*, *Kangarina*, *Pontocythere* and *Paracytheridea* are more common to the west and nearly absent to the east. Many other species, including *Cytheromorpha paracastanea* (Swain), *Pellucistoma magniventra* Edwards, and species of *Loxoconcha* and *Cytherura*, reach their maximum abundances close to the mouths of the Coatzacoalcos, Grijalva and San Pedro and San Pablo rivers.

Seasonally high evaporation on the Yucatan shelf produces a tongue of very saline water (>37°/oo) that may be traced across much of the eastern part of the Bay of Campeche (Fig. 2A). Where this saline tongue sinks to the bottom near the Terminos Lagoon and the San Pedro and San Pablo rivers, *Cytherella vermillionensis* reaches its highest abundance (20%) and seems to be a reliable indicator for this saline water mass.

These baythymetric and regional trends are superimposed on a somewhat variable local pattern of changing faunal composition and diversity. Nearshore, these local anomalies are thought to be due to the discharge of major rivers and lagoons, as evidenced by the occurrence of brackish-water species. These specimens may signify occasional periods of lowered salinity nearshore during rainy seasons, or they may have been transported seawards as sedimentary particles by river discharge. For one or two very rare species, it is suspected that the specimens have been reworked

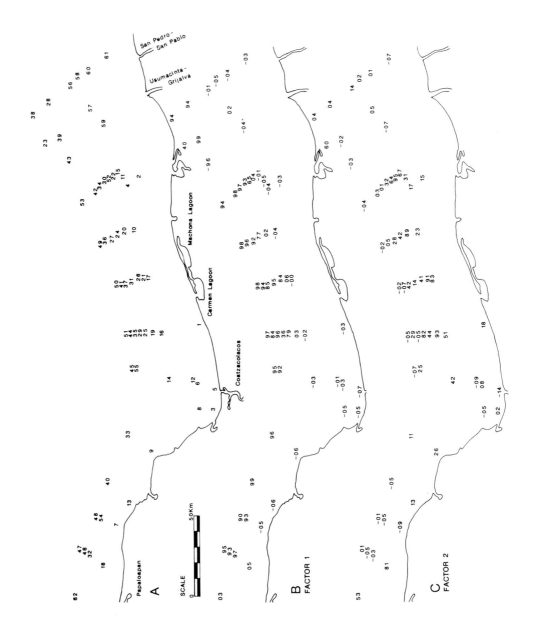

A

62
18
47
46
32
48
54
7
13
40
9
33
8
3
5
14
12
6
45
55
51
44
35
25
19
16
50
41
37
31
26
21
17
49
36
27
24
20
10
53
42
30
39
52
25
11
4
2
43
23
39
38
28
56 58 60
57
59
61
1
40
99
94
94
94
−96
−01 −05 −04 −03
Papaloapan
Coatzacoalcos
Carmen Lagoon
Machona Lagoon
Usumacinta-Grijalva
San Pedro-San Pablo
SCALE
50 Km
0

B   FACTOR 1

53
81
−01
−05
−09
13
99
−06
−06
96
−03
−05
−05
−01
−03
−07
95
92
97
84
96
36
79
−02
98
94
85
98
96
92 7
95
84
06
−00
02
−04
98 97
93
65
04
−01
−04 −05 −03
94
60
−96
02
−04
94
94
99
40
1
94
−07
−06
−05
03
95
93
97
90
93
05

C   FACTOR 2

53
81
−01
−05
−09
13
26
11
42
−05
02
−14
−09
08
18
51
93
44
82
−05
−29
−05
−07
25
−02
−07
−42
14
41
91
83
−02
05
28
42
89
23
01 2
34
95
67
17
31
15
−04
60
−02
−03
94
04
04
05
−07
14
02
01
−07

FACTOR 1
FACTOR 2

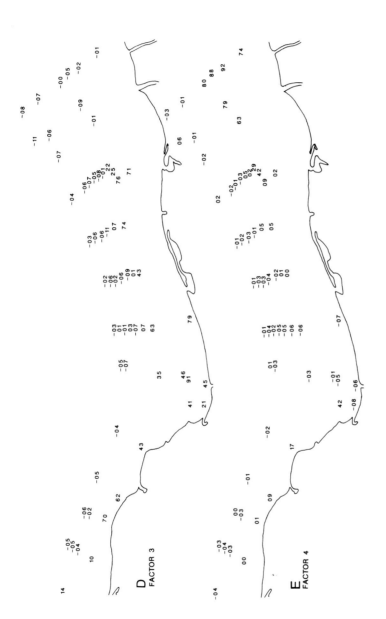

Fig. 3. (A) Sampling locations in the study area (see also Table 1). (B-E) Varimax-rotated factor loadings for Factors 1 (B), 2 (C), 3 (D) and 4 (E) on the original variables (stations). The factor loadings are rounded to two digits; thus the loading of factor 1 on station 1 = −0.03356 is given on the map as −03, and the loading of factor 3 on station 1 = 0.78703 is given as 79, etc.

Table 1. Number, station code, location and depth in metres of the 62 samples.

| No. | Stn. | Latitude | Longitude | depth | No. | Stn. | Latitude | Longitude | depth |
|-----|------|----------|-----------|-------|-----|------|----------|-----------|-------|
| 1 | 35 | 18°18.1' | 94°02.9' | 23.0 | 32 | J5 | 18°54.6' | 95°23.3' | 129.0 |
| 2 | R10 | 18°39.0' | 93°10.7' | 26.8 | 33 | 31 | 18°42.2' | 94°39.5' | 142.0 |
| 3 | A28 | 18°13.9' | 94°32.0' | 23.6 | 34 | R3 | 18°53.2' | 93°15.2' | 150.0 |
| 4 | R9 | 18°43.7' | 93°14.5' | 36.4 | 35 | O7 | 18°40.5' | 94°05.1' | 153.0 |
| 5 | A35 | 18°12.5' | 94°24.4' | 29.1 | 36 | Q7 | 18°50.0' | 93°33.4' | 156.0 |
| 6 | 33 | 18°19.0' | 94°23.0' | 37.6 | 37 | P3 | 18°44.4' | 93°48.4' | 157.0 |
| 7 | K6 | 18°44.0' | 95°12.0' | 41.0 | 38 | T11 | 19°14.7' | 92°50.0' | 177.0 |
| 8 | A29 | 18°19.8' | 94°31.8' | 34.9 | 39 | 84-3 | 19°06.0' | 92°58.4' | 178.0 |
| 9 | 30 | 18°33.5' | 94°47.0' | 45.0 | 40 | 29 | 18°48.5' | 94°57.4' | 182.0 |
| 10 | Q2 | 18°40.7' | 93°29.1' | 45.0 | 41 | P2 | 18°45.1' | 93°48.6' | 182.0 |
| 11 | R8 | 18°45.8' | 93°12.1' | 46.0 | 42 | R2 | 18°54.3' | 93°16.5' | 183.0 |
| 12 | A26 | 18°20.5' | 94°22.1' | 43.0 | 43 | 46 | 19°02.5' | 93°07.1' | 180.0 |
| 13 | 28 | 18°40.9' | 95°04.6' | 50.0 | 44 | O8 | 18°41.7' | 94°05.1' | 190.0 |
| 14 | A25 | 18°27.9' | 94°22.4' | 57.0 | 45 | 32 | 18°41.0' | 94°18.3' | 190.0 |
| 15 | R7 | 18°45.8' | 93°09.2' | 60.7 | 46 | J6 | 18°55.6' | 95°22.5' | 172.0 |
| 16 | O3 | 18°31.1' | 94°05.3' | 61.0 | 47 | 25 | 18°56.6' | 95°21.7' | 201.0 |
| 17 | P7 | 18°35.1' | 93°45.7' | 65.0 | 48 | 26 | 18°52.8' | 95°09.4' | 201.0 |
| 18 | J3 | 18°49.8' | 95°27.1' | 65.4 | 49 | Q9 | 18°52.3' | 93°34.2' | 209.0 |
| 19 | O4 | 18°34.5' | 94°05.3' | 80.0 | 50 | P1 | 18°45.7' | 93°48.8' | 204.0 |
| 20 | Q4 | 18°44.3' | 93°30.7' | 86.0 | 51 | O9 | 18°42.5' | 94°07.2' | 370.0 |
| 21 | P6 | 18°37.4' | 93°46.2' | 82.0 | 52 | R5 | 18°50.1' | 93°12.5' | 100.0 |
| 22 | R6 | 18°48.8' | 93°11.3' | 82.0 | 53 | 41 | 18°58.0' | 92°20.3' | 250.0 |
| 23 | 83-2 | 19°09.9' | 92°59.6' | 109.0 | 54 | K2 | 18°51.2' | 95°09.9' | 163.0 |
| 24 | Q5 | 18°46.2' | 93°31.5' | 100.0 | 55 | 1118 | 18°37.3' | 94°18.0' | 115.0 |
| 25 | O5 | 18°37.5' | 90°05.3' | 103.0 | 56 | T5 | 19°02.0' | 92°40.3' | 58.6 |
| 26 | P5 | 18°40.2' | 93°46.9' | 106.0 | 57 | 82-3 | 18°55.2' | 92°48.3' | 54.0 |
| 27 | Q6 | 18°47.9' | 93°32.2' | 118.0 | 58 | T3 | 18°59.2' | 92°37.9' | 39.4 |
| 28 | T9 | 19°08.7' | 92°46.6' | 120.0 | 59 | 44 | 18°51.3' | 92°53.1' | 36.0 |
| 29 | O6 | 18°39.5' | 94°05.2' | 124.0 | 60 | T2 | 18°56.5' | 92°35.7' | 28.5 |
| 30 | R4 | 18°51.4' | 93°13.7' | 121.0 | 61 | 48 | 18°49.9' | 92°29.0' | 17.2 |
| 31 | P4 | 18°41.7' | 93°47.5' | 123.0 | 62 | 22 | 18°58.1' | 95°36.9' | 62.0 |

fom Pliocene outcrops on the nearby shore. Seasonal changes in current patterns may also play a part in these local differences. Some stations off Carmen and Machona lagoons (mostly not included in this analysis) show high rates of carapace dissolution and sparse, pyritized assemblages. Perhaps local hydrographical conditions there favour seasonal low oxygen levels, as has been shown for parts of the southwest Louisiana shelf (Locklin & Maddocks, 1982, and references cited therein).

## BIOGEOGRAPHICAL AND BATHYMETRICAL RELATIONSHIPS

Table 2 gives the tentative geographical ranges of 61 species. These distributions are based solely on direct comparison with material from the Caribbean, Florida, Texas and Louisiana shelves, including types for Howe & van den Bold (1975), Kontrovitz (1976) and unpublished material. Our findings show strong similarity between the continental-shelf faunas of the southern and northern Gulf. A total of 90 of these species occur on the terrigenous and carbonate shelves of the

Table 2. List of the 99 ostracod species used in the factor analysis with known geographical distribution (1 = Caribbean, Florida, Louisiana and Texas; 2 = Florida, Louisiana and Texas; 3 = Lousiana and Texas) and bathymetric ranges given in metres.)

| | |
|---|---|
| 2*Actinocythereis* sp. A. (50–150) | *Malzella* spp. (20–150) |
| *Actinocythereis* sp. B. (20–65) | 3*Megacythere repexa* Garbett & Maddocks, 1979. (25–50) |
| *Ambocythere* sp. cf. *A.* sp. A of Cronin, 1983. (100–370) | *Microcytherura choctawhatcheensis* (Puri, 1954) (35–80) |
| *Argilloecia* spp. (50–370) | |
| 2*Aurila amygdala* (Stephenson, 1944)(20–60) | 3*Munseyella louisianensis*, Kontrovitz, 1976. (65–370) |
| Bairdiidae spp. (20-60) | |
| 2*Basslerites minutus* van den Bold, 1958. (20–120) | *Neocytherideis* spp. (20–50) |
| 2*Buntonia* n. sp. (80–190) | 1*Neomoceratina mediterranea* (Ruggieri, 1953) (20–50) |
| *Buntonia* spp. | |
| 2*Bythocythere* sp. A. | *Orionina* sp. A. (20–50) |
| 2*Cativella* sp. A. | 3*Paracypris* sp. A of Kontrovitz, 1976. (45–200) |
| 2*Bythoceratina* sp. A | 3*Paracypris?* sp. B. (20–80) |
| *Caudites* spp. (20–60) | 2*Paracytheridea altila* Edwards, 1944. (20–80) |
| 3*Cushmanidea* sp. A. (20–60) | 2*Paracytheridea rugosa* Edwards, 1944. (20–100) |
| *Cyprideis* spp. | 2*Paracytheroma texana*, Garbett & Maddocks, 1979. (20–60) |
| *Cythere?* sp. A. (80–150) | |
| 2*Cytherella vermilionensis* Kontrovitz, 1976. (20–65) | 1*Paracytheroma stephensoni* (Puri, 1954)(20–40) |
| 3*Cytherella* sp. A. (80–370) | 3*Paradoxostoma ensiforme* Brady, 1868. (20–90) |
| 3*Cytherella* sp. B. (60–200) | *Paradoxostoma* sp. A. (20–100) |
| *Cytherelloidea* spp. (20–100) | *Parakrithe* spp. (100–370) |
| 2*Cytheromorpha paracastanea* (Swain, 1955)(20–50) | 2*Pellucistoma magniventra* Edwards, 1944. (20–50) |
| 3*Cytheropteron* sp. aff. *C. hamatum* Sars, 1866. (20–250) | *Peratocytheridea* sp. A. (30–50) |
| | 2*Perissocytheridea* spp. (20–60) |
| 2*Cytheropteron morgani* Kontrovitz, 1976. (20–190) | 2*Pontocypris* sp. A. (20–120) |
| 1*Cytheropteron yorktownensis* (Malkin, 1953)(30–65) | 2*Pontocythere* sp. A. (20–80) |
| 3*Cytheropteron* sp. A. (100–200) | 2*Pontocythere* sp. B. (20–60) |
| 3*Cytheropteron* sp. B. (100–250) | *Pontocythere* sp. C. (20–50) |
| 2*Cytherura* sp. aff. *C. maya*, Teeter, 1975. (20–65) | *Pontocythere* spp. (20–100) |
| 1*Cytherura sablensis* (Benson & Coleman, 1963) (20–65) | 3*Propontocypris?* sp. A. |
| | *Proteoconcha* spp. (20–50) |
| 2*Cytherura sandbergi* Morales, 1966. | 1*Protocytheretta pumicosa* (Brady, 1869)(20–80) |
| 2*Cytherura* sp. aff.*C.sandbergi* Morales, 1966. (20–50) | 2*Protocytheretta* sp. A. (20–100) |
| 2*Cytherura* sp. A. (20–120) | 2*Pseudocythere* sp. A. (30–190) |
| 3*Cytherura* sp. B of Kontrovitz, 1978. (20–100) | 1*Pseudopsammocythere* ex. gr. *vicksburgensis* van den Bold, 1988. (20–180) |
| 2*Cytherura* sp. C. (20–65) | |
| *Cytherura* sp. D. (20–65) | 2*Pterygocythereis alophia* Hazel, 1983. (20–170) |
| *Cytherura* spp. (20–200) | 2*Pterygocythereis inexpectata* (Blake, 1933)(40-170) |
| 2*Echinocythereis margaritifera* (Brady, 1870)(30–200) | 2*Pterygocythereis* sp. A. (50–200) |
| 3*Echinocythereis spinireticulata* Kontrovitz, 1971. (45–370) | 3*Pumilocytheridea ayalai* Morales, 1966. (20–70) |
| | 2*Puriana convoluta* Teeter, 1975. (20–80) |
| *Eucytherura* spp. | 2*Puriana krutaki* Kontrovitz, 1976. (20–65) |
| 2*Henryhowella* ex. gr. *asperrima* (Reuss, 1850) (40–370) | 2*Puriana* sp. A. (20–80) |
| | Trachyleberidinae spp. |
| 1*Jugosocythereis pannosa* (Brady, 1869)(20–50) | *Triangulocypris* sp. A. (20–60) |
| 1*Kangarina* sp. cf. *K. ancycla* van den Bold, 1963. (30–190) | *Tuberculocythere?* sp. aff. *T.?* sp. A of Cronin, 1983. (60–370) |
| *Krithe* spp. (100–370) | *Tuberculocythere?* sp. C. |
| *Limnocythere* spp | *Xestoleberis* sp. A. (20–170) |
| 2*Loxoconcha moralesi* Kontrovitz, 1976. (20–50) | *Xestoleberis* sp. B. (20–170) |
| 2*Loxoconcha* sp. A. (20–100) | Ostracoda sp. A. (40–100) |
| 3*Loxoconcha* sp. B. (30–140) | Ostracoda sp. C. |
| 2*Loxoconcha* sp. C. (30–115) | Ostracoda sp. E. (80–200) |
| *Loxoconcha* sp. F. (20–100) | Ostracoda sp. F. (30–80) |
| *Loxoconcha* spp. (20–140) | Ostracoda sp. K. |
| 3*Macrocyprina skinneri* Kontrovitz, 1976. (20–80) | Ostracoda sp. L. |

northern Gulf of Mexico, and about 20 occur in carbonate environments of the Caribbean. Total species diversity on the Campeche shelf appears to be a little higher than in the northern Gulf of Mexico, although exact estimates are hampered by the fact that offshore faunas of the northern Gulf are almost as poorly known. Likely explanations for this higher diversity include the close proximity of reef faunas on the Yucatan platform and off Veracruz, higher nearshore salinities and higher temperatures.

The assemblages established by Kontrovitz (1976) on the Louisiana continental shelf were not defined in such a way as to be recognizable on the Campeche shelf, although many of the same species occur in Assemblages I and II of this study. An outer-shelf fauna (deeper than 100m) was not recognized by Kontrovitz, although he stated that species of *Krithe* and *Argilloecia* increase rapidly in abundance with depth. Species of those genera characterize our Assemblage III, but their ranges appear to begin at a greater (10 to 30m) depth. On the Texas continental shelf (Maddocks, unpublished data), Assemblages I and II are recognizable and characteristic of approximately the same depths.

The ostracod barren zone (10 to 40m) recognized on the continental shelf of the northern Gulf of Mexico (Benson & Coleman, 1963; van den Bold, 1977) is not an exception. In the study area, numerous samples collected between 15 and 50m are barren or have few ostracods; the coarser sediments, higher sedimentation rates, lower salinities and higher energy in the coastal zone may be responsible for these poor assemblages.

The bathymetrical ranges of species with well-defined trends are given in Table 2. Many ostracod species have approximately the same depth ranges on the Louisiana, Texas and Campeche shelves, for example: *Cytherella vermillionensis*, *Cytheromorpha paracastanea*, *Echinocythereis spinireticulata* Kontrovitz, *Loxoconcha moralesi*, *Loxoconcha* sp. A, *Paracytheridea altila* Edwards, *P. rugosa* Edwards, *Paradoxostoma* sp. A, *Pellucistoma magniventra*, *Protocytheretta pumicosa* (Brady), *Puriana convoluta*

Teeter, *P.* sp. A and species of *Pontocythere*. A few species, such as *Munseyella louisianensis*, *Pterygocythereis alophia*, *P. inexpectata* (Blake) and *Paracypris* sp. A, appear to range deeper in the southern than in the northern Gulf. *Henryhowella* ex. gr. *asperrima* and species of *Parakrithe* seem to first appear in deeper waters in the southern than in the northern Gulf.

## FACTOR ANALYSIS

A Q-mode factor analysis was applied to the relative-abundance data, using the subprogram 'Factor' of the Statistical Package for the Social Sciences (SPSS) (Kim, 1977) on the Burroughs 7800 computer of the UNAM. The data matrix consisted of 62 sampling stations (those yielding more than 100 specimens) and 99 species (those species that occurred in three or more samples and had at least one percent relative abundance in one of the samples). Initial factors were calculated by principal factoring with iteration (method PA2 of Kim, 1977). This method uses inferred factors and replaces the main vectors of the correlation matrix with communality estimates (the squared multiple correlations between each variable and the rest of the variables in the matrix). PA2 further performs an iteration procedure to improve the communality estimates. This procedure consists of the extraction of factors from the original correlation matrix, their replacement with the communality estimates, and the extraction of new factors from this reduced matrix. The new communality estimates are obtained from the factor variances. Six iterations were required for this analysis. Inspection of the eigenvalues and their cumulative percentages (the amount of information explained by the factors) and consideration of their meaning led to the decision to use four factors. The four factors explained 75.7% of the information in the data matrix; subsequent factors contributed little additional insight. To simplify and emphasize the factor structure, the factor matrix was rotated using the varimax (orthogonal) method, which simplifies the columns of the matrix and maximizes the variance of the squared loadings in each column. After rotation of the matrix, factor 1 explained 60.2% of the information in the data matrix, factor 2, 20.5%; factor 3, 9.9% and factor 4, 9.4%. The rotated factor loadings are mapped for each station in Figs 3B-E, and the transformation matrix used to recalculate these final factor loadings is given in Table 3.

The results of this factor analysis suggest that four factors are necessaary to explain the distribution of ostracods on the Campeche shelf. Three of the factors are associated with the bathymetry of the area, and a fourth factor reflects carbonate content of sediments and salinity. Factor 1 clearly shows the influence of bathymetry on faunal composition, and station loadings for this factor are conspicuously parallel to the bathymetrical gradient, with the highest loadings at the deepest stations. Factor 2 represents mid-shelf conditions with little terrigenous influence but rather sandy sediments. Factor 3 has high loadings for stations located near the mouths of rivers and lagoons, where influx of brackish-water species occurs. Factor 4 corresponds to the increased carbonate content of sediments on the eastern side of the study area and probably also reflects the high-salinity water mass reported for this area.

## CONCLUSIONS

Three bathymetrical assemblages exist on the Campeche shelf. Assemblage I, shallower than 60m, is represented by species of *Pontocythere*, *Cytherura* and *Loxoconcha*. Assemblage II, between 60 and 100m, is dominated by *Echinocythereis margaritifera* and species of *Cytheropteron*. Assemblage III, below 100m, is characterized by *Cytherella* sp. A and by species of *Krithe*, *Parakrithe* and *Argilloecia*.

Within each assemblage, systematic east-west changes in faunal composition and species diversity reflect the increased salinity and carbonate content of sediments eastward, *Loxoconcha moralesi*, *Cytheropteron* sp. aff. *C. hamatum*, *Paracytheroma texana*, *Neomonoceratina mediterranea* and *Paradoxostoma ensiforme*, for example, reach their highest abundance to the east, close to the Yucatan carbonate shelf. *Cytherella vermillionensis* is also common in the east, but its

Table 3. Transformation matrix showing correlations among the rotated factors and recalculated communalities.

|  | Factor 1 | Factor 2 | Factor 3 | Factor 4 |
|---|---|---|---|---|
| Factor 1 | 0.98 | 0.17 | −0.07 | −0.03 |
| Factor 2 | −0.11 | 0.87 | 0.47 | 0.10 |
| Factor 3 | 0.14 | −0.45 | 0.88 | −0.10 |
| Factor 4 | 0.05 | −0.13 | 0.04 | 0.99 |

highest abundance coincides with a high-salinity water tongue located between Terminos Lagoon and the San Pedro-San Pablo Rivers. Species of Bairdiidae, *Aurila*, *Kangarina*, *Pontocythere* and *Paracytheridea* occur more commonly in the west than in the east.

Proximity to rivers and lagoons causes local perturbation of these trends, with enhanced species diversity caused by influx of lagoonal species. *Paracytheroma paracastanea*, *Pellucistoma magniventra* and species of *Loxoconcha* and *Cytherura* occur more abundantly near river mouths.

Factor analysis suggests that four factors are necessary to explain the distribution of ostracods on the Campeche shelf. Factors 1, 2 and 3 are strongly associated with bathymetry and sediment composition, from terrrigenous sand for Factor 1 to soft muds for Factor 3. Factor 4 represents the influence of carbonate sediments and perhaps of a local high-salinity water mass.

Although a large number (90+) of species recognized in this study also occur in the northern Gulf of Mexico in similar bathymetrical zones, the Campeche shelf fauna appears to be more diverse than that of the northern Gulf. This may reflect proximity to the Yucatan carbonate platform with its coral reefs, as well as the higher salinities and temperatures. A few species, such as *Munseyella louisianensis*, *Pterygocythereis alophia*, *P. inexpectata* and *Paracypris* sp. A, appear to range deeper in the southern than in the northern Gulf, and *Henryhowella* ex. gr. *asperrima* and species of *Parakrithe* seem to first appear in deeper waters in the southern than in the northern Gulf. However, such comparisons are hindered at present by taxonomic questions and the relatively poor knowledge of bathymetrical distribution in the northern Gulf.

Morphological adaptations and other biological aspects of the relationships of ostracods to these nearshore environments remain poorly understood. Seasonal and geographical variation, ontogenetic differentiation, polymorphism, sexual dimorphism and other aspects of intraspecific population structure require more attention. The Campeche shelf, as an important microfaunistic corridor connecting the Yucatan carbonate province with the clastic province of the western Gulf of Mexico, will provide optimal material for these and other objectives of this long-term project.

## ACKNOWLEDGEMENTS

This study is part of an ongoing programme supported by the University of Mexico and CONACYT (PCCNCNA 031676). Additional laboratory facilities and support were provided by the University of Houston Department of Geosciences. Special gratitude is expressed to Dr Jorge Carranza-Fraeser, Chairman, and Dr Agustin Ayala-Castañares of the Institute of Marine Sciences and Limnology of the University of Mexico for their encouragement and support of this project. We thank M. Alatorre, S. P. Czitron, F. Ruiz and A. R. Padilla for access to their oceanographical data, Dr Lizarraga-Partida and E. Sainz for the oceanographical maps, R. Lecuanda and F. Ramos for the sedimentological map, A. Molina-Cruz and E. Sainz for their help with the factor analysis progam and W. A. van den Bold for the opportunity to examine his collections and for his constant willingness to discuss and advise. Special thanks are due to students of the Micropalaeontology Laboratory at ICMyL and the crew of the *Justo Sierra* for their aid in collecting the samples and shipboard preparation.

## REFERENCES

Benson, R. H. & Coleman, G. L. 1963. Recent marine ostracodes from the eastern Gulf of Mexico. *Paleont. Contr. Univ. Kans.*, Lawrence, Article 2, 1-52.

Bold, W. A. van den. 1977. Distribution of marine podocopid Ostracoda in the Gulf of Mexico and Caribbean, *In* Löffler, H. & Danielopol, D. *Aspects of Ecology and Zoogeography of Recent and Fossil Ostracoda*, 175-186. Dr W. B. Junk, The Hague.

Creager, J. S. 1958. *Bathymetry and sediments of the Bay of Campeche. Report of the Department of Oceanography and Meteorology*, Texas A&M University, 188 pp.

Cronin, T. M. 1979. Late Pleistocene marginal marine ostracodes from the southeastern Atlantic Coastal Plain and their paleoenvironmental implications. *Géogr. phys. quaternaire*, Montreal, 33(2), 121-173.

Garbett, E. C. & Maddocks, R. F. 1979. Zoogeography of Holocene cytheracean ostracodes in the bays of Texas. *J. Paleont.*, Tulsa, Ok., 53, 841-919.

Howe, H. V. & Bold, W. A. van den. 1975. Mudlump Ostracoda. *In* Swain, F. M. (Ed.), *Biology and Paleobiology of Ostracoda, Bull. Am. Paleont.*, Ithaca, 65(282), 303-315.

Hulings, N. C. 1967. Marine Ostracoda from the western North Atlantic between Cape Hatteras, North Carolina, and Jupiter Inlet, Florida. *Bull. mar. Sci.*, Coral Gables, 17, 629-659.

Kim, Jae-On. 1977. Factor analysis, 468-514. *In* Nie, N. H., Hull, C. H., Jenkins, J. G., Steinbenner, K. & Bent, D. H. (Eds), *SPSS, Statistical Package for the Social Sciences*, Mcgraw Hill, New York.

Kontrovitz, M. 1976. Ostracoda from the Louisiana continental shelf. *Tulane Stud. Geol. Paleont.*, New Orleans, 12, 49-100.

Krutak, P. R. 1971. The Recent Ostracoda of Laguna Mandinga, Veracruz, Mexico. *Micropaleontology*, New York, 17, 1-30.

Krutak, P. R. 1974. Standing crops of modern ostracods in lagoonal and reefal environments, Veracruz, Mexico. *West Indies Laboratory Special Publication* No. 6, 11-14, Faileigh Dickinson University, St. Croix, U.S. Virgin Islands.

Krutak, P. R. 1977. Change in ostracod standing crop, Laguna Mandinga, Veracruz, Mexico. *In* Frost, S. H., Weiss, M. P. & Saunders, J. B. (Eds), Reefs and Related cabonates, Ecology and Sedimentology. *Stud. Geol. am. Ass. Petrol. Geologists.*, Tulsa, 4, 209-218.

Krutak, P. R. 1982. Modern ostracodes of the Veracruz-Anton Lizado reefs, Mexico. *Micropaleontology*, New York, 28, 258-288.

Krutak, P. R. & Rickles, S. E. 1979. Equilibrium in modern coral reefs, western Gulf of Mexico-Role of ecology and ostracod microfauna. *Trans. Gulf Cst Ass. geol. Socs*, San Antonio, Tx., 29, 263-274.

Krutak, P. R., Rickles, S. E. & Gio-Argaez, R. 1980 Modern ostracod species diversity, dominance and biofacies patterns, Veracruz-Anton Lizardo Reefs, Mexico. *An. Centro Cienc. Mar Limnol.*, Mexico, 7, 181-198.

Lecuanda, R. S. & Ramos, F. 1985. Distribución de sedimentos en el sur del Golfo de México. *In* Licea, S., *et al.* (Eds), *Estudio Multidisciplinario en la Zona Económica Exclusiva del Golfo de México: Características Geológicas, Físicas, Químicas y Biológicas, Informe Técnico* No. 2, Instituto de Ciencias del mar y Limnología, Universidad Nacional Autónoma de México.

Lizarraga-Partida, M. L., Botello, A. V., Licea-Duran, S., Flores-Coto, C., Soto, L. A., & Yañez-Arancibia, A. 1984. *Informe de los resultados de la campaña oceanográfica "PROGMEX"*, 148 pp. Instituto de Ciencias del Mar y Limnología, México, Universidad Nacional Autónoma de México.

Locklin, J. A. & Maddocks, R. F. 1982. Recent foraminifera around petroleum production platforms on the southwest Louisiana shelf. *Trans. Gulf Cst Ass . geol. Socs*, Houston, Tx., 32, 377-397.

Machain-Castillo, M. L. 1987. Ostracode biofacies in the southern Gulf of Mexico. *Geological Society of America 22nd Annual Meeting, Abstracts with Programs*, 19, 753.

Maddocks, R. F. 1974. Ostracodes. *In* Bright, T. S. & Pequegnat, L. N. (Eds), *Biota of the West Flower Garden Bank*, 199-229. Flower Garden Ocean Research Center, Gulf Publishing Company, Houston, Texas.

Morales, G. A. 1966. Ecology, distribution and taxonomy of Recent Ostracoda of the Laguna de Terminos, Campeche, Mexico. *Boln Inst. Geol. Univ. nac., Méx.*, Mexico, 81, 1-103.

Nowlin, W. D., Jr. 1972. Winter circulation patterns and property distributions. *In* Capurro, L. R. & Reid, J. L. (Eds), *Contributions on the Physical Oceanography of the Gulf of Mexico. Texas A&M University , Oceanography Studies* 2, Gulf Publishing Company, Houston, Texas.

Padilla, A. R., Alatorre, M. A., Ruíz, F. & Czitron, S. P. R. 1986. Observaciones recientes de la estructura termohalina en el sur del Golfo de México. *Union Geofísica Mexicana, A. C. Reunión Anual, Morelia, Michoacán, México*, 434-440.

Puri, H. S. 1960. Recent Ostracoda from the west coast of Florida. *Trans. Gulf Cst Ass . geol. Socs*, Jackson, Mi., 10, 107-149.

Sandberg, P. A. 1964. Notes on some Tertiary and Recent brackish-water Ostracoda. *In* Puri, H. S. (Ed.), *Ostracodes as Ecological and Paleoecological Indicators* 496-514. *Pubbl. Staz. zool. Napoli*, Milano & Napoli, 33-Suppl.

DISCUSSION

Franciszek Adamczak: What kind of sedimentary structures did you find (if any) in the area investigated?

Ana María Pérez-Guzmán: This study is based on grab samples, and sedimentary structures were not recognized. Future studies in this project might have some information regarding this aspect.

# Seasonal distribution of Ostracoda on two species of marine plants and two holothurians in Okinawa, Japan

**Tomohide Nohara & Ryoichi Tabuki**

Department of Earth Sciences, University of the
Ryukyus, Okinawa, Japan.

## ABSTRACT

Twenty-nine species of live ostracods found on the
red alga *Jania adhaerens* Lamaroux were identified as belonging to 11 genera. Monthly individual abundance data show two prominent peaks in
February and September. Twenty-one species of
live ostracods belonging to 10 genera were found
on the angiosperm *Thalassia hemprichii* (Ehren-
berg) Ascherson. The monthly data show a single
prominent peak; seasonal change is characterized
by a maximum population density in February
and March and a low density in summer. On both
the red alga and the angiosperm, the dominant ostracod genera were *Xestoleberis* and *Paradoxostoma*. Forty-five species of live ostracods belonging to 20 genera were identified from the outer
surface of two species of holothurians, *Actinopyga*
sp. and *Holothuria* sp.; on the latter the population
density was lowest in summer.

## INTRODUCTION

The ecology of littoral ostracods has been studied
mainly in temperate areas (Theisen, 1966; Whatley
and Wall, 1969; Whatley, 1976.). Since May,

1984, the authors have been investigating the ecology of littoral ostracods living in the moat in the
inner area of a coral reef off Sesoko-jima, Okinawa. Earlier (Tabuki & Nohara, 1988), we
presented preliminary results of our studies on the
ostracods from this area. Here we present some
results of our studies on the biotic relationships of
ostracod species and their seasonal distribution on
the surface of a red alga and a sea grass. To provide
a comparison with these sessile plants, the epizoic
ostracods from two species of sea cucumbers were
also studied.

## SAMPLING LOCALITIES AND METHODS

Samples were collected for ostracods from the
moat off northwestern Sesoko-jima, Okinawa (Lat.
26°39'N, Long. 127°51'43"W). A general summary
of the climatic, physiographical, and oceanographical conditions of the sampling localities was
given in Tabuki & Nohara (1988).

Studies of the seasonal and areal variation of
ostracods were made in the littoral and sublittoral
zones during spring tides each month between

Table 1. List of living ostracod species from *Jania adhaerens* and number of specimens of each species collected at monthly intervals in 1985.

| species — date of sample '85 | 1 | 2 | 3 | 4 | 5 | 6 | 7 | 8 | 9 | 11 | 11 | 12 |
|---|---|---|---|---|---|---|---|---|---|---|---|---|
| | 25 | 21 | 28 | 25 | 21 | 21 | 19 | 31 | 30 | 1 | 26 | 26 |
| *Neonesidea* sp. | | | 1 | 2 | 4 | 4 | | | | 1 | 1 | 2 |
| *Propontocypris* sp. | | | | | | | 1 | | | | | |
| *Perissocytheridea inabai* | 1 | | | | | | | | | 1 | | |
| *Morkhovenia* sp. | | 1 | | | 2 | 2 | | | | 3 | | |
| *Callistocythere* sp. A | 1 | | 2 | 1 | 2 | | 4 | 2 | 7 | 11 | 2 | 6 |
| *Callistocythere* sp. B | | | | | 1 | | | | | | | |
| *Tenedocythere transoceanica* | | 1 | 1 | 1 | 2 | 4 | 32 | 8 | 12 | 9 | | 4 |
| *Tenedocythere?* sp. | | | | | | | | | | | | |
| *Semicytherura? miurensis* | 1 | | | | | | | | | | | |
| *Loxoconcha uranouchiensis* | | | | | 1 | 1 | | | | | 1 | |
| *Loxocorniculum* sp. A | 5 | | | | | 2 | 13 | 25 | 43 | 50 | 23 | 11 |
| *Loxocorniculum* sp. B | | | | | | | | 1 | | | | |
| *Loxocorniculum* sp. C | | | | | | | 4 | 3 | | 2 | 1 | |
| *Xestoleberis setouchiensis* | 2 | | | 12 | 2 | | 2 | 2 | | 1 | 3 | 1 |
| *Xestoleberis* cf. *X. hanaii* | 65 | 77 | 74 | 48 | 34 | 29 | 61 | 71 | 81 | 81 | 118 | 123 |
| *Xestoleberis* cf. *X. sagamiensis* | 11 | 9 | 14 | 5 | 3 | 5 | 10 | 4 | 2 | 13 | 21 | 4 |
| *Xestoleberis* sp. A | 125 | 124 | 134 | 79 | 158 | 83 | 76 | 63 | 89 | 52 | 38 | 58 |
| *Xestoleberis* sp. B | 5 | 34 | | | 3 | | | | | | | |
| *Xestoleberis* sp. C | | | | | 1 | | | | | | | |
| *Paradoxostoma affine* | 2 | | | 4 | 8 | 9 | 27 | 4 | | 12 | 5 | 11 |
| *Paradoxostoma lunatum* | 1 | | 5 | 2 | 1 | | | | | | 1 | |
| *Paradoxostoma* cf. *P. gibberum* | 2 | | 3 | | | 5 | 7 | 3 | 13 | 19 | | |
| *Paradoxostoma* sp. A | 1 | 5 | 7 | 9 | 4 | 10 | 17 | 2 | 2 | 3 | | 7 |
| *Paradoxostoma* sp. B | 40 | 15 | 87 | 150 | 45 | 161 | 109 | 48 | 54 | 47 | 20 | 35 |
| *Paradoxostoma* sp. C | | 2 | | | | | | | | 1 | | 1 |
| *Paradoxostoma* sp. D | 24 | 17 | | 7 | 3 | 2 | 8 | 44 | 35 | 48 | 70 | 41 |
| *Paradoxostoma* sp. E | 1 | 1 | | | | | 2 | | | 1 | | |
| *Paradoxostoma* sp. F | 4 | 9 | | | 12 | 5 | 1 | 18 | | | | |
| *Paradoxostoma* sp. G | 11 | | | | | 2 | | 1 | | | 1 | 1 |
| indet spp. | | | | | 1 | 1 | | | | | | 1 |
| total number of individuals | 309 | 264 | 327 | 309 | 284 | 325 | 380 | 339 | 339 | 355 | 304 | 309 |
| number of individuals per unit sample (50g) | 538 | 2809 | 1128 | 422 | 893 | 691 | 1377 | 1662 | 2493 | 1517 | 569 | 546 |

November 1984 to December 1985 and April 1987 to March 1988.

Two species each of marine plants and sea cucumbers were chosen for this study because they inhabit this locality all the year round. They are *Jania adhaerens* (red alga), *Thalassia hemprichii* (sea grass), *Holothuria* sp. (sea cucumber) and *Actinopyga* sp. (sea cucumber).

*Jania adhaerens* was collected monthly on a platform in the sublittoral zone from May 1984 to December 1985. This red alga appears hemi-spherical because of its complex intergrowth of filamentous thalli. Samples were collected together with the surrounding sea water in plastic bags and agitated in the bags with a 10% solution of formalin. Then the bags' contents were passed through a Taylor 425 mesh sieve. The residue in the sieve was dried and partitioned into samples which contained about 200 ostracods each. The ostracods were measured and their state of maturity ascertained.

*Thalassia hemprichii* is a green marine angiosperm. The fronds are about 5mm wide and from 5cm (in winter) to 10cm long (in summer). Although the fronds are short in winter because the edges wither, filamentous algae and calcareous particles are abundant on their surfaces, even in winter. This sea grass grows in shallower environments than *Jania* and is emergent during lowest tides. Samples were collected on a gravel substrate in the littoral zone from November 1984 to December 1985 and from April 1987 to March 1988. The sea grass was cut using scissors and put into a plastic bag in air. It is possible, therefore, that some ostracods escaped during sampling. After collection, these samples were treated in the same way as the *Jania* sample.

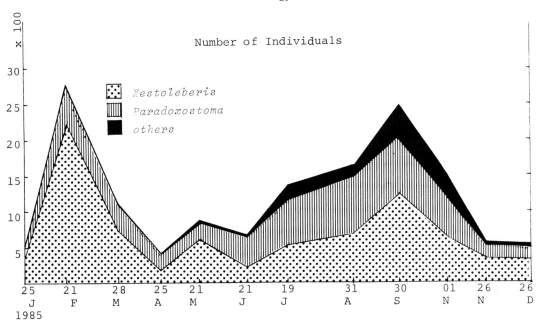

Fig. 1. Seasonal distribution of living Ostracoda collected from 50g (wet weight) of *Jania adhaerens*.

For comparison with these sessile plants, two mobile sea animals (*Holothuria* sp. and *Actinopyga* sp.) were studied with respect to their seasonal infestation by ostracods. Specimens of *Holothuria* sp. were collected on a sand or gravel substrate in the littoral zone from April 1987 to December 1987. They are black in colour and usually sand grains are attached to the exterior surface. Each month, 30 small individuals (19cm average length) were collected by hand and washed in a bucket. The washings were treated in the same way as the marine plants.

*Actinopyga* sp. was collected on a rocky substrate in the sublittoral zone from April 1987 to December 1987. They are brown in colour and smaller than *Holothuria* sp. Each month 20 individuals (17.5cm average length) were collected and processed the same way as the *Holothuria*.

## RESULTS

1) *Jania adhaerens*. Twenty-nine species of live ostracods belonging to 11 genera were identified in monthly samples of *Jania adhaerens* from January 1985 to December 1985 (Table 1). The genera *Xestoleberis* and *Paradoxostoma* were dominant

on the algal samples and comprised 80-90% of the ostracod population. The dominant species were *Xestoleberis* cf. *X. hanaii* Ishizaki, *Xestoleberis* sp. A and *Paradoxostoma* sp., and these 3 species comprised more than 60% of the population during any month.

The seasonal distribution of live ostracods per 50g (wet weight) of *Janaia adhaerens* shows two prominent peaks in February and September (Fig. 1). The dominant genera *Xestoleberis* and *Paradoxostoma* were most abundant then, and the 4 dominant species: *Xestoleberis* cf. *X. hanaii*, *Xestoleberis* sp. A, *Paradoxostoma* sp. B and *Paradoxostoma* sp. D also became abundant in the same months. However, such species as *Loxocorniculum* sp. A and *Tenedocythere* sp. increased in July, but were rarely found between February and May. In the case of these 2 species, adults made up 10% of the population. In *Paradoxostoma* sp. B and *Paradoxostoma* sp. D, adults comprised between 20 and 30% of the population. In *Xestoleberis* sp. A, the proportion of adults varied between 20 and 60%. However, in the dominant species *Xestoleberis* cf. *X. hanaii*, adults were found rarely, as reported earlier (Okubo, 1984). Some ostracods may live their entire lives on

Table 2.  List of living ostracod species from *Thalassia hemprichii* and number of specimens of each species collected at monthly intervals in 1984 and 1985.

| species \ date of sample | 11 09 | 12 21 | 1 25 | 2 22 | 3 28 | 4 25 | 5 21 | 6 21 | 7 19 | 8 31 | 9 25 | 11 01 | 11 26 | 12 25 |
|---|---|---|---|---|---|---|---|---|---|---|---|---|---|---|
| *Triebelina* sp. | | | | | | | 1 | | | | | | | |
| *Perissocytheridea inabai* | | | | | | 1 | | | | | | | | |
| *Morkhovenia* sp. | | | | | | | | 1 | | | | | | 1 |
| *Callistocythere* sp. A | | | | 2 | | 1 | | 1 | | | | 10 | 12 | |
| *Tenedocythere transoceanica* | | | | | | | | | | | | 3 | 2 | |
| *Radimella* sp. | | | | | | | | | | | | 2 | | 2 |
| *Loxoconcha uranouchiensis* | | | 2 | 1 | | | 2 | 7 | 3 | | | 4 | 2 | 2 |
| *Loxoconcha japonica* | 1 | | | | | | | | | | | | | |
| *Loxocorniculum* sp. A | 2 | 11 | 6 | 2 | | 3 | 9 | | 4 | 7 | 20 | 28 | 10 | 8 |
| *Xestoleberis setouchiensis* | | | | | | | 1 | | | | | | | 1 |
| *Xestoleberis* cf. *X. hanaii* | 46 | 82 | 101 | 127 | 123 | 109 | 114 | 6 | 12 | 29 | 57 | 85 | 62 | 75 |
| *Xestoleberis* cf. *X. sagamiensis* | | 1 | 4 | 7 | 1 | 5 | 1 | | | | | 1 | 3 | 2 |
| *Xestoleberis* sp. A | 5 | 4 | | | 3 | 1 | 2 | 1 | 1 | 1 | 2 | 44 | 15 | |
| *Xestoleberis* sp. B | | | 7 | 3 | | | 1 | | 1 | | | 4 | 2 | |
| *Paradoxostoma affine* | | | | 11 | | 3 | 1 | | | | 3 | 2 | 19 | 14 |
| *Paradoxostoma lunatum* | | | | | | | 1 | | | | | | | |
| *Paradoxostoma* sp. A | | | | | | | | | | | | | | 9 |
| *Paradoxostoma* sp. C | | 3 | 39 | 40 | 150 | 2 | 3 | | | | | | | 9 |
| *Paradoxostoma* sp. D | 67 | 35 | 92 | 18 | 30 | 46 | 23 | 7 | 6 | 13 | 26 | 67 | 57 | 77 |
| *Paradoxostoma* sp. F | 27 | 16 | 15 | | 1 | 35 | 21 | | 4 | 10 | 1 | | 54 | 17 |
| *Paradoxostoma* sp. G | | | 3 | | 1 | | | | | | | | | 1 |
| total number of individuals | 168 | 154 | 177 | 210 | 308 | 210 | 184 | 19 | 29 | 62 | 120 | 297 | 188 | 199 |
| number of individuals per unit sample (50g) | 92 | 237 | 128 | 538 | 466 | 47 | 90 | 6 | 9 | 41 | 62 | 330 | 121 | 429 |

*Jania adhaerens*, but the rarity of adults of *Xestoleberis* cf. *X. hanaii* implies a migration to other habitats, as pointed out by Whatley & Wall (1975).

*Xestoleberis* cf. *X. hanaii* showed seasonal changes in carapace length and height at each stage, as reported by Okubo (1984). These dimensions increased between January and May and decreased between July and September. Although it is known that population density in littoral Ostracoda increases in summer in the Japanese mainland (Okubo, 1984), in Okinawa it decreased from April to June.

b) *Thalassia hemprichii*. Twenty-one species of live ostracods belonging to 10 genera were found on this angiosperm sampled monthly from November 1984 to December 1985 (Table 2), and from April 1987 to March 1988. The genera *Xestoleberis* and *Paradoxostoma* comprised 12-59% of the population from November 1984 to December 1985.

The monthly record of individuals per 50g angiosperm shows only one prominent peak; sea-sonal change was characterized by maximum density in February and March and low density in summer (Figs 2 and 3).

The seasonal distribution of dominant species such as *Xestoleberis* cf. *X. hanaii* and *Paradoxostoma* sp. D is in accordance with the seasonal distribution pattern of the total population. However, *Paradoxostoma* sp. C became abundant in February and made up 48.7% of the population in March 1985.

For the *Thalassia* samples, the seasonal distribution of ostracods for 1984 to 1985 was different to the distribution of 1987 to 1988 (Fig. 3). This difference may be explained by the difference in water temperature in the two years. The scarcity of ostracods in summer may be partly caused by a higher water temperature in the moat than in other nearby environments (Fig. 3). Other factors such as salinity and dissolved oxygen content do not seem to be correlated with the seasonal distribution (Fig. 3). In the summer of 1987 the ostracod population decreased while the water temperature in the moat (35°C) became five degrees higher than the sea at the reef edge

Number of Individuals

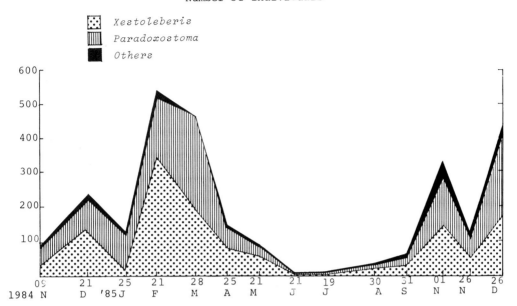

Fig. 2. Seasonal distribution of living Ostracoda collected from 50g (wet weight) of *Thalassia hemprichii*.

(30°C). Filamentous algae were scarce in summer and abundant in winter. Therefore, the low incidence of ostracods may be partly explained by the lack of filamentous algae on which they live. During the summer, the dominant genera *Xestoleberis* and *Paradoxostoma* may migrate to other habitats.

c) *Holothuroidea*. Thirty-five species of live ostracods belonging to 17 genera were recovered from the external surface of the sea cucumber *Holothuria* sp., sampled monthly from April to December 1987. Thirty-eight species of live ostracods belonging to 16 genera were identified from another sea cucumber, *Actinopyga* sp.

The genera *Xestoleberis* and *Paradoxostoma* were abundant on *Holothuria* sp. *Paradoxostoma* was represented by 10 species on *Holothuria* sp. and 13 species on *Actinopyga* sp. At the species level, *Xestoleberis* cf. *X. hanaii* occurred abundantly throughout the 9 months of the study, while *Paradoxostoma* sp. D and *Paradoxostoma* sp. F were found commonly from October to December. *Xestoleberis* and *Paradoxostoma* comprised from 57.2% (August) to 97% (December) of the

population. The number of individuals of the genus *Xestoleberis* was highest during April, May (81.5%) and June and then decreased to November (30.9%). Conversely, individuals of the genus *Paradoxostoma* were rather uncommon from April to August, but increased abruptly in abundance from August and reached a maximum in December (58.9%). *Xestoleberis*, therefore, was dominant until September, while *Paradoxostoma* became dominant after October. In total numbers, there were two peaks in May and October, and then an increase to December. The ostracod population became scarce in summer (July, August and September, Fig. 4). The seasonal distribution of *Xestoleberis* conforms to that of the total population. *Paradoxostoma* increased from September to December. The increase in total population in May might have been caused by the increase of *Xestoleberis* and the increase in October and December may similarly have been caused by *Xestoleberis* and *Paradoxostoma*.

On *Actinopyga* sp., *Xestoleberis* and *Paradoxostoma* comprised from 88.4% (December) to 34% (August) of the fauna. Two other genera,

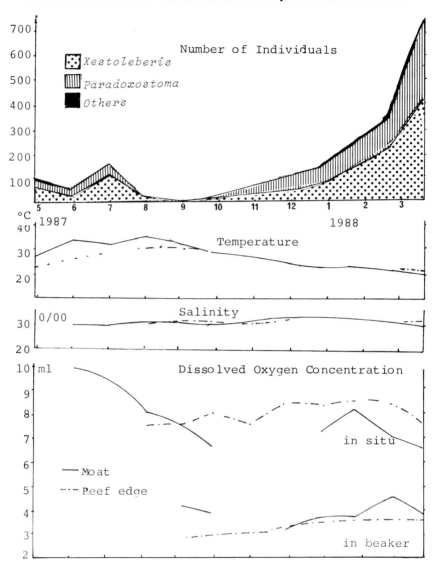

Fig. 3. Seasonal distribution of living Ostracoda collected from 50g (wet weight) of *Thalassia hemprichii* from Sesoko-jima, Okinawa.

*Loxocorniculum* and *Tenedocythere*, became more abundant than *Xestoleberis* and *Paradoxostoma* from July to September and made up 58.3% (*Loxocorniculum* 32.7% and *Tenedocythere* 25.6%) of the population in August. The peak in the seasonal distribution of the total population in July may have been caused by the increase of the genera *Xestoleberis*, *Loxocorniculum* and *Tenedocythere* (Fig. 5). The slight overall decrease from August to October may be due to the decrease of *Xestoleberis*

cf. *X. hanaii*. The increase from October to December may have been influenced by the increase of *Xestoleberis* cf. *X. hanaii* and *Paradoxostoma* sp. D. Adults of *Loxocorniculum* sp. A and *Tenedocythere transoceanica* Teeter were more abundant than juveniles externally on *Actinopyga* sp. The adults were more abundant than juveniles in *Paradoxostoma* sp. D, but *vice versa* in *Paradoxostoma* sp. F. However, in *Xestoleberis* cf. *X. hanaii* few adults were

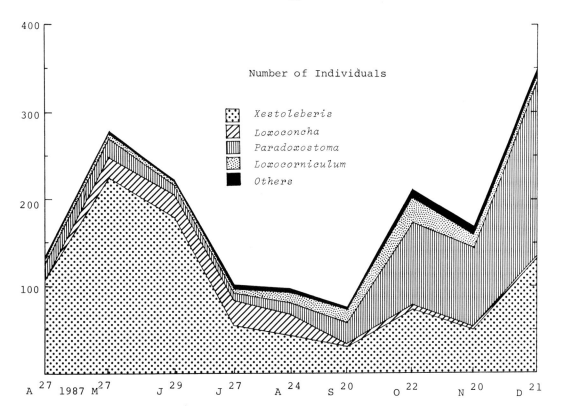

Fig. 4. Seasonal distribution of living Ostracoda collected from 30 individuals of *Holothuria* sp.

observed externally on either *Holothuria* sp. or *Actinopyga* sp.

For additional comparison, some ostracods inhabiting gravel were examined in December, 1987. Twenty-five species belonging to 10 genera were identified from gravel around a sea grass area and 21 species belonging to 7 genera were found from gravel on a reef flat area. During the same month, 17 species belonging to 8 genera were identified from *Holothuria* sp. and 21 species belonging to 7 genera from *Actinopyga* sp. There seemed to be no distinct difference in ostracod species diversity on holothurians or gravel, but slight differences were noticed in species composition. Though *Xestoleberis* cf. *X. hanaii* was dominant on sea cucumbers, other species such as *Leptocythere* sp., *Paradoxostoma* sp. and *Radimella* sp. dominated on gravel.

## CONCLUSIONS

Twenty-nine species of live ostracods found on the red alga, *Jania adhaerens*, were identified as belonging to 11 genera, the dominants being *Xestoleberis* and *Paradoxostoma*. Monthly data on individual abundance showed a pattern of two prominent peaks in February and September.

Twenty-one species of live ostracods belonging to 10 genera were found on the angiosperm, *Thalassia hemprichii*. The genera *Xestoleberis* and *Paradoxostoma* were again dominant. The monthly data of the individuals per unit angiosperm showed only one prominent peak, and seasonal change was characterized by a maximum population density in February and a low density in summer.

Thirty-five species of live ostracods belonging to 17 genera were found on the outer surface of a sea cucumber (*Holothuria* sp.). The monthly

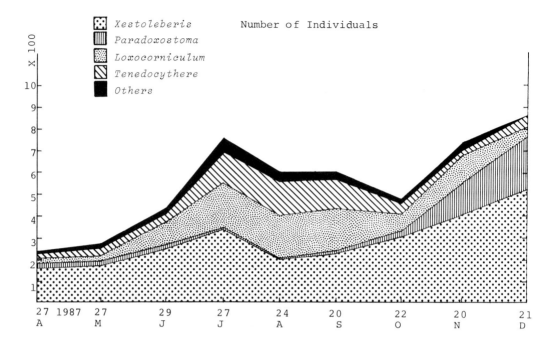

Fig. 5. Seasonal distribution of living Ostracoda collected from 20 individuals of *Actinopya* sp.

data indicate that the minimum population density occurs in summer. Thirty-eight species belonging to 16 genera were found externally on another sea cucumber (*Actinopyga* sp.).

Apparently ostracods on sea grass (*Thalassia* sp.) and sea cucumbers (*Holothuria* sp.) in the littoral zone decrease in numbers in summer and this may be in response to seasonal temperature maxima in the moat. On the other hand, ostracods on the alga (*Jania adhaerens*) and the sea cucumber (*Actinopyga* sp.) which live deeper than *Thalassia* and *Holothuria* sp. do not show a distinct population decrease in summer.

Individuals of *Xestoleberis* cf. *X. hanaii* were abundant in the cooler season and larger forms of this species were present in the cooler times of the seasons and smaller forms in the warmer months.

## ACKNOWLEDGEMENTS

We thank our former students for their help in the course of the present study. We also extend special

acknowledgement to Dr M. J. Grygier of the Smithsonian Institution, Washington, for reading the first draft of the manuscript and Professor R. C. Whatley, of the University College of Wales, Aberystwyth, for improving the manuscript.

## REFERENCES

Okubo, I. 1984. On the life history and the size of *Xestoleberis hanaii*. *Res. Bull. Shujitsu Coll.* **14**, 19-43.

Tabuki, R. & Nohara, T. 1988. Preliminary study on the ecology of ostracods from the moat of a coral reef off Sesoko Island, Okinawa, Japan. *In* Hanai, T., Ikeya, N. & Ishizaki, K. (Eds), *Evolutionary biology of Ostracoda, its fundamentals and applications*, proceedings of the Ninth International Symposium on Ostracoda, held in Shizuoka, Japan, 29 July - 2 August 1985, Developments in palaeontology and stratigraphy, **11**, 431-440, Kodansha Ltd., Tokyo and Elsevier, Amsterdam, Oxford, New York, Tokyo.

Theisen, B. T. 1966. The life history of seven species of ostracods, from a Danish brackish-water locality. *Meddr Dann. Fisk. og Havunders.*, København, n.s. 4, **8**, 215-270.

Whatley, R. 1976. Association between Podocopid Ostracoda and some animal substrates. *Abh. Verh. naturw. Ver.*

*Hamburg*, 18/19 (suppl.), 191-200.

Whatley, R. & Wall, D. R. 1969. A preliminary account of the ecology and distribution of Recent Ostracoda in the southern Irish Sea. *In* Neale, J. W. (Ed.), *The Taxonomy, Morphology and Ecology of Recent Ostracoda*, 268-298. Oliver & Boyd, Edinburgh.

Whatley, R. & Wall, D. R. 1975. The relationship between Ostracoda and algae in littoral and sublittoral marine environments. *In* Swain, F. M. (Ed.), *Biology and Paleobiology of Ostracoda. Bull. Am. Paleont.*, **65**(282), 173-203.

## DISCUSSION

Martin Angel: Where do the ostracods occur on the holothurians? Are they external or internal? How do you collect them?

Tomohide Nohara: We think that the ostracods live on the external surface of the holothurians. However, we have not yet observed exactly where they live. We collected the holothurians by hand, wearing plastic gloves and slowly moved them into a bucket. Then they were washed in the bucket to shake the ostraocds off the holothurians.

Rosalie Maddocks: Do any of the ostracod species that you found on holothurians live only on the holothurian, or are they found also on plants and sediment? Are any of the species ectoparasites or symbionts, or are they just living on the sediment and algae covering the holothurian?

Tomohide Nohara: The ostracod species living on the holothurians are also found on plants or sediments. We have no evidence as to whether the species are parasites or symbionts at present. We think that they live on the sediment and algae covering the holothurian, as well as directly on its external body surface.

# 28

# The Ostracoda of the Sekisei-sho area, Ryukyu Islands, Japan: a preliminary report on ostracods from coral reefs in the Ryukyu Islands

Ryoichi Tabuki & Tomohide Nohara

Department of Earth Sciences, College of Education, University of the Ryukyus, Okinawa, Japan.

## ABSTRACT

One hundred and forty-three ostracod species, both living and dead, belonging to 57 genera were collected in the Sekisei-sho area, which includes the largest coral reef in the Ryukyu Islands. The fauna is characterized by tropical and subtropical elements. There are some differences in species composition between the living and the dead assemblages, which can be explained by the fact that sampling was undertaken only on soft, mainly sandy bottoms by means of a soli-net sampler.

## INTRODUCTION

The ostracod faunas of coral reefs have been studied mainly in the Caribbean Sea, the Eastern Pacific and the Red and Java Seas (Kornicker, 1958; Hartmann, 1964, 1984; Allison & Holden, 1971; Maddocks, 1974; Teeter, 1975; Bonaduce et al., 1976; Krutak, 1982; Whatley & Watson, 1988). As a result of these studies, knowledge of the taxonomy, distribution, ecology and biogeography of ostracods indigenous to coral reefs and their associated environments has begun to accumulate. In coral reef areas from the Philippine Islands to the Ryukyu Islands of the Western Pacific, which are influenced by the warm Kuroshio Current, no faunal studies on ostracods have been undertaken except that by the authors (1988).

In October 1985, an oceanographical survey, that included sampling of surface sediments, was carried out in and around the Sekisei-sho area, Ryukyu Islands (Fig. 1). The present study was begun as part of this project, and preliminary results on the faunal characteristics of the ostracods are presented here.

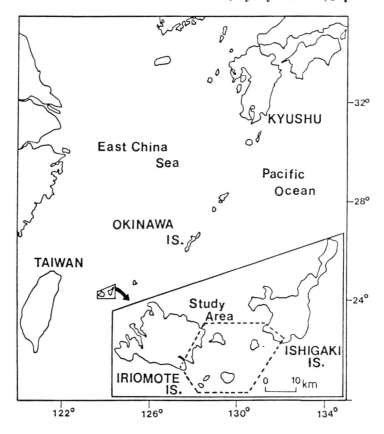

Fig. 1. Location map of study area.

## STUDY AREA

The Sekisei-sho area and its surroundings extend between two large islands, Ishigaki and Iriomote (Fig. 2). The area is characterized by a shallow, flat submarine topography that abruptly deepens towards oceanic areas to the north and south.

Yamamoto (1987) investigated the areal and vertical changes in water temperature and salinity in the study area, and recognized an oceanic water inflow into the inner part of the Sekisei-sho area, around the Kuro and Aragusuku Islands and through the channel between the Iriomote and Kohama Islands. Eguchi (1974) and Tatsuki & Fukuda (1974) observed that calcareous sands and gravels are distributed around the reefs in the area. Midorikawa & Ujiie (1987) studied in detail the calcareous sediments in the study area, and they concluded that gravels and medium to coarse-

grained sands are mainly distributed in the inner part of the Sekisei-sho area except for the central part between the Kuro and Kohama Islands, where finer sediments dominate; finer sediments also occur in the deeper areas to the south and north. Hatta & Ujiie (1987) discovered a difference in the benthonic foraminiferal species between the lagoonal area between the Kohama and Taketomi Islands and the area south of the lagoon.

## METHODS

In October 1985 surface sediment samples were collected from 30 locations by means of a soli-net sampler, which is a small sledge dredge of Ockelmann type, designed to collect surface sediments of one centimetre or less in thickness. The collection sites are plotted on Fig. 2 and listed in Table 1. Each sample was dried and 80g in dry weight

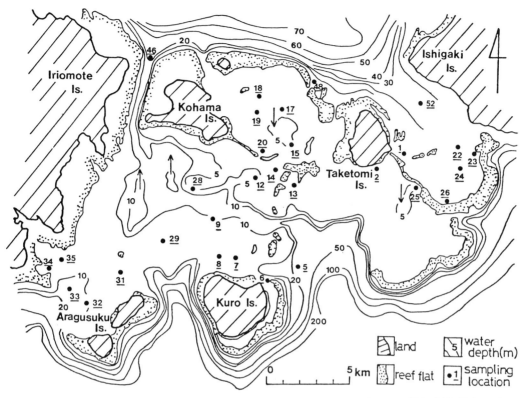

Fig. 2. The study area and sampling locations. The bathymetric chart is based upon chart nos 1206, 1285, 6513 and 6514 published by the Japanese Hydrographic Office.

was washed through 9 and 250 mesh sieves. The residue in the 250 mesh sieve was dried again and partitioned into unit samples, each of which contained more than 200 individual ostracods. Part of this residue was passed through 16, 32 and 60 mesh sieves to ascertain its grain-size distribution. Ostracod specimens were classified as 'living' specimens if they were judged to have been alive at the time of collection, that is, if they possessed a full set of appendages. Specimens were classified as 'dead' if they had partly preserved or no appendages.

## RESULTS AND DISCUSSION

One hundred and forty species belonging to 54 genera of Podocopida and 3 species of Myodocopida were identified, and only the former are discussed in this paper. There are many genera characteristic of tropical to subtropical, shallow water, and among them a group of genera, represented by *Paranesidea*, *Triebelina* and *Morkhovenia*, is known to appear characteristically or abundantly in coral reefs and their associated environments (Teeter, 1975; Bonaduce *et al.*, 1980; Whatley & Watson, 1988).

Table 1 presents data concerning the sediment samples and ostracod specimens. Many of the samples are of coarse sediments, mainly composed of gravels and medium to coarse-grained sands. Fine sediments, composed mainly of fine to medium-grained sands were collected from only 9 locations, which were situated at relatively deeper sites except for Station no. 8; 8 of these locations are in the inner part of the Sekisei-sho area. The fine sediment samples (except for that from Station no. 8) contained many dead specimens per unit sample. Few ostracods, however, were found in the coarse sediment samples, except at Station nos 5, 26, 46 and 48, all of which are located in the

Table 1.  Data from the sediment samples and ostracod specimens.

| Station no. | Depth (m) | Sediments | Weight of the residue (sand) | Ratio of the residue examined | Number of dead specimens | (A) | (B) | Number of living specimens | (C) |
|---|---|---|---|---|---|---|---|---|---|
| 1 | 3.0 | medium to coarse sand | 74.1 | 1 | 360 | 49 | 45 | 11 | 11 |
| 2 | 5.0 | medium to coarse sand | 78.3 | 1 | 265 | 34 | 33 | 7 | 7 |
| 5 | 49.0 | medium to coarse sand | 70.6 | 1/2 | 639 | 181 | 160 | 2 | 4 |
| 6 | 3.0 | gravelly coarse sand | 49.2 | 1 | 93 | 19 | 12 | 15 | 15 |
| 7 | 3.0 | coarse sand | 126.3** | 1 | 191 | 15 | 12 | 86 | 43 |
| 8 | 2.5 | fine to medium sand | 79.1 | 1 | 69 | 9 | 9 | 0 | 0 |
| 9 | 9.5 | fine sand | 9.8* | 1/2 | 265 | 541 | 402 | 1 | 2 |
| 12 | 8.0 | fine to medium sand | 75.4 | 1/16 | 557 | 1182 | 1114 | 1 | 16 |
| 13 | 2.0 | gravelly coarse sand | 58.1 | 1 | 132 | 23 | 17 | 6 | 6 |
| 14 | 3.0 | medium to coarse sand | 75.7 | 1 | 151 | 20 | 19 | 50 | 50 |
| 15 | 7.0 | fine to medium sand | 77.0 | 1/4 | 397 | 206 | 199 | 0 | 0 |
| 17 | 7.0 | medium to coars sand | 73.4 | 1 | 334 | 46 | 42 | 27 | 27 |
| 18 | 9.0 | fine to medium sand | 77.9 | 1/4 | 334 | 172 | 167 | 4 | 16 |
| 19 | 3.0 | medium to coarse sand | 149.7** | 1 | 290 | 19 | 18 | 48 | 24 |
| 20 | 6.0 | fine to medium sand | 76.1 | 1/4 | 490 | 258 | 245 | 1 | 4 |
| 22 | 2.5 | medium to coarse sand | 75.1 | 1 | 51 | 7 | 6 | 19 | 19 |
| 23 | 1.5 | sandy (coarse) gravel | 78.5** | 1 | 208 | 26 | 13 | 68 | 34 |
| 24 | 1.5 | coarse sand | 62.8 | 1 | 464 | 74 | 58 | 86 | 86 |
| 25 | 4.0 | medium to coarse sand | 78.7 | 1 | 109 | 14 | 14 | 20 | 20 |
| 26 | 1.0 | gravelly coarse sand | 48.8 | 1/4 | 268 | 220 | 134 | 57 | 228 |
| 28 | 3.0 | gravelly medium to coarse sand | 115.7** | 1 | 464 | 40 | 29 | 134 | 67 |
| 29 | 18.5 | fine to medium sand | 66.5 | 1/4 | 578 | 348 | 289 | 5 | 20 |
| 31 | 7.5 | fine to medium sand | 76.8 | 1/8 | 307 | 320 | 320 | 6 | 48 |
| 32 | 13.5 | medium to coarse sand | 78.7 | 1 | 29 | 4 | 4 | 1 | 1 |
| 33 | 9.5 | medium to coarse sand | 151.1** | 1 | 69 | 5 | 4 | 5 | 5 |
| 34 | 3.0 | medium to coarse sand | 72.1 | 1 | 313 | 43 | 39 | 4 | 4 |
| 35 | 4.5 | gravelly coarse sand | 107.6** | 1 | 641 | 60 | 40 | 9 | 9 |
| 46 | 27.0 | gravelly medium to coarse sand | 11.5* | 1 | 224 | 195 | 147 | 10 | 10 |
| 48 | 17.5 | medium to coarse sand | 74.9 | 1/8 | 494 | 528 | 494 | 4 | 32 |
| 52 | 28.0 | fine sand | 74.5 | 1/16 | 410 | 881 | 820 | 1 | 16 |

(A): Individual number of dead specimens per 10 grams of sands.
      "sands" is a fraction between 0.063 mm (250-mesh) and
      2 mm (9-mesh) in grain size of the sediments examined.
(B): Individual number of dead specimens per 10 grams of sediments.
(C): Individual number of living specimens per 80 grams of sediments.
* 13.2 grams of sediments from St. no. 9 and 15.2 grams from St. no.
  46 were used for the ostracod study.
** 160 grams of sediments were used for the study.

Table 2. List of the representative species of the dead ostracod assemblages from the Sekisei-sho area and its surroundings.

| Species name | Environment → Station no. → Depth (m) | LAGOON | | | | | LAGOON – OPEN SEA | | | | | | | | | | | | OPEN SEA | | | | | | | |
|---|---|---|---|---|---|---|---|---|---|---|---|---|---|---|---|---|---|---|---|---|---|---|---|---|---|---|
| | | 19 | 20 | 15 | 17 | 18 | 13 | 7 | 14 | 28 | 34 | 25 | 35 | 2 | 31 | 12 | 9 | 29 | 26 | 23 | 24 | 1 | 48 | 46 | 52 | 5 |
| | | 3.0 | 6.0 | 7.0 | 7.0 | 9.0 | 2.0 | 3.0 | 3.0 | 3.0 | 3.0 | 4.0 | 4.5 | 5.0 | 7.5 | 8.0 | 9.5 | 18.5 | 1.0 | 1.5 | 1.5 | 3.0 | 17.5 | 27.0 | 28.0 | 49.0 |
| *Neonesidea* sp. 1 | | | | | | | | | | | | | | | | | | | | | | | | | | |
| *N.* sp. 2 | | | | | | | | | | | | | | | | | | | | | | | | | | |
| *Paranesidea* sp. 1 | | | | | | | | | | | | | | | | | | | | | | | | | | |
| *Triebelina amicitiae* Keij | | | | | | | | | | | | | | | | | | | | | | | | | | |
| *T. sertata* Triebel | | | | | | | | | | | | | | | | | | | | | | | | | | |
| *Perissocytheridea inabai* Okubo | | | | | | | | | | | | | | | | | | | | | | | | | | |
| *Pontocythere subjaponica* (Hanai) | | | | | | | | | | | | | | | | | | | | | | | | | | |
| *Morkhovenia* sp. 1 | | | | | | | | | | | | | | | | | | | | | | | | | | |
| *M.* sp. 2 | | | | | | | | | | | | | | | | | | | | | | | | | | |
| *Keijia* cf. *demissa* (Brady) | | | | | | | | | | | | | | | | | | | | | | | | | | |
| *Callistocythere* sp. 1 | | | | | | | | | | | | | | | | | | | | | | | | | | |
| *C.* sp. 2 | | | | | | | | | | | | | | | | | | | | | | | | | | |
| *C.* sp. 3 | | | | | | | | | | | | | | | | | | | | | | | | | | |
| *Tenedocythere deltoides* (Brady) | | | | | | | | | | | | | | | | | | | | | | | | | | |
| *T. transoceanica* (Teeter) | | | | | | | | | | | | | | | | | | | | | | | | | | |
| *Radimella* cf. *virgata* Hu | | | | | | | | | | | | | | | | | | | | | | | | | | |
| *Neobuntonia* sp. 1 | | | | | | | | | | | | | | | | | | | | | | | | | | |
| *Jugosocythereis* cf. *Radimella elongata* Hu | | | | | | | | | | | | | | | | | | | | | | | | | | |
| *Trachyleberis* sp. 1 | | | | | | | | | | | | | | | | | | | | | | | | | | |
| *Pistocythereis bradyi* (Ishizaki) | | | | | | | | | | | | | | | | | | | | | | | | | | |
| *Loxoconcha uranouchiensis* Ishizaki | | | | | | | | | | | | | | | | | | | | | | | | | | |
| *L.* sp. 1 | | | | | | | | | | | | | | | | | | | | | | | | | | |
| *Loxocorniculum* sp. 1 | | | | | | | | | | | | | | | | | | | | | | | | | | |
| *L.* sp. 2 | | | | | | | | | | | | | | | | | | | | | | | | | | |
| *L.* sp. 3 | | | | | | | | | | | | | | | | | | | | | | | | | | |
| *Xestoleberis* cf. *hanaii* Ishizaki | | | | | | | | | | | | | | | | | | | | | | | | | | |
| *X.* cf. *sagamiensis* Kajiyama | | | | | | | | | | | | | | | | | | | | | | | | | | |
| *X.* sp. 1 | | | | | | | | | | | | | | | | | | | | | | | | | | |
| *X.* sp. 2 | | | | | | | | | | | | | | | | | | | | | | | | | | |
| *X.* sp. 3 | | | | | | | | | | | | | | | | | | | | | | | | | | |
| *X.* sp. 4 | | | | | | | | | | | | | | | | | | | | | | | | | | |
| *Ornatoleberis* sp. 1 | | | | | | | | | | | | | | | | | | | | | | | | | | |

%

■ 20 –   ⊠ 2 – 4
● 10 – 20   ⧄ 0 – 2
○ 4 – 10   □ 0

outer part of the Sekisei-sho area, which directly faces oceanic waters. Therefore, in the inner part of the Sekisei-sho area, dead specimens are concentrated in fine sediments at relatively deep sites, whereas, in the outer part of the area, a relationship between the abundance of dead specimens and the sediment grain-size or water depth is less clear. At present, the number of living specimens collected is not large enough to discuss the living ostracod assemblages in detail. However,

in contrast to the results based on dead specimens, large numbers of live specimens per unit sample were noted in some of the coarse sediment samples which were collected at relatively shallow sites, particularly Station nos 24, 26 and 28.

Table 2 is a list of representative species of the dead ostracod assemblages. Five locations, where fewer than 100 dead specimens were obtained are excluded from the list. The three major environments into which the sampling locations

Table 3. Species list of the living ostracod assemblages from the Sekisei-sho area and its surroundings.

| Environment | LAGOON | | | | | | | | LAGOON – OPEN SEA | | | | | | | | | | | | | | | OPEN SEA | | | | | | |
|---|---|---|---|---|---|---|---|---|---|---|---|---|---|---|---|---|---|---|---|---|---|---|---|---|---|---|---|---|---|---|
| Station no. | 19 | 20 | 15 | 17 | 18 | 13 | 8 | 6 | 7 | 14 | 28 | 34 | 25 | 35 | 2 | 31 | 12 | 9 | 29 | 26 | 23 | 24 | 22 | 1 | 33 | 32 | 48 | 46 | 52 | 5 |
| Depth (m) | 3.0 | 6.0 | 7.0 | 7.0 | 9.0 | 2.0 | 2.5 | 3.0 | 3.0 | 3.0 | 3.0 | 3.0 | 4.0 | 4.5 | 5.0 | 7.5 | 8.0 | 9.5 | 8.5 | 1.0 | 1.5 | 1.5 | 2.5 | 3.0 | 9.5 | 13.5 | 17.5 | 7.02 | 8.04 | 9.0 |
| *Neonesidea* sp. 1 | | | | | | | | | | | | | | | | | | | | | | | | | | | | | | 5 |
| *Paranesidea* sp. 1 | | | | | | | | | | | | | | | | | | | | | | | | | | | | 1 | | |
| *P.* sp. 2 | | | | | | | | | | | | | | | | | | | | | | | | | | | | 1 | | |
| *Macrocypris* cf. *decora* (Brady) | | | | 2 | | | | | | | | | | | | | | | | 1 | 1 | | | | | | | | | |
| *Propontocypris* sp. 1 | 3 | | | | 1 | | | | | | | | | | | | | | | | | | | | | | | | | |
| *P.* sp. 2 | | | | 2 | | | | | 3 | | 2 | 1 | | | | | | | | | | | | | | | | | | |
| *P.* sp. 3 | 1 | | | | | | | | | | 1 | | 3 | | | | | | | | | | | | | | | | | |
| *P.* sp. 4 | | | | | | | | | | | | | | | | | 1 | | | | | | | | | | | | | |
| *P.?* sp. | | | | 1 | | | | 1 | 11 | | 6 | | | | | | | | | | | | 7 | | | | | | | |
| *Perissocytheridea inabai* Okubo | | | | | | | | | | | | 1 | | | | | | | | | | | | | | | | | | |
| *Pontocythere subjaponica* (Hanai) | | | | | 1 | | | | | | 4 | | | | 5 | | | | | | | | | | | | | | | |
| *Parakrithella pseudadonta* (Hanai) | | | | | | | | | | | | | | | | | | | | 1 | | | | | | | | 1 | | |
| *Morkhovenia* sp. 1 | | | | | | | | | | | | | | | | 1 | | | 1 | 1 | | | | | | | | 1 | | |
| *M.* sp. 2 | 1 | | | | | | | | | | | | | | | | | 1 | | | | | | | | | | | | 1 |
| *Keijia* cf. *demissa* (Brady) | | | | 2 | | | | | 5 | 6 | 2 | 1 | | 2 | | 2 | | | | | | | | 1 | | | | 2 | | |
| *Callistocythere rugosa* Hanai | 1 | | | | | | | | | | | | | | | 2 | | | | | 1 | | | | | | | 1 | | |
| *C.* sp. 1 | | | | | | | | | | | | | | | | 2 | | | 3 | 1 | | 1 | 1 | 2 | | | 1 | | | |
| *C.* sp. 2 | 3 | | | 4 | | | | 5 | 18 | 6 | 8 | | 5 | | | 2 | | | 3 | 6 | | 3 | | | | | | 1 | | |
| *C.* sp. 3 | | | | | | | | | | | | | | 2 | | | | | | | | | | | | | | 2 | | |
| *Leptocythere* sp. 1 | | | | | 1 | | | | | | | | | | | | | | | | | | | 1 | | | 2 | | | |
| *Radimella* cf. *virgata* Hu | | | | | | | | | | | | | | | | | | | | | | | | | | | | | 1 | |
| *Coquimba* sp. 1 | | | 1 | | | | | | | | | | | | | 1 | | | | | | | | | | | | | | |
| *Cornucoquimba* sp. 1 | 1 | | | | | | | | | | | | | | | | | | | | | | | | | | | | | |
| *Trachyleberis* sp. 1 | | | | 1 | | | | | | | | | | | | | | | | | 1 | | | | | | | | | |
| *Keijiella* sp. 1 | 1 | | | | | | | | | | 1 | | | | 2 | | | | 1 | | | | | | | | | | | |
| *K.* sp. 2 | 1 | | | 1 | | | | 1 | 1 | 4 | | | | 1 | | | | | | | | | 1 | | | | 1 | | | |
| *Semicytherura* sp. 1 | 8 | | | | | | | 5 | 18 | 6 | 10 | 1 | 3 | | | | | 1 | 3 | 14 | 54 | 24 | 2 | 2 | | | 3 | | | |
| *S.* sp. 2 | | | | | | | | | | | | | | | | | | | | | | | | | | | | | | |
| *Loxoconcha uranouchiensis* Ishizaki | | | | | | | | | | | | | | | | | | | | | 1 | | | | | 1 | | | | |
| *Loxocorniculum* sp. 1 | | | | | | | | | 4 | 11 | 11 | | 1 | 1 | | | | | | | 1 | | 1 | | | | 1 | | | |
| *Xestoleberis* cf. *hanaii* Ishizaki | | | 5 | | | | | 4 | 14 | 6 | 18 | | 7 | | 2 | | | | 13 | 2 | 20 | 2 | 5 | 2 | | | 2 | | | |
| *X.* cf. *sagamiensis* Kajiyama | | | | | | | | | | 4 | 14 | | 1 | | | | | | | | | 14 | | | | | | | | |
| *X.* sp. 1 | 11 | | 9 | 2 | | | | 2 | 2 | 4 | 14 | | 1 | 1 | | | | | | | | | 1 | | | | | | | |
| *X.* sp. 2 | | | | | | | | | | 1 | 5 | | | 1 | | | | | | | | | | | | | | | | |
| *X.* sp. 3 | 8 | | | | | | | 2 | 2 | 3 | 3 | | | | | | | | 3 | 5 | 13 | | | 1 | | | | | | |
| *X.* sp. 4 | 7 | | | | | 3 | | 3 | 5 | 2 | 8 | | | | | | | | 18 | | 5 | 12 | | | | | | | | |
| *Ornatoleberis* sp. 1 | | | | | | | | | 27 | | 31 | | 1 | | | | | | | | | | | | | | | | | |
| *O.* sp. 2 | 1 | | | 1 | | | | 1 | 1 | 1 | 1 | | | | | | | | | 1 | 2 | | | 1 | | | | | | |
| *Paradoxostoma* sp. 1 | | | | | | 1 | | | | 2 | | | | | | | | | | | | | | | | | | | | |
| *P.* sp. 2 | | | | | | | | | 1 | | | | | | | | | | | | | | | | | | | | | |
| *P.* sp. 3 | | | | | | | | | | | | | | | | | | | | | | | | | | | | | | |
| *P.* sp. 4 | 2 | | 2 | | | | | | | | | | | | | | | | | | | | | | | | | | | |
| *P.* sp. 5 | | | | | | | | | | | | | | | | | | | | | | | | | | | | | | |
| Total number of living specimens | 48 | 1 | 0 | 27 | 4 | 6 | 0 | 15 | 86 | 50 | 34 | 4 | 20 | 9 | 7 | 6 | 1 | 1 | 5 | 57 | 68 | 86 | 19 | 11 | 5 | 1 | 4 | 10 | 1 | 2 |

are grouped lagoon, lagoon-open sea (transitional) and open sea, can be distinguished on the basis of other studies on topography, water masses, sediments and benthonic formaninifera carried out as part of the Sekisei-sho research project (Ujiie & Yada, 1987; Yamamoto, 1987; Midorikawa & Ujiie, 1987; Hatta & Ujiie, 1987).

The distribution patterns of most species seem to be independent of these environmental divisions. For example, *Xestoleberis* cf. *hanaii* Ishizaki, 1968, *X.* cf. *sagamiensis* Kajiyama, 1913 and *Loxoconcha uranouchiensis* Ishizaki, 1968 occur abundantly and ubiquitously, and *Loxoconcha* sp. 1, *Neonesidea* sp. 1, *Morkhovenia* sp. 1, *Tenedocythere deltoides* (Brady, 1890) and *T. transoceanica* (Teeter, 1975) are common at many localities. The occurrence, however, of some species can be correlated with certain environmental conditions. *Pistocythereis bradyi* (Ishizaki, 1968), *Trachyleberis* sp. 1 and *Perissocytheridea inabai* (Okubo, 1983) are characteristic of the lagoon and lagoon - open sea environments, while *Jugosocythereis* cf. *elongata* (Hu, 1979) seems to have a preference for water with an oceanic influence.

The species list for the living assemblages is given in Table 3. When it is compared with the list for the dead assemblages, some differences in species composition are evident. Living specimens of *Xestoleberis* cf. *hanaii* and *X.* cf. *sagamiensis*, both of which are dominant in the dead assemblages, are almost confined to three samples from the central part of the Sekisei-sho area (Station nos 7, 14 and 28), and *Loxoconcha* sp. 1, *Neonesidea* sp. 1, *Jugosocythereis* cf. *elongata*, *Tenedocythere deltoides* and *T. transoceanica*, all of which are commonly found in the dead assemblages, occur very rarely or are entirely lacking in the living assemblages. On the other hand, *Loxoconcha uranouchiensis*, *Callistocythere* sp. 3, *Xestoleberis* sp. 1, *X.* sp. 2, *X.* sp. 4 and *Ornatoleberis* sp. 1 are found more frequently in the living assemblages than their abundance in the dead assemblages would suggest.

Knowledge of the ecological distribution of each species is indispensible for understanding the different distribution patterns of dead and living specimens at the specific level. Since May 1984 the authors have been carrying out an ecological study on ostracods in the moat of a coral reef at Sesoko Island, Ryukyu Islands and, one of the preliminary results, is that ostracod species distribution is controlled by environmental factors such as water temperature, location within the moat, water depth and substrate (Tabuki & Nohara, 1988). Macroalgae, sea grasses and sandy bottoms are important substrates for ostracods. In addition to these, filamentous algae, which grow densely on some cobble to boulder-size gravels and partly on rocks, also play an important role in controlling ostracod distributions. The filamentous algae frequently trap sand grains and are covered with a thin sand layer; thus they are similar to a sandy bottom, although the population density of ostracods from filamentous algae is much higher than from sandy bottoms. At present, fourteen species are known to be common to the Sekisei-sho area and Sesoko Island, and they can be grouped into two substrate-related groups based on the ecological results from Sesoko Island:

1) the group represented by *Xestoleberis* cf. *hanaii*, *X.* cf. *sagamiensis* and *Tenedocythere transoceanica* (= *Hermanites* sp. of Tabuki & Nohara (1988)) lives by adhering to macroalgae and sea grasses.

2) the other group, comprising *Loxoconcha uranouchiensis*, *Callistocythere* sp. 3 and *Perissocytheridea inabai*, is composed of sandy bottom inhabitants.

In samples of filamentous algae, elements of both species groups are found. In the Sekisei-sho area, the former group is rare or absent as living individuals except for two small species, *Xestoleberis* sp. 2 and *X.* sp. 4, whereas the latter group occurs frequently as living specimens. This can be explained by the fact that the sampling of ostracods in the Sekisei-sho area was undertaken only on soft, mainly sandy bottom by means of a soli-net sampler, and most of the specimens of the species belonging to the first group might have been transported into the sandy bottom from their original habitats. However, there is a possibility that the soli-net sampler crossed vegetated areas during its operation, since Kida (1974) observed that algae and sea grasses are distributed on the sandy bottom in the Sekisei-sho area as well as on rock or gravel bottoms. When operating a soli-net sampler, one should pay attention to the distribution of vegetated areas in order to avoid mixing living ostracods from different habitats.

## CONCLUSIONS

Distribution patterns of ostracod genera or species in coral reefs in previous studies have been mainly based on dead specimens. A coral reef, however, is a high energy environment and consequently, transport of dead specimens and even occasionally of living ones, can take place. As a result, dead ostracod assemblages from soft, sandy bottoms become strongly mixed faunas composed of elements from various environments and habitats. On the other hand, the species diversity of ostracods in coral reefs is high (Teeter, 1975; Whatley & Watson, 1988), and this might imply that the coral reef provides various ecological niches for ostracods in both soft and hard bottom areas.

Tabuki & Nohara (1988) revealed that the ecological distibutions of ostracod species in the moat of a coral reef at Sesoko Island are controlled by some environmental factors such as water depth and substrate. The ecological results of Tabuki & Nohara (1988) suggest that the dead ostracod assemblages of the Sekisei-sho area have partly or mostly an allochthonous nature. In particular, those from the deeper sites, such as Station nos 5, 46 and 52, possibly contain many dead specimens derived from the different environments and habitats of the shallower aeas. Therefore, in order to understand the original ecological distributions of ostracod species, ostracod sampling should be planned so as to collect living specimens, covering separately the various habitats or ecological niches in both soft and hard bottoms of coral reefs.

## ACKNOWLEDGEMENTS

This study commenced as part of the research project supported by the Ministry of Education of Japan (Grant No. 60300021). The authors thank Professor H. Ujiie, University of the Ryukyus, for the opportunity to join this research project. We are deeply indebted to Dr M. J. Grygier, Smithsonian Institution, Washington, D.C., and Dr K. J. Lupardus, Okinawa Kokusai University, for reading and improving the manuscript. Thanks are also due to two anonymous reviewers of the present paper and Professor R. C. Whatley, University College of Wales, for their constructive suggestions.

## REFERENCES

Allison, E. C. & Holden, J. C. 1971. Recent Ostracodes from Clipperton Island, Eastern Tropical Pacific. *Trans. San Diego Soc. Nat. Hist.*, San Diego, **16**(7), 165-214.

Bonaduce, G., Masoli, M. & Pugliese, N. 1976. Benthic Ostracoda from the Gulf of Aqaba (Red Sea). *Pubbl. Staz. zool. Napoli*, Milano & Napoli, **40**, 372-428.

Bonaduce, G., Masoli, M., Minichelli, M. & Pugliese, N. 1980. Some New Benthic Marine Ostracod species from the Gulf of Aqaba (Red Sea). *Boll. Soc. paleont. italiana*, Modena, **19**, 143-178.

Eguchi, M. 1974. Scleractinia in the Proposed Underwater Park Site of the Kerama Islands and the Yaeyama Islands. *In General Survey for the Underwater Park and Park Planning in the Kerama Islands and Yaeyama Islands, Okinawa Prefecture*. Rep. Underwater Park Center, Okinawa Prefecture, Naha, 37-48. (In Japanese.)

Hartmann, G. 1964. Zur Kenntnis der Ostracoden des Roten Meeres. *Kieler Meeresforsch.*, Kiel, **20**, 35-127.

Hartmann, G. 1984. Zur Kenntnis der Ostracoden der polynesischen Inseln Huahinë (Gesellschaftsinseln) und Rangiroa (Tuamotu-Inseln). Mit Bemerkungen zur Verbreitung und Ausbreitung litoraler Ostracoden und einer Übersicht über die bislang auf den pazifischen Inseln gefundenen Arten. *Mitt. hamb. zool. Mus. Inst.*, Hamburg, **81**, 117-169.

Hatta, A. & Ujiie, H. 1987. Benthic Foraminiferal Assemblages in the Sekisei Lagoon. *The Earth Mon.*, Tokyo, **9**(3), 159-162. (In Japanese.)

Kida, W. 1974. Marine Algae in the Coral Reef of the Proposed Underwater Park Site of the Kerama Islands and the Yaeyama Islands. *In General Survey for the Underwater Park and Park Planning in the Kerama Islands and Yaeyama Islands, Okinawa Prefecture*. Rep. Underwater Park Center, Okinawa Prefecture, Naha, 63-84. (In Japanese.)

Kornicker, L. S. 1958. Ecology and Taxonomy of Recent Marine Ostracodes in the Bimini Area, Great Bahama Bank. *Publ. Inst. Marine Sci.*, Univ. Texas, Port Aransas, **5**, 194-300.

Krutak, P. R. 1982. Modern ostracodes of the Veracruz-Anton Lizardo reefs, Mexico. *Micropaleontology*, New York, **28**(3), 258-288.

Maddocks, R. 1974. *Biota of the West Flower Garden Bank, ostracodes*, 200-229. Flower Garden Ocean Research Center. Gulf Publishing Co., Houston.

Midorikawa, Y. & Ujiie, H. 1987. Substrates in the Sekisei-sho, Yaeyama Islands. *The Earth Mon.*, Tokyo, **9**(3), 152-158. (In Japanese.)

Tabuki, R. & Nohara, T. 1988. Preliminary Study on the Ecology of Ostracods from the Moat of a Coral Reef off Sesoko Island, Okinawa, Japan. *In* Hanai, T., Ikeya, N. & Ishizaki, K. (Eds), *Evolutionary biology of Ostracoda, its fundamentals and applications*, proceedings of the Ninth International Symposium on Ostracoda, held in Shizuoka, Japan, 29 July - 2 August 1985, Developments in palaeontology and stratigraphy, **11**, 429-437, Kodansha Ltd., Tokyo and Elsevier, Amsterdam, Oxford, New York, Tokyo.

Tatsuki, K. & Fukuda, T. 1974. Underwater View in the Proposed Underwater Park Site of the Kerama Islands and the Yaeyama Islands. *In General Survey for the Underwater Park and Park Planning in the Kerama Islands and Yaeyama Islands, Okinawa Prefecture*. Rep. Underwater Park Center, Okinawa Prefecture, Naha, 49-62. (In Japanese.)

Teeter, J. W. 1975. Distribution of Holocene Marine Ostracoda from Belize. *In* Wantland, K. F. & Pusey, W. C. (Eds), Belize shelf carbonate sediments, clastic sediments and ecology. *Am. Assoc. Petrol. Geol., Studies in Geology*, **2**, 400-499.

Ujiie, H. & Yada, S. 1987. The Sekisei-sho, a large Coral Reef in the Northwestern Active Margin of the Pacific. *The Earth Mon.*, Tokyo, **9**(3), 124-129. (In Japanese.)

Whatley, R. C. & Watson, K. 1988. A Preliminary Account of the Distribution of Ostracoda in Recent Reef and Reef Associated Environments in the Pulau Seribu or Thousand Island Group, Java Sea. *In* Hanai, T., Ikeya, N. & Ishizaki, K. (Eds), *Evolutionary biology of Ostracoda, its fundamentals and applications*, proceedings of the Ninth International Symposium on Ostracoda, held in Shizuoka, Japan, 29 July - 2 August 1985, Developments in palaeontology and stratigraphy, **11**, 399-411, Kodansha Ltd., Tokyo and Elsevier, Amsterdam, Oxford, New York, Tokyo.

Yamamoto, S. 1987. Water Mass Structure in the Sekisei Lagoon, Yaeyama Islands. *The Earth Mon.*, Tokyo, **9**(3), 135-140. (In Japanese.)

**Plate 1**

Fig. 1. *Neonesidea* sp. 4, lateral view of right valve. Station no. 5. X 50.

Fig. 2. *Neonesidea* sp. 1, lateral view of left valve. Station no. 5. X 60.

Fig. 3. *Paranesidea* sp. 2, lateral view of right valve. Station no. 5. X 50.

Fig. 4. *Triebelina sertata* Triebel, 1948, lateral view of left valve. Station no. 31. X 70.

Fig. 5. *Triebelina amicitiae* Keij, 1974, lateral view of left valve. Station no. 25. X 50.

Fig. 6. *Perissocytheridea inabai* Okubo, 1983, lateral view of left valve. Station no. 2. X 50.

Fig. 7. *Pontocythere subjaponica* (Hanai, 1959), lateral view of left valve. Station no. 5. X 70.

Fig. 8. *Parakrithella pseudadonta* (Hanai, 1959), lateral view of right valve. Station no. 52. X 70.

Fig. 9. *Morkhovenia* sp., lateral view of right valve. Station no. 46. X 100.

Fig. 10. *Keijia* cf. *demissa* (Brady, 1868), lateral view of left valve. Station no. 29. X 70.

Fig. 11. *Coquimba* sp. 2, lateral view of left valve. Station no. 29. X 80.

Fig. 12. *Neobuntonia* sp. 1, lateral view of right valve. Station no. 14. X 50.

Fig. 13. *Tenedocythere transoceanica* (Teeter, 1975), lateral view of right valve. Station no. 14. X 70.

Fig. 14. *Jugosocythereis* cf. *elongata* (Hu, 1979), lateral view of right valve. Station no. 5. X 50.

**Plate 2**

Fig. 1.  *Neocytheromorpha* sp., lateral view of right valve. Station no. 48. X 70.

Fig. 2.  *Radimella* cf. *virgata* Hu, 1979; lateral view of right valve. Station no. 46. X 70.

Fig. 3.  *Trachyleberis* sp. 1, lateral view of right valve. Station no. 18. X 50.

Fig. 4.  *Neocytheretta* sp., lateral view of left valve. Station no. 52. X 70.

Fig. 5.  *Cletocythereis* sp., lateral view of right valve. Station no. 52. X 70.

Fig. 6.  *Pistocythereis bradyi* (Ishizaki, 1968), lateral view of left valve. Station no. 52. X 50.

Fig. 7.  *Callistocythere rugosa* Hanai, 1957; lateral view of right valve. Station no. 17. X 100.

Fig. 8.  *Callistocythere* sp. 1, lateral view of right valve. Station no. 23. X 70.

Fig. 9.  *Loxoconcha uranouchiensis* Ishizaki, 1968; lateral view of right valve. Station no. 24. X 70.

Fig. 10.  *Loxocorniculum* sp. 2, lateral view of right valve. Station no. 15. X 70.

Fig. 11.  *Loxocorniculum* sp. 1, lateral view of right valve. Station no. 35. X 70.

Fig. 12.  *Gambiella* sp., lateral view of left valve. Station no. 15. X 100.

Fig. 13.  *Xestoleberis* cf. *hanaii* Ishizaki, 1968; lateral view of right valve. Station no. 31. X 70.

Fig. 14.  *Xestoleberis* sp. 1, lateral view of right valve. Station no. 7. X 100.

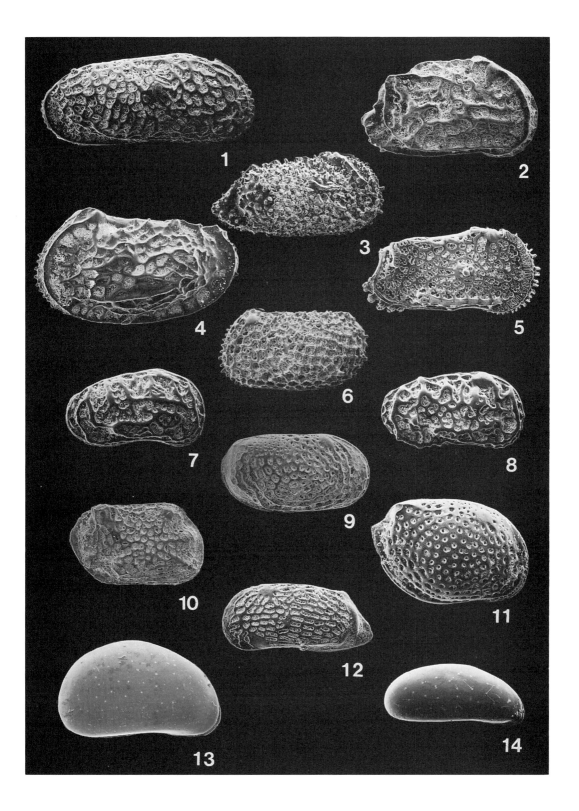

# MORPHOLOGY

# 29

# Morphological relationships of bioluminescent Caribbean species of *Vargula* (Myodocopa)

Anne C. Cohen & James G. Morin

Department of Biology, University of California at Los Angeles, Los Angeles, California 90024, U.S.A.

## ABSTRACT

In the Caribbean Sea, species-specific bioluminescent courtship displays are performed by over 50 species (not all described) of cypridinid ostracods having the diagnostic morphological characters of the genus *Vargula*. A preliminary cladistic analysis of 53 morphological characters on 28 taxa in 6 genera, using the Phylogenetic Analysis Using Parsimony (PAUP) computer program, and rooting with an outgroup, *Doloria levinsoni*, produced a preliminary cladogram. The results of the analysis suggest that (1) *Vargula* is polyphyletic, (2) all Caribbean species belong to 2 monophyletic groups, which are distantly related and almost exclusively Caribbean, and (3) the larger Caribbean group can be subdivided into 4 monophyletic groups. Comparison of luminescent characters with the cladogram shows that all Caribbean clades have luminescent displaying species, and suggests that two aspects of the bioluminescent behaviour (rapid flashing and even lateral displays) are each restricted by past evolution to different monophyletic groups, while another (vertical shortening displays) appears to be plesiomorphic.

## INTRODUCTION

The bioluminescent displays performed by the numerous ostracod species that we are studying, are species-specific indicators that have led us to discover that there are more than 50 such signalling species in the Caribbean Sea. These displays, which are apparently used for courtship, are among the most complex bioluminescent signals known in the sea (Morin, 1986). These signalling ostracods all have the diagnostic morphological characters of the genus *Vargula*. Prior to our research, only 4 species of *Vargula* had been described from the Gulf of Mexico and Caribbean: *V. magna* Kornicker, 1983; *V. bullae* Poulsen, 1962,

*V. parasitica* (Wilson, 1913) and *V. harveyi* Kornicker & King, 1965. We have located no living specimens of the first 3 species, but bioluminescence has been reported for *V. parasitica* (Harvey, 1924). We have collected and observed displays of *V. harveyi*. We also have described another 11 signalling species from Panama (Cohen & Morin, 1986; Morin & Cohen, 1988; Cohen & Morin, 1989) and partially described approximately 43 more (Morin & Cohen, in preparation) from various Caribbean locations. Each of the signalling species has a distinct reef habitat and apparently restricted geographical distribution.

The genus *Vargula* Skogsberg, 1920, is currently defined (Poulsen, 1962; Kornicker, 1975) only by a unique combination of characters and not by any synapomorphies (see Appendix 1: Glossary). The family Cypridinidae is defined by a synapomorphy: suckers on the male's 1st antennae, which are used to grasp the female during mating (Okada & Kato, 1949). Kornicker (1983) used differences in morphology of these suckers to divide the Cypridinidae into suprageneric taxa and assigned 13 genera (including *Vargula*) to the *Cypridina* Genus Group Cypridininae (Cypridinini). The signalling Caribbean species are clearly assignable to the *Cypridina* Genus Group (Cypridinini, Cypridininae), but their precise generic position is less certain.

To investigate the generic and specific affinities of the Caribbean bioluminescent species, we performed a preliminary cladistic analysis, using morphological characters. To resolve the taxonomic status of these species and because *Vargula* is a poorly defined genus, we included in our analysis not only the Caribbean signalling species, but also non-Caribbean species and genera sharing possibly unique character states (synapomorphies) with *Vargula*. Some character states are unique to Caribbean species, so we formed the hypothesis that a cladistic analysis would assign those species to monophyletic groups. Finally we have mapped some components of the species-specific bioluminescent display behaviour onto the morphologically derived cladogram in order to detect if these behavioural characters have been restricted by past evolution.

## MATERIAL

### Field Work

See Cohen & Morin (1986, 1989) and Morin & Cohen (1988) for details. The following is a summary of our methods: Specimens were collected as they displayed in the water column. They were captured by a diver using either (1) individually sealed and numbered discrete net traps (1 per display), or (2) repeated sweeps of a net through many displays. Specimens were collected in coral reef habitats (including sea grass, reef crest, spur and groove, walls, patch reefs, large and small sandy areas) of the San Blas Islands, Panama; Roatan, Honduras; Ambergris Cay, Carrie Bow Cay, and Glovers Reef, Belize; Grand Cayman; Discovery Bay and Long Bay, Jamaica; San Salvador Island, Bahamas; SW Puerto Rico; St. Croix, Virgin Islands; Barbados; Curacao; and Bonaire (Table 1).

### Morphological Characters

Morphological terminology follows Morin & Cohen (1988). Character states (Appendices 2 and 3) were determined from specimens or published descriptions (Table 2). Not all described species of *Vargula* have been included (there are 34 species described in the genus).

The completed cladistic analysis placed all of the published Caribbean species of *Vargula* into monophyletic groups, each defined by its own suite of diagnostic character states used in this analysis. We then used those diagnostic character states to assign unpublished (but partially described) Caribbean species in our collection, to the monophyletic groups of the cladistic analysis. The resulting information allows a preliminary estimate of the size, geographical distribution and luminescent behavioural characteristics of the groups.

### Behavioural Characters

Luminescent display behaviour is described in Morin (1986), Cohen & Morin (1986, 1989) and Morin & Cohen (1988, and unpublished data). Only the

Table 1. Distribution of the known *Vargula* species (described and undescribed) from the Caribbean by clade and geographical location. *V. magna* from the Gulf of Mexico is not listed.

| GROUP: LOCATION | F | A T | Z | H | U | Total |
|---|---|---|---|---|---|---|
| Panama | 2 V. graminicola V. shulmanae | 0 | 2 V. kuna V. mizonomma | 2 V. psammobia V. ignitula | 5 V. contragula V. lucidella V. micamacula V. noropsela V. scintilla | 11 |
| Honduras | 1 WLU* | 1 CMU | 0 | 2 RD, ODH | 3 HCO, IR, VSD | 7 |
| Belize | 3 VFF, LSU, GF | 1 M | 0 K | 3 MSH, BSD | 4 BCO, MWU ZZU, XD | 11 |
| Jamaica | 4 V. harveyi V. parasitica VF, VG | 3 Q, X V | 0 | 4 Z, WD OBD, A | 0 | 11 |
| Grand Cayman | 0 | 1 GSU | 0 | 0 | 0 | 1 |
| Puerto Rico | 0 | 2 WHU, BLU | 0 | 0 | 0 | 2 |
| Virgin Islands | 0 | 3 V. bullae D, SLD | 0 | 0 | 0 | 3 |
| Bahamas | 1 SWD | 2 OLD, U | 0 | 0 | 0 | 3 |
| Barbados | 0 | 2 TB, FWU | 0 | 0 | 0 | 2 |
| Curacao | 0 | 2 CSU, LD | 0 | 0 | 0 | 2 |
| Total | 11 | 17 | 2 | 11 | 12 | 53 |

* Letter abbreviations = unpublished species in our collections.

following very obvious components of the bioluminescent display behaviour were used in this preliminary comparison with the morphological analysis: display direction, general duration of pulses and regularity of spacing between pulses. Congruence between behavioural and morphological characters was examined by mapping the behavioural characters onto a morphological cladogram.

## Systematic Analysis

Cladistic analysis was performed with PAUP version 2.4.1 (Swofford, 1985; Platnick, 1987, 1988;

Fink, 1986; see Appendix 1, Glossary), using 28 taxa and 53 morphological characters (Appendices 2 and 3). This program offers a variety of algorithms and options. All of our analyses employed:

1) characters weighted to scale;
2) 10 trees retained at each step of analysis;
3) global branch swapping;
4) multiple parsimony;
5) additional consensus tree computed.

Other options were varied so that our analysis employed 3 different methods:

1) states were unordered for all characters and the cladogram was rooted with a single outgroup, *Doloria levinsoni*;

2) all states were unordered and the cladogram was rooted with multiple outgroups (6 taxa not presently assigned to *Vargula*);

3) states were hypothetically ordered in 27 characters and unordered in 26 characters, and the cladogram was rooted with *D. levinsoni*.

**Outgroups**

*Doloria levinsoni* was chosen as the outgroup for single outgroup analysis because:

1) from outgroup comparisons (below), we propose the hypothesis that it has the most plesiomorphic character states of the *Cypridina* Genus Group;

2) most character states used in our analysis have been described for *D. levinsoni*.

The following character states also occur in outgroups (Genus Groups, tribe, subfamily other than the *Cypridina* Genus Group), so we propose the hypothesis that the states are plesiomorphic in *D. levinsoni* compared to the states of the same characters in all or most other members of the *Cypridina* Genus Group:

1) The male endopodite of the 2nd antenna is 3-jointed and reflexed in *Doloria*. This character state is also present in the *Monopia* and *Codonocera* Genus Groups (2 of the 3 other Genus Groups), Gigantocypridinini (the other tribe), Azygocypridininae (the other subfamily), and other families of Myodocopida, but absent from all other genera of the *Cypridina* Genus Group.

Table 2. Sources used for determination of character states of each taxon used in cladistic analysis.

Source:

Type series of specimens:

*Vargula shulmanae* Cohen & Morin, 1986
*V. graminicola* Cohen & Morin, 1986
*V. parasitica* (Wilson, 1913)
*V. tsujii* Kornicker & Baker, 1977
*V. contragula* Cohen & Morin, 1986
*V. scintilla* Cohen & Morin, 1989
*V. noropsela* Cohen & Morin, 1989
*V. lucidella* Cohen & Morin, 1989
*V. micamacula* Cohen & Morin, 1989
*V. ignitula* Cohen & Morin, 1989
*V. psammobia* Cohen & Morin 1989
*V. kuna* Morin & Cohen, 1988
*V. mizonomma* Morin & Cohen, 1988
*V. norvegica* (Baird, 1860)

Other specimens:

*V. harveyi*\* Kornicker & King, 1965
*V. hilgendorfii*\*\* (Müller, 1890)
*V.* new species M\*
*Cypridina americana*\*\* (Müller, 1890)

Published descriptions:

*Vargula magna* Kornicker, 1984
*V. bullae* Poulsen, 1962
*V. plicata* Poulsen, 1962
*V. sutura* Kornicker, 1975
*V. hamata* Kornicker, 1975
*Sheina orri* Harding, 1966 (Kornicker, 1986a)
*Bathyvargula* (Poulsen, 1962)
*Cypridinodes* (Poulsen, 1962; Kornicker, 1975)
*Doloria levis* Skogsberg, 1920
*D. levinsoni* Kornicker, 1975

\* = In authors' collection (Los Angeles County Museum of Natural History); \*\* = on loan to authors (see Acknowledgements).

2) The basis of the mandible of *D. levinsoni* bears 4-5 dorsal bristles (1-2 of these are proximal to the midpoint of the dorsal margin). The Azygocypridininae have 3-14 dorsal bristles (all distal except in *A. rudjakovi* Kornicker, 1970, 2 are proximal), while all other cypridinids have only 3 such bristles.

3) The ventral mandibular basis of *D. levinsoni* bears a total of 6 a- and b-bristles. In the Azygocypridininae the number of a- and b-bristles is 5-11, but in all other cypridinids there are fewer than 6 (except *Rugosidoloria serrata* Kornicker, 1975).

4) The male of *D. levinsoni* has 25 terminal comb teeth on the 7th limb. In the Gigantocypridinini the number of terminal comb-teeth present in the male is 36-140, but there are less than 25 in all other cypridinids.

5) In males of *D. levinsoni*, the 7th limb has about 74 bristles of which 17-20 are distal (not terminal) comb bristles. In males the 7th limb has 18-19 (3 distal) in the *Monopia* Genus Group, 12 (1 distal) in the *Pterocypridina* Genus Group, 10-16 (2-4 distal) in the *Codonocera* Genus Group, about 120-300 in the Gigantocypridinini, 0 to about 150 in the Azygocypridininae, but males of all other members of the *Cypridina* Genus Group have fewer bristles (except *Cypridinodes favus* Brady, 1902).

6) The furca of *D. levinsoni* has 11 claws. The number of claws is 9-12 in the *Codonocera* Genus Group, 6-10 in the *Pterocypridina* Genus Group, 10-15 in the Gigantocypridinini, about 20 in the Azygocypridininae, but less than 11 in all other members of the *Cypridina* Genus Group (except *Skogsbergia costai* Kornicker, 1974, *Paradoloria dorsoserrata* (Müller, 1908), *Vargula hilgendorfii* and ?*V. danae* (Brady, 1880)).

7) The 6th limb of *D. levinsoni* has 24-29 end bristles. The 6th limb has about 24-30 in the *Codonocera* Genus Group, 26 in *Rugosidoloria*, 24-27 in the *Monopia* Genus Group, 27-30 in the Gigantocypridinini, 50 in Azygocypridininae, but less than 20 terminal bristles in all *Cypridina* Genus Group genera except *Macrocypridina*, some *Cypridinodes* species, *Paradoloria australis* Poulsen, 1962, *Paradoloria pellucida* (Kajiyama,

1912), *Paravargula hirsuta* (Müller, 1906b) and *Skogsbergia hesperida* (Müller, 1906a).

A comparison of characters between tribes, subfamilies, families, and orders, leads us to propose the hypothesis that the following taxa have the following character states which are plesiomorphic within the Cypridinidae (or Cypridininae) as indicated in parentheses:

1) The Azygocypridininae are the only members of the Myodocopida with a claw on the anterior edge of the furca, a state shared with the Halocyprida (and therefore a plesiomorphic state in the Cypridinidae); all furcal claws are ventral in the Cypridininae (apomorphic in the Cypridinidae).

2) Like most of the Philomedidae, the Azygocypridininae have 4-5 beta-bristles on the 4th limb (therefore plesiomorphic in the Cypridinidae), and the Gigantocypridinini have 3-4 (plesiomorphic in the Cypridininae). There are fewer beta-bristles in the other myodocopid families, but the Cylindroleberididae have a 4th limb highly modified for comb-feeding, and the other 2 families are probably less closely related to the Cypridinidae (Kornicker, 1986b). None of the Cypridinini have more than 3 beta-bristles on the 4th limb.

*Cypridina*, *Cypridinodes*, *Bathyvargula* and *Sheina* were included in the analysis because these are genera in the *Cypridina* Genus Group which share synapomorphies with *Vargula* (which lacks autapomorphies; see below).

1) *Cypridina* and *Vargula* are the only ostracods which are known to produce bioluminescence from the upper lip.

2) *Cypridinodes*, *Bathyvargula*, *Sheina* and *Vargula* are the only myodocopids with both a sclerotized, ventral, subterminal finger on the 2nd endopodial joint of the mandible and a pair of long tusks on the upper lip.

3) *Sheina* and some species of *Vargula* are the only cypridinids with a long ventroterminal mandibular bristle with a bulbous base.

Autapomorphies and extremely homoplastic characters were omitted from the computer analysis. Autapomorphies were not used in the computer analysis because:

1) they increase computer time, and

2) they are not shared between taxa, thus they

can show nothing about relationships between such taxa.

Autapomorphies for the various genera are: *Cypridina* (about 25 species):

1) valve shape oval, only moderately high (about 50-60% of length), with truncated (anterior straight or concave) rostrum and a short bluntly pointed keel (caudal process) located in middle of posterior margin;

2) upper lip with unique configuration of 2 anterior unpaired and 2 posterior paired tusks;

3) 4th limb with broadly based (proximal width > 1/2 length) exopodite abruptly narrowing distal to proximal bristle;

4) 5th limb with only 7-8 a- plus b-bristles on 2nd joint; and possibly

5) male 1st antenna with very long c- and f-bristles (also occurring in *Gigantocypris*).
*Cypridinodes* (about 9 species):

1) valve shape oval and high, usually with rather narrow snout-like rostrum with slight anterior concavity near tip and a small bluntly (truncated) pointed keel near middle of posterior margin;

2) upper lip with uniquely shaped tusks with longitudinal hair rows and toothed proximal posterior lobe;

3) mandible with only 2 well-developed claws;

4) 4th limb with short thin exopodite (1/8-1/3 length of 1st endopodial joint), and relatively long (length about 4X distal width) 1st endopodial joint (lacking cutting tooth and with curved flat claw-like tip on outer alpha-bristle).

*Bathyvargula* (probably 2 species) shares with *Metavargula* (probably 2 species) (Kornicker, 1979):

1) male 1st antenna with very broad flat filaments on sensory bristle of 5th joint and no rows of small suckers on distal filaments of b- and c-bristles;

2) mandible with sclerotized ventral finger of 2nd joint bearing marginal spines or teeth.

*Sheina* (1 species): Mandible with reduced male coxal endite (process with only 1-3 spines) and terminal endopodial bristle with bulbous base and terminal broad flattened pad.

Homoplastic characters have convergent or parallel states in groups lacking a common ancestor, and vary frequently not only within *Vargula*, but also within other cypridinid genera. Specific diagnostic characters, many quite homoplastic, of Caribbean signalling species are listed elsewhere in their published descriptions.

## RESULTS

### Morphological Analysis of Described Species

The three methods that were employed to manipulate the data in the PAUP program produced similar results but with some differences:

1) Using method 1 a single shortest tree (cladogram) (93.445 steps = character state changes; see Appendix 1: Glossary) was produced by combining unordered states for all characters and rooting with *Doloria levinsoni* as the outgroup (Fig. 1). In this cladogram, *Vargula norvegica*, the type species of *Vargula*, is the sister group to a clade including all the Caribbean species plus *V. tsujii*, *V. magna*, *V. plicata*, *Sheina orri*, *Cypridina americana*, *Bathyvargula* and *Cypridinodes* (i.e. all of the taxa to the left of *V. norvegica* in Fig. 1).

2) Method 2 (using multiple outgroups with unordered characters) resulted in a tree (93.445 steps) with midpoint rooting; this cladogram is not described further below.

3) Method 3 (using 27 hypothetically ordered and 26 unordered characters, and a single outgroup) produced 2 longer trees (both 96.979 steps).

The trees produced by all three methods assigned the Caribbean signalling species to 2 well supported and only distantly related monophyletic groups: (1) the exclusively Caribbean F-group and (2) the almost exclusively Caribbean A-group (Figs. 1 and 2). The A-group was composed of exclusively Caribbean monophyletic subgroups T, Z, H and U (with possible further subdivisions of the U-group), plus *Vargula tsujii* from California and *V. magna* from the Gulf of Mexico and the SE coast of the United States. Relationships among subgroups of the A-group differed slightly between the trees with unordered versus some ordered characters. In the ordered tree (method 3)

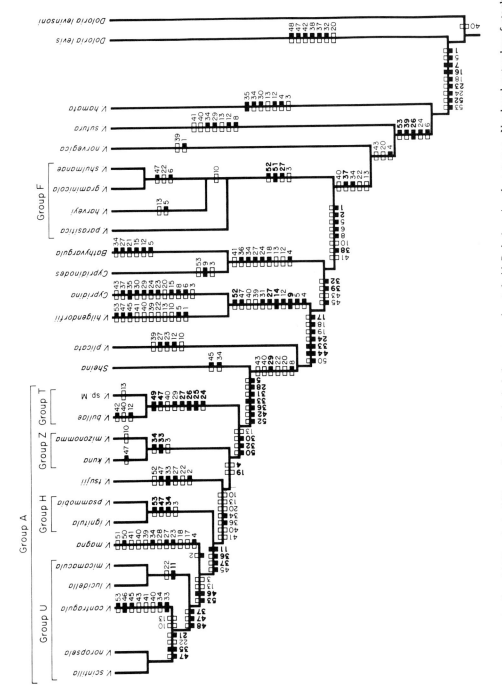

Fig. 1. Cladogram resulting from PAUP analysis of 53 unordered morphological characters and rooted with *D. levinsoni* as the outgroup. Numbered rectangles refer to changes in states of characters listed in Appendix 2; solid rectangles = no homoplasy; open rectangles = much homoplasy; half-filled rectangles = some homoplasy. Bold faced numbers refer to synapomorphies that best support each clade and were used in the analysis in Fig. 2. *Sheina = Sheina orri, Cypridina = Cypridina americana, V.* sp. M = *Vargula* new species M (Belize).

the U group was the sister group to the H plus M and T groups (and the H-group was the sister group of the M plus T-groups), while *V. magna* was the sister group of the rest of the A-group.

There were numerous homoplasies, but of varying degree, i.e. some were only slightly (half closed rectangles in Fig. 1) and others strongly inconsistent (open rectangles in Fig. 1) with the cladogram. For example, character no. 42 provides a good synapomorphy (reduction of bristles on inner 3rd joint on the 5th limb) for the A-group, but is somewhat homoplastic because the reduction also occurs in *Doloria levis* and is reversed in *V. magna*. Therefore, 2 grades of homoplasies are shown in Fig. 1: slight and strong. Only changes in character states are shown in Fig. 1. Because autapomorphies were excluded, lack of character state changes on terminal branches indicates agreement with the clade to which that branch belongs. Conversely, numerous character state changes on terminal branches indicate divergence from the clade to which that branch belongs.

The 3 methods produced slightly different arrangements of non-Caribbean taxa. The longer tree (method 3, not figured), constructed with some ordered characters differed from the shortest tree (method 1, Fig. 1) in the following ways: *Vargula hamata* and *V. sutura* exchanged positions; there was an unresolved trichotomy between *Cypridinodes*, *Bathyvargula*, and a clade consisting of the A-group plus *Sheina*, *V. plicata*, *Cypridina*, *V. hilgendorfii* and *V. magna*; *Sheina orri* was the sister group to *V. plicata* (and together they changed places with *V. hilgendorfii* and *Cypridina americana*).

## Assignment to Cladogram of Partially Described Caribbean Species

By applying the diagnostic synapomorphies (bold numbers in Fig. 1, asterisks in Appendix 2) to the 39 only partially described new luminescent species we have collected from the Caribbean, we have been able to assign each of the partially described species to one of the various Caribbean clades (F, A, T, Z, H and U) (Fig. 2). Table 1 shows the designation of all described and partially

described species from the Caribbean by clades and geographical location.

## Comparison with Major Bioluminescent Display Characters

Species with vertical shortening displays (both upward and downward) were found in all Caribbean species groups (Fig. 2). Species with rapid flashing displays belong only to the F-group: *V. graminicola* and 3 undescribed species. Species with evenly spaced lateral displays belonged only to the U- and H-groups: *V. noropsela*, *V. psammobia*, *V. contragula*, and 9 undescribed species from Honduras, Belize, Jamaica and the Bahamas.

## DISCUSSION

We propose the hypothesis that the genus *Vargula*, as presently composed and defined, is polyphyletic. *Vargula norvegica* is the type species of the genus. None of the species of *Vargula* included in this analysis were assigned to a monophyletic group which both included *V. norvegica* and excluded other genera. Fig. 1 shows that all of the species used in the analysis, except the 2 species of *Doloria*, are grouped into one large clade; this clade is the large branch of the tree to the left of *Doloria*. This all-embracing clade is defined by 9 synapomorphies, i.e. changes in the states of characters 1, 5, 7, 16, 18, 23, 24, 52, 53. This clade includes *Bathyvargula*, *Cypridinodes*, *Cypridina* and *Sheina*, as well as the *Vargula* species. *V. norvegica* is also part of a smaller, poorly defined (by only 3 homoplastic characters: 4, 20, 43) clade which includes all of the species to the left of *V. norvegica* in Fig. 1, including the other 4 genera. *V. norvegica* is not part of any of the separate clades (individual branches) to the left of it in Fig. 1. The other genera incorporated in the analysis all have unique diagnostic characters (probable synapomorphies listed in the Material section above) not shared with *V. norvegica* nor with any species of *Vargula*. Thus, *Vargula* as a genus should be broken up.

We further propose the hypothesis that the Caribbean bioluminescent signalling species

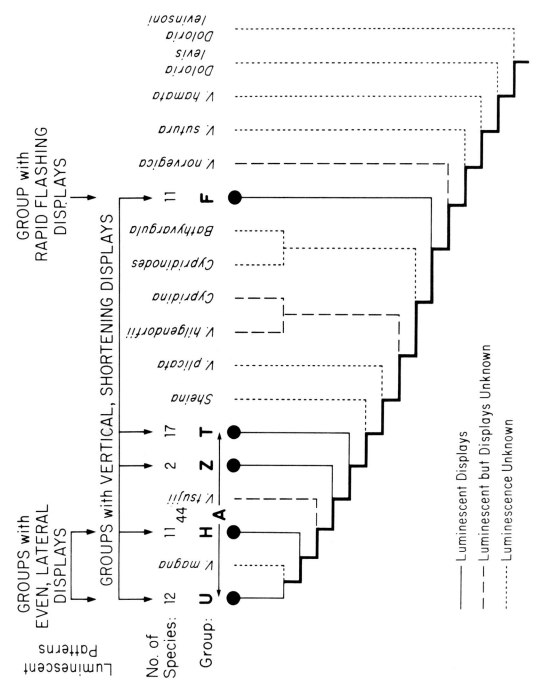

Fig. 2. Cladogram (based upon cladogram of Fig. 1) showing Caribbean monophyletic species groups, total number of species within each clade, and luminescent behavioural characters mapped onto morphological cladogram.

belong to 2, only distantly related monophyletic clades (F and A), that are not part of the genera *Vargula* (as traditionally conceived), *Cypridinodes Bathyvargula*, or *Cypridina*. Relationships within the A-group, as well as between the Caribbean groups and non-Caribbean taxa varied slightly in the 3 analyses. *V. tsujii* (NE Pacific) is the only strictly non-Caribbean species placed within the A-group. *V. magna* occurs within the greater Caribbean area (Gulf of Mexico and Carolinas). Although *V. magna* may be related to that Caribbean group, it had 7 strongly homoplastic character states when included in that group (Fig. 1). *V. contragula* also differed from its clade in many character states.

Monophyly of the various Caribbean groups was strongly supported in all of the cladograms (Figs 1 and 2). The basal tree divisions of all other taxa from *Doloria* and from *Doloria* plus *Vargula hamata*, and the clade encompassing Group-A, *Sheina*, and *V. plicata* were also well supported. Assignment of the Caribbean signalling species to other or new genera is postponed until we complete descriptions of more of the Caribbean species which have the diagnostic characters of these groups.

### Bioluminescent Display Behaviour

We propose the hypothesis that 2 behavioural character states follow the ancestry of the Caribbean signalling ostracods and hence that those characters have been constrained by evolutionary history (Fig. 2). Rapid flashing is restricted to the F-group and evenly spaced lateral patterns to the H- and U-groups. However, vertical shortening displays occur in all clades, which would suggest that this is a plesiomorphic character state. The A-group also includes 4 species which are not known to signal, but 3 of these have not been observed in their natural habitats: *V. parasitica* (known to be bioluminescent), *V. magna* and *V. bullae*. *Vargula tsujii*, from California, is known to be bioluminescent, but has not been observed to display. Based on their location between the bioluminescent Caribbean clades (Figs 1, 2), we predict that species of *Sheina*, *Bathyvargula*, and *Cypridinodes* will be bioluminescent or will have secondarily lost luminescence.

### Note on Distribution

Geographical distribution varies, but overlaps, among the monophyletic Caribbean groups (Table 1). A biogeographical analysis will follow completion of our taxonomic studies. Results to date indicate that the monophyletic F-group is exclusively Caribbean, especially the western sector. The monophyletic A-group contains only Caribbean species plus *Vargula tsujii* from California and Baja California. The H- and U-Groups are found primarily in the western Caribbean. The T-Group is the most widespread. Highest diversity occurs along the continental margins (mainland and nearshore islands) or larger islands. The lowest diversity is found in the eastern Caribbean and in small islands. Apparent sibling species pairs occur sympatrically but in different ecological niches, for example in Panama: *Vargula shulmanae* and *V. graminicola*, *V. kuna* and *V. mizonomma*, *V. psammobia* and *V. ignitula*, *V. noropsela* and *V. lucidella*, *V. micamacula* and *V. lucidella*.

### ACKNOWLEDGEMENTS

We thank the Los Angeles County Museum of Natural History for use of their facilities. We thank Shin-Ichi Hiruta (Hokkaido University of Education), Sheila Halsey and Geoffrey Boxshall (British Museum of Natural History), and Brad Myers for loan of specimens of *Vargula hilgendorfii*, *Cypridina norvegica*, and *C. americana*, respectively. Margaret Kowalczyk prepared the figures. This work was supported by the National Science Foundation (BSR-87-00324), the National Geographic Society and the American Philosophical Society.

# APPENDIX 1

## Glossary

This glossary is appended at the request of both referees and the editors:

**apomorphy** = derived character state. Of a homologous pair of character states, the apomorphic state is the state evolved from the plesiomorphic state. It is a single homologous character state that is present only in certain taxa and defines all of those particular taxa having it as a monophyletic group (e.g., 2 pairs of antennae in Crustacea; hairs in Mammalia; probably the unique [3 cups with tapetal layer] medial eye in Maxillopoda; a worm-like 7th limb in Myodocopida).

**autapomorphy** = derived character state unique to a single monophyletic taxon; it can define that taxa, but cannot show shared relationships with other taxa (e.g., a defining character state of *Vargula contragula* is the bulge on the anteroventral margin of the shell; since all of the other species in our analysis lack that character state, that character state cannot help in analyzing relationships of those species to *V. contragula*).

**character** = structure or function used to describe, compare and combine or separate taxa. It is expressed as states (e.g., the character 'colour' may be expressed as the states 'red' and 'blue').

**character state** = expression of a single homologous character (e.g., 'worm-like' and 'leg-like' are states of the homologous character '7th limb' of Ostracoda).

**clade** = monophyletic group (see below), a branch on a cladogram.

**cladistic analysis** = analysis of genealogical relationships of taxa that groups the taxa on the basis of uniquely shared character states.

**cladogram** = branching diagram (tree) resulting from cladistic analysis; pattern of distribution of shared derived characters.

**consensus tree** = cladogram combining common features of 2 or more trees (in PAUP, equally parsimonious trees); branching points that differ between trees are presented as unresolved polychotomies (i.e. more than 2 branches at the same disputed node).

**global branch swapping** = technique used to increase probability of finding the shortest tree in a cladistic analysis of a large data set; successive pairs of branches of the tree are exchanged.

**homoplasy** (homoplastic character states) = parallelism or convergence. 'A character [state] found in 2 or more species is homoplasous (nonhomologous) if the common ancestor of these species did not have the character in question, or if one character [state] was not the precursor of the other' (Wiley, 1981). Parallelism is the origin of seemingly similar character states more than once in closely related taxa (probably evolving twice from a similar genetic substrate; e.g., possibly secretion of bioluminescence from the upper lip of *Cypridina* and *Vargula*); convergence is the origin of seemingly similar character states more than once in distantly related taxa (probably evolving twice from different genetic substrates; e.g., eyes of cephalopods and fishes).

**ingroup** = a group of taxa whose phylogenetic relationships one wishes to analyze.

**midpoint rooting** = see root.

**monophyletic group** = 'a group of species that includes an ancestral species (known or hypothesized) and all of its descendants' (Wiley, 1981); a taxon defined by the possession of at least one character state unique to that group alone (examples of groups generally considered to be monophyletic are: Vertebrata, Crustacea, Ostracoda, Myodocopida, Cypridininae, *Gigantocypris*).

**multiple parsimony** = technique used to increase the probability of finding shortest tree in a cladistic analysis of a large data set; the computer program retains and considers more than one equally short tree.

**ordered characters** = characters in a cladistic analysis that are arranged in a transformation series, i.e. the states are arranged in an evolutionary order (e.g., if 'worm-like' is designated as state 0, 'leg-like' as state 1, and 'absent' as state 2 of the character '7th limb' and these states are ordered, the cladistic analysis assumes that absent 7th limbs evolved from leg-like 7th limbs which evolved from worm-like 7th limbs).

**outgroup** = taxon (or taxa) not belonging to ingroup (taxa that are target of cladistic analysis) and used (1) to root cladogram (tree) and (2) to polarize character states. Outgroup analysis is a method used to determine which states are apomorphic (derived) at the taxonomic level of the ingroup. Outgroup and ingroup must be related closely enough that they share clearly homologous characters. A character state that occurs in both outgroup and ingroup is plesiomorphic (primitive) in the ingroup, whereas a state of the same character that occurs only in the ingroup is apomorphic in the ingroup. (For example the Dantyinae could be used as the outgroup for an analysis of the Sarsiellinae, and the Platycopida might be used as an outgroup to the Podocopida.)

**paraphyletic** = group (taxon) composed of taxa sharing the same *immediate* common ancestor but not including all of the taxa sharing that common ancestor. 'The group is based on the common possession of plesiomorphic character' states (Humphries and Parenti, 1986). It is not defined by any synapomorphies (unique shared character not shared with any other taxon) (examples of groups generally considered to be paraphyletic are: reptiles, invertebrates, a definition of Ostracoda that excluded the Cypridacea).

**parsimony** (as an assumption in cladistic analysis) = assumption that evolution proceeded by fewer rather than more changes; therefore the shortest cladogram is the one most likely to show true phylogenetic relationships between taxa. The shortest cladogram (tree) is the one requiring the least number of character state changes (i.e. number of steps). [Note: however, after generating cladograms, character state changes required by the shortest tree and those required by longer trees can be compared and evaluated on the basis of biological probability (e.g., chances that a character state evolved more than once or was lost more than once), and a longer cladogram chosen as the best hypothesis of taxonomic relationships.]

**PAUP** = see Phylogenetic Analysis Using Parsimony.

**Phylogenetic Analysis Using Parsimony** = a computer software program for doing cladistic analyses, written by and available from Swofford, D. L., 607 East Peabody Drive, Champaign, Illinois 61620. This program is evaluated and recommended in Platnick, N. I. 1988. Programs for quicker relationships, *Nature*, London, **335**, 310; Platnick, N. I. 1987. An empirical comparison of microcomputer parsimony programs, *Cladistics*, **3**, 121-144; Fink, W. 1986. Microcomputers and phylogenetic analysis, *Science*, New York, **234**, 1135-1139.

**plesiomorphy** = primitive character state. 'Of a pair of homologues, the plesiomorphic character [state] is the [state] that arose earlier in time and gave rise to the later, apomorphic, character' state (Wiley, 1981). In outgroup analysis the plesiomorhic character state is present in both the ingroup and outgroup (e.g., 2 pairs of antennae is plesiomorphic in Podocopa, Myodocopa, all Ostracoda and Copepoda, etc. [i.e. plesiomorphic *within* the Crustacea, although apomorphic for the Crustacea compared to other Arthropoda]; hairs in dogs and kangaroos [plesiomorphic within mammals]; the medial eye in Podocopa and Myodocopa [plesiomorphic within Ostracoda], the worm-like 7th limb in Cypridinidae and Philomedidae [plesiomorphic within Myodocopida]).

**polyphyletic group** = group (taxon) composed of taxa not sharing the same *immediate* common ancestor and excluding the ancestor that is common to all of the taxa included in that group (examples of groups generally considered to be polyphyletic are: Protozoa; an ostracod taxon that is composed of only Cypridacea and Cyprinidacea; a grouping of all colonial animals or all winged animals or all shelled animals or all worm-like animals; un-natural groups).

**root** = taxon hypothesized to be more primitive and to have evolved earlier than all other taxa considered in a cladistic analysis and therefore placed in the most basal position in a cladogram (tree). See outgroup. In a PAUP, midpoint rooting occurs when the cladistic analysis places the designated outgroup(s) within the ingroup; this indicates that the ingroup is not a monophyletic group (e.g., midpoint rooting would probably occur if *Cypridina* and *Cypris* were analyzed as the ingroup with *Pterocypridina* and *Pontocypris* designated as outgroups; or if bats and bees were analyzed as the ingroup with buffaloes, bison and beetles as the outgroups).

**shortest tree** = see parsimony.

**sister groups** = opposing branches of a cladogram, 'two taxa that are more closely related to each other than either is to a third taxon' (Humphries & Parenti, 1986) (e.g., Myodocopa and Podocopa within the living Ostracoda [or Crustacea], if the Platycopida are included within the Podocopa).

**steps** = character state changes in a cladistic analysis; fractions result if all characters are weighted equally, but some characters have more states than others. The shortest cladogram (tree) is the one requiring the least number of character state changes to show relationships between taxa

(see parsimony).

**synapomorphy** = shared apomorphy; a unique derived character state 'shared by and defining a group of organisms within the context of a larger group' (Humphries & Parenti, 1986). The synapomorphy originated in the immediate common ancestor of the group defined.

**unordered characters** = characters in a cladistic analysis that are not arranged in a transformation series, i.e. the states are not arranged in an evolutionary order (e.g., if 'worm-like' is designated as state 0, 'leg-like' as state 1, and 'absent' as state 2 of the character '7th limb', unordered means that the cladistic analysis does not assume that absent 7th limbs evolved from leg-like 7th limbs which evolved from worm-like 7th limbs).

**weighted to scale** = all characters are weighted equally so that characters with more states have the same weight as those with fewer states (this results in fractions in the number of steps in a cladogram).

## APPENDIX 2

Characters and character states used in the cladistic analysis of *Vargula* species and related genera.

| Characters | | Character States | Ordered? |
|---|---|---|---|
| **Valve:** | | | |
| *1. | male L:H | 0  1.20-1.40mm<br>1  1.45-1.65<br>2  1.70-1.90 | + |
| *2. | male keel | 0  >half valve height<br>1  *circa* half valve height<br>2  <*circa* half valve height | - |
| 3. | spines on ridge<br>of caudal infold | 0  absent<br>1  male only<br>2  both sexes | - |
| **Furca:** | | | |
| *4. | which claws fused<br>to lamella | 0  none<br>1  2nd<br>2  2nd + 4th<br>3  male 2nd, 3rd, 4th<br>   female 2nd, 4th | + |
| *5. | total no. claws | 0  11-12<br>1  9<br>2  8<br>3  6-7 | + |
| **ANTERIOR OF BODY** | | | |
| *6. | sclerotized projections | 0  1 large rounded<br>1  1 large rounded with dent + 1<br>   tiny rounded & more ventral<br>2  1 large rounded + 1-2<br>   pointed & more ventral<br>3  none | - |
| **UPPER LIP** | | | |
| **7. | paired long tusks | 0  absent<br>1  present | + |
| *8. | anterior unpaired<br>area | 0  not ventrally extended<br>1  ventrally extended<br>2  unique multiple extensions | - |
| **9. | tusk expansions | 0  none<br>1  additional tusk-like<br>   long anterior nozzles<br>2  larger tusks with toothed<br>   posterior basal lobe | - |
| 10. | hairs on tusks | 0  tusk absent<br>1  few short<br>2  few long | - |

Appendix 2. Characters and character states used in the cladistic analysis of *Vargula* species and related genera - (continued)

| Characters | Character States | | | | | Ordered? |
|---|---|---|---|---|---|---|
| | 3 | some long | | | | |
| | 4 | many long | | | | |
| **COPULATORY ORGAN** | | | | | | |
| **11. posterior basal | 0 | absent | | | | - |
| tubercles | 1 | few rounded | | | | |
| | 2 | many pointed | | | | |
| **ANTENNA 1** | | | | | | |
| *12. sensory filaments | 0 | 10 | long prox. + 2 | shorter | | - |
| on bristle of 5th joint | 1 | 9 | long prox. + 2 | shorter | | |
| | 2 | 10 | long prox. + 3 | shorter | | |
| | 3 | 11 | long prox. + 2 | shorter | | |
| | 4 | 11 | long prox. + 2 | fatter | | |
| | 5 | 10-11 | long prox. + 3-4 | shorter | | |
| | 6 | 13 | long prox. + several shorter | | | |
| **ANTENNA 2 EXOPODITE** | | | | | | |
| 13. average number of | 0 | 15-25 | | | | - |
| ventral spines on | 1 | 11.5-14 | | | | |
| bristle of 2nd joint | 2 | 8.5-11 | | | | |
| | 3 | 5.5-8 | | | | |
| | 4 | 0 | | | | |
| **14. number of | 0 | 4 | | | | + |
| bristles on 9th joint | 1 | 3 | | | | |
| **ANTENNA 2 ENDOPODITE** | | | | | | |
| *15. number of joints | 0 | 3 | | | | + |
| | 1 | 1 | | | | |
| **16. male reflexed | 0 | yes | | | | + |
| | 1 | no | | | | |
| **MANDIBLE** | | | | | | |
| *17. terminal b-bristle | 0 | no bulbous base | | | | + |
| | 1 | with bulbous base | | | | |
| *18. terminal b-bristle | 0 | without proximal teeth/spines | | | | + |
| | 1 | with proximal teeth/spines | | | | |
| | 2 | with proximal bulb, but no teeth/spines | | | | |
| *19. distal d-bristle | 0 | short spines | | | | + |
| of basis | 1 | longer hairs only | | | | |
| 20. longest bristle | 0 | without proximal hairs | | | | - |
| of 1st joint | 1 | with proximal hairs | | | | |
| *21. c-bristles of basis | 0 | no hairs | | | | + |
| | 1 | some with hairs | | | | |
| 22. no. of dorsal bristles | 0 | 2 | | | | - |
| of 2nd joint reaching | 1 | 1-2 just or almost | | | | |
| beyond joint | 2 | none | | | | |

Appendix 2. Characters and character states used in the cladistic analysis of *Vargula* species and related genera - (continued).

| Characters | | Character States | Ordered? |
|---|---|---|---|
| *23. | sclerotized ventral finger on 2nd joint | 0   none<br>1   thin<br>2   thick | + |
| *24. | finger shape | 0   none<br>1   straight, pointed tip<br>2   slightly curved, narrow toothed tip<br>3   slightly curved, rounded tip<br>4   very curved, rounded tip | - |
| **25. | c- and d-bristles of basis | 0   not close<br>1   close together | + |
| **26. | portion of dorsal margin of basis lacking bristles | 0   less than half<br>1   proximal 2/3-3/4<br>2   proximal 4/5 | + |
| *27. | terminal claw arrangement and relative size | 0   LV=MV>D<br>1   D=LV<MV<br>2   D=MV<LV<br>3   D>LV=MV<br>4   MD>LD>LV & LV with bulge<br>5   LD=LV>MV & MV thin<br>6   big M>LV & D | - |
| *28. | tip of dorsal (D) claw | 0   evenly curved<br>1   bent | + |
| **4TH LIMB:** | | | |
| *29. | no. a-bristles | 0   4<br>1   3<br>2   5 | - |
| *30. | no. b-bristles | 0   3<br>1   2 | - |
| *31. | no. alpha-bristles | 0   2<br>1   1 | + |
| *32. | no. beta-bristles | 0   3<br>1   2<br>2   1 | + |
| **33. | size of teeth on b- & d-claws | 0   long with pointed tips<br>1   large with rounded tips<br>2   medium thin on b-, medium small on d-claws<br>3   medium-small on b-, small on d-claws<br>4   small | - |
| *34. | size and shape of tooth on 1st joint | 0   large & trifid<br>1   medium sized & bifid<br>2   large & bifid<br>3   very large & bifid | - |

Appendix 2. Characters and character states used in the cladistic analysis of *Vargula* species and related genera - (continued).

| Characters | | Character States | | Ordered? |
|---|---|---|---|---|
| | | 4 | large & with 4 cusps | |
| | | 5 | low & without cusps | |
| | | 6 | apparently absent | |
| **35. | no. c-bristles | 0 | 3 | + |
| | | 1 | 2 | |
| | | 2 | 4 (1 short) | |
| | | 3 | 4 long | |
| *36. | relative lengths of alpha- and beta-bristles | 0 | alpha>beta | + |
| | | 1 | alpha=beta | |
| | | 2 | beta>alpha | |
| *37. | ratio lengths cl- to unringed claw-like b-bristle | 0 | 64-75% | - |
| | | 1 | 100-150% | |
| | | 2 | <50% | |
| | | 3 | 175-200% | |
| **5TH LIMB:** | | | | |
| *38. | 4th & 5th joints fused? | 0 | no | + |
| | | 1 | yes | |
| *39. | no. of bristles on 4th & 5th joints | 0 | 4th=5-6, 5th=2 | + |
| | | 1 | 4th=3-4, 5th=2 | |
| | | 2 | 5 on fused joints | |
| | | 3 | 4 on fused joints | |
| | | 4 | 3 on fused joints | |
| | | 5 | 2 on fused joints | |
| 40. | no. of b-bristles on 2nd joint | 0 | 7 | - |
| | | 1 | 8 | |
| | | 2 | 8 or 9 | |
| | | 3 | 12 | |
| | | 4 | 6 | |
| | | 5 | 3 | |
| 41. | shape of distal tooth of protopodite | 0 | long with round end & nonterminal bumps | - |
| | | 1 | long, straight with round end | |
| | | 2 | with blunt bent tip | |
| | | 3 | small, finger-like | |
| | | 4 | absent | |
| *42. | no. of bristles on inner lobe of 3rd joint | 0 | 3 | + |
| | | 1 | 2 | |
| | | 2 | 1 | |
| 43. | process on end joint | 0 | none | - |
| | | 1 | several inner teeth or spines | |
| | | 2 | 1 or more rounded inner bumps | |
| | | 3 | process between bristles | |

Appendix 2. Characters and character states used in the cladistic analysis of *Vargula* species and related genera - (continued).

| Characters | | Character States | | Ordered? |
|---|---|---|---|---|
| **44. | size of teeth on  bristles of 2nd joint | 0<br>1 | larger and longer<br>thinner and shorter | + |
| *45. | no. of unringed claw-like a-bristles on 2nd joint | 0<br>1<br>2<br>3 | none<br>2<br>3<br>4 | - |
| **46. | relative lengths of peg bristle & longest tooth of of 1st joint | 0<br>1<br>2 | peg bristle<tooth<br>peg bristle>tooth<br>peg but only 3 teeth | - |
| *47. | number and evenness of cusp row on smallest tooth of 1st joint | 0<br>1<br>2<br>3<br>4<br>5<br>6<br>7 | 9 teeth in 3 bulges<br>9-15 with 2 proximal bulges<br>9-17 with 1-2  proximal bulges<br>7-18 with 1  proximal bulge<br>4-8 alternating long/short<br>15-19 even<br>5-13 even<br>7-8 with 1 larger distal tooth | - |
| *48. | hairs on end bristles of inner lobe of 3rd joint | 0<br>1 | no<br>yes | - |
| **49. | 2 bristles of outer lobe of 3rd  joint | 0<br>1 | both terminal<br>inner bristle lateral | + |
| 6TH LIMB: | | | | |
| *50. | posterior  projection of end joint | 0<br>1<br>2 | little<br>some projection<br>much projection | + |
| 7TH LIMB: | | | | |
| **51. | longest tooth of end comb central | 0<br>1 | yes<br>no | + |
| *52. | no. of male distal lateral comb bristles | 0<br>1<br>2<br>3 | 15-25<br>4-14<br>1-3<br>none | + |
| *53. | type of terminal peg (position is on edge opposite comb unless otherwise noted) | 0<br>1<br>2<br>3<br>4<br>5<br>6 | big, very stout, hooked, jaw-like<br>long, thin, straight<br>short, stout, bent<br>male short, stout, bent; female short, thin, straight<br>male short, thin, curved; female very long, thin, hooked<br>male long, thin, inside comb; female stout, hooked<br>long, thin, in middle of tip | - |

\*    = some homoplasies, good synapomorphy for some clades (half filled rectangles in Fig. 1);
\*\*   = no homoplasies, excellent synapomorphies (filled rectangles in Fig. 1).

**APPENDIX 3.** Character matrix (from Appendix 2) used in the cladistic analysis of *Vargula* species and related genera (character state 9 = missing).

```
Taxa                 Characters
                     1 2 3 4 5 6 7 8 9 10111213141516171819202122232425262728293031323334353637383940414243444546474849 50

D. levinsoni         0 9 0 0 0 0 0 0 0 9 9 0 0 0 0 0 0 0 0 0 9 0 0 0 0 0 0 0 0 0 0 9 0 0 0 0 9 0 0 0 0 0 9 9 9 0 0 9
D. levis             0 9 0 0 0 9 0 0 0 9 0 0 0 0 0 0 0 1 0 0 0 0 0 0 0 0 0 0 0 1 9 0 0 0 3 1 0 1 0 1 0 0 9 0 0 1 0 0
V. shulmanae         1 2 2 2 1 2 1 0 0 1 0 0 3 0 0 1 0 1 0 1 0 2 2 1 0 1 1 0 0 0 0 0 2 0 0 2 0 1 0 0 0 1 0 3 0 4 0 0 0
V. graminicola       1 2 2 2 1 2 1 0 0 1 0 0 3 0 0 1 0 1 0 1 0 2 2 1 0 1 1 0 0 0 0 0 2 0 0 2 0 1 0 0 0 1 0 3 0 4 0 0 0
V. parasitica        1 2 2 2 1 9 1 0 0 0 0 0 3 0 0 1 0 1 0 1 0 9 2 1 0 1 1 0 0 0 0 0 9 2 0 0 9 0 1 9 9 0 1 9 3 9 9 9 0 0
V. harveyi           1 2 2 2 2 1 1 0 0 1 0 0 2 0 0 1 0 1 0 1 0 1 2 1 0 1 1 0 0 0 0 0 2 0 0 2 0 1 0 0 0 1 0 3 0 3 0 0 0
V. norvegica         0 2 9 2 1 1 1 0 0 0 0 0 0 0 1 0 1 0 1 0 0 2 1 0 1 0 1 0 0 0 0 0 0 0 0 0 0 0 1 0 0 1 0 3 0 3 0 0 0
V. tsujii            2 1 0 1 1 0 1 0 0 4 0 0 1 0 0 1 1 2 0 0 0 1 2 3 0 1 2 1 1 1 1 2 2 2 2 0 2 2 1 5 2 4 1 0 1 1 0 2 0 0 1
V. kuna              2 0 1 2 1 0 1 0 0 4 0 0 1 0 0 1 1 2 1 0 0 2 2 3 0 1 0 1 1 1 1 2 3 0 0 2 2 1 5 2 4 1 0 1 1 0 2 0 0 1
V. mizonomma         2 0 1 2 1 0 1 0 0 2 0 0 1 0 0 1 1 2 1 0 0 2 2 3 0 1 0 1 1 1 1 2 3 0 0 2 2 1 5 2 4 1 0 1 1 0 3 0 0 1
V. contragula        2 0 0 1 1 0 1 0 0 4 2 0 2 0 0 1 1 2 0 1 0 2 2 3 0 1 0 1 1 1 1 2 1 3 0 0 1 1 5 2 4 1 0 1 0 2 6 1 0 1
V. scintilla         2 0 0 1 1 0 1 0 0 3 1 0 0 0 0 1 1 2 0 1 0 1 2 3 0 1 0 1 1 1 1 2 4 5 0 0 0 1 5 0 1 1 2 1 3 1 3 0 0 1
V. kuna              2 0 0 1 1 0 1 0 0 3 1 0 0 0 0 1 1 2 0 1 0 1 2 3 0 1 0 1 1 1 1 2 4 5 0 0 0 1 5 0 1 1 2 1 3 1 3 0 0 1
V. lucidella         2 0 0 1 1 0 1 0 0 3 2 0 0 0 0 1 1 2 0 1 1 0 2 3 0 1 0 1 1 1 1 2 4 5 3 0 1 1 5 0 1 1 2 1 3 1 5 1 0 1
V. noropsella        2 0 0 1 1 0 1 0 0 3 2 0 0 0 0 1 1 2 0 1 1 0 2 3 0 1 0 1 1 1 1 2 4 5 3 0 1 1 5 0 1 1 2 1 3 1 5 1 0 1
V. ignitula          2 0 1 1 1 0 1 0 0 3 0 0 2 0 0 1 1 2 0 1 0 2 2 3 0 1 0 1 1 1 1 2 4 1 0 1 2 1 5 0 1 1 0 1 1 0 1 0 0 1
V. psammobia         2 0 1 1 1 0 1 0 0 3 0 0 2 0 0 1 1 2 0 1 0 2 2 3 0 1 0 1 1 1 1 2 4 1 0 1 2 1 5 0 1 1 0 1 1 0 1 0 0 1
V. bullae            2 0 9 2 1 9 1 0 0 9 0 3 3 0 0 1 1 2 9 0 0 2 2 4 1 2 4 1 0 0 1 1 9 9 0 2 9 1 5 1 9 2 0 1 9 9 9 1 2
V. n. sp. M          2 0 9 2 1 0 1 0 0 4 0 0 0 0 0 1 1 2 1 0 0 2 2 4 1 2 4 1 0 0 1 1 4 2 0 2 2 1 5 3 4 1 0 1 1 0 6 0 1 2
V. plicata           2 0 9 2 3 9 1 1 0 1 9 3 3 0 0 1 1 2 9 1 0 1 1 3 0 1 5 0 0 0 0 1 9 9 9 9 1 3 0 9 0 3 9 9 9 3 0 0 9
V. magna             9 2 0 2 1 0 1 0 0 9 9 0 2 0 0 1 0 1 9 1 0 2 1 3 0 1 2 0 1 1 1 2 9 4 0 0 0 1 3 1 0 1 0 1 3 0 9 0 0 0
V. hamata            1 2 2 3 1 0 1 0 0 9 9 2 2 0 0 1 0 1 9 0 0 9 2 3 0 0 0 0 0 1 0 0 9 1 2 0 9 0 0 9 0 0 0 9 9 9 0 0 0
V. hilgendorfii      1 0 0 1 0 0 1 1 1 1 0 1 2 1 0 1 0 1 0 1 0 0 2 1 0 1 3 0 0 0 1 1 9 2 0 0 2 1 2 4 0 0 3 0 3 0 6 0 0 0
V. sutura            1 2 0 0 1 9 1 1 0 9 9 6 1 0 0 1 0 9 0 0 0 9 2 1 0 1 0 0 2 0 0 0 9 6 9 0 9 0 1 4 2 0 0 9 9 9 9 0 0 0
Bathyvargula         2 9 0 0 2 9 1 1 0 9 9 4 4 0 1 1 0 0 9 1 1 9 2 2 0 9 4 0 0 0 0 9 0 0 0 1 9 1 1 0 3 0 1 9 3 0 9 9 0 9
Cypridinodes         9 9 2 9 3 9 1 1 2 4 9 5 9 0 0 1 0 0 9 1 0 9 2 9 0 9 6 0 0 0 0 0 0 6 0 9 9 1 1 9 9 0 9 9 9 9 9 0 9
Cypridina americana  2 9 2 1 1 3 1 2 1 4 0 1 3 1 1 1 0 1 0 0 0 1 0 0 0 1 3 0 1 1 1 1 0 2 1 0 0 1 4 5 4 0 0 0 1 0 7 0 0 0
Sheina orri          2 9 9 2 3 0 1 0 0 9 9 0 3 0 0 1 1 2 1 0 0 2 2 3 0 1 0 0 1 9 0 1 5 3 9 0 9 1 9 2 9 0 9 9 2 9 9 0 0 9
```

# REFERENCES

Baird, W. 1860. Note upon the genus *Cypridina* Milne-Edwards, with a description of some new species. *Proc. zool. Soc. Lond.*, London, **28**, 199-202.

Brady, G. S. 1880. Report on the Ostracoda dredged by H.M.S. *Challenger* during the years 1873-1876. *In Rep. scient. Results Voy. H.M.S. Challenger*, Zool., **1**, 1-184, pls. 1-44.

Brady, G. S. 1902. On new or imperfectly known ostracoda, chiefly from a collection in the Zoological Museum, Copenhagen. *Trans. zool. Soc. Lond.*, London, **16**, 179-210, pls. 21-25.

Cohen, A. C. & Morin, J. G. 1986. Three new luminescent ostracods of the genus *Vargula* (Myodocopida, Cyprinidae) from the San Blas region of Panama. *Contr. Sci. Los Angeles*, Los Angeles County Museum, Los Angeles, No. **373**, 1-23.

Cohen, A. C. & Morin, J. G. 1989. Six new luminescent ostracodes of the genus *Vargula* (Myodocopida, Cyprinidae) from the San Blas region of Panama. *J. Crustacean Biol.*, Lawrence, Kansas, **9**(2), 297-340.

Fink, W. 1986. Microcomputers and phylogenetic analysis, *Science*, New York, **234**, 1135-1139.

Harding, J. P. 1966. Myodocopan ostracods from the gills and nostrils of fishes. *In* Barnes, H. (Ed.), *Some Contemporary Studies in Marine Science*, 369-374. George Allen and Unwin Ltd., London.

Harvey, E. N. 1924. Studies on bioluminescence, XVI: What determines the color of the light of luminous animals: *Am. J. Physiol.*, Boston, Mass., **70**, 619-623.

Humphries, C. J. & Parenti, L. R. 1986. *Cladistic Biogeography*. Oxford Monographs on Biogeography, No. **2**, 98 pp. Oxford Science Publications, Clarendon Press, Oxford.

Kajiyama, E. 1912. The Ostracoda of Misaki, Part 2. *Zool. Mag. Tokyo*, Tokyo Zoological Society, **24**, 609-619, pl. 9.

Kornicker, L. S. 1970. Ostracoda (Myodocopina) from the Peru-Chile Trench and the Antarctic Ocean. *Smithson. Contr. Zool.*, Washington, D.C., **32**, 1-42.

Kornicker, L. S. 1974. Revision of the Cypridinacea of the Gulf of Naples (Ostracoda). *Smithson. Contr. Zool.*, Washington, D.C., **178**, 1-64.

Kornicker, L. S. 1975. Antarctic Ostracoda (Myodocopina). *Smithson. Contr. Zool.*, Washington, D.C., **163**, 1-720.

Kornicker, L. S. 1979. The marine fauna of New Zealand: benthic Ostracoda (suborder Myodocopina). *N. Z. Oceanogr. Inst. Mem.*, Wellington, **82**, 1-58.

Kornicker, L. S. 1983. The ostracode family Cypridinidae and the genus *Pterocypridina*. *Smithson. Contr. Zool.*, Washington, D.C., **379**, 1-29.

Kornicker, L. S. 1984. Cypridinidae of the continental shelves of southeastern North America, the northern Gulf of Mexico, and the West Indies (Ostracoda: Myodocopina). *Smithson. Contr. Zool.*, Washington, D.C., **401**, 1-37.

Kornicker, L. S. 1986a. Redescription of *Sheina orri* Harding, 1966, a myodocopid ostracode collected on fishes off Queensland, Australia. *Proc. biol. Soc. Wash.*, Washington, D.C., **99**, 639-646.

Kornicker, L. S. 1986b. Sarsiellidae of the Western Atlantic and Northern Gulf of Mexico, and revision of the Sarsiellinae (Ostracoda: Myodocopina). *Smithson. Contr. Zool.*, Washington, D.C., **415**, 1-217.

Kornicker, L. S. & Baker, J. H. 1977. *Vargula tsujii*, a new species of luminescent Ostracoda from Lower and Southern California (Myodocopa: Cyprininae). *Proc. biol. Soc. Wash.*, Washington, D.C., **90**, 218-231.

Kornicker, L. S. & King, C. E. 1965. A new species of luminescent Ostracoda from Jamaica, West Indies. *Micropaleontology*, New York, **11**, 105-110.

Morin, J. G. 1986. "Fireflies" of the sea: Luminescent signaling in marine ostracode crustaceans. *Fla Ent.*, Gainesville, **69**, 105-121.

Morin, J. G. & Cohen, A. C. 1988. Two new luminescent ostracodes of the genus *Vargula* (Myodocopida, Cyprinidae) from the San Blas region of Panama. *J. Crustacean Biol.*, Lawrence, Kansas, **8**, 620-638.

Müller, G. W. 1890. Neue Cyprininiden. *Zool. Jb.*, Jena, **5**, 211-252.

Müller, G. W. 1906a. Die Ostracoden der Siboga-Expedition. *Siboga Exped.*, Leiden, **30**, 1-40.

Müller, G. W. 1906b. Ostracoda. *Deutsche Tiefsee-Expedition 1898-1899*, **8**, 1-154.

Müller, G. W. 1908. Die Ostracoden. *D. Südpol.-Exped.*, 1901-1903, Berlin, 10, Zoologie **2**, 52-178.

Okada, Y. & Kato, K. 1949. Studies on luminous animals in Japan. III. Preliminary report on the life history of *Cypridina hilgendorfi. Bull. biogeogr. Soc. Japan*, Tokyo, **14**, 21-25.

Platnick, N. I. 1988. Programs for quicker relationships, *Nature*, London, **335**(6188), 310.

Platnick, N. I. 1987. An empirical comparison of microcomputer parsimony programs, *Cladistics*, London, **3**(2), 121-144.

Poulsen, E. M. 1962. Ostracoda-Myodocopa. 1:Cypridiniformes-Cypridinidae. *Dana Rep.*, Copenhagen, **57**, 1-414.

Skogsberg, T. 1920. Studies on marine ostracodes, 1: Cypridinids, halocyprids, and polycopids. *Zool. Bidr. Upps.*, Stockholm, Suppl., **1**, 1-784.

Swofford, D. L. 1985. Manual. *Phylogenetic analysis using parsimony*, version 2.4. Illinois Natural History Survey, Champaign, Illinois.

Wiley, E. O. 1981. *Phylogenetics, the theory and practice of phylogenetic systematics*, 439 pp. John Wiley and Sons, New York, Chichester, Brisbane, Toronto, Singapore.

Wilson, C. B. 1913. Crustacean parasites of West Indian fishes and land crabs, with descriptions of new genera and species. *Proc. U.S. natn. Mus.*, Washington, **44**, 189-277.

# DISCUSSION

Richard Reyment: The light flashes exhibited by each of the purported species may be regarded as point processes (*sensu* Cox, D. R. & Lewis, D. A. W. 1966. *The Theory of Point Processes*. Methuen, London), each with its own characteristics. By measuring the times between events, it should be possible to characterize each category in mathematical terms.

James Morin & Anne Cohen: We agree; the signals can be analyzed mathematically. We are aware of the statistical work of Cox, D. R. & Lewis, D. A. W. (1966. *Statistical Analysis of Series of Events*. Methuen). In our case, *each* species produces very stereotyped and predictable signal trains. From our video data we are able to generate accurate statistical averages (e.g., pulse duration, interpulse distances and intervals, pulse number, train length, train duration, etc.; see Cohen & Morin, 1989), and analyze them empirically. These behavioural characteristics can also be treated in the same way as the morphological data in our computer analyses.

Dietmar Keyser: Are these luminescent species restricted to one area or the whole reef?

James Morin & Anne Cohen: Yes. That is, each species shows restrictions to specific microhabitats in reef systems (e.g., *only* along the edge of sand pockets in a reef), but also occurs throughout these reefs wherever these microhabitats occur.

Dietmar Keyser: Do the animals live in these sites or are they just gathering there?

James Morin & Anne Cohen: The data we have on a few species indicate that the males signal in the water column immediately above the benthonic habitat where they live by day.

# 30

# Morphological changes and function of the inner lamella layer of podocopid Ostracoda

Dietmar Keyser

Zoological Institute and Zoological Museum,
University of Hamburg, West Germany.

## ABSTRACT

Ostracods have a carapace which encases the whole body. It consists of an outer calcified and an inner non-calcified cuticle. The latter has always been regarded as a place where respiration takes place, especially since the vibratory plates of the maxillula and mandibula, as well as the maxillary palp, are situated directly beside these thin cuticular sheets.

Investigation of the inner epidermal layer shows clearly that it contains numerous mitochondria, indicating it to be highly active, using a lot of energy. Respiration itself does not require much energy so perhaps the inner epidermal layer also has other functions. According to the ionic composition of the medium, several cytheracean and cypridacean species develop an apical labyrinth beneath the inner cuticle. This can be explained as being evidence of the osmotic regulation function of the cells in that region. Such a labyrinth is also found around the isthmus of the animal and shows

a slightly thickened cuticle as well as many dark staining small mitochondria with densely packed cristae. At other places, where the labyrinth is absent, the mitochondria are very large and have only very short and lightly packed cristae, which leaves the centre part of the mitochondrium empty. They thus appear swollen. The appearance of these types of excretory cells are compared in some genera and their importance, in ecological terms for the species involved, is discussed.

## INTRODUCTION

Like all arthropods crustaceans possess an outer skeleton in their chitinous cuticle. This cuticle in contrast to that of insects and spiders is often strengthened with calcium and magnesium salts. Although the exposed surface especially shows such depositions of inorganic salts, the inner surfaces (for example, the inner parts of folds or invaginations) have only a very thin cuticle. In decapods the gills and adjacent areas are constructed similarly.

This functional separation is obvious in such specialized crustaceans as the ostracods. They possess a carapace which encloses the entire body, the carapace having on the outside of the fold a heavily calcified cuticle, while the inside of the fold, as well as the body itself, has only a thin one, which is not calcified. The functional separation is clear, the outside being a passive defence system and also a skeletal system for the adhesion of muscles and other organs, while the thinness and the flexibility of the inner cuticle and the body wall enable movement and also respiration.

As early as 1865, Claus thought that the main task of the inner lamellae was respiration. G.W. Müller (1894), Fassbinder (1912) and Hartmann (1966-1975) all refer to this, but only Fassbinder adds a peculiar observation upon finding many small dark spots on a clean part of the inner cuticle without the epidermal layer. He thought that these might be holes for the interchange of gases.

In the present study, the inner lamella and the epidermal layer have been investigated with a view to establishing their function.

## MATERIAL AND METHODS

The material for the study was collected in different brackish water habitats in the Baltic and the North Sea. Specimens from polluted freshwater channels in Hamburg and from freshwater lakes around the city were also used.

The following species were sectioned and investigated using both light and electron microscopes: *Cyprideis torosa* (Jones, 1850); *Leptocythere psammophila* Guillaume, 1976; *Leptocythere lacertosa* (Hirschmann, 1912); *Loxoconcha elliptica* Brady, 1868; *Hirschmannia viridis* (O. F. Müller, 1785), *Paradoxostoma variabile* (Baird, 1835); *Xestoleberis aurantia* (Baird, 1838); *Cytherissa lacustris* (G. O. Sars, 1863); *Candona compressa* (Koch, 1837); *Cypria ophthalmica* (Jurine, 1820); *Cyclocypris ovum* (Jurine, 1820) and *Ilyocypris tuberculata* Daday, 1900.

The material for light (LM) and electron microscopy (EM) was fixed live in 2.5% glutaraldehyde in phosphate-buffer after Sörensen or in 3% GA and 8% tannic acid with 0.5M Sörensens phosphate-buffer at pH 7.

After rinsing in buffer it was postfixed in 1% osmium tetroxide in phosphate buffer. The material was then decalcified in an aqueous solution of ethylenediaminetetra acetic acid (EDTA) washed in buffer, dehydrated in graded acetone, and embedded in Spurr's medium (Spurr, 1969).

Sections were cut on an OM/U II REICHERT ultramicrotome. Semi-thin sections were stained with toluidine-blue and pyronin according to Holstein & Wulfhenkel (1971). The ultra-thin sections were stained with uranyl acetate (Stempac & Ward, 1964) and lead citrate (Reynolds, 1963) and viewed with a ZEISS EM 9.

Specimens used for the study in the scanning electron microscope were brushed clean, air or critical-point dried, glued to a stub with double-sided sticky tape and gold coated. The micrographs were then taken using a CAMSCAM DV 4 scanning electron microscope.

## RESULTS

The inner surface of the carapace of the ostracods which faces the body of the animal is the inner lamella. It is an extension of the calcified outer lamella or shell of the ostracods. The selvage at the margin of the shell pinpoints the dividing line between outer and inner lamella (Fassbinder, 1912; Hartmann, 1967). We find proximal from the selvage the more or less extended calcified inner lamella, and distal from it the calcified part of the outer lamella. This sometimes reaches around the edge of the shell and thus places the selvage on the inner side of the shell. The calcification of the inner lamella ends abruptly at the line of the inner margin (inner edge line). The line of concrescence marks the inner line of the fused zone of the calcified parts of the inner and outer calcified layers (Fig. 1).

The inner lamella connects the calcified part to the isthmus fold through a thin cuticle (Plate 1, Figs 1 and 2; Fig. 1), which often has a thickness of less than 0.1µm. This cuticle consists of a clearly visible

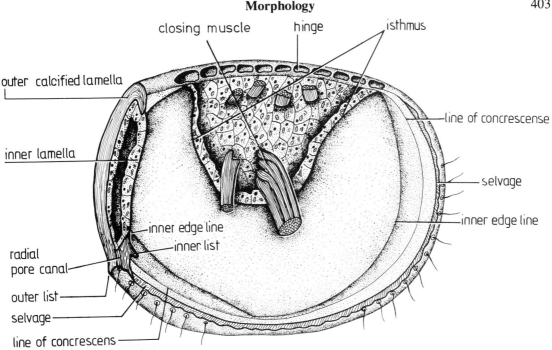

Fig. 1. Schematic representation of an ostracod valve with the body detached and its anterior end sectioned.

epicuticle and a thin exo/endocuticle, the latter two of which cannot be differentiated as such from one another. Internal structures of the layers are only occasionally observable as in, for example, *Ilyocypris tuberculata*. The layers occur again at the body cuticle, which is often thicker than the inner lamellar cuticle (Plate 1, Figs 5-7).

The fold of the carapace is lined on the inside with a continuous layer of epidermal cells (Plate 1, Fig. 5). Beneath the outer calcified lamella the thickness of the cells within the layer varies considerably, depending upon whether or not a nucleus is present in the observed area. In contrast the epidermal layer of the inner lamella is almost stable in thickness, varying only under different ecological conditions or at different times (moulting, feeding, etc.). It is indeed thicker during premoult, but thinner after ecdysis. Some cells possess filaments which extend from the outer lamella to a connection with a similar cell from the inner epidermal layer which also has these filaments (Plate 1, Fig. 6). Such filaments are connected to the inner part of the cuticle (outer and inner lamella) by hemidesmosomes, which connect the cell membrane to the cuticle by some

kind of sticky extracellular substance. In the middle, both cells (one from the outer, one from the inner lamellar epidermal layer) are held together by desmosomes, this being mainly the way in which single cells transmit forces over several cells, e.g., in muscles.

The epidermal cells of the outer and inner lamella are different in several ways. Epidermal cells of the outer lamella have, apart from the rounded or elongated nucleus, many dark staining vacuoles and only a few mitochondria as well as endoplasmic reticulum (Plate 1, Figs 5 and 6), while the epidermal cells of the inner lamella are characterized by the presence of many mitochondria and fewer vacuoles.

Prior to and during moulting, the number of large vacuoles with either dark- or weak-staining content is sometimes very high, although again only a few mitochondria and endoplasmic reticulae are present in the epidermal layer of the outer lamella (Plate 2, Fig. 8). Exclusive at moulting time is the large amount of glycogen, granules of which are distributed in the cell plasma.

The same is also true for the epidermal cells of the inner lamella. During moulting, large vacuoles with weak- or dark-staining contents are also found here, the amount of glycogen being raised. The number of mitochondria is always high and more or less restricted to the distal part of the cell, where at times they even occur in clusters. The endoplasmic reticulum is also always well developed (Plate 2, Fig. 8).

One feature is developed in only some species and then only in the central part of the inner epidermal layer. It begins just where the body lining bends towards the inner lamella and extends irregularly towards the margin. This is here called the apical labyrinth of the inner epidermal layer (Plate 1, Figs 3 and 4, 7; Plate 2, Fig. 12). It occurs in animals which live in brackish, limnic or other osmotically stressed regions, such as: *Cyprideis torosa*, *Cytherissa lacustris*, *Cypria ophthalmica*, *Candona compressa*, *Cyclocypris ovum* and *Ilyocypris tuberculata*. No indication of such a structure could be observed in *Xestoleberis aurantia*, *Leptocythere psammophila*, *Leptocythere lacertosa*, *Hirschmannia viridis*, *Loxoconcha elliptica* or *Paradoxostoma variabile*.

This apical labyrinth is an invagination of the plasma-membrane of the cell and occurs beneath the inner lamella cuticle extending often to the vicinity of the mitochondria (Plate 2, Fig. 11). All cells in this region have these invaginations. The mitochondria here are frequently different from those in cells outside this region, and those without such a labyrinth. The cells in the labyrinth region have mitochondria which appear denser, smaller and with well developed cristae, while adjacent cells without the labyrinth possess mitochondria which are 10 to 15 times larger, have a swollen appearance and have only a few short cristae, which leave the middle empty. The dividing line between these two cell types is very sharp (Plate 2, Fig. 10).

The underlying cuticle also shows a distinct differentiation. In the region of the labyrinth it is thicker and it does not stain quite as dark as in the non-labyrinth area (Plate 2, Fig. 11). The epicuticle is a very thin layer of about 0.03μm on top of the cuticle while the normal inner cuticle is 0.14μm thick. In the labyrinth region this is doubled in thickness, being about 0.3μm and is in the region

of a well adapted labyrinth due to a thickening of the innermost part of the cuticle. The outer part is more densely stained and smaller than the lightly built and weak staining inner part. In newly formed cuticle above small young cells, the cuticle has about the same thickness as the one in the labyrinth area but is very lightly stained over its whole thickness. The labyrinth in these young cells is only weakly developed.

In moulting animals the labyrinth persists during the process of moulting (Plate 2, Fig. 9). The cuticle prior to shedding is then in two distinct layers, the inner one having extra folds reaching into the cells of the inner epidermal layer to allow for the growth of the animal post-moulting.

Some individuals of *Cyprideis torosa* were examined which showed a growth of bacteria on the inner cuticle in the region of the labyrinth. This growth was sometimes developed as a distinct layer overlying the cuticle. Whether this was the reason why animals died after a few weeks in captivity, could not be determined.

Supporting fibres ('Stützzellen') with their filaments adhering to the cuticle are also developed in the region of the labyrinth. The filaments are connected to the cuticle as usual, but the cell surrounds this connecting site with a larger footlike extension than in adjacent areas with the small extensions which build the labyrinth.

The labyrinth extends from the isthmus-fold in varying degrees to the margin. It always begins very abruptly behind the fold of the body lining to the inner lamella. Often, the first cell is one with filaments or even muscles connected to it (Plate 1, Figs 3, 4 and 7).

## DISCUSSION

The inner lamella of ostracods has long been regarded as a place where respiration takes place (Claus, 1865; G. W. Müller, 1894; Bergold, 1910; Fassbinder, 1912). Respiration and osmoregulation are often connected with each other at the same site in many different groups of animals (crustaceans, fishes). However, this has never been proved in ostracods (Cannon, 1925; Weygoldt, 1960). This study produces several new facts to support such a connection:

1) The labyrinth is only present in animals, such as *Cyprideis* and all the limnic species, which live in a hypo-osmotic environment. It is absent in all marine species.

2) This structure exhibits a great extension of the surface of the plasma membrane of the epidermal cells, thus enabling a higher volume of water to be excreted (Plate 2, Figs 9 and 11).

3) The epidermal cells of the inner lamella always show a remarkably high number of mito-chondria (Okada, 1982), which is an indication of a high metabolic rate in this region (Plate 1, Fig. 7; Pl. 2, Fig. 12). Such a high rate cannot be explained as being necessary for respiration alone, but it would be required to furnish the energy needed to excrete water against the osmotic gradient.

4) The mitochondria of the cells in the labyrinth region are small and appear normal, while those of the adjacent cells are swollen (Plate 2, Fig. 10). This means that the water content in this region is higher than in cells with a labyrinth.

Both in isolation, but particularly in combination, these four lines of evidence, incontrovertibly demonstrate that osmo- and ionic regulation are important secondary functions of the inner lamella of ostracods which live in hypo-osmotic environments.

It is known that other crustaceans develop such structures as described here in the vicinity of the gills, where an osmoregulatory function has also been proved.

During investigations of the ultrastructural changes of the epidermal layer of the gill, Köhler-Günther & Bulnheim, Biologische Anstalt Helgoland, Hamburg, (pers. comm.) showed that there is a change in the shore-crab *Carcinus maenas*, when it is transferred from brackish to marine waters or the reverse. The same is true for *Cyprideis* from brackish oligohaline waters, which showed an extended labyrinth and a greater amount of mito-chondria, compared with few mitochondria when the animal was in polyhaline water. Whether this can help us in the understanding of the nature of noding in some brackish water ostracods is not yet clear and has to be studied. In all freshwater species so far investigated, the labyrinth was well developed and a large number of mitochondria were present.

As shown, the thin cuticle is also different in structure. This can be explained as being due to the necessity of facilitating an easy water exchange. This can only be achieved if the physical barrier of a tightly packed cuticle is somehow loosened, so that the pores within the cuticle fibrils became larger. That this is also neccessary during moulting, even with two cuticular sheets overlapping one another, is seen in Plate 2, Fig. 9.

In reflecting on these results, the question emerges of whether the possibility of developing such a labyrinth has something to do with the ability to take over new ecological niches, and to survive potentially fatal changes in the environment. The possession of such an ability certainly widens the ecological boundaries of a species and enables it to survive in places where others cannot follow.

## ACKNOWLEDGEMENTS

I would like to thank Miss M. Hänel, Hamburg, for the drawing and Dr D. Bürkel, Hamburg as well as Professor R. Whatley for reading the manuscript and for the most valuable suggestions.

## Plate 1

Abbreviations

| | | | |
|---|---|---|---|
| B | body | en | endocuticle |
| BC | body cavity | ex | exocuticle |
| I | isthmus | hep | hepatopancreas |
| IL | inner lamella (epidermal layer and cuticle) | l | labyrinth |
| OL | outer lamella (epidermal layer and cuticle) | m | mitochondrium |
| cut | cuticle | n | nervous element |
| ccut | calcified outer cuticle | s | supporting fibres ('Stützbalken') |
| ncut | newly formed cuticle | v | vacuole |

Fig. 1.   *Cypria ophthalmica*, male, with left valve detached. The internal organization of limbs can be seen. The area of the isthmus is torn open and in the marked square, a piece of the inner lamellar layer of the left valve can be seen.

Fig. 2.   *Cypria ophthalmica*, detail of Fig. 1 as marked. Connection between the isthmus region and the inner lamella, showing the borderline between the plain cuticle of the isthmus and the labyrinth of the inner lamella.

Fig. 3.   *Cypria ophthalmica*, detail of Fig. 2 as marked, showing the borderline in high magnification.

Fig. 4.   *Cypria ophthalmica*, detail of Fig. 2 as marked, showing the smooth cuticle on one edge and the labyrinth structure on the bottom of the torn cells.

Fig. 5.   *Cyprideis torosa*, section of the shell of an animal from polyhaline waters, showing the very wide calcified outer lamella. Note the few mitochondria in the cells of the inner lamella.

Fig. 6.   *Cypria ophthalmica*, section of the shell showing the connection of the two lamellae by the filaments in the epidermal cells (supporting cells). Note that mitochondria are found here in outer and inner epidermal cells near the filaments of the cells, producing energy for the filamental action.

Fig. 7.   *Cyprideis torosa*, section of the isthmus fold, showing the abrupt beginning of the labyrinth structure together with a high number of mitochondria.

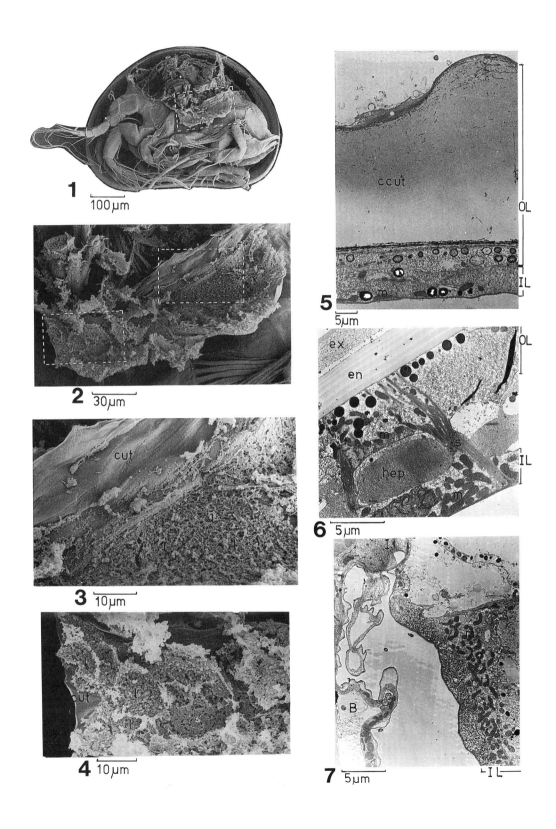

**1**  
100 μm

**2**  
30 μm

**3**  
cut  
10 μm

**4**  
cut  
10 μm

**5**  
ccut  
OL  
IL  
m  
5μm

**6**  
ex  
en  
m  
is  
hep  
m  
OL  
IL  
5μm

**7**  
I  
B  
IL  
5μm

**Plate 2**

Abbreviations

| | | | |
|---|---|---|---|
| B | body | en | endocuticle |
| BC | body cavity | ex | exocuticle |
| I | isthmus | hep | hepatopancreas |
| IL | inner lamella (epidermal layer and cuticle) | l | labyrinth |
| OL | outer lamella (epidermal layer and cuticle) | m | mitochondrium |
| cut | cuticle | n | nervous element |
| ccut | calcified outer cuticle | s | supporting fibres ('Stützbalken') |
| ncut | newly formed cuticle | v | vacuole |

Fig. 8. *Cyprideis torosa*, section of the shell of a moulting specimen, showing the large number of vacuoles and the high amount of glycogen in both the epidermal layers. Note that the new cuticle is folded beneath the old one.

Fig. 9. *Cyprideis torosa*, detail of Fig. 8, showing the labyrinth region and especially how the folding of the new cuticle does not interfere with the labyrinth.

Fig. 10. *Cyprideis torosa*, section of the inner lamella at the marginal border of the labyrinth, showing the dramatic change in the mitochondria in cells with labyrinth and cells without.

Fig. 11. *Cypria ophthalmica*, section of inner lamella, showing the labyrinth in close up, the underlying cuticle and a young cell with an incipient labyrinth and different cuticle.

Fig. 12. *Cyprideis torosa*, section of the shell showing outer and inner lamella with ā labyrinth.

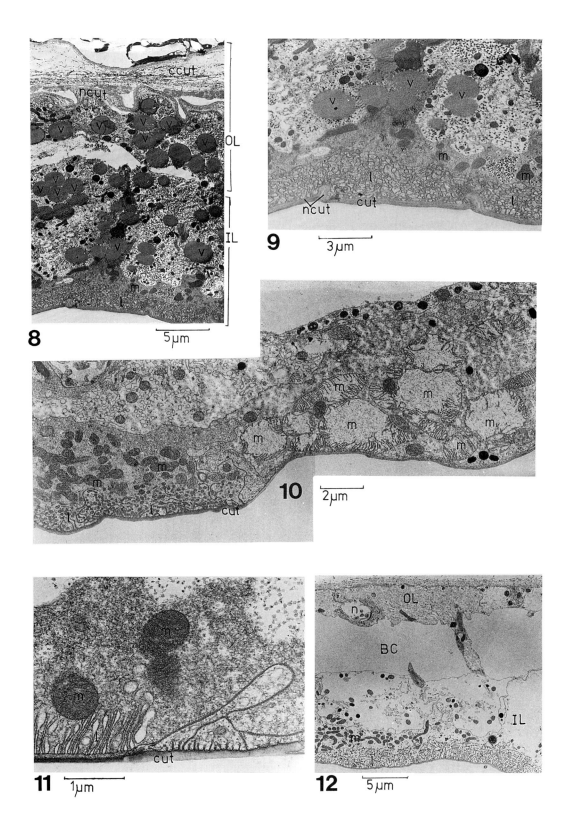

**8**
ccut
ncut
v
v v
v v v
v v
v
v v
v v
v
v
m
l
OL
IL
5μm

**9**
v
v v
m
m
l
l
ncut cut
3μm

**10**
m
m
m
m m
m
m
l l cut
2μm

**11**
m
m
cut
1μm

**12**
n
OL
BC
IL
m
l
5μm

## REFERENCES

Bergold, A. 1910. Beiträge zur Kenntnis des inneren Baues der Süßasser-Ostracoden. *Zool. Jb.*, Abteilung für Anatomie, Jena, **30**, 1-42.

Cannon, H. G. 1925. On the segmental excretory organs of certain fresh-water ostracods. *Phil. Trans. R. Soc.*, London, Series B, **214**, 1-27.

Claus, C. 1865. Über die Organisation der Cypridinen. *Z. wiss. Zool.*, Leipzig, **15**, 143-154.

Fassbinder, K. 1912. Beiträge zur Kenntnis des Süßwasser-ostracoden. *Zool. Jb.*, Abteilung für Anatomie und Ontogenie der Tiere, **32**, 533-576.

Hartmann, G. 1966-1975. Ostracoda, *In* Bronn, H. G. (Ed.), *Dr H. G. Bronns Klassen und Ordnungen des Tierreiches*, 5. Band I. Abt. 1-4 Lief., 1-786. Akademische Verlagsgesellschaft Geest & Portig KG, Leipzig.

Holstein, A. F., Wulfhenkel, U. 1971. Die Semidünschnitt-Technik als Grundlage für eine cytologische Beurteilung der Spermatogenese des Menschen, *Andrologie*, **3**, 65-69.

Müller, G. W. 1894. Die Ostracoden des Golfes von Neapel und der angrenzenden Meeresabschnitte, *Pubbl. Staz. zool. Napoli Monograph*, **21**, 1-404.

Okada, Y. 1982. Ultrastructure and Pattern of the Carapace of Bicornucythere. *In* Hanai, T. (Ed.), *Studies on Japanese Ostracoda*, 229-267. University of Tokyo.

Reynolds, E. S. 1963. The use of lead citrate at high pH as an electron opaque stain in electron microscopy. *J. Cell Biol.*, Rockefeller Institute, New York, **17**, 208.

Spurr, A. R. 1969. A low-viscosity epoxy resin embedding medium for electron microscopy. *J. Ultrastruct. Res.*, New York & London, **26**, 31-43.

Stempac, J. C., Ward, R. T. 1964. An improved staining method for electron microscopy. *J. Cell Biol.*, Rockefeller Institute, New York, **22**, 697.

Weygoldt, P. 1960. Embryologische Untersuchungen an Ostracoden. Die Entwicklung von *Cyprideis littoralis*. *Zool. Jb.*, Abteilung für Anatomie und Ontogenie, **78**, 369-426.

## DISCUSSION

Robin Whatley: Do you have any evidence that, as suggested by Kesling and, particularly in the case of *Krithe* by Peypouquet, that within the anterior part of the duplicature especially within the vestibulum of any ostracod, special cells occur associated with respiration and the *storage* of oxygen?

Dietmar Keyser: In our investigations we did not find such cells. To my knowledge there is no system of oxygen storage known in the animal kingdom besides the transportation of oxygen by haemoglobin or other similar substances. A storage device could also be the gas in the swim-bladder of fishes. There are no storage possibilities known in the cell itself.

Ken McKenzie: Was the TEM - photograph that you showed of a specimen about to go into moult indicative of early moult or later moult? I would have expected a thicker epicuticular plus endocuticular part if it was late in the moult cycle.

Dietmar Keyser: The specimen was just prior to ecdysis. Using the scheme of Drach (Drach, P. 1939. Mue et cycle d'intermue chez les crustacés décapodes. *Annls Inst. océanogr. Monaco*, **19**, 103-392) it was in the stage D2 or D3. Concerning the thickness of the layers, you have to realize that we are dealing with the cuticular layer of the inner lamella and this layer is always very thin.

# 31

# Observations on the ontogeny of the ocular sinus in a species of *Echinocythereis*

Mervin Kontrovitz

Department of Geosciences, Northeast Louisiana
University Monroe, U.S.A.

## ABSTRACT

This is the first report on the ontogeny of the ocular
sinus in podocopid ostracods. The ocular sinus of
each instar studied here (A-4 through adult) has a
unique set of features. In adults of *Echinocythereis
margaritifera* (Brady, 1870) the sinus is elongate
and tapers distally to a constriction at about mid-
length, beyond which it is expanded and bulbous.
A terminal (distal) concavity is the complement
of the inner eyespot surface. In the A-1 instar the
sinus resembles a low, wide cone with a slight
constriction and a well-developed terminal concavity. The sinus in the A-2 forms a short cone
with a shallow terminal concavity and no constriction; its length is only 35 to 45% of its diameter.
In the A-3, the sinus is low and hemispherical, with
a weak terminal concavity. The sinus in the A-4 is
low and mound-like, without a concavity. When
applied to fossils, this kind of study may be
useful in recognizing palaeoenvironments where
light was restricted; Lythgoe (1979) wrote that
adult animals that live in such environments
commonly retain juvenile ocular structures.

## INTRODUCTION

Little prior attention has been given to the ontogeny of ocular structures in podocopid ostracods.
Previous emphasis has been on adults, with only
brief mention of the ocular sinus in juveniles
(Kontrovitz, 1987). The present study is based
upon valves of *Echinocythereis margaritifera*
(Brady, 1870) and is the first to trace the ontogenetic development of the ocular sinus in a
podocopid taxon.

## MATERIAL AND METHODS

Specimens were collected from modern sediment
samples from 37 bottom localities on the continental shelf off Louisiana, northern Gulf of Mexico (Curtis, 1960; Kontrovitz, 1976), and from the
Quaternary Mississippi Mudlumps (Howe & van
den Bold, 1975). Carapaces from both sources are
alike and Hazel (1967) concluded that modern
forms '...found off the Mississippi Delta...' are
conspecific with the 'type' of *Cythereis margaritifera* Brady.

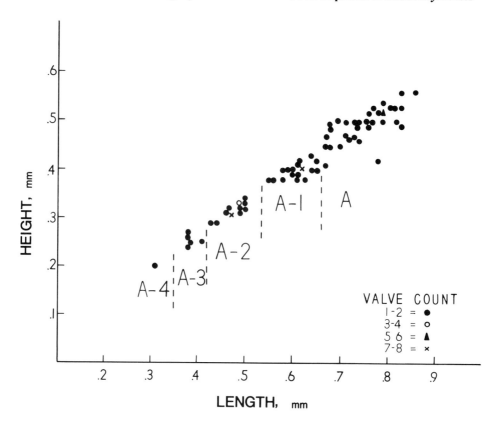

Fig. 1. Size distribution diagram of adult female and heteromorph left valves of *Echinocythereis margaritifera* (Brady, 1870) from the Louisiana Continental Shelf (Kontrovitz, 1976).

Six hundred and forty-one specimens of adults and juveniles were obtained from the modern sediments; 450 were taken from the mudlumps, reported to be about 15,000 to 15,500 years old (Howe & van den Bold, 1975).

The ocular sinuses of adult right and left valves are known to be different in each of several species of *Echinocythereis*. Also, there is a marked sexual dimorphism in the adult carapaces. Therefore, left valves of females and presumed females (heteromorphs) were used primarily to provide a means of comparison (Kontrovitz, 1987). See Whatley & Stephens (1977) for a discussion about the recognition of 'proto-males (technomorphs)' and 'proto-females (heteromorphs)'.

Specimens were considered to be either adults or juveniles based upon size, development of the hinge and inner lamella, and the degree of

calcification. Length and height measurements from all specimens were used to construct plots of valve size.

Internal moulds were made from 150 valves spanning the size range of all specimens. Valves were filled with the mounting medium Lakeside 70, heated to about 150 to 160°C, cooled to room temperature, and then dissolved in dilute (5%) HCl. The resulting moulds provide excellent representations of the ocular sinuses (Kontrovitz, 1982).

## RESULTS

### Adult shells

There is much intraspecific variation in the size of adults from the Louisiana continental shelf. Adult female left valves ($N = 52$) had lengths in the range

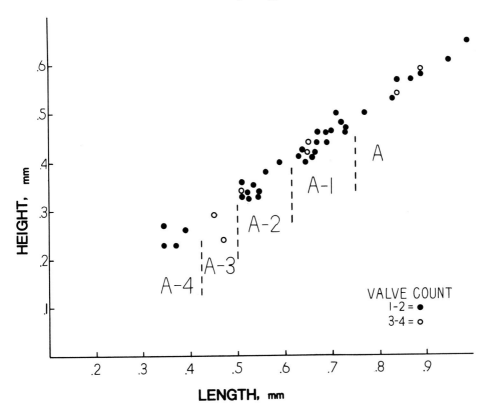

Fig. 2. Size distribution diagram of adult female and heteromorph left valves of *Echinocythereis margaritifera* (Brady, 1870) from the Mississippi Mudlumps (see Howe & van den Bold, 1975).

of 0.67 to 0.86mm, with a mean of 0.75mm and a standard deviation of 0.14. The A-1 moults were 0.55 to 0.65mm in length, A-2s were 0.43 to 0.50mm, A-3s were 0.38 to 0.41mm, and there was one A-4, which was 0.31mm in length (Fig. 1).

For mudlump specimens, adult female left valves ($N = 16$) were 0.77 to 0.99mm in length, with a mean of 0.85mm, and a standard deviation of 0.09. The A-1 moults were 0.63 to 0.73mm, A-2s were 0.51 to 0.59mm, A-3s were 0.45 to 0.47mm, and A-4's were 0.35 to 0.39mm (Fig. 2). A two-tailed $T$-Test for small samples indicates that the adult specimens from the continental shelf and mudlumps are not significantly different in size ($p = 0.05$; see Johnson, 1973, 268).

Only adults of this species have a robust hinge with a full complement of elements that are typical of the genus *Echinocythereis* (Puri, 1953).

For example, in the left valve, adults have a distinct anterior socket with a post-jacent antero-median tooth, and a distinct socket at the posterior cardinal angle. Juveniles (A-1 and earlier) have only a shallow elongate anterior groove, no post-jacent antero-median socket, and a long shallow groove at the posterior.

The 'width' of the inner lamella anteriorly is equal to about 9.0 to 12.5% of valve length, in all those specimens considered to be adults. In juveniles (heteromorphs) the 'width' of the inner lamella is about 4 to 6% of valve length for A-1, and less for earlier moults.

The degree of calcification is difficult to quantify, but as expected, the carapace of juveniles is much more fragile in all aspects, including the hinge and inner lamella.

There are slight differences in the central muscle

scars of adults compared to juveniles, but this too is difficult to quantify. In juveniles the adductor scars are smaller in proportion to valve size and are situated at mid-length, while in adults they are somewhat displaced toward the anterior. In juveniles the two frontal scars are slightly more separated from each other than in adults.

## Ocular sinuses in adults

To serve as a standard, the ocular sinus of an adult female left valve is described. In lateral view, the ocular sinus is elongate and constricted in diameter at some point giving a stalked appearance in all adults (Plate 1, Fig. 1). The base of the ocular sinus is represented by the portion of the mould that infills the ocular pit, as is seen in the valve interior. The diameter is about 75 to 100% of the length of the entire ocular sinus. From the base, the sinus tapers distally to a constriction that has a diameter of about 40 to 50% of that of the base. The constriction is situated at about mid-length of the sinus and is just proximal to the tapetal layer of the eye (Andersson & Nilsson, 1981). The length of the ocular sinus is the distance from its base to its distal end, perpendicular to the inner surface of the valve. In female adult left valves ($N = 52$), the sinuses were from 75 to 98µm in length (Fig. 3).

The ocular sinus is circular in cross-section from the base up to and including the first part of the expanded distal portion, beyond the constriction. The expanded portion is irregular in cross-section, depending upon the configuration of the terminal lobes.

The distal portion of an adult sinus is bulbous and lobed, representing the space occupied by the lens cells, retinal cells, and rhabdoms of the lateral eyecup (Andersson & Nilsson, 1981). Its diameter is about 85 to 105% of that of the base. There is always a terminal subcentral concavity that is the complement of the inner surface of the eyespot (Kontrovitz & Myers, 1984).

The terminal concavity is bordered at both its anterior and posterior by a lobe. The anterior lobe is broadly rounded, transects about 65 to 80 degrees of arc, and has an elevation of about 7 to

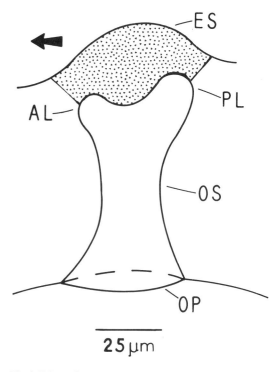

Fig. 3. Schematic drawing of a section through the ocular region of an adult female left valve of *Echinocythereis margaritifera* (Brady, 1870); OS = ocular sinus; OP = ocular pit; AL = anterior lobe; PL = posterior lobe; ES = eyespot (shown with a pattern).

12µm above the deepest part of the terminal concavity. The posterior lobe has an elevation of about 10 to 15µm; it may be rim-like or more broadly rounded, transecting about 120 to 130 degrees of arc. These lobes correspond to the anterior and posterior concavities of the inner surface of the eyespot (Kontrovitz & Myers, 1984). The lobes and the intervening low sloping surface lend a saddle-like appearance to the distal end of the ocular sinus.

The ocular sinuses of the adult right and left valve are not equal in size and proportions. The right sinus is shorter and thicker in appearance; the only opening into the sinus is through the ocular pit. In the left sinus there is an additional opening through an extension of the anterior hinge socket (Kontrovitz, 1987).

Although male left valves have greater length/height ratios than females, the ocular sinuses are similar in shape and are in proportion to valve size (length).

The valves of mudlump specimens are like those from the modern shelf. There are no differences in any characters including those related to ocular structures.

## Ocular sinuses in juveniles

In the A-1 moult, the ocular sinus is shaped like a low, thick, distally truncated cone (Plate 1, Fig. 2). The length of the sinus is about 65% of the diameter of the base, which is circular in cross-section. There is only one opening, through the ocular pit. Unlike in the adult, the left hinge of the juvenile does not have a deep anterior hinge socket that is open into the ocular sinus.

There is only a slight hint of a constriction, just proximal to the terminal lobes. The overall stoutness of the sinus and lack of a distinct constriction provide the best means of distinguishing the ocular sinuses of adults from those of A-1 and earlier stages. Indeed, this can be used to distinguish adult from juvenile valves, if one is willing to destroy the specimens to do so.

In the A-1 moult there is a terminal concavity rimmed with lobes at its anterior and posterior, much in the manner of the adult. The lobes may be low and faint or well-developed, in different specimens. In some, the anterior lobe is more rounded than the posterior. No differences have been noted between the right and left ocular sinuses in any of the pre-adult moult stages.

In the A-2 moults, the ocular sinus is low and the length is about 35 to 45% of the diameter of the base. The sinus is without constriction; it is a low rounded mound with a shallow, broad, subcentral concavity. The posterior lobe is rounded, while the anterior lobe is lower and narrower (Plate 1, Fig. 3). This indicates, that even at this stage, the inner surface of the eyespot is convex at its centre and concave at its margins (Kontrovitz & Myers, 1984).

Moults at the A-3 stage have an ocular sinus that is low and rounded, without a constriction. The length is only about 20 to 25% of the diameter of the base. There is a very shallow terminal

concavity with a round posterior lobe and a lower round anterior lobe (Plate 1, Fig. 4). It is presumed that the corresponding eyespot has a complementary surface.

The A-4 stage has a low mound-like ocular sinus that slopes off gently in all directions. There is neither a constriction nor a subcentral concavity at the distal end (Plate 1, Fig. 5). This indicates that the adjacent eyespot has only a broad concavity to mark its inner surface.

No earlier moults were found among the 1091 specimens secured for this study. It would be interesting to learn at what moult stage the ocular sinus is developed in this and other species of podocopid ostracods.

## DISCUSSION

Some ocular stuctures of fossil ostracods may be useful in interpreting water depths. Benson (1976, 1984) indicated a general decrease in eyespot size with increasing water depth and that at depths of 600 to 900m the eyespots 'disappear'. There have been attempts to apply this, but the reports do not quantify the relationships of eye characters and water depth (McKenzie & Peypouquet, 1984; Bonaduce & Danielopol, 1988).

Kontrovitz & Myers (1985, 1988) have shown that the eyes of podocopid ostracods would be of little use below depths of about 280m in the oceans, because of the design of the eye and the low light levels. Modifications to the eyespot-tapetum system could do little to enhance vision below that depth. It is most common '... to find that when conditions for vision become so difficult that investment in ... more specialized eyes no longer results in a worth-while return in visual information the eyes regress but do not entirely disappear' (Lythgoe, 1979, 104).

If Lythgoe (1979) is correct, it would be prudent to investigate the ontogeny of the ocular sinuses in many groups of modern ostracods. We may then be able to recognize fossil ostracods that lived near or beyond the limits of vision in the oceans. The adults may have juvenile ocular structures in the carapaces; even if the eyespots are absent at some water depth, the organisms may

retain some of the other structures, including the
ocular sinus.

**Plate 1**

Ocular region, internal moulds, left valves of *Echinocythereis margaritifera* from the Louisiana Continental Shelf; scanning
electron micrographs, about X320.

Fig. 1. Adult, female.
Fig. 2. Heteromorph, A-1.
Fig. 3. Heteromorph, A-2.
Fig. 4. Heteromorph, A-3.
Fig. 5. Heteromorph, A-4.

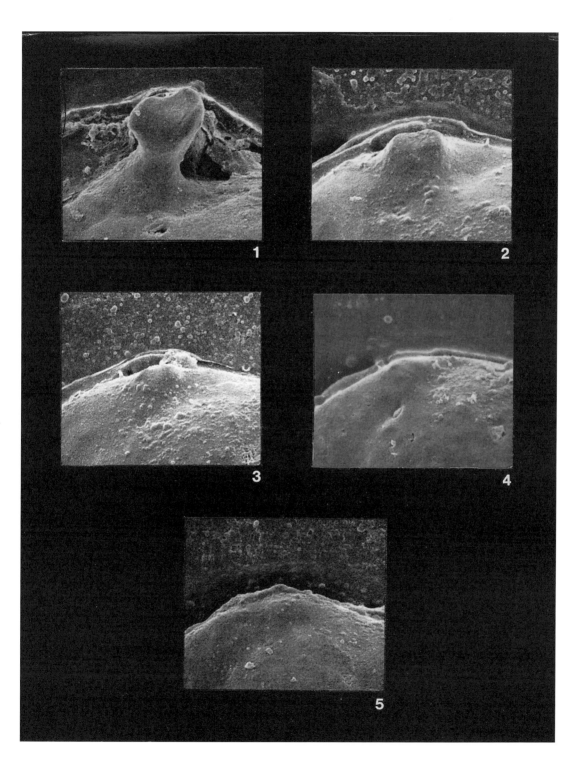

1
2
3
4
5

## REFERENCES

Andersson, A. & Nilsson, D. E. 1981. Fine structure and optical properties of an ostracode (Crustacea) nauplius eye. *Protoplasma*, Leipzig, **107**, 361-374.

Benson, R. H. 1976. The evolution of the ostracode *Costa* analysed by "Theta-Rho difference" (Discussion). In Hartmann, G. (Ed.), Evolution of Post-Paleozoic Ostracoda. *Abh. Verh. naturw. Ver. Hamburg*, N.F., Vereins in Hamburg, 18/19 (Suppl.), 127-139.

Benson, R. H. 1984. Estimating greater paleodepths with ostracodes, especially in past thermospheric oceans. *Palaeogeog. Palaeoclimatol. Palaeoecol.*, Amsterdam, **48**, 104-107.

Bonaduce, G. & Danielopol, D. L. 1988. To see and not to be seen: The evolutionary problems of the Ostracoda Xestoleberididae. *In* Hanai, T., Ikeya, N. & Ishizaki, K. (Eds), *Evolutionay biology of Ostracoda, its fundamentals and applications*, proceedings of the Ninth International Symposium on Ostracoda, held in Shizuoka, Japan, 29 July - 2 August 1985, Developments in palaeontology and stratigraphy, **11**, 375-398. Kodansha Ltd., Tokyo and Elsevier, Amsterdam, Oxford, New York, Tokyo.

Brady, G. S. 1870. Appendice (ostracodes), Vera Cruz at Carmen. *Fonds de la Mer*, Paris, **1**, 191-192.

Curtis, D. M. 1960. Relation of environmental energy levels and ostracod biofacies in east Mississippi delta region. *Bull. Am. Ass. Petrol. Geol.*, Chicago, **44**, 471-494.

Hazel, J. E. 1967. Classification and distribution of Recent Hemicytheridae and Trachyleberididae (Ostracoda) of northeastern North America. *U.S. Geol. Surv. Prof. Paper 564*, Washington, 1-72.

Howe, H. V. & Bold, W.A. van den. 1975. Mudlump Ostracoda. *In* Swain, F. M. (Ed.), Biology and Paleobiology of Ostracoda. *Bull. Am. Paleont.*, Ithaca, **65**(282), 303-315.

Johnson, R. 1973. *Elementary Statistics*, 480 pp. Duxbury Press, North Scituate, Massachusetts.

Kontrovitz, M. 1976. Ostracoda from the Louisiana Continental Shelf. *Tulane Stud. Geol. Paleont.*, New Orleans, **12**, 49-100.

Kontrovitz, M. 1982. A new method of producing internal molds in ostracode valves. *J. Paleont.*, Chicago, **56**, 818-820.

Kontrovitz, M. 1987. Ocular sinuses in some modern and fossil species of *Echinocythereis* (Ostracoda). *Micropaleontology*, New York, **33**, 93-96.

Kontrovitz, M. & Myers, J. H. 1984. A study of the morphology and dioptrics of some ostracode eyespots. *Trans. Gulf Cst Ass. geol. Socs*, **34**, 369-372.

Kontrovitz, M. & Myers, J. H. 1985. Ostracode vision and water depth: Paleoenvironmental implications. *Geol. Soc. Amer., Abstr. Progm.*, Boulder, **17**, 633.

Kontrovitz, M. & Myers, J. H. 1988. Ostracode eyes as paleoenvironmental indicators: Physical limits of vision in some podocopids. *Geology*, Boulder, **16**, 293-295.

Lythgoe, J. N. 1979. *The Ecology of Vision*, 244 pp. Clarendon, Oxford.

McKenzie, K. G. & Peypouquet, J.-P. 1984. Oceanic palaeoenvironment of Miocene Fyansford Formation from Fossil Beach near Mornington, Victoria, interpreted on the basis of Ostracoda. *Alcheringa*, Sydney, **8**, 291-303.

Puri, H. S. 1953. Contribution to the study of the Miocene of the Florida Panhandle. *Geol. Bull. Fla*, Tallahassee, **36**, 345 pp.

Whatley, R. C. & Stephens, J. M. 1977. Precocious sexual dimorphism in fossil and recent Ostracoda. *In* Löffler H. & Danielopol, D. L. (Eds), *Aspects of Ecology and Zoogeography of Recent and Fossil Ostracoda*, proceedings of the Sixth International Symposium on Ostracoda, Saalfelden, 69-91. Dr. W. Junk b.v. Publ., The Hague.

## DISCUSSION

Dietmar Keyser: In which direction does the elevated part of the eyestalk shadow the light?

Mervin Kontrovitz: The ostracod valves in Fig. 3 above indicate that light is gathered over a span of about 80 to 100 degrees of arc. That is, light is gathered over about 40 to 50 degrees on each side away from the centre of the eye tubercle.

Richard Benson: What do you think is the minimum diameter of the lens of the ocular tubercle where it is effective for sight or is this structure just vestigial?

Mervin Kontrovitz: I don't know, but *presume* it would be a diameter not much smaller than a 'rhabdom' or a retinal cell (Andersson & Nilsson, 1981). It would have to allow at least four photons per 100 milliseconds to pass; that is the calculated sampling time.

Roger Kaesler: How does the vision of immature instars differ from that of adults? The morphological difference suggests pronounced differences in vision.

Mervin Kontrovitz: We have passed light through the eye tubercles of adults, but not yet through those of juveniles. I suppose that juveniles, with less space for retinal cells and the like, would be less able to resolve images as compared to adults. Also, it seems logical that with smaller eyespots, useful vision in juveniles would be restricted to shallower waters.

Robin Whatley: Note that *Poseidonamicus ocularis* may have the juvenile eye in adult stages.

Mervin Kontrovitz: I would very much like to see the material that you mentioned here and at the Houston Symposium.

Ken McKenzie: Probably the eyestalk region is not only important for seeing. In other crustaceans it is known to be important for moulting, vitellogenesis and colour patterns. In juveniles which have eyestalks, the functions are probably located closer to what amounts in ostracods to the brain region. The discovery of a definite eyestalk in podocopid ostracods is most exciting for these reasons as well as in relation to the animal's capability for sight.

# PALAEOECOLOGY

# 32

# Biotope indicative features in Palaeozoic ostracods: a global phenomenon

**Gerhard Becker[1] & Martin J. M. Bless[2]**

[1]Geological-Palaeontological Institute, Frankfurt,
West Germany.
[2]The Natural History Museum, Maastricht,
The Netherlands.

## ABSTRACT

A review is given of the research on the palaeoecology of Devonian and Carboniferous ostracods, undertaken by the present authors over several years. Three biotope indicative assemblages are distinguished in terms of energy levels. In the present paper, the different terminologies on the subject used independently by both authors are correlated, and emphasis is given to ostracod communities characterizing marine, low-energy levels (Thuringian Assemblage), and especially to indicators for different biotopes of depth and shoreline. The existence of such, biotope indicative ostracod assemblages can be recognized on a worldwide scale as late as Permian, and even back to the Upper Ordovician.

## INTRODUCTION

Studies on the palaeoecology of Palaeozoic ostracods have been undertaken by the present authors for many years, with emphasis on Devonian assemblages (G.B.) and Carboniferous faunas (M.B.) respectively. This paper reviews our current understanding of biotope indicative assemblages, notably in Devonian to Permian sediments, whereby, we try to give a synopsis of our independently undertaken investigations. Special reference is made to some communities of the Thuringian Assemblage.

## OSTRACODA AND UNIFORMITARIANISM

The distribution of Recent ostracods is controlled by a number of ecological factors, and notably by temperature and salinity (primary factors *sensu* Neale, 1965). The resulting distribution patterns are modified by regional and local, partly interacting environmental conditions such as energy levels and substrate, food supply and competitors, predators, local temperature variations and depth differences, variations in light intensity, oxygen content, pH value (Van Morkhoven, 1962, 145-151; Hartmann, 1975, 575-594, 662-664;

Peypouquet, 1983, 26; Whatley, 1988, 104), but not by depth-dependent pressure.

In attempting to interpret fossil ostracod assemblages, we must follow uniformitarian principles (e.g. Neale, 1983; Whatley, 1983a, 54, 1988, 105-109). However, several morphologies occur in Palaeozoic ostracods which are unknown in modern faunas or whose function is not yet understood (e.g. most palaeocopid features). Moreover, the inferred relationship between a particular morphology and mode of life through geological time may be questionable, even for taxa with living descendants (Whatley, 1983a, 54). This applies especially to Palaeozoic ostracods. Other groups (e.g., Bairdiidae) are of 'considerable antiquity extending back to at least the Carboniferous' (Whatley & Watson, 1988, 409).

Therefore, reconstruction of the palaeoenvironment, in which these ostracods lived, is based first on interpretations of the sedimentary environment (wherein they have been buried but may not necessary have lived) and on the associated fauna and flora. These approaches may sometimes yield contradicting interpretations (Benson & Sylvester-Bradley, 1971, 65). Nevertheless, throughout the Palaeozoic, a limited number of ostracod assemblages existed, each of those characterized by the frequent presence of distinct morphologies (and often distinct taxa at the familial or generic levels), which have been observed repeatedly in comparable deposits with morphologically comparable associated faunas or floras (see also Whatley, 1983a, 56, 62; 1983b, 135).

## PALAEOZOIC OSTRACOD ASSEMBLAGES

### General

On the basis of carapace morphology, Becker (*in* Bandel & Becker, 1975, 61) distinguished, at first in the Devonian, different ostracod assemblages. Such communities are named (Eco-)Assemblages (Ökotypen) and considered by Becker to be indicative of different, distinct energy levels. There are three main Assemblages:

1) The Eifelian Assemblage (Eifeler Ökotyp) Fig. 1. Thick-shelled, frequently heavily lobed and ornamented ostracod faunas characterizing essentially high energy environments (in this connection, see Whatley *et al.*, 1982). Subassemblages are indicative of lagoon, back-reef, reef complex, open-marine and offshore environments (Becker, 1969, Figs 2-6; 1971, Fig. 4/1-6; see also Whatley & Watson, 1988).

Bless *et al.*, (1988, 348) introduced the term 'mixed marine ostracode assemblages' for faunas indicating the 'restricted shallow nearshore shelf'. Such communities, in which bairdiaceans are extremely rare or even absent, may represent a special subassemblage category within the Eifelian Assemblage.

2) The Thuringian Assemblage (Thüringer Ökotyp) Figs. 2-11. This term applies to rather thin-shelled, often smooth ostracods, frequently possessing long spines or fine thorn-like spines. Many of these are podocopids. They were first described by Gründel (1961, 1962) and Blumenstengel (1965a, 1979) from the Dinantian and late Devonian of eastern Thuringia (Ostthüringisches Schiefergebirge, East Germany) and were, by mistake, believed to be endemic (Jordan, 1964, 74). According to Becker (*in* Bandel & Becker, 1975, 61) the cosmopolitan Thuringian Assemblage is characteristic of marine, low-energy levels.

Bless (*in* Dreesen *et al.*, 1985, 332-336) distinguished, within faunas belonging to the Thuringian Assemblage *sensu* Becker, several categories of ostracods showing obvious similarities, both taxonomic and morphological.

*Category A*. Kirkbyacean ostracods. Reticulated palaeocopids with nodes and admarginal frills. Often with biotope indicative spines (Figs 3/8, 10/D). This category coincides with the 'amphivalente Kirkbyacea' (Becker, in press). Most probable nectobenthic forms (Hartmann, 1963, 48-50; Becker & Adamczak, in press).

*Category B*. Thuringian-type ostracods. Thin-shelled podocopid taxa, frequently with long spines, including *Discoidella*-type ostracods. Characteristic genera of the Thuringian Assemblage *sensu stricto* (Becker, 1982a, 162) are *Tricornina* Boucek *sensu lato* (e.g., Figs 2/4, 3/3, 9/5, 10/C), *Rectoplacera* Blumenstengel (Fig. 7/G), *Rectonaria* Gründel (e.g., Figs 2/9, 7/H)) and

*Processobairdia* Blumenstengel (Figs 3/10, 6/8, 7/T). Crawlers or swimmers (*Discoidella*-type ostracods). Species of the family Tricorninidae Blumenstengel often have extremely long spines which would certainly have hindered the animal crawling on the sediment surface. Therefore, the question arises as to whether these species were occasional swimmers. The conspicuously rounded ventral margin coincides with such a lifestyle (Hartmann 1964, 36). The Tricorninidae, however, were placed (Gründel & Kozur, 1972, 907) into the superfamily Cytheracea which contains only species unable to swim. Therefore, an inclusion into the superfamily Drepanellacea Ulrich & Bassler, as first proposed by Gründel (1966, 96), seems probably more correct.

*Category C. Microcheilinella*-like ostracods. Rather thick-shelled and inflated podocopids with pronounced ventral overlap, smooth or with conspicuous ornament (Figs 9/12, 10/P): crawlers or burrowers. According to Becker (1982a, 162), taxa with 'biotope indicative spines' (e.g., *Rectoplacera*) belong to Category B.

*Category D.* Heterogeneous palaeocopids. Usually characteristic of the Eifelian Assemblage *sensu lato. Kirkbyella* Coryell & Booth (Fig. 9/13) may be included. Taxa with biotope indicative spines or fine striate ornamentation are considered to be genuinely autochthonous within the Thuringian Assemblage: crawlers, burrowers and probably occasional swimmers.

*Category E.* Bairdiaceans and bairdiocypridaceans. Usually smooth-shelled podocopids. Also common in the offshore subassemblage of the Eifelian Assemblage. In the Thuringian Assemblage often with fine, biotope indicative spines, e.g., *Bairdia* McCoy *sensu lato* (Fig. 10/N-O), *Acratia* Delo (Fig. 6/14), *Ceratacratia* Blumenstengel (Figs 3/11, 7/R): crawlers.

*Category F.* Paraparchitaceans. Rather large, thick and smooth-shelled podocopids, partly with dorsal spines. Common to abundant in lagoonal or mixed-marine subassemblages of the Eifelian Assemblage. In thanatocoenoses within the Thuringian Assemblage (*Shishaella* Tschigova, Fig. 9/17).

*Category G.* Healdiaceans. Smooth-shelled podocopids (metacopids) with posterior ornamentation. Present in nearshore and offshore communities of the Eifelian Assemblage. Taxa with long spines, e.g., *Marginohealdia* Blumenstengel (Fig. 3/7) and *Timorhealdia* Bless (Fig. 11/7), are characteristic of the Thuringian Assemblage: mostly crawlers.

3) The Entomozoacean Assemblage (Entomozoen Ökotyp). This term applies to very thin-shelled ostracods with fingerprint ornament: mostly considered to be pelagic.

Blumenstengel (1965a, 208) first assumed the entomozoaceans to be planktonic, and later (1973, 76) to be benthonic. As for the whole group, Becker (1976, 216) and Gooday (1983, 760) questioned the benthonic mode of life and supposed that species with inflated carapaces were swimming forms. The question of whether the entomozoaceans may be considered to be meropelagic or holopelagic *sensu* Hartmann (1975, 575-576) and epipelagic or bathypelagic remains speculative, as well as the question of whether these animals may have lived nearshore or offshore (Becker & Bless 1987, 51, 54). Casier's suggestion (1987, 200), that they had a nectobenthonic, nearshore lifestyle, is not conclusive.

4) Mixed Assemblages *sensu* Becker (Mischfaunen). Mixtures between the Eifelian and the Thuringian assemblages and the Thuringian and Entomozoacean assemblages respectively are often reported (e.g., Becker, 1982, 164), indicating interdigitations of facies (Faziesüberschneidungen, Becker *in* Becker & Sánchez de Posada, 1977, 192). As for biocoenoses, mixtures between the Eifelian and the Entomozoacean assemblages had never been observed by the present authors, additional evidence that the entomozoaceans did not live in nearshore environments.

In the present paper, neither the Entomozoacean Assemblage nor the Eifelian Assemblage are dealt with further. For additional information, see Becker, 1976; Becker *et al.*, 1975; Becker & Bless, 1974; Bless, 1983; Bless *et al.*, 1976, 1981; Bless & Jordan, 1971; Bless & Sánchez de Posada, 1972; Rebske *et al.*, 1985; Sánchez de Posada & Bless, 1971; Wang, 1976, 1983, 1986, 1987.

## The Thuringian Assemblage

In connection with conodont studies in eastern Thuringia (Ostthüringisches Schiefergebirge), Gründel (1961, 1962) and Blumenstengel (1965a) discovered what were then completely new ostracod assemblages, in which spinose forms dominated both quantitatively and qualitatively. The most abundant taxa (represented by 100 or more specimens recorded by the original authors) are shown in Figs 2 and 3. Bairdiaceans and cytheraceans are the commonest elements. Kirkbyaceans and healdiaceans also occur frequently. The development of long, slender spines in genera, which otherwise are practically spineless (e.g., species of *Bairdia* sensu lato) or the development of extra long spines or flanges in species, which in the Eifelian Assemblage only display rather subduced spines (e.g., healdiaceans), is notable in this fauna.

Some taxa occur in both the Thuringian and Eifelian assemblages, notably bairdiaceans, but also many less abundant forms such as *Discoidella* Croneis & Gale, hollinellids, kirkbyaceans and pachydomellids. The Thuringian forms, however, are always thin-shelled. Comparison with the Eifelian Assemblage reveals some similarities with offshore shelf assemblages but practically none with lagoonal ones. This suggests a natural order of ecological niches ranging from lagoon through reef and shallow offshore shelf into the Thuringian (basinal) facies.

The original assemblages from eastern Thuringia as described by Gründel (1961, 1962) and Blumenstengel (1965a, 1979) were found in nodular limestones, alternating with or embedded in shales often with abundant entomozoacean ostracods (Mischfaunen *sensu* Becker). The rather monotonous associated macrofauna (mainly cephalopods, dacryoconarids, crinoids and trilobites) suggests a poorly aerated substrate. These features and the thin-shelled carapaces of the ostracods have often been interpreted as an indication of deep-water (basinal) facies (Becker, *in* Bandel & Becker, 1975, 61; Lethiers & Crasquin, 1985, 420), below the storm-wave base and the pterygocline *sensu* Liebau (1980, 189). This interpretation is supported by the frequent association with pelagic entomozoaceans.

Becker (1982, 163), and later Dreesen *et al.* (1985) and Bless (1987), however, emphasized that ostracod assemblages of the Thuringian Assemblage also occur in relatively shallow, open-marine sediments, presumably always deposited in rather calm, low-energy environments.

Faunas of the Thuringian Assemblage are not always silicified. Olempska (1979) discovered calcareous faunas of the Thuringian Assemblage in the late Devonian of Poland. Specimens belonging to the Thuringian Assemblage have also been observed as moulds in shales (e.g., Namurian of Meré, northern Spain; Sánchez de Posada, 1976; Becker, 1982). Conversely, many silicified ostracod faunas of the Eifelian Assemblage have been observed. Therefore, silicification is not a necessary criterion for the occurrence of Thuringian assemblages, as believed by Lethiers & Crasquin (1985, 420).

Thuringian Assemblages also occur in the nearby regions of Moravia (Dvorák *et al.*, 1986) and the Hartz Mountains (Gründel, 1972; Blumenstengel, 1974), and in more distant areas of Europe; i.e. Carnic Alps (Bandel & Becker, 1975), Belgium (Dreesen *et al.*, 1985), northern Spain (Becker, 1977, 1981a, 1981b, 1982), North Africa (Becker, 1987) and Asia, i.e. northeastern Siberia (Shilo *et al.*, 1984), Kazakhstan (Buschmina, 1977), Timor (Gründel & Kozur, 1975; Bless, 1987); Figs 4-11.

Mixtures of Thuringian and Eifelian Assemblages have been described from the late Devonian of Poland (Olempska, 1979), the Lower Devonian of Canada (Berdan & Copeland, 1973) and Japan (Kuwano, 1987). The degree of mixture between these assemblages may provide an index to geographical or bathymetrical positions.

A still unsolved problem is the apparently rapid lateral spreading of the Thuringian Assemblage (Becker, 1982, 171). The proposal of Kozur (1972) and Gründel & Kozur (1975) to equate the Thuringian Assemblage with modern psychrospheric faunas was questioned by Becker (1987, 97).

In summary, ostracods of the Thuringian

Assemblage (associated with entomozoaceans, cephalopods, trilobites, crinoids, primitive brachiopods, foraminifera and conodonts) are widespread, occurring from basinal facies to shallow-marine deposits. The main ecological parameters (notably normal salinity and low energy) may be superimposed on the other important factors (e.g., food supply), not easy to recognize in the fossil record. Careful sedimentological and palaeoecological analyses, however, can assist in reconstructing the original environment.

## EXAMPLES OF THURINGIAN ASSEMBLAGE ASSEMBLAGES

Recently, seven ostracod faunas of different Palaeozoic ages (Givetian to Lower Permian) from several continents (Western Europe, North Africa, Asia) have been reported by Shilo *et al.* (1984), Dreesen *et al.* (1985), Swennen *et al.* (1986), Bless (1987) and Becker (1987, 1988, in press). All these faunas are from low energy sediments, but they were deposited at different depths in dissimilar geological situations. They all belong to Becker's Thuringian Assemblage, having common morphological and taxonomical characters. Rare species within these faunas, unusual in the Thuringian Assemblage (and marked by an asterisk in Figs 5-10), give information on palaeo-depth and shoreline.

1) Late Givetian. Maider, eastern Morocco (Fig. 5).
A silicified faunule of very small and fragile specimens was derived from the uppermost beds of an outcrop near Ouihalane (northern Maider, Morocco). The bedrock is composed of banks of reef debris with reef builders no longer *in situ* (lumachelle limestones with corals and stromatolites *sensu* Massa, 1965). A late Givetian age is proved by conodonts (*varcus* Zone; Bultynck, 1985, 264). The depositional environment was offshore, at some distance from the reef and below wave base. The ostracods are in a down-slope mud flow containing reef debris and are clearly a thanatocoenosis.

The ostracod fauna is clearly dominated (both species and individuals) by cytheraceans (bythocytherids) belonging to the genera *Monoceratina* Roth, *Praebythoceratina* Gründel

and *Hercynocythere* Blumenstengel. The latter genus was previously known only from the Lower Hartz region (Blumenstengel, 1974). Bythocytherids occur in the shelf region and are less common in the deep sea (Whatley, 1988, Fig. 4). In the tropics they are often very common in inner shelf and sometimes intertidal environments, while in higher latitudes they are more common on the mid- and outer shelf (Whatley, pers. comm.). Rare associated forms of the Thuringian Assemblage are the kirkbyaceans (*Amphissites* Girty and *Villozona* Gründel) and the bairdiocypridacean genus *Baschkirina* Rozhdestvenkaya; the latter form with a weak posteroventral spine. Bairdiaceans are only represented by the genus *Bairdiacypris* Bradfield. Exotic taxa, from nearshore environments are species of *Kozlowskiella* Pribyl and *Eridoconcha* Ulrich & Bassler. However, since both genera are represented by more than one specimen, they may belong to the biocoenosis. This may be confirmed by the finding of juveniles (Becker, 1973, 68; Whatley, 1983a, 55-56, 1983b, 134).

2) Early Upper Famennian. Belgium (Fig. 6). From the basinal Upper Famennian nodular limestones of Baelen in eastern Belgium (Goé Nord, Member 10; Dreesen *et al.*, 1985, Fig. 5), a well preserved, silicified ostracod fauna was obtained. The depositional environment of the ostracod-bearing limestone was below wave base, but within the photic zone and probably thus only some tens of metres deep.

The fauna is dominated by bairdiaceans (*Bairdia* sensu lato, *Acratia* and *Processobairdia*), bairdiocypridaceans (*Bairdiocypris* Kegel or *Praepilatina* Polenova) and species of (partly spinose) *Microcheilinella*-like ostracods. These mostly smooth-shelled forms are associated with cytheraceans with lateral spines (*Tricornina* sensu lato, *Monoceratina* and *Berounella* Boucek) and kirkbyaceans (*Kirkbya* Jones) clearly representing the Thuringian Assemblage. The only exotic species seems to be the Beyrichiacean sp. 104, Becker & Bless, 1974.

3) Late Upper Famennian. Tafilalt, eastern Morocco (Fig. 7). Several silicified ostracod faunas had been detected in the nodule bearing late Upper Famennian near Erfoud (Tafilalt, eastern

Morocco) (Becker, 1987, Figs 1 and 2). The thin (condensed) Upper Devonian sequence (do I-VI) was deposited in a northwardly shallowing basin (Alberti, 1972, 151). The ostracod faunas were derived from its upper part (do VI), dated by conodonts (Alberti, 1970, 26).

The faunas are mixtures of cytheraceans (*Tricornina* sensu lato), 'rectonariids' (*Rectonaria*, *Orthonaria* Blumenstengel), bairdiaceans with dorsal spines and terminodorsal thorn-like spines (*Bairdia* sensu lato, *Acratia*, *Clinacratia* Blumenstengel, *Processobairdia* and *Ceratacratia*), bairdiocypridaceans with posteroventral spines or flanges (*Karinadomella* Becker, *Rectoplacera*, *Necrateria* Gründel, *Praepilatina*) and kirkbyaceans (*Amphissites*, *Villozona*), associated with heavily ornamented healdiaceans (*Marginohealdia* Blumenstengel, *Heterma* Gründel) and rare entomozoaceans (*Richterina* Gürich *sensu lato*). For most species, the number of individuals is low, only *Processobairdia olempskae* Becker, 1987, *Bairdia* (*Bairdia*) *nidensis* Olempska, 1979 and *B.* (*Bairdia*), gr. *hypsela* Rome, 1971 are represented by more than 20 specimens. This ostracod assemblage indicates a basinal environment.

4) Upper Famennian and Tournaisian. Northeastern Siberia (Fig. 8). Shilo *et al.* (1984) reported silicified ostracod faunas from Upper Famennian (Fa 2) and Middle Tournaisian (Tn 2) sequences in the Omolon Region (northeastern Siberia). 'Within this area sedimentary environments . . . [of] the Upper-Famennian-Tournaisian succession . . . ranging from rather deep subtidal through intertidal to supratidal occur' (Swennen *et al.*, 1986, 237). Some ostracod assemblages (associated with moravamminid foraminifera) are believed to be 'deep subtidal shelf indicators' (Swennen *et al.*, 1986, 247).

The ostracod assemblages reflect the observed variation in depositional environments. Shallow-shelf, nearshore assemblages of Upper Famennian age are characterized by *Serenida* Polenova, *Parapribylites* Pribyl and *Evlanovia* Egorov, possibly indicating the same (ecological niche) as *Hollinella* Coryell and *Copelandella* Bless & Jordan (Shilo *et al.*, 1984, 139, Fig. 3). Open-marine, unstable shelf environments are characterized by bairdiaceans, bairdiocypridaceans, kirkbyaceans and

*Microcheilinella* Geis. Relatively deep shelf environments are marked by the presence of cytheraceans associated with moravamminid foraminifera (Swennen *et al.*, 1986, 247). All the fauna is silicified.

5) Upper Visean. Southwestern Morocco (Fig. 9). The Upper Visean ostracod assemblage from southwestern Morocco (Dreesen *et al.*, 1985, Fig. 1) closely resembles the ostracod faunas from the Upper Givetian of northeastern Morocco.

Bairdiaceans (*Bairdia* sensu lato, *Bohlenatia* Gründel, *Bairdiacypris*, *Acratia*) and bairdocypridaceans (*Baschkirina*, *Bairdiocypris* and the peculiar *Michrocheilinella shiloi* Bless, 1984) are the dominants, kirkbyaceans (*Amphissites* and *Kirkbya*) are rather abundant, a species of the Cladocopina (*Discoidella*) occurs, and species of *Shishaella* and *Healdia* Roundy (and several hollinomorphs) are rare elements of this fauna. Cytheracean diversity (*Tricornina* sensu lato, *Saalfeldella* Gründel, *Monoceratina* sensu lato, and *Pseudomonoceratina* Gründel & Kozur) is high and in this respect, the Upper Visean fauna resembles the late Givetian fauna of northeastern Morocco. The rare elements of the fauna are probably allochthonous.

6) Upper Westphalian. Northern Spain (Fig. 10). A diverse ostracod fauna of exceptionally good preservation (Becker, 1978, plates 1-4) had been encountered during conodont studies in the Upper Westphalian Escalada Formation near Vega de Sebarga. This small village is situated in the valley of the River Ponga, a feeder of the River Sella in Asturias (northern Cantabrian Mountains, northern Spain); see Becker, 1978, Fig. 2. The light grey, thick-bedded limestones are rich in fusulinids indicating the *Fusulinella* Zone, Subzone B, Subdivision $B_1$ (Upper Moscovian). Macrofossils (corals, brachiopods) are rare. Microfacies studies in the area (see Becker, 1978, 44) revealed micritic bioarenites (fossiliferous wackestones and packstones), indicating shallow-water environments with very weak to moderately strong wave energy. The fauna must have lived in a rather calm environment (Becker, 1982, 163) because of the excellent preservation of the generally thin-shelled carapaces of the Thuringian Assemblage.

Bairdiaceans, often with delicate spines,

dominate both in diversity (*Bairdia* sensu lato, *Bohlenatia, Bairdiacypris, Acratia, Acanthoscapha* Ulrich & Bassler) and abundance. Other ostracod groups include abundant kirkbyaceans (*Amphissites, Kegelites* Coryell & Booth, *Kellettina* Swartz, *Kirkbya, Aurikirkbya* Sohn and *Coronakirkbya* Sohn, *Roundyella* Bradfield, *Kirkbyella*), rare bythocytherids (*Tricornina* and *Pseudomonoceratina*) and cladocopids (*Discoidella*). Bairdiocypridaceans are represented by the genera *Baschkirina, Praepilatina* (both with posteroventral spines) and the peculiar species *Microcheilinella shiloi*. Specimens of *Knoxiella*? Egorov, *Shishaella* and *Healdia* (of the Eifelian Assemblage) may be autochthonous (or at least parautochthonous). Rare specimens of the freshwater ostracod genus *Carbonita* are certainly allochthonous contaminants.

7) Lower Permian. Timor (Fig. 11). Rich and diverse ostracod faunas have been described from the Lower Permian of Timor (Gründel & Kozur, 1975; Bless, 1987). All assemblages have been prepared from limestone infilling the last chambers of ammonites and nautiloids. Characteristic of these assemblages is the high diversity (49 taxa reported from 10 samples) and the low number of specimens per taxon.

## CONCLUSIONS

All the faunas discussed belong to the Thuringian Assemblage and indicate low-energy environments. However, the actual taxa making up the assemblage differs from place to place. For example, bairdiaceans or cytheraceans may dominate in one regional assemblage while such groups as paraparchitaceans and kloedenellaceans may be associates elsewhere. All these characters are indicators for distinct positions of the biotopes with respect to depth, distance from sedimentary sources and shoreline (facies chain, 'Fazieskette' *sensu* Liebau, 1980, 175).

In reconstructing a bathymetric model, example 3 (late Upper Famennian of eastern Morocco) and example 6 (Upper Westphalian of northern Spain) represent the extremes of a basinal environment and a shallow-water niche respectively.

Low energy environments in a shallow near-shore realm are believed to be only short-lived. Any accident, of however short duration such as temperature change, changing of current directions or storms, can disturb the ecological equilibrium and can kill off the community and possibly destroy all evidence of its existence. The rare occurrence of fossil microfaunas from such biotopes may thereby be explained.

The other ostracod faunas (examples 1, 2, 4, 5 and 7) occupy more central positions in the model under discussion (Fig. 12). With the single exception of the late Givetian fauna from eastern Morocco (example 1), all these faunas are dominated by bairdiaceans and bairdiocypridaceans. This implies open-shelf conditions. Possibly, the relative depth and distance from the palaeo-shoreline may be indicated by the presence of rare, biotope indicative forms. Therefore, the Upper Famennian fauna from eastern Belgium (example 2, with *Processobairdia*) may have lived farther offshore than the Middle Tournaisian fauna from northeastern Siberia (example 4, with *Coryellina* Bradfield) and the Upper Visean fauna from southwestern Morocco (example 5, with *Shishaella*). However, the latter fauna is especially rich in cytheraceans and resembles, therefore, the late Givetian fauna (example 1); the rare elements may be allochthonous. With the exception of the basinal assemblage seen in example 3, this fauna is believed to have lived at the greatest depth.

The model discussed may be applicable to other geological intervals since ostracod faunas of the Thuringian Assemblage are known from the Upper Ordovician (Blumenstengel, 1965b-d; Knüpfer, 1968), through the Silurian (Blumenstengel, 1963), to the Triassic (Kozur, 1972). It is applicable to all continents; recently ostracods of the Thuringian Assemblage have been described from central Japan (Kuwano, 1987) and southern China (Wang, 1988). The same is true for the Eifelian Assemblage; 'certainly, these [Thuringian and Eifelian] Assemblages occur also in the Silurian [of Baltoscandia and Central Europe]' (Hansch, litt. comm.).

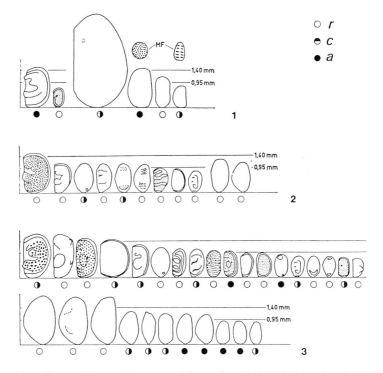

Fig. 1. Eifelian Assemblage. Characteristic nearshore ostracod faunas from the Middle Devonian of the Eifel region indicating lagoonal (1), reef (2) and offshore (3) environments. After Becker (1969, 1971).

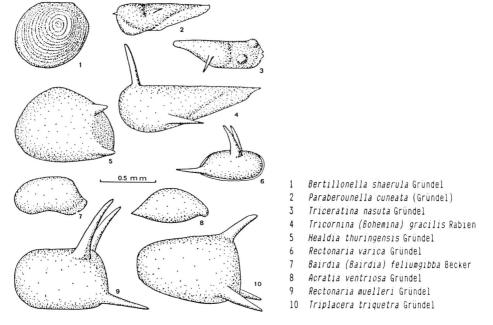

1  *Bertillonella shaerula* Gründel
2  *Paraberounella cuneata* (Gründel)
3  *Triceratina nasuta* Gründel
4  *Tricornina (Bohemina) gracilis* Rabien
5  *Healdia thuringensis* Gründel
6  *Rectonaria varica* Gründel
7  *Bairdia (Bairdia) feliumgibba* Becker
8  *Acratia ventriosa* Gründel
9  *Rectonaria muelleri* Gründel
10 *Triplacera triquetra* Gründel

Fig. 2. Thuringian Assemblage. Characteristic ostracod fauna (cladocopids, cytheraceans, healdiaceans, bairdiaceans and 'rectonariids') from the Dinantian of eastern Thuringia indicating basin facies. After Gründel (1961, 1962). Only species represented by at least 100 specimens are included.

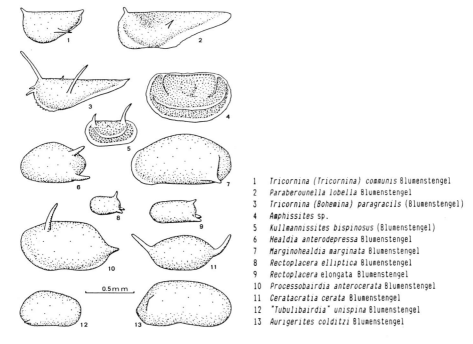

| | |
|---|---|
| 1 | *Tricornina (Tricornina) communis* Blumenstengel |
| 2 | *Paraberounella lobella* Blumenstengel |
| 3 | *Tricornina (Bohemina) paragracils* (Blumenstengel) |
| 4 | *Amphissites* sp. |
| 5 | *Kullmannissites bispinosus* (Blumenstengel) |
| 6 | *Healdia anterodepressa* Blumenstengel |
| 7 | *Marginohealdia marginata* Blumenstengel |
| 8 | *Rectoplacera elliptica* Blumenstengel |
| 9 | *Rectoplacera elongata* Blumenstengel |
| 10 | *Processobairdia anterocerata* Blumenstengel |
| 11 | *Ceratacratia cerata* Blumenstengel |
| 12 | *"Tubulibairdia" unispina* Blumenstengel |
| 13 | *Aurigerites colditzi* Blumenstengel |

Fig. 3. Thuringian Assemblage. Characteristic ostracod fauna (cytheraceans, kirkbyaceans, healdiaceans, 'rectonariids' and bairdiaceans) from the Upper Devonian of eastern Thuringia indicative of basin facies. After Blumenstengel (1965a). Only species represented by at least 100 specimens are included.

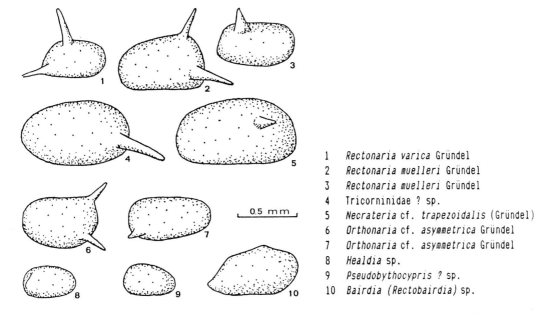

| | |
|---|---|
| 1 | *Rectonaria varica* Gründel |
| 2 | *Rectonaria muelleri* Gründel |
| 3 | *Rectonaria muelleri* Gründel |
| 4 | Tricorninidae ? sp. |
| 5 | *Necrateria* cf. *trapezoidalis* (Gründel) |
| 6 | *Orthonaria* cf. *asymmetrica* Gründel |
| 7 | *Orthonaria* cf. *asymmetrica* Gründel |
| 8 | *Healdia* sp. |
| 9 | *Pseudobythocypris* ? sp. |
| 10 | *Bairdia (Rectobairdia)* sp. |

Fig. 4. Thuringian Assemblage. Characteristic ostracod fauna ('rectonariids', healdiaceans and bairdiaceans) from the Dinantian of Bohemia indicative of basin facies. After Dvořák *et al.* (1986).

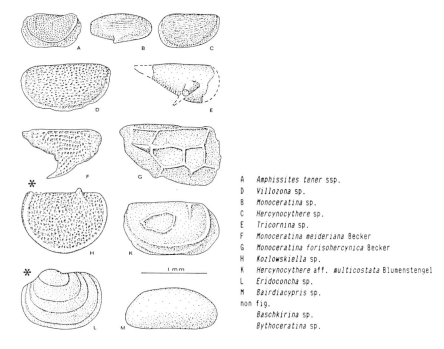

| | |
|---|---|
| A | *Amphissites tener* ssp. |
| D | *Villozona* sp. |
| B | *Monoceratina* sp. |
| C | *Hercynocythere* sp. |
| E | *Tricornina* sp. |
| F | *Monoceratina meideriana* Becker |
| G | *Monoceratina forisohercynica* Becker |
| H | *Kozlowskiella* sp. |
| K | *Hercynocythere* aff. *multicostata* Blumenstengel |
| L | *Eridoconcha* sp. |
| M | *Bairdiacypris* sp. |
| non fig. | |
| | *Baschkirina* sp. |
| | *Bythoceratina* sp. |

Fig. 5. Thuringian Assemblage. Characteristic ostracod assemblage from a deeper offshore environment from the late Givetian of eastern Morocco dominated by small and fragile cytheraceans (Category B of Bless); bairdiaceans (Category E) are very rare. After Becker (1988). Specimens marked by an asterisk are extremely rare and specifically biotope indicative forms (see Fig. 12).

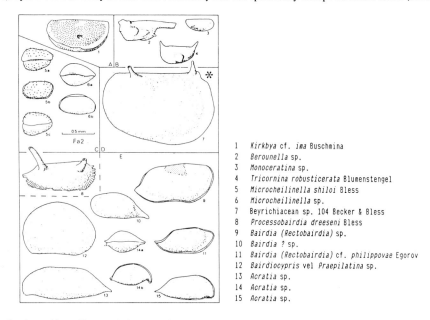

| | |
|---|---|
| 1 | *Kirkbya* cf. *ima* Buschmina |
| 2 | *Berounella* sp. |
| 3 | *Monoceratina* sp. |
| 4 | *Tricornina robusticerata* Blumenstengel |
| 5 | *Microcheilinella shiloi* Bless |
| 6 | *Microcheilinella* sp. |
| 7 | Beyrichiacean sp. 104 Becker & Bless |
| 8 | *Processobairdia dreeseni* Bless |
| 9 | *Bairdia (Rectobairdia)* sp. |
| 10 | *Bairdia* ? sp. |
| 11 | *Bairdia (Rectobairdia)* cf. *philippovae* Egorov |
| 12 | *Bairdiocypris* vel *Praepilatina* sp. |
| 13 | *Acratia* sp. |
| 14 | *Acratia* sp. |
| 15 | *Acratia* sp. |

Fig. 6. Thuringian Assemblage. Characteristic ostracod assemblage from a relatively deep open-marine environment from the Upper Famennian of eastern Belgium dominated by bairdiaceans (with *Processobairdia*), associated with cytheraceans and a spinose *Microcheilinella* species. After Dreesen et al. (1985). Specimens marked by an asterisk are specifically biotope indicative forms (see Fig. 12).

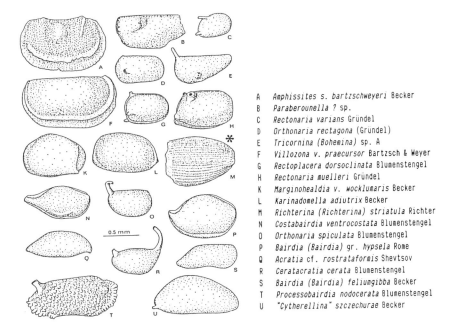

| | |
|---|---|
| A | *Amphissites s. bartzschweyeri* Becker |
| B | *Paraberounella ?* sp. |
| C | *Rectonaria varians* Gründel |
| D | *Orthonaria rectagona* (Gründel) |
| E | *Tricornina (Bohemina)* sp. A |
| F | *Villozona v. praecursor* Bartzsch & Weyer |
| G | *Rectoplacera dorsoclinata* Blumenstengel |
| H | *Rectonaria muelleri* Gründel |
| K | *Marginohealdia v. wocklumaris* Becker |
| L | *Karinadomella adiutrix* Becker |
| M | *Richterina (Richterina) striatula* Richter |
| N | *Costabairdia ventrocostata* Blumenstengel |
| O | *Orthonaria spiculata* Blumenstengel |
| P | *Bairdia (Bairdia)* gr. *hypsela* Rome |
| Q | *Acratia* cf. *rostrataformis* Shevtsov |
| R | *Ceratacratia cerata* Blumenstengel |
| S | *Bairdia (Bairdia) feliumgibba* Becker |
| T | *Processobairdia nodocerata* Blumenstengel |
| U | *"Cytherellina" szczechurae* Becker |

Fig. 7. Thuringian Assemblage. Characteristic basinal ostracod assemblage from the late Upper Famennian of eastern Morocco dominated by 'rectonariids' (Category B of Bless) and bairdiaceans (with *Processobairdia*, *Ceratacratia*; Category E), associated with cytheraceans (Category B) and entomozoaceans. After Becker (1987). Specimens marked by an asterisk are extremely rare and specifically biotope indicative forms (see Fig. 12).

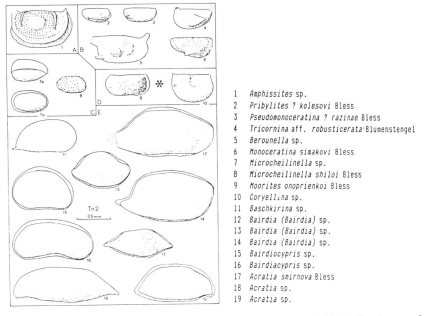

| | |
|---|---|
| 1 | *Amphissites* sp. |
| 2 | *Pribylites ? kolesovi* Bless |
| 3 | *Pseudomonoceratina ? razinae* Bless |
| 4 | *Tricornina* aff. *robusticerata* Blumenstengel |
| 5 | *Berounella* sp. |
| 6 | *Monoceratina simakovi* Bless |
| 7 | *Microcheilinella* sp. |
| 8 | *Microcheilinella shiloi* Bless |
| 9 | *Moorites onoprienkoi* Bless |
| 10 | *Coryellina* sp. |
| 11 | *Baschkirina* sp. |
| 12 | *Bairdia (Bairdia)* sp. |
| 13 | *Bairdia (Bairdia)* sp. |
| 14 | *Bairdia (Bairdia)* sp. |
| 15 | *Bairdiocypris* sp. |
| 16 | *Bairdiacypris* sp. |
| 17 | *Acratia smirnova* Bless |
| 18 | *Acratia* sp. |
| 19 | *Acratia* sp. |

Fig. 8. Thuringian Assemblage. Characteristic ostracod assemblage indicative of a deep, subtidal shelf environment from the Middle Tournaisian of northeastern Siberia dominated, by bairdiaceans, bairdiocypridaceans and cytheraceans, associated with a spinose *Microcheilinella* species. After Shilo *et al.* (1984), Dreesen *et al.* (1985). Specimens marked by an asterisk are specifically biotope indicative forms (cf. Fig. 12).

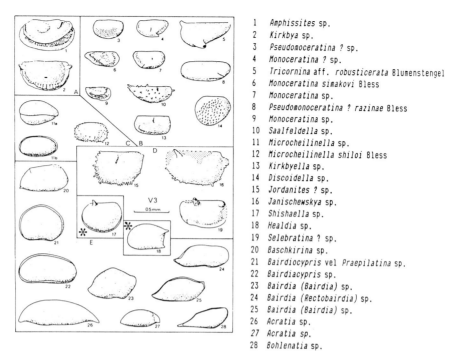

1 *Amphissites* sp.
2 *Kirkbya* sp.
3 *Pseudomoceratina ?* sp.
4 *Monoceratina ?* sp.
5 *Tricornina* aff. *robusticerata* Blumenstengel
6 *Monoceratina simakovi* Bless
7 *Monoceratina* sp.
8 *Pseudomonoceratina ? razinae* Bless
9 *Monoceratina* sp.
10 *Saalfeldella* sp.
11 *Microcheilinella* sp.
12 *Microcheilinella shiloi* Bless
13 *Kirkbyella* sp.
14 *Discoidella* sp.
15 *Jordanites ?* sp.
16 *Janischewskya* sp.
17 *Shishaella* sp.
18 *Healdia* sp.
19 *Selebratina ?* sp.
20 *Baschkirina* sp.
21 *Bairdiocypris* vel *Praepilatina* sp.
22 *Bairdiacypris* sp.
23 *Bairdia (Bairdia)* sp.
24 *Bairdia (Rectobairdia)* sp.
25 *Bairdia (Bairdia)* sp.
26 *Acratia* sp.
27 *Acratia* sp.
28 *Bohlenatia* sp.

Fig. 9. Thuringian Assemblage. Characteristic ostracod assemblage from an open-marine shelf environment from the Upper Visean of southwestern Morocco dominated by cytheraceans and bairdiaceans, associated with a spinose *Microcheilinella* species and cladocopids. After Dreesen *et al.* (1985). Specimens marked by an asterisk are specifically biotope indicative forms (cf. Fig. 12).

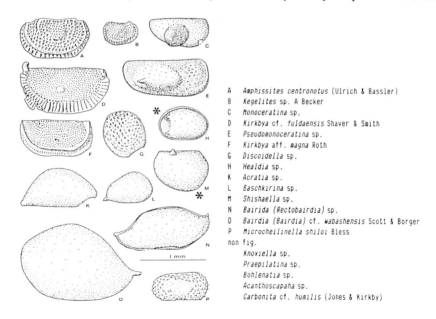

A *Amphissites centronotus* (Ulrich & Bassler)
B *Kegelites* sp. A Becker
C *Monoceratina* sp.
D *Kirkbya* cf. *fuldaensis* Shaver & Smith
E *Pseudomonoceratina* sp.
F *Kirkbya* aff. *magna* Roth
G *Discoidella* sp.
H *Healdia* sp.
K *Acratia* sp.
L *Baschkirina* sp.
M *Shishaella* sp.
N *Bairida (Rectobairdia)* sp.
O *Bairdia (Bairdia)* cf. *wabashensis* Scott & Borger
P *Microcheilinella shiloi* Bless
non fig.
*Knoxiella* sp.
*Praepilatina* sp.
*Bohlenatia* sp.
*Acanthoscapaha* sp.
*Carbonita* cf. *humilis* (Jones & Kirkby)

Fig. 10. Thuringian Assemblage. Characteristic ostracod assemblage of a wave protected, calm water niche in a shallow-water environment from the Upper Westphalian of northern Spain, dominated by bairdiaceans (Category E of Bless) and kirkbyaceans (Category A), associated with a spinose *Microcheilinella* species (Category C) and cladocopids (Category B). After Becker (in press). Specimens marked by an asterisk are extremely rare and specifically biotope indicative forms (see Fig. 12).

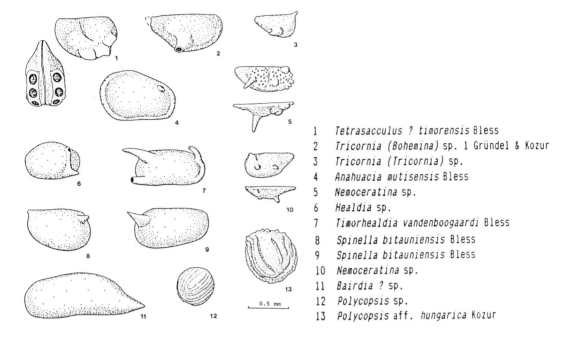

1    *Tetrasacculus ? timorensis* Bless
2    *Tricornia (Bohemina)* sp. 1 Gründel & Kozur
3    *Tricornia (Tricornia)* sp.
4    *Anahuacia mutisensis* Bless
5    *Nemoceratina* sp.
6    *Healdia* sp.
7    *Timorhealdia vandenboogaardi* Bless
8    *Spinella bitauniensis* Bless
9    *Spinella bitauniensis* Bless
10   *Nemoceratina* sp.
11   *Bairdia ?* sp.
12   *Polycopsis* sp.
13   *Polycopsis* aff. *hungarica* Kozur

Fig. 11. Thuringian Assemblage. Selected cytheraceans, 'rectonariids' and cladocopids (Category B of Bless), bairdiaceans (Category E) and healdiaceans (Category G) discovered in shallow shelf deposits in the Lower Permian of Timor. After Bless (1987).

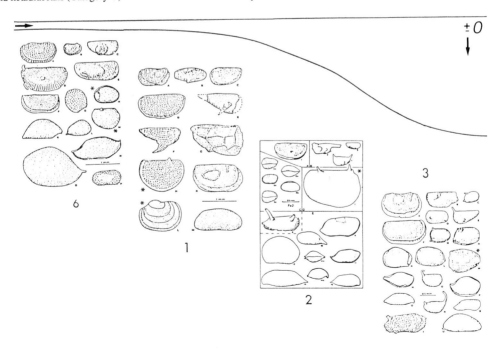

Fig. 12. Four subassemblages of the Thuringian Assemblage each representative of different water depths and distances from Palaeozoic shorelines (6 = nearshore niche, 1 = offshore, 2 = open-marine, 3 = basinal).

## REFERENCES

Alberti, H. 1970. Neue Trilobiten-Faunen aus dem Oberdevon Marokkos. *Göttinger Arb. Geol. Paläont.*, Gottingen, **5**, 15-29.

Alberti, H. 1972. Zur geologischen Entwicklung der Sahara Occidental an der Wende Devon/Karbon. *Newsl. Stratigr.*, **2**, 149-152.

Bandel, K. & Becker, G. 1975. Ostracoden aus paläozoischen-pelagischen Kalken der Karnischen Alpen (Silurium bis Unterkarbon). *Senckenberg. leth.*, Frankfurt a.M. **56**, 1-283.

Becker, G. 1969. Zur Paläokologie der Ostracoden. *Natur Mus.*, Frankfurt, **99**, 198-208.

Becker, G. 1971. Paleoecology of Middle Devonian ostracodes from the Eifel Region, Germany. *Bull. Cent. Rech. Pau - SNPA*, **5** suppl., 801-816.

Becker, G. 1973. Paläokolgische Analyse einer Ostracoden-Fauna aus dem Oberdevon von Belgien. *Neues Jb. Geol. Paläont. Abh.*, Stuttgart, **142**, 59-71.

Becker, G. 1976. Oberkarbonische Entomozoidae (Ostracoda) im Kantabrischen Gebirge (N-Spanien). *Senckenberg. leth.*, **57**, 201-223.

Becker, G. 1977. Thuringian ostracods from the Famennian of the Cantabrian Mountains (Upper Devonian, N. Spain). *In* Löffler, H. & Danielopol D. L. (Eds), *Aspects of ecology and zoogeography of Recent and fossil Ostracoda*, 459-474. Dr W. Junk b.v. Publishers, The Hague.

Becker, G. 1978. Flachwasser-Ostracoden aus dem hohen Westfal Asturiens (Kantabrisches Gebirge, N-Spanien). 1. Palaeocopida. *Senckenberg. leth.*, Frankfurt a.M., **59**, 37-69.

Becker, G. 1981a. Ostracoda aus cephalopoden-führendem Oberdevon im Kantabrischen Gebirge (N-Spanien). 1. Hollinacea, Primitiopsacea, Kirkbyacea, Healdiacea und Bairdiocypridacea. *Palaeontographica*, A, **173**, 1-63.

Becker, G. 1981b. *Vitissites* n. g. und die Herkunft der Amphissitidae (Ostracoda). *Senckenberg. leth.*, Frankfurt a.M., **62**, 173-191.

Becker, G. 1982a. Ostracoda aus cephalopoden-führendem Oberdevon im Kantabrischen Gebirge (N-Spanien). 2. Bairdiacea, Cytheracea und Entomozoacea. *Palaeontographica*, A, **178**, 109-182.

Becker, G. 1987. Ostracoda des Thüringer Ökotyps aus dem Grenzbereich Devon/Karbon N-Afrikas (Marokko, Algerien). *Palaeontographica*, A, **200**, 45-104.

Becker, G. 1988. ''Hercyn''-Ostracoden aus N-Afrika (Givetium; Marokko). *Neues Jb. Geol. Paläont. Abh.*, Stuttgart, **177**(1), 74-92.

Becker, G. In press. Flachwasser-Ostracoden aus dem hohen Westfal Asturiens (Kantabrisches Gebirge, N-Spanien). 2. Podocopida. *Senckenberg. leth.*, Frankfurt a.M., **68**.

Becker, G. & Adamczak, F. J. In press. On *Aurikirkbya wordensis* (Hamilton). Stereo-Atlas Ostracod Shells, London, **16**, 1-4.

Becker, G. & Bless, M. J. M. 1974. Ostracod stratigraphy of the Ardenno-Rhenish Devonian and Dinantian. *Internat. Symp. belg. micropaleont. limits from Emsian to Visean*, Bruxelles, **1**, 1-52.

Becker, G. & Bless, M. J. M. 1987. Cypridinellidae (Ostracoda) aus dem Oberdevon Hessens (Unterer Kellwasser-Kalk; Lahn-Dill-Gebiet und östliches Sauerland, Rechtsrheinisches Schiefergebirge). *Geol. Jb.*, Hessen, **115**, 29-56.

Becker, G., Bless, M. J. M. & Kullmann, J. 1975. Oberkarbonische Entomozoen-Schiefer im Kantabrischen Gebirge (Nordspanien). *Neues Jb. Geol. Paläont. Abh.*, Stuttgart, **150**, 92-110.

Becker, G. & Sánchez de Posada, L. C. 1977. Ostracoda aus der Moniello-Formation Asturiens (Devon; N-Spanien). *Palaeontographica*, A, **158**, 114-203.

Benson, R. H. & Sylvester-Bradley, P. C. 1971. Deep-sea ostracodes and the transformation of ocean to sea in the Tethys. *Bull. Cent. Rech. Pau*, SNPA, **5** suppl., 63-91.

Berdan, J. M. & Copeland, M. J. 1973. Ostracodes from the Lower Devonian formations in Alaska and Yukon Territory. Prof. Pap. U.S. geol. Surv., Washington, **825**, 1-47.

Bless, M. J. M. 1983. Late Devonian and Carboniferous ostracode assemblages and their relationship to the depositional environment. *Bull. Soc. belge Géol.*, **92**, 31-52.

Bless, M. J. M. 1987. Lower Permian ostracodes from Timor (Indonesia). *Proc. K. ned. Akad. Wet.*, Amsterdam, Section B, **90**(1), 1-30.

Bless, M. J. M., Boonen, P., Dusar, M. & Soille, P. 1981. Microfossils and depositional environment of the late Dinatian carbonates at Heibaart (northern Belgium). *Annls Soc. géol. Belg.*, Liège, **104**, 135-154.

Bless, M. J. M, Bouckaert, J., Bouzet, Ph., Conil, R., Cornet, P., Fairon-Demaret, M., Groessens, E., Longerstaey, P. J., Meessen, J. P. M. Th., Paproth, E., Pirlet, H., Streel, M., van Ameron, H. W. J. & Wolf, M. 1976. Dinantian rocks in the subsurface north of the Brabant and Ardenno-Rhenish massifs in Belgium, the Netherlands and the Federal Republic of Germany. *Meded. Rijks geol. Dienst*, Leiden, N.S., **27**, 82-195.

Bless, M. J. M. & Jordan, H. 1971. The new genus *Copelandella* from the Carboniferous - the youngest known beyrichiacean ostracodes. *Lethaia*, Oslo, **4**, 185-190.

Bless, M. J. M. & Sánchez de Posada, L. C. 1972. Sobre la aparición de ostrácodos nectónicos en la Cordillera-Cantábrica. *Breviora geol. astúr.*, Oviedo, **17**, 25-30.

Bless, M. J. M., Streel, M. & Becker, G. 1988. Distribution and paleoenvironment of Devonian to Permian ostracod assemblages in Belgium with reference to some late Famennian to Permian marine nearshore to ''brackish-water'' assemblages dated by microspores. *Annls Soc. géol. Belg.*, Liège, **110**, 347-362.

Blumenstengel, H. 1963. Zur Mikrofauna des Thüringer Ockerkalks (Silur). *Geologie*, **12**, 349-354.

Blumenstengel, H. 1965a. Zur Taxionomie und Biostratigraphie verkieselter Ostracoden aus dem Thüringer Oberdevon. *Freiberger Forsch.-H.*, C **183**, 1-127.

Blumenstengel, H. 1965b. Zur Ostracodenfauna eines Kalkgerölls aus dem Thüringer Lederschiefer (Ordovizium). *Freiberger Forsch.-H.*, C **182**, 63-78.

Blumenstengel, H. 1965c. Biostratigraphie des ostthüringisch-vogtländischen Paläozoikums. *Ber. geol. Ges. D.D.R.*, Berlin, **10**, 201-202.

Blumenstengel, H. 1965d. Ergebnisse der Ostracodenforschung aus dem Paläozoikum der Deutschen Demokratischen

Republik (Übersicht). *Ber. dt. Ges. geol. Wiss.*, Berlin, A, Geol. Paläont., **13**, 159-166.

Blumenstengel, H. 1973. Zur stratigraphischen und faziellen Bedeutung der Ostracoden im Unter- und Mittelharz. *Z. geol. Wiss. Berlin*, Themenh., **1**, 67-79.

Blumenstengel, H. 1974. Ostracoden aus dem Mitteldevon des Harzes (Blankenburger Zone). *Freiberger Forsch.-H.*, **C 298**, 19-43.

Blumenstengel, H. 1979. Die Ostracodenfauna der Wocklumeria-Stufe (Oberdevon) bei Saalfeld im Thüringer Schiefergebirge. *Z. geol. Wiss.*, Berlin, **7**, 521-557.

Bultynck, P. 1985. Lower Devonian (Emsian) - Middle Devonian (Eifelian and lowermost Givetian) conodont successions from the Ma'der and the Tafilalt, southern Morocco. *Cour. Forsch.-Inst. Senckenberg*, **75**, 261-286.

Buschmina, L. S. 1977. Novye vidy ostrakod iz nizhnego karbona tzentralnogo Kazakhstana. *Trudy Inst. Geol. Geofiz.*, Akad. Nauk SSSR, Sibirskoe Otdelenie, **345**, 84-94.

Casier, J.-G. 1987. Étude biostratigraphique et paléoécologique des ostracodes du recif de Marbre Rouge du Haumont à Vodelée (partie supèrieur du Frasnien, Bassin de Dinant, Belgique). *Rev. Paléobiol.*, **6**, 193-204.

Dreesen, R., Bless. M. J. M., Conil, R., Flajs, G. & Laschet, Ch. 1985. Depositional environment, paleoecology and diagenetic history of the "Mabre rouge à crinoides de Baelen" (late Upper Devonian, Verviers Synclinorium, eastern Belgium). *Annls Soc. géol. Belg.*, Liège, **108**, 311-359.

Dvořák, J., Friákova, O., Hladil, J., Kaldova, J., Kukal, Z. & Bless, M. J. M. 1986. A field trip to the Famennian of the Moravian Karst (CSSR). *Annls Soc. géol. Belg.*, Liège, **109**, 267-273.

Gooday, A. J. 1983. Entomozoacean ostracods from the Lower Carboniferous of south-western England. *Palaeontology*, London, **26**, 755-788.

Gründel, J. 1961. Zur Biostratigraphie und Fazies der Gattendorfia-Stufe in Mitteldeutschland unter besonderer Berücksichtigung der Ostracoden. *Freiberger Forsch.-H.*, **C 111**, 53-173.

Gründel, J. 1962. Zur Taxionomie der Ostracoden der Gattendorfia-Stufe Thüringens. *Freiberger Forsch.-H.*, **C 151**, 51-105.

Gründel, J. 1966. Zur Entwicklung und Taxionomie der Tricornidae (Ostracoda) in Mitteleuropa. *Paläont. Z.*, **40**, 89-102.

Gründel, J. 1972. Ostracoden (Crustacea) aus dem Visé des Harzes (Kulm-Fazies). *Freiberger Forsch.-H.*, **C 176**, 10-30.

Gründel, J. & Kozur, H. 1972. Zur Taxonomie der Bythocytheridae und Tricorninidae (Podocopida, Ostracoda). *Monatsber. dt. Akad. Wiss. Berl.*, **13**(10/12), 907-937.

Gründel, J. & Kozur, H. 1975. Psychrosphärische Ostracoden aus dem Perm von Timor. *Freiberger Forsch.-H.*, **C 304**, 39-45.

Hartmann, G. 1963. Zur Phylogenie und Sytematik der Ostracoden. *Z. Zool. Syst. EvolForsch.*, Frankfurt am Main, **1**, 1-154.

Hartmann, G. 1964. The problem of polyphyletic characters on ostracods and its significance to ecology and systematics. *Pubbl. Staz. zool. Napoli*, Milano & Napoli, **33** suppl.,
32-44.

Hartmann, G. 1975. Ostracoda. *In Dr H. G. Bronns Klassen und Ordnungen des Tierreichs*, **5**, Arthropoda. Abt. 1, Crustacea, Buch 2, Tl. 4, Lfg. 4, 569-786. Fischer.

Jordan, H. 1964. Zur Taxonomie und Biostratigraphie der Ostracoden des höchsten Silur und Unterdevon Mitteleuropas. *Freiberger Forsch.-H.*, **C 170**, 1-147.

Knüpfer, J. 1968. Ostracoden aus dem Oberen Ordovicium Thüringens. *Freiberger-Forsch-H.*, **C 234**, 5-29.

Kozur, H. 1972. Die Bedeutung triassischer Ostracoden für stratigraphische und paläökologische Untersuchungen. *Mitt. Ges. Geol. Bergbaust.*, **21**, 623-660.

Kuwano, Y. 1987. Early Devonian conodonts and ostracodes from Central Japan. *Bull. natn. Sci. Mus.*, Tokyo, Ser. C, **13**, 77-105.

Lethiers, F. & Crasquin, S. 1985. Reconnaissance des milieux profonds de la Paléotéthys à l'aide des ostracodes. *Bull. Soc. géol. Fr.*, Paris, **3**, 415-423.

Liebau, A. 1980. Paläobathymetrie und Ökofaktoren: Flachmeer-Zonierungen. *Neues Jb. Geol. Paläont. Abh.*, Stuttgart, **160**, 173-216.

Massa, D. 1965. Observations sur les séries siluro-dévoniennes des confins algéro-marocains du Sud. *Not. Mém. Comp. franç. Pétrol.*, **8**, 5-187.

Neale, J. 1965. Some factors influencing the distribution of Recent British Ostracoda. *Pubbl. Staz. zool. Napoli*, Milano & Napoli, **33**, suppl., 247-307.

Neale, J. 1983. The Ostracoda and uniformitarianism. 1. The later record: Recent, Pleistocene and Tertiary (Presidential address, 1981). *Proc. Yorks. geol. Soc.*, Leeds, **44**, pt. 3, no. 21, 305-326.

Olempska, E. 1979. Middle to Upper Devonian Ostracoda from the southern Holy Cross Mountains, Poland. *Palaeont. pol.*, Warszawa, **40**, 57-162.

Peypouquet, J.-P. 1983. Paléobathymetrie et paléohydrologie dans la coupe du Kef (Tunisie nord-orientale) entre le Maestrichtien et l'Ypresien sur la base des ostracodes. *Bull. Soc. paleont. italiana*, **22**, 21-29.

Rebske, W., Rebske, Ch., Bless, M.J.M., Paproth, E., & Steemans, Ph. 1985. Over enkele fossilien uit de Klerf-Schichten (onder Emsien) bij Waxweiler (Eifel, BRD) en hun leefmilieu. *Grondboor Hamer*, Nederlandse Geologisch Vereniging, **5**, 142-155.

Sánchez de Posada, L. C. 1976. Quelques remarques au sujet de la répartition des faunes d'ostracodes carbonifères de la Chaine Cantabrique. *Annls Soc. géol. N.*, Lille, **96**, 407-412.

Sánchez de Posada, L. C. & Bless, M.J.M. 1971. Una microfauna del Westfaliense C de Astúrias. *Revta esp. Micropaleont.*, Madrid, **3**, 193-204.

Shilo, N.A., Bouckaert, J., Afanasjeva, G. A., Bless, M. J. M., Conil, R., Erlanger, O. A., Gagiev, M. H., Lazarev, S. S., Onoprienko, Yu. I., Poty, E., Razina, T. P., Simakov, K. V., Smirnova, L. V., Streel, M. & Swennen, R. 1984. Sedimentological and paleontological atlas of the late Famennian and Tournaisian deposits in the Omolon Region (NE-USSR). *Annls Soc. géol. Belg.*, Liège, **107**, 137-247.

Swennen, R., Bless, M. J. M., Bouckaert, J., Razina, T. P. & Simakov, K. V. 1986. Evaluation of the transgression-regression events in the Upper Famennian-Tournaisian strata of the southeastern Omolon Area (NE-Siberia, USSR). *Annls Soc. géol. Belg.*, Liège, **109**, 237-248.

Van Morkhoven, F.P.C.M. 1962. *Post-Palaeozoic Ostracoda, their morphology, taxonomy and economic use*, Volume I, General, 204 pp., 79 figs, 8 tables, 1 enclosure. Elsevier, Amsterdam, London, New York.

Wang, Shang-qi. 1976. A new ostracode genus *Paramoelleritia* from the Devonian deposits in Guangxi. *Acta paleont. sin.*, Beijing, **9**, 231-239.

Wang, Shang-qi. 1983. Late Devonian pelagic ostracod sequences in Luofu region of Guangxi. *Bull. Nanjing Inst. Geol. Palaeont.*, Acad. sinica, **6**, 159-172.

Wang, Shang-qi. 1986. Discovery of Devonian entomozoids (Ostracoda) from southern Guizhou and its significance. *Acta micropalaeont. sin.*, **3**, 81-88.

Wang, Shang-qi. 1987. Entomozoids (Ostracoda) from Upper Devonian at Hangshan section of Longlin, Guangxi. *Acta palaeontol. sin.*, Beijing, **5**, 306-317.

Wang, Shang-qi 1988. Late Paleozoic ostracod associations from south China and their paleontological significances. *Acta palaeontol. sin.*, Beijing, **27**, 91-102.

Whatley, R. C. 1983a. The application of Ostracoda to palaeoenvironmental analysis. *In* Maddocks, R. F. (Ed.), *Applications of Ostracoda*, proceedings of the Eighth International Symposium on Ostracoda, July 26-29, 1982, 51-57. Univ. Houston Geos., Houston, Texas.

Whatley, R. C. 1983b. Some simple procedures for enhancing the use of Ostracoda in palaeoenvironmental analysis. *In* Costa, L. I. (Ed.), Proc. Symposium Biostratigraphy North Sea, Stavanger, 1981. *Norwegian Petrol. Directorate Bull.*, **2**, 129-146.

Whatley, R. C. 1988. Ostracoda and palaeogeography. *In* DeDeckker, P., Colin, J.-P. & Peypouquet, J.-P. (Eds), *Ostracoda in the Earth Sciences*, 103-123. Elsevier, Amsterdam, Oxford, New York, Tokyo.

Whatley, R. C., Trier, K. & Dingwall, P. M. 1982. Some preliminary observations on certain mechanical and biophysical properties of ostracod carapace. *In* Bate, R. H., Robinson, E. & Sheppard, L. M. (Eds), *Fossil and Recent ostracods*, 76-104. Ellis Horwood Ltd., Chichester for British Micropalaeontological Society.

Whatley, R. C. & Watson, K. 1988. A preliminary account of the distribution of Ostracoda in Recent reef and reef associated environments in the Pulau Seribu or the Thousand Island Group, Java Sea. *In* Hanai, T., Ikeya, N. & Ishizaki, K. (Eds), *Evolutionary biology of Ostracoda, its fundamentals and applications*, proceedings of the Ninth International Symposium on Ostracoda, held in Shizuoka, Japan, 29 July - 2 August 1985, Developments in palaeontology and stratigraphy, **11**, 399-411, Kodansha Ltd., Tokyo and Elsevier, Amsterdam, Oxford, New York, Tokyo.

# 33

# The colonization of subsurface habitats by the Loxoconchidae Sars and the Psammocytheridae Klie

**Dan L. Danielopol[1] & Gioacchino Bonaduce[2]**

[1]Limnological Institute, Austrian Academy of Sciences, A - 5310 Mondsee, Austria.
[2]Palaeontological Institute, University of Naples, Naples 80121, Italy

## ABSTRACT

Representatives of the genera *Tuberoloxoconcha* and *Psammocythere* are widely distributed in marine interstitial habitats. Recent species of the genera *Pseudolimnocythere*, *Kliella* and *Nannokliella*, live in subterranean freshwater. Their phylogenetic affinities with epigean ostracods are discussed. A comparative morphological study of the carapace and the walking legs of subterranean Pseudolimnocytherinae, with those of related epigean species, suggests that preadaptations of these structures facilitated the colonization of the interstitial environment. The wide salinity tolerance of the Pseudolimnocytherinae and their ability to colonize submarine springs in karstic coastal areas was an important aid in their invasion of the freshwater subterranean realm. It is suggested that representatives of both the Pseudolimnocytherinae (Loxoconchidae) and Psammocytheridae colonized freshwater habitats during the Upper Miocene. Regional geological events, such as the Messinian Salinity Crisis, could on one hand destroy a part of the marine ostracod fauna of the Mediterranean, yet on the other hand, facilitate the adaptation to freshwater of the interstitial fauna living in karstic springs. The modern marine interstitial Pseudolimnocytherinae and Psammocytheridae in the Mediterranean are recent immigrants which arrived after the reopening of the connection with the Atlantic in the Pliocene.

## INTRODUCTION

The evolution of subterranean organisms from epigean ones presents an intriguing and stimulating series of problems. Why should an epigean organism leave the earth's surface where it can easily find its resources in daylight, to adopt a troglobitic life style subject to more constrained conditions, e.g., no light, sometimes reduced space and limited food resources? Scientists in recent decades have described, from both marine and freshwater, interstitial and/or cavernous habitats, more than 300 species of ostracods which live exclusively in this peculiar environment (Danielopol & Hartmann, 1986). These ostracods display remarkable morphological and biological peculiarities. They have reduced ocular structures, are unpigmented, have a reduced chaetotaxy and some of the phanerae (like the sensorial setae) are very well developed. Many interstitial dwelling ostracods have a carapace of reduced size and of elongated shape (Danielopol & Hartmann, 1986).

For the marine interstitial Ostracoda, Hartmann (1973) considered such peculiarites as the minute size of the carapace or the reduced number of phanerae, adaptations to this special subsurface environment. One can ask, how the morphological characters of the subsurface ostracods developed. There are three possibilities:

1) Through a gradual morphological change of the animal in the subsurface habitats. Analogues exist for the surface dweller fauna (Mayr, 1982).

2) Through a few, but important, morphological and/or physiological changes, when the organisms switch from the epigean environmental situation to the subsurface one (for such examples see Westheide, 1987).

3) Through changes which occurred in another environment and which represent efficient functional solutions for the future occupation of the subsurface environment. We call such characteristics, 'preadaptations' (Osche, 1962). They become useful in the new hypogean environment and can be further improved through evolutionary changes.

Mayr (1976, 100) made the interesting distinction between preadaptation related to a functional shift and that for a habitat shift. We shall see here that some of the morphological preadaptations of the ostracods are useful for a change of habitat.

Maddocks (1976) considered the morphological characters mentioned by Hartmann (1973), as adaptations to the marine interstitial life, as evolutionary trends (for a similar position see also Wouters, 1987). Obviously, we deal with two different ways of defining the adaptation. Hartmann (1973) considers adaptation a character state, while Maddocks (1976) and Wouters (1987) see it as a process. In both cases one assumes that in order to define what is an adaptation to the subsurface environment for an ostracod one needs to know the phylogenetical affinities between closely related epigean and hypogean ostracods as well as their ecological requirements.

Unfortunately, until recently there have been few studies which explicitly consider these prerequisites when discussing either morphological trends or static adaptations of marine and/or freshwater hypogean ostracods (e.g., Hartmann, 1974; Maddocks, 1976; Gottwald, 1983; Carbonel et al., 1986, Marmonier & Danielopol, 1988). Maddocks (1976) and Danielopol (1976) showed that within the phylogenetic lineage *Anchistrocheles-Pussella* (Pussellinae) one notices a trend in the reduction of the carapace size and in the number of the setae of various limbs. A similar trend is visible in the case of the *Polycope* of the species group *loxobanosi* (Hartmann, 1977). There are minimal morphological changes of the carapace in size and shape when one compares epigean and hypogean species of *Kovalevskiella* (Carbonel et al., 1986) and of *Xestoleberis* of the group *arcturi* (Gottwald, 1983) or within one species when one looks at various populations of *Nannocandona faba* (Marmonier & Danielopol, 1988). In the case of the Pussellinae, one can suggest that the success in the colonization of the interstitial habitats is due to their gradual morphological adaption to this environment, whereas in the case of *Kovalevskiella, Xestoleberis* and *Nannocandona* colonization success is due to the preadaption features existing in the epigean dweller species.

Another interesting problem concerning the colonization of subterranean waters by ostracods is their passage from the marine environment to inland freshwater. Vandel (1965, 272) remarked: 'Strangely enough the majority of aquatic cavernicoles have not been derived from freshwater epigeous forms, but from marine forms.' Danielopol (1977) pointed out that in the case of ostracods only few representatives from the marine interstitial habitats directly colonized the freshwater subterranean realm. This should be the case with representatives of the Pseudolimnocytherinae and Psammocytheridae. When and how the marine ostracods of these latter groups crossed the salinity boundary and became further adapted to new freshwater, subterranean habitats remains unclear (Danielopol, 1980).

Considering the epigean Crustacea, Ekman (1913, 1914, 1918) showed that several glacial relicts, actually living in freshwater inland habitats of Northern Europe, could cross the salinity boundary due to their wide tolerance to low salt concentrations. He showed that there are at least two pathways of colonization of inland freshwaters by the North-European glacial forms, i.e. the active colonization of the estuaries and the passive colonization due to the damming of coastal embayments by glaciers. Thienemann (1950) and Segerstrale (1957) refined Ekmann's zoogeographical model, but many ecophysiological problems remain open to investigation, when one tries to explain the detailed distribution of various Crustacea of marine origin. The differential success in the distribution of several populations of *Gammarus duebeni* in Ireland and England for instance, were explained by Sutcliffe & Shaw (1968), only by careful examination of the water chemistry of the habitats and the osmoregulatory capacities of different populations living in these two geographical areas.

Considering the marine fauna which colonized subterranean freshwater habitats, one is confronted with various possibilities. Sket, in a series of publications, (e.g., 1970, 1986) argued that many invertebrates around the Mediterranean coasts actively migrated first into surface freshwater and from there colonized inland subterranean habitats. Stock (1980) proposed a passive model where the marine interstitial organisms during marine regressive phases were stranded along the coast and colonized, as a *refugium*, the freshwater subterranean environment. The ecophysiological aspects as well as the details of these two possible migration pathways have been little investigated. Few data exist on the migration of the marine ostracods which colonized the hypogean freshwater (Danielopol, 1980, Danielopol & Hart, 1985).

The present paper investigates the case histories of two ostracod groups, the Pseudolimnocytherinae and the Psammocytheridae, which successfully colonized both the marine and freshwater interstitial habitats. In order to reconstruct their history, we shall describe the phylogenetic relationships within these ostracod groups as well as their ecology and the geographical distribution of present and past ostracod species. We also propose for these groups a speculative model of the way they spread into the freshwater subterranean habitats along the Mediterranean coast. The merit of this model, in our opinion, is its plausibility as well as its possibility to be tested in future field and experimental observations.

## MATERIAL, SAMPLING SITES AND METHODS

The ostracods we examined were collected mostly by us or our students, i.e., the *Psammocythere* of the Mediterranean by G.B. and co-workers and *Tuberoloxoconcha* and *Pseudolimnocythere* by D.L.D. and colleagues. Data on this material have been reported in Bonaduce *et al.* (1980); Costa (1985) and Danielopol (1979, 1980). A new fossil *Pseudolimnocythere* species has been studied by Huber-Mahdi and Danielopol (unpublished). This material stems from the Badenien (Middle Miocene) of the Parathethys, at Hainburg, in the Vienna Basin. Besides this material, D.L.D. investigated the type material of *Psammocythere remanei*, Klie 1936. Additional information on the sampling sites, sampling methods and Ostracoda collected in southern France and southern Italy is given below.

The material of *Tuberoloxochoncha tuberosa* (Hartmann) was collected in southern France at Banyuls-sur-Mer, the type locality of this species (Hartmann, 1953). We dug 30cm diameter and 30-40cm deep holes following the Karaman and Chappuis method (Danielopol, 1978). The sampling site is located in a small beach (Ba-3), close to Les Elmes beach, at 0.2-0.5m distance from the sea.

At site Ba-2, a beach in front of La Baillaury, a temporary stream, enters the Bay of Banyuls, the sediment grain size is very similar to that at site

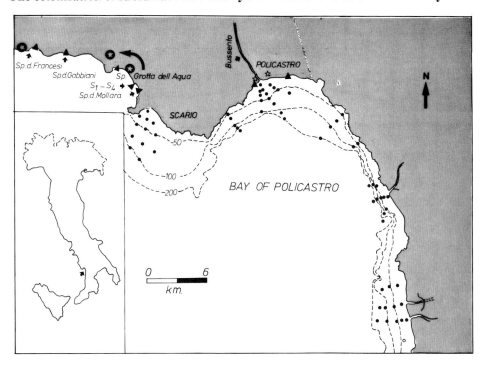

Fig. 1. The Bay of Policastro (southern Italy) and the sampling sites for ostracods (dots = benthonic samples collected by Bonaduce and co-workers (Costa, 1985); triangles = supra- and sublittoral sites from where interstitial and phytal ostracods have been sampled by Danielopol; white stars in black dots = sampling sites in caves; white star = sample taken from a well containing freshwater in the village of Policastro and from a Karaman-Chappuis hole in the Busento estuary; S1 - S4 location of the submarine karstic springs investigated during this study). The chart is redrawn with modifications from Costa (1985).

Ba-3 (see Delamare-Debouteville, 1960, Fig. 189). The salinity of the interstitial waters varies according to the flow of the stream (e.g., in May 1984 the salinity was 14%o). Here we found a single adult male of *Pseudolimnocythere* sp.

Our ostracod material from southern Italy was collected in the area of the Bay of Policastro (Fig. 1). A systematic survey of the offshore benthonic ostracods is given by Costa (1985) who also mentions the presence of *Psammocythere* sp. off Scario and Busento on sands, sandy pelitic and pelitic sediments between 50 and 100m. Where mixohaline water occurs, valves of *Tuberoloxocon-cha tuberosa* (Plate 1, Figs D-F) were collected from a pool in a littoral cave close to the Spiaggia dei Francesi (Fig. 1). The pool is 1m above sea level. Living and dead specimens of *Tuber-oloxoconcha* sp. (Danielopol, 1981) (Plate 1, Figs A-C) were found in the sandy sediments of subma-rine springs which discharge into the littoral area of

Molara beach in the northern part of the Scario area (Fig. 1, S1-S4) at a depth of 1-2m and some 10 to 20m offshore.

The area is karstic dominated by the cliffs of the Bulgheria Mountains. The rocks are Jurassic limestones (Di Stefano, 1984) containing both a well developed palaeokarst with filled caves and a modern karstic drainage system which dis-charges partly under the sea through diffuse springs. Small beaches exist in this area (Fig. 1). The sediment is mainly coarse sand, gravel and cobbles in the supra- and sublittoral (Fig. 2B). The salinity of the interstitial water which discharges in this area varies between 9 and 14%o, while the surface sea water is 33 to 35%o (Fig. 5). The submarine seepage flow of the karst water in one spring was 3.8ml. Besides *Tuberoloxoconcha* sp. in these springs we found a rich ostracod assem-blage. In spring No. 2 at a depth of 10 to 20cm *Loxoconcha stellifera* (Plate 1, Fig. G),*Callistocythere* aff. *pallida* (Plate 1, Fig. H), *Paradoxostoma* sp. and *Xestoleberis parva* (Plate 1, Fig. J) occur. Spring No. 3 in the superficial sandy sediments (0-5cm) yielded *Tuberoloxoconcha* sp., *Callistocythere* aff. *pallida*, *Cytherois* sp. and *Xestoleberis* sp. At 20 to 30cm depth one still finds as well as *Tuberoloxoconcha* sp., both *Xestoleberis* sp. and *Cytherois* sp. Spring No. 4 in the superficial sandy layer (0-5cm) contains *Tuberoloxoconcha* sp., *Aglaiocypris rara*, *Propontocypris* sp., *Loxoconcha stellifera*, *Loxocauda* sp., *Xestoleberis* sp., *Paradoxostoma* sp. and *Callistocythere* aff. *pallida*. In the deeper sediment layers (20-40m) besides the species mentioned above *Polycope* sp. and *Cytherois* sp. have been recovered.

With the exception of *Tuberoloxoconcha* sp. and *Polycope* sp., which are blind and seem to be exclusively interstitial, the other species are sighted and also occur on sediments covered with algae or sandy pelitic sediments.

We also investigated the interstitial habitats of several other beaches, notably Molara, Grotta dell'Aqua, Spiaggia dei Gabbiani, Spiaggia dei Francesi and the beach south of Busento (Fig. 1). At the Grotta dell'Aqua beach we found another interesting interstitial ostracod, *Xestoleberis delamarei* Hartmann (Plate 1, Fig. I). Samples from a drip-pool in the Grotta dell'Aqua (Fig. 1), from a Karaman-Chappuis hole in the estuary of the River Busento and from a well on the beach close to the river, contained only surface dwelling invertebrates of freshwater origin. The salinity of these habitats is given in Fig. 5. The sublittoral sediments off Busento beach, an area outside the karst zone, contain only fine silty sand (Fig. 2B) which is not inhabited by interstitial ostracods. For comparative studies, we also sampled phytal ostracods from algae growing on rocks around the submarine springs or close to the beaches. We found various species of *Xestoleberis*, e.g., *X. parva*, *X. pellucida*, *X. communis* and species of *Loxoconcha*, *Paradoxostoma* and *Semicytherura*.

Fig. 2A shows the various devices which we used for sampling ostracods, i.e. a plankton net for the phytal and superficial sandy sediments (the 0-5cm layer) and a Bou-Rouch pump (Danielopol, 1978) for the deeper layers in the spring areas. A Lee seepagemeter as described by Danielopol & Niederreiter (1987) was used to measure the amount of water discharging through the sediments (Fig. 2A).

## THE INTERSTITIAL LOXOCONCHIDAE AND PSAMMOCYTHERIDAE: THEIR PHYLOGENETICAL RELATIONSHIPS TO SURFACE DWELLERS

The Loxoconchidae Sars is a family with several hundred Recent and fossil species. The phylogenetic relationships of these species are poorly known. Athersuch and Horne (1984) reviewed a group of genera (with small carapace size) including *Hirschmannia* Elofson and *Elofsonia* Wagner. They also discussed several genera which have interstitial marine and freshwater representatives, i.e. *Nannocythere* Schäfer, *Tuberoloxoconcha* Hartmann and *Pseudolimnocythere* Klie. For this latter genus, Hartmann & Puri (1974) erected the subfamily Pseudolimnocytherinae. A small enigmatic loxoconchid species, '*Loxoconcha*' *helgolandica* Klie lives interstitially in the brackish water of Heligoland.

The Pseudolimnocytherinae lineage is characterized by small, elongated carapaces (between 320 and about 500µm length) with the anterior end higher than the posterior, and lacking ventroposterior sexual dimorphism in the valves. The hinge is hennodont with a smooth central bar in the left valve and a poorly developed tooth on the posterior cardinal area on the right valve (Plate 1, Fig. E). The antennula has elongated distal segments. Three genera are currently assigned to this subfamily: *Elofsonia*, *Tuberoloxoconcha* and *Pseudolimnocythere*. The Pseudolimnocytherinae probably stem from small loxoconchids, such as *Hirschmannia*, which have a carapace and antennula similar to the former mentioned genera. The hinge of *Hirschmannia* is gongylodont, which is typical of most of the Loxoconchidae, but with a smooth median element. Within the Pseudolimnocytherinae, *Elofsonia* species have an antennula with six segments instead of five segments in *Tuberoloxoconcha* and *Pseudolimnocythere*. This latter genus differs from the former two in having

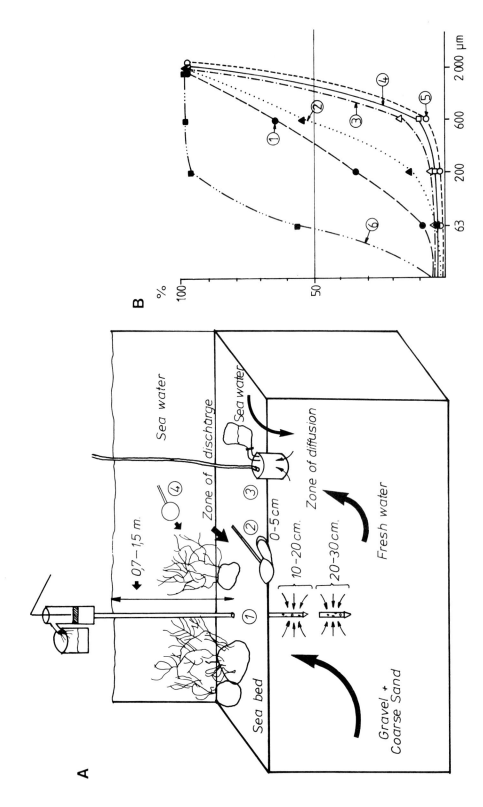

Fig. 2. A = sampling techniques used during the survey of the marine interstitial and phytal fauna around Scario; 1 - the Bou-Rouch pump technique; 2 - the skimming of sand procedure with a plankton net; 3 - the Lee-seepagemeter used to quantify the groundwater discharge to the sea; 4 - the procedure for collecting algae, i.e., isolated algal stems collected by hand and with a plankton net. B = grain-size analysis of sediments from submarine springs close to Molara beach (1, 3, 4, 5) from the sublittoral sand of Policastro (6) and a supralittoral area from Banyuls-sur-Mer, - site Ba-3, southern France (2); 1 - sediment sucked up by the Bou-Rouch pump at spring No. 2, close to the Molara beach; 3, 4, 5 surficial sediment (0-5cm depth) around the Molara springs Nos. 2, 3 and 4.

the sieve plates on the valves placed in funnel-like pores (Danielopol, 1980). *Elofsonia* and *Tuberoloxoconcha* species have sieve plates located at the surface of the carapace (Plate 1, Figs B and C). *Elofsonia*, therefore, appears to be more primitive than either *Tuberoloxoconcha* or *Pseudolimnocythere*. Danielopol (1980), in the Mediterranean, distinguished two species of *Tuberoloxoconcha*, *T. tuberosa*, with carapace slightly pitted around the periphery (Plate 1, Figs D, F) and a new species with a strongly ornate carapace (pitted on most of its surface, Plate 1, Fig. A). *Pseudolimnocythere* is known through two living species, *P. hypogea* (Klie, 1938) and *P. hartmanni* (Danielopol, 1979). A fossil species, *Pseudolimnocythere* sp. from Middle Badenian sediments of the Vienna Basin, differs from *P. hartmanni* mainly in the shape of the ventral margin of the valves which are more concave.

The Psammocytheridae Klie are cytherids with very primitive features, such as seven segments on the antennula, the subdistal segment is devoid of setae (as in the Bythocytheridae) and the endopodites of the maxilla and the thoracopods with four endopodial segments (Fig. 4B, C, E, F, H, L, J, M, N). Hartmann and Puri (1974) and Hiruta (1987) recognized the isolated position of this family within the Cytheracea. All the representatives of the genus *Psammocythere* (Klie, 1936) are Recent interstitial dwellers. An additional subapical antennular segment exists in other cytherid families, such as the Microcytheridae, Bythocytheridae and Entocytheridae. The latter family has antennular setae very similar to those of the Psammocytheridae. Within the Psammocytheridae, Danielopol (1977) recognized three genera: *Kliella* Schäfer, 1945; *Nannokliella* Schäfer, 1945 and *Psammocythere* Klie, 1936 which all have similar types of hemipenes with two clasping lobes (Fig. 4).

The marine interstitial *Psammocythere* has an elongate carapace with the dorsal margin parallel to the ventral (Fig. 4A), and the antennae have short distal claws (Fig. 4D). The maxillar endopodite of the male is transformed into a clasping organ (Fig. 4E); the freshwater interstitial *Kliella* and *Nannokliella* have small carapaces (Fig. 4G) more

rounded in dorsal view, the antennula has few long setae (Fig. 4H, L), the antenna has long distal claws (Fig. 4I), the maxilla is not sexually dimorphic and walking leg 3 is shorter than either 1 or 2 (Fig. 4J, M, N).

From all these data, one can differentiate two lineages within the Psammocytheridae: one represented by the Recent marine interstitial *Psammocythere* and a second by the present day freshwater genera *Kliella* and *Nannokliella*.

## ECOLOGICAL AND GEOGRAPHICAL DISTRIBUTION OF THE LOXOCONCHIDAE AND THE PSAMMOCYTHERIDAE

The Loxoconchidae is primarily a marine family mainly inhabiting shallow water. There are few deep sea Loxoconchidae (Whatley, 1983).

Important to our argument is a discussion of the ecology and biogeography of certain loxoconchid genera: *Elofsonia baltica*, widely distributed in shallow marine and estuarine environments in northern Europe, lives on different types of sediments, from sandy to pelitic, and on algae. It tolerates a wider range of salinity from 1 to 34°/∞ (Theisen, 1966, Olenska & Sywula, 1988). Athersuch and Horne (1984) include literature on the ecological distribution of this species, especially that of Whittaker. The development of *E. baltica* proceeds rapidly, within 1-2 months, and the species is very eurythermal (Theisen, 1966). By contrast, *Elofsonia pusilla* and *E. sustinensis* are more stenohaline.

There are other loxoconchids which display ecological similarities with *Elofsonia baltica*, e.g., *Hirschmannia viridis*, *Loxoconcha elliptica* and *Cytheromorpha fuscata*. All these species live both on algae and on or in sands and pelitic sediments. They have a wide salinity tolerance (2-35°/∞) living in brackish waters of estuaries and lagoons as well as in normal saline waters. *Cytheromorpha fuscata* also occurs in freshwater (Aladin, 1986; Theisen, 1966 and Neale & Delorme, 1985).

In our samples from the karstic springs at Scario, in interstitial habitats with a salinity of 14.4°/∞, we found *Loxoconcha stellifera* and *Loxocauda* sp. Bonaduce *et al.* (1988) and *L.*

*stellifera* in the infralittoral of the Tunesian shelf under a wide range of salinites.

The presence of a rich population of *Loxocauda* sp. in spring No. 4 at 20 to 40cm sediment depth at Scario suggests that this species can actively colonize interstitial habitats in coarse sands. *Loxocauda* sp. is a sighted species with an environmental range the same as that of *Loxocauda decipiens*, which lives on algae and sands. '*Loxoconcha*' *helgolandica* lives interstitially in the slightly saline water of Heligoland (Danielopol, 1980).

*Tuberoloxoconcha* is an amphiatlantic species which also occurs in the Mediterranean (Fig. 3 and chart in Danielopol, 1980). *Tuberoloxoconcha* sp. lives mainly in interstitial habitats on the northern coast of the Mediterranean and the Black Sea (Fig. 3; Danielopol, 1980). *Tuberoloxoconcha* sp. figured in Barbeito-González (1971) was found on sandy sediments around the island of Naxos down to 45m. Recently, Horne (pers. comm.) has found in northern Scotland, on both the eastern and western coasts, *Tuberoloxoconcha tuberosa* in Quarternary and Recent sediments. It is believed that this species was a phytal dweller during the Quaternary, but additional evidence is needed to support this contention.

*Tuberoloxoconcha tuberosa* lives in coarse sand (Fig. 2A) in habitats with normal salinity. Transported valves have been found in a marine cave in Scario, which suggests that this species also lives in littoral and sublittoral superficial sediment. We did not find *T. tuberosa* in the unsaturated zone of supralittoral sandy sediments. *Tuberoloxoconcha* sp. lives interstitially in waters with low salinities. At the Marina d'Orsei in Sardegna, it was found in waters with a salinity not less than $5°/oo$ (Danielopol, 1979, 1980), at Scario in the karstic spring area the salinity is around $8-15°/oo$, e.g., at the spring No. 4 the salinity was $14.4°/oo$. In southern Italy, *Tuberoloxoconcha* sp. was encountered in superficial coarse sand (0-5cm layer) and at 10-20cm at the spring No. 3, at 20-40cm depth at spring No. 4.

*Tuberoloxoconcha nana* is distributed in the Black and Azov Seas. It lives interstitially in supralittoral and sublittoral sands along the coasts of Bulgaria, Roumania and the Soviet Union (Fig. 3; Danielopol 1979, 1980). The salinity of the Black Sea on the Roumanian coast, near Constanta, where *T. nana* has been found, varies between 14 and $18°/oo$.

Recent *Pseudolimnocythere* species are known only from freshwater and brackish water habitats around the Mediterranean (Fig. 3; Danielopol, 1979, 1980). *Pseudolimnocythere* sp. is a species occurring in the Middle Miocene (Badenian) of the Vienna Basin in the Paratethys. From the ostracod assemblage and the palaeopopulation structure, in various samples of the sediments from Hainburg, it is suggested that this species lived in an estuary or lagoon with very low salinity (the species occur mixed with such freshwater taxa as *Darwinula*, *Candona* and *Ilyocypris*). It was transported post-mortem, at various times and in large numbers, to a marine shallow environment with *Cytherois*, *Xestoleberis* and other littoral species.

From the palaeogeographical standpoint it is interesting to notice that during the Middle Miocene, the Hainburg area was located on the western margin of the Central Paratethys. This basin was connected to the Mediterranean by a seaway located in the Istria and Drava zone in Yugoslavia (see Rögl & Steininger, 1983). This situation suggests that during the Middle Miocene, *Pseudolimnocythere* could have had a wider geographical distribution in the Mediterranean-Paratethys basins.

The Psammocytheridae have a worldwide distribution and all species live interstitially. *Psammocythere remanei* lives in coarse superficial *Amphioxus* sand in littoral/sublittoral areas around Heligoland (Klie, 1936, and Keyser, pers. comm. to DLD). *Psammocythere hartmanni* was found by one of us (G.B.) in coarse littoral sand from Malta (McKenzie, 1977). Other species of this genus have been collected by Bonaduce and co-workers in southern Italy (Bays of Taranto and Policastro) on sandy and sandy pelitic substrates (Bonaduce *et al.*, 1980, and Costa, 1985). In the Bay of Policastro *Psammocythere* occurs in normal salinity and in mixohaline waters (Costa, 1985). Gottwald (1983) described *Psammocythere santacruzensis* from interstitial habitats of the

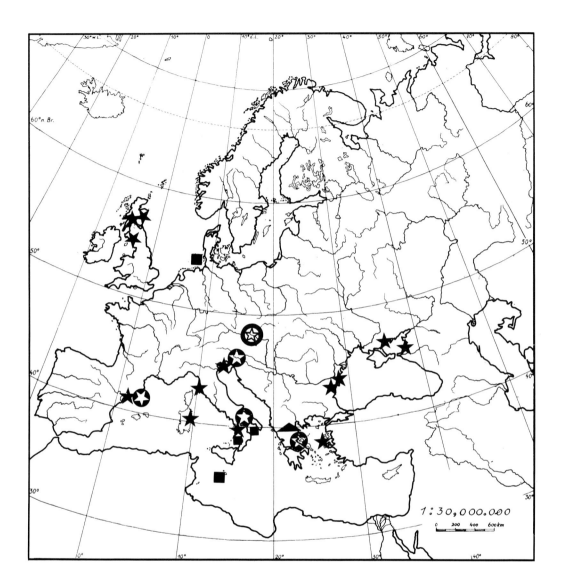

Fig. 3. The European distribution of Recent and fossil representatives of the genera *Tuberoloxoconcha* (black star), *Pseudolimnocythere* (white star) with Recent dwellers (black dot) and fossil Tertiary representatives (black ring), *Kliella* and *Nannokliella* (triangle), *Psammocythere* (square).

Galapagos Islands. A new species of *Psammocythere* has been found by Hiruta (1987) on Hokkaido Island in Japan. The species live both in saturated and unsaturated sands. *Kliella* and *Nannokliella* species were collected by Schäfer (1945) in deep Norton wells around Carla lake near Larissa in the Peloponese, Greece (Fig. 3; Danielopol, 1977 and 1980). The environment is freshwater and, from the faunal association and ostracod morphology, Danielopol (1977) postulated that Schäfer's material came from an interstitial freshwater environment.

## THE PREADAPTATION OF THE PSEU-DOLIMNOCYTHERINAE FAVOURING THE COLONIZATION OF SUBTERRANEAN HABITATS

Organisms which already have morphological and biological characters that can be useful for life in subsurface habitats are at an advantage in comparison with those that have to gradually evolve such characters. Cuenot (1909), Kosswig (1960, 1974), Wilkens (1972) and Rouch & Danielopol (1987) showed that preadapted and especially generalist species with large ecological tolerances have obvious advantages in actively colonizing the subterranean environment. Below we present arguments that several morphological and physiological peculiarities of the Pseudolimnocytherinae, i.e. the reduced carapace size, the shape of the distal claws of the walking legs and the wide salinity tolerance of the representatives of this ostracod group can partly explain its success in the colonization of subsurface habitats.

Considering the Psammocytheridae, we have no information on species living in epigean habitats, therefore, their reduced carapace size (length between 0.2 and 0.4mm) and the elongated distal claws of the walking limbs cannot be used for the purpose of the present discussion.

Remane (1933) and Elofson (1941) demonstrated that ostracods which live in sandy sediments have carapaces of smaller size than those living on plants or pelitic sediments. Williams (1972) evaluated the importance of the void dimensions of marine sediments for the inhabitation of various meiofaunal groups (including Ostracoda). He noticed that in coarse sediments the number of ostracods increased. The ostracods with short and elongated carapaces occurred in sediments with both large and small interstitial voids.

In the case of the Loxoconchidae, the carapace size varies between 0.3mm to more than 0.8mm length. One can classify the representatives of this family in two size groups, i.e. those species of smaller size (less than 0.5mm length) and those of larger size (more than 0.5mm length). Table 1 shows that the loxoconchid fauna found during 4 faunal surveys, covering a wide range of surface and interstitial habitats is dominated by species of the large size class.

All the Pseudolimnocytherinae have small carapace sizes, ranging between 0.3 and 0.5mm length. Phytal species of *Elofsonia* are 0.4 to 0.5mm, while those of the interstitial *Pseudolimnocythere* and *Tuberoloxoconcha* are 0.3 to 0.35mm long (Danielopol, 1980; Athersuch & Horne, 1984). The fossil epigean species *Pseudolimnocythere* sp. studied by Huber-Mahdi and Danielopol (unpublished) has the same size as the Recent living interstitial species of this genus.

The carapaces of many Loxoconchidae in the posteroventral area are rounded (e.g., *Loxoconcha stellifera*, Plate 1, Fig. G). In the Pseudolimnocytherinae the shape of the carapace is rectangular or elongate in both epigean dwellers, such as of *Elofsonia* (e.g., *E. baltica*, Athersuch & Horne, 1984) and interstitial dwellers such as *Tuberoloxoconcha* sp., (Plate 1, Fig. A) or *Pseudolimnocythere hartmanni* (Danielopol, 1980).

Most benthonic ostracods that walk on the substrate have slightly bent distal claws on the maxilla and thoracic legs. This characteristic is also functionally useful for those animals which walk within interstitial voids. In this microspace, the ostracods walk in the same way as they do on the surface of clastic sediments. Ostracod species which are adapted to live on algae, or those which live fixed on various animals or plants, have the distal claws strongly curved or hook-shaped. In the case of Loxoconchidae, Kamiya (1988) showed that *Loxoconcha japonica* which lives on *Zostera* has hook-shaped distal claws in contrast to *L. uranouchiensis*, a species living on sandy bottoms which has straight claws.

In the Pseudolimnocytherinae, both species such as *Elofsonia baltica* which live on sandy sediments and on algae and the interstitial *Tuberoloxoconcha* and *Pseudolimmnocythere* have walking legs with slightly bent claws (Athersuch & Horne, 1984; Hartmann, 1953; Danielopol, 1980). Therefore, we consider that the change of life-style from surface substrates to interstitial ones did not necessitate any change in the morphology

Table 1. Distribution (percentages) of loxoconchid species by carapace size according to various faunal surveys.

| Author(s) | Carapace length (mm) % from N species | | Total no. loxoconchid species (N) | Geographical area |
|---|---|---|---|---|
| | <0.5 | >0.5 | | |
| Barbeito-Gonzàles (1971) | 25 | 75 | 16 | Mediterranean Sea (Naxos) |
| Bonaduce et al. (1975) | 33 | 66 | 24 | Mediterranean Sea (Adriatic Sea) |
| Hartmann (1974) | 20 | 80 | 10 | Indian & Atlantic Oceans (South African Coasts) |
| Teeter (1975) | 20 | 80 | 10 | Caribbean Sea (Belize) |

of the walking limbs.

Aladin, in a series of papers on the osmoregulational abilities of the Ostracoda (see review Aladin, 1986) showed that loxoconchids that live in estuarine or strongly fluctuating salinity environments are euryhaline and can change their osmoregulation from hypo-osmosis in normal marine habitats to hyperosmosis in brackish water.

*Elofsonia baltica* is a pseudolimnocytherid which tolerates 1°/ₒₒ salinity waters (Theisen, 1966) and *Pseudolimnocythere* sp. in the Vienna Basin most probably lived in an environment with a strong freshwater character, as discussed in the previous section. The subterranean species *Pseudolimnocythere hypogea* lives in southern Italy in fresh and slightly brackish waters (Klie, 1938). Therefore, it is easy to understand the ability of interstitial *Tuberoloxoconcha* to colonize the karstic submarine springs from Scario. Even the crossing of the salinity boundary does not for a *Pseudolimnocythere* species seem to represent an exceptional challenge to its physiology.

There are indications that the Psammocytheridae also can tolerate lower salinities. For instance, Costa (1985) found *Psammocythere* sp. in the River Busento in Policastro Bay in mixohaline conditions.

In conclusion, the carapace size, the distal claws of the walking legs and the euryhalinity of Pseudolimnocytherinae are preadaptations which make the colonization of the interstitial environment relatively easy. Within this subfamily there was no necessity to evolve major morphological changes in order to become interstitial inhabitants and to further colonize a wide range of subsurface habitats.

## WHEN AND HOW PSEUDOLIMNOCYTHERINAE AND PSAMMOCYTHERIDAE COLONIZED FRESHWATER SUBTERRANEAN HABITATS ALONG THE MEDITERRANEAN COASTS

The present ecological and geographical distribution of the interstitial representatives of these two ostracod groups shows that there is a sharp separation between the marine interstitial species and the freshwater subterranean dwelling forms. *Tuberoloxoconcha* and *Psammocythere* species live in marine interstitial habitats, mainly in littoral and sublittoral areas, while *Pseudolimnocythere*, *Kliella* and *Nannokliella*, live mainly in freshwater interstitial and cavernous systems. This strong segregation in the Mediterranean area is, in our opinion, due to a special geological event: the Messinian Salinity Crisis.

During the Messinian stage of the Upper Miocene, an important regressional phase of the Mediterranean occurred. The connection between

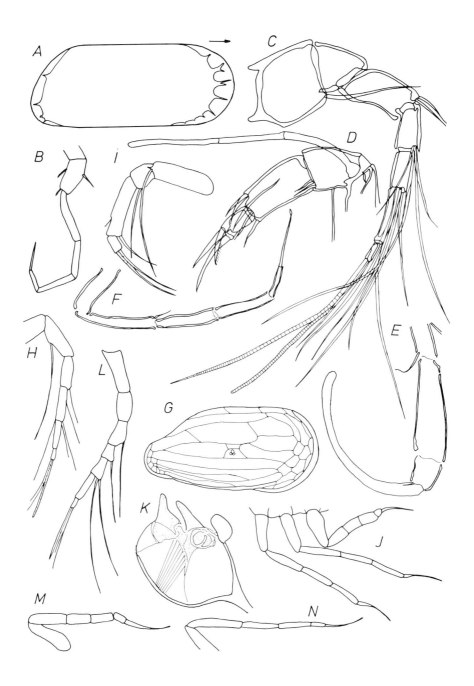

Fig. 4. Carapace and limbs of various Psammocytheridae (redrawn from various authors). A - D = *Psammocythere remanei* from Heligoland; A - right valve; B - the walking leg 1; C - antennula; D - antenna; E - F = *Psammocythere hartmanni* from Malta; E - male clasping organ, maxilla; F - walking leg 3; G, I - N = *Nannokliella dichtioconcha* from Greece; G - right valve; I - antenna; K - hemipenis; L - antennula; M - N - walking legs 3 and 2; H, J - *Kliella hyaloderma* from Greece; H - antennula; J - walking legs 1 - 3 ; A, B, G - N - female; C - F, K - male. A - B from Klie, 1936; C - D from Gottwald, 1983; E - F from McKenzie, 1977 and G - N from Schäfer, 1945.

the Mediterranean and the Atlantic was closed and thick evaporitic deposits formed under an arid climate (Hsü *et al.*, 1977, Cita *et al.*, 1978, Hsü, 1987). Along the coasts, limestone terrain underwent active karstification, rivers deepened their channels through erosion and developed canyons deep in the former sea bed. During this period a high number of organisms became extinct in the Mediterranean; drastic changes in the ostracod fauna have been noted by Benson (1975). With the reopening of the connection between the Atlantic and the Mediterranean in the Pliocene, a new stock of Atlantic ostracods penetrated into the Mediterranean.

It is very plausible that *Kliella, Nannokliella* and *Pseudolimnocythere*, which lived in shallow marine interstitial habitats before the Messinian Salinity Crisis, migrated into freshwater during this geological event (see below). Those populations located in deeper marine areas disappeared. The present day *Psammocythere* and *Tuberoloxoconcha* species are immigrants which penetrated from the Atlantic into the Mediterranean after the reopening of the sea connection between these areas in the Pliocene.

How did the migration from the marine habitats into the freshwater continental water happen? As mentioned above, there are currently two widely accepted possibilities. That of Sket (1970) and that of Stock (1980). Our present knowledge on the Pseudolimnocytherinae and Psammocytheridae do not fit into either of these models.

Should we accept the historical model of Sket (1970), i.e. a migration of organisms through surface running waters which discharge into the Mediterranean and subsequently a colonization of subterranean inland waters, then one would expect to find in the various drainage basins around the Mediterranean and elsewhere, e.g., along the Atlantic coast, ostracods of marine origin. As we saw, this is not the case with the two ostracod groups in discussion.

Following the regression model of Stock (1980), one would expect to find stranded marine interstitial faunas along wider areas than one can see in the case of the ostracod groups studied here, i.e. in both karstic and non-karstic areas. We saw that

with one exception, the unique specimen of *Pseudolimnocythere* from site Ba-2 at La Baillaury, Banyuls, all the other sampling sites where *Pseudolimnocythere, Kliella* and *Nannokliella* occur are located adjacent to or within karstic areas.

Below, we propose a highly speculative model on the pathway taken by *Pseudolimnocythere* and the representatives of the Psammocytheridae which colonized subterranean freshwater.

We consider that during the Upper Messinian stage interstitial Pseudolimnocytherinae and Psammocytheridae existed in both euhaline and mixohaline habitats along the Mediterranean coast. Various karstic springs seeped under the sea, as do the present day springs of Scario. We know (see data in Bouillin *et al.*, 1985) that the Jurassic Calabrian limestones like those at Aspromonte could have already been karstic during the Cretaceous or the Palaeogene. This suggests that submarine springs, like those existing at Scario, could have occurred since ancient times and, therefore, offer the possibility to be perennially inhabited by an interstitial fauna.

During the Messinian regressive phase, littoral springs became isolated from the sea but continued to remain active as long as water reserves in the karstified mountains remained sufficient. The water in these springs became completely freshened and during the arid period the water temperature increased. During short transgressive phases, these spring habitats turned once again to mixohaline conditions. Fig. 5 compares the present day situation of the coastal karstic area at Scario with a hypothetical situation during the Messinian regressive phase. Euryhaline ostracods living in these springs could because of very stressful conditions, change from an amphiosmotic regulation to a hyperosmotic one within several generations as in the genetic model proposed by Parsons (1987). This author showed that under severe stress conditions evolutionary change can be rapid. Parsons (1987) explains this as a result of several factors, e.g., fragmentation of large populations into smaller ones, the action of very strong directional selection within small populations, occurrence of changes dependent upon a few major genes and upon important mutational events and

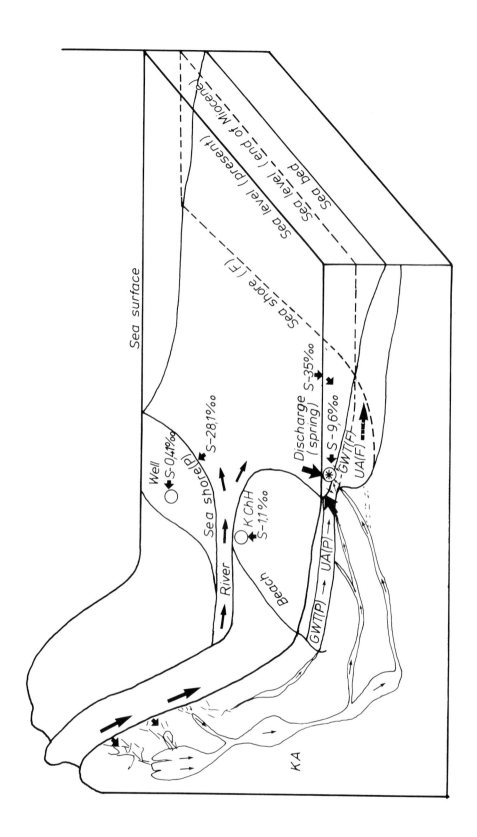

Fig. 5. Diagrammatic view of the groundwater flow through the karstic or porous (unconfined) aquifers in the Scario area during the present and (hypothetically) during the Late Miocene. S - salinity values measured during our investigations; KA - Karst system; UA - porous (unconfined) aquifer in sandy gravel sediments; GWT - groundwater table; U - present day situation; F - fossil (Late Miocene) situation; KChH - Karaman-Chappuis hole.

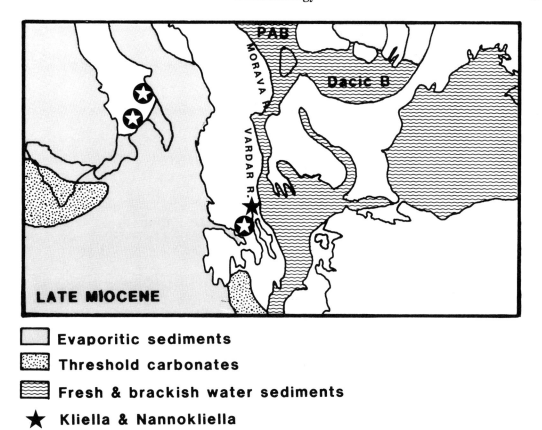

□ **Evaporitic sediments**

▨ **Threshold carbonates**

≋ **Fresh & brackish water sediments**

★ **Kliella & Nannokliella**

✪ **Pseudolimnocythere**

Fig. 6. Details of the palaeogeography of the Proto-Aegean Sea and the eastern Paratethys during the late Miocene (redrawn with modifications from Sonnenfeld, 1974).

recombination. All this produced accelerated evolutionary change which recalls the process of genetic assimilation described by Waddington (1959). In our case, if we consider that the submarine springs may have become isolated with their fauna during the Messinian Salinity Crisis, we obtain within a perennial habitat, stress conditions with respect to temperature and salinity. This could be partially responsible for an acceleration of evolutionary change to the new freshwater conditions.

Fig. 6 shows, on a palaeogeographical map, the present day distribution of the freshwater Psammocytheridae. The area around Lake Carla in Greece, where *Kliella* and *Nannokliella* species occur, is located not far from the suggested

palaeocoast during the Messinian, as reconstructed by Sonnenfeld (1974). The communication of the Paratethys with the Mediterranean through the Morava-Vardar connection allowed the existence of low salinity waters in the Proto-Aegean Sea, which contrasted with those existing in the southern part of the Mediterranean where evaporites formed. From such a brackish water environment, the marine Kliellinae could have rapidly adapted during a warm climate to the subterranean freshwaters of this karstic area. This applies also to the interstitial *Pseudolimnocythere* from the northern part of Euboea (Danielopol, 1980). Once the marine ostracods had adapted to the coastal feshwater subterranean environment, then they could actively begin to migrate further inland.

There are ecophysiological arguments which point out that the coastal karstic springs are optimal places for a potential migration towards the freshwater habitats. Marine waters have high concentrations of $Na^+/Cl^-$ ions, whereas in the freshwaters of karstic areas $Ca^{2+}/HCO_3^-$ ions are more abundant. Dalla Via (1986), who studied the adaptation to freshwater conditions of the Mediterranen decapod *Palaemonetes antenarius*, showed that osmotic change was possible in those areas where a balance between the critical values of $Na^+/Ca^{2+}$ could be achieved. A reduction in $Na^+$ ion uptake by the shrimp can be achieved when the $Ca^{2+}$ concentration in the water is high enough to permit a decrease of the animal's membrane permeability and, therefore, the loss of ions through the surface of the body is minimized. Such environmental situations occur on the Mediterranean karstic coasts (Sket, 1986). This should also be the case in the submarine springs at Scario.

Aladin (1984) studied the impact of temperature on the osmoregulatory ability of three ostracod species. He found that the osmotic concentration of the haemolymph and the salinity ranges of these ostracods increased at 24°C as compared with 10°C. It is possible, therefore, that during the warm climate of the Messinian, euryhaline marine ostracods might have more readily come to tolerate low salinities and even to adapt faster to freshwater habitats, than in periods with lower temperature regimes. We consider, therefore, that during the Upper Miocene, euryhaline Pseudolimnocytherinae and Psammocytheridae could have enjoyed favourable opportunities to migrate to freshwater subterranean habitats through the pathways of the karstic springs.

The importance of our model is due to the fact that it can be empirically tested. One can try to change the salinity tolerance of *Tuberoloxoconcha* and *Psammocythere* species following the experimental protocols used by Waddington (1959), i.e. to culture these ostracods under strong selective pressures of low salinity and high temperature conditions. New field data for Recent and fossil Pseudolimnocytherinae and Psammocytheridae could provide new information in order to corroborate or invalidate our model.

## CONCLUSIONS

1) The colonization of marine interstitial waters is a recurrent process.

2) Many ostracod species can actively penetrate into and live in deep coarse sediments if they are already preadapted.

3) Candidates for the interstitial mode of life are members of moderately specialized ostracod groups with small carapace sizes which previously lived on both phytal and sandy substrates.

4) The passage from a marine interstitial environment into a freshwater one is most easily accomplished by those groups which are already euryhaline and amphiosmotic.

5) Submarine karstic springs, which seep through coarse littoral sediments, are perennial habitats inhabitated by interstitial marine ostracods. These ostracods are potential candidates to migrate into inland waters. Once adapted to freshwater subterranean conditions they can actively spread further.

6) It is speculated that regional geological events, such as the Messinian Salinity Crisis, could on one hand destroy a part of the marine ostracod fauna of the Mediterranean, yet on the other hand facilitate the adaptation to freshwater of the interstitial fauna living in karstic springs.

7) The modern marine interstitial genera *Tuberoloxoconcha* (Loxoconchidae) and *Psammocythere*, in the Mediterranean are recent immigrants which arrived after the reopening of the connection with the Atlantic in the Pliocene.

8) The importance of our hypothetical explanation of the origin and distribution of the subterranean Pseudolimnocytherinae and Psammocytheridae lies in the possibility that it can be checked and improved through additional information. The discovery of new Recent and fossil Pseudolimnocytherinae and Psammocytheridae could corroborate or invalidate our model.

9) The fact that we believe the colonization process of the interstitial environment to be a recurrent one, allows us to experiment on Recent living interstitial ostracods in order to obtain a

better insight into their Tertiary history.

10) It is particularly important to devise experiments to simulate the thermal or the salinity stress to which those ostracods would have been submitted, when living in the littoral karstic areas during the Messinian Salinity Crisis.

11) Finally, we consider that our data support the active migration model of Rouch and Danielopol (1987).

## ACKNOWLEDGEMENTS

One of us (DLD) is much indebted to Drs N. Coineau and J. Soyer (Banyuls-sur-Mer) who facilitated his stay and study in 1981 at the Laboratoire Arago. Further he is grateful to Drs D. Horne (London), S. Hiruta (Kushiro), K. McKenzie (Waga Waga), D. Keyser (Hamburg), R. Rouch (Moulis), B. Sket (Ljubljana), Y. Tambareau (Toulouse), W. Geiger (Mondsee) and Professor R. C. Whatley (Aberystwyth) for useful discussions and information on the ostracods and their environment mentioned in this study. T. Huber-Mahdi and W. Piller (Vienna) offered unpublished information on the palaeoecology of *Pseudolimnocythere* sp. Dr C. Maybury and Professor R. C. Whatley improved the English of the manuscript.

**Plate 1**

Cytheridae from interstitial and phytal habitats around Scario.

Figs A-C.     *Tuberoloxoconcha* sp., Spring No. 2, near Molara beach, sublittoral.
Fig. A.        Female right valve.
Figs B-C.     Sieve pores, details of A.

Figs D-F.     *Tuberoloxoconcha tuberosa*, near Spiaggia dei Francesi, littoral cave.
Fig. D.        Right valve.
Fig. E.        Internal lateral view of right valve.
Fig. F.        Detail of D.

Fig. G.        *Loxoconcha stellifera*, left valve, Spring No. 2, near Molara beach, sublittoral.

Fig. H.        *Callistocythere* aff. *pallida*, female left valve, Spring No. 2, near Molara beach, sublittoral.

Fig. I.         *Xestoleberis delamarei*, female left valve, Grotta dell'Aqua beach, supralittoral habitat.

Fig. J.        *Xestoleberis parva*, female left valve. Spring No. 2, near Molara beach, sublittoral.

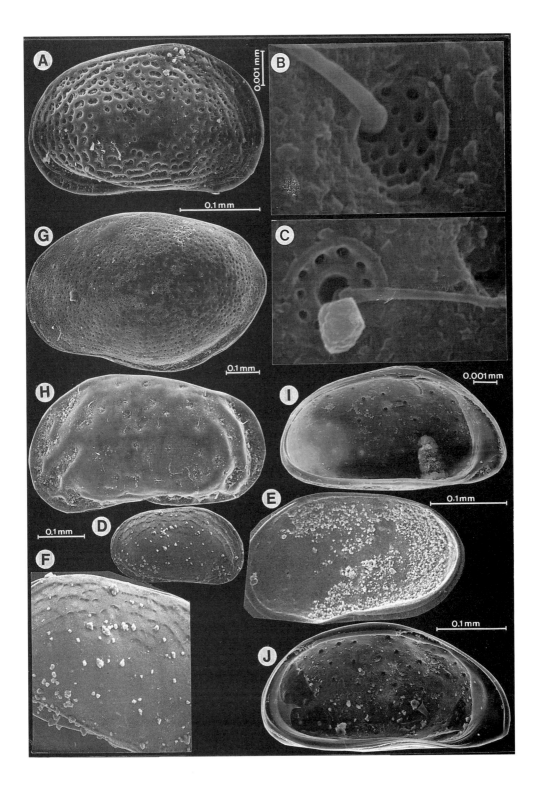

# REFERENCES

Aladin, N. V. 1984. The influence of temperature on the osmoregulation abilities of the Branchiopoda and Ostracoda. *Zool. Jn.*, Moscow, **63**, 1158-1163. (In Russian, English summary.)

Aladin, N. V. 1986. Haemolymph osmoregulation peculiarities in Ostracoda and Branchiopoda from thalassic and athalassic brackish water. *In* V. V. Khlebovich (Ed.), Hydrobiological investigation of estuaries. *Proc. Zool. Inst. USSR Acad. Sci.*, Leningrad, **141**, 75-97. (In Russian.)

Athersuch, J. & Horne, D. 1984. A review of some European genera of the family Loxoconchidae (Crustacea: Ostracoda). *Zool. J. Linn. Soc.*, London, **81**, 1-22.

Barbeito-Gonzàlez, P. 1971. Die Ostracoden des Küstenbereiches von Naxos (Griechenland) und ihre Lebensbereiche. *Mitt. hamb. zool. Mus. Inst.*, Hamburg, **67**, 255-326.

Benson, R. 1975. Changes in the ostracods of the Mediterranean with the Messinian Salinity Crisis. *Palaeogeogr. Paleoclimat. Paleoecol.*, Amsterdam, **20**, 147-170.

Bonaduce, G. Ciampo, G. & Masoli, M. 1975. Distribution of Ostracoda in the Adriatic Sea. *Pubbl. Staz. zool. Napoli*, Milano & Napoli, **40** Suppl., 1-304.

Bonaduce, G., Mascellaro, M., Masoli, M. & Pugliese, N. 1980. Ostracodi. *In* La sedimentazione recente del Golfo di Taranto. *Ann. Fac. Sc. Nautiche Napoli*, 49-50, app. 3, 1-96.

Bonaduce, G., Masoli, M. & Pugliese, M. 1988. Remarks on the benthic Ostracoda on the Tunisian Shelf. *In* Hanai, T., Ikeya, N., & Ishizaki, K. (Eds), *Evolutionary biology of Ostracoda, its fundamentals and applications*, proceedings of the Ninth International Symposium on Ostracoda, held in Shizuoka, Japan, 29 July - 2 August 1985, Developments in palaeontology and stratigraphy, **11**, 449-466, Kodansha Ltd., Tokyo and Elsevier, Amsterdam, Oxford, New York, Tokyo.

Bouillin, J. P., Majesté-Menjoulas, C., Olivier-Pierre, M. F., Tambareau, Y. & Villatte, J. 1985. Transgression de l'Oligocène inferieur (formation de Palizzi) sur un karst bauxitique dans les zones internes calabaro-peloritaines (Italie). *C. r. hebd. Seanc. Acad. Sci.*, Paris, **301**, Ser. 2, 415-429.

Carbonel, P., Colin, J.-P., Danielopol, D. L. & Londeix, L. 1986. Kovalevskiella (Ostracoda, Timiriaseriinae), genre à mode de vie benthique depuis l'Oligocène, son adaptation à la vie interstitielle. *Géobios*, Lyon, **19**, 677-687.

Cita, M. B., Wright, R. C., Ryan, W. B. F. & Longinelli, A. 1978. Messinian paleoenvironments. *In* Hsü, K. & Montadert, L. (Eds), *Init. Repts DSDP*, **42**, 1003-1035. U.S. Govt Printing Office, Washington.

Costa, R. 1985. *Le ostracofaune benthoniche delle facies di fondo della piattaforma continentale del Golfo di Policastro: aspetti applicativi*. Unpub. tesi di laurea in Micropaleontologia. Univ. di Trieste, 57 pp.

Cuenot, L. 1909. Le peuplement des places vides dans la nature et l'origine des adaptations. *Rev. gen. Sci.*, Paris, **20**, 8-14.

Dalla Via, G. J. 1986. Ecological zoogeography of *Palaemonetes antennarius* (Crustacea, Decapoda). *Arch. Hydrobiol.*, Stuttgart, **106**, 251-262.

Danielopol, D. L. 1976. Supplementary data on *Pussella botosaneanui* Danielopol (Ostracoda, Bairdiidae). *Vie Milieu*, Paris, 26, 2A, 261-273.

Danielopol, D. L. 1977. On the origin and diversity of European freshwater interstitial ostracods. *In* Löffler, H. & Danielopol, D. (Eds), *Aspects of ecology and zoogeography of Recent and fossil Ostracoda*, 295-305. W. Junk, The Hague.

Danielopol, D. L. 1978. *Introduction to groundwater ecology. Lecture notes for the UNESCO Training Course in Limnology*. 55 pp. Limnologisches Inst., Wien.

Danielopol, D. L. 1979. On the origin and the antiquity of the Pseudolimnocythere species (Ostracoda, Loxoconchidae). *Biologia gallo-hellen.*, Toulouse, **8**, 99-107.

Danielopol, D. L. 1980. An essay to assess the age of the freshwater interstitial ostracods of Europe. *Bijdr. Dierk*, Amsterdam & Leiden, **50**, 243-291.

Danielopol, D. L. & Hartmann, G. 1986. Ostracoda. *In* Botosaneanu, L. (Ed.), *Stygofauna Mundi*, 265-294. E. J. Brill - W. Backhuys, Leiden.

Danielopol, D. L. & Hart, C. W. 1985. Notes on the center of origin and of the antiquity of the Sphaeromicolinae, with description of Hobbsiella, new genus (Ostracoda: Entoytheridae). *Stygologia*, Leiden, **1**, 54-70.

Danielopol, D. L. & Niederreiter, R. 1987. Eine Übersicht über Geräte für Grundwasserforschung, gebaut am Institut für Limnologie der Österreichischen Akademie der Wissenschaften, Mondsee. Int. *Arbeitsgem. Donauforschung, 26, Arbeitstagung, Wissenschaftl. Kurzreferate*, 521-523, Passau.

Delamare-Debouteville, Cl. 1960. *Biologie des eaux souterraines littorales et continentales*, 740 pp. Hermann, Paris.

Di Stefano, G. 1984. Nuove osservazioni sulla geologia de M. Bulgheria in prov. di Salerno. *Boll. Soc. geol. ital.*, Roma, **13**, 191-198.

Ekman, S. 1913. Studien über die marinen Relikte der nordeuropäischen Binnengewässer. 2. Die Variationen der Kopfform bei *Limnocalanus grimaldii* (de Guerne) und *L. macrurus* G. O. Sars. *Int. Revue d. Ges. Hydrobiol. u. Hydrogr.*, Leipzig, **6**, 335-372.

Ekman, S. 1914. Studien über die marinen Relikte der nordeuropäischen Binnengewässer. 3. Über das Auftreten von *Limnocalanus grimaldii* (de Guerne) und *Mysis oculta* (Fabr.) im Meere, besonders im Ostseebecken. *Int. Revue d. Ges. Hydrobiol. u. Hydrogr.*, Leipzig, **6**, 493-517.

Ekman, S. 1918. Studien über die marinen Relikte der nordeuropäischen Binnengewässer. 4. und 5. *Int. Revue d. Ges. Hydrobiol. u. Hydrogr.*, **8**, Leipzig, 321-337.

Elofson, O. 1941. Zur Kenntnis der marinen Ostracoden Schwedens mit besonderer Berücksichtigung des Skageraks. *Zool. Bidrag Upps.*, Stockholm, **19**, 215-534.

Gottwald, J. 1983. Interstitielle Fauna von Galapagos. *Mikrofauna Meeresboden*, Wiesbaden, **90**, 1-187.

Hartmann, G. 1953. Ostracodes des eaux souterraines littorales de la Mediterranee et de Majorque. *Vie Milieu*, Paris, **4**, 238-253.

Hartmann, G. 1973. Zum gegenwärtigen Stand der Erforschung der Ostracoden interstitieller Systeme. *Annls Speleol.*, Paris, **28**, 417-426.

Hartmann, G. 1974. Zur Kenntnis des Eulitorals der

afrikanischen Westküste zwischen Angola und Kap der Guten Hoffnung und der afrikanischen Ostküste von Südafrika und Mocambique unter besonderer Berücksichtigung der Polychaeten und Ostracoden. Die Ostracoden des Untersuchungsgebietes. *Mitt. hamb. zool. Mus. Inst.*, Hamburg, **69**, 229-520.

Hartmann, G. & Puri, H. S. 1974. Summary of neontological and paleontological classification of Ostracoda. *Mitt. hamb. zool. Mus. Inst.*, Hamburg, **70**, 7-73.

Hiruta, S. 1987. A new species of marine interstitial Ostracoda of the genus *Psammocythere* from Kushiro, Hokkaido. *Zool. Sci.*, **4**, 1112.

Hsü, K. J. 1987. The desiccation of the Mediterranean Sea. *Endeavour*, London, N. S., **11**, 67-72.

Hsü, K. J., Montadert, L., Bernoulli, D., Cita, K. B., Erickson, A., Garrisson, R. E., Kidd, R. B., Meliere, F., Müller, C. & Wright, R. 1977. History of the Mediterranean Salinity Crisis. *Nature*, London, **267**(5610), 399-403.

Kamiya, T. 1988. Morphological and ethological adaptations of Ostracoda to microhabitats in Zostera beds. *In* Hanai, T., Ikeya, N., & Ishizaki, K. (Eds), *Evolutionary biology of Ostracoda, its fundamentals and applications*, proceedings of the Ninth International Symposium on Ostracoda, held in Shizuoka, Japan, 29 July - 2 August 1985, Developments in palaeontology and stratigraphy, **11**, 303-318, Kodansha Ltd., Tokyo and Elsevier, Amsterdam, Oxford, New York, Tokyo.

Klie, W. 1936. Ostracoden der Familie Cytheridae aus Sand und Schell von Helgoland. *Kieler Meeresforsch.*, Kiel, **1**, 49-72.

Klie, W. 1938. Ostracoden aus unterirdischen Gewässern in Süditalien. *Zool. Anz.*, Leipzig, **123**, 148-155.

Kosswig, K. 1960. Zur Phylogenese sogenannter Anpassungsmerkmale bei Höhlentieren. *Int. Revue Ges. Hydrobiol.*, Berlin, **45**, 493-512.

Kosswig, K. 1974. Form und vermeintliche Funktion. Zum Problem der Anfangsstadien sogenannter Konstruktiver Evolution. *Z. Zool. Syst. EvolForsch.*, Frankfurt am Main, **12**, 81-93.

McKenzie, K. G. 1977. Bonaducecytheridae, a new family of cytheracean Ostracoda, and its phylogenetic significance. *Proc. biol. Soc. Wash.*, Washington, **90**, 263-273.

Maddocks, R. F. 1976. Pussellinae are interstitial Bairdiidae (Ostracoda). *Micropaleontology*, New York, **22**, 194-214.

Marmonier, P. & Danielopol, D. L. 1988. Decouverte de *Nannocandona faba* Ekman (Ostracoda, Candoninae) en Basse Autriche, son origine et son adaption au milieu interstitiel. *Vie Milieu*, Paris, **38**, 35-48.

Mayr, E. 1976. Evolution and the diversity of life. *Selected essays*, 721 pp. Harvard Univ. Press, Cambridge, Mass.

Mayr, E. 1982. *The growth of biological thought*, 974 pp. Harvard Univ. Press, Cambridge, Mass.

Neale, J. W. & Delorme, L. D. 1985. *Cytheromorpha fuscata*, a relict Holocene marine ostracod from freshwater inland lakes of Manitoba, Canada. *Revta esp. Micropaleont.*, Madrid, **17**, 41-64.

Olenska, M. & Sywula, T. 1988. Ostracoda of the Gulf of Gdansk. 1. Species from the sandy bottom off Rewa. *Fragm. Faun.*, Warzawa, **31**, 445-457.

Osche, G. 1962. Das Präadaptationsphänomen und seine Bedeutung für die Evolution. *Zool. Anz.*, Jena, **169**, 14-49.

Parsons, A. 1987. Evolutionary rates under environmental stress. *Evolut. Biol.*, New York, **21**, 311-347.

Remane, A. 1933. Verteilung und Organisation der benthonische Mikrofauna der Kieler Bucht. *Wiss. Meeresunters.*, Kiel, **21**, 160-221.

Rögl, F. & Steininger, F. F. 1983. Vom Zerfall der Tethys zu Mediterran and Paratethys. Die neogene Palaeogeographie und Palinspastik des zirkum-mediterranen Raumes. *Ann. Naturhist. Mus. Wien*, Wien, **85/A**, 135-163.

Rouch, R. & Danielopol, D. L. 1987. L'origine de la faune aquatique souterraine; entre le paradigme du refuge et le modele de la colonisation active. *Stygologia*, Leiden, **3**, 345-372.

Schäfer, H. W. 1945. Grundwasser-Ostracoden aus Griechenland. *Arch. Hydrobiol.*, Stuttgart, **40**, 847-866.

Segerstrale, S. G. 1957. On the immigration of the glacial relicts of Northern Europe, with remarks on their prehistory. *Soc. Scient. Fennica. Comment. Biol.*, Helsingfors, **16**, 1-117.

Sket, B. 1970. Über Struktur und Herkunft der unteriridschen Fauna Jugoslawiens. *Biol. Vest.*, Ljubljana, **18**, 69-78.

Sket, B. 1986. Ecology of the mixohaline hypogean fauna along the Yugoslav coast. *Stygologia*, Leiden, **2**, 317-338.

Sonnenfeld, P. 1974. The Upper Miocene evaporite basins in the Mediterranean region - a study in paleooceanography. *Geol. Rundsch.*, Stuttgart, **63**, 1133-1172.

Stock, H. J. 1980. Regression model evolution as exemplified by the genus *Pseudoniphargus* (Amphipoda). *Bijdr. Dierk*, Amsterdam & Leiden, **50**, 105-144.

Sutcliffe, D. W. & Shaw, J. 1968. A reexamination of observations on the distribution of *Gammarus duebeni* Lilljeborg in relation to the salt content in fresh water. *J. anim. Ecol.*, London, **36**, 579-597.

Teeter. J. 1975. Distribution of Holocene marine Ostracoda from Belize. *Amer. Assoc. Petrol. Geol. Studies in Geology*, Tulsa, **2**, 400-499.

Theisen, B. F. 1966. The life history of seven species of ostracods from a Danish brackish-water locality. *Meddr. Danm. Fisk. og. Havunders.*, København, Ny serie, **4**, 215-270.

Thienemann, A. 1950. Verbreitungsgeschichte der Süsswassertierwelt Europas. *Die Binnengewässer*, Stuttgart, **18**, 809 pp.

Vandel, A. 1965. *Biospeleology. The biology of cavernicolous animals*, 524 pp. Pergamon Press, Oxford.

Waddington, C. H. 1959. Canalisation of development and genetic assimilation of acquired characters. *Nature*, London, **183**(4676), 1654-1655.

Westheide, W. 1987. Progenesis as a principle in meiofauna evolution. *J. nat. Hist.*, London, **21**, 843-854.

Whatley, R. C. 1983. Some aspects of the palaeobiology of Tertiary deep-sea Ostracoda from the S.W. Pacific. *J. micropalaeontol.*, London, **2**, 83-104.

Wilkens, H. 1972. Über Präadaptationen für das Höhlenleben, untersucht am Laichverhalten ober- und unterirdischer Populationen des *Astyanax mexicanus* (Pisces). *Zool. Anz.*, Jena, **188**, 1-11.

Williams, R. 1972. The abundance and biomass of the interstitial fauna of a graded series of shell-gravels in relation to the available space. *J. anim. Ecol.*, London, **41**, 623-646.

Wouters, K. 1987. *Comontocypris* gen. nov., a marine interstitial new genus of the family Pontocyprididae (Crustacea, Ostracoda). *Bull. Inst. r. Sci. nat. Belg., Biologie*, Bruxelles, **57**, 163-169.

## DISCUSSION

Robin Whatley: How are *Psammocythere* dispersed to such places as the Galapagos?

Dan Danielopol: We suspect that interstitial Ostracoda are transported passively onto volcanic islands where they settle in new sandy habitats. Evidence for possible passive dispersion of marine interstitial fauna on floating objects and weeds which also contain clastic sediments as well as with sand balast of ships has been presented by Gerlach, S. 1977. Means of meiofauna dispersal. *In* Sterrer, W. & Ax, P. (Eds), The meiofauna species in time and space. *Microfauna Meeresboden*, Wiesbaden, **66**, 89-103. Bonaduce and I will discuss this problem for the interstitial Xestoleberididae from the Stromboli Island in the Mediterranean.

# 34

# Adaptive strategies and evolutionary processes in Ostracoda: examples from the Eocene and the Eocene-Oligocene boundary of the northern Aquitaine Basin.

Odette Ducasse, Lucienne Rousselle & Olivier Bekaert

Department of Geology & Oceanography, University of Bordeaux I, France

## ABSTRACT

Seven polymorphic ostracod species have been studied over the Eocene-Oligocene boundary in the northern Aquitaine Basin. The influence of environment on these species, as shown by a gradual climatic deterioration during the Upper Eocene, is monitored. This microevolution is linked to the following adaptive strategies:

1) fluctuations in population size,

2) changes in the frequency and the number of morphotypes,

3) fluctuations in the degree of variation of some of their measured parameters.

Just before the sharp environmental cooling of Eocene-Oligocene times, polymorphism declines and variability decreases in most cases. The rare morphotypes which persist occur in abundance and

may give rise to innovations.

It is necessary to point out the difference between those features, the origin of which is genetic, and those which are merely an ecophenotypic response to the environment. This is particularly important as far as individuals and especially their ornamentation are concerned.

## INTRODUCTION

During our studies on the structure of ostracod populations since 1978 in the Aquitaine Palaeogene, we (O.D. and L.R.) have noticed that the space and time transformations that affect the studied species, indicate some adaptive modalities and evolutionary trends which can be related to

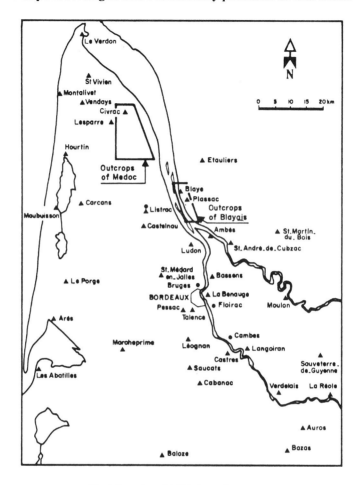

Fig. 1. Location of drill holes and outcrops.

environment. We have also observed that these phenomena are more evident in the margino-littoral area than in the epibathyal zone (Ducasse *et al.*, 1983).

Here we present the observations recorded for 7 species of ostracods traced through the Eocene and the Eocene-Oligocene boundary on the Aquitainian continental shelf in laguno-marine to shallow marine, more or less sheltered, unstable and, therefore, very 'demanding' environments. These seven species are:

*Hammatocythere oertlii* (Ducasse) (Ducasse *et al.*, 1978, 1988)
*Pokornyella ventricosa* (Bosquet) (Ducasse *et al.*, 1981, 1985, 1987)
*Quadracythere cassidea* (Reuss) (Ducasse *et al.*, 1985, 1987)
*Leguminocythereis inflata* Ducasse
*Leguminocythereis striatopunctata* (Roemer)
*Leguminocythereis pertusa* (Roemer) (Ducasse & Rousselle, 1988)
*Krithe rutoti* Keij (Ducasse & Rousselle, 1989).

The studied material comes from numerous boreholes and outcrops, the location of which are indicated in Fig. 1.

## THE ADAPTIVE STRATEGIES OF POPULATIONS

It is always difficult to assess the extent to which the structure of fossil populations is representative, but the following may be observed:

1) Fluctuations in population size, i.e., the increase or decrease in the number of individuals per unit area or volume.

2) Polymorphism, or the simultaneous existence

of several distinct morphotypes of a species. This polymorphism is more or less pronounced, but it is sometimes so reduced that only few, or even only a single morphotype may be present at any one time.

Polymorphism can be interpreted as a measure of genetic wealth that may be regarded as an advantageous adaptive phenomenon in unstable environments. High levels of polymorphism and population density are not observed in all species under the same environmental influences, but may correspond to their respective ecological optima (Sacchi, 1974; Keen, 1982).

As Fig. 2 shows, *Pokornyella ventricosa*, *Leguminocythereis striatopunctata* and *Krithe rutoti* are abundant and very polymorphic during the Lower and Middle Eocene, in marine environments under a warm and wet climate. Conversely, *Hammatocythere oertlii* is most strongly polymorphic during the Upper Eocene in a lagoonal to shallow marine, more or less sheltered environment, which became very unstable during climatic deterioration.

Decline in both polymorphism and population density is observed at the time and space boundaries of species and often immediately precedes their extinction. The modulation of polymorphism (the number, nature and proportions of morphotypes) and the number of individuals of each morphotype are related to environmental changes. The phenomenon induces microevolution and the different structural stages in the populations of the various species which constitute good local stratigraphical indices.

3) Fluctuations in the variation of morphotypes. A comparative study of the variation of some measurable parameters (length, height, width) shows that these values often tend to decrease until the morphotypes (*Q. cassidea* 'apostolescui') or the species (*P. ventricosa*, *L. striatopunctata*) become extinct. This phenomenon, which can be quantified and is often associated with the decline of polymorphism, may indicate a strengthening of the genetic heterogeneity and a resurgence of the homeostasy of populations when faced with inimical environmental conditions (Falconer, 1961). The variations remain almost the same if the morphotypes are tolerant (great longevity and broad

adaptive possibilities) as seen, for example, in the *Q. cassidea* 'helmeted' morphotype.

## THE EVOLUTIONARY AND/OR ADAPTIVE MODIFICATIONS OF INDIVIDUALS

Within species and/or morphotypes some morphological polarities were recorded through time:

1) Variations of the size of individuals: size can be shown to decrease in time in *P. ventricosa* and, on the contrary, increase in *Q. cassidea*, *L. striatopunctata*, *L. pertusa* and *K. rutoti*.

2) Variations in shape: this can be assessed by the sphericity indices length/height and length/width, and an example is the elongation of the outline of *P. ventricosa* and *K. rutoti*.

3) The modifications of ornamentation: within species, this manifests itself in, for example, the development of the longitudinal costae in *P. ventricosa*, the disappearance of ribs and the development of reticulation in *L. striatopunctata*, the weakening of ornamentation in *Q. cassidea* and becoming smooth in *L. pertusa*. Within some tolerant morphotypes this can also be observed, i.e., changes in the *H. oertlii* 'common' and *L. inflata* 'perennial' morphotypes.

Also during the Middle Eocene, a period of high sea level and a subtropical climate, the architectural features of the carapaces are often pronounced, sometimes coated, for example in the *Q. cassidea* 'helmeted', *L. inflata* 'corpulent' and *L. pertusa* 'aquitaine' morphotypes.

In the middle and late Upper Eocene and the early Oligocene, in margino-littoral more unstable environments and during climatic oscillations, the ornamentation tends to disappear, e.g., in the *Q. cassidea* 'helmeted', *L. inflata* 'late' and *L. pertusa* 'erasa' morphotypes.

It is often difficult to separate the evolutionary modifications of morphology from those that indicate a direct effect of the environment. In 1963, Mayr pointed out the necessary distinction that needs to be made between polymorphism which concerns genetically fixed characters and polyphenism which is related to an ecophenotypic effect. It is also important to note that not all

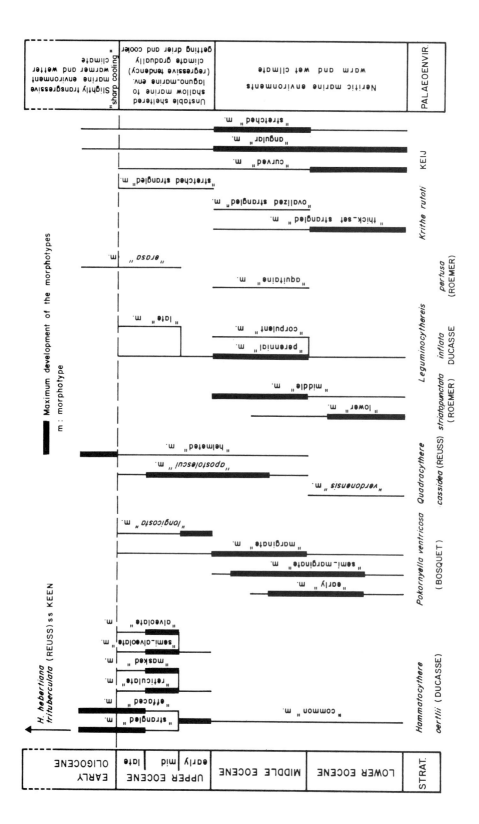

Fig. 2. Evolution of some polymorphic ostracod species in the northern Aquitaine Basin during the Eocene; the effects of the Eocene-Oligocene 'crisis'.

species exhibit the same level of phenotypic behaviour.

## THE EOCENE-OLIGOCENE BOUNDARY CRISIS

The Eocene-Oligocene boundary corresponds to important palaeoceanographical and climatic changes throughout the world. In Aquitaine, deposits of Upper Eocene age were generally deposited during a marine regression compared with those of Middle Eocene age, which were deposited during transgressive episodes. The last regression of the epoch was the largest and took place in northern Aquitaine in the late Upper Eocene. In the basal Oligocene a rather weak transgression took place.

A climatic deterioration occurred gradually through the Upper Eocene and a cooling associated with drought toward the end of the Eocene precedes a sharp drop of temperature at the Eocene-Oligocene boundary. In the basal Oligocene, environmental conditions became warmer and wetter again (Cavelier, 1979; Chateauneuf, 1980; Pomerol, 1985).

The effects of this 'crisis' were strongly felt on the northern Aquitaine continental shelf and are marked by a substantial turnover of the ostracod fauna (Ducasse *et al.*, 1985). Numerous species became extinct, among them such important elements as *P. ventricosa* and *L. inflata*; others persisted and modified their population structure with a resultant decline in the level of polymorphism and the appearance in flood proportions associated with monomorphism as seen in *Q. cassidea*. Some new species appear, for example *H. oertlii* gave rise to a new Oligocene species *H. hebertiana* by selective conservation and isolation of its 'strangled' and 'effaced' morphotypes (Ducasse *et al.*, 1988).

## CONCLUSIONS

It is necessary to emphasize the importance of the intraspecific variations observed in the 7 studied species. Their ability to exist as a number of distinct morphotypes makes it possible to relate their microevolution directly to the fluctuating features of a shallow marine area under the influence of global climatic perturbation.

Polymorphism is itself an adaptive strategy because it is a consequence of a strong genetic heterogeneity (connected with homeostasy).

It should be noted that the comparative study of variation reveals that morphotypes are capable of strengthening their homeostasy by possessing a genetic structure, a major feature of which results in a reduction of the range about the mean, in terms of their carapace dimensions.

Moreover, morphotypes seem to have some degree of independence within polymorph populations. This is evidenced by their environmental preferences, (e.g., *H. oertlii* 'alveolate' and 'masked' morphotypes, *L. inflata* 'corpulent' and 'late' morphotypes, *K. rutoti* 'ovalized strangled' morphotype) or their tolerance (e.g., the *H. oertlii* 'common', *P. ventricosa* 'marginate', *Q. cassidea* 'helmeted' and *L. inflata* 'perennial' morphotypes). Speciation by selective conservation and isolation of morphotypes also documents this.

## REFERENCES

Cavelier, C. 1979. La limite Eocène-Oligocène en Europe occidentale. *Mém. Univ. L. Pasteur*, Strasbourg, **54**, 280 pp.

Chateauneuf, J. J. 1980. Palynostratigraphie et paléoclimatologie de l'Eocène supérieur et de l'Oligocène du Bassin de Paris. *Mém Bur. Rech. géol. minièr.*, Paris, **116**, 360 pp.

Ducasse, O., Bekaert, O., Ringeade, M. & Rousselle, L. 1987. Polymorphisme et évolution chez les Ostracodes du Paléogène aquitain: phénomènes micro-évolutifs chez *Pokornyella ventricosa* (Bosquet) et *Quadracythere cassidea* (Reuss) à l'Eocène et à l'Oligocène inférieur dans le Nord du Bassin d'Aquitaine. *Trav. C.R.M.*, n° **8**, Villefranche-sur-mer, Fr., 20-38.

Ducasse, O., Lété, C. & Rousselle, L. 1985. Contribution à l'étude paléontologique d'une crise paléogène: populations d'ostracodes à la limite Eocène-Oligocène dans le Médoc (Gironde). *Bull. Inst. Geol. Bassin d'Aquitaine*, Bordeaux, **38**, 141-175.

Ducasse, O., Lété, C. & Rousselle, L. 1988. Polymorphism and speciation. Medoc Ostracods at the Eocene/Oligocene Boundary (Aquitaine, France). *In* Hanai, T., Ikeya, N. & Ishizaki, K. (Eds), *Evolutionary biology of Ostracoda, its fundamentals and applications*, proceedings of the Ninth International Symposium on Ostracoda, held in Shizuoka, Japan, 29 July - 2 August 1985, Developments in palaeontology and stratigraphy, **11**, 939-947, Kodansha Ltd., Tokyo and Elsevier, Amsterdam, Oxford, New York,

Tokyo.

Ducasse, O. & Rousselle, L. 1988. Le genre *Leguminocythereis* (Ostracodes) dans le Paléogène nord-aquitain: espèces et populations, histoire évolutive locale. *Géobios*, Lyon, **21**, fasc. 2, 137-167.

Ducasse, O. & Rousselle, L. 1989. *Krithe rutoti* Keij (Ostracodes): structure des populations et évolution a l'Eocène et au moment de la crise Eocène-Oligocène sur le plateau continental nord-aquitain. *Annls Paléont.*, Paris, **75**(1), 1-22.

Ducasse, O., Rousselle, L. & Peypouquet, J. P. 1983. Processes of evolution in marginal-coastal and bathyal Ostracods, Palaeogene of Aquitaine, France. *In* Maddocks, R. F. (Ed.), *Applications of Ostracoda*, proceedings of the Eighth International Symposium on Ostracoda, July 26-29, 1982, 605-611. Univ. Houston Geos., Houston, Texas.

Falconer, D. S. 1961. *Introduction to quantitative genetics*, x + 365 pp. Oliver & Boyd, Edinburgh & London.

Keen, M. C. 1982. Intraspecific variation in Tertiary ostracods. *In* Bate, R. H., Robinson, E. & Sheppard, L. M. (Eds), *Fossil and Recent Ostracods*, 381-405. Ellis Horwood Ltd., Chichester for the British Micropalaeontological Society.

Mayr, E. 1963. *Animal species and evolution*, xvi + 797 pp. Belknap Press of Harvard Univ. Press, Cambridge, Mass.

Pomerol, Ch. 1985. La transition Eocène-Oligocène est-elle un phénomène progressif ou brutal? *Bull. Soc. géol. Fr.*, Paris, **8**(1, 2), 263-267.

Sacchi, C. F. 1974. Le polymorphyisme dans le regne animal. *Mém. Soc. zool. Fr.*, Paris, **37**, 61-101.

# 35

# Biofacies of early Permian Ostracoda: response to subtle environmental change

**Roger L. Kaesler, Jonathan C. Sporleder &
James A. Pilch**

Department of Geology, Museum of Invertebrate
Paleontology & Paleontological Institute,
The University of Kansas, U.S.A.

## ABSTRACT

A thin stromatolite bed and associated limestone facies crop out along a 260km long, north-south belt in the Midcontinent of the United States. The stromatolite bed was deposited in shallow-water, restricted-marine environments where stromatolites were protected by high salinity and frequent desiccation from browsing organisms that might otherwise have destroyed them. Lithological evidence suggests that the southern part of the outcrop belt, where water was probably deeper, was the site of lower energy, more nearly normal marine environments than occurred to the north. In the southernmost part of the study area, the stromatolite bed is absent.

Thirty-five localities were sampled. Fourteen samples that yielded ostracods from the stromatolite bed are discussed here. In general, the samples contained fewer specimens than were found in a previous study to the north. The number of species in the samples is low, especially in the north, where *Paraparchites humerosus* and *Sansabella bolliaformis* predominate. Assemblages from northern localities typically have few individuals of species characteristic of nearshore, terrigenous rocks: *Cavellina_nebrascensis*, *Bairdia beedei*, *Hollinella bassleri*, *Pseudobythocypris pediformis*, and *Healdia simplex*. To the south more individuals of these species occur. Nevertheless, one can extract most of the useful palaeoenvironmental information by focusing attention on the two most abundant species and on lithological evidence.

## INTRODUCTION

Fossil Ostracoda are most readily obtainable from such fine-grained terrigenous rocks as shales and mudstones, which are poorly understood petrographically. Over the past two decades, petrographical analyses of limestones from a great many depositional environments have noted that ostracods comprise at least a few percent of many carbonate rocks. Nevertheless, these ostracods in limestones can almost never be identified, even to the generic level. They are seen only rarely on bedding planes and are nearly always impossible to extract for study. On the other hand, the petrographical study of limestones is quite an advanced science. It is indeed unfortunate that students of the Ostracoda, intent on deriving a comprehensive palaeoecology, have been unable to tap the wealth of palaeoenvironmental knowledge developed by carbonate petrographers.

We have managed to extract ostracods from limestone by means of crushing rather than using chemical methods. The limestones we studied have been analyzed in detail petrographically (Sporleder, in press). Our intention here is to demonstrate some relationships between the ostracod assemblages and the distribution of limestone microfacies. We hope to show that most of the useful palaeoenvironmental information can be obtained by studying only a very few species of Ostracoda that are dominant in the assemblages.

The stromatolitic limestone bed that is the focus of this study is from the lower part of the Americus Limestone Member of the Foraker Limestone. Kaesler & Denver (1988) studied ostracods from northern outcrops of the same bed. The stromatolite bed is typically a peloid algal boundstone with encrusting foraminifera, *Spirorbis* worm tubes and ostracods. It was deposited in a restricted, nearshore environment in the very early Permian (Wolfcampian). It is typically only 10 to 20cm thick, yet it crops out along a belt that extends for 200km across Kansas in the Midcontinent of the United States. Fig. 1 shows a generalized stratigraphical section of the stromatolite and associated facies and indicates palaeoenvironmental conditions (Sporleder, in press). Correlative limestone beds extend at least another 50km farther south than the stromatolite before passing into terrigenous rocks in northern Oklahoma. The datum for the schematic cross section of Fig. 2 is the base of the upper limestone of the Americus, a bed of remarkably uniform thickness and lithology (Harbaugh and Demirmen,

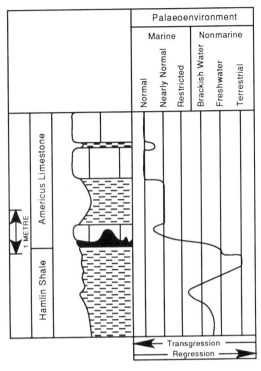

Fig. 1. Generalized stratigraphical section of the Americus Limestone Member of the Foraker Limestone (Permian, Wolfcampian). The stromatolite bed of this study lies at the base of the Americus (after Sporleder, in press).

1964).

## MATERIAL AND METHODS

Large hand specimens of the stromatolitic limestone and its southern lateral equivalents from 35 localities were crushed to a size of about 7mm, processed with Amosol (a petroleum-naptha-based, mineral-spirit solvent), washed through 20-mesh and 100-mesh sieves, and picked for ostracods. Fourteen samples (Fig. 3) yielded 1323 specimens, although ostracods were rare in some samples. Altogether 2370 ostracods were found, but more than one thousand of them were from samples of beds other than the stromatolite, and are not considered further in our study. Although 10 species or species groups were recovered, only 8 species occurred in the stromatolite bed (Table 1).

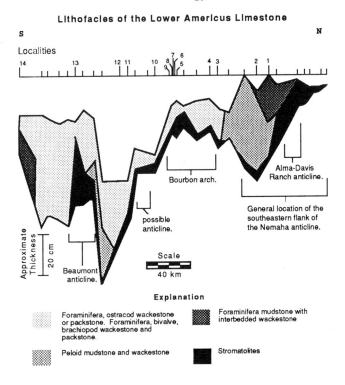

Fig. 2. Diagrammatic cross section of the lower Americus Limestone showing generalized lithology, thickness and locations of all localities sampled. The stromatolite bed yielded ostracods at the numbered localities.

Most of the ostracods are smooth. One might suspect that such highly ornamented species as *Hollinella bassleri* (Knight, 1928), for example, might be under-represented in our samples because they might have been retained in the rock matrix during crushing, making them unavailable for study. We have considered and rejected the possibility that our samples may be biased in this way. We have examined numerous fragments of limestone and have found few ornamented forms, yet numerous smooth ones embedded in them. Moreover, we have found as high a proportion of ornamented as smooth forms that are free of limestone matrix in comparison to their apparent abundance in the sample.

## Statistical considerations

Much of our analysis is based on ratios and percentages of species in our samples. In spite of the fact that ratios and percentages have been applied widely in biology and palaeontology and are frequently the most appropriate measure to use, they are fraught with statistical difficulties (Sokal & Rohlf, 1981, 17-19). One serious drawback is that the same ratio can be obtained from different pairs of numerators and denominators and might, therefore, be the result of quite different biological processes. Another difficulty is that ratios and percentages are inherently less accurate than the raw data from which they are computed. These two drawbacks result in values that are dependent on sample sizes, so that confidence limits around them may be quite broad.

Fig. 4 plots two hypothetical samples: A with 50% species 1 and 40% species 2 and B with 70% species 1 and 20% species 2. The 95% confidence limits based on sample sizes of 50, 100, and 200 are shown. Note that if samples of size 50 had been collected, these two quite different samples would have had overlapping confidence limits making them statistically indistinguishable. With sample sizes of 100 the two samples appear to be only barely different.

Fig. 5 is presented to help the reader assess the precision of percentages of species reported. The ordinate of the figure shows the number of specimens in each the 14 samples of the stromatolite bed. Next to these are shown the 95%

confidence bands for percentages of 0, 25, 50, 75, and 100 from samples of sizes 10 to 300. Note that if samples contain less than about 35 individuals the confidence bands are quite wide and overlap extensively. (Confidence limits were set using Rohlf & Sokal, 1969, Table W, 208-214.)

Synergistic and antagonistic environmental effects, especially in the marine environment, often obscure the kinds of trends that permit straightforward environmental interpretation. In palaeoecology the problem of interpretation is compounded by the taphonomic overprint, which introduces mixing and differential preservation of faunas. Moreover, with the notable exception of grain size of substrate, most environmental parameters are simply not recorded in the rock record. In the search for meaning, one must often be content with finding only very weak trends and relationships. We have computed a number of correlation coefficients (Table 1) and linear regressions, many of which have values quite close to zero. In all instances, we have tested them for statistical significance.

## RESULTS AND DISCUSSION

Table 2 lists species of Ostracoda found in samples of the stromatolite bed. Two species dominate the assemblages: *Paraparchites humerosus* Ulrich & Bassler, 1906, and *Sansabella boliaformis* (Ulrich & Bassler, 1906). Their relative abundances in comparison to all other species are shown in a ternary plot (Fig. 6). Kaesler & Denver (1988) found that the same two species were dominant in samples of northern outcrops of the same bed. In their study, however, *Bairdia*

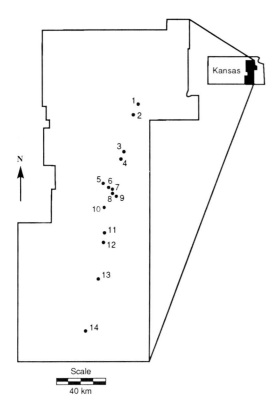

Fig. 3. Map of east-central Kansas showing the fourteen localities sampled that yielded ostracods (after Sporleder, in press).

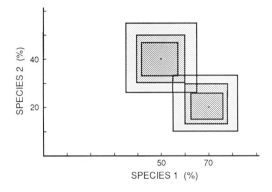

Fig. 4. Percentages of two species in each of two hypothetical samples with 95 percent confidence limits for sample sizes of 50 (light shading), 100 (intermediate shading) and 200 (dark shading).

*beedei* Ulrich & Bassler, 1906, was much more common; and the number of species in most of their samples was lower.

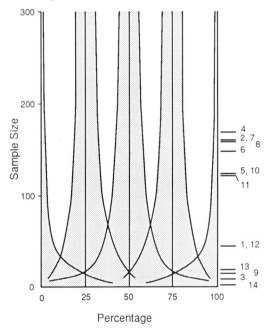

Fig. 5. Numbers of ostracods from samples of the stromatolite bed and 95-percent confidence bands about percentages of 0, 25, 50, 75 and 100 for sample sizes from 10 to 300.

## Geographical effects

In general, samples from the southern part of the study area (stations 12, 13, and 14) have a smaller proportion of *P. humerosus* than samples from the northern part (stations with low numbers) (Fig. 6). The interpretation of this trend, however, must be tempered by consideration of the smaller numbers of ostracods in samples of southern outcrops (Fig. 5).

The most striking trend among the ostracods is the increasing northward dominance of *P. humerosus* (Fig. 7). North of locality 10 (Fig. 3), it comprises up to 90% of the ostracod fauna. Southwards, it declines in dominance to as little as 5% at locality 13 and is altogether absent from the sample collected at locality 14. Its decreasing dominance coincides with a trend towards increasingly more open-marine conditions. The stromatolite is absent south of locality 13, possibly the result of browsing pressure from aquatic herbivores. The decreased proportion of *P. humerosus* to the south supports previous contentions that species of *Paraparchites* were able to withstand high salinity (Dewey, 1987, 1988) and brief intervals of subaerial exposure and that they fed on the cyanobacteria that formed the stromatolite (Kaesler & Denver, 1988).

*Sansabella bolliaformis*, the subdominant species, shows no such pronounced trend toward increased dominance. The slope of the linear regression line (Fig. 7), although suggestive, is not significantly different from zero. Moreover, the apparent weak trend is probably only an artefact of the compositional nature of the data. *S. bolliaformis* and *P. humerosus* together comprise 94% of the

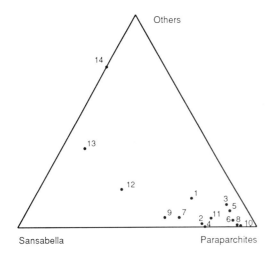

Fig. 6. Ternary diagram showing relative abundances of *Paraparchites humerosus*, *Sansabella bolliaformis* and the six other species in samples of the stromatolite bed.

ostracod fauna. Given this fact and the decreased proportion of *P. humerosus* southward, it is inevitable that *S. bolliaformis* should increase in relative abundance in the same direction. Nevertheless, the relationships between these two species and others in the study are not simple. The abundance of *P. humerosus* is strongly negatively correlated with that of *S. bolliaformis* ($P < 0.01$ - Table 1).

**Structural and stratigraphical effects**

Early in the late Carboniferous, parts of the Mid-continent and some western and southwestern parts of the United States were affected by block faulting of the Precambrian basement. Following erosion, sedimentary rocks were draped over these buried ridges forming faulted anticlines, a process that was facilitated by differential compaction. Some of these features were of major importance and are now the sites of important reserves of petroleum. Others were small, and it is likely that a few remain to be discovered.

By Wolfcampian time, these tectonic features had been eroded flat. Nevertheless, differential compaction of sediments and perhaps minor faulting continued so that they exerted a subtle but detectable influence on depositional environments (Fisher, 1980). At sites of buried anticlines the

lower Americus Limestone is characterized by such high-energy deposits as grainstone, packstone and some wackestone, whereas over the synclines it comprises largely wackestone and mudstone. Moreover, the stromatolite bed is usually thicker over buried anticlines, suggesting subaerial exposure that prevented browsing herbivores from destroying the stromatolite-building cyanobacteria.

Fig. 8 shows the distribution of total thickness of the lower beds of the Americus Limestone. The diagram is based only on the fourteen localities at which ostracods were found in the stromatolite bed. Sporleder (in press) presents a more extensive analysis of his much larger data set. Where the beds are thickest, they overlie the buried Beaumont anticline. The central part of the study area, where the beds are thinnest, is on the southern flank of the Bourbon arch, the site of deposition of a foraminiferal ostracod packstone.

**Palaeocommunities**

The number of species does not change significantly and shows no geographical trend in the area, although an index of species diversity that considers evenness would probably increase southward because of the diminishing importance of the dominant species, *P. humerosus*. The ostracod assemblage of the stromatolite bed is remarkable, however, in the relative abundances of individuals of species. As was noted previously by Kaesler and Denver (1988), species that are characteristic of nearshore terrigenous mudstones and shales, although present, are uncommon in the stromatolite bed. These include *Cavellina nebrascensis* (Geinitz, 1866); *Bairdia beedei* Ulrich and Bassler, 1906; *Pseudobythocypris pediformis* (Knight, 1928); and *Healdia simplex* Roundy, 1926.

One measure of the palaeoenvironment is given by the sample size, the number of ostracods found in each sample (Fig. 8). The measure is crude because additional effort could be expended at any locality to obtain more ostracods. (Larger samples could be collected and more material picked so that one could presumably get any

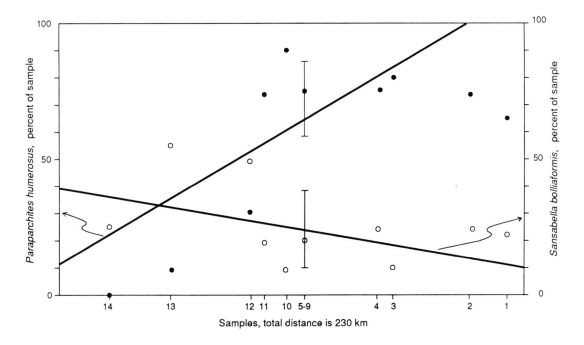

Fig. 7. Plots of abundances of *Paraparchites humerosus* and *Sansabella bolliaformis* in percent against sample numbers, showing linear regression lines. Distances on the abscissa are scaled in proportion to distances from an arbitrary southern datum, the Kansas-Oklahoma border.

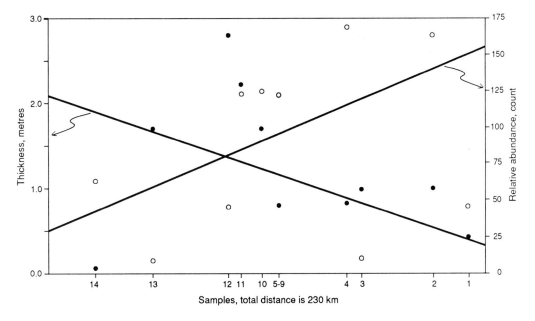

Fig. 8. Plots of thicknesses of the lower beds of the Americus Limestone and of number of ostracods found in samples of approximately equal weight against sample numbers, showing linear regression lines. Distances on the abscissa are scaled in proportion to distances from an arbitrary southern datum, the Kansas-Oklahoma border.

number of ostracods for study.) Our samples, however, were of approximately equal weight, so that a comparison of the numbers of ostracods found in the samples has some palaeoenvironmental meaning. The variation across the area is complex. The largest numbers of individuals occur in samples from the middle part of the area, and many fewer ostracods were found southwards toward the more open marine environments.

At least three processes may contribute to this pattern, but it is not possible to assess their relative importance. First, ostracods living in restricted marine environments may have proliferated in the absence of competition from other species. If *P. humerosus* fed on cyanobacteria, the very presence of the stromatolite bed and the small numbers of such aquatic herbivores as gastropods suggest that food was not a limiting resource for *P. humerosus*. The abundance of the cyanobacteria may, therefore, have accounted for the abundance of ostracods.

Second, many groups of organisms other than ostracods are typically more abundant in open marine settings than in restricted environments. The apparent paucity of the ostracod fauna to the south may have resulted simply from dilution of the ostracods by other elements of the fauna. This process seems unlikely to explain fully the pattern, however, because large numbers of foraminifera were found throughout the study area, even where ostracods are abundant.

Third, although crushing has enabled us to remove ostracods from limestone deposited in both fresh water and under restricted marine conditions, we have not had similar success with marine limestone. This is true even for quite argillaceous limestone that is known to contain abundant ostracods. Freshwater limestone is deposited as low-Mg calcite rather than as high-Mg calcite or aragonite and is likely to be less intensively recrystallized during diagenesis. Consequently, ostracods may be less firmly bound in such rocks and, therefore, more likely to be freed during crushing. Whether this explanation applies to limestone from a restricted marine environment depends on the mineralogy of the grains that comprise the limestone. It would apply to the

rocks we studied only if the carbonate grains accreted by the cyanobacteria to form the stromatolites were composed of some mineral other than aragonite. Sandberg (1983, 19) has demonstrated '...an oscillating Phanerozoic temporal trend...' in mineralogy of nonskeletal carbonate grains. Unfortunately for this interpretation, the Wolfcampian was a time when the carbonate curve was above the aragonite threshold. According to Sandberg's model, Wolfcampian nonskeletal carbonate is expected to have been precipitated as high-Mg calcite or aragonite rather than as low-Mg calcite. The source of the grains that the stromatolites comprise, skeletal or nonskeletal, is unknown.

## CONCLUSIONS

1) *Paraparchites humerosus* and *Sansabella bolliaformis* are the dominant and subdominant species in the ostracod assemblages associated with the stromatolite bed at the base of the lower Americus Limestone.

2) *Paraparchites humerosus* has a much lower relative abundance in the south where the environment of deposition was less restricted and more nearly normal marine.

3) *Sansabella bolliaformis* may have had greater relative abundance where more nearly normal marine conditions prevailed, but the data do not show a statistically significant geographical trend.

4) Six other species that occurred in association with the dominant and subdominant species comprise only 6% of the ostracod fauna.

5) The distribution of *P. humerosus* and *S. bolliaformis* is controlled by geographical, stratigraphical and structural effects as well as by other biota in the palaeocommunity.

6) The number of species changes little over the study area, but the distribution of individuals among species is more nearly even in the south where the dominance of *P. humerosus* is less pronounced.

## ACKNOWLEDGEMENTS

We are grateful to Jane Priesner of The University of Kansas Paleontological Institute, who prepared all the illustrations, and to A. J. Rowell, Jack D. Keim, and Bryon Wiley for assistance with photography. Specimens have been reposited with The University of Kansas Museum of Invertebrate Paleontology.

## REFERENCES

Dewey, C. P. 1987. Paleoecology of a hypersaline Carboniferous ostracod fauna. *J. micropalaeontol*., London, **6**(2), 29-33.

Dewey, C. P. 1988. Lower Carboniferous ostracode assemblages from Nova Scotia. *In* Hanai, T., Ikeya, N. & Ishizaki, K. (Eds), *Evolutionary biology of Ostracoda, its fundamentals and applications*, proceedings of the Ninth International Symposium on Ostracoda, held in Shizuoka, Japan, 29 July - 2 August 1985, Developments in palaeontology and stratigraphy, **11**, 685-696. Kodansha Ltd., Tokyo and Elsevier, Amsterdam, Oxford, New York, Tokyo.

Fisher, W. L. 1980. *Variation in stratigraphy and petrology of the uppermost Hamlin Shale and Americus Limestone related to the Nemaha structural trend in northeastern Kansas*. Unpub. doctoral dissertation, The University of Kansas, Lawrence, Kansas, 166 pp.

Geinitz, H. B. 1866 (1867). Carbonformation und Dyas in Nebraska. *Verhandlungen der Kaiserlichen Leopoldino-Carolinischen deutschen Akademie der Naturforscher*, Dresden, 91 pp.

Harbaugh, J. W. & Demirmen, F. 1964. Application of factor analysis to petrologic variations of Americus Limestone (Lower Permian), Kansas and Oklahoma. *Kans. State Geol. Surv. Spec. Distrib. Publ.*, Lawrence, Kansas, **15**, 1-40.

Kaesler, R. L. & Denver, L. E. II. 1988. Distribution and diversity of nearshore Ostracoda: environmental control in the Early Permian. *In* Hanai, T., Ikeya, N. & Ishizaki, K. (Eds), *Evolutionary biology of Ostracoda, its fundamentals and applications*, proceedings of the Ninth International Symposium on Ostracoda, held in Shizuoka, Japan, 29 July - 2 August 1985, Developments in palaeontology and stratigraphy, **11**, 671-683. Kodansha Ltd., Tokyo and Elsevier, Amsterdam, Oxford, New York, Tokyo.

Knight, J. B. 1928. Some Pennsylvanian ostracodes from the Henrietta Formation of eastern Missouri, part 1. *J. Paleont.*, **2**, 229-267.

Rohlf, F. J. & Sokal, R. R. 1969. *Statistical Tables*, 253 pp. W. H. Freeman & Co., San Francisco.

Roundy, P. V. 1926. Mississippian formations of San Saba County, Texas, part 2, the microfauna. *U.S. Geol. Surv, Prof. Pap.*, Washington, D.C., **146**, 5-23.

Sandberg, P. A. 1983. An oscillating trend in Phanerozoic non-skeletal carbonate mineralogy. *Nature*, London, **305**(5929), 19-22.

Sokal, R. R. & Rohlf, F. J. 1981. *Biometry*. 2nd edition, 859 pp. W. H. Freeman & Co., San Francisco.

Sporleder, J. C. In press. Structural control of the distribution of subtidal to supratidal paleoenvironments of the Americus Limestone. *Kans. State Geol. Surv., Geology Series 4*, Lawrence, Kansas.

Ulrich, E. O. & Bassler, R. S. 1906. New American Paleozoic Ostracoda. Notes and descriptions of Upper Carboniferous genera and species. *U.S. Nat. Mus. Proc.*, Washington, D.C., **30**, 149-164.

# 36

# Palaeoecology of Upper Carboniferous Ostracoda from the Lake Murray Formation, southern Oklahoma

Larry W. Knox

Department of Earth Sciences,
Tennessee Technological University, U.S.A.

## ABSTRACT

Ostracods from 81 continuous channel samples from part of a thick shale sequence in the lower part of the Lake Murray Formation (Middle Pennsylvanian, Ardmore Basin, Oklahoma) were analyzed in order to determine their environment of deposition.

Dark grey to olive-grey mudshales representing deposition in prodeltaic offshore environments in water depths of some 100m yield an ostracod assemblage of moderate abundance and diversity. The assemblage was dominated by Genus A sp. A; *Pseudobythocypris tomlinsoni* (Harlton, 1929); Genus A sp. B; and *Healdia colonyi* Coryell and Booth, 1933. During regression, pale orange mudshales were deposited in an offshore environment in water depths of approximately 50m.

These shallower water sediments contain an assemblage of abundant ostracods with equitable species distribution and high diversity. In order of decreasing abundance, the assemblage includes *Healdia colonyi*; *Fabalicypris warthini* (Bradfield, 1935); *Healdia formosa* Harlton, 1928; *Healdia oblonga* Bradfield, 1935; *Cavellina* sp. A; *Pseudobythocypris tomlinsoni*; *Monoceratina* n. sp. Bradfield 1935; *Shivaella brazoensis* (Coryell and Sample, 1932); *Amphissites centronotus* (Ulrich and Bassler, 1906); *Hollinella* sp. A; *Amphissites girtyi* Knight, 1928a; and *Kirkbya* cf. *K*. *punctata* Kellett, 1933. Yellow-grey to medium grey clayshales were deposited in a distant offshore marine environment (not significantly influenced by deltaic sedimentation) at depths of 50-100m. These offshore clayshales contain an ostracod assemblage that is moderately abundant, but of low species diversity. The assemblage is dominated by *Pseudobythocypris tomlinsoni*; Genus A sp. A; *Healdia asper* Cooper, 1946; *Fabalicypris warthini*; and *Orthobairdia oklahomaensis* (Harlton, 1927).

## INTRODUCTION

A number of palaeoecological studies of North American Pennsylvanian ostracods have been published in recent years. Notable among these are Brondos and Kaesler (1976), Haack and Kaesler (1980), Kaesler (1983), and most recently by Melnyk and Maddocks (1988). These studies have largely been concerned with interbedded carbonate and shale sequences from the midcontinent of North America; noticeably lacking have been studies of basins in which shales are the predominant rock type.

The purpose of this paper is to present results of a palaeoecological study of ostracods from the Middle Pennsylvanian portion of a very thick sequence of terrigenous shales at a single section near Ardmore, Oklahoma (NW1/4,NW1/4,SW1/4, sec. 15, T3S, R3E) (Fig. 1). A depositional history of the sequence, independent of the ostracods, was established by analysis of the invertebrate macrofauna, lithostratigraphical sequence, clay mineralogy, sediment colour, organic carbon content and distribution of sediment size.

## MATERIAL AND METHODS

A continuous series of 1/3m channel samples was taken from the base of unit 9 to the middle of unit 11 at the Dutton Ranch locality (Fig. 2). Unit 10 (a sandstone) was not sampled. In order to efficiently compare ostracod abundances, exactly 400 grams of each sample, or in some cases every other sample, was processed by boiling with sodium carbonate and washing over a 75µm (200 mesh) sieve. Each dry residue was separated into three size fractions: a) coarser than 840µm (20 mesh), b) less than 840µm and greater than 300µm (50 mesh), and c) less than 300µm. The

less than 840µm and greater than 300µm fraction was picked for all ostracods. The total number of equivalent ostracod carapaces of each species was determined by adding the larger total of either the left or right valve count to the number of carapaces without regard to sex.

Colour was determined on freshly broken dry samples by comparison with the Geological Society of America Rock-Color Chart. Clay mineralogy was determined by X-ray diffraction of nine samples, and total organic carbon (total carbon minus carbonate carbon) was determined for ten samples. Sediment size distribution of 15 samples was determined by pipette analysis.

## STRATIGRAPHY

The Ardmore Basin is a geographically narrow structural basin located in the northern part of the southern Oklahoma Aulacogen (Fig. 1). Lower and Middle Pennsylvanian rocks at the Dutton Ranch locality (Fig. 2) comprise fine-grained terrigenous basinal sediments and differ markedly from shallow-shelf carbonates deposited just 16km to the northeast (Tennant *et al.*, 1982). Tennant (1981) suggested an Atokan age for the Lake Murray Formation based on its stratigraphical position. A primitive species of the fusulinid genus *Beedeina* occurs 7.5m above the base of unit 11 (Wahlman, pers. comm., 1989). The Atokan-Desmoinesian series boundary has generally been placed at the lowest level of the standard Zone of *Beedeina*. The boundary in the study interval has been tentatively placed at that level (Fig. 2). However, the ostracod species *Amphissites girtyi* occurs near the base of unit 9, and elsewhere in the midcontinent this species is interpreted to be exclusively of Desmoinesian age (Shaver and Smith, 1974). Therefore, the series boundary may occur as low as the base of unit 9.

## DEPOSITIONAL ENVIRONMENT

### Macrofauna

A model for the succession of macrofaunal communities related to water depth has recently been

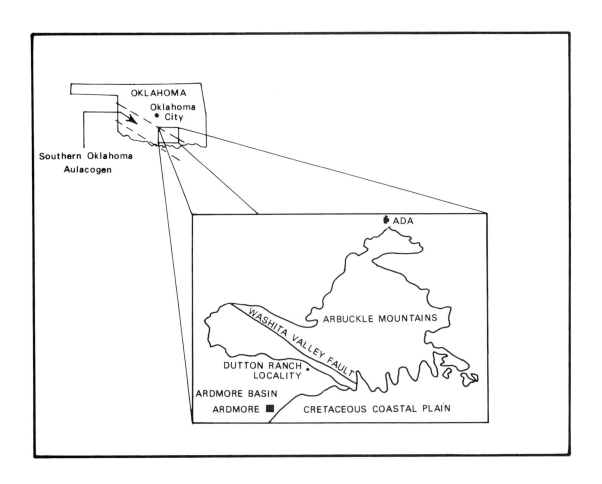

Fig. 1. Locality map.

proposed for North American Pennsylvanian cyclothems of the midcontinent (Boardman *et al.*, 1984). The macrofauna from units 9 and 11 of the Lake Murray Formation consist in large part of crinoid plates, chonetid brachiopods and small bivalves and gastropods. Mapes (pers. comm., 1988) has examined the macrofauna and has suggested that, from near the base of unit 9 to the top of the sampled part of unit 11, the fauna belongs to the stenotopic crinoid-brachiopod-fusulinid-coral-sponge community, representing water depths of approximately 50-100m. An exception is the interval from 0-5.5m above the base of unit 11

that contains abundant fusulinids interpreted as representing maximum water depths of 5-10m.

**Clay mineralogy**

The clay mineralogy of 12 shale samples from units 9 and 11 was determined by X-ray diffraction. Typical clay mineral percentages are illustrated in Fig. 2. The percentage of kaolinite is highest in the lower part of unit 9 (24%), this decreases to 15% in the upper part of unit 9, and averages only 2% in unit 11. Smectite is lowest in the lower part of unit 9 (5%); this

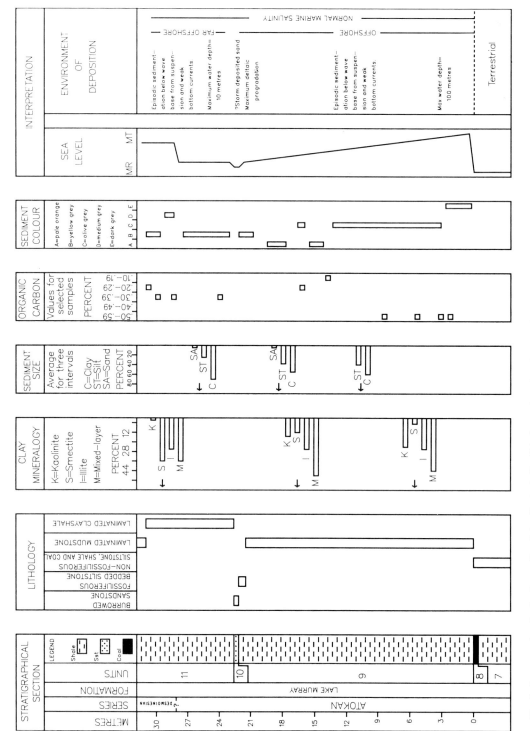

Fig. 2.  Summary data for the Dutton Ranch locality including lithology, clay mineralogy, sediment size distribution, organic carbon percentage, sediment colour and interpretation of environment of deposition.

increases to 12% in the upper part of unit 9 and to 26% in unit 11.

In marine basins, kaolinite is considered to be concentrated near shore and to be a good indicator of palaeogeography in basins as old as Pennsylvanian (Brown *et al.*, 1977). The near shore concentration may result because kaolinite flocculates at lower salinities (Edzwald and O'Melia, 1975) and/or because kaolinite floccules are larger than those of other clay minerals (Gibbs, 1977) and, therefore, less likely to be transported seawards. The presence of kaolinite in significant quantities in unit 9 suggests proximity to the palaeo-shoreline and its virtual absence in unit 11 suggests an offshore marine environment.

### Sediment size and lithology

Ten shale samples from unit 9 averaged 39% silt and 58% clay (mudshale); five samples from unit 11 averaged 25% silt and 71% clay (clayshale) (Fig. 2). The somewhat larger sediment size in unit 9 suggests a nearer shore environment than unit 11, since in most cases energy levels decrease seawards.

In thin section most of units 9 and 11 consist of regularly to discontinuously laminated (laminae 2-5mm thick) mudshale or clayshale (Fig. 2). This type of lamination suggests episodic deposition from suspension, weak bottom currents, and relatively still water well below wave base. Infaunal animals, recognized mostly as trace fossils, are often common enough to obliterate pre-existing sedimentary structures in Upper Carboniferous rocks elsewhere in North America (Miller and Knox, 1985). Lamination in the Lake Murray Formation suggests that the upper part of the sediment column was too low in oxygen to support infaunal organisms, which would have disrupted or destroyed the lamination by burrowing through the sediment (Howard, 1975).

### Organic carbon and sediment colour

Within the studied interval organic carbon varies from 0.52-0.58% in the lower part of unit 9 to 0.16-0.30% in the upper part of unit 9 and 0.20-0.38%

in unit 11 (Fig. 2). Sediment colour and percentage of organic carbon are important features of the shales in the study interval because they are closely associated with each other as well as with distinctive assemblages of ostracods. Shales with highest organic carbon percentages are dark grey to medium olive-grey in colour (Fig. 2) and are associated with ostracods of cluster A (Fig. 5). Sediments with intermediate organic carbon values are yellow-grey (Fig. 2) and yield ostracods of cluster D (Fig. 5). The lowest organic carbon percentages occur in pale orange shales (Fig. 2) with ostracods of cluster B (Fig. 5).

Colour is probably the most obvious feature of shales, and in some cases, may have environmental significance. Potter *et al.*, (1979) suggested that shales seem to partition into two colour series: a red to greenish-grey series based on a decreasing $Fe^{3+}/Fe^{2+}$ ratio, and a greenish-grey to black series based on increasing carbon content. The amount of organic carbon is a partly independent control of colour because even small amounts of organic matter favour the reduced form $Fe^{2+}$. X-ray diffraction revealed that pyrite is absent in shales of the study interval. Given the absence of disseminated pyrite, the only other material likely to be responsible for the dark grey colour of these shales is probably their high organic carbon content (Blatt *et al.*, 1980, p. 404).

Berner (1984) proposed that the amount of organic matter in sediments is controlled by three environmental factors: rate of production of organic matter in surface waters of the basin, rate of sedimentation of non-organic particles and rate of oxidation of organic matter in the sediment column. It is not possible to conclude which of these factors, or combinations of them, controlled the content of organic carbon in the study interval. However, within the sequence environmental gradients associated with these factors were steep enough to control carbon content and, presumably, to produce significantly different assemblages of ostracods.

### Interpretation of environment of deposition

Depositional history of the Lake Murray sequence

was interpreted by analysis of invertebrate macro-faunas, sedimentary structures, stratigraphical sequence, mineralogy, distribution of sediment size, and organic carbon content.

Initially, a rapid transgression flooded terrestrial environments at the base of unit 9. The rapid nature of the transgression is suggested by the presence of stenotopic crinoid-brachiopod communities within 1m above the coal, and the absence of typical nearshore shallow water molluscan communities. The rapid transgression may have occurred as a result of tectonic movements on the Washita Valley fault bordering the aulacogen on the Northeast, or because of glacial-eustatic sea level change typical of Pennsylvanian cyclothems of the midcontinent. The remainder of unit 9 represents a shallowing-upward sequence indicated by increasing grain size and the presence of shallow water fusulinid faunas immediately above and below unit 10. The shallowing was apparently the result of sedimentary infilling of the marine basin by progradation of mud from a nearby delta that supplied significant quantities of kaolinitic clay to the basin. Maximum water depths (near 100m) may have occurred during deposition of the lowest few metres of unit 9 and minimum depths (near 50m) near the top of the unit. These estimates depend on three assumptions: crinoid-brachiopod communities were typical of 50-100m water depths; no change of eustatic sea level occurred after initial transgression; and basinal infilling decreased water depth some 35m, based on a 30% post-depositional compaction (Lineback, 1966) of unit 9.

In unit 9, Cluster A ostracods occur in samples representing greatest water depths, and Cluster B ostracods in samples representing minimum water depths. The presence of crinoids and brachiopods within a few centimetres above and below the sandstone suggests that the upper part of unit 9 and the lower part of unit 11 were never subaerially exposed during deposition. The laminated nature of the sediments and their lack of erosional features suggest that sedimentation was episodic and that deposition occurred below wave base from suspension and weak bottom currents, perhaps as density flows. Most of the macrofossil taxa

consists of immature individuals; their small size may represent catastrophic kills due to episodic sedimentation. Alternatively, the muddy substrate may have lacked the cohesiveness necessary to support mature benthonic organisms (Walker and Diehl, 1986).

In unit 11, from 0-5.5m above the base an abundant fusulinid fauna suggests water depths not exceeding 5-10m. Above this point the presence of crinoid-brachiopod communities imply a second transgression resulting in water depths again in the 50-100m range. Ostracods of Cluster D occur in samples representing the deep water environment of unit 11. The laminated nature of the rocks of unit 11, lack of kaolinitic clay, fine-grained sediment, and absence of traction transport structures (flaser bedding, ripple marks, cross bedding, etc.) imply episodic sedimentation below wave base in a distant offshore marine basin, not strongly influenced by deltaic sedimentation.

## NATURE OF THE OSTRACOD FAUNA

### Transported *versus* non transported ostracod assemblages

Analysis of the population age structure of ostracod species from the studied interval suggests that small specimens were sometimes subject to post mortem transport. As a consequence, if presence-absence data are utilized without respect to population structure, it is likely that life and death assemblages will be confused. The method of Whatley (1983) was used to distinguish between life and death assemblages of ostracods in this study. Population age structures for *Amphissites girtyi* and *Pseudobythocypris tomlinsoni* from samples 18 and 53 of unit 9 illustrate the transport problem (Fig. 3). Using Whatley's criteria, population age structures of both species in sample 53 represent a low energy life assemblage. In sample 18, specimens of *P. tomlinsoni* were not transported after death, but the truncated instar distribution (large individuals missing) of *A. girtyi* indicates that specimens of this species were post mortem transported and represent a low energy death assemblage. Visual inspection of sample

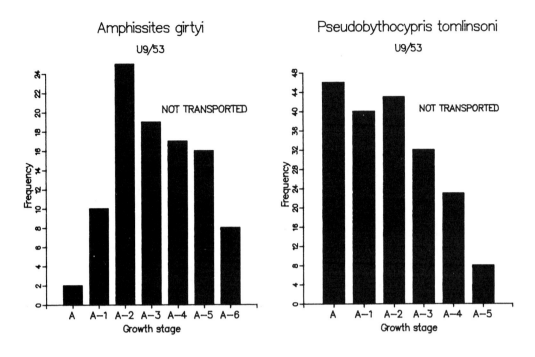

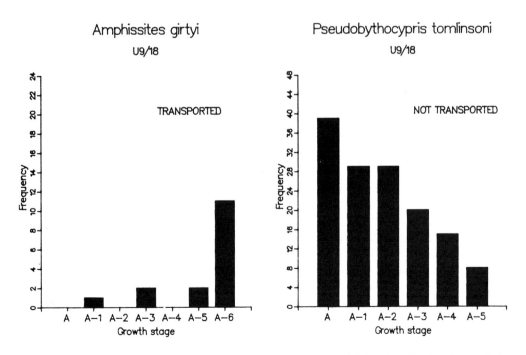

Fig. 3. Population age structure for *Amphissites girtyi* and *Pseudobythocypris* from unit 9. Numbers after the oblique stroke indicate sampled interval above base of unit.

residues smaller than 300µm suggests that post mortem transportation occurred commonly for several species. In order to eliminate, in large part, the effects of post mortem transportation, only ostracod frequency counts in the size fraction less than 840µm and greater than 300µm were used in this study.

**Abundance of ostracods**

The number of equivalent ostracod carapaces for each 1/3m channel sample is shown in Fig. 4. The average number of carapaces in sample groups segregated by colour is as follows: a) 117

in olive-grey mudshales of unit 9, b) 199 in pale orange mudshales of unit 9, and c) 175 in olive-grey to yellow-grey and medium dark grey clayshales of unit 11. In order to emphasize environmental effects related to colour, a few samples of intermediate colours are not included in these averages. The two samples with the most ostracods were from unit 9 (samples 52 and 53), which had 437 and 944 equivalent carapaces, respectively.

**Cluster analysis**

Four major groups of samples (significantly different at $P<0.05$) were identified with Q-mode

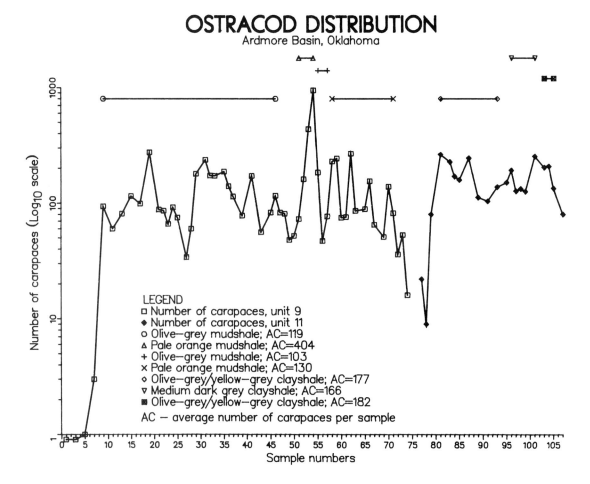

Fig. 4. Number of equivalent ostracod carapaces for each 1/3m sample. Samples 1 to 74 are from unit 9; samples 77 to 107 are from unit 11.

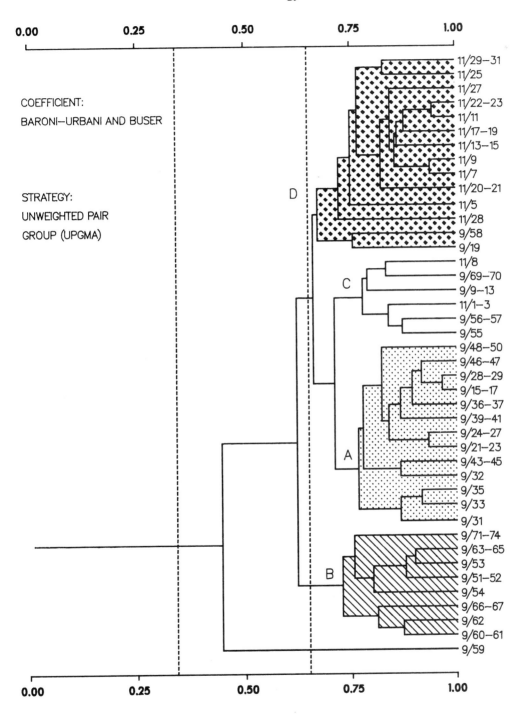

Fig. 5. Q-mode cluster dendrogram for the Dutton Ranch locality. Numbers at the right indicate unit number (before the oblique stroke) and numbers of single or combined intervals (after the oblique stroke). Coefficient values to the right of the dashed line between values 0.5 and 0.75 are positively correlated at P<0.05. Between the dashed lines the coefficient values are corrleated at below the 95% level.

cluster analysis (Fig. 5). Stratigraphically adjacent samples were combined in order to produce equivalent ostracod carapace counts of some 200 per interval. Cluster A consisted exclusively of sampled intervals of olive-grey to dark grey laminated mudshales from the lower part of unit 9. These samples were interpreted as representing prodeltaic sediments deposited in offshore environments under what are thought to be the deepest water conditions represented in the study interval. Two sample intervals from the lowest part of unit 9, the lowest and third lowest intervals, cluster with groups C and D, respectively.

Group B consisted exclusively of sample intervals of pale orange mudshales from the upper part of unit 9 representing regression due to deltaic progradation. This interval represents a time when such environmental factors as sediment size, substrate stability, oxygen content, and water depth were at an optimum for many species of ostracods, as witnessed by their abundance and diversity.

Group C consisted of a scattering of sample intervals including the lowest interval of unit 9, two samples of olive-grey mudshales that occur between pale orange mudshale intervals in the upper part of unit 9 and mudshales immediately below and above the sandstone (unit 10). Why these samples cluster at significantly different levels is not clear; possibly they comprise ostracod species that were tolerant of unstable environmental conditions suggested by juxtaposition of different sediment types.

Finally, cluster D samples include 12 of the 14 intervals from unit 11, the third lowest interval of unit 9, and one interval from between pale orange mudshales near the top of unit 9. These samples are primarily yellow-grey clayshales interpreted as representing deposition in distant offshore environments not strongly influenced by deltaic sedimentation. Water depth was moderate, probably in the range of 50-100m.

### Ostracod assemblages

Summary data for the four assemblages of ostracods in each cluster are shown in Tables 1

and 2. The ostracod assemblage from samples of cluster A has moderate species diversity, has relatively small numbers of specimens (low incidence) per sample, and is dominated by Genus A sp. A, *Pseudobythocypris tomlinsoni*, and *Healdia colonyi*. The cluster B ostracod assemblage has high species diversity, relatively high incidence, and an equitable species distribution. The assemblage from cluster D has moderate species diversity, moderate numbers of specimens per sample, and is dominated by *Pseudobythocypris tomlinsoni*, Genus A sp. A, *Healdia asper*, *Fabalicypris warthini*, and *Orthobairdia oklahomaensis*.

Palaeoenvironmental trends are apparent for only one morphological group. Reticulate species belonging to the genera *Amphissites* and *Kirkbya* occur almost exclusively in cluster B samples representing the shallowest water depths of the study interval.

### SUMMARY

The following sequence of events is suggested for deposition of the study interval at the Dutton Ranch locality. A rapid transgression flooded a coal-forming environment, resulting in marine water depths in the range of 50-100m, initially close to the maximum of that range. Kaolin-rich sediments, dark grey to olive-grey in colour, began to accumulate in pro-deltaic environments. Sedimentation was episodic, resulting in juvenile macrofaunal populations by frequent catastrophic kills. Regression occurred as a result of basin infilling. During the regressive phase, a moderate diversity assemblage of ostracods was established, dominated by species of the family Healdiidae, as well as *Pseudobythocypris tomlinsoni*.

Near the point of maximum regression, probably in water depths of some 50m, pale orange mudshales were deposited in a well oxygenated pro-deltaic environment. The deep water ostracod assemblage was rapidly replaced by a high abundance, high diversity, equitable assemblage of ostracods characterized by *Healdia colonyi*, *Fabalicypris warthini*, *Healdia formosa*, *Healdia oblonga*, *Cavellina* sp. A, *Pseudobythocypris*

Table 1. Percentage of most abundant ostracod species occurring in each cluster group. A dash (-) indicates less than one per cent.

| | CLUSTER | | | |
| | A | B | C | D |
| --- | --- | --- | --- | --- |
| HOLLINIDAE | | | | |
|    *Hollinella* sp. A | 1.1 | 4.3 | 2.1 | - |
| KIRKBYIDAE | | | | |
|    *Kirkbya* cf. *Kirkbya punctata* | - | 3.1 | - | 0 |
| AMPHISSITIDAE | | | | |
|    *Amphissites girtyi* | 0 | 3.5 | - | - |
|    *Amphissites centronotus* | - | 4.8 | 0 | 0 |
| PARAPARCHITIDAE | | | | |
|    *Shivaella brazoensis* | 1.2 | 5.1 | 1.2 | 1.3 |
| BAIRDIACEA | | | | |
|    *Bairdia pompilioides* Harlton, 1928 | - | 2.3 | 1.2 | 1.7 |
|    *Fabalicypris warthini* | - | 11.1 | 6.0 | 7.6 |
|    *Orthobairdia oklahomaensis* | 0 | 2.5 | - | 7.1 |
| BYTHOCYTHERIDAE | | | | |
|    *Monoceratina* n. sp. Bradfield | - | 5.5 | 1.0 | 1.0 |
| HEALDIIDAE | | | | |
|    Genus A sp. A | 31.7 | 4.0 | 16.3 | 17.1 |
|    Genus A sp. B | 8.1 | - | 3.1 | 2.5 |
|    *Healdia asper* | - | - | - | 15.3 |
|    *Healdia colonyi* | 21.8 | 19.1 | 25.2 | 3.7 |
|    *Healdia formosa* | 5.7 | 6.9 | 1.7 | - |
|    *Healdia nucleolata* Knight, 1928b | 2.2 | 2.8 | 3.0 | 1.4 |
|    *Healdia oblonga* | 1.8 | 6.5 | 5.2 | - |
|    *Healdia simplex* Roundy, 1926 | - | 1.1 | 1.4 | - |
| BAIRDIOCYPRIDIDAE | | | | |
|    *Pseudobythocypris tomlinsoni* | 24.5 | 6.1 | 26.0 | 34.4 |
| CAVELLINIDAE | | | | |
|    *Cavellina* sp. A | - | 6.4 | 3.3 | 4.6 |
| TOTALS | 98.1 | 95.1 | 96.7 | 97.7 |

Table 2. Summary data for cluster groups.

| CLUSTER | A | B | C | D |
| --- | --- | --- | --- | --- |
| Number of intervals per cluster | 13 | 8 | 6 | 14 |
| Number of ostracods in cluster | 2662 | 2616 | 999 | 3359 |
| Average number of ostracods per interval | 205 | 327 | 166 | 240 |
| Number of species in cluster | 21 | 35 | 22 | 25 |

*tomlinsoni, Monoceratina* n. sp. Bradfield, *Shivaella brazoensis, Amphissites centronotus, Hollinella* sp. A, *Amphissites girtyi,* and *Kirkbya* cf. *K. punctata,* in order of decreasing abundance.

After deposition of the basal 5.5m of unit 11, a second major transgression resulted in water depths in the seaward end of the 50-100m range and abandonment of the delta. In this distant offshore marine environment an ostracod assemblage similar to that of the deep water phase of unit 9 was established that was dominated by *Pseudobythocypris tomlinsoni,* Genus A sp. A, *Healdia asper, Fabalicypris warthini,* and *Orthobairdia oklahomaensis.*

## CONCLUSIONS

1) Colour appears to be significant in the interpretation of environment of deposition in thick, basinal shale sequences of Pennsylvanian age. Ostracod assemblages are closely correlated with different rock colours, which apparently reflect steep gradients of such environmental factors as organic production, sedimentation rate, and the amount of oxygen near the sediment-water interface.

2) Determination of ostracod assemblages may allow distinction of shallow versus deep marine water within the 50-100m range.

## ACKNOWLEDGEMENTS

Penny Baxter and Helen King did the X-ray diffraction of the clay minerals. Pat Patzer spent many hours picking a significant portion of the ostracod residues. Alan Horowitz of Indiana University supplied the program for cluster analysis. Robin Whatley and two anonymous reviewers made constructive comments on an earlier draft of this paper. Their assistance is gratefully acknowledged.

## REFERENCES

Berner, R. S. 1984. Sedimentary pyrite formation: an update. *Geochim. cosmochim. Acta,* London, **48**, 605-615.

Blatt, H., Middelton, G. V., & Murray, R. C. 1980. *Origin of sedimentary rocks,* 2nd ed., 782 pp. Prentice-Hall, New Jersey.

Boardman, D. R., Mapes, R. H., Yancey, T. E. & Malinky, J. M. 1984. A new model for the depth-related allogenic community succession within North American Pennsylvanian cyclothems and implications on the black shale problem. *In* Hyne, N. J. (Ed.), Limestones of the Mid-Continent. *Tulsa geol. Soc. Spec. Pub.,* Tulsa, **2**, 141-182.

Bradfield, H. H. 1935. Pennsylvanian Ostracoda of the Ardmore Basin, Oklahoma. *Bull. Am. Paleont.,* Ithaca, **22**(73), 1-173.

Brondos, M. D. & Kaesler, R. L. 1976. Diversity of assemblages of late Paleozoic Ostracoda. *In* Scott, R. W. & West, R. R. (Eds), *Structure and Classification of Paleocommunities,* 213-234. Dowden, Hutchinson & Ross, Stroudsburg.

Brown, L. F. Jr, Bailey, S. W., Cline, L. M. & Lister, J. S. 1977. Clay mineralogy in relation to deltaic sedimentation patterns of Desmoinesian cyclothems in Iowa-Missouri. *Clays Clay Miner.,* Oxford, **25**, 171-186.

Cooper, C. L. 1946. Pennsylvanian ostracodes of Illinois. *Ill. Geol. Sur. Bull.,* Urbana, **70**, 1-177.

Coryell, H. N. & Booth, R. T. 1933. Pennsylvanian Ostracoda - a continuation of the study of the Ostracoda fauna from the Wayland Shale, Graham, Texas. *Am. Midl. Nat.,* Notre Dame, Ind., **14**, 258-279.

Coryell, H. N. & Sample, C. H. 1932. Pennsylvanian Ostracoda - a study of the Ostracoda fauna of the East Mountain Shale, Mineral Wells Formation, Mineral Wells, Texas. *Am. Midl. Nat.,* Notre Dame, Ind., **13**, 245-281.

Edzwald, J. K. & O'Melia, C. R. 1975. Clay distributions in recent estuarine sediments. *Clays Clay Miner.,* Oxford, **23**, 39-44.

Gibbs, R. J. 1977. Clay mineral segregation in the marine environment. *J. sedim. Petrol.,* Menasha, **47**, 237-243.

Haack, R. C. & Kaesler, R. L. 1980. Upper Carboniferous ostracode assemblages from a mixed carbonate-terrigenous-mud environment. *Lethaia,* Oslo, **13**, 147-156.

Harlton, B. H. 1927. Some Pennsylvanian Ostracoda of the Glenn and Hoxbar Formations of southern Oklahoma and of the upper part of the Cisco Formation of northern Texas. *J. Paleont.,* **1**, 203-212.

Harlton, B. H. 1928. Pennsylvanian ostracods of Oklahoma and Texas. *J. Paleont.,* **2**, 132-141.

Harlton, B. H. 1929. Some Upper Mississippian (Fayetteville) and Lower Pennsylvanian (Wapanucka-Morrow) Ostracoda of Oklahoma and Arkansas. *Am. J. Sci.,* New Haven, **18**, 254-270.

Howard, J. D. 1975. The sedimentological significance of trace fossils. *In* Frey, R.W. (Ed.), *The study of trace fossils,* 131-146. Springer-Verlag, New York.

Kaesler, R.L. 1983. Ostracoda from Pennsylvanian subsurface shales: environmental indicators. *In* Maddocks, R. F. (Ed.), *Applications of Ostracoda,* proceedings of the Eighth International Symposium on Ostracoda, July 26-29, 1982, 116-130. Univ. Houston Geos., Houston, Texas.

Kellett, B. 1933. Ostracodes of the Upper Pennsylvanian and the Lower Permian strata of Kansas, Pt. 1: the Aparchitidae, Beyrichiidae, Glyptopleuridae, Kloedenellidae, Kirkbyidae, and Youngiellidae. *J. Paleont.,* **7**, 59-108.

Knight, J. B. 1928a. Some Pennsylvanian ostracodes from the Henrietta Formation of eastern Missouri, part 1. *J. Paleont.,*

**2**, 229-267.

Knight, J. B. 1928b. Some Pennsylvanian ostracodes from the Henrietta Formation of eastern Missouri, part 2. *J. Paleont.*, **2**, 318-337.

Lineback, J. A. 1966. Deep-water sediments adjacent to the Borden Siltstone (Mississippian) delta in southern Illinois. *Ill. geol. Surv. Circ.*, **401**, 1-48.

Melnyk, D. H. & Maddocks, R. F. 1988. Ostracode biostratigraphy of the Permo-Carboniferous of central and north-central Texas, Part I: Paleoenvironmental framework. *Micropaleontology*, New York, **34**, 1-20.

Miller, M. F. & Knox, L. W. 1985. Biogenic structures and depositional environments of a Lower Pennsylvanian coal-bearing sequence, northern Cumberland Plateau, Tennessee, U.S.A. *In* Curran, H. A. (Ed.), Biogenic structures: their use in interpreting depositional environments. *Soc. econ. Paleont. Miner. Spec. Publs*, Tulsa, **35**, 67-97.

Potter, P.E., Maynard, J. B. & Pryor, W. A. 1979. *Sedimentology of shale*, 306 pp. Springer-Verlag, New York.

Roundy, P. V. 1926. Mississippian formations of San Saba County, Texas, part 2, the microfauna. *U. S. geol. Surv. Prof. Pap.*, Washington, **146**, 5-23.

Shaver, R. H. & Smith, S. G. 1974. Some Pennsylvanian kirkbyacean ostracods of Indiana and midcontinent series terminology. *Indiana geol. Surv. Rep. Prog.*, Bloomington, **31**, 1-59.

Tennant, S. H. 1981. *Lithostratigraphy and depositional environments of the upper Dornick Hills Group (Lower Pennsylvanian) in the northern part of the Ardmore Basin, Oklahoma*. Unpub. M.Sc. thesis, University of Oklahoma, Norman, 291 pp.

Tennant, S. H., Sutherland, P. K., & Grayson, R. C. Jr. 1982. Stop descriptions - northern Ardmore Basin. *In* Sutherland, P. K. (Ed.), Lower and Middle Pennsylvanian stratigraphy in south-central Oklahoma. *Okla. geol. Surv. Guidebook*, Norman, **20**, 38-40.

Ulrich, E. O. & Bassler, R. S. 1906. New American Paleozoic Ostracoda. Notes and descriptions of Upper Carboniferous genera and species. *U. S. natn. Mus. Proc.*, Washington, **30**, 149-164.

Walker, K. R. & Diehl, W. W. 1986. The effect of synsedimentary substrate modification on the composition of paleocommunities: paleoecologic succession revisited. *Palaios*, Tulsa, **1**, 65-74.

Whatley, R. C. 1983. The application of Ostracoda to palaeoenvironmental analysis. *In* Maddocks, R. F. (Ed.), *Applications of Ostracoda*, proceedings of the Eighth International Symposium on Ostracoda, July 26-29, 1982, 51-77. Univ. Houston Geos., Houston, Texas.

# 37

# Ostracods in late Pleistocene and Holocene sediments from the Fram Strait, eastern Arctic

Nasser Mostafawi

Geological-Palaeontological Institute and Museum, Kiel, West Germany

## ABSTRACT

The samples from a 3.49m long core taken from 1444m water depth in Fram Strait were studied for their ostracods (see Fig. 1 for the core's drilling site). Most of the species observed are well-known from Pleistocene and Recent sediments. Their palaeoecology can, therefore, be interpreted by comparison with the Recent representative of the species. Ostracods were found only in the 0.250-0.500mm fraction and in insufficient numbers to be used for a stratigraphical study. A list of the samples studied and the number of ostracods they contained is given in Table 1. In samples 264-266cm and 290-292cm two valves of *Sarsicytheridea punctillata* (Brady, 1865) were encountered. Samples from the upper part of the core contain a few specimens of *Krithe* sp.; *Cytheropteron hamatum* Sars, 1869; *Cytheropteron arcticum* Neale & Howe, 1973 and *Cytheropteron* sp. Only in samples 25-27cm and 33-35cm were three specimens of *Acanthocythereis dunelmensis* (Norman, 1865) found. The most abundant species in the samples are *Muellerina abyssicola* and *Krithe* sp. In this core we are dealing with a mixed assemblage of shallow and deep water species. The occurrence of sporadic specimens of the shallow water species is interpreted as being due to downslope transport.

The extremely low abundance of the microfauna in the core samples can probably be attributed to a combination of low benthonic productivity combined with dilution by ice rafted detritus.

## THE RESULTS AND THEIR INTERPRETATION

For the interpretation of the results two assumptions are made. Firstly that the late-glacial communities were identical to modern communities. This assumption seems justified because the late-glacial ostracod faunas discussed below contain assemblages, which are, for the most part, similar

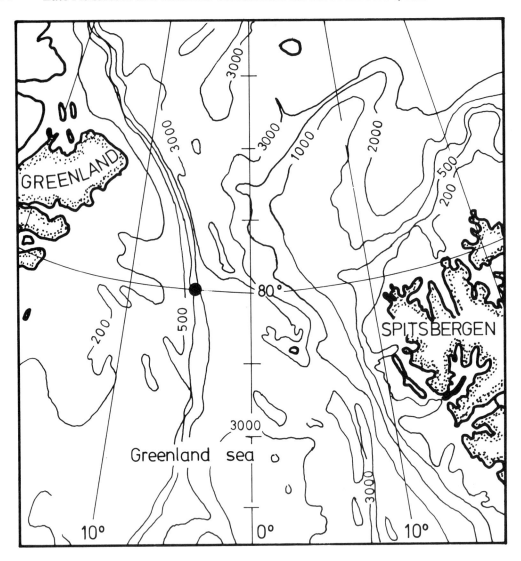

Fig. 1. Location map of the studied core.

to modern assemblages. Applying this assumption most of the ostracod species encountered are characteristic of littoral, outer-shelf marine faunas in the northern hemisphere, and have been an increasingly prolific constituent of such ostracod faunas since the Pleistocene, e.g., *Cytheropteron hamatum*, *C. arcticum*, *Acanthocythereis dunelmensis* and *Sarsicytheridea punctillata*. The presence of *Muellerina abyssicola* and *Krithe* sp. in this core may be taken as evidence of deep water, because their relatively great abundance (73.9 to 100% of the total fauna, adult and juvenile) in samples 246-248cm and 252-254cm and good preservation suggests that they are autochthonous. *Muellerina abyssicola* is a widespread amphiatlantic species, as can be seen on the distribution map given by Hazel (1970). It seems to live within broad ranges of water depth between 111-1628m (Hazel, 1970) and 400-1380m (Benson *et al.*, 1983). However, this species occurs mainly in depths of less than 500m. The second assumption, therefore, is that the ostracods in this core are not reworked

Table 1. Ostracod distribution in core GIK 21308-4 (80°01.0'N, 04° 49.9'W). ( ) = broken or abraded valves.

| Depth in core (cm) | M. abyssicola | K. sp. | C. hamatum | C. arcticum | C. sp. | A. dunelmensis | S. punctillata |
|---|---|---|---|---|---|---|---|
| 25 - 27 | | | | | | 1 (1) | |
| 33 - 35 | | | | | | (1) | |
| 80 - 82 | 1 | 1 | | (1) | (1) | | |
| 110 -112 | 1(1) | 1 | (1) | | | | |
| 128 -130 | 2 | (1) | (1) | (1) | | | |
| 150 -152 | 1(2) | 1 | (1) | | | | |
| 205 -207 | 2 | (1) | | (1) | | | |
| 225 -227 | 2 | 1 | (1) | | (1) | | |
| 246 -248 | 8(2) | 5(1) | 1(1) | 2(1) | 1 | | |
| 252 -254 | 4 | 34(2) | | | | | |
| 264 -266 | | | | | | | (1) |
| 290 -292 | | | | | | | 1 |

from older sediments. This asssumption seems justified because *Muellerina abyssicola* and *Krithe* sp. typical for the water depth of this core dominate the samples and are well preserved. Whereas the specimens of *Cytheropteron hamatum, C. arcticum, C.* sp., *Acanthocythereis dunelmensis* and *Sarsicytheridea punctillata* are in part broken or abraded. *Acanthocythereis dunelmensis* and *Sarsicytheridea punctillata* are cold and shallow water (10-100m) amphiatlantic species, which seem to tolerate reduced salinities (Athersuch, 1982; Rosenfeld, 1977). *Cytheropteron arcticum* is characteristic of the Arctic region and occurs at depths between 14-261m (Neale & Howe, 1975). *Cytheropteron hamatum* lives in the Norwegian and Barents Sea at depths in the range 192-373m and most of the stations where it occurs are at depths of less than 300m (Sars, 1922-1928; Neale & Howe, 1975). Only Whatley & Masson (1979) regarded this species as a deep water form, because they found some dead specimens in a sample from the area of the Rosemary Bank in the N.E. Atlantic at a depth of 1100m.

The fauna can be used to give information about conditions in both shallow and relatively deep water. Therefore, and because of the fragmentation and abrasion of the valves of the present material, the shallow water species must have been carried downslope into deep water.

It is of interest to note that the foraminifera encountered in the core samples also indicate a mixed assemblage. The foraminiferal assemblage consists predominantly of benthonic forms (*Elphidium, Cibicides* and *Cassidulina*).

Although the identification of transported shells can be difficult, there is little doubt that transported ostracods are present in most of the core samples. Currents on the ocean bottom may be important transportation agents. The transport of benthonic mico-organisms, as well as certain larger organisms, by currents, is widely known. However, icebergs are also transporting agents of great capacity and their wide distribution in this area indicates that they must have contributed considerable material to the sediments.

## ACKNOWLEDGEMENTS

I wish to express my gratitude to the following people: U. Schuldt and W. Reimann who helped with the photographs, M. Petersen for typing the manuscript, J. Thiede for stimulating discussions. I am grateful to the German Ministry of Research and Technology for financial support.

**Plate 1**

All the material is from core 21308-4, Fram Strait, deposited in the Forschungsinstitut Senckenberg (SMF; catalogue-Nr. Xe 14051-14057). Magnification X70, except where stated.

Figs 1-2.     *Muellerina abyssicola* (Sars, 1865).
Fig. 1.        Exterior left valve, sample 246-248cm; Xe 14051.
Fig. 2.        Exterior right valve, the same sample; Xe 14051.

Figs 3-5.     *Krithe* sp.
Fig. 3.        Exterior left valve, sample 252-254cm Xe 14052.
Fig. 4.        Exterior right valve, the same sample; Xe 14052.
Fig. 5.        Interior left valve, the same sample; Xe 14052.

Fig. 6.        *Sarsicytheridea punctillata* (Brady, 1865). Exterior left valve, sample 290-292cm; Xe 14054.

Fig. 7.        *Acanthocythereis dunelmensis* (Norman, 1865). Exterior right valve, juvenile, sample 25-27cm; Xe 14053.

Fig. 8.        *Cytheropteron hamatum* Sars, 1869. Exterior right valve X80, sample 246-248cm; Xe 14056.

Fig. 9.        *Cytheropteron* sp. Exterior left valve, X90, sample 246-248cm; Xe 14057.

Figs 10-11. *Cytheropteron arcticum* Neale and Howe, 1973.
Fig. 10.      Exterior left valve, X90, sample 246-248cm; Xe 14055.
Fig. 11.      Exterior right valve, X90, the same sample; Xe 14055.

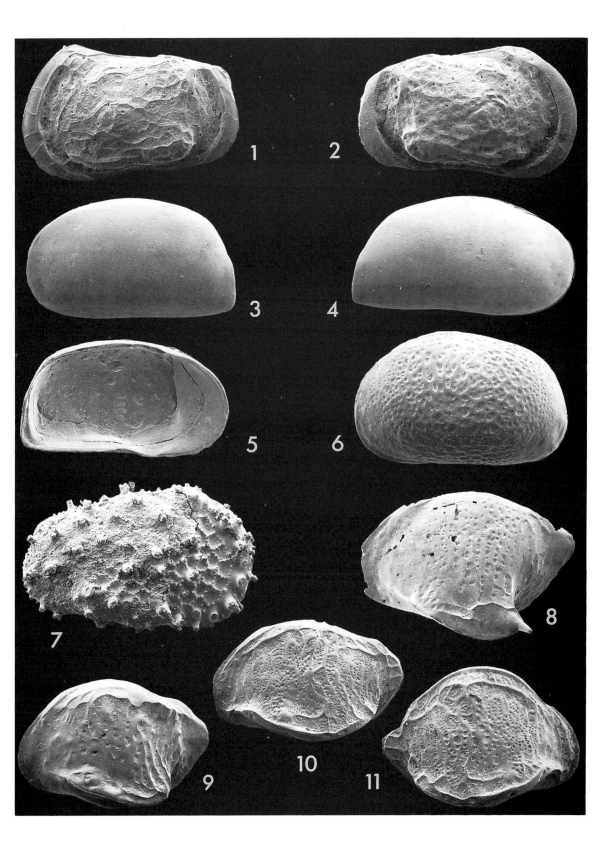

# REFERENCES

Athersuch, J. 1982. Some ostracod genera formerly of Cytherideidae Sars. *In* Bate, R. H., Robinson, E. & Sheppard, L. M. (Eds), *Fossil and Recent ostracods*, 231-275. Ellis Horwood Ltd., Chichester for British Micropalaeontological Society.

Benson, R. H., DelGrosso, R. M. & Steineck, P. L. 1983. Ostracods distribution and biofacies, Newfoundland continental slope and rise. *Micropaleontology*, New York, **29**(4), 430-453.

Hazel, J. E., 1970. Ostracod zoogeography in the southern Nova Scotian and northern Virginian faunal provinces. *U.S. Geol. Surv. Prof. Paper*, Washington D.C., **529-E**, 21 pp.

Neale, J. W. & Howe, H. V. 1975. The marine Ostracoda of Russian Harbour, Novya Zemlya and other high latitude faunas. *In* Swain, F. M. (Ed.), *Biology and Palaeobiology of Ostracoda. Bull. Am. Paleont.*, Ithaca, **65**(282), 381-431.

Rosenfeld, A. 1977. Die rezenten Ostracoden-Arten in der Ostsee. *Meyniana*, Kiel, **29**, 11-49.

Sars, G. O. 1922-1928. An account of the Crustacea of Norway. 9: *Ostracoden*, Bergen Museum, 277 pp.

Whatley, R. C. & Masson, D. G. 1979. The ostracod genus *Cytheropteron* from the Quarternary and Recent of Great Britain. *Revta esp. Micropaleont.*, Madrid, **11**(2), 222-227.

# 38

# The palaeoecology of Ostracoda in late Pleistocene sediments from Borehole 85 GSC 1 in the western Beaufort Sea

Qadeer Siddiqui & John Milne

Department of Geology, Saint Mary's University, Halifax, Nova Scotia, Canada B3H 3C3.

## ABSTRACT

Borehole 85 GSC 1, on the outer shelf of the western Beaufort Sea, was found to penetrate three late Pleistocene sedimentary units, informally designated from top to bottom as Unit I (12.22m), Unit II (27.4m) and Unit III (12.6m) respectively.

Unit II was not fossiliferous. Units I & III contained both coastal (marginal marine) and marine ostracod assemblages. Since all the species found are living today, the palaeoecology of the fossil assemblages has been interpreted from knowledge of their modern occurrence.

Units I and III contain three ostracod assemblages: a coastal assemblage represented by *Pteroloxa cumuloidea* Swain, 1963; a tolerant (eurythermal, euryhaline) assemblage including *Cytheropteron montrosiense* Brady, Crosskey & Robertson, 1874; *Cytheropteron sulense* Lev, 1972; *Heterocyprideis sorbyana* (Jones, 1856); *Normanicythere leioderma* (Norman, 1869); *Paracyprideis pseudopunctillata* Swain, 1963 and a fully marine assemblage consisting of *Bythocythere* sp., *Krithe glacialis* Brady, Crosskey & Robertson, 1874 and *Rabilimis mirabilis* (Brady, 1868).

Unit I differs from Unit III in that *Heterocyprideis sorbyana*, *Normanicythere leioderma* and *Paracyprideis pseudopunctillata* are missing from the tolerant assemblage, and *Rabilimis mirabilis* is missing from the fully marine assemblage.

Units I and III contain a large number of valves of *Pteroloxa cumuloidea*; an instar diagram based on 53 valves is included.

## INTRODUCTION

In 1984-1985, the Atlantic Geoscience Centre of the Geological Survey of Canada conducted an extensive shallow water seismic and drilling programme on the Natsek Plain, part of the Yukon continental shelf, in the western Beaufort Sea

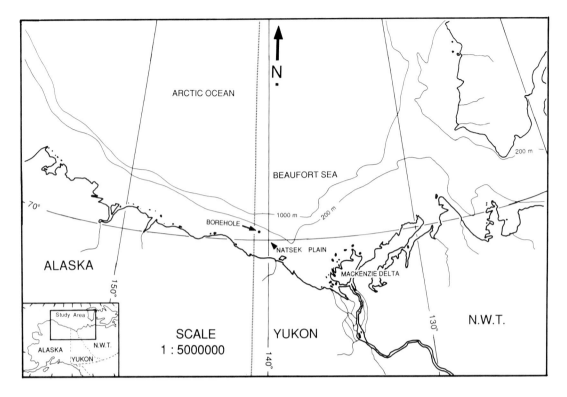

Fig. 1. Location of borehole 85 GSC 1.

(Fig. 1). In September 1985 a 52.5m core was collected from Borehole 85 GSC 1 (lat. 70° 08' 23.9"N, long. 140°28'17.9"W) in 44.5m of water depth for sedimentological and micropalaeontological study (Fig. 2). This core is mainly late Pleistocene in age, based on thermoluminescence dating (pers. comm.).

The specimens illustrated in Plates 1 and 2 of this paper will be deposited in the palaeontological collections of the Geological Survey of Canada, Ottawa.

## LITHOSTRATIGRAPHY

Three distinct sedimentary units have been identified in the borehole (Fig. 2) by means of seismic stratigraphy, lithology, grain size and x-ray analysis (Milne, 1987). The uppermost part of the core is a thin veneer of muddy sand with gravel, which extends from the sediment/water interface down to 0.28m; presumably this is Holocene in age.

Below the presumed Holocene sediments is Unit I, extending from 0.28 to 12.5m, and consisting of finely laminated silty clay with sporadic layers of sand, pebbles and plant detritus.

Unit II extends from 12.5 to 39.9m; it consists of homogeneous clay with silty bands which become commoner with depth. As determined by clay mineralogy, this unit is composed mostly of rock flour with a small amount of illite, chlorite and kaolinite. These sediments were probably deposited in a lacustrine environment (Milne, 1987).

Unit III extends from 39.9 to 52.5m. It consists of sandy silty clay with a high percentage of pebbles in the upper part; the lower part is extensively laminated with a high percentage of plant detritus.

## OSTRACOD ASSEMBLAGES AND THEIR PRESENT DISTRIBUTION

In a recent paper on the ostracod fauna from 5

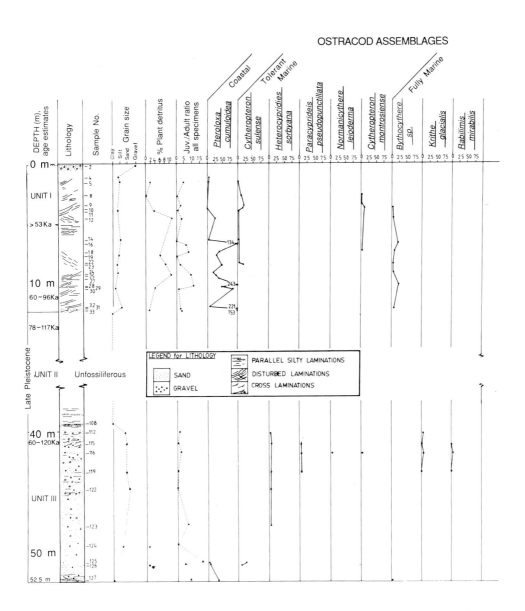

Fig. 2. Ostracod distribution and simplified stratigraphy of borehole 85 GSC 1.

wells in the eastern Beaufort Sea, Siddiqui (1988) identified 5 assemblages, 4 marine and one fresh-water. These assemblages were recognized by grouping the species whose modern preferred habitats were similar; most of the species are still living, so the preferences were known. There were so many species in this extensive study that only the commonest or most typical were selected as indicators of their respective assemblages; the dis-tributions of these species were then charted. The core from Borehole 85 GSC 1 is short and contains only 9 ostracod species; 8 of these species

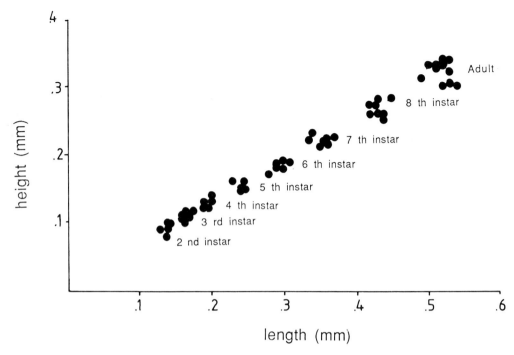

Fig. 3. Discontinuous size distribution of *Pteroloxa cumuloidea.*

were also found in the 5 wells series, the exception being *Bythocythere* sp. Only 3 of the assemblages described previously were represented in this core, the freshwater and deep marine ones being missing.

The marine assemblages of 85 GSC 1 and the ostracod species found in them are as follows:

A. Coastal (marginal marine), with reduced salinity:

*Pteroloxa cumuloidea* Swain, 1963

B. Tolerant marine (eurythermal, euryhaline), from locations with variable salinity and temperature:

*Cytheropteron montrosiense* Brady, Crosskey & Robertson, 1874

*Cytheropteron sulense* Lev, 1972

*Heterocyprideis sorbyana* (Jones, 1856)

*Normanicythere leioderma* (Norman, 1869)

*Paracyprideis pseudopunctillata* Swain, 1963

C. Fully marine, with typical marine salinity at all times and sufficient water depth to maintain a steady temperature:

*Bythocythere* sp.

*Krithe glacialis* Brady, Crosskey & Robertson, 1874

*Rabilimis mirabilis* (Brady, 1868)

The stratigraphical distribution of these species in borehole 85 GSC 1 is shown in Fig. 2 and is explained below.

*Unit I*: 32 samples were examined, of which 26 contained ostracods. *Pteroloxa cumuloidea,* representative of the coastal assemblage, occurred abundantly in most samples, particularly in the lower half of the unit. The tolerant assemblage is represented by the two species of *Cytheropteron,* occurring only in the upper part of the unit. The fully marine assemblage is represented by *Bytho-cythere* sp. and *Krithe glacialis,* the former ranging from the lower to the middle part of the unit, and *K. glacialis* occurring in only one sample, in its upper part. Most of the ostracods in this unit are associated with silt mixed with a high percentage of plant detritus; this environment was evidently particularly suitable for *P. cumuloidea.*

*Unit II*: All 75 samples examined were devoid of microfossils.

*Unit III*: 21 samples were examined, of which 14 contained ostracods, all three assemblages being represented. The coastal species *Pteroloxa cumuloidea* occurs only in the lower part of the unit. The tolerant assemblage, represented by all five of its characteristic species, ranges through most of the unit. The fully marine assemblage, including *Bythocythere* sp., *Krithe glacialis* and *Rabilimis mirabilis*, is confined to the upper part.

## OSTRACOD DISTRIBUTION AND PALAEOECOLOGY

The high juvenile/adult ratio of the ostracods in Units I and III, mostly of *P. cumuloidea*, suggests that these may have been buried *in situ*. All instars from the second to the adult are present, agreeing with Whatley's description of a low energy biocoenosis (Whatley, 1983, 1988). In spite of the fact that ostracods characteristic of different habitats are found in some of the same samples, the distribution shown in Fig. 2 shows an inverse relationship between *P. cumuloidea* and the tolerant group on one hand, and the fully marine group on the other. The mixture may be attributed either to these species overlapping in part of their ranges, or, more likely, to some mixing and reworking, either by ice rafting or by ice gouging (Briggs, 1983; Reimnitz *et al.*, 1987).

This material provides some interesting data on the little known species *Pteroloxa cumuloidea*, described by Swain (1963) from the Gubik Formation (late Pliocene and Quaternary) of the Arctic coastal plain in Alaska. It also occurs in Quaternary sediments of the MacKenzie Delta. Its known geological range is Pleistocene to Recent and it thrives in coastal waters of the Beaufort Sea (Siddiqui, 1988).

An instar diagram based on 53 valves is included (Fig. 3); the specimens are believed to represent all instars from the second to the adult.

## ACKNOWLEDGEMENTS

We wish to thank Drs S. Blasco and P. Hill of the Atlantic Geoscience Centre for access to the material on which this study was based.

**Plate 1**

Figs 1-2. *Pteroloxa cumuloidea* Swain, 1963.

Fig. 1.    Male right valve, length 610μm, X108.
Fig. 2.    Female left valve, length 573μm, X115.

Figs 3-4. *Cytheropteron montrosiense* Brady, Crosskey & Robertson, 1874.

Fig. 3.    Male right valve, length 647μm, X96.
Fig. 4.    Female right valve, length 518μm, X123.

Figs 5-6. *Cytheropteron sulense* Lev, 1972.

Fig. 5.    Male left valve, length 610μm, X108.
Fig. 6.    Female right valve, length 518μm, X125.

Fig. 7.    *Heterocyprideis sorbyana* (Jones, 1856). Female right valve (specimen lost).

Fig. 8.    *Normanicythere leioderma* (Norman, 1869). Male left valve, length 980μm, X69.

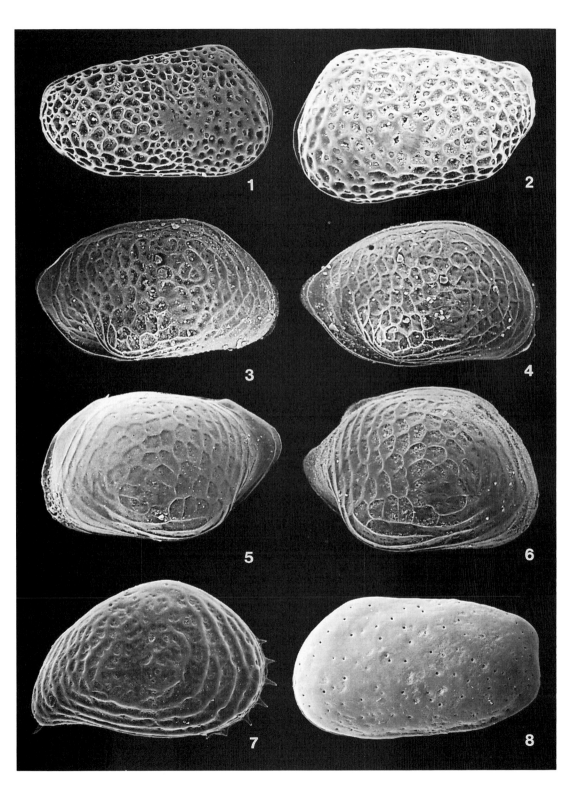

**Plate 2**

Fig. 1. *Paracyprideis pseudopunctillata* Swain, 1963, male right valve, length 864μm, X72.

Figs 2-3. *Bythocythere* sp.

Fig. 2. Male right valve, length 536μm, X117.
Fig. 3. Male left valve, length 490μm, X126.

Fig. 4. *Krithe glacialis* Brady, Crosskey & Robertson, 1874, female left valve, length 888μm, X71.

Fig 5-6. *Rabilimis mirabilis* (Brady, 1868).

Fig. 5. Female right valve, length 1184μm, X52.
Fig. 6. Female left valve, length 1193μm, X51.

Additional Figs   Adult-2nd instars, left valves. *Pteroloxa cumuloidea* Swain, 1963.  The bar represents 1mm.

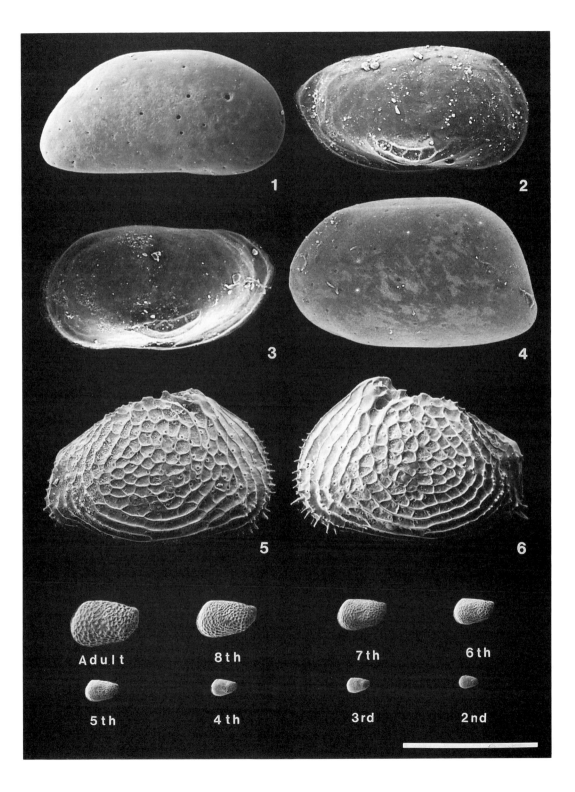

Adult 8th 7th 6th

5th 4th 3rd 2nd

## REFERENCES

Briggs, W. M. Jr. 1983. Ice rafting of ostracodes and other microfauna, Beaufort Sea, Alaska. Abstracts with programs, 96th Annual Meeting, *Geol. Soc. Am.*, Boulder, Co., 532.

Lev, O. M. 1972. Binomical and palaeogeographical condition of the marine Neogene - Quaternary basins of the USSR North, according to Ostracoda. *In New tectonics and palaeogeography of the Soviet Arctic related to the mineral resources evaluation*, 15-20. Inst. Arctic Geol., Leningrad.

Milne, J. 1987. *The litho-, bio- and seismic stratigraphy of Quaternary deposits on the Yukon continental shelf.* Honours thesis, St. Mary's University, Halifax, Nova Scotia, Canada, 159 pp.

Reimnitz, E., Kempema, E. W. & Barnes, P. W. 1987. Anchor ice, seabed freezing and sediment dynamics in shallow Arctic seas. *J. geophys. Res.*, Richmond, Va., **92**(13), 14671-14678.

Siddiqui, Q. A. 1988. The Iperk Sequence (Plio-Pleistocene) and its Ostracod Assemblages in the Eastern Beaufort Sea. *In* Hanai, T., Ikeya, N. & Ishizaki, K. (Eds), *Evolutionary biology of Ostracoda, its fundamentals and applications*, proceedings of the Ninth International Symposium on Ostracoda, held in Shizuoka, Japan, 29 July - 2 August 1985, Developments in palaeontology and stratigraphy, **11**, 533-570, Kodansha Ltd., Tokyo and Elsevier, Amsterdam, Oxford, New York, Tokyo.

Swain, F. M. 1963. Pleistocene Ostracoda from the Gubik Formation, Arctic coastal plain, Alaska. *J. Paleont.*, **37**(4), 798-834.

Whatley, R. C. 1983. The Application of Ostracoda to Palaeoenvironmental Analysis. *In* R. F. Maddocks, (Ed.) *Applications of Ostracoda*, proceedings of the Eighth International Symposium on Ostracoda, July 26-29, 1982, 51-77. Univ. Houston, Geos., Houston, Texas.

Whatley, R.C. 1988. Population structure of ostracods: some general principles for the recognition of palaeo-environments. *In* De Deckker, P., Colin, J.-P. & Peypouquet, J.-P. (Eds). *Ostracoda in the Earth Sciences*, 103-124. Elsevier, Amsterdam, Oxford, New York, Tokyo.

## DISCUSSION

William Austin: Is there any faunal evidence for the onset of glaciation at the top of your Unit III?

Qadeer Siddiqui: No, the same fauna occurs to the top of Unit III.

David Horne: How large is *Pteroloxa*?

Qadeer Siddiqui: It is approximately 600μm.

David Horne: I don't regard this as small for a Loxoconchid - 600μm is a fairly 'normal' size.

Robin Whatley: In my opinion, *Pteroloxa* is congeneric with *Roundstonia* Neale, and is not, therefore, an endemic genus. I would welcome your comments on this.

Qadeer Siddiqui: I was not suggesting that *Pteroloxa* was an endemic genus, but that Swain's two species are; they have not been found, as far as I know, outside Alaska and the MacKenzie Delta as fossils; *P. cumuloidea* is living now in the Beaufort Sea. The genus *Pteroloxa* certainly needs further investigation.

# 39

# Ostracoda from Holocene calcareous tufa deposits in southern Belgium: a palaeoenvironmental analysis

**Anne Van Frausum[1] &**
**Karel Wouters[2]**

[1]Instituut voor Aardwetenschappen, University of Leuven, Leuven, Belgium.
[2]Koninklijk Belgisch Instituut voor Natuurwetenschappen, Brussels, Belgium.

## ABSTRACT

Holocene calcareous tufa deposits of some 6m thick are exposed in the valley of the river Fonds de Ry at Treignes (S. Belgium). They have previously been dated as Boreal, Atlantic and Subboreal deposits by means of pollen analysis (Geurts, 1976). The tufa was sampled in detail, and the contained ostracods were quantitatively studied. Species diversity was relatively high with 25 species present. Factor analysis of the quantitative data resulted in the recognition of three major ostracod associations:

Association 1: *Cyclocypris ovum, Cyclocypris laevis, Prionocypris zenkeri, Eucypris pigra, Candona candida* and *Pseudocandona marchica,* occurring in the Boreal part of the section, and probably characteristic of stagnant water bodies and/or slowly running waters.

Association 2: *Pseudocandona brevicornis, Psychrodromus olivaceus, Eucypris pigra, Cryptocandona vavrai* and *Pseudocandona albicans* occurs in the Atlantic in alternation with Association 1. It probably characterizes an environment with running water and intensive spring activity.

Association 3: *Pseudocandona zschokkei, Ilyocypris bradyi, Ilyocypris inermis, Potamocypris zschokkei* and *Eucypris pigra* occurs exclusively in the Subboreal part of the section, and represents a fauna of spring dwelling ostracods which often lived interstitially.

The transitions between ostracod associations coincide with the pollen zones and with the transitions between the associations of land molluscs. It seems reasonable to assume that, apart from local ecological conditions, the ostracod fauna was

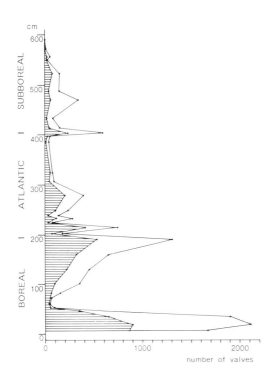

Fig. 1. Total number of adult and juvenile valves per 100g of dry sediment for each investigated level. Adult valves = shaded, juvenile valves = white.

also influenced by major climatic fluctuations. This is a good indication that ostracods can probably be used as climatic indicators.

## INTRODUCTION

Several deposits of fossiliferous calcareous tufa occur in Belgium. They are mostly of Postglacial age. One of these tufa deposits, in the neighbourhood of Treignes, has been investigated for Ostracoda. A tentative reconstruction of the palaeoenvironment has been made, based upon the association of the species encountered.

## GEOLOGICAL SETTING

There are several calcareous tufa deposits in Belgium. A survey is given by Gullentops & Mullenders (1972). Some of the deposits are Recent,

the more elaborate, however, are fossil. The most important tufas have already been dated by means of pollen analysis (Mullenders et al., 1963; Geurts, 1976).

One of them is situated at Treignes, in southern Belgium, near the French border. The tufa deposits are located in the valley of the river 'Fonds de Ry'. The total thickness of the deposit amounts to 10.60m (Geurts, 1976). The oldest beds belong to the Preboreal. The deposits rest upon limestone and schists of the Frasnian in the north, on limestone of the Givetian in the centre and on schists and limestone of the Couvinian in the south.

## SAMPLING AND ANALYSIS

Upstream, the 'Fonds de Ry' is still forming a small amount of calcareous tufa. Downstream, however, the river has incised a small valley into the fossil deposits.

The samples have been taken at a place where a vertical section with a height of 6m is made by the 'Fonds de Ry'. On the basis of sedimentological data, 47 levels could be distinguished. 44 samples were taken, 33 of them being examined for ostracods. Each of the 44 samples was investigated for its $CaCO_3$ content. The juvenile as well as the adult valves were counted. The species were identified, and for each specimen it was noted whether it was a left valve, a right valve, or a carapace.

The results of the counting are shown graphically. Fig. 1 reveals the total number of adult and juvenile valves per 100g of dry sediment for each investigated level. Fig. 2 shows the number of adult valves of each species per 100 adults of each level. By means of factor analysis the observed data could be interpreted more precisely.

## RESULTS

The following 25 species of ostracods were found:
*Ilyocypris bradyi* Sars, 1890
*Ilyocypris inermis* Kaufmann, 1900
*Candona candida* (O. F. Müller, 1785)
*Candona neglecta* (Sars, 1887)
*Pseudocandona zschokkei* (Wolf, 1919)

*Pseudocandona marchica* (Hartwig, 1899)
*Pseudocandona hartwigi* (G. W. Müller, 1900)
*Pseudocandona albicans* (Brady, 1864)
*Pseudocandona brevicornis* (Klie, 1925)
*Pseudocandona* sp. cf. *P. breuili* (Paris, 1920)
*Fabaeformiscandona fabaeformis* (Fischer, 1851)
*Cryptocandona vavrai* (Kaufmann, 1900)
*Candonopsis* sp.
*Nannocandona faba* Ekman, 1914
*Cyclocypris laevis* (O. F. Müller, 1785)
*Cyclocypris ovum* (Jurine, 1820)
*Eucypris pigra* (Fischer, 1851)
*Prionocypris zenkeri* (Chyzer, 1858)
*Herpetocypris reptans* (Baird, 1835)
*Psychrodromus olivaceus* (Brady & Norman, 1889)
*Scottia pseudobrowniana* Kempf, 1971
*Cavernocypris subterranea* (Wolf, 1919)
*Potamocypris fulva* (Brady, 1868)
*Potamocypris zschokkei* (Kaufmann, 1900)
*Potamocypris villosa* (Jurine, 1820)

By comparing the association of species in each level, the total section could be divided into three units. As the section had already been dated by pollen analysis (Mullenders *et al.*, 1963; Geurts, 1976), the division in units could be compared with the climatic periods of the Postglacial. It was found that each unit corresponds almost exactly to a climatic period.

Unit 1 extends from the actual water level of the river up to a height of 1.95m. This interval represents the Boreal (pollen zones IV, V and VI). The most important species of Unit 1 are: *Cyclocypris ovum, Cyclocypris laevis, Prionocypris zenkeri, Eucypris pigra, Candona candida* and *Pseudocandona marchica*.

Unit 2 extends from 1.95m up to 4.05m and represents the Atlantic period (pollen zone VIIa). The most important species are: *Pseudocandona brevicornis, Psychrodromus olivaceus, Eucypris pigra, Cryptocandona vavrai* and *Pseudocandona albicans*.

Unit 3 starts at 4.05m and ends at 6.00m, where soil cover begins and represents the Subboreal (pollen zones VIIb and VIII). The most important species are: *Pseudocandona zschokkei, Ilyocypris* sp., *Potamocypris zschokkei* and *Eucypris pigra*.

The palaeoenvironmental reconstructions were based upon known ecological data for Recent Ostracoda. For this purpose, the publications of Nüchterlein (1969) and Hiller (1972) were mainly used.

In Unit 1 the most common species are *Cyclocypris ovum* and *Cyclocypris laevis. Cyclocypris ovum* can be found in the most diverse environments, while *Cyclocypris laevis* has a preference for running water. *Prionocypris zenkeri* is mostly found in rather cold, running water. *Eucypris pigra* and *Candona candida* are cold water species and are found in groundwaters, springs, spring brooks and stagnant waters. *Pseudocandona marchica* is a polythermophilic species.

In Unit 2, with the exception of *Pseudocandona albicans*, all species have a preference for cold water. *Pseudocandona brevicornis* appears fairly suddenly; it is almost exclusively found in springs. This species has not been recorded previously in Belgium. *Psychrodromus olivaceus* is mostly found in springs and spring brooks. *Eucypris pigra* is a cold water species which is mostly found in association with springs. *Cryptocandona vavrai* occurs predominantly in springs, groundwater and small rivers. *Pseudocandona albicans* is a mesothermophilic species and has a preference for running waters.

Unit 3 is dominated by *Pseudocandona zschokkei*. This species is typical of groundwaters (Klie, 1938). It is also found living in small water bodies associated with springs. *Pseudocandona zschokkei* has already been found in the Condat Tufa in the Dordogne, which is probably of Interglacial age (Preece, Thorpe & Robinson, 1986). *Ilyocypris bradyi* has a preference for cold water and is often found in association with springs and small running waters. *Ilyocypris inermis* lives almost exclusively in spring brooks. *Potamocypris zschokkei* is found in the cold water environments of springs, spring brooks and groundwaters. *Eucypris pigra* is only found in association with springs.

It is interesting to note the presence of *Nannocandona faba* in Unit 1 and Unit 2. This is the first record for Belgium of this remarkable species. According to Ekman (1914) it probably has a northern origin. Marmonier & Danielopol (1988)

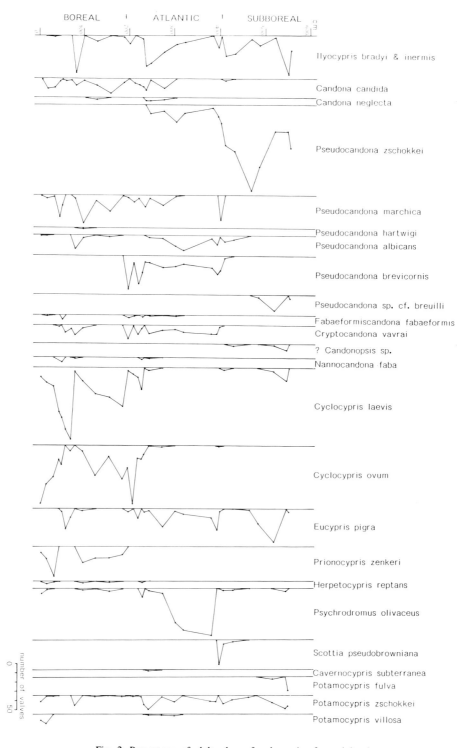

Fig. 2. Percentages of adult valves of each species for each level.

discuss the colonization of the interstitial environment by this species.

Considering all the ecological data, there seems to be an evolution in the associations of ostracods throughout time. During the Boreal (Unit 1), the ostracods lived in rather diverse environments, with a slight preference for slow flowing waters. During the Atlantic (Unit 2) the environment was such that it favoured faunas typical of springs and spring brooks. Finally, during the Subboreal, the ostracod fauna seems to include elements which are from the groundwater environment.

## DISCUSSION

The ostracods of tufa deposits in Belgium have not previously been studied. Elsewhere in Europe, however, several tufas have been investigated. In eastern Europe Absolon (1973); Diebel & Pietrzeniuk (1975, 1977, 1978, 1980, 1984) and Pietrzeniuk (1985) are the most important authors of studies on ostracods of tufas. Most of these tufa deposits are interglacial in age, the ostracod associations are not the same as those of the tufa deposit of Treignes. In Italy, Devoto (1965) studied lacustrine tufas from the Lower Liri Valley. In England, Scotland and Ireland the biostratigraphy of Postglacial tufa deposits have been investigated by Robinson (1979), Preece, Benett & Robinson (1984), Preece & Robinson (1984) and Preece, Coxon & Robinson (1986). The species of sediments of Flandrian age are largely similar to those found in Treignes. In Preece & Robinson (1984) and Preece, Thorpe & Robinson (1986), several sites were investigated for Mollusca and Ostracoda. The Mollusca proved to be useful as biostratigraphal indicators. The ostracods, however, merely reflected local changes in the depositional environments.

It is striking that the units deduced from the ostracod assemblages in the studied section at Treignes coincide with the climatic periods derived from pollen analysis. They also coincide with the units that could be distinguished by Gullentops (1974) in her study of terrestrial gastropods. Further investigation of tufas in Belgium should make

clear whether this reflects a local or a more general ecological phenomenon.

For each investigated sample, the juveniles were counted. However, they were not used in the factor analysis, as it is difficult to interpret their numbers. Without measuring the valves, it is difficult to establish the stage of the juveniles. Furthermore, it is not always possible to determine the species to which they belong. This is especially so for the very young stages.

The contribution of the juveniles to the total number of valves varies from 17% to 82%. It mostly fluctuates around 60%. According to Whatley (1983), the ostracod population age structure can be used as an indicator of the water energy of the environment of deposition. In the tufaceous sediments of Treignes, however, a correlation between the nature of the sediment and the percentage of juveniles could not be detected. Here, the contribution of juveniles to the total number of valves seems to be more dependent on the species present rather than on the type of sediment. The high percentage of juveniles, however, indicates that the ostracods probably represent a biocoenosis.

On the basis of the sedimentological characteristics, it was not possible to explain the presence or absence of particular ostracod associations. Absolon (1973), however, pointed out that, by comparing 40 vertical sections of tufa deposits, there is a correlation between different types of tufa deposits and specific ostracod associations. Knowing the relationship between the palaeoenvironment and sedimentological characteristics, he was able to deduce the palaeoecology of the contributing ostracods. Taking these palaeoecological data into account, the palaeoenvironment of the tufa deposits of Treignes can be inferred. The sediments of Unit 1 were deposited in ponds and smaller stagnant water bodies. Unit 2 should probably represent a mixture of spring, river and marsh deposits.

On the basis of the ecological and palaeoecological data of the ostracods of the studied section, it is possible to reconstruct the local palaeoenvironmental conditions. During the Boreal, the valley of the 'Fonds de Ry' was probably characterized by slowly running water and small ponds. The Atlantic had a wetter climate than the Boreal and this is reflected in an intensified spring activity and by the presence of spring brooks and dispersed marshy areas. During the Subboreal, the amount of surface water was apparently much lower in the study area, as is indicated by the presence of subterranean species.

The ostracod associations in the tufa help to distinguish the various hydrological regimes, which correspond to vegetation changes. These vegetation changes also influence the gastropod fauna. This is a natural consequence: hydrology affects ostracods, but also trees and shrubs, and the snails which feed on them. So ostracods can be used as climatic indicators.

Other tufa deposits in Belgium are now being investigated to corroborate the results obtained in Treignes.

As a concluding remark, it has to be stressed that the knowledge of the synecology of recent freshwater ostracods is too inadequate to fully interpret fossil associations. All conclusions presented in this paper are based upon autoecological data.

## ACKNOWLEDGEMENTS

The first author acknowledges a grant from the Belgian Institute for the Encouragement of Scientific Research in Industry and Agriculture (I.W.O.N.L.).

## REFERENCES

Absolon, A. 1973. Ostracoden aus einigen Profilen spät- und postglacialer Karbonatablagerungen in Mitteleuropa. *Mitt. bayer. St.Paläont. Hist. Geol.*, München, **13**, 47-94.

Devoto, G. 1965. Lacustrine Pleistocene in the Lower Liri Valley. *Geologica romana*, Rome, **4**, 291-368.

Diebel, K. & Pietrzeniuk, E. 1975. Ostracoden aus dem holozänen Travertin von Bad Langensalza. *Quartärpaläont.*, Berlin, **1**, 27-55.

Diebel, K. & Pietrzeniuk, E. 1977. Ostracoden aus dem Travertin von Taubach bei Weimar. *Quartärpaläont.*, Berlin, **2**, 119-137.

Diebel, K. & Pietrzeniuk, E. 1978. Die Ostrakodenfauna des eeminterglazialen Travertins von Burgtonna in Thüringen. *Quartärpaläont.*, Berlin, **3**, 87-91.

Diebel, K. & Pietrzeniuk, E. 1980. Pleistozäne Ostracoden vom Fundort des *Homo erectus* bei Bilzingsleben. *Ethnogr.-archäol. Z.*, Berlin, **21**, 26-35.

Diebel, K. & Pietrzeniuk, E. 1984. Jungpleistozäne Ostrakoden aus Sedimenten der Parkhöhlen von Weimar. *Quartärpaläont.*, Berlin, **5**, 285-319.

Ekman, S. 1914. Kenntnis der Schwedischen Susswasser-Ostracoden. *Zool. Bidr. Upps.*, Stockholm, **3**, 1-36.

Geurts, M.-A. 1976. Genèse et stratigraphie des travertins de fond de vallée en Belgique. *Acta Geogr. Lovan.*, Louvain-la-Neuve, **16**, 1-66.

Gullentops, A. 1974. *Bijdrage tot de biogeografische en kwartair geologische studie van Gastropodenassociaties in het bekken van de Fonds de Ry (prov. Namen)*. Unpubl. Lic. thesis, University of Leuven, 1-146.

Gullentops, F. & Mullenders, W. 1972. Age et formation de dépôts de tuf calcaire holocène en Belgique. *Congr. Coll. Univ. Liège*, Liège, **67**, 113-135.

Hiller, D. 1972. Untersuchungen zur Biologie und zur Ökologie limnischer Ostracoden aus der Umgebung von Hamburg. *Arch. Hydrobiol.*, Stuttgart, Suppl. **40**(4), 400-497.

Klie, W. 1938. Ostracoda Muschelkrebse. *Tierwelt Dtl.*, Jena, **34**, 1-230.

Marmonier, P. & Danielopol, D. 1988. Découverte de *Nannocandona faba* Ekman (Ostracoda, Candoninae) en Basse Autriche. Son origine et son adaptation au milieu interstitiel. *Vie Milieu*, Université de Paris, **38**(1), 35-48.

Mullenders W., Duvigneaud, J. & Coremans, M. 1963. Analyse pollinique de dépôts de tuf calcaire et de tourbe à Treignes (Belgique). *Grana palynol.*, Stockholm, **4**, 439-448.

Nüchterlein, H. 1969. Süsswasserostracoden aus Franken. Ein Beitrag zur Systematik und Ökologie der Ostracoden. *Int. Revue ges. Hydrobiol.*, Berlin, **54**, 223-287.

Pietrzeniuk, E. 1985. Ostrakoden aus dem holozänen Travertin von Weimar. *Z. geol. Wiss. Berlin*, **13**, 207-233.

Preece, R. C., Benett, K. D. & Robinson, J. E. 1984. The Biostratigraphy of an Early Flandrian Tufa at Inchrory, Glen Avon, Banffshire. *Scott. J. Geol.*, Geological Societies of Edinburgh & Glasgow, Edinburgh, **20**(2), 143-159.

Preece, R. C., Coxon, P. & Robinson, J. E. 1986. New biostratigraphic evidence of the Post-glacial colonization of Ireland and for Mesolithic forest disturbance. *J. Biogeogr.*, Hull, **13**, 487-509.

Preece, R. C. & Robinson, J. E. 1984. Late Devensian and Flandrian Environmental History of the Ancholme Valley, Lincolnshire: Molluscan and Ostracod Evidence. *J. Biogeogr.*, Hull, **11**, 319-352.

Preece, R. C., Thorpe, P. M. & Robinson, J. E. 1986. Confirmation of an interglacial age for the Condat tufa (Dordogne, France) from biostratigraphic and isotopic data. *J. Quat. Sci.*, London, **1**, 57-65.

Robinson, J.E. 1979. The Ostracod Fauna of the Interglacial Deposits at Sugworth, Oxfordshire. *Phil. Trans. R. Soc.*, London, B, **289**, 99-106.

Whatley, R. C. 1983. The Application of Ostracoda to Palaeoenvironmental Analysis. *In* Maddocks, R. F. (Ed.), *Applications of Ostracoda*, proceedings of the Eighth International Symposium on Ostracoda, July 26-29, 1982, 51-77. Univ. Houston Geos., Houston, Texas.

# DISCUSSION

Dan Danielopol: Did *Pseudocandona zschokkei* live in an epigean environment during the Subboreal? Where did the
species come from in your tufaceous material? Was it
from subterranean waters, discharged more abundantly
during the Subboreal or from other epigean habitats in the
surrounding area?

Anne Van Frausum & Karel Wouters: We do not think that
*Pseudocandona zschokkei* lived in an epigean environment
during the Subboreal, but it is difficult to point out whether
the species came from subterranean waters discharged
outside the study area, or whether it lived in the hypogean
waters after deposition of the sediments.

# PALAEOGEOGRAPHY, ZOOGEOGRAPHY AND PALAEOZOOGEOGRAPHY

# 40

# Cretaceous halocyprid Ostracoda

**Jean-Paul Colin[1] & Bernard Andreu[2]**

[1]Esso Rep., 33321 Bègles, France and c/o
Exxon Production Research Company, Houston,
Texas, U.S.A.
[2]Résidence Joffre, 59640 Jeumont, France.

## ABSTRACT

Two groups of probably planktonic Ostracoda, seemingly related to some Recent halocyprids (Myodocopida, Halocypridacea) have been reported from various mid-Cretaceous localities, essentially Albian to Lower Cenomanian, in the palaeo-South Atlantic (Gabon, Congo, Ivory Coast, Brazil), the South Tethyan margins (Morocco, Israel, Iran) and Western and Central Europe (southern England, Switzerland, Czechoslovakia). The first group is characterized by the presence of a well developed anterior rostrum and rostral incisure. The second, more common group, lacks the rostrum and, therefore, is only questionably attributed to the halocyprids. The systematic position, biostratigraphical potential, the palaeoecology and the palaeobiogeography of these forms are discussed.

## INTRODUCTION

Halocyprid ostracods are essentially planktonic (Angel, 1983), with non-calcified or weakly calcified carapaces which, therefore, are very seldom preserved fossils (Van Morkhoven, 1962; 1963). However, as we will show in this paper, several forms apparently belonging to this group were reported from mid-Cretaceous sediments of the South Atlantic, western and central Europe and along the margins of the Tethys.

Halocyprid ostracods belong to the order Myodocopida, suborder Myodocopina and superfamily Halocypridacea (Sylvester-Bradley, 1961).

Myodocopid ostracods have their origin in the early Palaeozoic (Entomozoidae). In the Mesozoic there are infrequent records of the Cypridinacea and the Polycopidae: *Triadocypris* Weitschat (Cypridinacea) in the Triassic of Svalbard (Weitschat, 1983), *Cypridina* Milne-Edwards (Cypridinidae) in the Middle Jurassic (Dépêche, 1984), the late Cretaceous (Bosquet, 1847; Van Veen, 1936; Van Morkhoven, 1962; 1963) and the Eocene (Keij, 1957; Van Morkhoven, 1963; Ducasse *et al.* 1985), *Philomedes* Liljeborg (Philomedidae) in the early Cretaceous (Donze, 1965; Neale, 1976) and *Pokornyopsis* Kozur (Thaumatocyprididae) in the Jurassic (Kornicker & Sohn, 1976) and the Triassic (Colin, personal observations in the Triassic of Minorca, Balearic Islands, Spain).

The first mention of forms allegedly related to the Recent genus *Conchoecia* in the Cretaceous was made by Pokorný (1964) in the Coniacian of Czechoslovakia. The species named *Conchoecia cretacea* (Plate 1, Fig. 1), has a distinct rostrum and incisure and caudal projection at mid-shell height, and faint longitudinal striations. All known species of *Conchoecia* have rostral incisure and caudal projection above mid-shell height.

Subsequently, species tentatively assigned to this genus were reported from various mid-Cretaceous localities (Albian to Upper Cenomanian): Kaye (1965) in southern England, Grosdidier (1973) in Iran (Persian Gulf), Grosdidier (1979) in Gabon, Oertli (*in* Charollais *et al.*, 1977) in Switzerland, Rosenfeld & Raab (1984) in Israel, Koutsoukos & Dias-Brito (1987) and De Azevedo *et al.* (1987a, 1987b) in Brazil and Andreu (in preparation) in Morocco.

## SYSTEMATICS

Cretaceous halocyprids generally occur in small numbers and, therefore, with the exception of Pokorný's species, have all been left in open nomenclature. They are characterized by a very thin and fragile carapace and an ornamentation consisting of distinct to faint longitudinal to concentric striations. They are often poorly preserved, crushed or strongly deformed.

Based on the presence or absence of a rostrum and rostral incisure, two distinct morphological groups have been recognized:

1) *Typical halocyprids* with a straight dorsal margin, the presence of a well developed anterior rostrum and rostral incisure, usually above mid-height of valve, and longitudinal or concentric striations. These are represented by:

*Conchoecia cretacea* Pokorný, 1964: Coniacian of Czechoslovakia (Plate 1, Fig. 1); *Conchoecia* sp. A Kaye (1965): Upper Albian of southern England; *Conchoecia* GA E 1 Grosdidier (1979): Lower Albian of Gabon (Plate 1, Figs 2-4);

2) While those members of the first group can only doubtfully be ascribed to *Conchoecia*, members of the second group are only doubtfully halocyprids.

They are often found associated with members of the previous group, as in southern England (Kaye, 1965) and in Gabon (Grosdidier, 1979), and are characterized by an arched dorsal margin and a faint longitudinal striation, somewhat similar to that of the Palaeozoic Entomozoacea, as previously noted by Kaye (1965). These forms are represented by:

*?Conchoecia* sp. B Kaye (1965): Upper Albian of southern England; '*Conchoecia*' IR O 27 (Grosdidier, 1973), originally identified as belonging to the non-marine limnocytherid genus *Timiriasevia*: Lower Albian of Iran (Persian Gulf); '*Conchoecia* GA D 31 Grosdidier (1979): Lower Albian of Gabon (Plate 1, Fig. 5); Gen. aff. '*Conchoecia*' sp. 215 Rosenfeld & Raab (1984): Upper Albian of Israel (Plate 1, Figs 7-8); Gen. aff. '*Conchoecia*' sp. 150 Rosenfeld & Raab (1984): Upper Albian-Lower Cenomanian of Israel (Plate 1, Fig. 6); *Conchoecia*? sp. Oertli (in Charollais *et al.*, 1977): Upper Albian of Switzerland (Plate 1, Fig. 9). '*Conchoecia*' sp. Andreu (in preparation): Upper Albian-Lower Cenomanian of Morocco (Plate 1, Figs 10-12).

Although more common than the typical rostrate halocyprids, the systematic position of these forms is very problematical. Clearly none of them belongs within the genus *Conchoecia*. They certainly do not seem attributable to any Recent halocyprid genus, although they bear some resemblance to species of the recently described genus *Deeveya* Kornicker & Iliffe (1985) from caves in the Bahamas. The genus, however, is probably new. More material needs to be studied since the internal characters and especially the muscle scars have never been observed, and until these characters have been described, even their attribution to the Halocypridacea must be in doubt.

The length:height plot diagram (Fig. 1) shows that essentially three groups can be differentiated:

1) The dominant group is characterized by a length:height ratio of about 1.40, by discontinuous faint longitudinal striae (10 to 15), and is represented

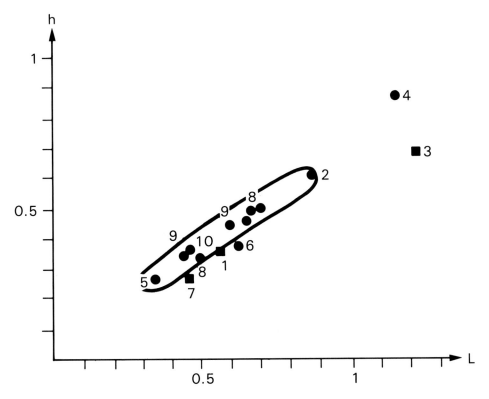

Fig. 1. Length:height plot of various species of Cretaceous halocyprids. 1 = *Conchoecia* sp. A Kaye, 1965; 2 = ?*Conchoecia* sp. B Kaye, 1965; 3 = *Conchoecia cretacea* Pokorný, 1964; 4 = Gen. aff. '*Conchoecia*' sp. 215 Rosenfeld & Raab, 1984; 5 = Gen. aff. '*Conchoecia*' sp. 150 Rosenfeld & Raab, 1984; 6 = *Conchoecia*? sp. Oertli, 1977; 7 = *Conchoecia* GA E 1 Grosdidier, 1979; 8 = '*Conchoecia*' GA D 31 Grosdidier, 1979; 9 = '*Conchoecia*' IR O 27 (Grosdidier, 1973); 10 = '*Conchoecia*' sp. Andreu (this paper).

by the following species; ?*Conchoecia* sp. B Kaye, Gen. aff. '*Conchoecia*' sp. 150 Rosenfeld & Raab, '*Conchoecia*' IR O 27 (Grosdidier), '*Conchoecia*' sp. Andreu (in preparation) and '*Conchoecia*' GA D 31 Grosdidier;

2) The second group, at present only represented by a single species, *Conchoecia*? Oertli (in Charollais *et al.* 1977), is very similar in shape and ornamentation, but the carapace is more elongate (length:height = 1.65).

3) The third group, also monospecific with Gen. aff. '*Conchoecia*' sp. 215 Rosenfeld & Raab, has a much higher and larger carapace (length:height = 1.42; length = 1.18mm) and has some 15, deeper longitudinal striae on the surface of the valves.

## GEOGRAPHICAL DISTRIBUTION (Fig. 2)

### West Africa

Grosdidier (1973) reported two species, '*Conchoecia*' GA E 1 (without rostrum) and *Conchoecia* GA E 1 (with rostrum and rostral incisure) from the Lower Albian, Lower Madiéla Formation of Gabon, in which they are associated with Radiolaria and planktonic Foraminiferida (*Hedbergella* spp.). They characterize the base of Grosdidier's ostracod zone MAb, which correlates with the calpionellid *Colomiella recta* Zone of Chevalier & Fischer (1982). Another ostracod present in this zone is the cytherid *Sergipella transatlantica* Krömmelbein (1967), a species also known from Brazil.

In West Africa, similar faunas have been found in the Ivory Coast and in the Congo Republic

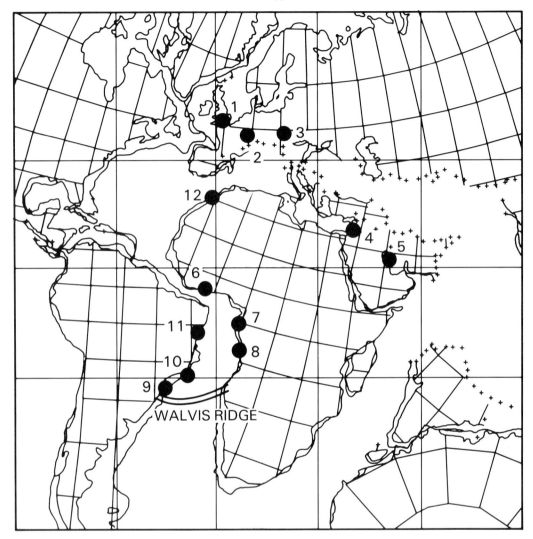

Fig. 2. Geographical distribution of Cretaceous halocyprids (plotted on an earliest Cenomanian map, after Smith & Briden, 1977). 1 = Southern England (Kaye, 1965); 2 = Switzerland (Charollais *et al.*, 1977); 3 = Czechoslovakia (Pokorný, 1964); 4 = Israel (Rosenfeld & Raab, 1984); 5 = Iran (Grosdidier, 1973); 6 = Ivory Coast (unpublished); 7 = Gabon (Grosdidier, 1979); 8 = Congo (unpublished); 9-10 = Santos and Campos basins (De Azevedo *et al.* 1987a, 1987b; Koutsoukos & Dias-Brito, 1987; 11 = Morocco (Andreu, this paper); 12 = Sergipe Basin (Koutsoukos pers. comm.).

(Colin, pers. obs.)

## Morocco

A species related to the group of ?*Conchoecia* sp. B of Kaye (without rostrum) has been found by Andreu (in preparation) in sediments dated as uppermost Albian (Vraconian) and Lower Cenomanian by the planktonic foraminifera *Rotalipora* *gandolfi* and *Rotalipora* aff. *montsalvensis* from the Agadir Basin, Morocco. This species was found in green marls, with a relatively rich and diverse microfauna. Planktonic foraminifera increase in proportion from the base to the top of the sampled section and obviously outnumber the benthonic foraminifera at the top. The environment was probably inner to outer neritic, with good connection to the open sea.

## Brazil

In Brazil, the genus *Conchoecia* has been reported (but not illustrated) from the Lower to Middle Albian of the Campos and Santos-Florianopolis basins (Guaruja Formation) by Koutsoukos & Dias-Brito (1987) and De Azevedo *et al.* (1987 a-b) and in the lowermost Albian (Riachuelo Formation, Angico Member) and Lower Cenomanian (Cotinguiba Formation, Sapucari Member, *Pseudotissotia* zone) of the Sergipe Basin (Koutsoukos pers. comm.). In the Florianopolis and Santos basins *Conchoecia* is known from the Lower to Middle Albian, in outer neritic facies associated with the planktonic foraminifera *Favusella* and *Hedbergella*, the benthonic foraminifera *Dorothia*, *Gavelinella* and *Lenticulina*, and radiolarians (Koutsoukos & Dias-Brito, 1987). In the Sergipe Basin, halocyprids may represent up to 6-10% of the total microfauna and the environment interpreted as middle-deep neritic (Koutsoukos, pers. comm.).

In the central-eastern part of the Campos Basin, *Conchoecia* occurs in sediments of the same age and is also associated with *Favusella*, *Hedbergella*, but also with the benthonic foraminifera *Trocholina*, *Lenticulina*, and calcisphaerulids (*Pithonella* spp.). Although interpreted by Koutsoukos & Dias-Brito (1987) as an inner neritic assemblage, we believe this type of assemblage most likely to typify the outer shelf. Koutsoukos & Dias-Brito (1987) concluded that the palaeowater depths were probably shallow (not more than 50m), but this does not conflict with our interpretation. In both basins, the Lower to Middle Albian carbonate sediments represent the first pre-oceanic marine deposition on the floor of the spreading South Atlantic.

## Europe: Switzerland

Oertli, in Charollais *et al.* (1977), reported and illustrated *Conchoecia*? sp., which lacks a rostrum, from the Middle Albian, *Anahoplites intermedius* Ammonite Zone, of the Swiss Jura. The lithology is a sandy, glauconitic, clayey marl (the clay fraction represents 57%, with 50% of illite).

Associated microfossils include the podocopid ostracods (*Cornicythereis*, *Cytherella*, *Neocythere*, *Pontocyprella*, *Protocythere*, *Rehacythereis* and *Schuleridea*) and benthonic foraminifera represented by the arenaceous genera *Dorothia*, *Haplophragmoides*, and *Tritaxia*, and the calcareous genus *Lenticulina*. No planktonic foraminifera were recorded.

## Europe: England

Kaye (1965), described and illustrated two species assigned to the genus *Conchoecia* from the Upper Albian, *H. orbignyi* Ammonite Zone, of Kent, southern England. One of these, *Conchoecia* sp. A, possesses a rostrum, while ?*Conchoecia* sp. B lacks this structure.

## Europe: Czechoslovakia

Pokorný (1964), described the first reported Cretaceous halocyprid named *Conchoecia*? *cretacea* from the Coniacian of Czechoslovakia. This species has a rostrum and rostral incisure.

## The Middle East: Iran

We agree with Rosenfeld & Raab (1984) who attributed *Timiriasevia* IR 0 27 of Grosdidier (1973, 4, Figs 31b-c) from the Albian of the Persian Gulf, to the genus '*Conchoecia*' (without rostrum).

## The Middle East: Israel

Rosenfeld & Raab (1984), illustrated two species of the non-rostrate group, from the Albian to Upper Cenomanian of Israel:

Gen. aff. '*Conchoecia*' sp. 215;
Gen. aff. '*Conchoecia*' sp 150.

These forms are associated with species of the benthonic genera *Monoceratina*, *Polycope*, *Krithe*, *Cytheropteron*, *Paracypris*, '*Bairdia*' and dwarf specimens of other ostracod genera. The authors interpreted the palaeoenvironment as deep bathyal, between 500 and 1000m.

## BIOSTRATIGRAPHY

With the exception of Pokorný's species *Conchoecia*? *cretacea* from the Coniacian of Czechoslovakia, all the records discussed above are restricted to the Albian to Lower Cenomanian stratigraphical interval. In fact, only 2 of the 10 reported species (Fig. 3), seem to extend into the Lower Cenomanian; these are Gen. aff. '*Conchoecia*' 159 Rosenfeld & Raab (1984) and '*Conchoecia*' sp. Andreu (in preparation) from Morocco. These halocyprids can, therefore, be considered as valuable indicators for this Albian to Lower Cenomanian interval in sediments otherwise devoid of age diagnostic microfossils. In the South Atlantic they have been successfully used as local stratigraphical markers by Grosdidier (1979) in Gabon (MAb Ostracod Zone of the Madiéla Formation) and in southern Brazil by De Azevedo *et al.* (1987a) and Koutsoukos & Dias-Brito (1987) for the Lower to Middle Albian. In Israel, Gen. aff. '*Conchoecia*' sp. 215 is considered as a good marker for the Albian *Monoceratina shimonensis* Assemblage Zone by Rosenfeld & Raab (1984).

## PALAEOECOLOGY

The available literature precludes the accurate reconstruction of the environment in which these Cretaceous halocyprids lived. They have been found in sedimentary environments ranging from shallow marine carbonate facies (as in the Albian Madiéla Formation of Gabon (Grosdidier, 1979) and the Macae Formation of the Campos Basin of Brazil (De Azevedo *et al.*, 1987b), to argillaceous shelf environments (as in the Albian of southern England (Kaye, 1965) and Switzerland (Charollais *et al.*, 1977), to deep water basinal facies (as in Israel (Rosenfeld & Raab, 1984).

In the South Atlantic, these halocyprids are often associated with Radiolaria, e.g., in the Albian Madiéla Formation of Gabon (Grosdidier, 1979) and in the Lower Albian of the Florianopolis and Santos basins of Brazil (Koutsoukos & Dias-Brito, 1987). In both examples, halocyprids occur in the first marine transgressive sediments of the Albian, which overlie lagoonal to non-marine sediments of

Neocomian to Aptian age. Grosdidier (1979) interprets the presence of halocyprids with Radiolaria as a suggestion of relatively strong oceanic influences. Koutsoukos & Dias-Brito (1987) characterized the Lower Albian phase in the Santos and Campos basins of southern Brazil as representing the early pre-oceanic phase of a shallow (less than 50m) tropical sea and that the impoverished and low diversity fauna resulted from hypersaline conditions which prevailed at the time.

In Israel, Rosenfeld & Raab (1984), on the basis of association of these halocyprids with the genera *Monoceratina*, *Krithe*, '*Bairdia*', *Polycope*, *Cytheropteron*, *Paracypris* and dwarf specimens of other ostracod genera, propose a deep water environment, between about 500 and possibly 1000m. Dwarfism may indicate some restricted circulation conditions.

## PALAEOBIOGEOGRAPHY

The occurrence of apparently closely related species of halocyprids since the Lower Albian in the South Atlantic (Brazil, Gabon, Congo, Ivory Coast) and in the Tethys (Israel, Iran), once more poses the problem of the age of the first marine connection between the South and the North Atlantic. Where did Cretaceous halocyprids originate from: the South Atlantic or the Tethys ?

In the South Atlantic they are only known from north of the Rio Grande-Walvis Ridge, a region which was poorly connected with the southern South Atlantic during the mid-Cretaceous. From the Albian to the Santonian the Walvis Ridge acted as an important faunal barrier (Benson, 1984; Reyment & Dingle, 1987, Dingle, 1988) and conditioned the development of anoxic conditions in the Angola Basin from the Albian to the Santonian (Maillot, 1983; Zimmermann *et al.* 1987). Very few data are known on Albian ostracods from the South Atlantic (Benson, 1984; Tambareau, 1982). However, Dingle (1984) did not report any halocyprids in his very detailed study of the mid-Cretaceous ostracods from offshore southern Africa and the Falkland Plateau.

Many authors have discussed the age of the first connection between the North and South

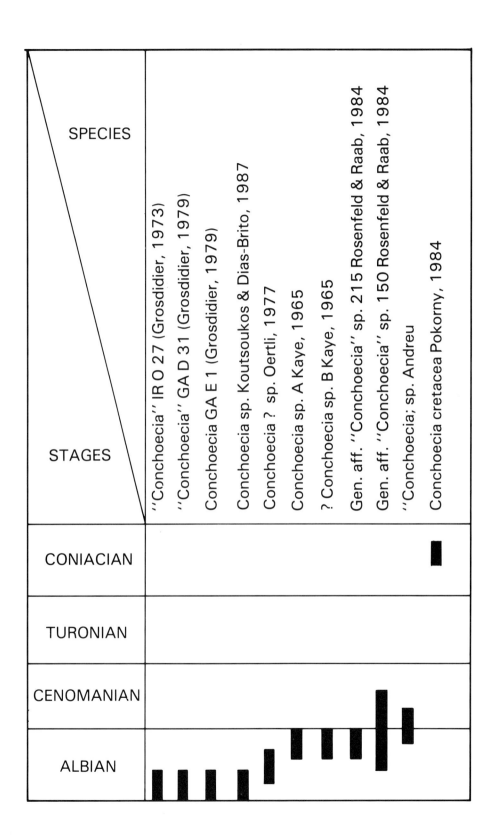

Fig. 3. Stratigraphical distribution of Cretaceous halocyprids.

Atlantic. A detailed review can be found in Dias-Brito (1987). Ages given in the literature range from the Aptian (Nairn & Stehli, 1973; Wiedmann & Neugebauer, 1978) to the Turonian (Reyment & Mörner, 1977), with a strong predilection for the Lower to Middle Albian (Kennedy & Cooper, 1975; Moullade & Guérin, 1982; Reyment & Dingle, 1987).

The presence of halocyprids since the base of the Albian in the South Atlantic and in the Tethys strongly favours a connection as early as the Lower Albian. This hypothesis is strongly supported by recent evidence of the co-occurrence of planktonic foraminifera and nannofossil associations (Nannoconus truitti) by Dias-Brito (1987). These observations are also in perfect agreement with the recent discovery by Chevalier & Fischer (1982) of the Tethyan planktonic foraminifera *Biglobigerinella barri/sigali* and *Globigerinelloides algerianus* in the Upper Aptian (Madiéla Formation) and of the calpionellid *Colomiella recta* in the Lower Albian of Gabon. The genus *Colomiella* is well known from Tethyan Aptian to Lower Albian sediments (Mexico, Tunisia, Spain, southern France, Somalia, Iran). Caron (1978) recognized the Tethyan planktonic foraminiferal zones of the Upper Aptian-basal Albian *Globigerinelloides algeriana-Ticinella bejouaensis* in DSDP sites 363-364 offshore Angola.

Connection through a trans-Saharian seaway, as suggested by Scheibnerova (1978) is not likely to have been possible before the uppermost Cenomanian-Lower Turonian.

Therefore, the Cretaceous halocyprids probably originated in the Tethys during the Lower Albian or possibly in the Upper Aptian. The generally accepted surface oceanic current circulation models for this period, support this interpretation (Berggren & Hollister, 1977). The Tethyan origin of this group of ostracods is in agreement with the myodocopid phylogeny postulated by Kornicker & Sohn (1976) and by Neale (1983). They propose a common ancestor for the Halocypridacea and the Thaumatocypridacea. During the Permian and the Mesozoic, the latter group had a Tethyan distribution (genera *Thaumatotoma* Kornicker & Sohn and *Pokornyopsis* Kozur).

The geographical distribution of Cretaceous halocyprids is, therefore, strongly controlled by the opening of the South Atlantic and the main Albian eustatic sea-level rise (Tambareau, 1982; Babinot & Colin, 1988).

## ACKNOWLEDGEMENTS

The authors would like to express their sincere gratitude to Drs A. Rosenfeld (Jerusalem, Israel), E. Koutsoukos (Rio de Janeiro, Brazil) and H. J. Oertli (Pau, France) for the loan of material. The senior author is also grateful to the management of Exxon Production Research Company (Houston, Texas, U.S.A.) for the permission to publish this paper. We also wish to thank Dr G. Tronchetti (Marseille) for the identification of the foraminifera from Morocco.

## REFERENCES

Angel, M. V. 1983. A review of the progress of research on Halocyprid and other oceanic planktonic ostracods. *In* Maddocks, R. F. (Ed.), *Applications of Ostracoda*, proceedings of the Eighth International Symposium on Ostracoda, July 26-29, 1982, 529-548. Univ. Houston Geos., Houston, Texas.

Babinot, J.-F. & Colin, J.-P. 1988. Paleobiogeography of Tethyan Cretaceous marine ostracods. *In* Hanai, T., Ikeya, N. & Ishizaki, K. (Eds), *Evolutionary biology of Ostracoda, its fundamentals and applications*, proceedings of the Ninth International Symposium on Ostracoda, held in Shizuoka, Japan, 29 July - 2 August 1985, Developments in palaeontology and stratigraphy, **11**, 823-839, Kodansha Ltd., Tokyo and Elsevier, Amsterdam, Oxford, New York, Tokyo.

Benson, R. H. 1984. Estimating greater paleodepths with ostracodes, especially in the thermospheric oceans. *Palaeogeogr. Palaeoclimat. Palaeoecol.*, Amsterdam, **48**, 107-141.

Berggren, W. A. & Hollister, C. D. 1977. Plate tectonics and paleocirculation-commotion in the ocean. *Tectonophysics*, Amsterdam, **38**, 11-48.

Bosquet, J. 1847. Description des Entomostracés fossiles des terrains tertiaires de la France et de la Belgique. *Mém. Acad. r. Belgique*, Bruxelles, **24**, 1-142.

Caron, M. 1978. Cretaceous planktonic foraminifers from DSDP Leg 40, Southeastern Atlantic ocean. *In* Bolli, H. M., Ryan, W. E. F. *et al.* (Eds), *Init. Rep. D.S.D.P.*, Washington, **40**, 561-678.

Charollais, J., Moullade, M., Oertli, H. J. & Rapin, F. 1977. Découverte de microfaunes de l'Albien Moyen et supérieur dans la vallée de Joux (Jura vaudois, Suisse). *Géobios*,

Lyon, **10**(5), 683-695.

Chevalier, J. & Fischer, M. 1982. Présence de *Colomiella* Bonet (Calpionellidea) dans le Crétace inférieur (Madiéla) du Gabon. *Cah. Micropaléontol.*, Paris, **2**, 29-34.

De Azevedo, R. L. M., Gomide, J. & Viviers, M. C. 1987a. Geohistoria da Bacia de Campos, Brasil do Albiano ao Maastrichtiano. *Revta bras. Geociênc.*, São Paulo, **17**(2), 139-146.

De Azevedo, R. L. M., Gomide, J., Viviers, M. C. & Hashimoto, A. T. 1987b. Bioestratigrafia do Cretaceo marinho da Bacia de Campos, Brasil. *Revta bras. Geociênc.*, São Paulo, **17**(2), 147-153.

Dépêche, F. 1984. Les ostracodes d'une plate-forme continentale au Jurassique: recherches sur le Bathonien du Bassin parisien. *Mém. Sci. Terre, Univ. Pierre et Marie Curie*, Paris, 1-405.

Dias-Brito, D. 1987. A Bacia de Campos no Mesocretaceo: uma contribuicao a palaeoceanografia do Atlantico Sul primitivo. *Revta bras. Geociênc.*, São Paulo, **17**(2), 162-167.

Dingle, R. V. 1984, Mid-Cretaceous ostracoda from Southern Africa and the Falkland Plateau. *Ann. S. Afr. Mus.*, Cape Town, **93**(3), 97-211.

Dingle, R. V. 1988. Marine ostracode distributions during the early breakup of southern Gondwanaland. *In* Hanai, T., Ikeya, N., & Ishizaki, K. (Eds), *Evolutionary biology of Ostracoda, its fundamentals and applications*, proceedings of the Ninth International Symposium on Ostracoda, held in Shizuoka, Japan, 29 July - 2 August 1985, Developments in palaeontology and stratigraphy, **11**, 841-854, Kodansha Ltd., Tokyo and Elsevier, Amsterdam, Oxford, New York, Tokyo.

Donze, P. 1965, Espèces nouvelles d'ostracodes des couches de base du Valanginien de Berrias (Ardêche). *Trav. Lab. Géol. Univ. Lyon, N.S.*, **12**, 87-107.

Grosdidier, E. 1973. Associations d'ostracodes du Crétacé d'Iran. *Rev. Inst. fr. Pétrole*, Paris, **28**(2), 131-168.

Grosdidier, E. 1979. Principaux ostracodes marins de l'intervalle Aptien-Turonien du Gabon (Afrique occidentale). *Bull. Centres Rech. Explor.-Prod., Elf-Aquitaine*, Pau, **3**(1), 1-35.

Kaye, P. 1965. Some new British Albian Ostracoda. *Bull. Br. Mus. nat. Hist.*, Geology, London, **11**(5), 217-254.

Keij, A. 1957. Eocene and Oligocene Ostracoda of Belgium. *Mém. Inst. r. Sci. nat. Belg.*, Bruxelles, **136**, 1-210.

Kennedy, W. J. & Cooper, M. 1975. Cretaceous ammonite distribution and the opening of the South Atlantic. *J. geol. Soc. London*, London, **131**, 283-288.

Kornicker, L. S. & Iliffe, T. M. 1985. Deeveyinae, a new subfamily of ostracoda (Halocyprididae) from a marine cave on the Turks and Caicos Islands. *Proc. biol. Soc. Wash.*, Washington, **98**, 476-493.

Kornicker, L. S. & Sohn, I. G. 1976. Phylogeny, ontogeny and morphology of Recent and fossil Thaumatocyprididae (Myodocopa, Ostracoda). *Smithson. Contr. Zool.*, Washington, D.C., **29**, 1-126.

Koutsoukos, E. M. & Dias-Brito, D. 1987. Paleobatimetria da margem continental do Brasil durante o Albiano. *Revta bras. Geociênc.*, São Paulo, **17**(2), 86-91.

Krömmelbein, K. 1967. Ostracoden aus der marinen 'Kuesten-Kreide' Brasiliens - 2: *Sergipella transatlantica* n.g., n.sp., und *Aracajuia benderi* n.g., n.sp., aus der Ober-Aptium/Albian. *Senckenberg. leth.*, Frankfurt a.M., **48**(6), 525-533.

Maillot, H. 1983. Les paléoenvironnements de l'Atlantique Sud: Apports de la géochimie sédimentaire. *Soc. géol. Nord, Pub.*, Lille, **9**, 1-316.

Moullade, M. & Guérin, S. 1982. Le problème des relations de l'Atlantique Sud et de l'Atlantique central au Crétacé Moyen: nouvelles données microfauniques d'après les forages D.S.D.P. *Bull. Soc. géol. Fr.*, Paris, **24**(3), 511-517.

Nairn, A. E. M. & Stehli, F. G. 1973. A model for the South Atlantic. *In* Nairn A. E. M. & Stehli, F. G. (Eds), *The ocean basins and margins*, **1**, 211-517. Plenum Press, New York.

Neale J. W. 1976. On *Philomedes donzei* Neale sp. nov. *Stereo-Atlas Ostracod Shells*, Welwyn Garden City, **3**(2), 9-12.

Neale, J. W. 1983. Geological history of the Cladocopina. *In* Maddocks, R. F. (Ed.), *Applications of Ostracoda*, proceedings of the Eighth International Symposium on Ostracoda, July 26-29, 1982, 612-626. Univ. Houston Geos., Houston, Texas.

Pokorný, V. 1964. *Conchoecia*? *cretacea* n. sp., first fossil species of the family Halocyprididae (Ostracoda, Crustacea). *Acta Univ. Carol., Geol.*, Praha, **2**, 176-179.

Reyment, R. A. & Dingle, R. V. 1987. Palaeogeography of Africa during the Cretaceous period. *Palaeogeogr. Palaeoclimat. Palaeoecol.*, Amsterdam, **59**, 93-116.

Reyment, R. A. & Mörner, N. A. 1977. Cretaceous transgressions exemplified by the South Atlantic. *Palaeontol. Soc. Japan*, Tokyo, **21**, 247-262.

Rosenfeld, A. & Raab, M. 1984. Lower Cretaceous ostracodes from Israel and Sinai. *Geol. Surv. Israel spec. Publ.*, Jerusalem, **4**, 85-134.

Scheibnerova, V. 1978. Aptian and Albian benthic foraminifera of Leg 40, sites 363 and 464, Southern Atlantic. *In* Bolli, H. M. Ryan, W. B. F. *et al.* (Eds), *Init. Repts DSDP*, **40**, 741-757. U.S. Govt Printing Office, Washington.

Smith, A. G. & Briden, J. C. 1977. *Mesozoic and Cenozoic paleocontinental maps*, 1-63. Cambridge University Press, Cambridge.

Sylvester-Bradley, P. C. 1961. Order Myodocopida Sars, 1866. *In* Moore, R. C. (Ed.), *Treatise on Invertebrate Paleontology, Pt. Q, Arthropoda*, **3**, 387-406. Univ. Kansas Press.

Tambareau, Y. 1982. Les ostracodes et l'histoire géologique de l'Atlantique Sud au Crétacé. *Bull. Centres Rech. Explor.-Prod., Elf-Aquitaine*, Pau, **6**(1), 1-37.

Van Morkhoven, F. P. C. M. 1962. *Post-Palaeozoic Ostracoda, their morphology, taxonomy and economic use*, Volume **I**, General, 204 pp., 79 figs, 8 tables, 1 enclosure. Elsevier, Amsterdam, London, New York.

Van Morkhoven, F. P. C. M. 1963. *Post-Palaeozoic Ostracoda, their morphology, taxonomy and economic use*, Volume **II**, Generic descriptions, 478 pp., 763 figs, 4 tables. Elsevier, Amsterdam, London, New York.

Van Veen, J. E. 1936. Die Cypridinidae der Maastrichter Tuffkreide und des Kunrader Korallenkalkes von Sued-Limburg. *Natuurh. Maandbl.*, Maastricht, **25**(11/12), 169-170.

Weitschat, W. 1983. Ostracoden (O. Myodocopida) mit

**Plate 1**

Fig. 1.     *Conchoecia cretacea* Pokorný, 1964. Carapace, right view. Coniacian, Czechoslovakia.

Figs 2-4.     *Conchoecia* GA E 1 Grosdidier, 1979. Albian (Madiela Formation) of Gabon.
Fig. 2.     Carapace, left view.
Fig. 3.     Carapace, right view.
Fig. 4.     Carapace, dorsal view.

Fig. 5.     '*Conchoecia*' GA D 31 Grosdidier, 1979. Albian (Madiela Formation) of Gabon. Carapace, right view.

Fig. 6.     Gen. aff. '*Conchoecia*' sp. 150 Rosenfeld & Raab, 1984. Albian of Israel. Carapace, right view.

Figs 7-8.     Gen. aff. '*Conchoecia*' sp. 215 Rosenfeld & Raab, 1984. Albian of Israel.
Fig. 7.     Carapace, left view.
Fig. 8.     Carapace, right view.

Fig. 9.     *Conchoecia* ? sp. Oertli (in Charollais et al. 1977). Albian of Switzerland. Carapace.

Figs 10-12. '*Conchoecia*' sp. Andreu. Albian of Morocco.
Fig. 10.     Carapace, left view.
Fig. 11.     Carapace, right view.
Fig. 12.     Carapace, left view.

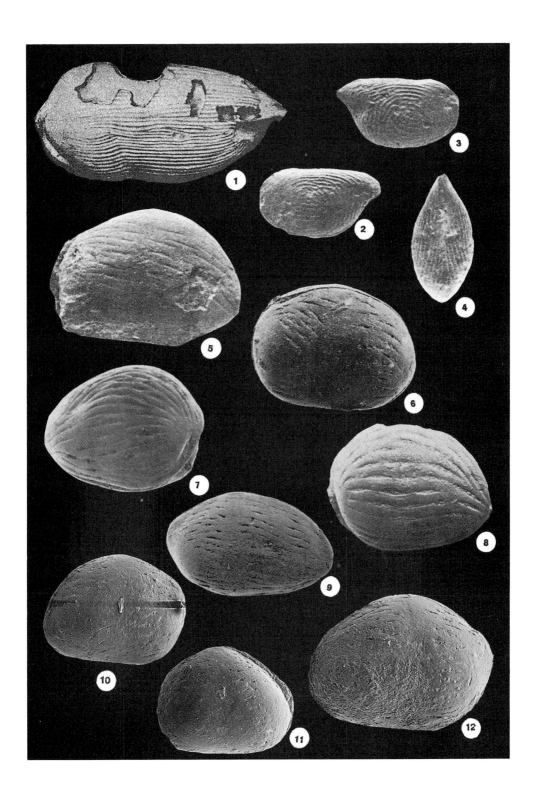

weichkoerper-erhaltung aus der unter-Trias von Spitzbergen. *Palaeont. Z.*, Stuttgart, **57**(3/4), 309-323.

Wiedmann, J. & Neugebauer, J. 1978. Lower Cretaceous ammonites from the South Atlantic Leg 40 (DSDP), their stratigraphic value and sedimentological properties. *In* , H., Ryan, W. B. F. *et al.* (Eds), *Init. Repts DSDP*, **40,** 709-734. U.S. Govt Printing Office, Washington.

Zimmermann, H. B., Boersma, A. & McCoy, F. W. 1987. Carbonaceous sediments and palaeoenvironment of the Cretaceous South Atlantic Ocean. *In* Brooks, J. & Fleet, A. J. (Eds), *Marine Petroleum Source Rocks. Geol. Soc. spec. Publ.*, **26**, 271-286. Blackwell Scientific, Oxford.

# 41

# Palaeogeographical significance of ostracod biofacies from Mississippian strata of the Black Warrior Basin, northwestern Alabama: a preliminary report.

Chris P. Dewey[1], T. Mark Puckett[2] & Hugh B. Devery[1]

[1]Department of Geology and Geography, Mississippi State University, U.S.A.
[2]Department of Geology, University of Alabama, U.S.A.

## ABSTRACT

The Black Warrior Basin of Mississippi and Alabama is bounded by the Appalachian and Ouachita orogenic belts. During Chesterian time the basin comprised a carbonate shelf and parts of two prograding clastic wedges derived from the southwest and northeast.

Shallow marine ostracods from the outcrop belt of the Black Warrior Basin have been used to examine the palaeoenvironmental relationships between the southwesterly derived clastic wedge (Floyd-Pride Mountain-Parkwood interval) and the carbonate shelf (Monteagle-Bangor formations).

At least three ostracod assemblages are recognized based upon the genera they contain: Bairdiacean-kirkbyacean faunas, indicative of carbonate offshore conditions; a Binodicope-quasillitacean fauna, indicative of clastic offshore conditions; Kloedenellacean dominated faunas, indicative of clastic nearshore conditions; and mixed assemblages.

Salinity fluctuation, associated with clastic progradation, was probably the main factor controlling the distribution of the ostracod assemblages. Reduced salinities on the shelf at the distal end of the clastic wedge restricted the distribution of such normal marine shelf indicators as the bairdiaceans, kirkbyaceans, quasillitaceans, hollinomorphs and binodicopes but allowed the development of nearshore kloedenellacean faunas far from any palaeo-shoreline.

## INTRODUCTION

The Black Warrior Basin of Mississippi and Alabama (Fig. 1) is a triangular region bounded to the southeast by the Appalachian tectonic belt and to

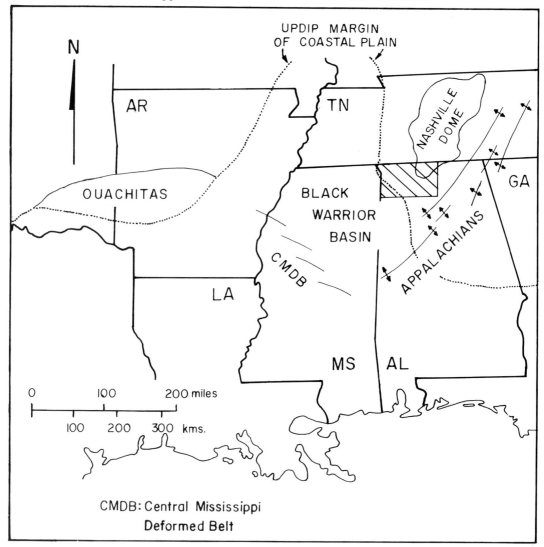

Fig. 1. Location map showing study area in northwestern Alabama.

the southwest by the Ouachita tectonic belt and its subsurface extension, the central Mississippi deformed belt. The northern boundary of the basin is marked by the Nashville Dome (Mellen, 1947; Thomas, 1988).

During Chesterian time, the Black Warrior Basin was situated near the southern continental margin of North America (Fig. 2) on a palaeogeographical feature known as the Alabama promontory (Thomas, 1988). Deposition of the Monteagle and Bangor Limestones (Fig. 3) occurred on the east Warrior platform (Thomas, 1972) and was

synchronous with deposition of prograding clastic wedges. Chesterian clastic wedges include the Floyd Shale-Pride Mountain-Hartselle-Parkwood wedge derived from a southwestern source, and the Pennington wedge derived from the northeast (Thomas, 1974; Mack et al., 1983).

An alternative model for the provenance of the sandstones within the Floyd-Parkwood interval suggests that they were derived from a cratonic, northwesterly source, possibly the Ozark dome (summarized in Cleaves & Bat, 1988). This model is based upon isolith maps of the sandstone

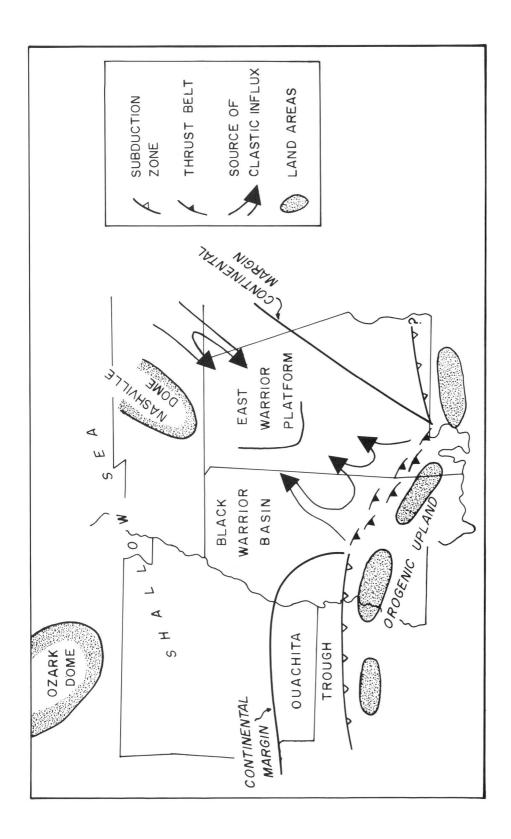

Fig. 2. Possible palaeotectonic setting of the Black Warrior Basin during Chesterian time (based upon Wickham *et al.*, 1976; Craig & Conner, 1979; Mack *et al.*, 1983).

intervals in the Floyd (Lewis and Evans sands) and Parkwood (Carter sand) in the subsurface of Mississippi and Alabama.

Using compositional evidence, Mack *et al.* (1983) suggest that the southwestern clastic source was provided by the initiation of tectonic collision involving the Alabama promontory with either a magmatic arc and subduction complex, or a microcontinent and associated continental-margin arc components to the southwest. To the north and northwest of the colliding fragment, in the Ouachita embayment, the deep water Ouachita trough was receiving turbidite and hemipelagic sediments of the Stanley-Jackfork sequence (Morris, 1974; Sutherland & Manger, 1979). The deep water facies in western Mississippi include black siliceous claystones and dark cherts, which are similar to some lithologies in the Stanley (Thomas, 1974), and may indicate the easternmost extent of the Ouachita trough. The Black Warrior Basin may, therefore, have formed part of the southern continental margin of North America and also a foreland basin to the Ouachita orogen with subduction of the North American plate occurring southward under a magmatic arc or microcontinental fragment and associated arc (Wickham *et al.*, 1976; Mack *et al.*, 1983).

## STRATIGRAPHY

### Pride Mountain - Monteagle Interval

The Pride Mountain Formation (Fig. 3) is an extensive, lithologically complex tongue of the Floyd Shale that prograded onto the East Warrior platform. It was the first major clastic progradation into the Black Warrior Basin and consists primarily of clay shales and three sandstone units together with minor limestones (Thomas, 1972; Di Giovanni, 1984). Provenance of the sandstones within the Pride Mountain has been a matter of debate. Cleaves & Broussard (1980) suggested that they were deposited by delta systems with a northwestern source, whereas Thomas (1974) and Mack *et al.*, (1983) suggested a southwesterly source. Holmes (1981) and Di Giovanni (1984) suggested that the Pride Mountain sandstones

were deposited as northwest-southeast trending sand bodies on a shallow marine shelf, parallel to depositional strike.

To the northeast, the Pride Mountain interfingers with the carbonate facies of the Monteagle which is interpreted to be an upward shoaling oolitic tidal-bar sequence (Thomas, 1972; Handford, 1978; Maples & Archer, 1986). Thus, the lateral transition from the Pride Mountain to Monteagle represents the interaction of clastic progradation with carbonate depositional processes on the shelf.

### Upper Floyd - Bangor Interval

The Floyd Shale (Fig. 3) is a grey-buff weathering clay shale, that is interpreted as a lagoonal or prodeltaic deposit. Upper deposits of the Floyd prograded northeastward onto the East Warrior platform during lower Bangor deposition (Thomas, 1972). Sediments of the Bangor Limestone record at least six shallow marine shelf depositional cycles, each of which began with a transgressive facies and shoaled upward into inter- and supratidal sediments (Scott, 1978). The lower Bangor carbonate shelf grades southwestward into the Floyd shale and indicates the changeover from transgressive carbonate processes to progradational clastic processes.

## OSTRACOD PALAEOECOLOGY

### General

Ostracods have been used widely as indicators of distance from shore, bathymetry and salinity in Carboniferous palaeoenvironments. This has led to the recognition of three types of ostracod fauna (Bless, 1983):

1) *Carbonita*-type faunas, indicative of freshwater conditions (Bless & Pollard, 1973),

2) Bairdiacean, palaeocope, paraparchitacean and kloedenellacean faunas, indicative of conditions which range from brackish, through hypersaline marginal marine, out to distal shelf environments. An extensive literature exists concerning the distribution of these faunas (see, for

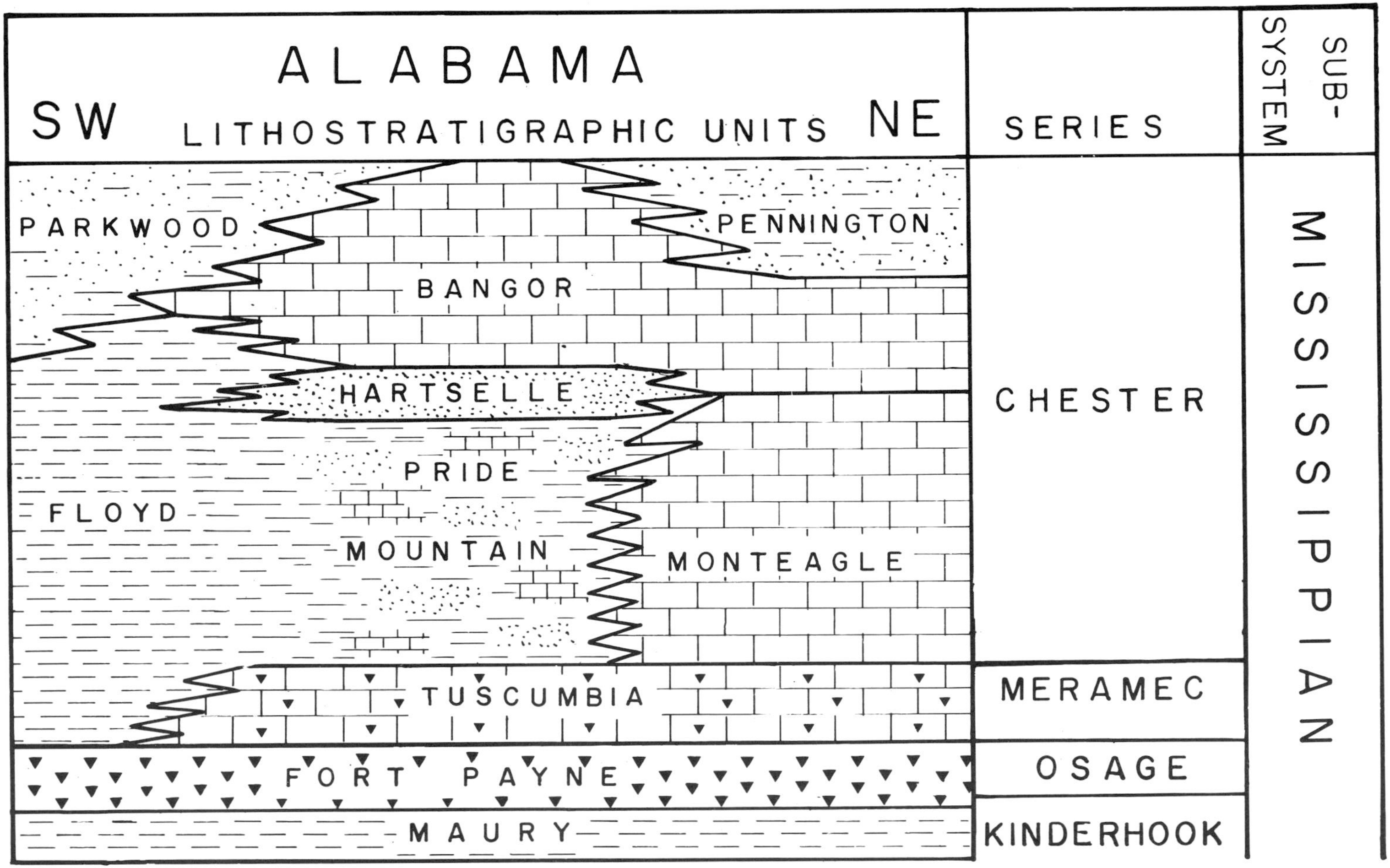

Fig. 3. Mississippian lithostratigraphy of Alabama (redrawn from Thomas, 1972).

example; Van Ameron *et al.*, 1970; Haack & Kaesler, 1980; Kaesler, 1982; Crasquin, 1984; Dewey, 1987; Melnyk & Maddocks, 1988).

3) Entomozoacean-tricorninid-type faunas, indicative of deep basinal environments (Bandel & Becker, 1975; Becker *et al.*, 1975).

In the Black Warrior Basin, we have begun to examine the effects of progradation at the fringes of a carbonate platform, and also the effects of clastic *versus* carbonate substrates upon the distribution of shallow marine ostracod faunas (Devery, 1987; Puckett, 1987), (Table 1). Three ostracod assemblages have been recognized in this study, all of which can be related to shelf environments. It is important that the ostracod assemblages can be determined by reference to distinct morphotypes that are obvious at relatively high taxonomic levels. We recognize that the assemblages obscure the fine palaeoecological detail that is available at the species level and it is expected that more precise palaeoecological interpretations will be produced as further taxonomic work is completed (Plates 1 and 2). The primary intent, however, was to produce a palaeoecological tool that is useful to a non-specialist.

**Pride Mountain Interval**

In northwestern Alabama, the Pride Mountain consists of shales with interbedded sandstones and limestones. Due to the lateral facies variations, it is difficult to correlate units between sections. Ostracod faunas range from highly abundant to sparse, and from excellent preservation to poor. As a result of these factors, the ostracod palaeoecology of the Pride Mountain has been difficult to interpret.

Two distinct assemblages and one mixed assemblage have been defined from the Pride Mountain in northern Alabama (Puckett, 1987).

*Assemblage I.* This assemblage is characterized by high percentages of bairdiaceans and kirkbyaceans. Dominant bairdiaceans include *Bairdia*, *Bairdiolites* and *Orthobairdia* and the dominant kirkbyacean is *Polytylites*, although *Aurikirkbya* and *Kirkbya* are also present. The assemblage also contains healdiaceans, a few

individuals of the hollinomorph *Tetrasacculus* and rare paraparchitaceans. The Kloedenellacea are conspicuous by their absence. The assemblage is derived from bioclastic limestones that contain a diverse fauna of macro-invertebrate suspension feeders.

*Assemblage II.* This assemblage represents a mixed group of faunas that may contain elements of Assemblage I and III. Samples that yield Assemblage II-type faunas were collected from fossiliferous grey shales. The unifying character of this assemblage is that it is not dominated by kloedenellaceans or by bairdiaceans and kirkbyaceans, although members of these superfamilies are common in some samples. The dominant palaeocope is the binodicope, *Mammoides*, although rare individuals of *Hollinella* and *Tetrasacculus* are also present. Another important group of ostracods in this assemblage is the Metacopina, including *Healdia* and the quasillitacean, *Graphiadactyllis*.

*Assemblage III.* This assemblage is dominated by members of the Kloedenellacea, primarily *Sansabella*, *Nufferella*, *Sargentina* and *Oliganisus*. The kloedenellacean faunas are typically found in grey, sparsely fossiliferous shales, and normally have faunas with low diversity and low abundance. One exception to this is found in a sample that is dominated by the genus *Sargentina*. This fauna is associated with *Amphissites*, *Polytylites*, *Healdia*, *Graphiadactyllis* and kloedenellaceans, and was found at the extreme top of a bioclastic limestone unit below a sparsely fossiliferous shale.

**Upper Floyd-Bangor Interval**

The Bangor Limestone interfingers with distal deposits of the Floyd Shale, resulting in outcrops characterized by a mudstone lithofacies at one extreme and an oolitic and skeletal grainstone lithofacies at the other. Frequently, outcrops are a combination of these extremes, where the amount of mudstone and the lateral persistency and thickness of the limestones are variables. The development of these lithofacies is a function of the interactions between clastic influx due to

Table 1. Distribution of representative ostracods from assemblages in the Pride Mountain and Bangor intervals.

| Interval | | Pride Mountain | | | Bangor | |
|---|---|---|---|---|---|---|
| Assemblage | | I | II | III | A | B |
| **Ostracods** | | | | | | |
| Amphissities rugosus | | | | | | |
| Amphissites centronotus | | R | | | | |
| Polytylites superus | | | | | | |
| Polytylites quincollinus | | | | | | |
| Kirkbya spp. | | | | | | |
| Hollinella radiata | | | R | | | |
| Tetrasacculus mirablilis | | R | R | | R | |
| Coryellina ventricornis | | | | | R | |
| Pseudoparaparchites sp | | | | | R | |
| Bairdia spp. | | | | | | |
| Orthobairdia sp. A. | | | | | | |
| Bairdiacypris curvis | | | | | | |
| Bairdiolites spp. | | | | | | |
| Mammoides dorsospinosa | | | | | | |
| Graphiadactyllis fayettevillensis | | | | | | |
| Healdia spp. | | | | | | |
| Cavellina parva | | | | | | |
| Paracavellina elliptica | | | | | | |
| Nufferella wellsi | | | | | | |
| Sansabella spp. | | | | | R | |
| Knoxiella sp. A. | | | | | | |
| Sargentina alleni | | | | | | |
| Geffenina johnsoni | | | | | | |
| Oliganisus geisi | | | | | | |
| Glyptopleura spp. | | R | | | | |
| Glyptopleurina bulbosa | | R | | | | |

progradation and activity of the carbonate factory during times of low progradation and/or transgression.

In general, ostracods are much more abundant in the Bangor interval than in the Pride Mountain, although the assemblages are similar. Two distinct ostracod assemblages have been defined (Devery, 1987) from the Bangor interval.

*Assemblage A.* This assemblage is characterized by the abundance of bairdiaceans including *Bairdia*, *Orthobairdia*, *Bairdiolites* and *Bairdiacypris* together with such kirkbyaceans as *Polytylites*, *Amphissites* and *Kirkbya*. *Tetrasacculus*, *Coryellina* and *Pseudoparaparchites*, are exclusive to this assemblage. Unlike the bairdiacean-kirkbyacean Assemblage I

of the Pride Mountain, kloedenellaceans are a consistent part of this fauna, and such genera as *Glyptopleura*, *Glyptopleurina* and *Oliganisus* may be common. *Sansabella* also occurs in some samples, but is the dominant kloedenellacean in only two samples. The presence of glyptopleurids in association with bairdiaceans and kirkbyaceans, combined with the absence of several kloedenellaceans that are typical of Assemblage B, make this a distinctive association between true bairdiacean-kirkbyacean and kloedenellacean faunas. It is possible that further work may reveal a subdivision of the assemblage according to the composition and abundance of the kloedenellacean component. Other components of Assemblage A include the platycope *Cavellina* and the metacope *Healdia*, but these forms are also found in Assemblage B. Assemblage A occurs mostly in calcareous mudstones and the interbedded mudstone and limestone lithofacies. Associated benthonic, filter-feeding, macro-invertebrates (Waters, 1978) are similar to those found in Assemblage I in the Pride Mountain.

*Assemblage B*. This assemblage is the most widespread in the Floyd-lower Bangor interval and is characterized by the abundance of kloedenellaceans including *Sansabella* as the dominant genus and *Geffenina*, *Nufferella* and *Knoxiella* as accessory components. Ostracods assigned to this assemblage occur in the calcareous mudstone lithofacies. The assemblage is very similar to Assemblage III in the Pride Mountain interval.

## DISCUSSION

The palaeoenvironmental setting of the Pride Mountain can be interpreted as a clastic-dominated shelf where carbonate deposition occurred as local aggradational events, except where the Pride Mountain grades into the Monteagle carbonate shoal in northeastern Alabama. The Pride Mountain was the earliest progradational event of the Floyd wedge and represents clastic inundation of the platform. Conversely, the Bangor represents a carbonate platform environment (Thomas, 1972) onto which fine clastic material prograded northeastward at the distal end of the Floyd wedge (Thomas, 1974).

Ostracod faunas in the Pride Mountain show the effects of deposition on a normal marine salinity clastic-dominated shelf and grade into more restricted faunas where the effects of progradation were strongest. True bairdiacean-kirkbyacean dominated faunas are rare and, together with the binodicope-quasillitacean fauna, probably represent the only palaeoenvironments in which offshore conditions were attained during deposition of the Pride Mountain. Since most ostracod faunas in the Pride Mountain can be related to the kloedenellacean Assemblage, it is likely that the widespread occurrence of nearshore

faunas on the shelf can be related to the effects of progradation.

Ostracods from the Floyd and Bangor interval show the effects of a normal marine salinity carbonate platform oscillating back and forth into mud-dominated environments associated with the distal portions of the Floyd wedge. The fact that the bairdiacean-kirkbyacean faunas in the Bangor often contain glyptopleurid kloedenellaceans suggests a more proximal rather than a distal shelf setting. The distribution of kloedenellacean faunas dominated by the genus *Sansabella* suggests a close proximity to shoreline.

All the evidence gathered to date shows that Black Warrior Basin was being influenced by relatively shallow water or nearshore conditions.

Implied in this discussion, is that a shoreline existed to the southwest during both Bangor and Pride Mountain deposition. Palaeogeographical reconstructions suggest that the open ocean was to the south (Mack *et al.*, 1983; Dewey, 1985), however the development of a tectonic upland (Fig. 2) that was a source of clastic sediment (Mack *et al.*, 1983) may have prevented a connection to the open ocean. Since the ostracod faunas are taxonomically similar to typical midcontinent Chesterian ostracods (Cooper, 1941) it is likely that the Black Warrior Basin connected northward between the Ozark and the Nashville domes to the midcontinental epeiric seaway (Craig & Conner, 1979). To the southwest, shallow water, nearshore conditions would have been associated with clastic progradation from the developing orogenic upland.

The palaeoenvironmental parameters controlling ostracod distribution in the Black Warrior Basin, therefore, can be related to the interaction of two depositional systems: the effects of clastic progradation onto the shelf and the effects of carbonate-producing transgressive cycles on the shelf.

In the broad sense, substrate selection controlled the occurrence of bairdiacean-kirkbyacean and binodicope-quasillitacean faunas. The distribution of kloedenellacean, healdiacean and cytherellacean ostracodes, however, cannot be simply related to substrate, since several species occur in both clastic and carbonate substrates.

Furthermore, changes in ostracod diversity and abundance occur in some sections where there is no noticeable change in lithology or macrofaunal content. Physical parameters of the palaeoenvironment that may vary during progradation include type and amount of terrigenous input, salinity, dissolved oxygen and organic content, all of which may have been significant controls on the distribution of ostracods.

In both the Pride Mountain and Bangor intervals, deteriorating palaeoenvironmental conditions, as indicated by a reduction in macrofaunal content and diversity, resulted in the development of low diversity kloedenellacean faunas that may be of high or low abundance. The restriction of the stenohaline bairdiaceans, palaeocopes and quasillitaceans suggests that it was the effects of reduced salinities at the distal end of the progradational wedge that permitted the occurrence of the 'nearshore' kloedenellaceans. Transitional faunas, therefore, represent intermediate conditions in which such nearshore kloedenellaceans as *Sansabella*, are replaced by offshore genera such as *Glyptopleura* and *Oliganisus*. Thus, it would be possible to develop 'nearshore' kloedenellacean faunas relatively far from shore.

## CONCLUSIONS

At least three distinct ostracod assemblages can be recognized in the Black Warrior Basin:

A bairdiacean-kirkbyacean dominated assemblage, indicative of open shelf offshore conditions of normal salinity. In the Black Warrior Basin this assemblage is best developed on carbonate substrates and includes Assemblage I from the Pride Mountain and Assemblage A from the Bangor.

A binodicope-quasillitacean fauna, found only in fossiliferous shales, suggesting an affinity to open shelf offshore conditions. The absence of this assemblage in the Bangor Limestone and its distribution in the Pride Mountain (Assemblage II), indicate a preference for clastic substrates.

Kloedenellacean-dominated faunas, associated mostly with sparsely fossiliferous, fine clastic substrates and calcareous mudstones. Abundances of ostracods in the fauna (Assemblage III from the

Pride Mountain and Assemblage B from the Bangor) can vary considerably.

In addition to the three distinct ostracod assemblages, mixed faunas also occur in northern Alabama. Ostracod faunas responded gradually to changing palaeoenvironmental conditions across the coenocline, except where such changes were abrupt. Intermediate faunas can contain species from end-member assemblages such as the bairdiacean-kikbyacean fauna or the kloedenellacean fauna. Glyptopleurid kloedenellaceans seem to be restricted to intermediate faunas. Healdiacean and cytherellacean ostracods are widely distributed across the coenocline and offer the greatest potential for biostratigraphical use.

Ostracod assemblages in the Black Warrior Basin developed in response to the interactions of carbonate-producing, transgressive events and clastic-producing, progradational events. Research to date suggests that the main controls upon ostracod distribution patterns were substrate and salinity although, at this stage, palaeo-oxygen and organic content cannot be eliminated as contributing factors.

## AKNOWLEDGEMENTS

Acknowledgment is made to the Donors of the Petroleum Research Fund, administered by the American Chemical Society, to the Mississippi Mineral Resources Institute and also to Mississippi State University for financial support of this research.

## REFERENCES

Bandel, K. & Becker, G. 1975. Ostracoden aus paläozoischen pelagischen Kalken der Karnischen Alpen (Silurium bis Unterkarbon). *Senkenberg. leth.*, Frankfurt a.M., **56**, 1-83.

Becker, G., Bless, M. J. M. & Kullmann, J. 1975. Oberkarbonische Entomozoen-Schiefer im Kantabrischen Gebirge (Nordspanien). *Neues Jb. Geol. Paläont.* Abh., Stuttgart, **150**, 92-110.

Bless, M. J. M. 1983. Late Devonian and Carboniferous ostracode assemblages and their relationship to the depositional environment. *Geologie*, Berlin, **92**, 31-53.

Bless, M. J. M. & Pollard, J. E. 1973. Paleoecology and ostracode faunas of Westphalian ostracode bands from Limburg, the Netherlands and Lancashire, Great Britain. *Meded.*

**Plate 1**

Fig. 1. *Polytylites superus* (Croneis & Gale, 1939); left aspect, length 1.0mm, Bangor.

Fig. 2. *Kirkbya elongata* Cooper, 1941; left aspect, length 0.75mm, Bangor.

Fig. 3. *Polytylites quincollinus* (Harlton, 1924); left aspect, length 1.0mm, Pride Mountain.

Fig. 4. *Amphissites centronotus* (Ulrich & Bassler, 1906); left aspect, length 1.2mm, Pride Mountain.

Fig. 5. *Hollinella radiata* (Jones & Kirkby, 1886); left aspect, male, length 1.8mm, Pride Mountain.

Fig. 6. *Pseudoparaparchites* sp. A; right aspect, length 0.8mm, Bangor.

Fig. 7. *Amphissites rugosus* Girty, 1910; left aspect, length 1.0mm, Pride Mountain.

Fig. 8. *Tetrasacculus mirabilis* (Croneis & Gale, 1939); right aspect, male, length 0.8mm, Pride Mountain.

Fig. 9. *Tetrasacculus mirabilis* (Croneis & Gale, 1939); right aspect, female, length 0.9mm, Pride Mountain.

Fig. 10. *Bairdiolites elongatus* Croneis & Funkhouser, 1938; right aspect, length 0.8mm, Bangor.

Fig. 11. *Bairdia golcondensis* Croneis & Gale, 1939; right aspect, length 1.0mm, Pride Mountain.

Fig. 12. *Orthobairdia* sp. A; right aspect, length 1.2mm, Pride Mountain.

Fig. 13. *Bairdiacypris curvis* (Cooper, 1941); right aspect, length 1.25mm, Bangor.

Fig. 14. *Healdia* sp. A; left aspect, length 0.66mm, Pride Mountain.

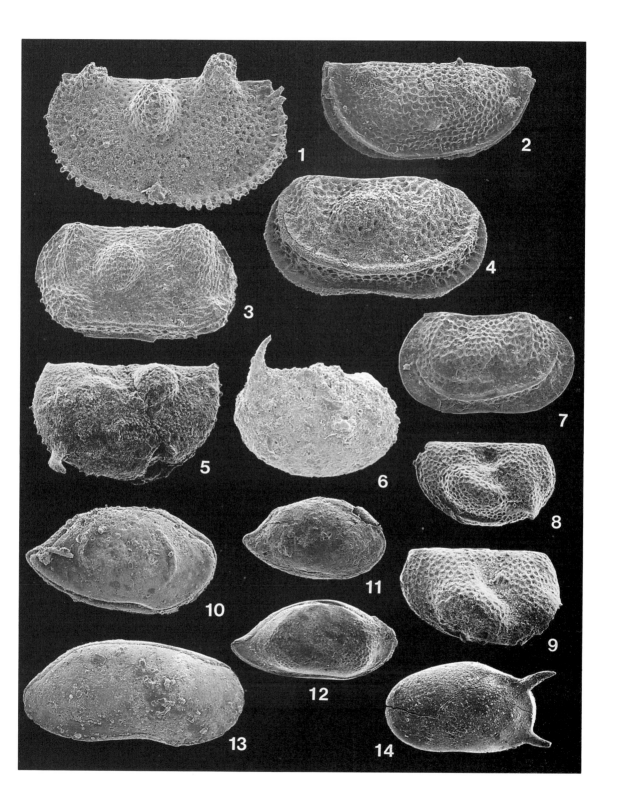

**Plate 2**

Fig. 1.  *Geffenina johnsoni* Coryell & Sohn, 1938; left aspect, length 1.0mm, Bangor.

Fig. 2.  *Glyptopleura berniciana* Robinson, 1978; left aspect, length 0.95mm, Bangor.

Fig. 3.  *Nufferella wellsi* Coryell & Sohn, 1938; right aspect, length 0.93mm, Pride Mountain.

Fig. 4.  *Glyptopleurina bulbosa* Croneis & Gale, 1939; left aspect, length 0.8mm, Bangor.

Fig. 5.  *Paracavellina elliptica* Cooper, 1941; right aspect, length 1.1mm, Pride Mountain.

Fig. 6.  *Sansabella truncata* Cooper, 1941; left aspect, length 0.97mm, Pride Mountain.

Fig. 7.  *Oliganisus geisi* Croneis & Gutke, 1939; left aspect, length 1.35mm, Bangor.

Fig. 8.  *Cavellina parva* Cooper, 1941; right aspect, length 0.96mm, Pride Mountain.

Fig. 9.  *Knoxiella* sp. A; left aspect, length 0.8mm, Pride Mountain.

Fig. 10.  *Coryellina ventricornis* (Jones & Kirkby, 1886); postero-right aspect, length 0.78mm, Bangor.

Fig. 11.  *Mammoides dorsospinosa* Sohn, 1961; left aspect, length 0.9mm, Pride Mountain.

Fig. 12.  *Sargentina alleni* Coryell & Johnson, 1939; right aspect, length 1.0mm, Pride Mountain.

Fig. 13.  *Graphiadactyllis fayettevillensis* (Harlton, 1929); left aspect, length 1.12mm, Pride Mountain.

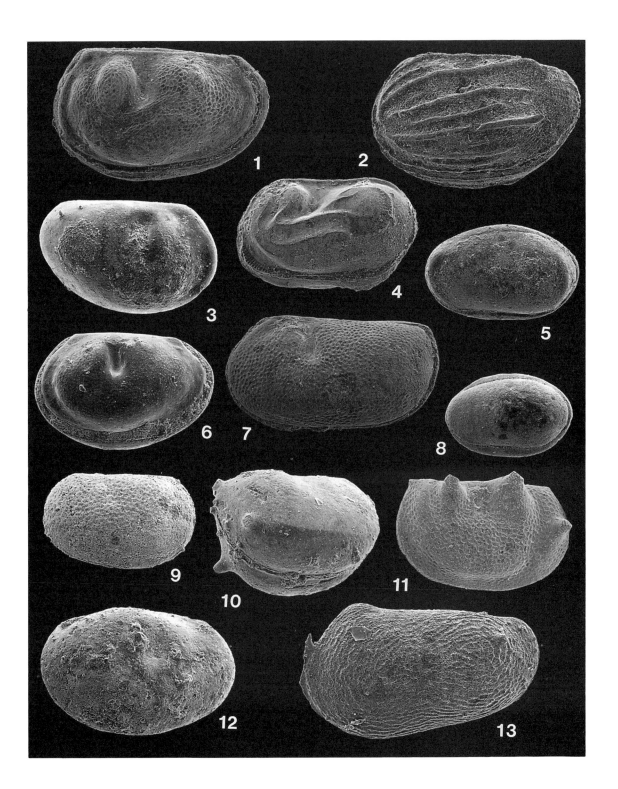

*Rijks geol. Dienst.*, Leiden, N.S., **24**, 21-53.

Cleaves, A. W. & Broussard, M. C. 1980. Chester and Pottsville depositional systems, outcrop and subsurface in the Black Warrior Basin of Mississippi and Alabama. *Trans. Gulf Cst Ass. geol. Socs*, **30**, 49-60.

Cleaves, A. W. & Bat, D. T. 1988. Terrigenous clastic facies distributions and sandstone diagenesis, subsurface Lewis and Evans format units (Chester Series), on the northern shelf of the Black Warrior Basin. *Trans. Gulf Cst Ass. geol. Socs*, **38**, 177-186.

Cooper, C. L. 1941. Chester ostracodes of Illinois. *State Geol. Surv. of Illinois, Rept. of Invest.*, Urbana, Illinois, **77**, 5-101.

Craig, L. C. & Connor, C. W. 1979. Paleotectonic investigations of the Mississippian System in the United States. *U.S. Geol. Surv. Prof. Pap.* **1010**, 559 pp. 15 pls.

Crasquin, S. 1984. L'écozone a Bairdiacea et Paraparchitacea (Ostracoda) au Dinantien. *Géobios*, Lyon, **17**(3), 341-348.

Devery, H. B. 1987. *Analysis of the microfauna, facies variation and stratigraphy of selected outcrops of the Bangor Limestone (Chesterian; Mississippian) in Colbert, Franklin and Lawrence Counties, northwest Alabama.* Unpub. M.Sc. thesis, Mississippi State University, 206 pp.

Dewey, C. P. 1985. The palaeobiogeographic significance of Lower Carboniferous crustaceans (ostracodes and peracarids) from western Newfoundland and central Nova Scotia, Canada. *Palaeogeogr. Palaeoclimat. Palaeoecol.*, Amsterdam, **49**, 175-188.

Dewey, C. P. 1987. Palaeoecology of a hypersaline Carboniferous ostracod fauna. *J. micropalaeont.*, London, **6**, 29-33.

Di Giovanni, M. 1984. *Stratigraphy and deposition of the lower Pride Mountain Formation (Upper Mississippian) in the Colbert County area, Alabama.* Unpub. M.Sc. thesis, University of Alabama, Tuscaloosa, 143 pp.

Haack, R. C. & Kaesler, R. L. 1980. Upper Carboniferous ostracode assemblages from a mixed carbonate-terrigenous-mud environment. *Lethaia*, Oslo, **13**, 147-156.

Handford, C. R. 1978. Monteagle Limestone (Upper Mississippian) - Oolitic tidal bar sedimentation in southern Cumberland Plateau. *Amer. Assoc. Petrol. Geol. Bull.*, Tulsa, Oklahoma, **62**, 644-656.

Holmes, J. W. 1981. *The depositional environment of the Mississippian Lewis Sandstone in the Black Warrior Basin of Alabama.* Unpub. M.Sc. thesis, University of Alabama, Tuscaloosa, 172 pp.

Kaesler, R. L. 1982. Ostracoda as environmental indicators in Late Pennsylvanian subsurface shales. *Third North Amer. Paleont. Conv. Procs*, Boulder, Colorado, **1**, 275-280.

Mack, G. H., Thomas, W. A. & Horsey, C. A. 1983. Composition of Carboniferous sandstones and tectonic framework of southern Appalachian-Ouachita Orogen. *J. sedim. Petrol.*, Menasha, **53**, 931-946.

Maples, C. G. & Archer, A. W. 1986. Shoaling-upward sequences and facies dependent trace fossils in the Monteagle Limestone (Mississippian) of Alabama. *S. East. Geol.*, Duke University, Durham, North Carolina, **27**, 35-43.

Mellen, F. F. 1947. Black Warrior Basin, Alabama and Mississippi. *Amer. Assoc. Petrol. Geol. Bull.*, Tulsa,

Oklahoma, **31**, 1801-1816.

Melnyk, D. H. & Maddocks, R. F. 1988. Ostracode biostratigraphy of the Permo-Carboniferous of central and north-central Texas, Part 1: Paleoenvironmental framework. *Micropaleontology*, New York, **34**, 1-20.

Morris, R. C. 1974. Carboniferous rocks of the Ouachita Mountains, Arkansas: A study of facies patterns along the unstable slope and axis of a flysch trough. *Geol. Soc. Amer., Spec. Pap.*, Boulder, Colorado, **148**, 241-279.

Puckett, T. M. 1987. *Biofacies analysis of ostracodes from the Pride Mountain interval in Colbert, Lawrence and Jefferson Counties, northern Alabama.* Unpub. M.Sc. thesis, Mississippi State University, 202 pp.

Scott, G. L. 1978. Deposition, facies patterns and hydrocarbon potential of Bangor Limestone (Mississippian) northern Black Warrior Basin, Alabama and Mississippi. *In* Moore, W. H. (Ed.), *Mississippian Rocks of the Black Warrior Basin.* 17th Mississippi Geol. Soc. Field Trip, 34-54. Mississippi Geological Society.

Sutherland, P. K. & Manger, W. L. 1979. Comparison of Ozark Shelf and Ouachita Basin facies for Upper Mississippian and Lower Pennsylvanian Series in eastern Oklahoma and western Arkansas. *Oklahoma Geol. Surv. Guidebook* **19**, 81 pp.

Thomas, W. A. 1972. Mississippian stratigraphy of Alabama, *Ala. Geol. Surv. Monogr.* **12**, 121 pp.

Thomas, W. A. 1974. Converging clastic wedges in the Mississippian of Alabama. *Geol. Soc. Amer. Spec. Pub.*, Boulder, Colorado, **148**, 187-207.

Thomas, W. A. 1988. The Black Warrior Basin. *In* Sloss, L. L. (Ed.), *Sedimentary Cover - North American Craton*, Geological Society of America, The Geology of North America, Boulder, Colorado, **D2**, 471-492.

Van Ameron, H. W., Bless, M. J. M. & Winkler Prins, C. F. 1970. Some paleontological and stratigraphical aspects of the Upper Carboniferous Sama Formation (Asturias, Spain). *Meded. Rijks geol. Dienst.*, Leiden, N.S., **21**, 9-79.

Waters, J. A. 1978. *The paleontology and paleoecology of the lower Bangor Limestone (Chesterian, Mississippian) in northwestern Alabama.* Unpub. Ph.D. dissert., Indiana University, 193 pp.

Wickham, J., Roeder, D. & Briggs, G. 1976. Plate tectonic models for the Ouachita foldbelt. *Geology*, Boulder, Colorado, **4**, 173-176.

## DISCUSSION

Ken McKenzie: Do you think that the carapaces which proved empty of sediment when broken open are indicative of rapid burial?

Chris Dewey: No, if rapid sedimentation had been the case, we would not expect to see the diverse filter feeding macrofauna.

Ken McKenzie: Was there any associated pyritization?

Chris Dewey: Not in these particular sediments.

# 42

# Pandemic ostracod communities in the Tethyan Triassic

**Edith Kristan -Tollmann**

A-1180 Wien, Scheibenbergstrasse 53, Austria.

## ABSTRACT

By means of some examples of ostracod faunas, and of some characteristic ostracod species, it can be demonstrated that surprisingly, many benthonic ostracods of the Triassic were widespread throughout the Tethyan realm. It was also surprising to find that some of the Triassic species have long stratigraphical ranges. Taxa which extended through the whole of the Triassic, from the Werfenian (Scythian) or Anisian to the end of the Rhaetian, are now known.

## INTRODUCTION

It has been recently demonstrated in a series of papers that many groups of Triassic macro- and micro-organisms are surprisingly widespread throughout the Tethyan realm (Kristan-Tollmann, 1986; 1987; 1988a; 1988b; 1988c; Kristan-Tollmann & Tollmann, 1983; Tollmann & Kristan-Tollmann, 1985). It is now possible to illustrate some examples from the Triassic ostracod faunas of the Tethys. Until recently, very few complete Triassic ostracod faunas had been analysed from all the different parts of this vast area, some 13000km long. In this first survey, two facts can be recognized:

1) Throughout the entire area the same ostracod faunal associations are found in the same palaeoenvironments. The percentage of similar ostracod species in such faunas from distant sites can be very high; up to 100%. A smaller percentage of identical species is generally a consequence of a change in environmental factors. Ostracod faunas, hitherto classified as 'endemic', will surely continue to be appraised as such as long as faunas of analogous ecological sites in the Tethyan realm remain unknown (for examples see below).

2) Apart from complete ostracod faunas, it is apparent that some Triassic species achieved not only a very far reaching regional distribution, but also a very large stratigraphical range and facies distribution (for examples see below).

## EXAMPLES OF TRIASSIC, PAN-TETHYAN OSTRACODS

### Ostracod Associations

*Rhaetian ostracods from the Zlambach Marls of Central Iran.* The first example is the rich ostracod fauna of the Rhaetian Marls from the 'Salt Spring' at Bagerabad, 60km northeast of Isfahan, Iran (Kristan-Tollmann *et al.*, 1980). This fauna comprises 25 species, of which 21 were already known from the Tethyan Triassic of the Eastern Alps, and only 4 species were new. In fact, one of these new species from Iran, *Mostlerella dizluense* (Kristan-Tollmann), has been found in three localities in the Northern Calcareous Alps of Austria (Kristan-Tollmann, 1982). The ostracod fauna from the 'Salt Spring' locality are from a near-reef facies, demonstrated not only by the characteristics of the fauna, but also by the intercalation of small patch reefs between the marls exposed in the field, and by the nature of the other micro- and macrofauna. A comparison of this fauna with that from the Plackles summit of the Hohe Wand Mountains in Lower Austria, also supports the interpretation of its being in close proximity to a reef environment. Characteristic species from both localities are:

*Hiatobairdia subsymmetrica* Kristan -Tollmann
*Hiatobairdia labrifera* Kristan -Tollmann
*Carinobairdia triassica* Kollmann
*Carinobairdia alpina* Kollmann
*Dicerobairdia bicornuta kollmanni* Kristan -Tollmann
*Kerocythere (K. ) hartmanni* (Bolz & Kozur)
*Kerocythere (Rekocythere ) mostleri* (Bolz & Kozur)
*Mostlerella dizluense* (Kristan -Tollmann).

Only the common species *Cytherella acuta* Urlichs and the rare *Triceratina fortenodosa* (Urlichs) had been originally described from the Rhaetian Kössen Beds of Kössen in the Tyrol, Austrian Alps. Photographic illustrations of the most important ostracods from the 'Salt Spring' locality in Iran have been given recently by Kristan-Tollmann (1988)b.

*Rhaetian ostracods from the Kioto Marls, Kumaun, Indian Himalayas, India.* A small ostracod fauna from the Rhaetian Kioto Marls from Kumaun in the Indian Himalayas, provides a good example of an equivalent to the Zlambach Marls, type 'Zwicshenkögel' in the Salzkammergut of Northern Austria. This fauna represents a transition, from the marls of an open marine basin, to the intercalated Hallstatt Limestone. This fauna from the Kioto Marls is totally dominated by representatives of the Hungarellinae (Kristan-Tollmann & Gupta, 1987):

*Triadohealdia trigonia* Kristan -Tollmann
*Torohealdia opisthocostata* Kristan -Tollmann

*Torohealdia tuberosa* Kristan -Tollmann
*Signohealdia robusta* Kristan -Tollmann.

All these species were originally described from the westernmost sector of the Tethys. It is also worthy of note, that all the other elements of the fauna of the Kioto Marls, i.e., foraminifera, holothurian sclerites and both attached and planktonic crinoids, are identical to those of the Zlambach Marls. These marly facies of the basin are intercalated at the margins of the uppermost part of the Hallstatt Limestone and are exposed in many parts of the Salzkammergut in the Northern Calcareous Alps of Austria.

*Upper Triassic ostracods from the shelf off Northwestern Australia.* In the Sahul Shoals No. 1 Borehole on the shelf off Northwestern Australia, south of Timor, marine Triassic was reached at 1880-1890m. The ostracod fauna from this site consists mainly of species already known from the Upper Triassic of the westernmost Tethys (Kristan-Tollmann, 1986). The most common species is *Cytherella acuta* Urlichs, first described from the Rhaetian Kössen Marls of Kössen in the Austrian Tyrol and later found as a common taxon in the Rhaetian of the 'Salt Spring' fauna near Bagerabad, Iran (see above and Kristan-Tollmann *et al.*, 1980). Other characteristic Tethyan ostracods from this well are: *Nodobairdia mammilata* Kollmann and *Tethyscythere austriaca* (Kozur & Bolz).

## Some widespread ostracod species

*Judahella (Judahella) tuberculifera* (Gümbel)

This taxon provides a very impressive example of a species widely distributed in a regional sense, as well as stratigraphically, and in various facies. It is found in the Alpine Tethyan and also in the Germanic Epicontinental Sea, ranging from the Upper Werfenian to the Rhaetian and it can be traced through the Tethys from Europe to China (Kristan-Tollmann, 1986).

*Judahella (Judahella ) andrusovi* Kozur & Bolz

*J. (J.) andrusovi* like *J. (J.) tuberculifera* belongs to those Triassic ostracods which show the widest stratigraphical and regional distribution. It represents one of the few Triassic ostracods to occur in both the Alpine and the Germanic-Sephardic facies region during the Middle and Upper Triassic. *J. (J.) andrusovi* was first described from the Kössen Beds (Rhaetian) outcropping in the Ampel Brook (Ampelsbach) north of Aachenwald in the Austrian Tyrol. It is also known from the Kössen Beds of Weisslofer Brook, near Kössen in the Tyrol. The author has found it at three sites in the Zlambach Marls (Rhaetian) from Austria: Stambach near Bad Goisern, Salzkammergut, Upper Austria; Grünbachgraben on the eastern side of the Untersberg, Salzburg; Plackles Peak on the Hohe Wand near Wiener Neustadt, Lower Austria. Additionally it has been recorded from the Rhaetian marls from the Salt Well section near Dizlu, northeast of Bagerabad, Central Iran (Kristan-Tollmann *et al.*, 1980) and I have specimens from the Carnian (Raibl Beds) of the Zsámbék 14 Borehole in the Transdanubian Mountains, Hungary. Basha (1982) noted its occurrence in the Carnian Hisban Limestone from Wadi Hisban in the Jordan Valley west of Amman, Jordan. Finally this speices is present in the Leidapo section, 30km south of Guiyang in Southern China. This series, consisting of an alternation of limestone and marls, belongs despite its Cassian Beds facies type, to the Upper Anisian Leidapo Subformation of the Anisian Qingyan Formation.

*Gruendelicythere ampelsbachensis* Kozur & Bolz

This taxon is certainly restricted to the Tethys and is found from Western Europe to Southern China in different formations of Upper Anisian and Lower Carnian to Rhaetian age.

*Nodobairdia mammilata* Kollmann

Similarly, this species, which was first described from the Lower Carnian of the Southern Tyrol of Italy, was later recorded from the Rhaetian Zlambach Marls of the Northern Calcareous Alps, Austria; from the 'Salt Spring' near Bagerabad, Iran and from the Upper Triassic of the Sahul Shoals No. 1 well, off Northwestern Australia (Kristan-Tollmann, 1986). Recently, the author has noted this species in faunas from the Reifling Limestone of Anisian age from the Schneeberg in the Northern Calcareous Alps, Austria and the Raibl Beds of Carnian age from the Zsámbék 14 Borehole in the Transdanubian Mountains, Hungary.

*Mostlerella nodosa* Kozur

From our current knowledge of the species, *M. nodosa* seems to be restricted to the Carnian stage. However, its regional distribution is wide and also its occurrence in many different facies regions is striking. At present it is known from the following: Upper Carnian - marls from Veszprem Quarry near the slaughter house, Hungary; Lower to Middle Carnian - Opponitz Beds from the Stiegengraben near Göstling/Ybbs, Lower Austria; Veszprem football field, Hungary; the Raibl Beds of the Zsámbék 14 Borehole of the Transdanubian Mountains, Hungary; the Cassian Marls of Pralongia (Ruones meadows), southeast of Corvara, Italy; the Hisban Limestone Formation of the Wadi Hisban of the Jordan Valley near Amman, Jordan and the Mohilla Formation of the Devora 2A Borehole in Northern Israel (Gerry *et al.*, this volume).

*Ptychobairdia kuepperi* Kollmann

The pan-Tethyan distribution of this species is demonstrated for the first time. Hitherto, *P. kuepperi* has been known from many sites in the Zlambach Marls of the Eastern Alps, Austria, with a stratigraphical range from Sevatian (Upper Triassic) to Lower Liassic. Its occurrence in Sevatian and Liassic formations from Timor, can now be documented. The sites of the outcrops of these beds are located along the Meto River, south of Soë, in the central part of Western Timor. At this locality, the Sevatian is represented by the Pötschen Limestone (formally Aitutu Formation) which contains many typically alpine taxa, for example the commonly occurring species:

*Ptychobairdia kuepperi* Kollmann
*Anisobairdia cincta* Kollmann
*Anisobairdia gibba* Kristan -Tollmann
*Anisobairdia? fastigata* (Bolz)
*Triadohealdia alexandri* Kristan -Tollmann
*Triadohealdia opisotruncata* Kristan -Tollmann
*Torohealdia amphicrassa* Kristan -Tollmann
*Torohealdia opisthocostata* Kristan -Tollmann etc.

Some of these species exhibit some minor differences in comparison with the type material from the Eastern Alps. A detailed study will show to what extent a separation into subspecies may be necessary.

## ACKNOWLEDGEMENTS

This study was done under the support of the IGCP-Projects 203 and 272. Furthermore I would like to express my best thanks to the two referees and to one of them, for correcting the English of the manuscript.

## REFERENCES

Basha, S. H. S. 1982. Microfauna from the Triassic Rocks of Jordan. *Revue Micropaléont.*, Paris, **25**, 3-11.

Kristan-Tollmann, E. 1982. Bemerkungen zur triadischen Ostracoden-Gattung *Mostlerella.Neues.Jb.Geol.Paläont.*, Monatshefte, 1982, 560-572.

Kristan-Tollmann, E. 1986. Observations on the Triassic of the southeastern margin of the Tethys - Papua/New Guinea, Australia and New Zealand. *Neues. Jb. Geol. Paläont.* Monatshefte, 1986, 201-222.

Kristan-Tollmann, E. 1987. Triassic of the Tethys and its relations with the Triassic of the Pacific Realm. *In* McKenzie, K.G. (Ed.), *Shallow Tethys 2. Proceedings of the International Symposium on Shallow Tethys 2, Wagga Wagga* , 169-186. A. A. Balkema, Rotterdam, Boston.

Kristan-Tollmann, E. 1988a. A comparison of late Triassic agglutinated foraminifera of western and eastern Tethys. *In* Gradstein, F. M. & Rögl, F. (Eds), Second Workshop on Agglutinated Foraminifera. *Abh.Geol.B -A.*, Wien, **41**, 245-253.

Kristan-Tollmann, E. 1988b. Unexpected microfaunal communities within the Triassic Tethys. *In* Audley-Charles, M. G. & Hallam, A. (Eds), *Gondwana and Tethys. Spec. Publs geol. Soc.*, London, **37**, 213-223.

Kristan-Tollmann, E. 1988c. Unexpected communities among the crinoids within the Triassic Tethys and Panthalassa. *In* Burke, R. D., Mladenov, P. V., Lambert, P. & Parsley, R. L. (Eds), *Echinoderm Biology*. Proc. Sixth Int. Echinoderm Conf., 133-142. A. A. Balkema, Rotterdam.

Kristan-Tollmann, E. & Gupta, V. J. 1987. Remarks on the microfauna of the Rhaetian Kioto-Marls from Kumaun, Himalaya.*Neues.Jb.Geol.Paläont.*,Monatshefte,467-492.

Kristan-Tollmann, E. & Tollmann, A. 1983. Überregionale Züge der Tethys in Schichtfolge und Fauna am Beispiel der Trias zwischen Europa und Fernost, speziell China. *SchrReihe erdwiss. Komm. österr. Akad. Wiss.*, Wien, **5**, 177-230.

Kristan-Tollmann, E., Tollmann, A. & Hamedani, A. 1980. Beiträge zur Kenntnis der Trias von Persien II. *Mitt. öst. geol. Ges.*, Wien, **73**, 163-235.

Tollmann, A. & Kristan-Tollmann, E. 1985. Paleogeography of the European Tethys from Paleozoic to Mesozoic and the Triassic Relations of the Eastern Part of Tethys and Panthalassa. *In* Nakazawa, K. & Dickins, J. M. (Eds), *The Tethys*, 3-22. Tokai Univ. Press.

## DISCUSSION

Avraham Honigstein: Are you able to show an example of a common ostracod species occurring in different stratigraphical levels, but from the same palaeoenvironmental area?

Edith Kristan-Tollmann: There are a lot of examples of common ostracods which have wide stratigraphical ranges within the same palaeoenvironment in the Triassic, e.g., we find *Judahella tuberculifera* , *J. andrusovi* and *Gruendelicythere ampelsbachensis* in the same palaeoenvironment (Cassian facies) from the Upper Anisian to the Carnian and *J. tuberculifera* occurs in a facies of comparable type (the Zlambach Marls) which ranges from the Upper Anisian to the Rhaetian.

Kenneth McKenzie: You showed slides of the same species occurring on both the northern flank of Tethys and in Timor and North West Australia. Does this mean that North West Australia was relatively near to the northern shore of the Tethys in the Triassic?

Edith Kristan-Tollmann: According to our present knowledge there is no difference in the ostracod faunas of the northern and southern border zone of the Tethys.

Dick Van Harten: Is there in your opinion a general relation between the geographical distribution of species and their duration in time?

Edith Kristan-Tollmann: Despite our present limited knowledge of Triassic ostracods within the Tethyan area, one can state that there are species which are pan-Tethyan and have long stratigraphical ranges, while others, with the same wide palaeozoogeographical distribution, have relatively short stratigraphical ranges. In relation to endemic ostracod faunas in the Tethyan area; I believe that there will be very few found, after more detailed studies of this vast region have been undertaken. Today, therefore, a statement on the stratigraphical ranges of possible endemic faunas and species cannot be given.

# 43

# Phylogeny and historical biogeography of the Megalocypridinae Rome, 1965; with an updated checklist of the subfamily

**Koen Martens[1] & August Coomans[2]**

[1]Koninklijk Belgisch Instituut voor Natuurwetenschappen, Hydrobiology, Vautierstraat 29, 1040 Brussels, Belgium. [2]Rijksuniversiteit Gent, Laboratoria voor Morfologie en systematiek der Dieren, K.L. Ledeganckstraat 35, 9000 Gent, Belgium.

## ABSTRACT

Nineteen soft part and valve morphology characters of the different genera of the Megalocypridinae were analysed using outgroup comparison and the ontogenetic method, and an attempt was made to reconstruct the phylogeny of the subfamily. However, it proved impossible to decide upon an absolute polarization within the group and two alternative hypotheses are proposed. Some comments on the historical zoogeography of the subfamily are also offered. It is thought that the Megalocypridinae originated as a distinct group in the Upper Cretaceous-Lower Palaeogene and that most of the generic evolution occurred in East Africa.

## INTRODUCTION

Rome (1965a, 1965b) erected the subfamily Megalocypridinae to include the genera *Megalocypris* Sars, 1898, *Hypselecypris* Rome, 1965 and *Apatelecypris* Rome, 1965. Martens (1986) transferred *Sclerocypris* Sars, 1924 (syn.: *Bharatcypris* Battish, 1978) to the Megalocypridinae and erected 2 new genera: *Madagascarcypris* and *Eundacypris*. Martens (1987a, 1987b, 1987c, 1988) described new species and discussed aspects of the morphology of the genus *Sclerocypris*. Finally, Wouters *et al.* (in press) transferred *Tanganyikacypris* Kiss, 1961 to the Megalocypridinae, but retained Tanganyikacypridini DeDeckker & Wouters, 1983 as a tribe of the subfamily. A second genus of the

Tanganyikacypridini from Lake Tanganyika was also reported, but left in open nomenclature by Wouters *et al.* (in press).

The present contribution offers a cladistic approach to the classification of the Megalocypridinae. For this we rely on a set of morphological characters, most of which have been extensively described and illustrated in Martens (1986, 1988).

## PHYLOGENY

### Introduction

We have chosen the cladistic approach of Hennig (Wiley 1981, Von Vaupel Klein 1984) to analyse the generic phylogeny of the Megalocypridinae. Using this method, genealogical relationships are traced by *character-analysis*, which consists mainly of two steps: (1) to distinguish between homologies (same origin) and homeoplasies (same appearance but independent origin), and (2) to polarize the available alternative character states, i.e. decide which character state is most primitive (plesiomorphic) and which is relatively advanced or derived (apomorphic). Different methods exist to distinguish between apomorphic and plesiomorphic character states: (1) *outgroup comparison* (see below), (2) *ontogeny* (see below), (3) *geological age* (absolute and relative age of the occurrence of an apomorphic character state), (4) *progression rule* (relying on biogeography), (5) *commonality principle* (the most widely occurring character state in a certain group is the most plesiomorphic).

The most frequently used methods are the outgroup comparison and the ontogenetic method. The commonality principle lacks a theoretical justification and is, therefore, used less and less frequently.

### Outgroup comparison

Sets of characters are studied and compared between the taxonomic group (X), for which the internal phylogeny is to be established, and an outgroup (Y). Although mostly the adelphotaxon (= the group with which X shares a common ancestor) is chosen as an outgroup, in theory any

related group can be chosen (Wiley, 1981). Subsequently, a list of characters common to both groups X and Y is established. For each character, the different character states (A, A', A",...) in the various taxa of the groups X and Y are monitored. For example, the character 'valve ornamentation' has the different character states 'smooth', 'pitted', 'spiny', etc.

There are 5 possible situations which can occur in an outgroup comparison:

1) the character state A occurs in all taxonomic ingroups (TIG) and in all taxonomic outgroups (TOG): A is not relevant for the present discussion.

2) A occurs in one or more TIG and in all TOG: A is a plesiomorphic character state.

3) A occurs in a few TIG, the alternative A' in some TIG and in all TOG: A is an apomorphic character state.

4) A occurs in all TIG and in no TOG: A is a synapomorphic character state for the group X, but is irrelevant for the Linnean hierarchy in group X.

5) A occurs in one TIG and in no TOG: A is a synapomorphy for that particular TIG.

Especially characters of categories 2) and 3) are used in character analyses.

### Ontogeny

This method largely relies on the concept of recapitulation: the ontogeny reflects the phylogeny. When used with some care, it can yield important results. The following (simplified) rule is applied: character states occurring in later juvenile stages are more apomorphic than those found in earlier juvenile stages.

To discuss the advantages and difficulties experienced with both techniques is beyond the scope of the present chapter. For this, we refer to the comprehensive review in Wiley (1981).

### Transformation series

Characters can have more than two different character states. In such cases, a transformation series or *morphocline* exists, e.g., the degree of inward

Table 1. Characters and character states used for the cladistic analysis of the genera of Megalocypridinae. Characters which are discussed in the text are marked with an *. A = apomorphy, P = plesiomorphy.

| No. | Character | Character State 1 | Character State 2 |
|-----|-----------|-------------------|-------------------|
| 1 | pattern of seminal tubes | type 1 | type 2 |
| 2* | penultimate segment of T1 | divided (P) | fused (A) |
| 3* | Müller-organ | present | absent |
| 4 | tooth bristles on 3rd endite of Mx1 | smooth | toothed |
| 5* | flagellated claws on A1 | absent (P) | present (A) |
| 6 | 1 extra spine on 3rd end. of Mx1 | absent (P) | present (A) |
| 7 | number of setae on 3rd segm. of T1 | 1 (P) | 2 (A) |
| 8* | inward displacement of ant. selvage | (see discussion) | |
| 9 | caudal part of ovarium reflexed | dors. (A) | ventr. (P) |
| 10 | inward displacement of ventr. selvage of RV | weak (P) | strong (A) |
| 11 | shape of terminal palp segment of Mx1 | dilated (P) | curved (A) |
| 12 | suppl. branches of furcal attachment | absent (P) | present (A) |
| 13 | caudal inner list on LV | smooth (P) | striated(A) |
| 14 | caudal inner margin on both valves | normal (P) | sinuous(A) |
| 15 | number of lobes in medial shield of hemipenis | 1 (P) | 2 (A) |
| 16 | labyrinth in hemipenis | normal (P) | very (A) elongated |
| 17 | apical armature on term. segm. of A2 | present (P) | absent (A) |
| 18* | valve margins in adults | denticul.(P) | smooth (A) |
| 19* | prominent inner list on RV | present (P) | absent (A) |

displacement of the selvage in the Megalocypridinae (see below). Such a series comprises various degrees of primitive and advanced character states.

## PHYLOGENY OF THE MEGALOCYPRIDINAE

### Outgroups

Two genera which are morphologically closely related to the Megalocypridinae are *Cypris* O. F. Müller, 1776 and *Chlamydotheca* Saussure, 1858. To use the latter genus as an outgroup, however, complicates matters as the taxonomy of *Chlamydotheca* is in urgent need of revision. In our opinion, at least the species of the *C. unispinosa* -group belong in a separate genus. The present contribution, however, is not the place to erect new taxa and we will, therefore, refer to both *Chlamydotheca*

s.s. and the *C. unispinosa* -group as *Chlamydotheca* s.l. Notwithstanding these difficulties, *Cypris* and *Chlamydotheca* s.l. are used as outgroups as demonstrated below.

### Character analysis

Nineteen morphological characters, both carapace and soft part, are used for the present analysis; they are listed in Table 1. The polarization of some of these characters is discussed below.

*Character 2: penultimate segment of T1 divided or not*
In Cypridinae, this segment is not divided, while both other groups (*Chlamydotheca* s.l. and Megalocypridinae) have two separate segments. In this case, we can use two criteria to determine which character state is plesiomorphic.
1) All juvenile stages of both *Chlamydotheca*

s.l. and the Megalocypridinae have this segment divided.

2) Segment 4a carries an apical seta in the latter two groups. This seta is still present in the same place in the Cypridinae, i.e. halfway along the fused segment 4a+b. This is a morphological relict of an ancient separation.

According to these criteria, the fused segment is the apomorphic condition.

*Character 3: Müller-organ*
Martens (1986, 1987a) mentioned the presence of bladder-like organs between the A1 and the A2 in all Megalocypridinae and proposed to call it Müller-organs, after G.W. Müller who was the first to describe this organ. Müller-organs are found in all groups of *Chlamydotheca* s.l. and in all Megalocypridinae, but are absent in *Cypris*. They are also present in some Eucypridinae, but little is known about their occurrence in other Cyprididae. The presence or absence, morphology and function of this organ could prove to be of great significance in a phylogenetic analysis of the Cyprididae as a whole, but no polarization is possible to date.

*Character 5: flagellated claws on A1*
This character is the main synapomorphy uniting all Megalocypridinae, for the following reasons:
1) These structures do not occur in any other group of the Cyprididae.
2) They occur in the final 3 instar stages and in the adult of the Megalocypridinae, but are absent in all earlier juveniles, where normal, short setae occur on the A1.
3) No juvenile stage of *Chlamydotheca* s.l. has flagellated claws.

This clearly indicates that this character is an apomorphy within the Megalocypridinae and not a plesiomorphic character which has become secondarily reduced in *Chlamydotheca* s.l..

*Character 8: degree of inward displacement of selvage*
This is the most important transformation series in the present analysis and we will, therefore, discuss all arguments in full. This character shows 4

different states in the Megalocypridinae. They will here be defined by the position of the anterior selvage of the right valve (RV).

1) Inward displacement over more than half the distance between valve margin (vm) and inner margin (im).
2) Selvage inwardly displaced over *circa* 1/3 of this distance.
3) Selvage scarcely displaced (less than 1/20 of the said distance).
4) No anterior selvage.

All groups in *Chlamydotheca* s.l. and *Cypris* have a strongly inwardly displaced selvage (8-1).

The problem, then, is to determine which of these character states is the plesiomorphic and which the most apomorphic. There are alternative hypotheses, both accepting the fact that the selvage is an ancient structure, which has disappeared and re-appeared several times, but which has been present through most of the phylogenetic history of the Cyprididae, at least up to level 00 (Figs 1 and 2).

The first hypothesis accepts that a selvage was present in the entire phylogenesis of the taxa here considered and that the direct ancestor of the Megalocypridinae and *Chlamydotheca* s.l. had a prominent selvage on both valves. This character was then retained in the *Chlamydotheca*-line, but the selvage became gradually reduced in the Megalocypridinae. In this scheme, *Sclerocypris* would be an example of the group with the most plesiomorphic character state and *Eundacypris* would display the most apomorphic condition (Fig. 1).

The second hypothesis assumes a reduction of inward displacement of the selvage between level 00 and 0 (Fig. 2), suggesting that the direct ancestor of *Chlamydotheca* s.l. and the Megalocypridinae lacked a selvage. A subsequent re-introduction of this structure is then postulated to have occurred in both groups, *simultaneously* and *independent* of one another. According to this hypothesis, *Eundacypris* and *Apatelecypris* would have the more plesiomorphic character state, while *Sclerocypris* would represent the completely apomorphic condition.

To decide between the two hypotheses, outgroup comparison offers little help. Both are equally

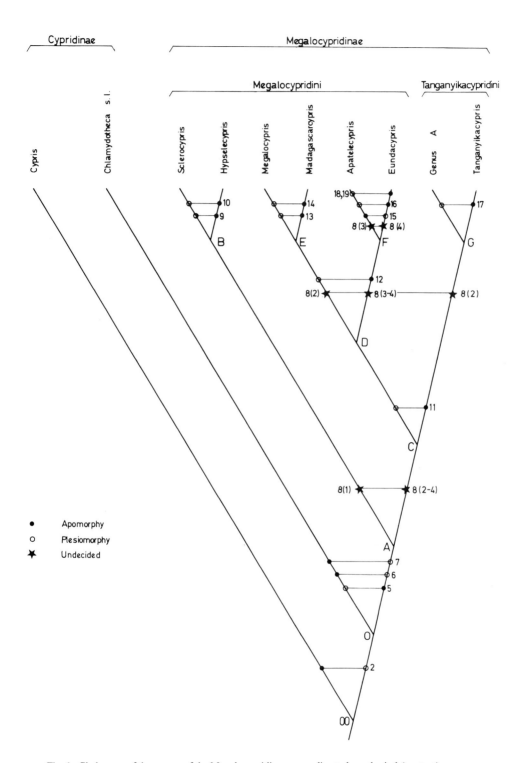

Fig. 1. Cladogram of the genera of the Megalocypridinae, according to hypothesis 1 (see text).
Characters used are numbered as in Table 1.

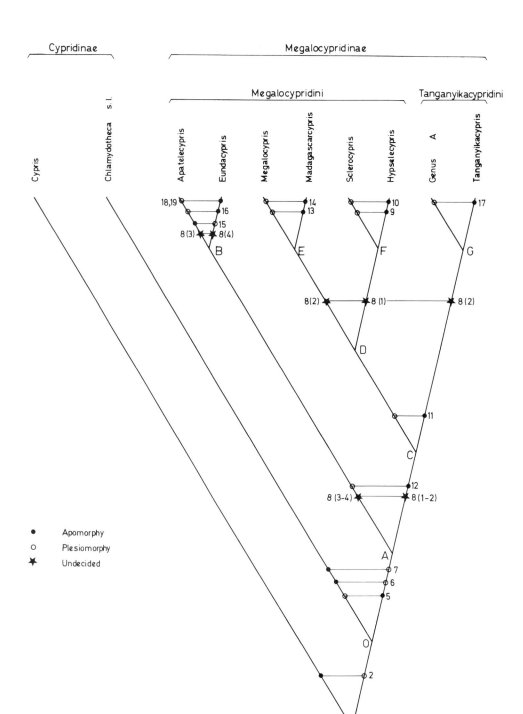

Fig. 2. Cladogram of the genera of the Megalocypridinae, according to hypothesis 2 (see text).
Characters used are numbered as in Table 1.

acceptable, except when the principle of greatest parsimony is applied. The first hypothesis indeed requires two steps less (a reduction and a simultaneous re-introduction in the two groups) and is the more parsimonious.

Ontogeny is also of little help. Juvenile stages of all three groups (Megalocypridinae and the two outgroups) have very similar marginal carapace structures, with selvages hardly or not at all displaced and with large inner lists on a relatively narrow calcified part of the inner lamella on both valves. If this juvenile morphology represents a plesiomorphic condition, and not a supplementary adaptation of the juvenile (a so-called juvenile apomorphy), then this plesiomorphic character state goes back at least to the common ancestor of the three taxa under consideration and is, therefore, of little use in the present discussion.

None of the two major methods can thus be used to polarize the present transformation series. Although we are able to deduce a relative arrangement of the genera involved, we will have to look for other features to attempt construction of the absolute polarity.

*Characters 18 & 19*
Both features in adult *Apatelecypris* (the crenulated ventro-caudal valve margin and the large inner list on the RV) also occur in nearly all Megalocypridinid instars. They are, therefore, plesiomorphic character states. This indicates that *Apatelecypris* is either an ancient genus (hypothesis 2) or is paedomorphic (hypothesis 1).

**The fossil record**

Fossils of Megalocypridinae s.s. are rare. The only specimens reported belong to *Sclerocypris* and are known only from Quaternary deposits (*Sclerocypris* cf. *clavularis* from Lake Turkana (Carbonel & Peypouquet 1979 and Cohen *et al.*, 1983); *S. jenkinae* from Lake Elmenteita (Cohen & Nielsen, 1986); *S. bicornis* as *Chlamydotheca* spec. from Egypt (Boukhari & Guernet, 1985). No other fossil Megalocypridinae or Megalocypridinae-like animals have, to the best of our knowledge, been reported from Africa (Grekoff 1957, 1958, 1960,

among others). Thus, the palaeontological record is of little help in the present analysis.

**Discussion**

The character analysis does not allow one to decide upon the absolute polarity of the genera in the subfamily. We therefore present both major possibilities in two cladograms (Figs 1 and 2). Neither of these can immediately be excluded from the discussion and there are arguments *pro* and *contra* for both.

The first hypothesis (Fig. 1), with *Sclerocypris* being the oldest genus, is corroborated by the following arguments:

1) The more parsimonious hypothesis with regard to character 8 (selvage) can be accepted.

2) The old (but not always accepted) axiom that the most widespread and radiated (also outside Africa) taxon (*Sclerocypris*) is also the oldest, is applicable.

However, if we accept this scheme, then we must consider *Apatelecypris* paedomorphic. This is not so if the second hypothesis (Fig.2), with *Sclerocypris* and *Hypselecypris* being the most recently developed taxa, is accepted. Then the two plesiomorphic character states (18 and 19) could indeed indicate that *Apatelecypris* is an ancient genus.

**ZOOGEOGRAPHY**

Representatives of the Megalocypridinae have so far been reported from Africa (south of the Tropic of Cancer) and India. Most species are known from South (mainly the South-West) and East Africa.

*Sclerocypris* occurs on most of the continent (excluding Central and West Africa and the Cape Province); *Hypselecypris* is restricted to eastern Zaire; *Madagascarcypris* seems to be endemic to Madagascar; *Megalocypris* species are confined to the Cape Province of South Africa, *Apatelecypris* and *Eundacypris* occur in Namibia, while all Tanganyikacypridini are endemic to Lake Tanganyika. The distribution of these genera in Africa is shown in Fig. 4.

It should be noted that 1/3 of the African

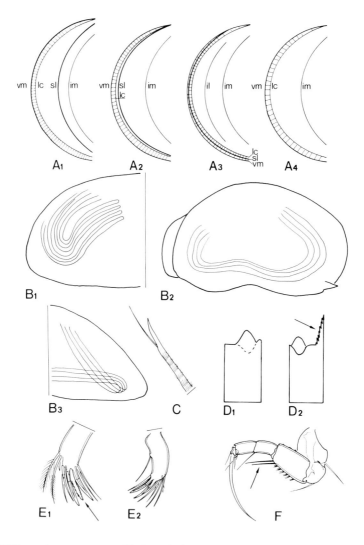

Fig. 3. Illustration of different character states used in the analysis.

A. Schematic representation of the four character states of the inward displacement of the anterior selvage in the RV (ch. 8).

A1. sl displaced over more than 1/2 the distance between vm and im (*Sclerocypris* and *Hypselecypris*).

A2. sl displaced over c. 1/3 of the said distance (*Megalocypris*, *Madagascarcypris* and Tanganyikacypridini).

A3. sl scarcely inwardly displaced (*Apatelecypris*).

A4. no anterior sl (*Eundacypris*). (il= inner list. im= inner margin. lc= line of concrescence. sl= selvage. vm= valve margin.)

B. Schematic representation of patterns of seminal tubes.

B1. type 1, with lower branches extending dorsally of the central adductor muscles (Megalocypridinae).

B2. type 2a, with lower branches extending ventrally of the central adductor muscles (*Cypris latissima*, after G. W. Müller, 1900).

B3. type 2b (*Pseudocypris acuta*), after G. W. Müller, 1914).

C. Flagellated claw on A1 (Megalocypridinae).

D. Schematic representation of the terminal segment of A2 in males of Megalocypridinae.

D1. without apical armature (*Tanganyikacypris*).

D2. with apical armature (other Megalocypridinae)

E. Mx1, third endite.

E1. *Chlamydotheca* spec., showing extra spine on third endite.

E2. Megalocypridinae and juvenile *Chlamydotheca*, without the extra spine.

F. T1 of adult *Chlamydotheca* spec., showing 2 apical setae on the third segment.

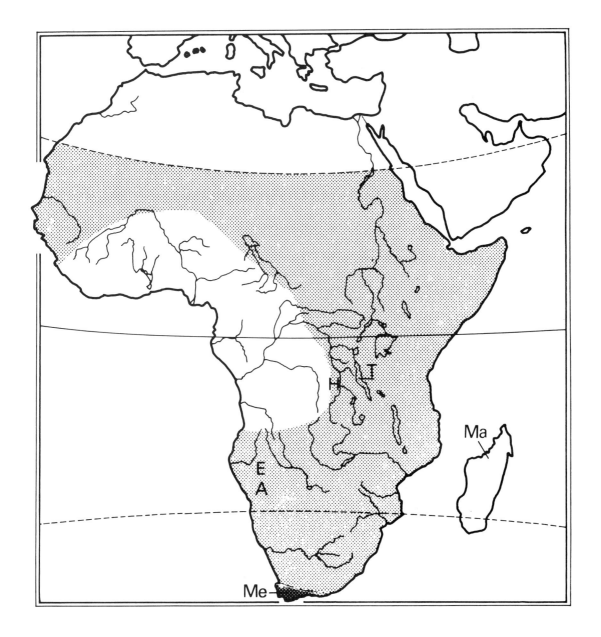

Fig. 4. Distribution of the Megalocypridinae in Africa. Rastered area = distribution of the genus *Sclerocypris* . H = *Hypselecypris*. A = *Apatelecypris* . E = *Eundacypris* . Ma = *Madagascarcypris* . Me = *Megalocypris* . T = Tanganyikacypridini.

species have been only reported from their type locality. Furthermore, because the phylogeny of this group remains unresolved, little can be said with certainty about its historical zoogeography. However, the following statements can be put forward:

1) The quest for the ancestor of the Megalocypridinae depends largely on the Linnean hierarchy of the genera within this subfamily. If *Sclerocypris* is taken to be the oldest genus, then a Gondwana-taxon is indicated as the ancestor from which probably also *Chlamydotheca* s.l. emerged. Should *Apatelecypris* be more closely related to the ancestral form, then we are looking for a completely different ancestor.

2) Whichever hypothesis is preferred, the Megalocypridinae s.s. probably did not evolve as such until after the separation of South America and Africa, *circa* 85 MY BP. The presence of a true megalocypridinid in South America would, of course, immediately nullify this hypothesis.

3) Again, independent from which hierarchy is preferred, there are a number of reasons to believe that most of the generic evolution occurred in East Africa. In spite of the fact that *Megalocypris* is to date restricted to the Cape Province, it is most likely that this genus also originated in or near the Rift. For this, we rely on the sole assumption that all Megalocypridinae with a type 2) selvage have a monophyletic origin.

a) Tanganyikacypridini are to be regarded as East African relicts of this phyletic line.

b) *Madagascarcypris* probably speciated through vicariance when Madagascar drifted to its present position some 65 MY BP. There are three possible positions for this island in the Cretaceous, but the one immediately next to East Africa is most favoured to date (Bosellini, 1986; Thyson, 1986; Wild, 1975; Owen, 1983).

c) The ancestor to the phyletic line with the type (2) selvage, evolved from a *Sclerocypris*-like form (hypothesis 1) or *vice versa* (hypothesis 2). Either way, the missing link between both phyletic lines could be represented by the Somalian *Sclerocypris pardii* Martens, 1987, which combines aspects of both *Megalocypris* and *Sclerocypris*.

The southward migration of *Megalocypris* could be correlated with the northward drift of the African continent and the corresponding shifts of climate belts. Another, perhaps more plausible, hypothesis is that the areal distribution of *Megalocypris* expanded through most of the southern part of the continent and that the present day distribution is but a relict of this former extent. Either way, fossil evidence (or even living relict populations) should then be found in the region between East Africa and the Cape. Furthermore, *Megalocypris* is not the only Cape element to have its supposed origin in East Africa. Various examples from phytogeography have also been reported (Taylor, 1978).

4) *Sclerocypris* species in India could have been present when the Indian subcontinent left Africa (*circa* 65 MY BP) or may have been introduced more recently *via* Arabia, Iran, etc.. However, no fossil or recent evidence of this is available. Both dispersions are possible if *Sclerocypris* is an ancient taxon (hypothesis 1), but if *Sclerocypris* is a more recent genus (hypothesis 2), then the first possibility (the 'Noah's Ark' principle of the Indian subcontinent) should most likely be excluded.

## CONCLUSIONS

Despite combining the data from the character analysis (based on outgroup comparison and the ontogenetic method) with the palaeontological record and present day distributions, it still proved impossible to decide upon a Linnean hierarchy of genera within the Megalocypridinae. Evidence from the different disciplines is indeed contradictory to some extent.

## ACKNOWLEDGEMENTS

The basic research for the present paper was conducted while one of us (KM) was research assistant with the National Fund for Scientific Research. Dr F. Fiers and Dr K. Wouters (Brussels) offered valuable comments on earlier versions of the manuscript. Dr K. Roche (Ghent) kindly corrected the English.

# APPENDIX

## Checklist of the Megalocypridininae

Family    CYPRIDIDAE Baird, 1845
Subfamily MEGALOCYPRIDINAE Rome, 1965
Tribe     MEGALOCYPRIDINI Rome, 1965
Genus     *Sclerocypris* Sars, 1924
    syn. *Bharatcypris* Bhattish, 1978
*S. bicornis* (G. W. Müller, 1900)
    syn. *Eucypris elongata* Spandl, 1924 syn. nov.
*S. clavularis* Sars, 1924 (type species)
    syn. *Eucypris capensis* Daday, 1910
*S. coomansi* Martens, 1986
*S. dayae* Martens, 1988
*S. dedeckkeri* Martens, 1988
*S. dumonti* Martens, 1988
*S. exserta exserta* Sars, 1924
*S. exserta makarikarensis* Martens, 1988
*S. flabella* (Vavra, 1897)
*S. jenkinae* Klie, 1933
*S. devexa* (Daday, 1910)
*S. jaini* Bhattia & Mannikeri, 1975
    syn. *S . indica* Bhattish & Mannikeri, 1975 nec
        Deb, 1973)
*S. longisetosa* Martens, 1988
*S. major* Sars, 1924
*S. multiformis* (Kiss, 1960)
    syn. *Eucypris serratamarginata* Kiss, 1960
*S. pardii* Martens, 1987
*S. rajasthaniensis* Deb, 1973
    syn. *S. indica* Deb, 1973
    syn. *Bharatcypris mackenziei* Bhattish, 1978
*S. rothschildi* (Daday, 1910)
*S. sarsi* Martens, 1987
    (nom. nov. pro *S. tuberculata* (Sars, 1924)
    nec (Methuen, 1910)
*S. tuberculata* (Methuen, 1910)
    syn. *Afrocypris biconica* Klie, 1933
*S. venusta* (Vavra, 1897)
*S. virungensis* Martens, 1988
*S. woutersi* Martens, 1988
*S. zelaznyi zelaznyi* Martens, 1988
*S. zelaznyi etoshensis* Martens, 1988

Remark: *Candonocypris dentatus* Victor & Michael, 1975 is probably a juvenile of a third Indian

*Sclerocypris* species.

Genus  *Hypselecypris* Rome, 1965
*H. wittei* Rome, 1965 (type species)

Genus  *Megalocypris* Sars, 1898
*M. durbani* (Baird, 1862)
    syn. *M. hodgsoni* (nom. nud. in Sars, 1898)
*M. hispida* Sars, 1924
*M. princeps* Sars, 1898 (type species)

Genus  *Madagascarcypris* Martens, 1986
*M. voeltzkowi* (G. W. Müller, 1898) (type species)

Genus  *Apatelecypris* Rome, 1965
*A. schultzei* (Daday, 1913) (type species)
    syn. *Megalocypris brevis* Sars, 1924

Genus  *Eundacypris* Martens, 1986
*E. superba* (Sars, 1924) (type species)

Tribe    Tanganyikacypridini DeDeckker &
    Wouters, 1983
Genus  *Tanganyikacypris* Kiss, 1961
*T. matthesi* Kiss, 1961 (type species)
*T. stappersi* Wouters *et al.* (in press)

Genus  Genus A aff. *Tanganyikacypris*

# REFERENCES

Bosellini, A. 1986. East African continental margins. *Geology*, **14**, 76-78.
Boukhari, M. & Guernet, C. 1985. Ostracoda of the Pleistocene of Fayoum: systematic view and new observations about the characteristics of an old lake. *Revue Micropaléont.*, Paris, **28**, 32-40.
Carbonel, P. & Peypouquet, J.-P. 1979. Les ostracodes des séries du Bassin de l'Omo. *Bulletin de l'institut de géologie du Bassin d'Aquitaine*, **25**, 167-199.
Cohen, A. S., Dussinger, R. & Richardson, J. 1983. Lacustrine paleochemical interpretations based on eastern and southern African ostracodes. *Palaeogeogr. Palaeoclimat . Palaeoecol.*, Amsterdam, **43**, 129-151.
Cohen, A. S. & Nielsen, C. 1986. Ostracodes as Indicators of Palaeohydrochemistry in lakes: a late Quaternary example from Lake Elmenteita, Kenya. *Palaios*, **1**, 601-609.
Grekoff, N. 1957. Ostracodes du Bassin du Congo. I. Jurassique supérieur et Crétace inférieur du nord du bassin. *Annls Mus. r. Congo Belge*, Bruxelles, Tervuren, Serie in 8°, Sciences Géologiques, **19**, 97 pp., 6 pls.

Grekoff, N. 1958. Ostracodes du Bassin du Congo. III. Tertiaire. *Annls Mus. r. Congo Belge*, Bruxelles, Tervuren, Serie in 8°, Sciences Géologiques, **22**, 36 pp., 3 pls.

Grekoff, N. 1960. Ostracodes du Bassin du Congo. II. Crétace. *Annls Mus. r. Congo Belge*, Bruxelles, Tervuren, Serie in 8°, Sciences Geologiques, **35**, 70 pp., 10 pls.

Martens, K. 1986. Taxonomic revision of the subfamily Megalocypridinae Rome, 1965 (Crustacea, Ostracoda). *Verhandelingen van de Koninklijke Akademie voor Wetenschappen, Letteren en Schone Kunsten van België, Klasse der Wetenschappen*, **48**(174), 81 pp., 64 pls.

Martens, K. 1987a. On *Sclerocypris pardii* n.sp. (Crustacea, Ostracoda) from Somalia. *Monitore zool. ital.*, Firenze, **22**, 59-72.

Martens, K. 1987b. Further additions to the description of *Sclerocypris devexa* (Daday, 1910) (Crustacea, Ostracoda), with a note on the position of this species. *Biologisch Jaarboek Dodonaea*, **5**(1), 86-91.

Martens, K. 1987c. Homology and functional morphology of the sexual dimorphism in the antenna of *Sclerocypris* Sars, 1924 (Crustacea, Ostracoda, Megalocypridinæ). *Bijdr. Dierk.*, Amsterdam & Lieden, **57**, 183-190.

Martens, K. 1988. Seven new species and two new subspecies of *Sclerocypris* Sars, 1924 from Africa, with new records of some other Megalocypridinae (Crustacea, Ostracoda). *Hydrobiologia*, Den Haag, **162**, 243-273.

Owen, H. G. 1983. *Atlas of continental displacement 200 million years to the present*, 160 pp. Cambridge University Press.

Rome, D. R. 1965a. Crustacea, Ostracoda. *In* Hanström, B., Brinck, P. & Rudebeck, G. (Eds), Results of the Lund University Expedition in 1950-1951. *S. Afr. anim. Life*, Stockholm, **11**, 9-58.

Rome, D. R. 1965b. *Ostracodes. Parc National de l' Upemba*, **69**, 71 pp. Mission G.F. De Witte.

Taylor, H. C. 1978. Capensis. *In* Werger M. J. A. (Ed.), *Biogeography and ecology of southern Africa*, 171-230, Dr W. Junk Publ., The Hague.

Thyson, P. D. 1986. *Climatic changes and variability in Southern Africa*, 220 pp. Oxford Univ. Press.

Von Vaupel Klein, J. C. 1984. A primer of a phylogenetic approach to the taxonomy of the genus *Euchirella* (Copepoda, Calanoida). *Crustaceana*, Supplement, Leiden, **9**, 194 pp.

Wild, H. 1975. Phytogeography and the Gondwanaland position of Madagascar. *Boissiera*, Genève, **24**, 107-117.

Wiley, E. O. 1981. *Phylogenetics, the theory and practice of phylogenetic systematics*, 439 pp. J. Wiley & Sons, N.Y.

Wouters, K., Martens, K. & DeDeckker, P. (In press). On the systematic position of *Tanganyikacypris* Kiss, 1961, with a description of *T. stappersi* sp. n. (Crustacea, Ostracoda), *In* Malz, H. (Ed.), First European Ostracod Meeting, Vol. 1.

## DISCUSSION

Joe Hazel: All cladograms based on only Recent material are working hypotheses that can be proved or disproved only by palaeontological or, less telling, biogeographic evidence. Would you agree?

Koen Martens: We most certainly do not agree, because your statement implies that soft bodied taxa which lack fossils cannot be subject to cladistic analyses. However, especially for such animal and plant groups, cladism offers a scientific method to deduce phylogenetic relationships. With regard to groups that do have a fossil record (e.g., ostracods), we still disagree. Every new character and every new taxon (in- or outgroup) included in a new analysis constitutes a valuable falsification of a former cladogram. Fossils form only a limited group of all possible test cases and they can furthermore hamper character analyses because most fossils constitute hard parts only. Cladistic character analyses, however, should deal with both soft and hard part morphology.

Robert Ross: Do you have evidence of absence of Megalocypridinae in described Cretaceous and Tertiary assemblages across Asia or even South America or are you still in need of a fossil record from these areas?

Koen Martens: It is of course logically impossible to prove the absence of a taxon in a certain region or period. One can only have evidence of its presence there and then by actually finding it. But it is certainly true that the more collections of a given period and region are described without finding a certain group, the greater the probability that the said group is indeed absent.

Relatively few Cretaceous and Tertiary ostracods from South America and Africa have been described but none of these appear sufficiently close to the Megalocypridinae to be lodged in this subfamily. But especially for Africa itself, a larger fossil record would be most helpful to unravel the history of the group. For the same period in Asia, we have more information thanks to a large number of papers by Chinese palaeontologists which appeared in the last decade. None of these taxa looks megalocypridinid either, except perhaps for some species of the genus *Limnocypridea*, which have a carapace shape which is superficially similar to that of *Apatelecypris schultzei*.

Dan Danielopol: The correlation between Recent Megalocypridinae and fossil triangular and trapezoidal Cypridinae causes you difficulties because these ostracod shapes occur convergently. Due to constructional constraints, ostracods with a high carapace have to strengthen their dorsal margin. One of the solutions is to produce dorsal tubercles on one of the valves (in the triangular carapace shape) or dorsal bars or they can strengthen the cardinal areas.

Koen Martens: This could very well be the case, still we would feel more comfortable about this when the structure of the valve margins of *Limnocypridea* species would prove sufficiently different to allow total rejection of any form of phylogenetical relationship between *Limnocypridea* and *Apatelecypris*.

# 44

# Non-marine Cretaceous ostracods from Argentina and their palaeobiogeographical relationships

**Eduardo A. Musacchio**

National University of Patagonia San Juan
Bosco, Comodoro Rivadavia, Argentina.

## ABSTRACT

The present paper is based on a fossil register which includes 86 species of previously described, non-marine ostracods belonging to 24 genera. Some of the genera are worldwide, such as *Cypridea* sensu lato, *Ilyocypris* and *Darwinula* and some exhibit remarkable similarities at the species level (as do the associated charophytes) to contemporary assemblages in the Northern Hemisphere, particularly during the Neocomian, Aptian and Maastrichtian. Similarly, the genera and the faunal sequence of the assemblages seem to exhibit similarities with Asian faunas. Gondwanian species with '*Limnocythere*'-like carapaces are well represented by *Looneyellopsis* and *Wolburgiopsis*. Likewise, species with '*Ilyocypris*'-like carapaces are well represented by *Neuquenocypris*. Some affinities, both at the generic and species level, with early Cretaceous Brazilian microfaunas can be established, although only in a limited number of cases.

## INTRODUCTION

In the past 18 years new, well preserved southern South American ostracod assemblages, mainly from Patagonia, have been described. The early Cretaceous faunas include several species of *Cypridea* Bosquet, 1852 *sensu lato* and clavatoracean charophytes as in many other 'Wealden' facies worldwide (Musacchio, 1979). In late Cretaceous assemblages, however, specimens of '*Ilyocypris*' Brady & Norman, 1889 *sensu lato* are abundant, together with other taxa with '*Ilyocypris*'-like carapaces (Musacchio, 1973, 1989 and Uliana & Musacchio, 1978). The lowermost early Tertiary fauna found in the Neuquén Basin approximates to that of the terminal Cretaceous but it is much less diverse (Musacchio & Moroni, 1983).

In southern South America, non-marine deposits were laid down in different palaeoenvironments and representatives of the early and late Cretaceous are widely distributed. Three sedimentary basins: the Neuquén Basin, the Gulf of San Jorge and the Northwest Argentinian Basin and their allied areas have received the most micropalaeontologic attention (Fig. 1). The Neuquén Basin has the greatest importance in chronological terms. This was, in part, a back-arc basin during the uppermost Jurassic and early Cretaceous. Here,

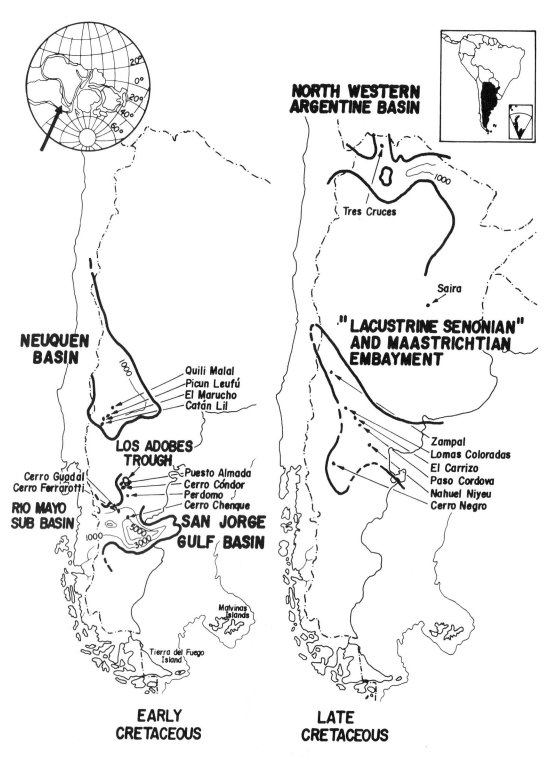

Fig. 1. Sedimentary basins in Argentina with localities studied.

the 'Andico' stratigraphical megasequence (Groeber, 1953) includes marine levels with ammonoids (Leanza & Wiedmann, 1980) and non-marine levels with charophytes and ostracods. In the late Cretaceous the 'Riográndico' megasequence (Groeber, 1953) is continental at its base. However, at its top, which embraces the Cretaceous-Tertiary boundary, there are a number of marine episodes, with ammonoids and planktonic foraminifera (Bertels, 1970; Camacho & Riccardi, 1978). Five zones based on non-marine calcareous microfossils taking into account the multiple control offered by the different groups of fossils mentioned above, have been previously proposed (Fig. 2, after Musacchio, 1989) The use of a global scale for marine parts of the sequence seems to be possible in some of these zones (Musacchio, 1979).

The present study is an attempt to distinguish pandemic genera and species with large or definable geographical ranges, from taxa endemic to South America or found, up to now, only in Patagonia. There are, however, many taxa whose biogeographical relationships are still far from being understood with accuracy.

## REPRESENTATIVE GONDWANIAN TAXA

Two unrelated groups of ostracods include several conspicuous or peculiar species up to now not recorded outside South America. The first is a group of small 'Limnocythere'-like ostracods belonging to the genera Looneyellopsis Krömmelbein & Weber, 1971 and Wolburgiopsis Musacchio, 1978. The second group includes several 'Ilyocypris'-like species belonging to the cypridacean genus Neuquenocypris (Musacchio, 1973), Notwithstanding their many endemic taxa, neither group seems to be completely restricted to South America.

## 'LIMNOCYTHERE'-LIKE GENERA

In many assemblages, both of early and late Cretaceous age a number of small cytheraceans are abundant. The carapace of several species resemble the genus Limnocythere Brady, 1868. The two first species of the group to be described were 'Wolburgia' plastica Musacchio, 1970 and

'Wolburgia' chinamuertensis Musacchio, 1970; both from late Hauterivian to early Barremian sediments of the Neuquén Basin. The generic assignment, particularly for the first species, was based on resemblance to Wolburgia atherfieldensis Anderson, 1966 from the Wealden of England. Krömmelbein & Weber, 1971 subsequently described Looneyellopsis brasiliensis from the Itaparica and Lower Candeias stratigraphical units of the Reconcavo-Tucano Basin of Brazil.

To date, twelve South American species described under Looneyellopsis Krömmelbein & Weber, 1971; Wolburgiopsis Musacchio, 1978 and Huillicythere Musacchio, 1978 seem to represent a natural grouping of cytheraceans. The presence of a robust hinge of merodont type (Plate 1, Figs 3a-b) distinguishes the later genus from the Limnocytheridae sensu stricto.

## THE GENUS 'NEUQUENOCYPRIS'

The genus 'Neuquenocypris' (Musacchio, 1973, 17; nom. transl. Musacchio, 1989) includes ilyocypridids of medium to large size with trapezoidal to rectangular carapaces of reversed overlap, which are frequently rimmed and strongly ornamented by hollow spines; the inner lamella lacks lists and the surface of the valves in most of the species, exhibits a 'mosaic'-like texture. This genus is similar to Ilyocypris Brady & Norman, 1889; Pelocypris Klie, 1939; C+yprideamorphella Mandelstam, 1956 and Parailyocypris Hou, 1956.

To date, the first record of Neuquenocypris comes from the Flabellochara harrisi assemblage of Aptian age. It is highly diverse during the late Cretaceous and remains so into the Palaeocene.

Outside South America, Ilyocypris colloti from the late Cretaceous of southern France resembles Neuquenocypris (see Babinot, 1975).

## RELATIONSHIPS WITH BRAZILIAN FAUNAS

Some species belonging to early Cretaceous associations from the Gulf of San Jorge Basin are comparable to Looneyellopsis brasiliensis Krömmelbein & Weber, 1971; Reconcavona? ultima

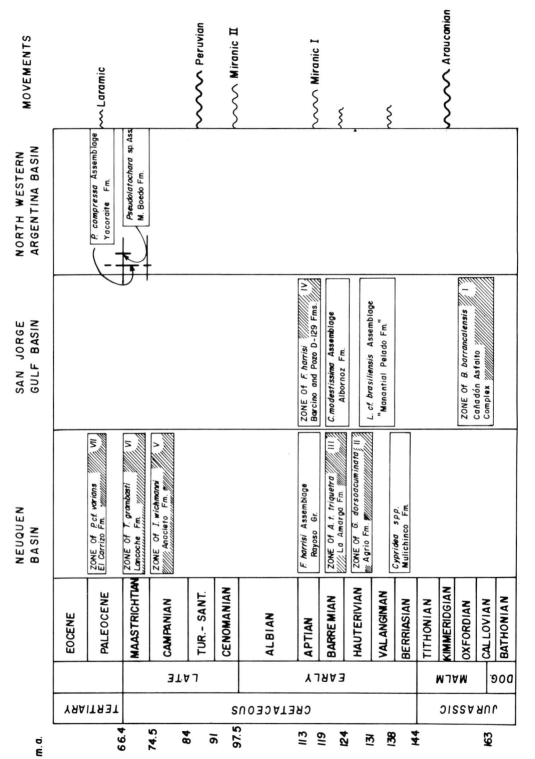

Fig. 2. Biostratigraphical sketch of the continental Cretaceous in Argentina based on calcareous microfossils.

Krömmelbein & Weber, 1971 and *Pattersoncypris angulata angulata* (Krömmelbein & Weber) [Hechem *et al.*, 1987]. These genera were originally described from Brazil. The genus *Petrobrasia* Krömmelbein, 1965 also occurs in Patagonia. Although the number of taxa common between Brazil and southern and western Argentina are limited in number, nonetheless together, they exhibit a coherent palaeogeographical distribution.

## PANDEMIC OSTRACODS

The following worldwide taxa are well represented in southern South America: *Cypridea* Bosquet, 1852 *sensu stricto*; *Cypridea* (*Ulwellia*) Anderson, 1940; *Rhinocypris* Anderson, 1941; *Ilyocypris* Brady & Robertson, 1885 and *Darwinula* Brady & Robertson 1885.

The Neocomian assemblages known from the Neuquén and Gulf of San Jorge basins are characterized by different species of *Cypridea* (*Ulwellia*). They are, respectively well represented by the assemblages of the La Amarga Formation of Hauterivian to early Barremian age (Musacchio, 1971; 1979) and the Puesto Albornoz Formation (Musacchio, 1987) of the same age. Several species (and morphotypes) belonging to this subgenus are shared by both basins, suggesting a free interchange at that time.

At the species level, most of the *Cypridea* (*Ulwellia*) taxa already described from Patagonia seem to be endemic to the region. However, a few species from other areas are somewhat similar, particularly to *Cypridea* (*Ulwellia*) *ludica* and *Cypridea* (*Ulwellia*) *cymerata* Musacchio, 1971. For example, both *Cypridea loango* Grosdidier, 1967 from 'Wealden' sediments of the West African continental margin, and *Cypridea* (*Ulwellia*) *subacuminata sichuaensis* Ye, 1982, from Lower Cretaceous sediments of the Sichuan Basin, China, resemble the two closely related Patagonian species mentioned above. The two species: *cymerata* and *ludica*, described from the 'Andico' megasequence of Neuquén, constitute a stable super-species, while different morphotypes occur in the Valanginian Hauterivian and early Barremian of the Neuquén Basin, as well as in

Neocomian sediments of the Gulf of San Jorge Basin (Musacchio, 1971; 1978; 1987).

*Cypridea* (*Ulwellia*) *modestissima* Musacchio, 1971 is included in this subgenus only with reservations. The poorly developed beak and the obsolescent alveolus resemble *Cypridea* (*Pseudocypridina*) Roth, 1933. Notwithstanding this, however, the species exhibits reversed overlap. The texture of the calcite forming the valves with its 'mosaic'-like design, differs from that of other species of *Cypridea* (*Ulwellia*) described from Patagonia.

*Cypridea* sensu stricto with normal overlap and *Rhinocypris* (or *Ilyocypris* sensu lato) first appear in Patagonia in the Aptian. *Cypridea* sensu stricto replaces *Cypridea* (*Ulwellia*) which dominated the Neocomian faunas.

In the Argentine late Cretaceous, however, *Ilyocypris* sensu stricto increases in abundance at the expense of *Cypridea*, which becomes very rare. *Ilyocypris wichmanni* Musacchio, 1973 *sensu lato*, a particularly well represented super-species, possesses a swollen posterodorsal area as in *Ilyocypris minor* Grekoff, 1956 and *Ilyocypris makunguensis* Marlière (in Grekoff, 1956), both from Cretaceous sediments of Zaire.

At least four species of *Darwinula* are presently recognized from early to late Cretaceous deposits in Patagonia (Musacchio, 1970; 1973; 1987; 1989) and *Cypridopsis pasocordobensis* Musacchio, 1973 occurs in the late Cretaceous of Neuquén.

## NORTH AMERICAN SIMILARITIES

The presence of morphological affinities between some ostracod species from southern South America (and some of the associated charophytes) and species from the Northern Hemisphere has been previously noted (Babinot, 1975; 1980; Musacchio & Chebli, 1975; Wang, 1978; Wang & Lu, 1982).

During the Aptian, in particular, an interchange between North and South America took place. From the Los Adobes sub-basin of the extra-Andean region of the Province of Chubut, three species of *Cypridea* sensu stricto and two of the allied charophytes from the *Flabellochara harrisi* assemblage, show varying degrees of similarity with species previously described mainly from Aptian sediments in the Rocky Mountains region (Musacchio in Musacchio & Chebli, 1975). They are: *Cypridea diminuta* Vanderpool, 1926; *Cypridea americana* Musacchio, 1975 and *Cypridea craigi* Musacchio, 1975. Of these, *Cypridea americana* strongly resembles *Cypridea hudsoni* Craig, 1961 (*nomen nudum* = *Cypridea* D Peck & Craig, 1962) and *Cypridea craigi* resembles *Cypridea trispinosa* Craig, 1961 (*nomen nudum* = *Cypridea* A Peck & Craig, 1962).

Data on Maastrichtian non-marine ostracods from North America is unknown to the author, but the charophytes *Platychara compressa* (Knowlton, 1899) and *Tolypella grambasti* Musacchio, 1978, both occur in South and North America. Finally, *Perissocytheridea informalis* Musacchio, 1978 has been discovered in Senonian (pre-Maastrichtian) sediments in central-west Argentina. To the author's knowledge, this is the earliest record of the genus, which was hitherto thought to be confined to the Cainozoic.

## ASIAN AFFINITIES

The sequence of Patagonian non-marine microfaunas from the mid part of the Upper Jurassic to the lowermost Tertiary, exhibits some features in common with those of Asia. In particular, the complete and well studied faunas from this interval in China and Mongolia (Chen, *et al.*, 1982a; 1982b; Gou & Cao, 1983; Hao *et al.*, 1983; Hou, 1982; Szczechura, 1978; 1981) exhibit certain similarities.

The Jurassic microfauna known from Cerro Cóndor in the Province of Chubut (Callovian-Oxfordian) includes species belonging to the genera *Bisulcocypris*, '*Metacypris*', *Timiriasevia* and *Darwinula* (see Musacchio, 1987). Similar generic assemblages occur in the Chinese Upper Jurassic of Gansu Province (Qi, 1985; Li, 1985).

Neocomian assemblages from Patagonia are characterized by the subgenus *Cypridea* (*Ulwellia*) and also include *Clinocypris*, *Dryelba*, *Darwinula* and the charophytes *Atopochara trivolvis*

*triquetra* Grambast, 1966 and *Mesochara* cf. *stipitata* (Wang, 1965).

*Cypridea* (*Ulwellia*) numbers decrease in the late Neocomian and the genus is eventually replaced by *Cypridea* sensu stricto and *Rhinocypris* associated with the charophytes *Flabellochara harrisi* (Peck, 1941) and *Porochara mundula* (Peck, 1941) in the Aptian.

No early Upper Cretaceous microfaunas are yet known in Patagonia, but the well developed Senonian ostracods (pre-Maastrichtian) suggest an increase in endemism. *Ilyocypris* is particularly well represented. At the end of the Cretaceous the ostracod fauna is very diverse and includes different endemic species of *Neuquenocypris*. Some species of this genus resemble the Asian genera: *Parailyocypris* and *Ilyocyprimorpha*. Also encountered are species whose morphology resembles that of the Talicyprideinae Hou, 1982 and *Timiriasevia* Mandelstam, 1947, which were abundant in China at this time.

The uppermost Cretaceous and the lowermost Tertiary charophytes from Patagonia and North Argentina compare well with coeval Chinese microfloras (Wang, 1978; Wang & Lu, 1982).

## FAUNAL INTERCHANGE

These similarities between non-marine animal and plant genera and species from Patagonia, west central Argentina and those from the Northern Hemisphere are not thought to be the function of random morphological convergence. On the contrary, they appear in coeval assemblages or in assemblages of closely similar ages, and seem to involve genuine biological relationships. Examples of comparable taxa, from both hemispheres, can be chosen from late Hauterivian-early Barremian, Aptian and Maastrichtian assemblages.

Similarly, the Callovian-Oxfordian faunas of non-marine ostracods also exhibit generic pandemism (Ballent, this volume). The only Palaeocene assemblage with *Peckichara* cf. *meridionalis* (Massieux, 1981), known from various localities (Musacchio & Moroni, 1983 and unpublished new information) exhibits a greatly reduced diversity and is clearly a relict Maastrichtian assemblage.

Other charophytes in their distribution strengthen the hypothesis of interchange between the two hemispheres in the Cretaceous (Musacchio & Chebli, 1975; Wang, 1978).

## THE DISPERSION OF THE OSTRACODS

At the present day many non-marine ostracods exhibit pandemic distributions. There are many cases where both latitudinal and longitudinal dysjunct distribution are thought to be due to the agency of migratory birds. However, during the Cretaceous Period, and owing to the low-grade of specialization of birds, other mechanisms of dispersion have been previously proposed.

The viability of desiccative eggs is well known in the literature. Helmdach (1979) has emphasized the effect of winds, not only with the dispersion of faunas, but moreover controlling the population dynamics and the evolutionary rate of the widely distributed species. Insects and flying reptiles have also been considered as dispersion agents of eggs (Sohn, 1969).

The palaeogeographical reconstructions illustrated by Barron (1987) for parts of the Lower and late Cretaceous would allow of terrestrial interchange between South and North America. The hypothesis of the existence of a land connection, which facilitated the interchange, must not be rejected.

Palaeogeographical changes are accelerated during orogeny. The 'initial miranic movements' (Stipanicic & Rodrigo, 1970) were proposed as a series of tectonic phases in the upper part of the early Cretaceous in the southern Andes. The Patagonian sequence with the *Flabellochara harrisi* assemblage, of presumably Aptian age, is postorogenic both in the Neuquén and the Gulf of San Jorge basins, as in the case of the Rocky Mountains with regard to the Comanchean movements. These events, not necessarily synchronous, could have palaeogeographical implications in both the Northern and Southern Hemisphere sectors of the 'Cordilleras'.

Faunal similarities with Asia, by the same token, could be related to the provision of a land bridge between Alaska and Siberia at this time.

However, details of the exact routes and dating of the respective events are still far from being well known.

## CONCLUSIONS

1) Distinct endemic non-marine ostracods flourished in Patagonia in both the early Cretaceous and late Cretaceous. Good examples are represented by '*Limnocythere*'-like cytheraceans, from the Valanginian to the Maastrichtian and by the cypridacean genus *Neuquenocypris* in the Aptian to Palaeocene. Brazilian affinities can be seen in the early Cretaceous, but only in a limited number of taxa.

2) The Patagonian assemblages during the Cretaceous also suggest interchange of pandemic taxa of the Cytheracea and Cypridacea, at both generic and specific levels.

A. *Cypridea* (*Ulwellia*) assemblages are well represented in the Neocomian.

B. *Cypridea* sensu stricto assemblages are well represented from the Aptian onwards, when the number of species of *Cypridea* (*Ulwellia*) is decreasing.

C. Endemism seems to increase after the early Cretaceous. In the uppermost Cretaceous, however, various genera are common to the Northern Hemisphere.

D. Asian affinities, both in the early and late Cretaceous are emphasized, suggesting an hypothetical bipolar-latitudinal design for many taxa. The distribution of the associated charophytes reinforces this hypothesis.

E. The only Palaeocene non-marine microfauna known from various different localities in the Neuquén Basin resembles those of the Maastrichtian but with reduced diversity.

## ACKNOWLEDGEMENTS

The author is indebted to Professor Robin Whatley (University College of Wales) who, kindly, revised and criticized the original manuscript and also to two anonymous referees, who made many valuable suggestions. However, the author is responsible for any failures in the present chapter.

**Plate 1**

Representative Gondwanian taxa.

Figs 1a-b. *Wolburgiopsis neocretacea* (Bertels, 1972). Uppermost Cretaceous.
Fig. 1a.    Carapace, right lateral view.
Fig. 1b.    Dorsal view (length = 0.47mm).

Figs 2a-b. *Looneyellopsis chinamuertensis* (Musacchio, 1970). Late Hauterivian-early Barremian.
Fig. 2a.    Carapace, dorsal view.
Fig. 2b.    Carapace right lateral view (holotype: length = 0.49mm).

Figs 3a-b. *Huillicythere* sp. Neocomian.
Fig. 3a.    Right valve hinge (approximately X270).
Fig. 3b.    Left valve hinge (approximately X203).

Fig. 4.      *Pattersoncypris* cf. *angulata angulata* (Krömmelbein & Weber, 1971). Aptian. Carapace right lateral view (length = 0.81mm).

Fig. 5.      *Neuquenocypris pecki* (Musacchio, 1978). Uppermost Cretaceous. Carapace left lateral view (holotype: length = 0.97mm).

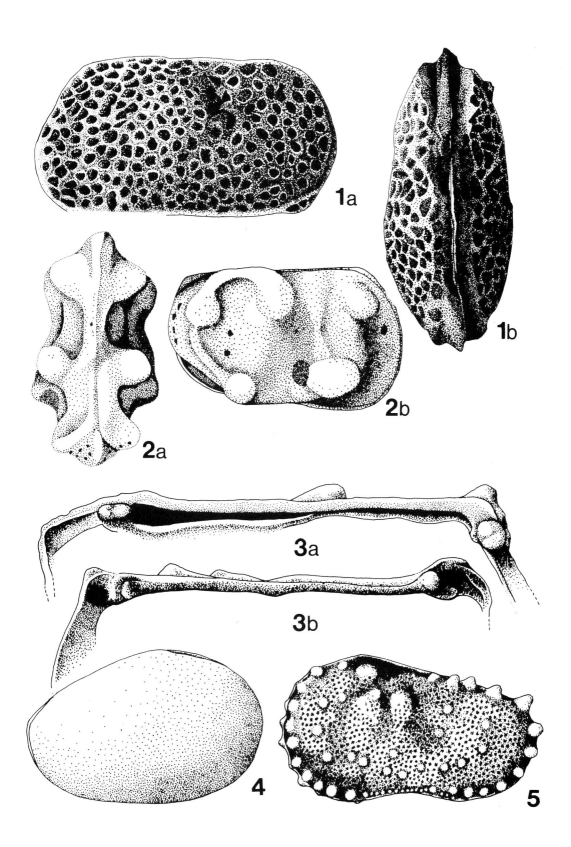

1a

1b

2a

2b

3a

3b

4

5

**Plate 2**

Species of Pandemic taxa.

Fig. 1.    *Cypridea (Cypridea) diminuta* Vanderpool, 1926. Aptian. Carapace right lateral view (length = 0.82mm).

Fig. 2.    *Cypridea (Ulwellia) ludica* Musacchio, 1971. Late Hauterivian-early Barremian. Carapace left lateral view (length = 1.11mm).

Figs 3a-b. *Cypridea (Cypridea) americana* Musacchio, 1975. Late Hauterivian-early Barremian.
Fig. 3a.    Carapace left lateral view.
Fig. 3b.    Carapace dorsal view (holotype: length = 0.97mm).

Fig. 4.    *Cypridea (Cypridea) craigi* Musacchio, 1975. Aptian. Carapace left lateral view (length = 1.01mm).

Figs 5a-b. *Rhinocypris* sp. Aptian.
Fig. 5a.    Carapace right lateral view (length = 0.65mm).
Fig. 5b.    Carapace dorsal view (length = 0.58).

Fig. 6.    '*Cypridea*' sp. Uppermost Cretaceous. Carapace left lateral view (length = 0.88mm).

Fig. 7.    *Ilyocypris wichmanni* Musacchio, 1973 *sensu lato*. Carapace left lateral view (length = 0.69mm).

Fig. 8.    *Altanicypris*? sp. Uppermost Cretaceous. Carapace right lateral view (length = 1.1mm).

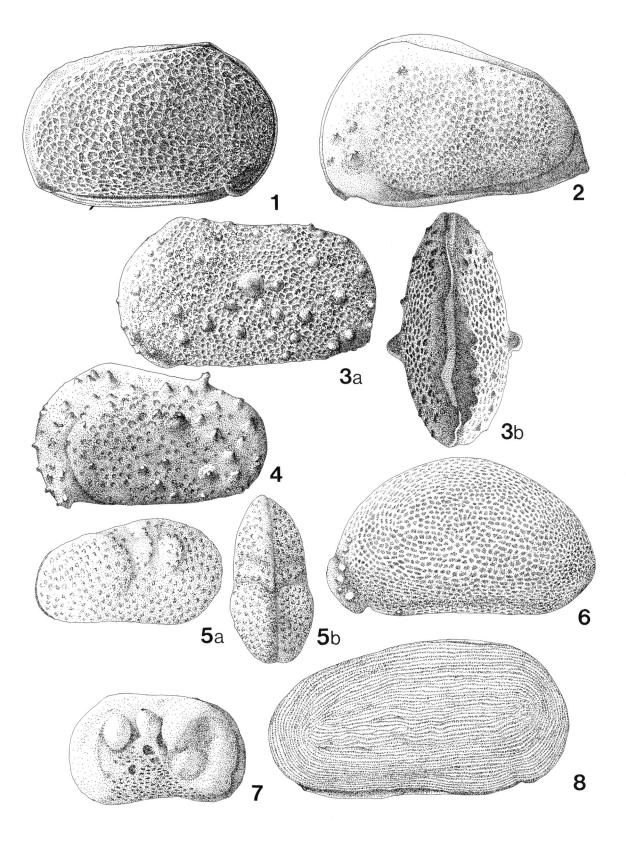

Likewise, the author wishes to thank Professor Florencia Murtagh de Perea (UNPSJB) for her help in organizing the English manuscript, Miss Ageda Pichiello for the drawings of the ostracods and Mr Andrés Blachakis for making the text figures. Dr Monique Feist (University of Languedoc) kindly shared with the author unpublished information concerning the presence of *Tolypella grambasti* in Alaska. Finally, the author is greatly indebted to the Argentine Oil Co. C.A.P.S.A. for the financial support which made posssible his attendance at the Symposium.

# REFERENCES

Babinot, J.-F. 1975. Études préliminaires sur les ostracodes du Valdonien-Fuvelien (Campanien continental) du Bassin d'Aix-en-Provence (Bouches du Rhône), France. *Paleob. Cont.*, **6**(1), 1-21.

Babinot, J.-F. 1980. *Les ostracodes du Crétacé supérieur de Provence*, Ph.D. thesis, RCP 510, Vols 1-3, Université de Provence.

Barron, J. 1987. Cretaceous plate tectonics reconstructions. *Palaeogeogr. Palaeoclimat. Palaeoecol.*, Amsterdam, **59**, 3-29.

Bertels, A. 1970. Los foraminíferos planctónicos de la Cuenca Cretácico-Terciaria en la Patagonia septentrional (Argentina), con consideraciones sobre la estratigrafía del Fortín General Roca (Provincia de Río Negro). *Rev. Asoc. Paleont. Argent.*, **7**(1), 1-50.

Camacho, H. & Riccardi, A. C. 1978. Invertebrados, megafauna. In Relatorio Geología y Recursos Naturales del Neuquén. *7th Congr. Geol. Argent.*, 137-144. Buenos Aires.

Chen, P. J., Li, W. B., Chen, J. H., Ye, Ch. H., Wang, Z. Shen., Y. B. & Sun, D. L. 1982a. Sequence of fossil biotic groups of Jurassic and Cretaceous in China. *Scientia sin.*, Ser. B, **25**(9), 1011-1020.

Chen, P. J., Li, W. B., Chen, J. H., Ye, C. H., Wang, Z., Shen, Y. B & Sun D. L. 1982b. Stratigraphical classification of Jurassic and Cretaceous in China. *Scientia sin.*, Ser. B, **25**(11), 1227-1248.

Gou, Y. S. & Cao, M. Z. 1983. Stratigraphic and biogeographic distribution of the *Cypridea*-bearing faunas in China. In Maddocks, R. F. (Ed.), *Applications of Ostracoda*, proceedings of the Eighth International Symposium on Ostracoda, July 26-29, 1982, 381-393. Univ. Houston Geos., Houston, Texas.

Grekoff, N. 1956. Ostracodes du Basin du Congo. I. Jurassique supérieur et Crétacé Inférieur du Nord du Basin. *Annls Mus. r. Congo Belge*, Sciences Géologiques, Bruxelles,Tervuren, **8**(17), 1-97.

Groeber, P. 1953. Mesozoico. In *Geografía de la República Argentina. Sociedad de Estudios Geográficos (GAEA)*, **2**(1), 9-541. Buenos Aires.

Hao, Y. Ch., Su, D. Y. & Li, Y. 1983. Late Mesozoic non-marine ostracods in China. In Maddocks, R. F. (Ed.), *Applications of Ostracoda*, proceedings of the Eighth International Symposium on Ostracoda, July 26-29, 1982, 372-380. Univ. Houston Geos., Houston, Texas.

Hechem, J. J., Figari, C. & Musacchio, E. A. 1987. Cuenca del Golfo San Jorge. El hallazgo del la Formación Pozo D-129 en superficie: información estratigráfica y paleontológica. *Petrotecnia*, **8**(11), 13-15.

Helmdach, F. F. 1979. Möglichkeiten der Verbreitung nichtmariner Ostrakoden-populationen un deren Auswirkung auf die Phylogenie und Stratigraphie. *Neues Jb. Geol. Paläontol. Mh.*, Stuttgart, **6**, 378-384.

Hou, Y. T. 1982. On the taxonomy and origin of cristid-ostracods. *Acta palaeont. sin.*, Beijing, **21**(1), 73-82.

Krömmelbein, K. & Weber, P. 1971. Ostracoden des "Nordost-Brasilianischen Wealden". *Beih. geol. Jb.*, Hannover, **115**, 1-93.

Leanza, H. & Wiedmann, J. 1980. Ammoniten des Valangin und Hauterive (Unterkreide) von Neuquén und Mendoza, Argentinien. *Eclog. geol. Helv.*, Lausanne, Basel, **73**(3), 941-981.

Li, Z. W. 1985. Middle Jurassic Ostracods from the Lonjiagou Formation of Gahai, Gansu. *Acta micropalaeont. sin.*, **2**(3), 257-260.

Musacchio, E. A. 1970. Ostrácodos de las superfamilias Cytheracea y Darwinulacea de la Formación La Amarga (Cretácico Inferior de la Provincia de Neuquén, Argentina. *Rev. Asoc. Paleont. Argent.*, **7**(4), 301-316.

Musacchio, E. A. 1971. Hallazgo del género *Cypridea* en Argentina y consideraciones estratigráficas sobre la Formación La Amarga (Cretácico Inferior) en la Provincia de Neuquén, Argentina. *Rev. Asoc. Paleont. Argent.*, **8**(2), 105-125.

Musacchio, E. A. 1973. Charóphytas y ostrácodos no-marinos del Grupo Neuquén (Cretácico Superior) en algunos afloramientos de las Provincias de Río Negro y Neuquén. *Rev. Mus. La Plata*, n.s., Pal., **48**(8), 1-32.

Musacchio, E. A. 1978. Ostrácodos del Cretácico Inferior en el Grupo Mendoza. Cuenca del Neuquén, Argentina. *7th Congr. Geol. Argent. Actas.*, Buenos Aires, **2**, 459-473.

Musacchio, E. A. 1979. Datos paleobiogeográficos de algunas asociaciones de foraminíferos, ostrácodos, y carófitos del Jurásico medio y el Cretácico inferior de Argentina. *Rev. Asoc. Paleont. Argent.*, **16**(3-4), 247-271.

Musacchio, E. A. 1981. Five Jurassic and Cretaceous non-marine, Ostracodal and Charophytal Associations (calcareous microfossils) from the San Jorge Gulf Basin, Argentina. *Zbl. Geol. Palaont.*, Teil I, H 7/8, 839-851.

Musacchio, E. A. 1989. In press. Biostratigraphy of the Non-Marine Cretaceous of Argentine Based on Calcareous Microfossils. *3rd Int. Cretaceous Symp.*, Tübingen.

Musacchio, E. M. & Chebli, G. 1975. Ostrácodos no marinos y carófitos del Cretácico inferior de las Provincias de Neuquén y Chubut, Argentina. *Rev. Asoc. Paleont. Argent.*, **12**(1), 7-96.

Musacchio, E. A. & Moroni, A. M. 1983. Charophyta y Ostracoda no marinos eoterciarios de la Formación El Carrizo en la Provincia de Río Negro. *Rev. Asoc. Paleont. Argent.* **20**(1-2), 21-33.

Peck, R. E. & Craig, W. W. 1962. Lower Cretaceous non-marine ostracods and charophytes of Wyoming and adjacent areas. *Wyoming Geol. Ass. Guidebook*, 33-43.

Qi, H. 1985. Upper Jurassic non-marine ostracods from Jingyuan, Gansu Province. Acta micropalaeont. sin., 2(3), 280-282.

Sohn. I. 1969. Non marine ostracods of early Cretaceous from Pine Valley quadrangle, Nevada. *U.S. Geol. Surv. Prof. Pap.* 643B, 1-9.

Szczechura, J. 1978. Results of the Polish Mongolian Palaeontological Expedition. Part VIII, Freshwater ostracods from the Nemegt Formation (Upper Cretaceous) of Mongolia. *Palaeont. pol.*, Warszawa, 38, 65-121.

Szczechura, J. 1981. The taxonomy of *Cypridea* Bosquet, 1852, and similar ostracods. *Neues Jb. Geol. Paläont. Abh.*, Stuttgart, 161(2), 254-269.

Stipanicic, P. & Rodrigo, F. 1970. El diastrofismo eo y mesocretácico en Argentina y Chile, con referencia a los movimientos jurásicos de la Patagonia. *4th Congr. Geol. Argent. Actas*, Buenos Aires, 2, 353-368.

Uliana, M. A. & Musacchio, E. A. 1978. Microfósiles calcáreos no-marinos del Cretácico superior en Zampal. Provincia de Mendoza, Argentina. *Rev. Asoc. Paleont. Argent.*, 15(1-2), 111-135.

Wang, Z. 1978. Cretaceous charophytes from the Yangtze-Han River Basin with a note on the classifications of Porocharaceae and Characeae. *Mem. Nanjing Inst. Geol. Paleont. Acad. Sin.*, 5(9), 61-92.

Wang, Z. & Lu, H. N., 1982. Classification and evolution of Clavatoraceae, with notes on its distribution in China. *Bull. Nanjing Inst. Geol. Paleont. Acad. Sin.*, 6(4), 77-104.

## DISCUSSION

Jean-Paul Colin: The geographical distribution of your genera: *Wolburgiopsis* and *Looneyellopsis* is much wider than you suggest. Representatives of these genera are present in various localities in Africa in the early Cretaceous, e.g., Zaire (see Grekoff's work), Chad and Ethiopia.

Ken McKenzie: Since you record many pandemic genera and species; how do you account for their common occurrence on the separate continental masses (even in the late Cretaceous and Devonian)?

Eduardo Musacchio: Several agents of dispersal for the eggs of Cretaceous ostracods (e.g., winds, insects and flying reptiles) have been discussed by various authors (see text). In addition the land connections between South America and other continents of the Northern Hemisphere must have aided the effectiveness of the aereal mechanisms of dispersion. Recent palaeogeographical reconstructions by Barron (1987) for the Cretaceous suggest the possibility of direct physical communication of the continents during parts of the period.

## POST SCRIPTUM

Professor Robin Whatley has kindly shared with the author unpublished information which allows a better understanding of the dispersion strategies of continental Cretaceous ostracods. Whatley (Whatley, R. C. In Press. The reproductive and dispersal strategies of Cretaceous non-marine Ostracoda: the key to pandemism. *Proceedings of the First International Symposium on non-marine correlations, ICGP 245, Urumqi, China 1987*) has recently considered the means by which many non-marine cypridaceans achieved a pandemic distribution due mainly to the evolution in the late Jurassic of the major dispersal strategies in the Cypridacea, namely parthenogenetic reproduction and a desiccation resistant egg which can be wind blown or transported by other animals. The widespread distribution of the Darwinulacea and Limnocytheridae is due to their long occupation of the non-marine environments (since the Devonian and Permian respectively), while the Cypridacea only entered freshwater permanently in the Bathonian.

# 45

# A preliminary study of the brackish and marine Ostracoda of the Pembrokeshire Coast, S. W. Wales

**Klaus Trier**

Glandwr, Mathry, Haverfordwest, Dyfed,
SA62 5HG, U.K.

## ABSTRACT

Forty species of living marine and brackish water ostracods are listed from intertidal and sublittoral algal collections taken around the 110-mile coastline of Pembrokeshire, S.W. Wales. Eighteen other species, found only as dead valves, came from sand scrapes taken at low water. These preliminary results, from collections accumulated over a 5-year period, are compared to faunas known from Cardigan Bay, the Bristol Channel and southern England.

## INTRODUCTION

The object of this study was to record the coastal ostracod fauna of the former county of Pembrokeshire, now part of Dyfed, S.W. Wales, and to compare the results with those of earlier workers in Cardigan Bay (Wall, 1969), to the north, in the Bristol Channel and Severn Estuary (Horne, 1980) and along the Dorset coast of S. England (Whittaker, 1972). The areas are shown in Fig. 1.

In September 1966, the Field Studies Council published the 'Dale Fort Marine Fauna' which included 10 species of ostracods either tow-netted or extracted from algae, in the vicinity of Dale Fort, Pembrokeshire (Fig. 1). They were listed as: *Cypridina* sp., *Philomedes globosus* (Lilljeborg), *Cythere lutea* (O. F. Müller), *Cythere albomaculata* Baird, *Leptocythere macallana* (Brady & Robertson), *Hemicythere villosa* (Sars), *Loxoconcha impressa* (Baird), *Xestoleberis aurantia* (Baird), *Sclerochilus contortus* (Norman) and *Paradoxostoma* sp. In the present study only the myodocopid species *Cypridina* sp. and *P. globosus* and the cytheracean *L. macallana* were not found; *C. albomaculata* is referred to the genus *Heterocythereis* and it and *L. impressa* (as *L. rhomboidea* (Fischer)) appears among the 58 marine and brackish species listed in Table 1, relating to my collections around the Pembrokeshire Coast.

The writer was unable to find any additional reference to ostracods specifically from the

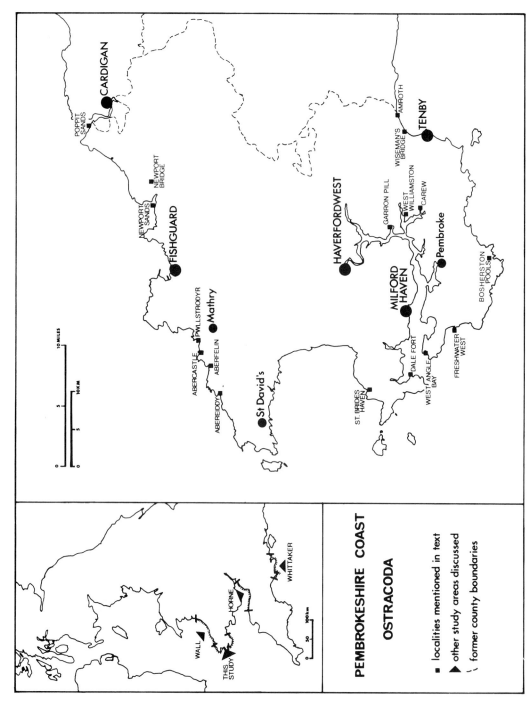

Fig. 1. Map of Pembrokeshire showing the sampling localities mentioned in the text. The inset shows the study area in terms of the comparative studies of Wall (1969), Whittaker (1972) and Horne (1980).

Pembrokeshire area. Encouraged by my friend, the late P. C. Sylvester-Bradley, two investigations were made. The first, aimed at collecting ostracods from sediment, took place between October 1974 and September 1976. Thirty-three sites were visited and sediment scraped from above the low water line. After the tide has receded one may frequently distinguish this line by its almost white colour in contrast to underlying sediment deposits that contain a high proportion of shell fragments, apparently separated by the action of the water. These scrapings were the source of material removed for examination. A total of 37 species were recorded from this survey and these are listed in column E of Table 1. Of these, 18 species were never found alive in the second investigation which followed. Most of the dead ostracods were single valves and many appeared to have come from the sublittoral, either washed from seaweeds or from sediment; some may even be sub-Recent. They are not considered further below unless otherwise stated.

The second investigation was more rewarding and aimed at obtaining material representative of the living fauna of the approximately 110 miles of coastline between Poppit Sands (at the mouth of the Teifi Estuary) on the old Cardiganshire-Pembrokeshire boundary to the north, and Amroth, near the old Pembrokeshire-Carmarthenshire boundary in the south (Fig. 1). Fifty-five sites were visited and from these 161 samples of seaweed were collected between May 1976 and July 1979. Of these, 151 were taken from intertidal rockpools and 10 by divers at three sublittoral sites in June 1977 and August 1978. The selection of sites was made with the object of reasonable equidistance along the coast, the known existence of rockpools containing algae, and ready accessibility to ensure collections could be made at several adjacent sites near the low watermark and with minimum delay while the day's tide was nearly at its lowest ebb. In some instances collections were made in tidal streams and these accounted for the brackish species listed in Table 1 (e.g., *Cyprideis torosa* (Jones), *Leptocythere lacertosa* (Hirschmann), *L. porcellanea* (Brady), *L. castanea* (Sars) and *Loxoconcha elliptica* Brady).

Whenever possible, at least one sample of algae was collected at the low watermark or just below, one at the mid and one at the high watermark. A team of divers collected 10 samples of algae at three sites in the vicinity of Abereiddy (Fig. 1) on the dates given above.

The algae collected during my surveys involved a wide variety of types, as follows: Greens - *Ulva lactuca*, *Cladophora* spp., *Enteromorpha* spp., *Codium* sp., *Bryopsis* sp. and *Chaetomorpha* spp.; Browns - *Fucus* spp. (with brown filamentous epiphytes including *Ectocarpus* spp.), *Dictyota* sp., *Laminaria saccharina* and *L. digitata* holdfasts; Reds - *Corallina officinalis*, *Chondrus* spp., *Polyides* sp., *Plumaria* spp., *Porphyra* spp., *Griffithsia* spp. and *Phycodrys* sp. The results showed that the best concentrations of ostracods were on the filamentous weeds and *Corallina*.

Pembrokeshire inlets vary greatly in character between those which are largely composed of shingle, those with many rockpools and the popular bathing beaches, some of which have rockpools at one or other end. Following severe south-westerly gales, a sandy beach may change to rocky ledges or *vice versa* and remain so for several months. Almost uniformly, inlets are surrounded by steeply rising cliffs and not infrequently access on foot is confined to a short period between high tides.

The method used to extract the ostracods was simple and based on a similar technique first described by Whatley & Wall (1975). The algae were uprooted and placed in a plastic container and topped up with seawater at the collection point. On return to base, usually within seven hours, the algae was tipped into a 150μm sieve to remove most of the seawater. Using a jet of 10% buffered formalin in seawater the contents of the sieve was flushed into a clean container topped up to cover the algae with the same solution, then left to soak for about half an hour. It was then returned to the 150μm sieve to separate the solids, washed with freshwater and left to dry. The resultant solids were passed through a 950μm, 410μm and 300μm sieve, all material being retained for examination under a stereo-microscope. All the material on which this study was based has now been presented

Table 1. Recent ostracods of the Pembrokeshire Coast, live and dead. Nomenclature follows Athersuch *et al*., 1989.

| Species | Live | | | | Dead | |
|---|---|---|---|---|---|---|
| | A | B | C | D | E | F |
| *Propontocypris pirifera* (G.W. Müller) | 33 | 0.35 | 4 | 12.20 | 2 | 0.07 |
| *Cythere lutea* O.F. Müller | 962 | 10.32 | 479 | 49.79 | 115 | 4.29 |
| *Eucythere declivis* (Norman) | | | | | 2 | 0.07 |
| *Leptocythere pellucida* (Baird) | | | | | 56 | 2.09 |
| *Leptocythere castanea* (Sars) | 44 | 0.47 | 4 | 9.09 | | |
| *Leptocythere lacertosa* (Hirschmann) | 47 | 0.50 | 13 | 27.66 | | |
| *Leptocythere porcellanea* (Brady) | 47 | 0.50 | 11 | 24.44 | | |
| *Leptocythere psammophila* Guillaume | | | | | 31 | 1.16 |
| *Leptocythere tenera* (Brady) | 1 | 0.01 | | | 25 | 0.93 |
| *Callistocythere badia* (Norman) | 20 | 0.21 | 12 | 60.00 | | |
| *Cyprideis torosa* (Jones) | 70 | 0.75 | 60 | 85.71 | | |
| *Cuneocythere semipunctata* (Brady) | | | | | 2 | 0.07 |
| *Neocytherideis subulata* (Brady) | | | | | 10 | 0.37 |
| *Sahnicythere retroflexa* (Klie) | | | | | 6 | 0.22 |
| *Pontocythere elongata* (Brady) | | | | | 282 | 10.53 |
| *Carinocythereis whitei* (Baird) | | | | | 32 | 1.19 |
| *Hiltermannicythere emaciata* (Brady) | | | | | 2 | 0.07 |
| *Hemicythere villosa* (Sars) | 80 | 0.86 | 4 | 5.00 | 205 | 7.65 |
| *Aurila convexa* (Baird) | 141 | 1.51 | 3 | 2.13 | 331 | 12.36 |
| *Aurila arborescens* (Brady) | 31 | 0.33 | 2 | 6.45 | | |
| *Heterocythereis albomaculata* (Baird) | 2378 | 25.50 | 1515 | 63.71 | 136 | 5.08 |
| *Urocythereis britannica* Athersuch | | | | | 1 | 0.04 |
| *Loxoconcha rhomboidea* (Fischer) | 344 | 3.69 | 73 | 21.22 | 1141 | 42.94 |
| *Loxoconcha elliptica* Brady | 46 | 0.49 | 18 | 39.13 | 4 | 0.15 |
| *Bonnyannella robertsoni* (Brady) | 3 | 0.03 | | | 70 | 2.61 |
| *Elofonsia baltica* (Hirschmann) | 17 | 0.18 | 7 | 41.18 | | |
| *Hirschmannia viridis* (O.F. Müller) | 875 | 9.39 | 563 | 64.34 | | |
| *Palmoconcha guttata* (Norman) | | | | | 24 | 0.90 |
| *Palmoconcha laevata* (Norman) | 3 | 0.30 | | | 51 | 1.90 |
| *Paracytheridea cuneiformis* (Brady) | 1 | 0.01 | | | 4 | 0.15 |
| *Hemicytherura cellulosa* (Norman) | 76 | 0.82 | 6 | 7.89 | 2 | 0.07 |
| *Hemicytherura clathrata* (Sars) | | | | | 14 | 0.52 |
| *Hemicytherura hoskini* Horne | 37 | 0.40 | 2 | 5.41 | 1 | 0.04 |
| *Microcytherura fulva* (Brady & Robertson) | | | | | 2 | 0.07 |
| *Semicytherura nigrescens* (Baird) | 420 | 4.50 | 26 | 6.19 | 1 | 0.04 |
| *Semicytherura angulata* (Brady) | 3 | 0.03 | | | 2 | 0.07 |
| *Semicytherura sella* (Sars) | 7 | 0.08 | 2 | 28.57 | | |
| *Semicytherura striata* (Sars) | 5 | 0.05 | | | 1 | 0.04 |
| *Semicytherura tela* Horne & Whittaker | 35 | 0.38 | 5 | 14.29 | | |
| *Cytheropteron latissimum* (Norman) | | | | | 11 | 0.41 |

Table 1. Recent ostracods of the Pembrokeshire Coast, live and dead. Nomenclature follows Athersuch *et al.*, 1989 - (continued).

| Species | Live | | | | Dead | |
|---|---|---|---|---|---|---|
|  | A | B | C | D | E | F |
| *Cytheropteron nodosum* Brady |  |  |  |  | 13 | 0.49 |
| *Xestoleberis aurantia* (Baird) | 393 | 4.22 | 254 | 64.63 |  |  |
| *Bythocythere bradyi* (Sars) |  |  |  |  | 32 | 1.19 |
| *Bythocythere robinsoni* Athersuch *et al.* |  |  |  |  | 28 | 0.96 |
| *Sclerochilus contortus* (Norman) |  |  |  |  | 12 | 0.45 |
| *Sclerochilus abbreviatus* Brady & Robertson | 3 | 0.03 | 1 | 33.33 |  |  |
| *Paradoxostoma abbreviatum* Sars | 20 | 0.21 | 15 | 75.00 |  |  |
| *Paradoxostoma bradyi* Sars | 76 | 0.82 | 21 | 27.63 | 10 | 0.37 |
| *Paradoxostoma ensiforme* Brady | 2 | 0.02 | 2 | 100.00 | 12 | 0.45 |
| *Paradoxostoma hibernicum* Brady | 314 | 3.37 | 106 | 33.76 |  |  |
| *Paradoxostoma normani* Brady | 15 | 0.16 | 7 | 46.67 |  |  |
| *Paradoxostoma porlockense* Horne & Whittaker | 94 | 1.01 |  |  |  |  |
| *Paradoxostoma pulchellum* Sars | 26 | 0.28 | 9 | 34.62 |  |  |
| *Paradoxostoma sarniense* Brady | 349 | 3.74 | 153 | 43.84 |  |  |
| *Paradoxostoma trieri* Horne & Whittaker | 152 | 1.63 | 91 | 59.87 |  |  |
| *Paradoxostoma variabile* (Baird) | 1922 | 20.64 | 1286 | 66.84 | 5 | 0.19 |
| *Cytherois fischeri* (Sars) | 168 | 1.80 | 112 | 66.67 |  |  |
| *Cytherois pusilla* Sars | 63 | 0.68 | 11 | 17.46 |  |  |
| TOTAL | 9323 | 100.00 | 4877 |  | 2678 | 100.00 |

Column A = Number of live specimens found, adults and juveniles.
Column B = Number of specimens as a percentage of total live fauna (9323 specimens).
Column C = Number of juveniles alone.
Column D = Number of juveniles as percentage of column A.
Column E = Number of dead specimens found.
Column F = Number of specimens as percentage of total dead fauna (2678 specimens).

to the Institute of Earth Studies, University College of Wales, Aberystwyth.

## DISCUSSION

Table 1 lists under column A the number of live specimens (adults and juveniles) of the 40 species of marine and brackish ostracods found along the Pembrokeshire Coast, in association with intertidal and sublittoral algae. The number of males, females and juveniles of each species is given.

Column B, then, shows the percentage each species represents out of the total specimens found in this survey, a grand total of 9323. Column C shows the quantity of juveniles of each species. Column D shows, for each species, the percentage of juveniles of the total ostracod count in column A. Columns E and F record the dead assemblages. Column E shows the number of dead specimens of each species found; column F the percentage each species represents out of the total 2678 specimens counted. The scope of the present paper precludes

Table 2. Summary of Recent live ostracods obtained from sublittoral algae, collected by divers at 3 sites on the Pembrokeshire Coast, June 1977 and August 1978.

| Species | A | B | C |
|---|---|---|---|
| *Propontocypris pirifera* (G. W. Müller) | 25 | 1 | 78.8 |
| *Cythere lutea* O. F. Müller | 39 | 271 | 32.2 |
| *Leptocythere tenera* (Brady) | 1 | | 100.0 |
| *Callistocythere badia* (Norman) | | 2 | 10.0 |
| *Hemicythere villosa* (Sars) | 2 | | 2.5 |
| *Aurila convexa* (Baird) | 11 | | 7.8 |
| *Heterocythereis albomaculata* (Baird) | 45 | 441 | 20.4 |
| *Loxoconcha rhomboidea* (Fischer) | 28 | 53 | 23.5 |
| *Bonnyannella robertsoni* (Brady) | 3 | | 100.0 |
| *Hirschmannia viridis* (O. F. Müller) | | 16 | 1.8 |
| *Palmoconcha laevata* (Norman) | 3 | | 100.0 |
| *Paracytheridea cuneiformis* (Brady) | 1 | | 100.0 |
| *Hemicytherura cellulosa* (Norman) | 12 | | 15.8 |
| *Hemicytherura hoskini* Horne | 16 | | 43.2 |
| *Semicytherura sella* (Sars) | 1 | | 14.3 |
| *Semicytherura striata* (Sars) | 1 | | 20.0 |
| *Semicytherura tela* Horne & Whittaker | 6 | 1 | 20.0 |
| *Xestoleberis aurantia* (Baird) | 9 | 1 | 2.5 |
| *Paradoxostoma abbreviatum* Sars | | 1 | 5.0 |
| *Paradoxostoma bradyi* Sars | 23 | 2 | 32.9 |
| *Paradoxostoma ensiforme* Brady | | 2 | 100.0 |
| *Paradoxostoma variabile* (Baird) | 9 | 59 | 3.5 |
| TOTAL | 235 | 850 | |

Column A = Adults.
Column B = Juveniles.
Column C = Adults and juveniles, as a percentage of total collected (Table 1), littoral and sublittoral.

anything more than the most basic statistics. No attempt, for instance, is made to differentiate between the various collections and types of algae on a seasonal basis. This would have involved a detailed ecological project, whereas my aim was, merely out of interest, to make a preliminary study of the phytal ostracods of the coast adjacent to my home and to list them.

A brief reference to the dead assemblages (column E, Table 1) shows that 18 species were not represented by live specimens, namely *Eucythere declivis* (Norman), *Leptocythere pellucida* (Baird), *L. psammophila* Guillaume, *Cuneocythere semipunctata* (Brady), *Neocytherideis subulata* (Brady), *Sahnicythere retroflexa* (Klie), *Pontocythere elongata* (Brady), *Carinocythereis whitei* (Baird), *Hiltermannicythere emaciata* (Brady), *Urocythereis britannica* Athersuch, *Palmoconcha guttata* (Norman), *Hemicytherura clathrata* (Sars), *Microcytherura fulva* (Brady & Robertson), *Cytheropteron latissimum* (Norman), *C. nodosum* Brady, *Bythocythere bradyi* Sars, *B. robinsoni* Athersuch *et al.*) and *Sclerochilus contortus*. Most of the species are sublittoral benthonic species living in sediment down to about 200 feet and would not be expected in algal samples; *H. clathrata* does not appear to live in Britain today (J. Whittaker, pers. comm.) and is probably sub-Recent. Conversely, 21 living phytal species were not found dead in the sediment scrapes.

The predominant living species in the Pembrokeshire collections were *Heterocythereis albomaculata* (making up 25.5% of the total number of ostracods collected between 1976 and 1979), *Paradoxostoma variabile* (Baird) (20.64%), *Cythere lutea* (10.32%) and *Hirschmannia viridis* (O. F. Müller) (9.39%). Of the rest, only *Semicytherura nigrescens* (Baird) (4.5%) and *Xestoleberis aurantia* (4.22%) exceed 4%, leaving 25.43% attributable to the remaining 34 species. The total number of juveniles found was almost exactly 52% of all live ostracods.

Table 2 lists the 22 species collected from sublittoral algae by divers at sites at Abereiddy, Pwllstrodyr and Abercastle. All species, except five, were represented among the faunas collected between high and low watermark. These exceptions were *Leptocythere tenera* (Brady), *Bonnyannella robertsoni* (Brady), *Palmoconcha laevata* (Norman), *Paracytheridea cuneiformis* (Brady) and *Paradoxostoma ensiforme* Brady, but their numbers were very low. Of interest, however, was the very large number of juveniles, compared to adults, which were found of *Cythere lutea*, *Heterocythereis albomaculata* and *Paradoxostoma variabile*. In the case of 4 other species, only juveniles were found. In others, notably *C. lutea* and *Loxoconcha rhomboidea*, these sublittoral samples accounted for 57% and 73% respectively of all juveniles of these species found. Seasonal factors undoubtedly are responsible. The samples were collected in June and August when the juvenile population of these species is probably at its highest. Other species (e.g., *P. variabile*) appear to have most juveniles in winter, witness the December 2nd 1978 algal collection from Abercastle at low water which yielded 301 specimens, all but 13 being *P. variabile* of which 211 were juveniles.

In conclusion, it was found that the most

productive, and hence the principal intertidal collecting sites were (from north to south) (Fig. 1):

| Site | National Grid Reference |
|---|---|
| Newport Sands | SN 053 406 |
| Pwllstrodyr | SM 866 338 |
| Abercastle | SM 853 337 |
| Aberfelin | SM 833 324 |
| Abereiddy | SM 796 312 |
| St. Brides Haven | SM 802 110 |
| West Angle | SM 852 032 |
| Freshwater West | SR 884 994 |
| Wiseman's Bridge | SN 145 060 |

Not all sites collected were marine. Later in the survey 5 brackish sites were investigated at West Williamston, Carew, Newport Bridge (Nevern Estuary), Garron Pill and the tidal stream between Bosherston Pools and Broadhaven. In total 7 species were recorded, namely *Leptocythere castanea*, *L. lacertosa*, *L. porcellanea*, *Cyprideis torosa*, *Loxoconcha elliptica*, *Elofsonia baltica* (Hirschmann) and *Cytherois fischeri* (Sars); they are included in Table 1.

Table 3 gives a comparison of the Pembrokeshire ostracod faunas with those reported from Cardigan Bay by Wall (1969), the Bristol Channel by Horne (1980) and the Dorset Coast of S. England by Whittaker (1972). All these areas have seen the most detailed systematic and ecological studies undertaken of the British coastal Ostracoda known to the author to date, and in all cases it has been possible to verify these records and compare them with the present material. The 40 live species recorded herein compare well with the 45 species recorded live by Wall (1969) in Cardigan Bay, 43 by Horne (1980) and 44 by Whittaker (1972) from the Bristol Channel and the Dorset Coast, respectively. Wall had many sublittoral benthonic species in his live records, whereas the other surveys, like my own, concentrated on shallower phytal collections.

Records of *Aurila arborescens* (Brady) and *Paradoxostoma trieri* Horne & Whittaker (its type-area) are of interest. Neither were found in the other surveys listed above although the latter

Table 3. A comparison of the Cardigan Bay, Pembrokeshire Coast, Bristol Channel and Dorset Coast ostracod faunas, live and dead. Nomenclature follows Athersuch *et al.*, 1989.

| Species | Cardigan Bay[1] Live | Dead only | Pembs. Coast[2] Live | Dead only | Bristol Coast[3] Live | Dead only | Dorset Coast[4] Live | Dead only |
|---|---|---|---|---|---|---|---|---|
| *Propontocypris trigonella* (Sars) | X | | | | | | | |
| *Propontocypris pirifera* (G. W. Müller) | | | X | | | | X | |
| *Cythere lutea* O. F. Müller | X | | X | | X | | X | |
| *Eucythere declivis* (Norman) | | X | | X | | | | |
| *Eucythere argus* (Sars) | | X | | | X | | | |
| *Leptocythere pellucida* (Baird) | X | | | X | | | | X |
| *Leptocythere baltica* Klie | X | | | X | | | X | |
| *Leptocythere castanea* (Sars) | X | | X | | X | | X | |
| *Leptocythere ciliata* Hartmann | | | | | X | | | |
| *Leptocythere lacertosa* (Hirschmann) | X | | X | | X | | X | |
| *Leptocythere macallana* (Brady & Robertson) | | X | | | | | X | |
| *Leptocythere porcellanea* (Brady) | X | | X | | X | | X | |
| *Leptocythere psammophila* Guillaume | X | | | X | X | | X | |
| *Leptocythere tenera* (Brady) | X | | X | | X | | | X |
| *Callistocythere littoralis* (G. W. Müller) | | X | | | | | X | |
| *Callistocythere badia* (Norman) | X | | X | | | | X | |
| *Callistocythere murrayi* Whittaker | | | | | | | X | |
| *Cyprideis torosa* (Jones) | X | | X | | X | | X | |
| *Cuneocythere semipunctata* (Brady) | X | | | X | X | | | |
| *Neocytherideis subulata* (Brady) | X | | | X | | | | X |
| *Sahnicythere retroflexa* (Klie) | X | | | X | | | | X |
| *Pontocythere elongata* (Brady) | X | | | X | X | | | |
| *Carinocythereis carinata* (Roemer) | | | | | | | | X |
| *Carinocythereis whitei* (Baird) | X | | | X | | | | |
| *Hiltermannicythere emaciata* (Brady) | X | | | X | | | | |
| *Pterygocythereis jonesii* (Baird) | | X | | | | | | |
| *Robertsonites tuberculatus* (Sars) | | X | | | | | | |
| *Basslerites teres* (Brady) | | | | | | | X | |
| *Hemicythere villosa* (Sars) | X | | X | | X | | X | |
| *Aurila convexa* (Baird) | X | | X | | X | | X | |
| *Aurila arborescens* (Brady) | | | X | | | | | |
| *Aurila woutersi* Horne | | | | | X | | | |
| *Heterocythereis albomaculata* (Baird) | X | | X | | X | | X | |
| *Urocythereis britannica* Athersuch | X | | | X | | | | X |
| *Loxoconcha rhomboidea* (Fischer) | X | | X | | X | | X | |
| *Loxoconcha elliptica* Brady | X | | X | | X | | X | |
| *Bonnyannella robertsoni* (Brady) | X | | X | | X | | | X |
| *Elofsonia baltica* (Hirschmann) | X | | X | | X | | X | |

Table 3. A comparison of the Cardigan Bay, Pembrokeshire Coast, Bristol Channel and Dorset Coast ostracod faunas, live and dead. Nomenclature follows Athersuch *et al.*, 1989 - (continued).

| Species | Cardigan Bay[1] Live | Dead only | Pembs. Coast[2] Live | Dead only | Bristol Channel[3] Live | Dead only | Dorset Coast[4] Live | Dead only |
|---|---|---|---|---|---|---|---|---|
| *Elofsonia pusilla* (Brady & Robertson) | | X | | | | | X | |
| *Hisrchmannia viridis* (O. F. Müller) | X | | X | | X | | X | |
| *Palmoconcha guttata* (Norman) | X | | | X | | | | X |
| *Palmoconcha laevata* (Norman) | X | | X | | X | | X | |
| *Sagmatocythere multifora* (Norman) | | X | | | | | | |
| *Paracytheridea cuneiformis* (Brady) | | X | X | | X | | | X |
| *Cytherura gibba* (O. F. Müller) | | | | | | | X | |
| *Hemicytherura cellulosa* (Norman) | X | | X | | X | | X | |
| *Hemicytherura clathrata* (Sars) | | X | | X | | | | |
| *Hemicytherura hoskini* Horne | | | X | | X | | | |
| *Microcytherura fulva* (Brady & Robertson) | | X | | X | | | | X |
| *Semicytherura nigrescens* (Baird) | | X | X | | X | | X | |
| *Semicytherura acuticostata* (Sars) | | X | | | X | | | |
| *Semicytherura angulata* (Brady) | | X | X | | X | | | |
| *Semicytherura cornuta* (Brady) | | X | | | | | X | |
| *Semicytherura producta* (Brady) | | X | | | | | | |
| *Semicytherura sella* (Sars) | X | | X | | | | X | |
| *Semicytherura simplex* (Brady & Norman) | | X | | | | | | X |
| *Semicytherura striata* (Sars) | X | | X | | X | | X | |
| *Semicytherura tela* Horne & Whittaker | | | X | | X | | X | |
| *Cytheropteron latissimum* (Norman) | X | | | X | | | | |
| *Cytheropteron depressum* (Brady & Norman) | X | | | | | | | X |
| *Cytheropteron dorsocostatum* Whatley & Masson | | X | | | | | | |
| *Cytheropteron nodosum* Brady | X | | | X | X | | | |
| *Cytheropteron subcircinatum* Sars | | X | | | | | | |
| *Xestoleberis nitida* (Liljeborg) | | | | | | | X | |
| *Xestoleberis aurantia* (Baird) | | | X | | | | X | |
| *Xestoleberis rubens* Whittaker | | | | | | | X | |
| *Bythocythere bradyi* Sars | X | | | | X | X | | |
| *Bythocythere intermedia* Elofson | | X | | | | | | |
| *Bythocythere robinsoni* Athersuch *et al.* | X | | | | X | | | |
| *Pseudocythere caudata* (Sars) | | X | | | | | X | |
| *Sclerochilus contortus* (Norman) | X | | | | X | | | X |
| *Sclerochilus abbreviatus* Brady & Robertson | | | X | | | | X | |
| *Sclerochilus gewemuelleri* Dubowsky | | | | | X | | X | |
| *Sclerochilus schornikovi* Athersuch & Horne | | | | | X | | | |
| *Sclerochilus truncatus* (Malcomson) | | X | | | | | | |
| *Paradoxostoma abbreviatum* Sars | X | | X | | X | | | |

Table 3.  A comparison of the Cardigan Bay, Pembrokeshire Coast, Bristol Channel and Dorset Coast ostracod faunas, live and dead.  Nomenclature follows Athersuch et al., 1989 - (continued).

| Species | Cardigan Bay[1] Live | Dead only | Pembs. Coast[2] Live | Dead only | Bristol Channel[3] Live | Dead only | Dorset Coast[4] Live | Dead only |
|---|---|---|---|---|---|---|---|---|
| *Paradoxostoma bradyi* Sars | X | | X | | X | | X | |
| *Paradoxostoma ensiforme* Brady | X | | X | | X | | | X |
| *Paradoxostoma fleetense* Horne & Whittaker | | | | | | | X | |
| *Paradoxostoma hibernicum* Brady | X | | X | | X | | | |
| *Paradoxostoma normani* Brady | X | | X | | X | | X | |
| *Paradoxostoma porlockense* Horne & Whittaker | | | X | | X | | | |
| *Paradoxostoma pulchellum* Sars | | | X | | | | X | |
| *Paradoxostoma robinhoodi* Horne & Whittaker | X | | | | X | | | |
| *Paradoxostoma sarniense* Brady | | | X | | | | X | |
| *Paradoxostoma trieri* Horne & Whittaker | | | X | | | | | |
| *Paradoxostoma variabile* (Baird) | X | | X | | X | | X | |
| *Cytherois fischeri* (Sars) | | | X | | X | | X | |
| *Cytherois pusilla* Sars | | | X | | | | X | |
| *Cytherois stephanidesi* Klie | | | | | | | X | |
| *Paracytherois* sp. | X | | | | X | | | X |

Key: [1]Wall, 1969; [2]this study; [3]Horne, 1980; [4]Whittaker, 1972.

has since been reported as far apart as W. Scotland and S. W. France (Horne & Whittaker, 1985). Thirty-one specimens of *A. arborescens* were collected from littoral algae at West Angle Bay in October 1977. This is the first and only live record from the British Isles, (although Whatley, pers. comm. has found dead specimens in Cardigan Bay), but it is known extensively from the Mediterranean under the name of *A. woodwardii* (Brady). Evidently it was more widespread in the Quaternary of the British Isles and was originally described from the Nar Valley Clay of Norfolk by Brady (1865) (see Athersuch *et al.*, 1985, for a redescription). As the West Angle Bay location is near the international oil refinery at Milford Haven it is tempting to suggest that such an exotic species could have been introduced. On the other hand, the location of Pembrokeshire, like Cornwall to the south, at the southwestern tip of Britain, may

indicate Gulf Stream influence and 'southern' species at their northernmost limits. Certainly in Cornwall, species of 'southern' affinity, like *Xestoleberis labiata* Brady & Robertson, occur which are found nowhere else in British waters today.

The distribution of *Xestoleberis aurantia* also warrants discussion. It is common in the Pembrokeshire and Dorset areas, but like other species of *Xestoleberis* it was absent from the extensive Cardigan Bay (Wall, 1969) and Bristol Channel collections (Horne, 1980). Other species of similar distribution were *Propontocypris pirifera* (G. W. Müller), *Sclerochilus abbreviatus* Brady & Robertson, *Paradoxostoma pulchellum* Sars, *P. sarniense* Brady and *Cytherois pusilla* Sars. On the other hand, *Hemicytherura hoskini* Horne and *Paradoxostoma porlockense* Horne & Whittaker were found both in Pembrokeshire and the

Bristol Channel, but not in Cardigan Bay nor the Dorset Coast. What this signifies in terms of ecology it is not possible at present to say, save to reinforce the patchiness of distribution of a species within its overall geographical range. Local conditions, exposure, tidal regimes and competition from other species, however, may all have some bearing on these distribution patterns.

## ACKNOWLEDGEMENTS

I will always be grateful to the late Peter Sylvester-Bradley for his lively, stimulating and sometimes humerous introduction to the joys of ostracod collecting and to Robin Whatley and his students for the many hours spent in identifying species and correcting my errors. The University College of Wales diving team provided much useful material as did my son, Mike Trier, from his snorkelling expeditions. I was always pleased to welcome visiting ostracodologists to my home for the enthusiasm and encouragement they brought. Among these John Whittaker produced the final draft of this manuscript and compared my records with those of his own studies and of Wall and Horne. Lastly, I owe to my wife for her skilful navigation along the many miles of often unmarked Pembrokeshire country lanes and for her support in a project that caused her inconvenience and sometimes doubts in its sanity, now, perhaps, redeemed.

## REFERENCES

Athersuch, J., Horne, D. J. & Whittaker, J. E. 1985. G. S. Brady's Pleistocene ostracods from the Brickearth of the Nar Valley, Norfolk, U.K. *J. micropalaeontol.*, London, 4(2), 153-158.

Athersuch, J., Horne, D. J. & Whittaker, J. E. 1989. *Marine and brackish water ostracods (Superfamilies Cypridacea and Cytheracea). In* Kermack, D. M. & Barnes, R. S. K. (Eds), Synopses of the British Fauna (New Series), no. 43. Linnean Society of London and E. J. Brill, Leiden.

Brady, G. S. 1865. On undescribed fossil Entomostraca from the Brickearth of the Nar. *Ann. Mag. nat. Hist.*, London, ser. 3, 16, 189-191.

Horne, D. J. 1980. *Recent Ostracoda from the Severn Estuary and Bristol Channel.* Unpub. Ph.D. thesis, University of Bristol.

Horne. D. J. & Whittaker, J. E. 1985. A revision of the genus *Paradoxostoma* Fischer (Crustacea; Ostracoda) in British waters. *Zool. J. Linn. Soc.*, London, 85, 131-203.

Wall, D. R. 1969. *The taxonomy and ecology of Recent and Quaternary Ostracoda from the southern Irish Sea.* Unpub. Ph.D. thesis, University of Wales, 555 pp., 45 pls, 39 figs, 2 tables.

Whatley, R. C. & Wall. D. R. 1975. The relationship between Ostracoda and algae in littoral and sublittoral marine environments. *Bull. Am. Paleont.*, Ithaca, 65, 173-203.

Whittaker, J. E. 1972. *The taxonomy, ecology and distribution of Recent brackish and marine Ostracoda from localities along the coast of Hampshire and Dorset.* Unpub. Ph.D. thesis, University of Wales, 643 pp., 68 pls, 81 figs, 20 tables.

# EDUCATIONAL

# 46

# Publish on Ostracoda:
# what, when, how, where?
# Introduction to a workshop

**Henri J. Oertli**

F - 64320 Bizanos, France

## ABSTRACT

The art of preparing a publication is seldom taught. The younger scientist often faces editorial problems for which some advice could be useful and prevent the loss of valuable time. This workshop introduction considers and offers recommendations of the 'what, how, when and where?' of paper preparations for submission to a scientific journal. Addressed are such questions as: *What* kind of papers should you produce (being aware that publishing is in your personal interest, and will often be required by your employer); *how* should you prepare it (you must be convinced that it will be a masterpiece...); *when* (and possibly with whom) should you publish; and *where* should your paper appear (who needs to see your results)?

Voluminous papers do prove to be a problem. Some may split into several less voluminous articles; but in other cases this seems undesirable and illogical. A table of publishers who accept large papers is given, with or without material conditions.

## INTRODUCTION

Rudyard Kipling once published a mini-poem, which inspired the title of this workshop, a poem that seems at first of restricted interest; but on considering it more closely, it is of profound truth. Here it is:

> I keep six honest serving-men
> They taught me all I knew;
> Their names are What and Why and When,
> And How and Where and Who.

'Publish or perish' is the famous scientists' incantation; it encapsulates very well the fact that the scientist, however clever he may be, who refuses (or is too lazy) to publish will not find recognition. He most probably has an employer: this should oblige him to contribute to publicizing the activity of his laboratory or institution. Also his career will, at least partly, be influenced by the number and quality of his publications.

'Number' and 'quality' are of course delicate and difficult to define. It should be evident that every publication has to be a kind of mature fruit, that numerous papers on strictly the same subject should be avoided, that an article has to be written with the utmost discipline and conviction to give the best possible arguments and solutions.

## WHAT?

Quite naturally this section is closely linked to the following: 'When?'; but first let us consider some aspects of the 'purely what?'.

You will want to publish research results. They may be based on new or revised ostracod material, on a bibliographic and/or taxonomic revision or be the result of 'palaeonto-philosophical thoughts'. It may be a monographical work, on one subject or several linked ones, or a single result which is considered as particularly interesting. Be sure that you bring new facts that are valuable, and, if possible, a long-awaited contribution. Of course, this obliges you to carry out an extensive bibliographic research. Also it will be most useful to consult colleagues in your field, as well as the last issues of *Cypris*, in order to be more or less sure that a similar study is not in progress elsewhere. Present your paper in a way that will allow colleagues to achieve the same or similar result (I refer here especially to sampled localities or figured specimens; taxonomic revisions, on the contrary, may result in controversy, which sometimes helps progress by stimulation). If you have found surprising results (perhaps a relict species, or a stratigraphically abnormal occurrence, or an amazingly heterogeneous association), be absolutely sure that they are not due to an error in your work or somebody else's; it's best to be cautious by resampling.

## WHEN?

Several possibilities of when to publish occur. In the first case the choice of time is free. The scientist will present his paper when he thinks he knows all there is to know about the chosen subject. This can, however, be unrealistic since some scientists never feel ready, and if they are lacking advice, supervision or encouragement from colleagues or superiors, a dramatic blockage may result.

An example of the second case is where a team has finished its study (or is obliged to do so by an imposed time limit), and will/must publish. Teamwork can be a good challenge, thanks to mutual criticism and encouragement, but this presupposes a team that really collaborates and progresses (which means that it must have an energetic leader).

Third case: publication of a partial result of a thesis (or another voluminous study). This is highly justified and should be encouraged, firstly because many theses never see the 'light of day' (or if they do, only long after they have been completed), and secondly because the writing of such papers, i.e. the fight for a logical, accurate presentation of the data, may reveal important gaps in the research; gaps that there is still time to fill in (the more and better while the subject is still 'fresh').

Fourth case: a paper is to be presented at a symposium. If the dates are known long in advance, you should have time to finish a study, perhaps one that otherwise would have been delayed. Which means that this incentive may be useful and welcome. However, we all know that symposia can also constitute a danger for the paper's quality, since often these articles are written in a hurry or deal with a subject of very limited (if any) interest, or constitute simply a 'warming up' of an already published subject. This concerns what I would call 'free ticket papers'; the employer pays for symposium participation only if a paper is presented.

Senior scientists are also exposed to a certain amount of danger, especially when their notoriety obliges them to produce at an abnormally high rate. However, they have the great advantage of more experience, of more stuff - and more staff (exceptions exist!). I admit that this is not yet an absolute guarantee...!

Fifth case: the advancement of the study is not really sufficient for publication, but has yielded very exciting results that the scientist would like to publicize as soon as possible, the more so if he/she perhaps knows or supposes that similar studies have begun elsewhere. In this case, it is normal that preliminary results are published.

## HOW?

Special attention should be given at the very beginning of a study leading to a publication, to the observational phase. It is essential that at this early stage the scientist looks at his material with the future publication in mind. This means careful and extensive observation and notes (that should prevent him from too frequent later re-examining, re-measuring) as well as a very complete bibliographical study. Indeed, you should not be obliged, near the completion of your work, to search again in the library after having already seen the papers, simply because you have forgotten their exact content or omitted to note the complete references. (By the way, note the complete, i.e. non-abbreviated journal titles, as some editors require them quoted in full.)

For the observational data as well as the consulted references, the card system proves helpful. Other methods may be preferred, (e.g., if a computer is your friend...). Except for field work, the notebook is impracticable; it does not allow later grouping of observations and consideration.

At a very early stage in your work, a kind of table of contents should be laid down, which will roughly follow the editors' *leitmotiv* IMRaD: Introduction, Materials/Methods, Results and Discussion.

### Introduction

State of the art before undertaking your research. Definition of the gap you want to fill in, and how you intend to do this.

### Materials/Methods

With what type of material and using which methods have you carried out your work? New techniques? Errors? In short, what has been done in order to get results?

### Results

Results of your study, not a mixture of known and new. Be short, do not 'tell everything'. Discuss also your problems and results that do not correspond to your initial hypothesis.

### Discussion/Conclusions

Significance of your results; comparison between existing studies and yours. Here too, be modest and cautious. Could your results be interpreted differently?

It is helpful to begin using folders corresponding to the different 'Contents' sections to include all your notes and ideas (which are generated during the day, but often also at night...) and leave these in an easily accessible place. Never use both sides of a sheet, otherwise you lose the facility to cut and paste paragraphs and chapters.

Once you have the impression that 'the fruit is mature', you will begin the painful task of writing the first draft. Begin it during an optimistic phase, try to write big parts without interruption (be it in pen or pencil or directly into a word processor). Don't stop for stylistic/orthographic or other uncertainties but leave a blank, to be filled in. Make provisional drafts of figures.

Having put down all your knowledge, you will begin the revision, which is best done after some days or even weeks, in order to distance yourself from the text. Be critical with yourself, revise every sentence and ask whether it expresses in a simple manner what you want to say. Would a foreigner understand easily? Does it follow on in a logical manner from the preceding sentence? Every sentence should be a little masterpiece! This will involve you in several rewritings, but that is normal!

Once you feel satisfied with yourself (or become completely allergic to your text), submit it to several people, preferably to friends who have distinct knowledge in your field and whom you know for their helpful criticism, but also to people who have some scientific background, but without being specialized. Often, the latter will be particularly useful in detecting passages difficult (or impossible!) to understand. Sentences that need two or more readings before they are understood must be rewritten.

Table 1.  List of publishers of palaeontological journals and books  accepting voluminous papers.

| Journal | Publishing place | Conditions |
|---|---|---|
| Paleontological Society of America Memoirs | Columbus/Ohio | Full page charge  (140 $/page). |
| Journal of Paleontology | Lawrence/Kansas | Less than 50 pp. Author must be SEPM member. 'Substantial page charges'. |
| Ecological Monographs (Ecological Society of America) | Lawrence/Kansas | 35 $/page for the first 16 pp., nothing beyond. Editor encourages submission of papers with significant ecological orientation. |
| Bulletin of American Paleontology | Ithaca/New York | For papers of more than 150 pp. |
| Palaeontographica americana | Ithaca/New York | For papers of more than 400 pp. For both: Authors contribute to the illustrations: 80-100 $/plate, 30-35 $/figure. |
| Schweizerische paläontologische Abhandlungen | Basel | Author must be Swiss, or the material must be from Switzerland, or it must be deposited in a Swiss Museum. Considerable printing contribution (no definite figures). |
| Palaeontologica polonica | Warsaw | Would accept two ostracod monographs (in good English!) to be published in one volume. Two reviews of competent ostracodologists to be sent with MS. 'No financial requirement. We are going to offer such a volume as a Polish contribution to the improvement of international circulation of palaeontological information'. |
| Bollettino della Società paleontologica italiana | Modena | Normally up to 50 manuscript pages, but larger monographs sometimes considered. Author's charge: 20% real cost of pages and illustrations. |
| Geobios | Lyon | For papers of more than 40 printed pages (illustrations included): page contribution. 75 reprints must be bought. |
| Elsevier, Earth Science Department | Amsterdam | 'In principle, we are interested in publishing monographs and review papers on Ostracoda. It will, however, depend on several factors if we actually decide to publish or not. Inform your audience that they should not hesitate submitting a proposal to us'. |
| Zitteliana and Mitteilungen der bayerischen Staatssammlung für Paläontologie und historische Geologie | Munich | Paper must have a connection with the Bavarian State Collection. Monographs need printing contribution. |
| Palaeontographica, A | Stuttgart | No page charges. Up to 15 plates. 50 free reprints. |
| Abhandlungen der senckenbergischen naturforschenden Gesellschaft | Frankfurt | Monographs possible, but no theses. Material must be deposited in the Senckenberg Museum. Contribution more or less obligatory, but in certain cases help through the Deutsche Forschungsgemeinschaft. Contact Heinz Malz. |
| Senckenbergiana lethaea (or S. biologica, S. marina) | Frankfurt | Same as above (usually less voluminous papers). |
| Courier Forschungs-Institut Senckenberg | Frankfurt | Mostly (but not exclusively) for studies performed at Senckenberg. Deposit of (some) material desired, but not obligatory. Theses possible. Author must contribute. Contact Heinz Malz. |
| Crustaceana Supplement | Leiden | Contribution dependent upon the size of the manuscript and the number of illustrations it contains. If manuscript on floppy disks or camera-ready copy: no subsidy required. |
| Special Papers in Palaeontology | London | Publication of larger papers (than in its sister journal: Palaeontology). Contribution to costs if possible, but not obligatory. |

N.B. Plate sizes and other author's instructions:  write to the editors concerned (addresses: see the journals, or Bowker - *vide* reference).

Anyway, be prepared to receive very critical views on what you had thought to be first class; this will be a hard experience!

## WHERE?

To be published in a major journal is a guarantee of a wide distribution for your paper, but this needs very serious preparation and top quality presentation. Usually only outstanding contributions will be accepted.

Be sure to follow exactly the guidelines for authors. It can be useful to attach, when sending your paper, informal referee's reports from two internationally known specialists. Also, before even mailing your paper, it is wise to address a letter to the editor, together with a first class abstract, asking if he thinks your paper might meet the journal's interests. Do not be discouraged if he refuses, you have a choice of numerous others.

If the contacted editor agrees to look at your paper, you must be aware that this is by no means the overall acceptance. The paper will be read carefully by referees. The result of this is that, in the best case, you have to retouch certain paragraphs and better explain this or that. In the worst case, it is a complete refusal. Usually, you will need to rewrite certain parts of the paper.

You may have fewer difficulties with minor and lesser known journals. The quality of printing may be as good, and publishing quicker, but as they have a more restricted circulation, do not forget to order a large number of reprints.

A big problem are voluminous papers whose splitting would destroy the paper's construction and coherence (this is not always the case for big papers!), and publication often seems impossible. In order to clear up the situation in this particular field, I sent some months ago a circular letter to all journal editors who I knew had sometimes published large papers on Ostracoda, asking if they would be prepared to consider for publication papers of approximately 100 pp. text, 20-40 figures/tables and 10-20 full-page plates - and what would be their conditions?

From 34 letters mailed, I received 18 replies. As far as they can be used, they are laid out in Table 1. It shows that places exist where publication of voluminous studies is still possible, partly even without author's contributions.

Also, the 'bible' of serial and non-serial publications (Bowker) shows that in earth sciences there exist some 350 titles of serials (i.e. regular publications) and 841 irregular serials; for palaeontology alone, the values are 42 and 48 respectively. Not included here are biological journals. Therefore, there is no lack of publication places!

## ACKNOWLEDGEMENTS

My sincerest thanks go to the two anonymous referees, as well as to Caroline Maybury and Robin Whatley, for their helpful remarks.

## REFERENCE

*The Bowker International Serials Database: Ulrich's International Periodicals Dictionary.* 1987-88. 26th ed., 2 vol. Irregular Serials and Annuals. 1987-88. 13th ed., 1 vol. Bowker, New York/London.

# 47

# Video recording in the study of living Ostracoda: techniques and preliminary results

I. G. Sohn

U.S. Geological Survey, Room E-308,
National Museum of Natural History,
Washington D.C., 20560, U.S.A.

## ABSTRACT

Instrumentation, techniques and the culture of os- tracods used in making a video recording are described. Potential applications of video record- ing are suggested. The 26 minute video shown at the Symposium was edited from more than seven hours of recording. It showed the movements of appendages and the brooding of eggs and instars of the non-swimming *Darwinula stevensoni* (Brady & Robertson, 1870); also the feeding behaviour and body functions of *Darwinula* and of a swim- ming cyprid, probably *Cypridopsis vidua* (O. F. Müller, 1776) sensu Kesling, 1951. The absence in the above taxa of statocysts for orientation is in- ferred.

## INTRODUCTION

Because the first circular announcing the 10th International Symposium on Ostracoda stated that video facilities would be available, I investigated the possibility of making a pilot video recording on living Ostracoda. The necessary equipment was available at the Smithsonian Institution. The non-swimming *Darwinula stevensoni* is an excel- lent subject for video observation. At high mag- nification swimming ostracods normally move too fast for recording and observation, except when they are feeding or resting.

Specimens used in making the video are deposited in the collections of the Department of Invertebrate Zoology, National Museum of Natu- ral History with the following catalogue numbers: *Darwinula stevensoni* recorded in September, USNM 193635a; in March USNM 193635b; and in June, USNM 193635c.

## INSTRUMENTATION

The video was made with the same equipment that Dr Kurt Fredriksson, Department of Mineral Sciences, National Museum of Natural History, used to record melting experiments with meteorites

(Fredriksson & Wlotzka, 1988). The following equipment was used: a 1965 petrographic microscope (Zeiss W1) equipped with an Optovar (Zeiss) magnification changer, a 2.5 objective, and a 6.3 eyepiece, a solid state video camera (Hitachi KPC 100U), a C-mount adapter, and a Betamax (Beta) format 1/2 inch video cassette recorder (VCR) equipped with a microphone. For the final segment in August 1988 a video home system (VHS) format 1/2 inch tape VCR was substituted for the Beta format.

Although the microscope is equipped with transmitted as well as reflected light sources, transmitted light was used in order to record, through the darwinulid valves, the brooding of eggs and retained instars within the carapace.

The photographs from the video (Plates 1-3) were made from the original 1/2 inch beta format recording with a Beta Hi-Fi VCR (Sanyo VCR 300), a Still Video Recorder (Sony MVR A770) that records on a 2" floppy disc frames from the video. Prints from the floppy disc can be made with a colour video printer (Sony Mavigraph UP500) or converted to black and white photographic negatives with a Polaroid Freeze Frame.

A microphone attachment recorded narrated observations and other data essential for documentation, study and/or editing of the tape. In addition, these data also helped in preparing the script for the voice in the edited video.

Home video recorders use 1/2 inch video tapes (high quality tapes yield better results) and professional equipment uses 3/4 inch tapes. Three different television systems are currently in use: in the Western Hemisphere (with few exceptions) - National Television System Committee (NTSC); outside the Western Hemisphere - Phase Alteration Line (PAL) or Séquentiel Couleur à Mémoire (SECAM). The voltage frequencies (Hz) vary among countries: with few exceptions, 60Hz is used in the Western Hemisphere; Africa, most of Asia, Europe and Oceania use 50Hz. Australia uses PAL and 50Hz.

Information on specifications can be obtained from TV and VCR dealers in each country. In order to view a video made on one of the TV systems using a TV having a different system, the cassette has to be commercially duplicated. The cost for this service varies among providers.

Additional information on video recording is available in the following publications.

Lenk, J. D. 1983. *Complete guide to videocassette recorder operation and servicing*, 365 pp. Prentice-Hall, New Jersey.

Pasternak, B. 1983. *Video cassette recorders: buying, using & maintaining*, 143 pp. Tab Books Inc., Blue Ridge Summit, Pa. 17214.

Sachorn, J. L. 1986. *Video magazine's guide to component TV*, 199 pp. MacGraw-Hill, New York.

Spottswood, R. *The Focal encyclopedia of film and television techniques*, 1969. 1100 pp. Focal Press, London.

## MAINTAINING AND EXTRACTING LIVE OSTRACODS

Methods of collecting living ostracods were summarized by Keyser (1988). Most ostracodologists have their own favourite tools for collecting.

I maintained living marine and freshwater ostracods, including the original water, mud and plants, at room temperature in covered glass containers for many years. Distilled water has been added to both marine and non-marine aquaria to compensate for evaporation and to maintain the original water chemistry of the culture. A mark was made at the water level on the outside of the container and water added periodically to that mark. Although complete drying of the aquaria should be avoided, evaporation of as much as 80% or more of the water column has not, so far as can be determined, harmed the cultures.

Except for an occasional piece of chicken egg shell and an annual small addition of ground lime slurry to maintain a high pH, a small quantity of dry fish food was added infrequently. With time, most plastic containers release toxic substances into the culture that may kill the animals. Valves of dead ostracods kept in plastic containers in water may become decalcified after several weeks.

The standard methods of extracting live ostracods from collections or aquaria is to transfer a small quantity of the sample into a shallow picking dish, examine with a microscope and transfer living ostracods with a pipette or brush into a second receptacle. A more efficient method is to take advantage of the non-wetting film that covers the ostracod carapace. Ostracods that are exposed to air float on the surface; living specimens struggle to break the surface tension of the water in order to sink.

By washing the sample through a sieve, raising the sieve in order to expose the ostracods to air, and gently lowering the sieve into a basin with water, the ostracods are concentrated on the surface of the water. Non-swimming ostracods are concentrated on the water surface by slightly tipping the picking receptacle so that the specimens are exposed to air and float on the water.

## METHODS USED

Three segments in the life cycle from eggs and instars within the marsupium of adults to released instars were recorded for *Darwinula stevensoni* over a period of 10 months. The morphology of the soft parts and their functions were documented.

Because the video was to record *Darwinula*, a non-swimming taxon, a glass flat-bottom cell culture slide was used in September 1987 in order to avoid repeated adjustments of the focus. Glass slides that have concave cells might have caused the animal to move out of the focus of the microscope. During the March and June 1988 recording sessions the cover of a plastic box, 2.5cm square and 5mm deep, placed on a glass microscope slide for ease of manipulation, was substituted for the glass slide.

The miniature 'aquarium' provided more room for several specimens of *Darwinula* and a few young of swimming cyprids (Plates 1-3). Both species are from a collection made in 1969 in a freshwater pond in Long Island, New York (Sohn, 1987, 151). Fragments of the detritus in the aquarium, upon which both the cyprids and darwinulids were observed feeding, and several dead cyprids were included with the ostracods in the plastic box cover.

Home video cassettes can be used to record at 3 speeds: 2 hours (SP), 4 hours (LP), and 6 hours (EP) and the VCR can be programmed for each of the above speeds. This information is noted on each cassette and each VCR model. The slowest speeds produce best results.

A digital counter records the number of revolutions of the sprocket in the VCR as an aid to finding any given segment on the tape; however, these numbers differ with each model of the VCR. Most VCRs have 'Pose' mode and 'Still' (Freeze Frame) capabilities.

The video recordings made in September 1987 and March 1988 were with a Beta system VCR. This video cassette was then duplicated on an RCA VCR by use of a VCR Dubbing Kit. Instructions are in the Owner's Manual. Some of the fidelity is lost each time a video is transcribed; consequently, only the master tape should be used for duplication. The tab provided at the side of the cassette should be removed from the master tape after the recording is completed in order to prevent the accidental erasure of the video.

A stage micrometer divided in 0.01mm was recorded through the microscope and the camera was adjusted for the image on the monitor screen to be 1mm wide. The Optivar magnification changer on the microscope was used to enlarge the width of the screen to 0.8mm, 0.6mm and 0.5mm. Scales of a 1mm wide screen and a 0.5mm wide screen were copied on graph paper from the projections on the screen of my home Zenith 21" diagonal TV receiver. These scales were then used to measure specimens, eggs and instars.

## ADVANTAGES OF VIDEOS FOR THE STUDY OF OSTRACODS

Because of the availability of television monitors in schools and laboratories, videos can be used for classroom lectures and for study by research workers. Important segments can be reviewed in slow motion and photographs can be made for study and publication (Plates 1-3).

Some applications of video recording designed for ostracod research are illustrated by the following examples.

Although the present study was to learn how to make a video of *Darwinula stevensoni* in order to document the brooding of eggs and instars in the marsupium, the sanitation strategies of *Darwinula* and the cyprid are a fortuitous bonus.

In September 1987 and June 1988 the darwinulid carapaces were empty of eggs, consequently, movement of the posterior process was clearly visible within the carapace. Movements of the respiratory seta of the mandibles, maxillae, and first thoracopods were synchronized and were recorded moving rhythmically in the closed and open carapaces. The antennulae, antennae and the second and third thoracopods were extruded from the carapace; their movements were not synchronized (Plate 3). In March 1988 the carapaces of all but one of the studied specimens contained from one to four eggs and/or instars. The ambulatory function of the dorsally geniculate antennulae and the ventrally bent antennae were recorded (Plate 3); the antennae are also sensory and grasping appendages. The closing and opening of the terminal claws of the antennae in a manner similar to that of the prehensile movement of fingers of the hominid hand were seen (Plate 3, Fig. 2c).

Darwinulids are known to crawl in the substrate. In the video they were seen crawling forward, usually oriented with the ventral margin up, using the geniculate antennulae to pull the carapace along; and less often they moved with the dorsal margin up, using the antennae and second and third thoracopods.

Specimens were recorded resting on their sides with the valves open or closed. Gravid darwinulids used the posterior process and the second and third thoracopods to manipulate the eggs and/or the instars within the marsupium.

The laterally flattened posterior process could be seen moving anteroventrally, probably creating currents that bring food to the instars and oxygen to the eggs and instars. The antennulae and antennae of instars were recorded extending from their carapaces inside the mothers' marsupia.

The released instars were very active; moving continuously among the debris, or resting on their backs with their valves parted and continuously moving their appendages. Several growth stages were recorded: the smallest instars (carapace length 0.17-0.24mm) did not yet have posterior legs (thoracopods), the larger instars (carapace length 0.25-0.35mm) already had posterior legs.

Two growth stages of *Cypridopsis vidua* sensu Kesling, 1951 were recorded as they fed on detritus and plant material.

By using its antennae and posterior legs the young cypridopsid, greatest length of 0.25mm and height of 0.17mm, crawled along the margins as well as the top and bottom surfaces of a fragment of plant material 2 x 1.3mm and 0.05mm thick that was suspended within the water column. The antennae appear to move food towards the mouth. This specimen stopped, reared on its legs (Plate 1, Fig.1), parted the valves and evacuated (Plate 1, Fig. 2). The specimen then crawled away from the excrement that consists of a ball of loosely packed strands that resembles a miniature tumbleweed (Plate 2, Figs 1-2).

The evacuation strategy of *Darwinula* differs from that of *Cypridopsis*. *Darwinula* did not extrude the thoracopods; it ejected a loosely packed excrement with sufficient force for the excrement to float away from between the parted valves. (Plate 1, Fig. 3). These differences were previously unknown (Sohn, 1988).

Video recording of taxa of additional categories of Ostracoda may reveal similar sanitation strategies and/or other distinctions among groups.

A larger cypridopsid, 0.48mm long, 0.32mm wide and 0.29mm high, was recorded feeding. The dorsal margin of this specimen is less convex than that of the younger instar. Its feeding habit was similar to that of the younger instar.

An exuvia of a moulted cypridopsid with a shell length of 0.57mm was recorded as it floated on the surface. The chitinous cover of the appendages and the chitinous covers of the hair from which the next growth stage had removed its hair and appendages could be seen (Plate 2, Fig. 3). This exuvia resembles the organic residue of decalcified valves of extant ostracods (Sohn, 1958).

The few dead carapaces recorded in the video suggest that bacteria and protozoans feed on and destroy ostracod shells. Future video recording of the destruction of dead ostracod carapaces may contribute information to the field of palaeocology.

Illustrations and descriptions of dissections do not impart the information obtained by video recording. The video showed that the antennulae and antennae serve an ambulatory function. As previously described, the terminal claws of the antennae open and close in a manner similar to that of the prehensile movement of the fingers of a hominid hand for use in feeding.

The swimming, crawling, feeding strategy and body functions of *Cypridopsis vidua* (O. F. Müller, 1776) sensu Kesling, 1951 were documented (Plates 1-2). *Darwinula* has elongated, coloured patches in the gut close to the dorsal margin of both valves and *Cypridopsis* has colour markings on the valves; these colours differ in different environments. The colour of those patches in *Darwinula* vary with the available food (Sohn, 1987, 160). By using reflected light, the colour of ostracods from different environments can be video recorded.

*Cypridopsis* was observed crawling on the top, bottom and along the margins of a submerged plant fragment; this can be explained by the gripping ability of the extruded appendages. However, because *Darwinula* crawls either with the ventral margin up or with the dorsal margin up, these similarities suggest that these ostracods do not have statocysts for orientation.

## ACKNOWLEDGEMENTS

I thank Dr Kurt Frediksson, Department of Mineral Sciences, National Museum of Natural History, for the use of his video recording equipment. Mr Donald Hales, Human Studies Film Archives, Smithsonian Institution, made available equipment for transfer of home video tape to professional tape (1/2" to 3/4" width) so that Mr Peter R. C. Erikson, Smithsonian Office of Telecommunications, could edit this 26 minute video from approximately 7 hours of video tape recordings. Mr Dane A. Penland, Photographer Special Assignments, Smithsonian Printing and Photographic Services, made still video images from the original (Beta version) of the video which he converted to black and white photographs. My colleague, Dr L. S. Kornicker, National Museum of Natural History, provided advice and encouragement. He and Dr C. W. Hart, Jr, National Museum of Natural History, reviewed this manuscript.

**Plate 1**

All movements are on a plane normal to the microscope.

Figs 1-2. *Cypridopsis vidua* (O. F. Müller, 1776) *sensu* Kesling, 1951. Young instar (greatest length 0.25mm and height of 0.17mm) feeding on floating decayed plant material, in the process of evacuating.

Fig. 1.   Specimen reared on its thoracopods.
Fig. 2.   Process of evacuation.
Fig. 3.   *Darwinula stevensoni* (Brady & Robertson, 1870). Left view of gravid female (greatest length 0.57mm) containing two eggs and one nauplius in the marsupium in the process of evacuating, and right view of a female (greatest length 0.60mm), with an empty marsupium, presumed to have released the instars. The excrement was in the process of leaving the carapace, it was forced out of the carapace and floated away to disappear from view beneath the anteroventral part of the larger female.

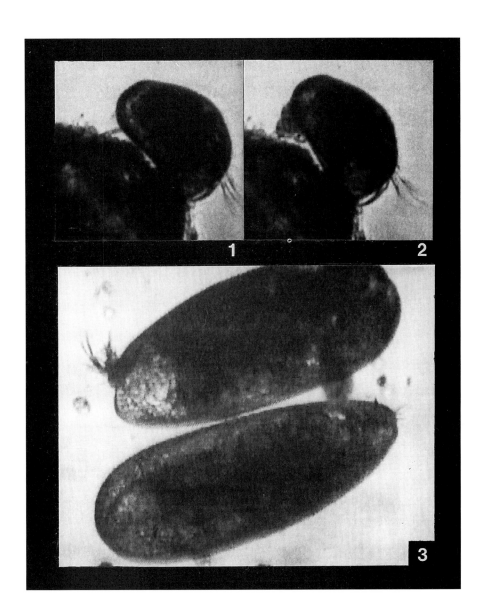

**Plate 2**

All movements are on a plane normal to the microscope.

Figs 1-3.  *Cypridopsis vidua* (O. F. Müller) sensu Kesling,  1951.  Sequential photographs continued from Plate 1, Figs 1-2.

Figs 1-2.  Unlike the darwinulid, the specimen crawled away  from the excrement.
Fig.  3.      Exuvia (greatest length 0.57mm) floating on the  surface of  the water. Note the chitinous covers of the hair  along  the posterior and of the antennulae and antennae.

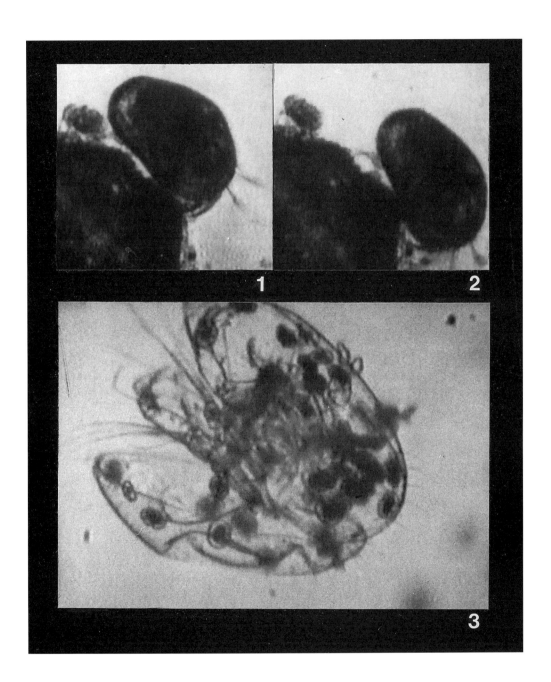

**Plate 3**

All movements are on a plane normal to the microscope.

Figs 1-2. *Darwinula stevensoni* (Brady & Robertson, 1870).

Fig. 1.   Photographs of the right side of a gravid female with two eggs and probably a nauplius within the marsupium; the left and right antennulae (a, b), right and left antennae (c, d), right and left second thoracopods (e, f) and right and left third thoracopods (g, h) extruded from the carapace.

Fig. 2.   Sequential photograph. Note that the antennulae are geniculate dorsally, that the right and left appendages move independently of each other; the grasping movement of the right antenna and the closed left antenna (c, d). The distal spine of the left third thoracopod can be seen extruding from the posterior of the carapace (h).

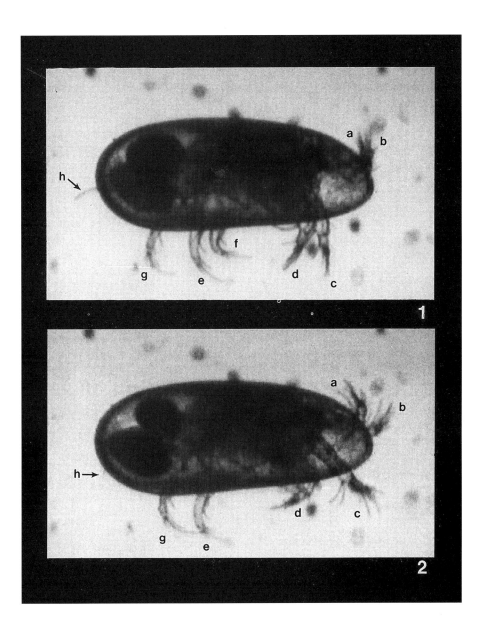

## REFERENCES

Fredriksson, K. & Wlotzka, F. 1988. Experiments with a Leitz 1750°C Heating Stage (abs.). *Meteoritics*, **23**(3), 269.

Keyser, D. 1988, Ostracoda. *In* Higgins, R. P. & Thiel, H. (Eds), *Introduction to the study of Meiofauna*, 488 pp. Smithsonian Institution Press, Washington, D.C.

Sohn, I. G. 1958. Chemical constituents of ostracodes; some applications to paleontology and paleoecology. *J. Paleont.*, Tulsa, Okla., **32**(4), 730-736.

Sohn, I. G. 1987. The ubiquitous ostracode *Darwinula stevensoni* (Brady and Robertson, 1870), redescription of the species and lectotype designation. *Micropaleontology*, New York, **33**(2), 150-163, 3 pls.

Sohn, I. G. 1988. Darwinulocopina (Crustacea: Podocopa), a new suborder proposed for nonmarine Paleozoic to Holocene Ostracoda. *Proc. Wash. Biol. Soc.*, Washington, D.C., **101**(4), 817-824.

# Taxonomic index

Note: Page numbers in *italics* refer to pages on which figs/tables appear. Numerous species may be included under pages referred to by the generic name only.

# Subject index

Note: Page numbers in *italics* refer to pages on which figs/tables appear.

Deep sea *contd.*
  xylophile fauna 307, 307–19, 309
Deep Sea Drilling Project (DSDP) 42, *43*, 44, 54, 130, 212, 213, 300, 312
  North Atlantic sites 130, *288*, 300, *302*
  Pacific sites 50, 288, *289*, 290, *302–3*, 312
Deltas 35
Denmark 245
Depth, diversity and evolution 82–3
Derbyshire 163, 164
Dessication-resistant egg 15, 18
Devonian 59–70, *103*, 162, 163, 569
  Czechoslovakia, ostracod fauna 233–7
  entomozocean genera 101, 102, *103*, *109*, 110, 167, *168*
  eustatic sea-level changes 110, *168*
  global events 101, 102, *109*, 110, 167, *168*
  ostracod assemblages 422
  palaeoecology, biotope indicative features 421–36
  Thuringian Assemblage 422, 424, 426, *429*
  transgressions 59, 60, 64, 68, 69, 101, 104, 107
Diatoms 335
Dinantian *162*, 162, 422, *429*
Dinarides 263
Dinaro–Hellenic area *27*
Dinoflagellates 124, 254, 256
Dinosaurs 7
Dissolved Organic Matter (DOM) 278
Diversity 3, 5
  Cainozoic deep-sea Ostracoda 41, *47*, 71–9, 82–5, 287, 290–2, *293*, *294–6*
  Cainozoic shallow water 77–9
  controls and factors affecting 82–4
  Devonian ostracod faunas of Canada 59, 64
  Mesozoic 11, *12*, *13*

psychrosphere formation *47*, 49
reef intervals *102*
transgressions and 5, 11, 14, 83, 167
Turonian 25, 32
Dominican Republic 221, *222*, 226, 228
Dorset *578–80*, 581
Dover, England 123, 124, *125*, *126*, 127, *128*, 130
Drake Passage 5, 6, 9, 46, 84, 293, 300
Dysaerobic communities 161, 164, 168, 169

East Antarctic Ice Cap 41, 46, 50
East China Sea 139, 146, 147, 149
Echinoids 90
Ecological niches 83
Ecology 333, 355
Ecosystem *274*
Ecozones 33
Educational aspects 583–602
Eifelian Assemblage 422, 423, 424, *428*
Eifelian fauna, Canada 59, 60, 64, 68
Elephant Island 278
El Niño 9, 107, 339
Emsian 59, 110
England 36, 245, 248, 515
  Cretaceous halocyprids 516, 519, 520
  deep-sea genera, origins 296, *297–8*
  Eocene–Oligocene sea-level changes 155–7
  Namurian entomozoacean ostracods 161, *162*, 164, *165*, *171*
  tufa deposits 509
  Turonian ostracods 123–37
  Wealden, *see* Wealden
Entomozoacean ostracods 423, 425, 426, 516
  global event affecting 101–12
  Namurian 161–71, *162*, *165*, *171*

Environmental changes 4, 18, 465–73
Environmental factors, evolution and diversity 82–3
Eocene 44, 49, 73, 114, *115*, 154
  adaptive strategies in 459–64
  deep-sea ostracods 84, 287, 296, 300
  diversity 49, 290, *291*, 292
  extinctions and originations 19, *21*, 73, 80, 81, 158
  Gulf Coast ostracods, changes 113–21
  halocyprid ostracods 515
  ostracod zones 114–18
  psychrospheric species 84
  sea-level changes 83, 114, 154, 155–7, 463
  species and generic diversity 49, 74, 76, 78–9, 85
  terminal extinction event 153, 158–9
  xylophile fauna 313
Eocene–Oligocene boundary 80, 84, 153–60, *157*
  adaptive strategies in 459–64
  ostracod fauna changes 157–8, 463
Eocene–Oligocene Boundary Crisis 463
Epigean organisms 437, 438, 439
Ethiopia 569
Euphausiids 278
Europe 87
  Cainozoic Ostracoda 71–87
  Cenomanian–Turonian Oceanic Anoxic Event 123
  Cretaceous halocyprid ostracods 515, 519
  entomozoacean ostracods 103, 110
  Eocene extinction event 158
  Lower Jurassic ostracods 212
  species and generic diversity 74–9
  Thuringian Assemblage 424, 425
  Triassic ostracods 542, 543
  tufa deposits 509